PROJECT

MANAGEMENT

A Systems Approach to Planning, Scheduling, and Controlling

TENTH EDITION

HAROLD KERZNER, Ph.D.

Senior Executive Director for Project Management
The International Insitute for Learning New York,
New York

WILEY
John Wiley & Sons, Inc.

Library of Congress Cataloging-in-Publication Data:

Kerzner, Harold.
    Project management : a systems approach to planning, scheduling, and conrolling/Harold Kerzner.—10th ed.
        p.   cm.
    Includes index.
    ISBN 978-0-470-27870-3 (cloth : acid-free paper)   1. Project management.   I. Title.
    HD69.P75K47 2009
    658.4′04—dc22

                                                            2008049907

Printed in the United States of America.

10 9 8 7 6 5 4 3 2 1

To
Dr. Herman Krier,
my Friend and Guru,
who taught me well the
meaning of the word "persistence"

Contents

Preface _____

Project management has evolved from a management philosophy restricted to a few functional areas and regarded as something nice to have to an enterprise project management system affecting every functional unit of the company. Simply stated, project management has evolved into a business process rather than merely a project management process. More and more companies are now regarding project management as being mandatory for the survival of the firm. Organizations that were opponents of project management are now advocates. Management educators of the past, who preached that project management could not work and would be just another fad, are now staunch supporters. Project management is here to stay. Colleges and universities are now offering graduate degrees in project management.

The text discusses the principles of project management. Students who are interested in advanced topics, such as some of the material in Chapters 21 to 24 of this text, may wish to read one of my other texts, *Advanced Project Management: Best Practices in Implementation* (Hoboken, NJ: Wiley, 2004), and *Project Management Best Practices: Achieving Global Excellence* (Hoboken, NJ: Wiley, 2006). John Wiley & Sons will also be introducing a five-book series on project management best practices, co-authored by Frank Saladis and Harold Kerzner, to accompany the above two books.

This book is addressed not only to those undergraduate and graduate students who wish to improve upon their project management skills but also to those functional managers and upper-level executives who serve as project sponsors and must provide continuous support for projects. During the past several years, management's knowledge and understanding of project management has matured to the point where almost every company is using project management in one form or another. These companies have come to the realization that project management and productivity are related and that we are now managing our business as though it is a series of projects. Project management coursework is now consuming more of training budgets than ever before.

General reference is provided in the text to engineers. However, the reader should not consider project management as strictly engineering related. The engineering examples are the result of the fact that project management first appeared in the engineering disciplines, and we should be willing to learn from their mistakes. Project management now resides in every profession, including information systems, health care, consulting, pharmaceutical, banks, and government agencies.

The text can be used for both undergraduate and graduate courses in business, information systems, and engineering. The structure of the text is based upon my belief that project management is much more behavioral than quantitative since projects are managed by people rather than tools. The first five chapters are part of the basic core of knowledge necessary to understand project management. Chapters 6 through 8 deal with the support functions of managing your time effectively, conflicts, and other special topics. Chapters 9 and 10 describe factors for predicting success and management support. It may seem strange that 10 chapters on organizational behavior and structuring are needed prior to the "hard-core" chapters of planning, scheduling, and controlling. These first 10 chapters are needed to understand the cultural environment for all projects and systems. These chapters are necessary for the reader to understand the difficulties in achieving cross-functional cooperation on projects where team members are working on multiple projects concurrently and why the people involved, all of whom may have different backgrounds, cannot simply be forged into a cohesive work unit without friction. Chapters 11 through 20 are more of the quantitative chapters on planning, scheduling, cost control, estimating, procurement, and quality. Chapters 21 through 24 are advanced topics and future trends. Chapter 25 is a capstone case study that can be related to almost all of the chapters in the text.

The changes that were made in the 10th edition include:

- A chapter on the business of scope changes
- A chapter on managing crises projects
- A chapter on the Iridium Project, which serves as a capstone case
- An appendix on using the book to study for the PMP® exam
- A section on understanding the collective belief on a project
- A section on the need for an exit champion
- A section on project financing
- A section on managing virtual teams
- A section on rewarding project teams
- A section on the need for an enterprise project management system
- A section on kickoff meeting
- A section on breakthrough projects
- A section on project audits
- A section on managing intellectual property

- A section on the problems associated with project scheduling
- A section on schedule compression myths
- A section on human behavior education
- A section on dysfunctional team behavior
- A section on validating project assumptions
- Existing sections from the 9th edition with expanded information include: the new breed of project manager; additional scheduling problems; a discussion on the difference between active and passive involvement by the sponsor; the need for challenging the decisions of the sponsor; information needed for effective estimating; managing stakeholder expectations; the project war room; power and authority; the management reserve; and Six Sigma. The chapters on risk management and procurement were restructured to be in better alignment with the PMBOK® Guide, 4th edition.

The text contains more than 25 case studies, more than 125 multiple choice questions, and nearly 400 discussion questions. In addition, there is a supplemental workbook (*Project Management Workbook to Accompany Project Management,* tenth edition) that contains more than 600 multiple choice questions, additional case studies, challenging problems, and crossword puzzles. There is also a separate book of cases (*Project Management Case Studies,* third edition) that provides additional real-world examples.

This text, the workbook, and the book of cases are ideal as self-study tools for the Project Management Institute's PMP® Certification exam. Because of this, there are tables of cross references on each chapter's opening page in the textbook detailing the sections from the book of cases, the workbook, and the Guide to the Project Management Body of Knowledge (PMBOK® Guide) that apply to that chapter's content. The left-hand margin of the pages in the text has side bars that identify the cross-listing of the material on that page to the appropriate section(s) of the PMBOK® Guide. At the end of most of the chapters is a section on study tips for the PMP® exam, including more than 125 multiple choice questions.

This textbook is currently used in the college market, in the reference market, and for studying for the PMP® Certification exam. Therefore, to satisfy the needs of all markets, a compromise had to be reached on how much of the text would be aligned to the PMBOK® Guide and how much new material would be included without doubling the size of the text. Some colleges and universities use the textbook to teach project management fundamentals without reference to the PMBOK® Guide. The text does not contain all of the material necessary to support each section of the PMBOK® Guide. Therefore, to study for the PMP® Certification exam, the PMBOK® Guide must also be used together with this text. The text covers material for almost all of the PMBOK® Guide knowledge areas but not necessarily in the depth that appears in the PMBOK® Guide.

An instructor's manual is available only to college and university faculty members by contacting your local Wiley sales representative or by visiting the Wiley website at www.wiley.com/kerzner. This website includes not only the instructor's manual but also 500 PowerPoint slides that follow the content of the book and help organize and execute classroom instruction and group learning. Access to the instructor's material can be provided only through John Wiley & Sons Publishers, not the author.

One-, two-, and three-day seminars on project management and the PMP® Certification Training using the text are offered by contacting Lori Milhaven, Executive Vice President, the International Institute for Learning, at 800-325-1533, extension 5121 (email address: lori.milhaven@iil.com).

The problems and case studies at the ends of the chapters cover a variety of industries. Almost all of the case studies are real-world situations taken from my consulting practice. Feedback from my colleagues who are using the text has provided me with fruitful criticism, most of which has been incorporated into the tenth edition.

The majority of the articles on project management that have become classics have been referenced in the textbook throughout the first 11 chapters. These articles were the basis for many of the modern developments in project management and are therefore identified throughout the text.

Many colleagues provided valuable criticism. In particular, I am indebted to those industrial/government training managers whose dedication and commitment to quality project management education and training have led to valuable changes in this and previous editions. In particular, I wish to thank Frank Saladis, PMP®, Senior Consultant and Trainer with the International Institute for Learning, for his constructive comments, recommendations, and assistance with the mapping of the text to the PMBOK® Guide as well as recommended changes to many of the chapters. I am indebted to Dr. Edmund Conrow, PMP®, for a decade of assistance with the preparation of the risk management chapters in all of my texts.

To the management team and employees of the International Institute for Learning, thank you all for 20 years of never-ending encouragement, support, and assistance with all of my project management research and writings.

Harold Kerzner
The International Institute for Learning
2009

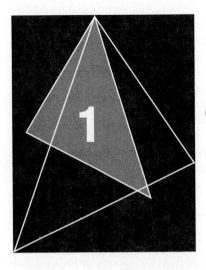

Overview

Related Case Studies (from Kerzner/*Project Management Case Studies,* 3rd Edition)	Related Workbook Exercises (from Kerzner/*Project Management Workbook and PMP®/CAPM® Exam Study Guide,* 10th Edition)	PMBOK® Guide, 4th Edition, Reference Section for the PMP® Certification Exam
• Kombs Engineering • Williams Machine Tool Company* • Hyten Corporation • Macon, Inc. • Continental Computer Corporation • Jackson Industries	• Multiple Choice Exam	• Integration Management • Scope Management • Human Resource Management

1.0 INTRODUCTION

Executives will be facing increasingly complex challenges during the next decade. These challenges will be the result of high escalation factors for salaries and raw materials, increased union demands, pressure from stockholders, and the possibility of long-term high inflation accompanied by a mild recession and a lack of borrowing power with financial institutions. These environmental conditions have existed before, but not to the degree that they do today.

---

*Case Study also appears at end of chapter.

In the past, executives have attempted to ease the impact of these environmental conditions by embarking on massive cost-reduction programs. The usual results of these programs have been early retirement, layoffs, and a reduction in manpower through attrition. As jobs become vacant, executives pressure line managers to accomplish the same amount of work with fewer resources, either by improving efficiency or by upgrading performance requirements to a higher position on the learning curve. Because people costs are more inflationary than the cost of equipment or facilities, executives are funding more and more capital equipment projects in an attempt to increase or improve productivity without increasing labor.

Unfortunately, executives are somewhat limited in how far they can go to reduce manpower without running a high risk to corporate profitability. Capital equipment projects are not always the answer. Thus, executives have been forced to look elsewhere for the solutions to their problems.

Almost all of today's executives are in agreement that the solution to the majority of corporate problems involves obtaining better control and use of existing corporate resources, looking internally rather than externally for the solution. As part of the attempt to achieve an internal solution, executives are taking a hard look at the ways corporate activities are managed. Project management is one of the techniques under consideration.

The project management approach is relatively modern. It is characterized by methods of restructuring management and adapting special management techniques, with the purpose of obtaining better control and use of existing resources. Forty years ago project management was confined to U.S. Department of Defense contractors and construction companies. Today, the concept behind project management is being applied in such diverse industries and organizations as defense, construction, pharmaceuticals, chemicals, banking, hospitals, accounting, advertising, law, state and local governments, and the United Nations.

The rapid rate of change in both technology and the marketplace has created enormous strains on existing organizational forms. The traditional structure is highly bureaucratic, and experience has shown that it cannot respond rapidly enough to a changing environment. Thus, the traditional structure must be replaced by project management, or other temporary management structures that are highly organic and can respond very rapidly as situations develop inside and outside the company.

Project management has long been discussed by corporate executives and academics as one of several workable possibilities for organizational forms of the future that could integrate complex efforts and reduce bureaucracy. The acceptance of project management has not been easy, however. Many executives are not willing to accept change and are inflexible when it comes to adapting to a different environment. The project management approach requires a departure from the traditional business organizational form, which is basically vertical and which emphasizes a strong superior–subordinate relationship.

1.1 UNDERSTANDING PROJECT MANAGEMENT

PMBOK® Guide, 4th Edition
1.2 What Is a Project?
1.3 What Is Project Management?

In order to understand project management, one must begin with the definition of a project. A project can be considered to be any series of activities and tasks that:

- Have a specific objective to be completed within certain specifications
- Have defined start and end dates
- Have funding limits (if applicable)
- Consume human and nonhuman resources (i.e., money, people, equipment)
- Are multifunctional (i.e., cut across several functional lines)

Project management, on the other hand, involves five process groups as identified in the PMBOK® Guide, namely:

- Project initiation
  - Selection of the best project given resource limits
  - Recognizing the benefits of the project
  - Preparation of the documents to sanction the project
  - Assigning of the project manager
- Project planning
  - Definition of the work requirements
  - Definition of the quality and quantity of work
  - Definition of the resources needed
  - Scheduling the activities
  - Evaluation of the various risks
- Project execution
  - Negotiating for the project team members
  - Directing and managing the work
  - Working with the team members to help them improve
- Project monitoring and control
  - Tracking progress
  - Comparing actual outcome to predicted outcome
  - Analyzing variances and impacts
  - Making adjustments
- Project closure
  - Verifying that all of the work has been accomplished
  - Contractual closure of the contract
  - Financial closure of the charge numbers
  - Administrative closure of the papework

Successful project management can then be defined as having achieved the project objectives:

- Within time
- Within cost
- At the desired performance/technology level
- While utilizing the assigned resources effectively and efficiently
- Accepted by the customer

The potential benefits from project management are:

- Identification of functional responsibilities to ensure that all activities are accounted for, regardless of personnel turnover
- Minimizing the need for continuous reporting
- Identification of time limits for scheduling
- Identification of a methodology for trade-off analysis
- Measurement of accomplishment against plans

- Early identification of problems so that corrective action may follow
- Improved estimating capability for future planning
- Knowing when objectives cannot be met or will be exceeded

Unfortunately, the benefits cannot be achieved without overcoming obstacles such as:

- Project complexity
- Customer's special requirements and scope changes
- Organizational restructuring
- Project risks
- Changes in technology
- Forward planning and pricing

Project management can mean different things to different people. Quite often, people misunderstand the concept because they have ongoing projects within their company and feel that they are using project management to control these activities. In such a case, the following might be considered an appropriate definition:

> Project management is the art of creating the illusion that any outcome is the result of a series of predetermined, deliberate acts when, in fact, it was dumb luck.

Although this might be the way that some companies are running their projects, this is not project management. Project management is designed to make better use of existing resources by getting work to flow horizontally as well as vertically within the company. This approach does not really destroy the vertical, bureaucratic flow of work but simply requires that line organizations talk to one another horizontally so work will be accomplished more smoothly throughout the organization. The vertical flow of work is still the responsibility of the line managers. The horizontal flow of work is the responsibility of the project managers, and their primary effort is to communicate and coordinate activities horizontally between the line organizations.

PMBOK® Guide, 4th Edition
1.6 Project Management Skills

Figure 1–1 shows how many companies are structured. There are always "class or prestige" gaps between various levels of management. There are also functional gaps between working units of the organization. If we superimpose the management gaps on top of the functional gaps, we find that companies are made up of small operational islands that refuse to communicate with one another for fear that giving up information may strengthen their opponents. The project manager's responsibility is to get these islands to communicate cross-functionally toward common goals and objectives.

The following would be an overview definition of project management:

> Project management is the planning, organizing, directing, and controlling of company resources for a relatively short-term objective that has been established to complete specific goals and objectives. Furthermore, project management utilizes the systems approach to management by having functional personnel (the vertical hierarchy) assigned to a specific project (the horizontal hierarchy).

PMBOK® Guide, 4th Edition
2.4.2 Organizational Structures

MANAGEMENT GAPS   FUNCTIONAL GAPS: DEPARTMENTIZATION   OPERATIONAL ISLANDS

FIGURE 1–1. Why are systems necessary?

The above definition requires further comment. Classical management is usually considered to have five functions or principles:

- Planning
- Organizing
- Staffing
- Controlling
- Directing

You will notice that, in the above definition, the staffing function has been omitted. This was intentional because the project manager does not staff the project. Staffing is a line responsibility. The project manager has the right to request specific resources, but the final decision of what resources will be committed rests with the line managers.

We should also comment on what is meant by a "relatively" short-term project. Not all industries have the same definition for a short-term project. In engineering, the project might be for six months or two years; in construction, three to five years; in nuclear components, ten years; and in insurance, two weeks. Long-term projects, which consume resources full-time, are usually set up as a separate division (if large enough) or simply as a line organization.

Figure 1–2 is a pictorial representation of project management. The objective of the figure is to show that project management is designed to manage or control company resources on a given activity, within time, within cost, and within performance. Time, cost, and performance are the constraints on the project. If the project is to be accomplished for an outside customer, then the project has a fourth constraint: good customer relations. The reader should immediately realize that it is possible to manage a project internally within time, cost, and performance and then alienate the customer to such a degree that no further business will be forthcoming. Executives often select project managers based on who the customer is and what kind of customer relations will be necessary.

Projects exist to produce deliverables. The person ultimately assigned as the project manager may very well be assigned based upon the size, nature, and scope of the deliverables. Deliverables are outputs, or the end result of either the completion of the project or the end of a life-cycle phase of the project. Deliverables are measurable, tangible outputs and can take such form as:

- **Hardware Deliverables:** These are hardware items, such as a table, a prototype, or a piece of equipment.

FIGURE 1–2. Overview of project management.

- **Software Deliverables:** These items are similar to hardware deliverables but are usually paper products, such as reports, studies, handouts, or documentation. Some companies do not differentiate between hardware and software deliverables.
- **Interim Deliverables:** These items can be either hardware or software deliverables and progressively evolve as the project proceeds. An example might be a series of interim reports leading up to the final report.

Another factor influencing the selection of the project manager would be the stakeholders. Stakeholders are individuals or organizations that can be favorably or unfavorably impacted by the project. As such, project managers must interface with these stakeholders, and many of the stakeholders can exert their influence or pressure over the direction of the project.

Some stakeholders are referred to as "active" or "key" stakeholders that can possess decision-making authority during the execution of the project. Each stakeholder can have his or her own set of objectives, and this could place the project manager in a position of having to balance a variety of stakeholder interests without creating a conflict-of-interest situation for the project manager.

Each company has its own categorization system for identifying stakeholders. A typical system might be:

- Organizational stakeholders
  - Executive officers
  - Line managers
  - Employees
  - Unions

- Product/market stakeholders
  - Customers
  - Suppliers
  - Local committees
  - Governments (local, state, and federal)
  - General public
- Capital market stakeholders
  - Shareholders
  - Creditors
  - Banks

1.2 DEFINING PROJECT SUCCESS

In the previous section, we defined project success as the completion of an activity within the constraints of time, cost, and performance. This was the definition used for the past twenty years or so. Today, the definition of project success has been modified to include completion:

- Within the allocated time period
- Within the budgeted cost
- At the proper performance or specification level
- With acceptance by the customer/user
- With minimum or mutually agreed upon scope changes
- Without disturbing the main work flow of the organization
- Without changing the corporate culture

The last three elements require further explanation. Very few projects are completed within the original scope of the project. Scope changes are inevitable and have the potential to destroy not only the morale on a project, but the entire project. Scope changes *must* be held to a minimum and those that are required *must* be approved by both the project manager and the customer/user.

Project managers must be willing to manage (and make concessions/trade-offs, if necessary) such that the company's main work flow is not altered. Most project managers view themselves as self-employed entrepreneurs after project go-ahead, and would like to divorce their project from the operations of the parent organization. This is not always possible. The project manager must be willing to manage within the guidelines, policies, procedures, rules, and directives of the parent organization.

All corporations have corporate cultures, and even though each project may be inherently different, the project manager should not expect his assigned personnel to deviate from cultural norms. If the company has a cultural standard of openness and honesty when dealing with customers, then this cultural value should remain in place for all projects, regardless of who the customer/user is or how strong the project manager's desire for success is.

As a final note, it should be understood that simply because a project is a success does not mean that the company as a whole is successful in its project management endeavors. Excellence in project management is defined as a continuous stream of successfully managed projects. Any project can be driven to success through formal authority and strong executive meddling. But in order for a continuous stream of successful projects to occur, there must exist a strong corporate commitment to project management, and this commitment *must be visible.*

1.3 THE PROJECT MANAGER–LINE MANAGER INTERFACE _____

PMBOK® Guide, 4th Edition
1.6 Project Management Skills

We have stated that the project manager must control company resources within time, cost, and performance. Most companies have six resources:

- Money
- Manpower
- Equipment
- Facilities
- Materials
- Information/technology

Actually, the project manager does *not* control any of these resources directly, except perhaps money (i.e., the project budget).[1] Resources are controlled by the line managers, functional managers, or, as they are often called, resources managers. Project managers must, therefore, negotiate with line managers for all project resources. When we say that project managers control project resources, we really mean that they control those resources (which are temporarily loaned to them) *through line managers.*

Today, we have a new breed of project manager. Years ago, virtually all project managers were engineers with advanced degrees. These people had a command of technology rather than merely an understanding of technology. If the line manager believed that the project manager did in fact possess a command of technology, then the line manager would allow the assigned functional employees to take direction from the project manager. The result was that project managers were expected to manage people.

Most project managers today have an understanding of technology rather than a command of technology. As a result, the accountability for the success of the project is now viewed as shared accountability between the project manager and all affected line managers. With shared accountability, the line managers must now have a good understanding of project management, which is why more line managers are now becoming PMP®S. Project managers are now expected to focus more so on managing the project's deliverables rather than providing technical direction to the project team. Management of the assigned resources is more often than not a line function.

Another important fact is that project managers are treated as though they are managing part of a business rather than simply a project, and as such are expected to make sound business decisions as well as project decisions. Project managers must understand business principles. In the future, project managers may be expected to become externally certified by PMI® and internally certified by their company on the organization's business processes.

---

1. Here we are assuming that the line manager and project manager are not the same individual. However, the terms *line manager* and *functional manager* are used interchangeably throughout the text.

In recent years, the rapid acceleration of technology has forced the project manager to become more business oriented. According to Hans Thamhain,

> The new breed of business leaders must deal effectively with a broad spectrum of contemporary challenges that focus on time-to-market pressures, accelerating technologies, innovation, resource limitations, technical complexities, social and ethical issues, operational dynamics, cost, risks, and technology itself as summarized below:

- High task complexities, risks and uncertainties
- Fast-changing markets, technology, regulations
- Intense competition, open global markets
- Resource constraint, tough performance requirements
- Tight, end-date-driven schedules
- Total project life-cycle considerations
- Complex organizations and cross-functional linkages
- Joint ventures, alliances and partnerships, need for dealing with different organizational cultures and values
- Complex business processes and stakeholder communities
- Need for continuous improvements, upgrades and enhancements
- Need for sophisticated people skills, ability to deal with organizational conflict, power, and politics
- Increasing impact of IT and e-business[2]

Dr. Thamhain further believes that there are paradigm shifts in technology-oriented business environments that will affect the business leaders of the future, including project managers. According to Dr. Thamhain, we are shifting from…

- … mostly linear work processes to highly dynamic, organic and integrated management systems
- …efficiency toward effectiveness
- …executing projects to enterprise-wide project management
- …managing information to fully utilizing information technology
- …managerial control to self-direction and accountability
- …managing technology as part of a functional speciality ot management of technology as a distinct skill set and professional
- …status[3]

Another example of the need for the project manager to become more actively involved in business aspects has been identified by Gary Heerkens. Heerkens provides several revelations of why business knowledge has become important, a few of which are[4]:

- It really doesn't matter how well you execute a project, if you're working on the wrong project!

2. H. J. Thamhain, *Management of Technology*, (Hoboken, NJ: Wiley, 2005), pp. 3–4.
3. See note 2; Thamhain; p. 28.
4. G. Heerkens, *The Business-Savvy Project Manager* (New York: McGraw-Hill, 2006), pp. 4–8.

- There are times when spending more money on a project could be smart business—even if you exceed the original budget!
- There are times when spending more money on a project could be smart business—even if the project is delivered after the original deadline!
- Forcing the project team to agree to an unrealistic deadline may not be very smart, from a business standpoint.
- A portfolio of projects that all generate a positive cash flow may not represent an organization's best opportunity for investment.

It should become obvious at this point that successful project management is strongly dependent on:

- A good daily working relationship between the project manager and those line managers who directly assign resources to projects
- The ability of functional employees to report vertically to line managers at the same time that they report horizontally to one or more project managers

These two items become critical. In the first item, functional employees who are assigned to a project manager still take technical direction from their line managers. Second, employees who report to multiple managers will always favor the manager who controls their purse strings. Thus, most project managers appear always to be at the mercy of the line managers.

Classical management has often been defined as a process in which the manager does not necessarily perform things for himself, but accomplishes objectives through others in a group situation. This basic definition also applies to the project manager. In addition, a project manager must help himself. There is nobody else to help him.

If we take a close look at project management, we will see that the project manager actually works for the line managers, not vice versa. Many executives do not realize this. They have a tendency to put a halo around the head of the project manager and give him a bonus at project termination, when, in fact, the credit should go to the line managers, who are continually pressured to make better use of their resources. The project manager is simply the agent through whom this is accomplished. So why do some companies glorify the project management position?

To illustrate the role of the project manager, consider the time, cost, and performance constraints shown in Figure 1–2. Many functional managers, if left alone, would recognize only the performance constraint: "Just give me another $50,000 and two more months, and I'll give you the ideal technology."

The project manager, as part of these communicating, coordinating, and integrating responsibilities, reminds the line managers that there are also time and cost constraints on the project. This is the starting point for better resource control.

Project managers depend on line managers. When the project manager gets in trouble, the only place he can go is to the line manager because additional resources are almost always required to alleviate the problems. When a line manager gets in trouble, he usually goes first to the project manager and requests either additional funding or some type of authorization for scope changes.

To illustrate this working relationship between the project and line managers, consider the following situation:

Project Manager (addressing the line manager): "I have a serious problem. I'm looking at a $150,000 cost overrun on my project and I need your help. I'd like you to do the same amount of work that you are currently scheduled for but in 3,000 fewer man-hours. Since your organization is burdened at $60/hour, this would more than compensate for the cost overrun."

Line Manager: "Even if I could, why should I? You know that good line managers can always make work expand to meet budget. I'll look over my manpower curves and let you know tomorrow."

The following day . . .

Line Manager: "I've looked over my manpower curves and I have enough work to keep my people employed. I'll give you back the 3,000 hours you need, but remember, *you owe me one!*"

Several months later . . .

Line Manager: "I've just seen the planning for your new project that's supposed to start two months from now. You'll need two people from my department. There are two employees that I'd like to use on your project. Unfortunately, these two people are available now. If I don't pick these people up on your charge number right now, some other project might pick them up in the interim period, and they won't be available when your project starts."

Project Manager: "What you're saying is that you want me to let you sandbag against one of my charge numbers, knowing that I really don't need them."

Line Manager: "That's right. I'll try to find other jobs (and charge numbers) for them to work on temporarily so that your project won't be completely burdened. Remember, you owe me one."

Project Manager: "O.K. I know that I owe you one, so I'll do this for you. Does this make us even?"

Line Manager: "Not at all! But you're going in the right direction."

When the project management–line management relationship begins to deteriorate, the project almost always suffers. Executives must promote a good working relationship between line and project management. One of the most common ways of destroying this relationship is by asking, "Who contributes to profits—the line or project manager?" Project managers feel that they control all project profits because they control the budget.

The line managers, on the other hand, argue that they must staff with appropriately budgeted-for personnel, supply the resources at the desired time, and supervise performance. Actually, both the vertical and horizontal lines contribute to profits. These types of conflicts can destroy the entire project management system.

The previous examples should indicate that project management is more behavioral than quantitative. Effective project management requires an understanding of:

- Quantitative tools and techniques
- Organizational structures
- Organizational behavior

Most people understand the quantitative tools for planning, scheduling, and controlling work. It is imperative that project managers understand totally the operations of each line organization. In addition, project managers must understand their own job description, especially where their authority begins and ends. During an in-house seminar on engineering project management, the author asked one of the project engineers to provide a description of his job as a project engineer. During the discussion that followed, several proj-ect managers and line managers said that there was a great deal of overlap between their job descriptions and that of the project engineer.

Organizational behavior is important because the functional employees at the interface position find themselves reporting to more than one boss—a line manager and one project manager for each project they are assigned to. Executives must provide proper training so functional employees can report effectively to multiple managers.

1.4 DEFINING THE PROJECT MANAGER'S ROLE _____

PMBOK® Guide, 4th Edition
2.3 Stakeholders
2.3.8 Functional Managers
2.3.6 Project Managers
Chapter 4 Project
   Integration Management

The project manager is responsible for coordinating and integrating activities across multiple, functional lines. The integration activities performed by the project manager include:

- Integrating the activities necessary to develop a project plan
- Integrating the activities necessary to execute the plan
- Integrating the activities necessary to make changes to the plan

These integrative responsibilities are shown in Figure 1–3 where the project manager must convert the inputs (i.e., resources) into outputs of products, services, and ultimately profits. In order to do this, the project manager needs strong communicative and interpersonal skills, must become familiar with the operations of each line organization, and must have knowledge of the technology being used.

An executive with a computer manufacturer stated that his company was looking externally for project managers. When asked if he expected candidates to have a command of computer technology, the executive remarked: "You give me an individual who has

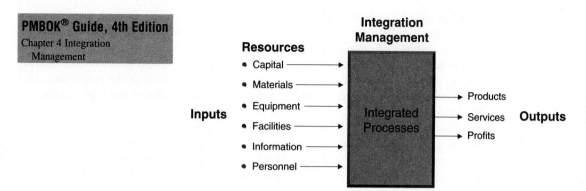

FIGURE 1–3. Integration management.

good communicative skills and interpersonal skills, and I'll give that individual a job. I can teach people the technology and give them technical experts to assist them in decision making. But I cannot teach somebody how to work with people."

The project manager's job is not an easy one. Project managers may have increasing responsibility, but very little authority. This lack of authority can force them to "negotiate" with upper-level management as well as functional management for control of company resources. They may often be treated as outsiders by the formal organization.

In the project environment, everything seems to revolve about the project manager. Although the project organization is a specialized, task-oriented entity, it cannot exist apart from the traditional structure of the organization. The project manager, therefore, must walk the fence between the two organizations. The term *interface management* is often used for this role, which can be described as managing relationships:

- Within the project team
- Between the project team and the functional organizations
- Between the project team and senior management
- Between the project team and the customer's organization, whether an internal or external organization

To be effective as a project manager, an individual must have management as well as technical skills. Because engineers often consider their careers limited in the functional disciplines, they look toward project management and project engineering as career path opportunities. But becoming a manager entails learning about psychology, human behavior, organizational behavior, interpersonal relations, and communications. MBA programs have come to the rescue of individuals desiring the background to be effective project managers.

In the past, executives motivated and retained qualified personnel primarily with financial incentives. Today other ways are being used, such as a change in title or the promise of more challenging work. Perhaps the lowest turnover rates of any professions in the world are in project management and project engineering. In a project environment, the project managers and project engineers get to see their project through from "birth to death." Being able to see the fruits of one's efforts is highly rewarding. A senior project manager in a construction company commented on why he never accepted a vice presidency that had been offered to him: "I can take my children and grandchildren into ten countries in the world and show them facilities that I have built as the project manager. What do I show my kids as an executive? The size of my office? My bank account? A stockholder's report?"

The project manager is actually a general manager and gets to know the total operation of the company. In fact, project managers get to know more about the total operation of a company than most executives. That is why project management is often used as a training ground to prepare future general managers who will be capable of filling top management positions.

1.5 DEFINING THE FUNCTIONAL MANAGER'S ROLE

PMBOK® Guide, 4th Edition
Chapter 9 Human Resources
  Management
9.1.2 HR Planning: Tools and
  Techniques

Assuming that the project and functional managers are not the same person, we can identify a specific role for the functional manager. There are three elements to this role:

- The functional manager has the responsibility to define *how* the task will be done and *where* the task will be done (i.e., the technical criteria).
- The functional manager has the responsibility to provide sufficient resources to accomplish the objective within the project's constraints (i.e., *who* will get the job done).
- The functional manager has the responsibility for the deliverable.

In other words, once the project manager identifies the requirements for the project (i.e., what work has to be done and the constraints), it becomes the line manager's responsibility to identify the technical criteria. Except perhaps in R&D efforts, the line manager should be the recognized technical expert. If the line manager believes that certain technical portions of the project manager's requirements are unsound, then the line manager has the right, by virtue of his expertise, to take exception and plead his case to a higher authority.

In Section 1.1 we stated that all resources (including personnel) are controlled by the line manager. The project manager has the right to request specific staff, but the final appointments rest with line managers. It helps if project managers understand the line manager's problems:

- Unlimited work requests (especially during competitive bidding)
- Predetermined deadlines
- All requests having a high priority
- Limited number of resources
- Limited availability of resources

- Unscheduled changes in the project plan
- Unpredicted lack of progress
- Unplanned absence of resources
- Unplanned breakdown of resources
- Unplanned loss of resources
- Unplanned turnover of personnel

Only in a very few industries will the line manager be able to identify to the project manager in advance exactly what resources will be available when the project is scheduled to begin. It is not important for the project manager to have the best available resources. Functional managers should not commit to certain people's availability. Rather, the functional manager should commit to achieving his portion of the objective within time, cost, and performance even if he has to use average or below-average personnel. If the project manager is unhappy with the assigned functional resources, then the project manager should closely track that portion of the project. Only if and when the project manager is convinced by the evidence that the assigned resources are unacceptable should he confront the line manager and demand better resources.

The fact that a project manager is assigned does not relieve the line manager of his functional responsibility to perform. If a functional manager assigns resources such that the constraints are not met, then *both* the project and functional managers will be blamed. One company is even considering evaluating line managers for merit increases and promotion based on how often they have lived up to their commitments to the project managers. Therefore, it is extremely valuable to everyone concerned to have all project commitments *made visible to all.*

Some companies carry the concept of commitments to extremes. An aircraft components manufacturer has a Commitment Department headed by a second-level manager. The function of the Commitment Department is to track how well the line managers keep their promises to the project managers. The department manager reports directly to the vice president of the division. In this company, line managers are extremely careful and cautious in making commitments, but do everything possible to meet deliverables. This same company has gone so far as to tell both project and line personnel that they run the risk of being discharged from the company for burying a problem rather than bringing the problem to the surface *immediately*.

In one automotive company, the tension between the project and line managers became so combative that it was having a serious impact on the performance and constraints of the project. The project managers argued that the line managers were not fulfilling their promises whereas the line managers were arguing that the project managers' requirements were poorly defined. To alleviate the problem, a new form was created which served as a contractual agreement between the project and the line managers who had to commit to the deliverables. This resulted in "shared accountability" for the project's deliverables.

Project management is designed to have shared authority and responsibility between the project and line managers. Project managers plan, monitor, and control the project, whereas functional managers perform the work. Table 1–1 shows this shared responsibility. The one exception to Table 1–1 occurs when the project and line managers are the same person. This situation, which happens more often than not, creates a conflict of interest.

16   OVERVIEW

PMBOK® Guide, 4th Edition
2.4.2 Organizational Structure
Figure 2–7

TABLE 1–1. DUAL RESPONSIBILITY

	Responsibility	
Topic	Project Manager	Line Manager
Rewards	Give recommendation: Informal	Provide rewards: Formal
Direction	Milestone (summary)	Detailed
Evaluation	Summary	Detailed
Measurement	Summary	Detailed
Control	Summary	Detailed

If a line manager has to assign resources to six projects, one of which is under his direct control, he might save the best resources for his project. In this case, his project will be a success at the expense of all of the other projects.

The exact relationship between project and line managers is of paramount importance in project management where multiple-boss reporting prevails. Table 1–2 shows that the relationship between project and line managers is not always in balance and thus, of course, has a bearing on who exerts more influence over the assigned functional employees.

PMBOK® Guide, 4th Edition
2.4.2 Organizational Structure
Figure 2–7

TABLE 1–2. REPORTING RELATIONSHIPS

		Project Manager (PM)/Line Manager (LM)/Employee Relationship			
Type of Project Manager	Type of Matrix Structure*	PM Negotiates For	Employees Take Technical Direction From	PM Receives Functional Progress From	Employee Performance Evaluations Made By
Lightweight	Weak	Deliverables	LMs	Primarily LMs	LMs only with no input from PM
Heavyweight	Strong	People who report informally to PM but formally to LMs	PM and LMs	Assigned employees who report to LMs	LMs with input from PM
Tiger teams	Very strong	People who report entirely to PM full-time for duration of project	PM only	Assigned employees who now report directly to PM	PM only

*The types of organizational structures are discussed in Chapter 3.

1.6 DEFINING THE FUNCTIONAL EMPLOYEE'S ROLE

Once the line managers commit to the deliverables, it is the responsibility of the assigned functional employees to achieve the functional deliverables. For years the functional employees were called subordinates. Although this term still exists in textbooks, industry prefers to regard the assigned employees as "associates" rather than subordinates. The reason for this is that in project management the associates can be a higher pay grade than the project manager. The associates can even be a higher pay grade than their functional manager.

In most organizations, the assigned employees report on a "solid" line to their functional manager, even though they may be working on several projects simultaneously. The employees are usually a "dotted" line to the project but solid to their function. This places the employees in the often awkward position of reporting to multiple individuals. This situation is further complicated when the project manager has more technical knowledge than the line manager. This occurs during R&D projects.

The functional employee is expected to accomplish the following activities when assigned to projects:

- Accept responsibility for accomplishing the assigned deliverables within the project's constraints
- Complete the work at the earliest possible time
- Periodically inform both the project and line manager of the project's status
- Bring problems to the surface quickly for resolution
- Share information with the rest of the project team

1.7 DEFINING THE EXECUTIVE'S ROLE

In a project environment there are new expectations of and for the executives, as well as a new interfacing role.[5] Executives are expected to interface a project as follows:

- In project planning and objective-setting
- In conflict resolution
- In priority-setting
- As project sponsor[6]

Executives are expected to interface with projects very closely at project initiation and planning, but to remain at a distance during execution unless needed for priority-setting and conflict resolution. One reason why executives "meddle" during project execution is that they are not getting accurate information from the project manager as to project status. If project managers provide executives with meaningful status reports, then the so-called meddling may be reduced or even eliminated.

---

5. The expectations are discussed in Section 9.3.
6. The role of the project sponsor is discussed in Section 10.1.

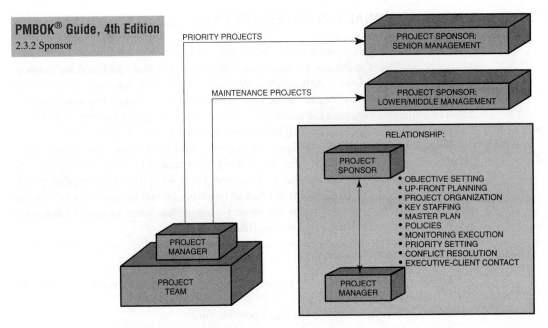

FIGURE 1–4. The project sponsor interface.

1.8 WORKING WITH EXECUTIVES

Success in project management is like a three-legged stool. The first leg is the project manager, the second leg is the line manager, and the third leg is senior management. If any of the three legs fail, then even delicate balancing may not prevent the stool from toppling.

The critical node in project management is the project manager–line manager interface. At this interface, the project and line managers must view each other as equals and be willing to share authority, responsibility, and accountability. In excellently managed companies, project managers do not negotiate for resources but simply ask for the line manager's commitment to executing his portion of the work within time, cost, and performance. Therefore, in excellent companies, it should not matter who the line manager assigns as long as the line manager lives up to his commitments.

Since the project and line managers are "equals," senior management involvement is necessary to provide advice and guidance to the project manager, as well as to provide encouragement to the line managers to keep their promises. When executives act in this capacity, they assume the role of project sponsors, as shown in Figure 1–4,[7] which also shows that sponsorship need not always be at the executive levels. The exact person appointed as the project sponsor is based on the dollar value of the project, the priority of the project, and who the customer is.

7. Section 10.1 describes the role of the project sponsor in more depth.

The ultimate objective of the project sponsor is to provide behind-the-scenes assistance to project personnel for projects both "internal" to the company, as well as "external," as shown in Figure 1–4. Projects can still be successful without this commitment and support, as long as all work flows smoothly. But in time of crisis, having a "big brother" available as a possible sounding board will surely help.

When an executive is required to act as a project sponsor, then the executive has the responsibility to make effective and timely project decisions. To accomplish this, the executive needs timely, accurate, and complete data for such decisions. Keeping management informed serves this purpose, while the all-too-common practice of "stonewalling" prevents an executive from making effective project decisions.

It is not necessary for project sponsorship to remain exclusively at the executive levels. As companies mature in their understanding and implementation of project management, project sponsorship may be pushed down to middle-level management. Committee sponsorship is also possible.

1.9 THE PROJECT MANAGER AS THE PLANNING AGENT

PMBOK® Guide, 4th Edition
Chapter 9 Project Human
Resources Management

The major responsibility of the project manager is planning. If project planning is performed correctly, then it is conceivable that the project manager will work himself out of a job because the project can run itself. This rarely happens, however. Few projects are ever completed without some conflict or trade-offs for the project manager to resolve.

In most cases, the project manager provides overall or summary definitions of the work to be accomplished, but the line managers (the true experts) do the detailed planning. Although project managers cannot control or assign line resources, they must make sure that the resources are adequate and scheduled to satisfy the needs of the project, not vice versa. As the architect of the project plan, the project manager must provide:

- Complete task definitions
- Resource requirement definitions (possibly skill levels)
- Major timetable milestones
- Definition of end-item quality and reliability requirements
- The basis for performance measurement

These factors, if properly established, result in:

- Assurance that functional units will understand their total responsibilities toward achieving project needs.
- Assurance that problems resulting from scheduling and allocation of critical resources are known beforehand.
- Early identification of problems that may jeopardize successful project completion so that effective corrective action and replanning can be taken to prevent or resolve the problems.

Project managers are responsible for project administration and, therefore, must have the right to establish their own policies, procedures, rules, guidelines, and directives—provided these policies, guidelines, and so on, conform to overall company policy. Companies with mature project management structures usually have rather loose company guidelines, so project managers have some degree of flexibility in how to control their projects. However, project managers cannot make any promises to a functional employee concerning:

- Promotion
- Grade
- Salary
- Bonus
- Overtime
- Responsibility
- Future work assignments

These seven items can be administered by line managers only, but the project manager can have indirect involvement by telling the line manager how well an employee is doing (and putting it in writing), requesting overtime because the project budget will permit it, and offering individuals the opportunity to perform work above their current pay grade. However, such work above pay grade can cause severe managerial headaches if not coordinated with the line manager, because the individual will expect immediate rewards if he performs well.

Establishing project administrative requirements is part of project planning. Executives must either work with the project managers at project initiation or act as resources later. Improper project administrative planning can create a situation that requires:

- A continuous revision and/or establishment of company and/or project policies, procedures, and directives
- A continuous shifting in organizational responsibility and possible unnecessary restructuring
- A need for staff to acquire new knowledge and skills

If these situations occur simultaneously on several projects, there can be confusion throughout the organization.

1.10 PROJECT CHAMPIONS

Corporations encourage employees to think up new ideas that, if approved by the corporation, will generate monetary and nonmonetary rewards for the idea generator. One such reward is naming the individual the "project champion." Unfortunately, the project champion often becomes the project manager, and, although the idea was technically sound, the project fails.

TABLE 1–3. PROJECT MANAGERS VERSUS PROJECT CHAMPIONS

Project Managers	Project Champions
• Prefer to work in groups	• Prefer working individually
• Committed to their managerial and technical responsibilities	• Committed to technology
• Committed to the corporation	• Committed to the profession
• Seek to achieve the objective	• Seek to exceed the objective
• Are willing to take risks	• Are unwilling to take risks; try to test everything
• Seek what is possible	• Seek perfection
• Think in terms of short time spans	• Think in terms of long time spans
• Manage people	• Manage things
• Are committed to and pursue material values	• Are committed to and pursue intellectual values

Table 1–3 provides a comparison between project managers and project champions. It shows that the project champions may become so attached to the technical side of the project that they become derelict in their administrative responsibilities. Perhaps the project champion might function best as a project engineer rather than the project manager.

This comparison does not mean that technically oriented project managers-champions will fail. Rather, it implies that the selection of the "proper" project manager should be based on *all* facets of the project.

1.11 THE DOWNSIDE OF PROJECT MANAGEMENT

Project management is often recognized only as a high-salaried, highly challenging position whereby the project manager receives excellent training in general management.

For projects that are done for external sources, the project manager is first viewed as starting out with a pot of gold and then as having to manage the project so that sufficient profits will be made for the stockholders. If the project manager performs well, the project will be successful. But the personal cost may be high for the project manager.

There are severe risks that are not always evident. Some project management positions may require a sixty-hour workweek and extensive time away from home. When a project manager begins to fall in love more with the job than with his family, the result is usually lack of friends, a poor home life, and possibly divorce. During the birth of the missile and space programs, companies estimated that the divorce rate among project managers and project engineers was probably twice the national average. Accepting a project management assignment is not always compatible with raising a young family. Characteristics of the workaholic project manager include:

- Every Friday he thinks that there are only two more working days until Monday.
- At 5:00 P.M. he considers the working day only half over.
- He has no time to rest or relax.
- He always takes work home from the office.
- He takes work with him on vacations.

1.12 PROJECT-DRIVEN VERSUS NON–PROJECT-DRIVEN ORGANIZATIONS

PMBOK® Guide, 4th Edition
2.4.2 Organizational Systems
2.2 Project-Based and
     Non–Project-Based

On the micro level, virtually all organizations are either marketing-, engineering-, or manufacturing-driven. But on the macro level, organizations are either project- or non–project-driven. The PMBOK® Guide uses the terms *project-based* and *non–project-based,* whereas in this text the terms *project-driven* and *non–project-driven* or *operational-driven* are used. In a project-driven organization, such as construction or aerospace, all work is characterized through projects, with each proj-ect as a separate cost center having its own profit-and-loss statement. The total profit to the corporation is simply the summation of the profits on all projects. In a project-driven organization, everything centers around the projects.

In the non–project-driven organization, such as low-technology manufacturing, profit and loss are measured on vertical or functional lines. In this type of organization, projects exist merely to support the product lines or functional lines. Priority resources are assigned to the revenue-producing functional line activities rather than the projects.

Project management in a non–project-driven organization is generally more difficult for these reasons:

- Projects may be few and far between.
- Not all projects have the same project management requirements, and therefore they cannot be managed identically. This difficulty results from poor understanding of project management and a reluctance of companies to invest in proper training.
- Executives do not have sufficient time to manage projects themselves, yet refuse to delegate authority.
- Projects tend to be delayed because approvals most often follow the vertical chain of command. As a result, project work stays too long in functional departments.
- Because project staffing is on a "local" basis, only a portion of the organization understands project management and sees the system in action.
- There is heavy dependence on subcontractors and outside agencies for project management expertise.

Non–project-driven organizations may also have a steady stream of projects, all of which are usually designed to enhance manufacturing operations. Some projects may be customer-requested, such as:

- The introduction of statistical dimensioning concepts to improve process control
- The introduction of process changes to enhance the final product
- The introduction of process change concepts to enhance product reliability

If these changes are not identified as specific projects, the result can be:

- Poorly defined responsibility areas within the organization
- Poor communications, both internal and external to the organization

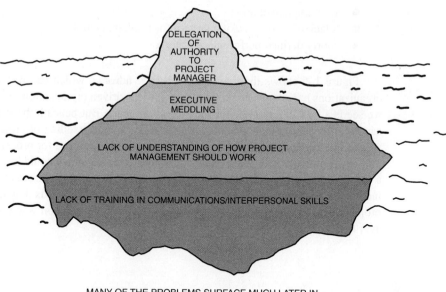

MANY OF THE PROBLEMS SURFACE MUCH LATER IN
THE PROJECT AND RESULT IN A MUCH HIGHER COST
TO CORRECT AS WELL AS INCREASE PROJECT RISK

FIGURE 1–5. The tip-of-the-iceberg syndrome for matrix implementation.

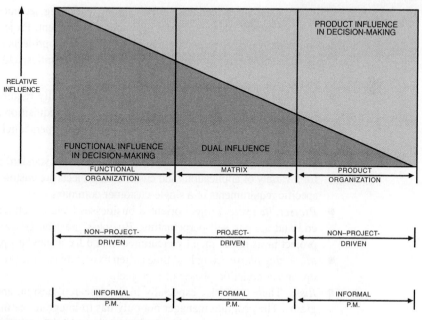

FIGURE 1–6. Decision-making influence.

- Slow implementation
- A lack of a cost-tracking system for implementation
- Poorly defined performance criteria

Figure 1–5 shows the tip-of-the-iceberg syndrome, which can occur in all types of organizations but is most common in non–project-driven organizations. On the surface, all we see is a lack of authority for the project manager. But beneath the surface we see the causes; there is excessive meddling due to lack of understanding of project management, which, in turn, resulted from an inability to recognize the need for proper training.

In the previous sections we stated that project management could be handled on either a formal or an informal basis. As can be seen from Figure 1–6, informal project management most often appears in non–project-driven organizations. It is doubtful that informal project management would work in a project-driven organization where the project manager has profit-and-loss responsibility.

1.13 MARKETING IN THE PROJECT-DRIVEN ORGANIZATION

PMBOK® Guide, 4th Edition
1.4.3 Projects and Strategic
Planning

Getting new projects is the lifeblood of any project-oriented business. The practices of the project-oriented company are, however, substantially different from traditional product businesses and require highly specialized and disciplined team efforts among marketing, technical, and operating personnel, plus significant customer involvement. Projects are different from products in many respects, especially marketing. Marketing projects requires the ability to identify, pursue, and capture one-of-a-kind business opportunities, and is characterized by:

- *A systematic effort.* A systematic approach is usually required to develop a new program lead into an actual contract. The project acquisition effort is often highly integrated with ongoing programs and involves key personnel from both the potential customer and the performing organization.
- *Custom design.* While traditional businesses provide standard products and services for a variety of applications and customers, projects are custom-designed items to fit specific requirements of a single-customer community.
- *Project life cycle.* Project-oriented businesses have a well-defined beginning and end and are not self-perpetuating. Business must be generated on a project-by-project basis rather than by creating demand for a standard product or service.
- *Marketing phase.* Long lead times often exist between the product definition, start-up, and completion phases of a project.
- *Risks.* There are risks, especially in the research, design, and production of programs. The program manager not only has to integrate the multidisciplinary tasks and project elements within budget and schedule constraints, but also has to

manage inventions and technology while working with a variety of technically oriented prima donnas.
- *The technical capability to perform.* Technical ability is critical to the successful pursuit and acquisition of a new project.

In spite of the risks and problems, profits on projects are usually very low in comparison with commerical business practices. One may wonder why companies pursue project businesses. Clearly, there are many reasons why projects are good business:

- Although immediate profits (as a percentage of sales) are usually small, the return on capital investment is often very attractive. Progress payment practices keep inventories and receivables to a minimum and enable companies to undertake projects many times larger in value than the assets of the total company.
- Once a contract has been secured and is being managed properly, the project may be of relatively low financial risk to the company. The company has little additional selling expenditure and has a predictable market over the life cycle of the project.
- Project business must be viewed from a broader perspective than motivation for immediate profits. Projects provide an opportunity to develop the company's technical capabilities and build an experience base for future business growth.
- Winning one large project often provides attractive growth potential, such as (1) growth with the project via additions and changes; (2) follow-on work; (3) spare parts, maintenance, and training; and (4) being able to compete effectively in the next project phase, such as nurturing a study program into a development contract and finally a production contract.

Customers come in various forms and sizes. For small and medium businesses particularly, it is a challenge to compete for contracts from large industrial or governmental organizations. Although the contract to a firm may be relatively small, it is often subcontracted via a larger organization. Selling to such a diversified heterogeneous customer is a marketing challenge that requires a highly sophisticated and disciplined approach.

The first step in a new business development effort is to define the market to be pursued. The market segment for a new program opportunity is normally in an area of relevant past experience, technical capability, and customer involvement. Good marketers in the program business have to think as product line managers. They have to understand all dimensions of the business and be able to define and pursue market objectives that are consistent with the capabilities of their organizations.

Program businesses operate in an opportunity-driven market. It is a common mistake, however, to believe that these markets are unpredictable and unmanageable. Market planning and strategizing is important. New project opportunities develop over periods of time, sometimes years for larger projects. These developments must be properly tracked and cultivated to form the bases for management actions such as (1) bid decisions, (2) resource commitment, (3) technical readiness, and (4) effective customer liaison. This strategy of winning new business is supported by systematic, disciplined approaches, which are illustrated in Figure 1–7.

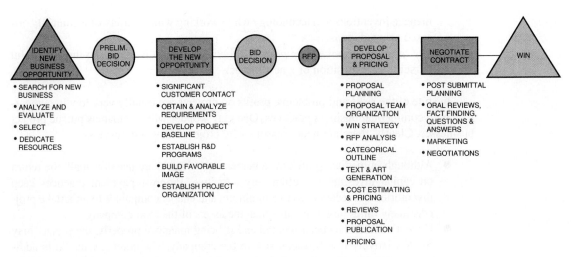

FIGURE 1–7. The phases of winning new contracts in project-oriented businesses.

1.14 CLASSIFICATION OF PROJECTS

The principles of project management can be applied to any type of project and to any industry. However, the relative degree of importance of these principles can vary from project to project and industry to industry. Table 1–4 shows a brief comparison of certain industries/projects.

For those industries that are project-driven, such as aerospace and large construction, the high dollar value of the projects mandates a much more rigorous project management approach. For non–project-driven industries, projects may be managed more informally than formally, especially if no immediate profit is involved. Informal project management is similar to formal project management but paperwork requirements are kept at a minimum.

TABLE 1–4.   CLASSIFICATION OF PROJECTS/CHARACTERISTICS

	Type of Project/Industry					
	In-house R&D	Small Construction	Large Construction	Aerospace/ Defense	MIS	Engineering
Need for interpersonal skills	Low	Low	High	High	High	Low
Importance of organizational structure	Low	Low	Low	Low	High	Low
Time management difficulties	Low	Low	High	High	High	Low
Number of meetings	Excessive	Low	Excessive	Excessive	High	Medium
Project manager's supervisor	Middle management	Top management	Top management	Top management	Middle management	Middle management
Project sponsor present	Yes	No	Yes	Yes	No	No
Conflict intensity	Low	Low	High	High	High	Low
Cost control level	Low	Low	High	High	Low	Low
Level of planning/scheduling	Milestones only	Milestones only	Detailed plan	Detailed plan	Milestones only	Milestones only

1.15 LOCATION OF THE PROJECT MANAGER

The success of project management could easily depend on the location of the project manager within the organization. Two questions must be answered:

- What salary should the project manager earn?
- To whom should the project manager report?

Figure 1–8 shows a typical organizational hierarchy (the numbers represent pay grades). Ideally, the project manager should be at the same pay grade as the individuals with whom he must negotiate on a daily basis. Using this criterion, and assuming that the project manager interfaces at the department manager level, the project manager should earn a salary between grades 20 and 25. A project manager earning substantially more or less money than the line manager will usually create conflict. The ultimate reporting location of the project manager (and perhaps his salary) is heavily dependent on whether the

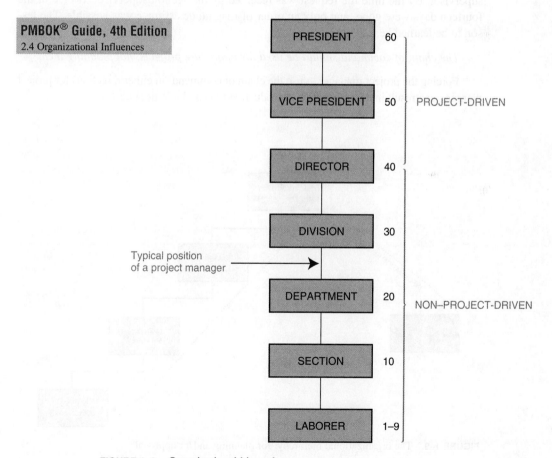

PMBOK® Guide, 4th Edition
2.4 Organizational Influences

FIGURE 1–8. Organizational hierarchy.

organization is project- or non–project-driven, and whether the project manager is responsible for profit or loss.

Project managers can end up reporting both high and low in an organization during the life cycle of the project. During the planning phase of the project, the project manager may report high, whereas during implementation, he may report low. Likewise, the positioning of the project manager may be dependent on the risk of the project, the size of the project, or the customer.

Finally, it should be noted that even if the project manager reports low, he should still have the right to interface with top executives during project planning although there may be two or more reporting levels between the project manager and executives. At the opposite end of the spectrum, the project manager should have the right to go directly into the depths of the organization instead of having to follow the chain of command downward, especially during planning. As an example, see Figure 1–9. The project manager had two weeks to plan and price out a small project. Most of the work was to be accomplished within one section. The project manager was told that all requests for work, even estimating, had to follow the chain of command from the executive down through the section supervisor. By the time the request was received by the section supervisor, twelve of the fourteen days were gone, and only an order-of-magnitude estimate was possible. The lesson to be learned here is:

The chain of command should be used for approving projects, not planning them.

Forcing the project manager to use the chain of command (in either direction) for project planning can result in a great deal of unproductive time and idle time cost.

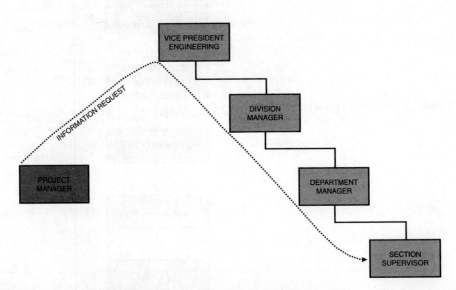

FIGURE 1–9.   The organizational hierarchy: for planning and/or approval?

1.16 DIFFERING VIEWS OF PROJECT MANAGEMENT

Many companies, especially those with project-driven organizations, have differing views of project management. Some people view project management as an excellent means to achieving objectives, while others view it as a threat. In project-driven organizations, there are three career paths that lead to executive management:

- Through project management
- Through project engineering
- Through line management

In project-driven organizations, the fast-track position is in project management, whereas in a non–project-driven organization, it would be line management. Even though line managers support the project management approach, they resent the project manager because of his promotions and top-level visibility. In one construction company, a department manager was told that he had no chance for promotion above his present department manager position unless he went into project management or project engineering where he could get to know the operation of the whole company. A second construction company requires that individuals aspiring to become a department manager first spend a "tour of duty" as an assistant project manager or project engineer.

Executives may dislike project managers because more authority and control must be delegated. However, once executives realize that it is a sound business practice, it becomes important, as shown in the following letter[8]:

> In order to sense and react quickly and to insure rapid decision-making, lines of communication should be the shortest possible between all levels of the organization. People with the most knowledge must be available at the source of the problem, and they must have decision-making authority and responsibility. Meaningful data must be available on a timely basis and the organization must be structured to produce this environment.
>
> In the aerospace industry, it is a serious weakness to be tied to fixed organization charts, plans, and procedures. With regard to organization, we successfully married the project concept of management with a central function concept. What we came up with is an organization within an organization—one to ramrod the day-to-day problems; the other to provide support for existing projects and to anticipate the requirements for future projects.
>
> The project system is essential in getting complicated jobs done well and on time, but it solves only part of the management problem. When you have your nose to the project grindstone, you are often not in a position to see much beyond that project. This is where the central functional organization comes in. My experience has been that you need this central organization to give you depth, flexibility, and perspective. Together, the two parts permit you to see both the woods and the trees.
>
> Initiative is essential at all levels of the organization. We try to press the level of decision to the lowest possible rung of the managerial ladder. This type of decision-making provides motivation and permits recognition for the individual and the group at all levels. It stimulates action and breeds dedication.

---

8. Letter from J. Donald Rath, Vice President of Martin-Marietta Corporation, Denver Division, to J. E. Webb, of NASA, October 18, 1963.

With this kind of encouragement, the organization can become a live thing—sensitive to problems and able to move in on them with much more speed and understanding than would be normally expected in a large operation. In this way, we can regroup or reorganize easily as situations dictate and can quickly focus on a "crisis." In this industry a company must always be able to reorient itself to meet new objectives. In a more staid, old-line organization, frequent reorientation usually accompanied by a corresponding shift of people's activities, could be most upsetting. However, in the aerospace industry, we must be prepared for change. The entire picture is one of change.

1.17 CONCURRENT ENGINEERING: A PROJECT MANAGEMENT APPROACH

In the past decade, organizations have become more aware of the fact that America's most formidable weapon is its manufacturing ability, and yet more and more work seems to be departing for Southeast Asia and the Far East. If America and other countries are to remain competitive, then survival may depend on the manufacturing of a quality product and a rapid introduction into the marketplace. Today, companies are under tremendous pressure to rapidly introduce new products because product life cycles are becoming shorter. As a result, organizations no longer have the luxury of performing work in series.

Concurrent or simultaneous engineering is an attempt to accomplish work in parallel rather than in series. This requires that marketing, R&D, engineering, and production are all actively involved in the early project phases and making plans even before the product design has been finalized. This concept of current engineering will accelerate product development, but it does come with serious and potentially costly risks, the largest one being the cost of rework.

Almost everyone agrees that the best way to reduce or minimize risks is for the organization to plan better. Since project management is one of the best methodologies to foster better planning, it is little wonder that more organizations are accepting project management as a way of life.

1.18 STUDYING TIPS FOR THE PMI® PROJECT MANAGEMENT CERTIFICATION EXAM

This section is applicable as a review of the principles or to support an understanding of the knowledge areas and domain groups in the PMBOK® Guide. This chapter addresses some material from the PMBOK® Guide knowledge areas:

- Integration Management
- Scope Management
- Human Resources Management

Understanding the following principles is beneficial if the reader is using this textbook together with the PMBOK® Guide to study for the PMP® Certification Exam:

- Definition of a project
- Definition of the triple constraint
- Definition of successful execution of a project
- Benefits of using project management
- Responsibility of the project manager in dealing with stakeholders and how stakeholders can affect the outcome of the project
- Responsibility of the project manager in meeting deliverables
- The fact that the project manager is ultimately accountable for the success of the project
- Responsibilities of the line manager during project management staffing and execution
- Role of the executive sponsor and champion
- Difference between a project-driven and non–project-driven organization

Be sure to review the appropriate sections of the PMBOK® Guide and the glossary of terms at the end of the PMBOK® Guide.

Some multiple-choice questions are provided in this section as a review of the material. There are other sources for practice review questions that are specific for the PMP® Exam, namely:

- *Project Management IQ®* from the International Institute for Learning (iil.com)
- *PMP® Exam Practice Test and Study Guide,* by J. LeRoy Ward, PMP, editor
- *PMP® Exam Prep,* by Rita Mulcahy
- *Q & As for the PMBOK® Guide,* Project Management Institute

The more practice questions reviewed, the better prepared the reader will be for the PMP® Certification Exam.

In Appendix C, there are a series of mini–case studies called Dorale Products that reviews some of the concepts. The minicases can be used as either an introduction to the chapter or as a review of the chapter material. These mini–case studies were placed in Appendix C because they can be used for several chapters in the text. For this chapter, the following are applicable:

- Dorale Products (A) [Integration and Scope Management]
- Dorale Products (B) [Integration and Scope Management]

Answers to the Dorale Products minicases appear in Appendix D.

The following multiple-choice questions will be helpful in reviewing the above principles:

1. The triple constraints on a project are:
   A. Time, cost, and profitability
   B. Resources required, sponsorship involvement, and funding
   C. Time, cost, and quality and/or scope
   D. Calendar dates, facilities available, and funding

2. Which of the following is not part of the definition of a project?
   A. Repetitive activities
   B. Constraints
   C. Consumption of resources
   D. A well-defined objective

3. Which of the following is usually not part of the criteria for project success?
   A. Customer satisfaction
   B. Customer acceptance
   C. Meeting at least 75 percent of specification requirements.
   D. Meeting the triple-constraint requirements

4. Which of the following is generally not a benefit achieved from using project management?
   A. Flexibility in the project's end date
   B. Improved risk management
   C. Improved estimating
   D. Tracking of projects

5. The person responsible for assigning the resources to a project is most often:
   A. The project manager
   B. The Human Resources Department
   C. The line manager
   D. The executive sponsor

6. Conflicts between the project and line managers are most often resolved by:
   A. The assistant project manager for conflicts
   B. The project sponsor
   C. The executive steering committee
   D. The Human Resources Department

7. Your company does only projects. If the projects performed by your company are for customers external to your company and a profit criterion exists on the project, then your organization is most likely:
   A. Project-driven
   B. Non–project-driven
   C. A hybrid
   D. All of the above are possible based upon the size of the profit margin.

ANSWERS

1. C
2. A
3. C
4. A
5. C
6. B
7. A

PROBLEMS

1–1 In the project environment, cause-and-effect relationships are almost always readily apparent. Good project management will examine the effect in order to better understand the cause and possibly prevent it from occurring again. Below are causes and effects. For each one of the effects, select the possible cause or causes that may have existed to create this situation:

Effects

1. Late completion of activities
2. Cost overruns
3. Substandard performance
4. High turnover in project staff
5. High turnover in functional staff
6. Two functional departments performing the same activities on one project

Causes

a. Top management not recognizing this activity as a project
b. Too many projects going on at one time
c. Impossible schedule commitments
d. No functional input into the planning phase
e. No one person responsible for the total project
f. Poor control of design changes
g. Poor control of customer changes
h. Poor understanding of the project manager's job
i. Wrong person assigned as project manager
j. No integrated planning and control
k. Company resources are overcommitted
l. Unrealistic planning and scheduling
m. No project cost accounting ability
n. Conflicting project priorities
o. Poorly organized project office

(This problem has been adapted from Russell D. Archibald, *Managing High-Technology Programs and Projects,* New York: John Wiley, 1976, p. 10.)

1–2 Because of the individuality of people, there always exist differing views of what management is all about. Below are lists of possible perspectives and a selected group of organizational members. For each individual select the possible ways that this individual might view project management:

Individuals

1. Upper-level manager
2. Project manager
3. Functional manager
4. Project team member
5. Scientist and consultant

Perspectives

a. A threat to established authority
b. A source for future general managers
c. A cause of unwanted change in ongoing procedures

      d. A means to an end
      e. A significant market for their services
      f. A place to build an empire
      g. A necessary evil to traditional management
      h. An opportunity for growth and advancement
      i. A better way to motivate people toward an objective
      j. A source of frustration in authority
      k. A way of introducing controlled changes
      l. An area of research
      m. A vehicle for introducing creativity
      n. A means of coordinating functional units
      o. A means of deep satisfaction
      p. A way of life

1–3 Consider an organization that is composed of upper-level managers, middle- and lower-level managers, and laborers. Which of the groups should have first insight that an organizational restructuring toward project management may be necessary?

1–4 How would you defend the statement that a project manager must help himself?

1–5 Will project management work in all companies? If not, identify those companies in which project management may not be applicable and defend your answers.

1–6 In a project organization, do you think that there might be a conflict in opinions over whether the project managers or functional managers contribute to profits?

1–7 What attributes should a project manager have? Can an individual be trained to become a project manager? If a company were changing over to a project management structure, would it be better to promote and train from within or hire from the outside?

1–8 Do you think that functional managers would make good project managers?

1–9 What types of projects might be more appropriate for functional management rather than project management, and vice versa?

1–10 Do you think that there would be a shift in the relative degree of importance of the following terms in a project management environment as opposed to a traditional management environment?

      a. Time management
      b. Communications
      c. Motivation

1–11 Classical management has often been defined as a process in which the manager does not necessarily perform things for himself, but accomplishes objectives through others in a group situation. Does this definition also apply to project management?

1–12 Which of the following are basic characteristics of project management?

      a. Customer problem
      b. Responsibility identification
      c. Systems approach to decision-making
      d. Adaptation to a changing environment

e. Multidisciplinary activity in a finite time duration

f. Horizontal and vertical organizational relationships

1–13 Project managers are usually dedicated and committed to the project. Who should be "looking over the shoulder" of the project manager to make sure that the work and requests are also in the best interest of the company? Does your answer depend on the priority of the project?

1–14 Is project management designed to transfer power from the line managers to the project manager?

1–15 Explain how career paths and career growth can differ between project-driven and non–project-driven organizations. In each organization, is the career path fastest in project management, project engineering, or line management?

1–16 Explain how the following statement can have a bearing on who is ultimately selected as part of the project team:

"There comes a time in the life cycle of all projects when one must shoot the design engineers and begin production."

1–17 How do you handle a situation where the project manager has become a generalist, but still thinks that he is an expert?

CASE STUDY

WILLIAMS MACHINE TOOL COMPANY

For 85 years, the Williams Machine Tool Company had provided quality products to its clients, becoming the third largest U.S.-based machine tool company by 1990. The company was highly profitable and had an extremely low employee turnover rate. Pay and benefits were excellent.

Between 1980 and 1990, the company's profits soared to record levels. The company's success was due to one product line of standard manufacturing machine tools. Williams spent most of its time and effort looking for ways to improve its bread-and-butter product line rather than to develop new products. The product line was so successful that companies were willing to modify their production lines around these machine tools rather than asking Williams for major modifications to the machine tools.

By 1990, Williams Company was extremely complacent, expecting this phenomenal success with one product line to continue for 20 to 25 more years. The recession of the early 1990s forced management to realign their thinking. Cutbacks in production had decreased the demand for the standard machine tools. More and more customers were asking for either major modifications to the standard machine tools or a completely new product design.

The marketplace was changing and senior management recognized that a new strategic focus was necessary. However, lower-level management and the work force, especially engineering, were strongly resisting a change. The employees, many of them with over 20 years of employment at Williams Company, refused to recognize the need for this change in the belief that the glory days of yore would return at the end of the recession.

By 1995, the recession had been over for at least two years yet Williams Company had no new product lines. Revenue was down, sales for the standard product (with and without modifications) were decreasing, and the employees were still resisting change. Layoffs were imminent.

In 1996, the company was sold to Crock Engineering. Crock had an experienced machine tool division of its own and understood the machine tool business. Williams Company was allowed to operate as a separate entity from 1995 to 1996. By 1996, red ink had appeared on the Williams Company balance sheet. Crock replaced all of the Williams senior managers with its own personnel. Crock then announced to all employees that Williams would become a specialty machine tool manufacturer and that the "good old days" would never return. Customer demand for specialty products had increased threefold in just the last twelve months alone. Crock made it clear that employees who would not support this new direction would be replaced.

The new senior management at Williams Company recognized that 85 years of traditional management had come to an end for a company now committed to specialty products. The company culture was about to change, spearheaded by project management, concurrent engineering, and total quality management.

Senior management's commitment to product management was apparent by the time and money spent in educating the employees. Unfortunately, the seasoned 20-year-plus veterans still would not support the new culture. Recognizing the problems, management provided continuous and visible support for project management in addition to hiring a project management consultant to work with the people. The consultant worked with Williams from 1996 to 2001.

From 1996 to 2001, the Williams Division of Crock Engineering experienced losses in 24 consecutive quarters. The quarter ending March 31, 2002, was the first profitable quarter in over six years. Much of the credit was given to the performance and maturity of the project management system. In May 2002, the Williams Division was sold. More than 80% of the employees lost their jobs when the company was relocated over 1,500 miles away.

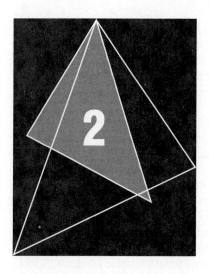

2

Project Management Growth: Concepts and Definitions

Related Case Studies (from Kerzner/*Project Management Case Studies*, 3rd Edition)	Related Workbook Exercises (from Kerzner/*Project Management Workbook and PMP®/CAPM® Exam Study Guide*, 10th Edition)	PMBOK® Guide, 4th Edition, Reference Section for the PMP® Certification Exam
• Goshe Corporation • MIS Project Management at First National Bank • Cordova Research Group • Cortez Plastics • L. P. Manning Corporation • Project Firecracker • Apache Metals, Inc. • Haller Specialty Manufacturing	• Multiple Choice Exam	• Integration Management • Scope Management

2.0 INTRODUCTION

PMBOK® Guide, 4th Edition
Chapter 4 Integration Management

The growth and acceptance of project management has changed significantly over the past forty years, and these changes are expected to continue well into the twenty-first century, especially in the area of multinational project management. It is interesting to trace the evolution and growth of project management from the early days of systems management to what some people call "modern project management."

The growth of project management can be traced through topics such as roles and responsibilities, organizational structures, delegation of authority and decision-making, and especially corporate profitability. Twenty years ago, companies had the choice of whether or not to accept the project management approach. Today, some companies foolishly think that they still have the choice. Nothing could be further from the truth. The survival of the firm may very well rest upon how well project management is implemented, and how quickly.

2.1 GENERAL SYSTEMS MANAGEMENT

Organizational theory and management philosophies have undergone a dramatic change in recent years with the emergence of the project management approach to management. Because project management is an outgrowth of systems management, it is only fitting that the underlying principles of general systems theory be described. Simply stated, general systems theory can be classified as a management approach that attempts to integrate and unify scientific information across many fields of knowledge. Systems theory attempts to solve problems by looking at the total picture, rather than through an analysis of the individual components.

General systems theory has been in existence for more than four decades. Unfortunately, as is often the case with new theory development, the practitioners require years of study and analysis before implementation. General systems theory is still being taught in graduate programs. Today, project management is viewed as applied systems management.

In 1951, Ludwig von Bertalanffy, a biologist, described so-called open systems using anatomy nomenclature. The body's muscles, skeleton, circulatory system, and so on, were all described as subsystems of the total system (the human being). Dr. von Bertalanffy's contribution was important in that he identified how specialists in each subsystem could be integrated so as to get a better understanding of the interrelationships, thereby contributing to the overall knowledge of the operations of the system. Thus, the foundation was laid for the evolution and outgrowth of project management.

In 1956, Kenneth Boulding identified the communications problems that can occur during systems integration. Professor Boulding was concerned with the fact that subsystem specialists (i.e., physicists, economists, chemists, sociologists, etc.) have their own languages. He advocated that, in order for successful integration to take place, all subsystem specialists must speak a common language, such as mathematics. Today we use the PMBOK® Guide, the Project Management Body of Knowledge, to satisfy this need for project management.

General systems theory implies the creation of a management technique that is able to cut across many organizational disciplines—finance, manufacturing, engineering, marketing, and so on—while still carrying out the functions of management. This technique has come to be called systems management, project management, or matrix management (the terms are used interchangeably).

2.2 PROJECT MANAGEMENT: 1945–1960

During the 1940s, line managers used the concept of over-the-fence management to manage projects. Each line manager, wearing the hat of a project manager, would perform the work necessitated by their line organization, and when completed, would throw the "ball" over the fence in hopes that someone would catch it. Once the ball was thrown over the fence, the line managers would wash their hands of any responsibility for the project because the ball was no longer in their yard. If a project failed, blame was placed on whichever line manager had the ball at that time.

The problem with over-the-fence management was that the customer had no single contact point for questions. The filtering of information wasted precious time for both the customer and the contractor. Customers who wanted firsthand information had to seek out the manager in possession of the ball. For small projects, this was easy. But as projects grew in size and complexity, this became more difficult.

Following World War II, the United States entered into the Cold War. To win a Cold War, one must compete in the arms race and rapidly build weapons of mass destruction. The victor in a Cold War is the one who can retaliate with such force as to obliterate the enemy.

The arms race made it clear that the traditional use of over-the-fence management would not be acceptable to the Department of Defense (DoD) for projects such as the B52 Bomber, the Minuteman Intercontinental Ballistic Missile, and the Polaris Submarine. The government wanted a single point of contact, namely, a project manager who had total accountability through all project phases. The use of project management was then mandated for some of the smaller weapon systems such as jet fighters and tanks. NASA mandated the use of project management for all activities related to the space program.

Projects in the aerospace and defense industries were having cost overruns in excess of 200 to 300%. Blame was erroneously placed upon improper implementation of project management when, in fact, the real problem was the inability to forecast technology. Forecasting technology is extremely difficult for projects that could last ten to twenty years.

By the late 1950s and early 1960s, the aerospace and defense industries were using project management on virtually all projects, and they were pressuring their suppliers to use it as well. Project management was growing, but at a relatively slow rate except for aerospace and defense.

Because of the vast number of contractors and subcontractors, the government needed standardization, especially in the planning process and the reporting of information. The government established a life-cycle planning and control model and a cost monitoring system, and created a group of project management auditors to make sure that the government's money was being spent as planned. These practices were to be used on all government programs above a certain dollar value. Private industry viewed these practices as an over-management cost and saw no practical value in project management.

2.3 PROJECT MANAGEMENT: 1960–1985

The growth of project management has come about more through necessity than through desire. Its slow growth can be attributed mainly to lack of acceptance of the new management techniques necessary for its successful implementation. An inherent fear of the unknown acted as a deterrent for managers.

Between the middle and late 1960s, more executives began searching for new management techniques and organizational structures that could be quickly adapted to a changing environment. The table below and Figure 2–1 identify two major variables that executives consider with regard to organizational restructuring.

Type of Industry	Tasks	Environment
A	Simple	Dynamic
B	Simple	Static
C	Complex	Dynamic
D	Complex	Static

Almost all type C and most type D industries have project management–related structures. The key variable appears to be task complexity. Companies that have complex tasks and that also operate in a dynamic environment find project management mandatory. Such industries would include aerospace, defense, construction, high-technology engineering, computers, and electronic instrumentation.

Other than aerospace, defense, and construction, the majority of the companies in the 1960s maintained an informal method for managing projects. In informal project management, just as the words imply, the projects were handled on an informal basis whereby the

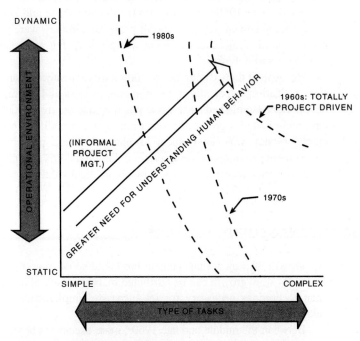

FIGURE 2–1.   Matrix implementation scheme.

authority of the project manager was minimized. Most projects were handled by functional managers and stayed in one or two functional lines, and formal communications were either unnecessary or handled informally because of the good working relationships between line managers. Many organizations today, such as low-technology manufacturing, have line managers who have been working side by side for ten or more years. In such situations, informal project management may be effective on capital equipment or facility development projects.

By 1970 and again during the early 1980s, more companies departed from informal project management and restructured to formalize the project management process, mainly because the size and complexity of their activities had grown to a point where they were unmanageable within the current structure. Figure 2–2 shows what happened to one such construction company. The following five questions help determine whether formal project management is necessary:

- Are the jobs complex?
- Are there dynamic environmental considerations?
- Are the constraints tight?
- Are there several activities to be integrated?
- Are there several functional boundaries to be crossed?

If any of these questions are answered yes, then some form of formalized project management may be necessary. It is possible for formalized project management to exist in only one functional department or division, such as for R&D or perhaps just for certain types of projects. Some companies have successfully implemented both formal and informal project management concurrently, but these companies are few and far between. Today we realize that the last two questions may be the most important.

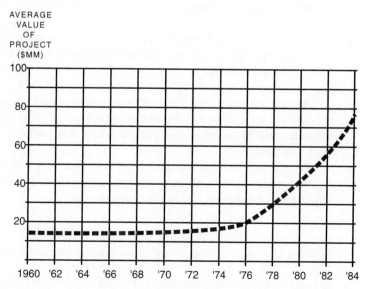

FIGURE 2–2. Average project size capability for a construction company, 1960–1984.

The moral here is that not all industries need project management, and executives must determine whether there is an actual need before making a commitment. Several industries with simple tasks, whether in a static or a dynamic environment, do not need project management. Manufacturing industries with slowly changing technology do not need project management, unless of course they have a requirement for several special projects, such as capital equipment activities, that could interrupt the normal flow of work in the routine manufacturing operations. The slow growth rate and acceptance of project management were related to the fact that the limitations of project management were readily apparent, yet the advantages were not completely recognizable. Project management requires organizational restructuring. The question, of course, is "How much restructuring?" Executives have avoided the subject of project management for fear that "revolutionary" changes must be made in the organization. As will be seen in Chapter 3, project management can be achieved with little departure from the existing traditional structure.

Project management restructuring has permitted companies to:

- Accomplish tasks that could not be effectively handled by the traditional structure
- Accomplish onetime activities with minimum disruption of routine business

The second item implies that project management is a "temporary" management structure and, therefore, causes minimum organizational disruption. The major problems identified by those managers who endeavored to adapt to the new system all revolved around conflicts in authority and resources.

Three major problems were identified by Killian[1]:

- Project priorities and competition for talent may interrupt the stability of the organization and interfere with its long-range interests by upsetting the normal business of the functional organization.
- Long-range planning may suffer as the company gets more involved in meeting schedules and fulfilling the requirements of temporary projects.
- Shifting people from project to project may disrupt the training of new employees and specialists. This may hinder their growth and development within their fields of specialization.

Another major concern was that project management required upper-level managers to relinquish some of their authority through delegation to the middle managers. In several situations, middle managers soon occupied the power positions, even more so than upper-level managers.

Despite these limitations, there were several driving forces behind the project management approach. According to John Kenneth Galbraith, these forces stem from "the imperatives of technology." The six imperatives are[2]:

---

1. William P. Killian, "Project Management—Future Organizational Concepts," *Marquette Business Review,* Vol. 2, 1971, pp. 90–107.

2. Excerpt from John Kenneth Galbraith, *The New Industrial State,* 3rd ed. Copyright © 1967, 1971, 1978, by John Kenneth Galbraith. Reprinted by permission of Houghton Mifflin Company. All rights reserved.

- The time span between project initiation and completion appears to be increasing.
- The capital committed to the project prior to the use of the end item appears to be increasing.
- As technology increases, the commitment of time and money appears to become inflexible.
- Technology requires more and more specialized manpower.
- The inevitable counterpart of specialization is organization.
- The above five "imperatives" identify the necessity for more effective planning, scheduling, and control.

As the driving forces overtook the restraining forces, project management began to mature. Executives began to realize that the approach was in the best interest of the company. Project management, if properly implemented, can make it easier for executives to overcome such internal and external obstacles as:

- Unstable economy
- Shortages
- Soaring costs
- Increased complexity
- Heightened competition
- Technological changes
- Societal concerns
- Consumerism
- Ecology
- Quality of work

Project management may not eliminate these problems, but may make it easier for the company to adapt to a changing environment.

If these obstacles are not controlled, the results may be:

- Decreased profits
- Increased manpower needs
- Cost overruns, schedule delays, and penalty payments occurring earlier and earlier
- An inability to cope with new technology
- R&D results too late to benefit existing product lines
- New products introduced into the marketplace too late
- Temptation to make hasty decisions that prove to be costly
- Management insisting on earlier and greater return on investment
- Greater difficulty in establishing on-target objectives in real time
- Problems in relating cost to technical performance and scheduling during the execution of the project

Project management became a necessity for many companies as they expanded into multiple product lines, many of which were dissimilar, and organizational complexities grew. This growth can be attributed to:

- Technology increasing at an astounding rate
- More money invested in R&D

- More information available
- Shortening of project life cycles

To satisfy the requirements imposed by these four factors, management was "forced" into organizational restructuring; the traditional organizational form that had survived for decades was inadequate for integrating activities across functional "empires."

By 1970, the environment began to change rapidly. Companies in aerospace, defense, and construction pioneered in implementing project management, and other industries soon followed, some with great reluctance. NASA and the Department of Defense "forced" subcontractors into accepting project management. The 1970s also brought much more published data on project management. As an example[3]:

> Project teams and task forces will become more common in tackling complexity. There will be more of what some people call temporary management systems as project management systems where the men [and women] who are needed to contribute to the solution meet, make their contribution, and perhaps never become a permanent member of any fixed or permanent management group.

The definition simply states that the purpose of project management is to put together the best possible team to achieve the objective, and, at termination, the team is disbanded. Nowhere in the definition do we see the authority of the project manager or his rank, title, or salary.

Because current organizational structures are unable to accommodate the wide variety of interrelated tasks necessary for successful project completion, the need for project management has become apparent. It is usually first identified by those lower-level and middle managers who find it impossible to control their resources effectively for the diverse activities within their line organization. Quite often middle managers feel the impact of a changing environment more than upper-level executives.

Once the need for change is identified, middle management must convince upper-level management that such a change is actually warranted. If top-level executives cannot recognize the problems with resource control, then project management will not be adopted, at least formally. Informal acceptance, however, is another story.

As project management developed, some essential factors in its successful implementation were recognized. The major factor was the role of the project manager, which became the focal point of integrative responsibility. The need for integrative responsibility was first identified in research and development activities[4]:

> Recently, R&D technology has broken down the boundaries that used to exist between industries. Once-stable markets and distribution channels are now in a state of flux. The industrial environment is turbulent and increasingly hard to predict. Many complex facts

3. Reprinted from the October 17, 1970, issue of *BusinessWeek* by special permission, © 1970 by McGraw-Hill, Inc., New York, New York 10020. All rights reserved.

4. Reprinted by permission of *Harvard Business Review*. From Paul R. Lawrence and Jay W. Lorsch, "New Management Job: The Integrator," *Harvard Business Review*, November–December 1967, p. 142. Copyright © 1967 by the Harvard Business School Publishing Corporation; all rights reserved.

about markets, production methods, costs and scientific potentials are related to investment decisions.

All of these factors have combined to produce a king-size managerial headache. There are just too many crucial decisions to have them all processed and resolved through regular line hierarchy at the top of the organization. They must be integrated in some other way.

Providing the project manager with integrative responsibility resulted in:

- Total accountability assumed by a single person
- Project rather than functional dedication
- A requirement for coordination across functional interfaces
- Proper utilization of integrated planning and control

Without project management, these four elements have to be accomplished by executives, and it is questionable whether these activities should be part of an executive's job description. An executive in a Fortune 500 corporation stated that he was spending seventy hours a week acting as an executive and as a project manager, and he did not feel that he was performing either job to the best of his abilities. During a presentation to the staff, the executive stated what he expected of the organization after project management implementation:

- Push decision-making down in the organization
- Eliminate the need for committee solutions
- Trust the decisions of peers

Those executives who chose to accept project management soon found the advantages of the new technique:

- Easy adaptation to an ever-changing environment
- Ability to handle a multidisciplinary activity within a specified period of time
- Horizontal as well as vertical work flow
- Better orientation toward customer problems
- Easier identification of activity responsibilities
- A multidisciplinary decision-making process
- Innovation in organizational design

2.4 PROJECT MANAGEMENT: 1985–2009

By the 1990s, companies had begun to realize that implementing project management was a necessity, not a choice. The question was not how to implement project management, but how fast could it be done?

Table 2–1 shows the typical life-cycle phases that an organization goes through to implement project management. In the first phase, the Embryonic Phase, the organization recognizes the apparent need for project management. This recognition normally takes

TABLE 2–1. LIFE-CYCLE PHASES FOR PROJECT MANAGEMENT MATURITY

Embryonic Phase	Executive Management Acceptance Phase	Line Management Acceptance Phase	Growth Phase	Maturity Phase
• Recognize need	• Visible executive support	• Line management support	• Use of life-cycle phases	• Development of a management cost/ schedule control system
• Recognize benefits	• Executive understanding of project management	• Line management commitment	• Development of a project management methodology	• Integrating cost and schedule control
• Recognize applications	• Project sponsorship	• Line management education	• Commitment to planning	• Developing an educational program to enhance project management skills
• Recognize what must be done	• Willingness to change way of doing business	• Willingness to release employees for project management training	• Minimization of "creeping scope"	
			• Selection of a project tracking system	

place at the lower and middle levels of management where the project activities actually take place. The executives are then informed of the need and assess the situation.

There are six driving forces that lead executives to recognize the need for project management:

- Capital projects
- Customer expectations
- Competitiveness
- Executive understanding
- New project development
- Efficiency and effectiveness

Manufacturing companies are driven to project management because of large capital projects or a multitude of simultaneous projects. Executives soon realize the impact on cash flow and that slippages in the schedule could end up idling workers.

Companies that sell products or services, including installation, to their clients must have good project management practices. These companies are usually non–project-driven but function as though they were project-driven. These companies now sell solutions to their customers rather than products. It is almost impossible to sell complete solutions to customers without having superior project management practices because what you are actually selling is your project management expertise.

There are two situations where competitiveness becomes the driving force: internal projects and external (outside customer) projects. Internally, companies get into trouble when the organization realizes that much of the work can be outsourced for less than it would cost to perform the work themselves. Externally, companies get into trouble when they are no longer competitive on price or quality, or simply cannot increase their market share.

Executive understanding is the driving force in those organizations that have a rigid traditional structure that performs routine, repetitive activities. These organizations are quite resistant to change unless driven by the executives. This driving force can exist in conjunction with any of the other driving forces.

New product development is the driving force for those organizations that are heavily invested in R&D activities. Given that only a small percentage of R&D projects ever make it into commercialization where the R&D costs can be recovered, project management becomes a necessity. Project management can also be used as an early warning system that a project should be cancelled.

Efficiency and effectiveness, as driving forces, can exist in conjunction with any other driving forces. Efficiency and effectiveness take on paramount importance for small companies experiencing growing pains. Project management can be used to help such companies remain competitive during periods of growth and to assist in determining capacity constraints.

Because of the interrelatedness of these driving forces, some people contend that the only true driving force is survival. This is illustrated in Figure 2–3. When the company recognizes that survival of the firm is at stake, the implementation of project management becomes easier.

The speed by which companies reach some degree of maturity in project management is most often based upon how important they perceive the driving forces to be. This is illustrated generically in Figure 2–4. Non–project-driven and hybrid organizations move quickly to maturity if increased internal efficiencies and effectiveness are needed. Competitiveness is the slowest path because these types of organizations do not recognize that project management affects their competitive position directly. For project-driven organizations, the path is reversed. Competitiveness is the name of the game and the vehicle used is project management.

Once the organization perceives the need for project management, it enters the second life-cycle phase of Table 2–1, Executive Acceptance. Project management cannot be

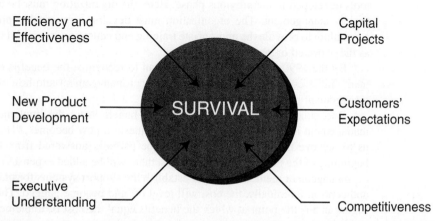

FIGURE 2–3. The components of survival. *Source:* Reprinted from H. Kerzner, *In Search of Excellence in Project Management.* New York: Wiley, 1998, p. 51.

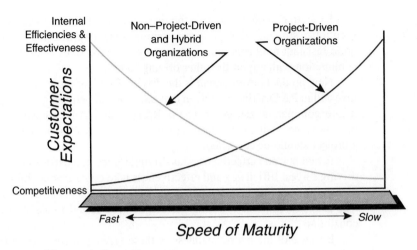

FIGURE 2–4. The speed of maturity.

implemented rapidly in the near term without executive support. Furthermore, the support must be visible to all.

The third life-cycle phase is Line Management Acceptance. It is highly unlikely that any line manager would actively support the implementation of project management without first recognizing the same support coming from above. Even minimal line management support will still cause project management to struggle.

The fourth life-cycle phase is the Growth Phase, where the organization becomes committed to the development of the corporate tools for project management. This includes the project management methodology for planning, scheduling, and controlling, as well as selection of the appropriate supporting software. Portions of this phase can begin during earlier phases.

The fifth life-cycle phase is Maturity. In this phase, the organization begins using the tools developed in the previous phase. Here, the organization must be totally dedicated to project management. The organization must develop a reasonable project management curriculum to provide the appropriate training and education in support of the tools, as well as the expected organizational behavior.

By the 1990s, companies finally began to recognize the benefits of project management. Table 2–2 shows the benefits of project management and how our view of project management has changed.

Recognizing that the organization can benefit from the implementation of project management is just the starting point. The question now becomes, "How long will it take us to achieve these benefits?" This can be partially answered from Figure 2–5. In the beginning of the implementation process, there will be added expenses to develop the project management methodology and establish the support systems for planning, scheduling, and control. Eventually, the cost will level off and become pegged. The question mark in Figure 2–5 is the point at which the benefits equal the cost of implementation. This point can be pushed to the left through training and education.

<table>
<tr><td>**PMBOK® Guide, 4th Edition**
1.5 Project Management in
Operations Management</td><td colspan="2">**TABLE 2–2. BENEFITS OF PROJECT MANAGEMENT**</td></tr>
</table>

Past View	Present View
• Project management will require more people and add to the overhead costs.	• Project management allows us to accomplish more work in less time, with fewer people.
• Profitability may decrease.	• Profitability will increase.
• Project management will increase the amount of scope changes.	• Project management will provide better control of scope changes.
• Project management creates organizational instability and increases conflicts.	• Project management makes the organization more efficient and effective through better organizational behavior principles.
• Project management is really "eye wash" for the customer's benefit.	• Project management will allow us to work more closely with our customers.
• Project management will create problems.	• Project management provides a means for solving problems.
• Only large projects need project management.	• All projects will benefit from project management.
• Project management will increase quality problems.	• Project management increases quality.
• Project management will create power and authority problems.	• Project management will reduce power struggles.
• Project management focuses on suboptimization by looking at only the project.	• Project management allows people to make good company decisions.
• Project management delivers products to a customer.	• Project management delivers solutions.
• The cost of project management may make us noncompetitive.	• Project management will increase our business.

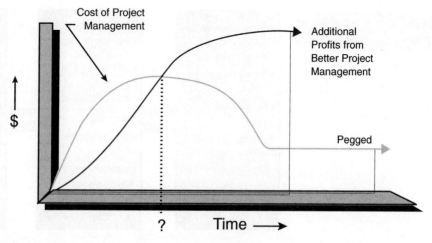

FIGURE 2–5. Project management costs versus benefits.

2.5 RESISTANCE TO CHANGE

Why was project management so difficult for companies to accept and implement? The answer is shown in Figure 2–6. Historically, project management resided only in the project-driven sectors of the marketplace. In these sectors, the project managers were given the responsibility for profit and loss, which virtually forced companies to treat project management as a profession.

In the non–project-driven sectors of the marketplace, corporate survival was based upon products and services, rather than upon a continuous stream of projects. Profitability was identified through marketing and sales, with very few projects having an identifiable P&L. As a result, project management in these firms was never viewed as a profession.

In reality, most firms that believed that they were non–project-driven were actually hybrids. Hybrid organizations are typically non–project-driven firms with one or two divisions that are project-driven. Historically, hybrids have functioned as though they were non–project-driven, as shown in Figure 2–6, but today they are functioning like project-driven firms. Why the change? Management has come to the realization that they can most effectively run their organization on a "management by project" basis, and thereby achieve the benefits of both a project management organization and a traditional organization. The rapid growth and acceptance of project management during the last ten years has taken place in the non–project-driven/hybrid sectors. Now, project management is being promoted by marketing, engineering, and production, rather than only by the project-driven departments (see Figure 2–7).

A second factor contributing to the acceptance of project management was the economy, specifically the recessions of 1979–1983 and 1989–1993. This can be seen from Table 2–3. By the end of the recession of 1979–1983, companies recognized the benefits of using project management but were reluctant to see it implemented. Companies

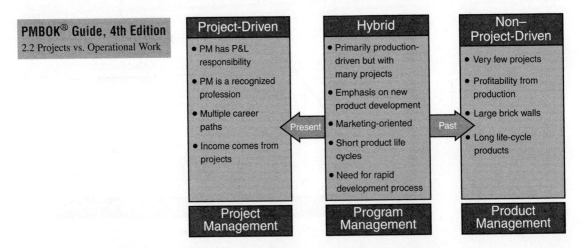

FIGURE 2–6. Industry classification (by project management utilization).

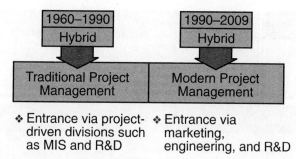

FIGURE 2–7. From hybrid to project-driven.

returned to the "status quo" of traditional management. There were no allies or alternative management techniques that were promoting the use of project management.

The recession of 1989–1993 finally saw the growth of project management in the non–project-driven sector. This recession was characterized by layoffs in the white collar/management ranks. Allies for project management were appearing and emphasis was being placed upon long-term solutions to problems. Project management was here to stay.

The allies for project management began surfacing in 1985 and continued throughout the recession of 1989–1993. This is seen in Figure 2–8.

- *1985:* Companies recognize that they must compete on the basis of quality as well as cost. Companies begin using the principles of project management for the implementation of total quality management (TQM). The first ally for project management surfaces with the "marriage" of project management and TQM.
- *1990:* During the recession of 1989–1993, companies recognize the importance of schedule compression and being the first to market. Advocates of concurrent engineering begin promoting the use of project management to obtain better scheduling techniques. Another ally for project management is born.

TABLE 2–3. RECESSIONARY EFFECTS

	Characteristics				
Recession	Layoffs	R&D	Training	Solutions Sought	Results of the Recessions
1979–1983	Blue collar	Eliminated	Eliminated	Short-term	• Return to status quo • No project management support • No allies for project management
1989–1993	White collar	Focused	Focused	Long-term	• Change way of doing business • Risk management • Examine lessons learned

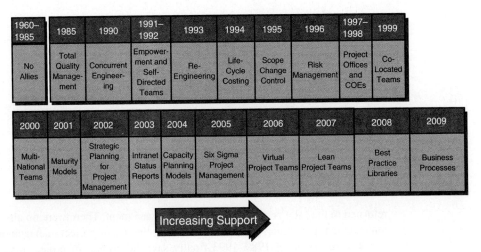

1960–1985	1985	1990	1991–1992	1993	1994	1995	1996	1997–1998	1999
No Allies	Total Quality Management	Concurrent Engineering	Empowerment and Self-Directed Teams	Re-Engineering	Life-Cycle Costing	Scope Change Control	Risk Management	Project Offices and COEs	Co-Located Teams

2000	2001	2002	2003	2004	2005	2006	2007	2008	2009
Multi-National Teams	Maturity Models	Strategic Planning for Project Management	Intranet Status Reports	Capacity Planning Models	Six Sigma Project Management	Virtual Project Teams	Lean Project Teams	Best Practice Libraries	Business Processes

Increasing Support

FIGURE 2–8. New processes supporting project management.

- *1991–1992:* Executives realize that project management works best if decision-making and authority are decentralized, but recognize that control can still be achieved at the top by functioning as project sponsors.
- *1993:* As the recession of 1989–1993 comes to an end, companies begin "re-engineering" the organization, which really amounts to elimination of organizational "fat." The organization is now a "lean and mean" machine. People are asked to do more work in less time and with fewer people; executives recognize that being able to do this is a benefit of project management.
- *1994:* Companies recognize that a good project cost control system (i.e., horizontal accounting) allows for improved estimating and a firmer grasp of the real cost of doing work and developing products.
- *1995:* Companies recognize that very few projects are completed within the framework of the original objectives without scope changes. Methodologies are created for effective change management.
- *1996:* Companies recognize that risk management involves more than padding an estimate or a schedule. Risk management plans are now included in the project plans.
- *1997–1998:* The recognition of project management as a professional career path mandates the consolidation of project management knowledge and a centrally located project management group. Benchmarking for best practices forces the creation of centers for excellence in project management.
- *1999:* Companies that recognize the importance of concurrent engineering and rapid product development find that it is best to have dedicated resources for the duration of the project. The cost of overmanagement may be negligible compared to risks of undermanagement. More organizations begin to use colocated teams all housed together.
- *2000:* Mergers and acquisitions create more multinational companies. Multinational project management becomes a major challenge.

- *2001:* Corporations are under pressure to achieve maturity as quickly as possible. Project management maturity models help companies reach this goal.
- *2002:* The maturity models for project management provide corporations with a basis to perform strategic planning for project management. Project management is now viewed as a strategic competency for the corporation.
- *2003:* Intranet status reporting comes of age. This is particularly important for multinational corporations that must exchange information quickly.
- *2004:* Intranet reporting provides corporations with information on how resources are being committed and utilized. Corporations develop capacity planning models to learn how much additional work the organization can take on.
- *2005:* The techniques utilized in Six Sigma are being applied to project management, especially for continuous improvement to the project management methodology. This will result in the establishment of categories of Six Sigma applications some of which are nontraditional.
- *2006:* Virtual project teams and virtual project management offices will become more common. The growth of virtual teams relies heavily upon trust, teamwork, cooperation, and effective communication.
- *2007:* The concepts of lean manufacturing will be applied to project management.
- *2008:* Companies will recognize the value of capturing best practices in project management and creating a best practices library or knowledge repository.
- *2009:* Project management methodologies will include more business processes to support project management.

As project management continues to grow and mature, it will have more allies. In the twenty-first century, second and third world nations will come to recognize the benefits and importance of project management. Worldwide standards for project management will be established.

If a company wishes to achieve excellence in project management, then it must go through a successful implementation process. This is illustrated in Situation 2–1.

Situation 2–1: The aerospace division of a Fortune 500 company had been using project management for more than thirty years. Everyone in the organization had attended courses in the principles of project management. From 1985 to 1994, the division went through a yearly ritual of benchmarking themselves against other aerospace and defense organizations. At the end of the benchmarking period, the staff would hug and kiss one another, believing that they were performing project management as well as could be expected.

In 1995, the picture changed. The company decided to benchmark itself against organizations that were not in the aerospace or defense sector. It soon learned that there were companies that had been using project management for fewer than six years but whose skills at implementation had surpassed the aerospace/defense firms. It was a rude awakening.

Another factor that contributed to resistance to change was senior management's preference for the status quo. Often this preference was based upon what was in the executives' best interest rather than the best interest of the organization. It was also common for someone to attend basic project management programs and then discover that the organization would not allow full implementation of project management, leading to frustration for those in the lower and middle levels of management. Consider Situation 2–2:

> **Situation 2–2:** The largest division of a Fortune 500 company recognized the need for project management. Over a three-year period, 200 people were trained in the basics of project management, and 18 people passed the national certification exam for project management. The company created a project management division and developed a methodology. As project management began to evolve in this division, the project managers quickly realized that the organization would not allow their "illusions of grandeur" to materialize. The executive vice president made it clear that the functional areas, rather than the project management division, would have budgetary control. Project managers would *not* be empowered with authority or critical decision-making opportunities. Simply stated, the project managers were being treated as expediters and coordinators, rather than real project managers.

Even though project management has been in existence for more than forty years, there are still different views and misconceptions about what it really is. Textbooks on operations research or management science still have chapters entitled "Project Management" that discuss only PERT scheduling techniques. A textbook on organizational design recognized project management as simply another organizational form.

All companies sooner or later understand the basics of project management. But companies that have achieved excellence in project management have done so through successful implementation and execution of processes and methodologies.

2.6 SYSTEMS, PROGRAMS, AND PROJECTS: A DEFINITION

In the preceding sections the word "systems" has been used rather loosely. The exact definition of a system depends on the users, environment, and ultimate goal. Business practitioners define a system as:

> A group of elements, either human or nonhuman, that is organized and arranged in such a way that the elements can act as a whole toward achieving some common goal or objective.

Systems are collections of interacting subsystems that, if properly organized, can provide a synergistic output. Systems are characterized by their boundaries or interface conditions. For example, if the business firm system were completely isolated from the environmental system, then a *closed system* would exist, in which case management would have complete control over all system components. If the business system reacts with the environment, then the system is referred to as *open*. All social systems, for example, are categorized as open systems. Open systems must have permeable boundaries.

If a system is significantly dependent on other systems for its survival, then it is an *extended system*. Not all open systems are extended systems. Extended systems are ever-changing and can impose great hardships on individuals who desire to work in a regimented atmosphere.

Military and government organizations were the first to attempt to define the boundaries of systems, programs, and projects. Below are two definitions for systems:

- *Air Force Definition:* A composite of equipment, skills, and techniques capable of performing and/or supporting an operational role. A complete system includes related facilities, equipment, material services, and personnel required for its operation to the degree that it can be considered as a self-sufficient unit in its intended operational and/or support environment.
- *NASA Definition:* One of the principal functioning entities comprising the project hardware within a project or program. The meaning may vary to suit a particular project or program area. Ordinarily a "system" is the first major subdivision of project work (spacecraft systems, launch vehicle systems).

PMBOK® Guide, 4th Edition
1.4.2 Program Management
Definition

Programs can be construed as the necessary first-level elements of a system. Two representative definitions of programs are given below:

- *Air Force Definition:* The integrated, time-phased tasks necessary to accomplish a particular purpose.
- *NASA Definition:* A relative series of undertakings that continue over a period of time (normally years) and that are designed to accomplish a broad, scientific or technical goal in the NASA long-range plan (lunar and planetary exploration, manned spacecraft systems).

Programs can be regarded as subsystems. However, programs are generally defined as time-phased efforts, whereas systems exist on a continuous basis.

Projects are also time-phased efforts (much shorter than programs) and are the first level of breakdown of a program. A typical definition would be:

- *NASA/Air Force Definition:* A project is within a program as an undertaking that has a scheduled beginning and end, and that normally involves some primary purpose.

As shown in Table 2–4, the government sector tends to run efforts as programs, headed by a program manager. The majority of the industrial sector, on the other hand, prefers to describe efforts as projects, headed by a project manager. Whether we call our undertaking project management or program management is inconsequential because the same policies, procedures, and guidelines tend to regulate both. For the remainder of this text, programs and projects will be discussed interchangeably. However, the reader should be aware that projects are normally the first-level subdivision of a program and that programs are more ongoing than projects. This breakdown will be discussed in more detail in Chapter 11.

TABLE 2–4. DEFINITION SUMMARY

Level	Sector	Title
System*	—	—
Program	Government	Program managers
Project	Industry	Project managers

*Definitions, as used here, do not include in-house industrial systems such as management information systems or shop floor control systems.

Once a group of tasks is selected and considered to be a project, the next step is to define the kinds of project units. There are four categories of projects:

- *Individual projects:* These are short-duration projects normally assigned to a single individual who may be acting as both a project manager and a functional manager.
- *Staff projects:* These are projects that can be accomplished by one organizational unit, say a department. A staff or task force is developed from each section involved. This works best if only one functional unit is involved.
- *Special projects:* Often special projects occur that require certain primary functions and/or authority to be assigned temporarily to other individuals or units. This works best for short-duration projects. Long-term projects can lead to severe conflicts under this arrangement.
- *Matrix or aggregate projects:* These require input from a large number of functional units and usually control vast resources.

Project management may now be defined as the process of achieving project objectives through the traditional organizational structure and over the specialties of the individuals concerned. Project management is applicable for any ad hoc (unique, one-time, one-of-a-kind) undertaking concerned with a specific end objective. In order to complete a task, a project manager must:

PMBOK® Guide, 4th Edition
1.3 What Is Project Management?

- Set objectives
- Establish plans
- Organize resources
- Provide staffing
- Set up controls
- Issue directives
- Motivate personnel
- Apply innovation for alternative actions
- Remain flexible

The type of project will often dictate which of these functions a project manager will be required to perform.

2.7 PRODUCT VERSUS PROJECT MANAGEMENT: A DEFINITION

> **PMBOK® Guide, 4th Edition**
> 4.1.1 Inputs to Project
>   Charter
> 4.1.1.1 Product Scope and Project
>   Scope and Chapter 5 Introduction

For all practical purposes, there is no major difference between a project and a program other than the time duration. Project managers focus on the end date of their project from the day they are assigned as project manager. Program managers usually have a much longer time frame that project managers and never want to see their program come to an end. In the early years of project management with the Department of Defense serving as the primary customer, aerospace and defense project managers were called program managers because the intent was to get follow-on government contracts each year.

But what about the definition of product management or product line management? Product managers function closely like program managers. The product manager wants his or her product to be as long-lived as possible and as profitable as possible. Even when the demand for the product diminishes, the product manager will always look for spin-offs to keep a product alive.

Although the PMBOK® Guide does not differentiate between project and program scope, it does differentiate between project and product scope:

- *Project scope* defines the work that must be accomplished to produce a deliverable with specified features or functions. The deliverable can be a product, service, or other result.
- *Product scope* defines the features or functions that characterize the deliverable.

Figure 2–9 shows the relationship between project and product management. When the project is in the R&D phase, a project manager is involved. Once the product is developed and introduced into the marketplace, the product manager takes control. In some situations, the project manager can become the product manager. Product and project management can, and do, exist concurrently within companies.

Figure 2–9 shows that product management can operate horizontally as well as vertically. When a product is shown horizontally on the organizational chart, the implication is that the product line is not big enough to control its own resources full-time and therefore shares key functional resources. If the product line were large enough to control its own resources full-time, it would be shown as a separate division or a vertical line on the organization chart.

Also shown in Figure 2–9 is the remarkable fact that the project manager (or project engineer) is reporting to a marketing-type person. The reason is that technically oriented project leaders get too involved with the technical details of the project and lose sight of when and how to "kill" a project. Remember, most technical leaders have been trained in an academic rather than a business environment. Their commitment to success often does not take into account such important parameters as return on investment, profitability, competition, and marketability.

To alleviate these problems, project managers and project engineers, especially on R&D-type projects, are now reporting to marketing so that marketing input will be included in all R&D decisions because of the high costs incurred during R&D. Executives must exercise caution with regard to this structure in which both product and project

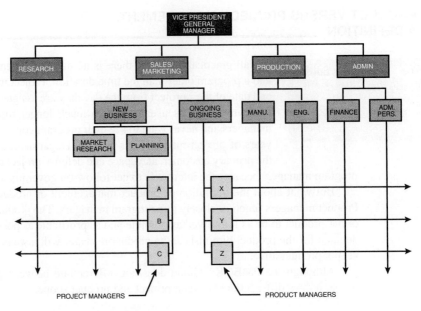

FIGURE 2–9. Organizational chart.

managers report to the marketing function. The marketing executive could become the focal point of the entire organization, with the capability of building a very large empire.

2.8 MATURITY AND EXCELLENCE: A DEFINITION

Some people contend that maturity and excellence in project management are the same. Unfortunately, this is not the case. Consider the following definition:

> Maturity in project management is the implementation of a standard methodology and accompanying processes such that there exists a high likelihood of repeated successes.

This definition is supported by the life-cycle phases shown in Table 2–1. Maturity implies that the proper foundation of tools, techniques, processes, and even culture, exists. When projects come to an end, there is usually a debriefing with senior management to discuss how well the methodology was used and to recommend changes. This debriefing looks at "key performance indicators," which are shared learning topics, and allows the organization to maximize what it does right and to correct what it did wrong.

The definition of excellence can be stated as:

> Organizations excellent in project management are those that create the environment in which there exists a *continuous* stream of successfully managed projects and where success is measured by what is in the best interest of *both* the company and the project (i.e., customer).

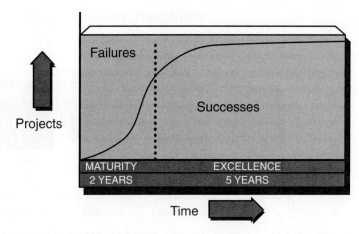

FIGURE 2–10. The growth of excellence.

Excellence goes well beyond maturity. You must have maturity to achieve excellence. Figure 2–10 shows that once the organization completes the first four life-cycle phases in Table 2–1, it may take two years or more to reach some initial levels of maturity. Excellence, if achievable at all, may take an additional five years or more.

Figure 2–10 also brings out another important fact. During maturity, more successes than failures occur. During excellence, we obtain a continuous stream of successful projects. Yet, even after having achieved excellence, there will still be some failures.

> Executives who always make the right decision are not making enough decisions. Likewise, organizations in which all projects are completed successfully are not taking enough risks and are not working on enough projects.

It is unrealistic to believe that all projects will be completed successfully. Some people contend that the only true project failures are the ones from which nothing is learned. Failure can be viewed as success if the failure is identified early enough so that the resources can be reassigned to other more opportunistic activities.

2.9 INFORMAL PROJECT MANAGEMENT: A DEFINITION

Companies today are managing projects more informally than before. Informal project management does have some degree of formality but emphasizes managing the project with a minimum amount of paperwork. Furthermore, informal project management is based upon guidelines rather than the policies and procedures that are the basis for formal project

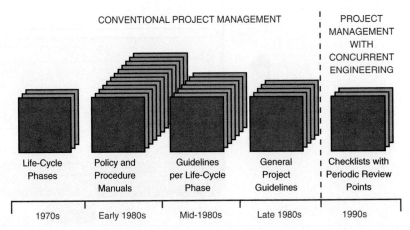

CONVENTIONAL PROJECT MANAGEMENT ┊ PROJECT MANAGEMENT WITH CONCURRENT ENGINEERING

| Life-Cycle Phases | Policy and Procedure Manuals | Guidelines per Life-Cycle Phase | General Project Guidelines | Checklists with Periodic Review Points |

| 1970s | Early 1980s | Mid-1980s | Late 1980s | 1990s |

FIGURE 2–11. Evolution of policies, procedures, and guidelines. *Source:* Reprinted from H. Kerzner, *In Search of Excellence in Project Management.* New York: Wiley, 1998, p. 196.

management. This was shown previously to be a characteristic of a good project management methodology. Informal project management mandates:

- Effective communications
- Effective cooperation
- Effective teamwork
- Trust

These four elements are absolutely essential for effective informal project management.

Figure 2–11 shows the evolution of project documentation over the years. As companies become mature in project management, emphasis is on guidelines and checklists. Figure 2–12 shows the critical issues as project management matures toward more informality.

As a final note, not all companies have the luxury of using informal project management. Customers often have a strong voice in whether formal or informal project management will be used.

2.10 THE MANY FACES OF SUCCESS

Historically, the definition of success has been meeting the customer's expectations regardless of whether or not the customer is internal or external. Success also includes getting the job done within the constraints of time, cost, and quality. Using this standard definition, success is defined as a point on the time, cost, quality/performance grid. But how many projects, especially those requiring innovation, are accomplished at this point?

Very few projects are ever completed without trade-offs or scope changes on time, cost, and quality. Therefore, success could still occur without exactly hitting this singular point. In this regard, success could be defined as a cube, such as seen in Figure 2–13. The singular point of time, cost, and quality would be a point within the cube, constituting the convergence of the critical success factors (CSFs) for the project.

Another factor to consider is that there may exist both primary and secondary definitions of success, as shown in Table 2–5. The primary definitions of success are seen through

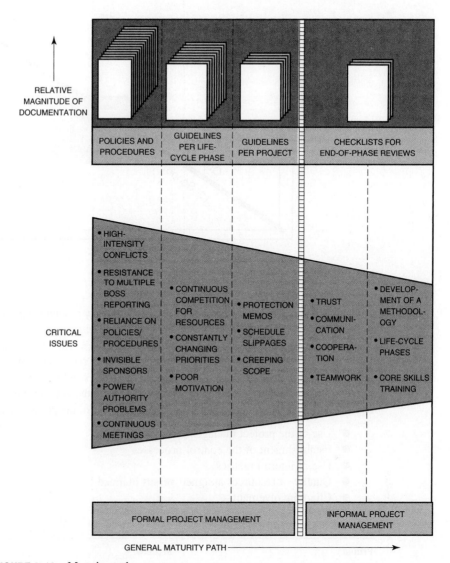

FIGURE 2–12. Maturity path.

the eyes of the customer. The secondary definitions of success are usually internal benefits. If achieving 86 percent of the specification is acceptable to the customer and follow-on work is received, then the original project might very well be considered a success.

The definition of success can also vary according to who the stakeholder is. For example, each of the following can have his or her own definition of success on a project:

- Consumers: safety in its use
- Employees: guaranteed employment
- Management: bonuses
- Stockholders: profitability
- Government agencies: compliance with federal regulations

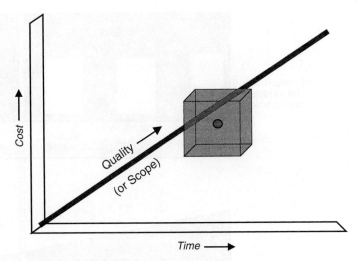

FIGURE 2–13. Success: point or cube?

It is possible for a project management methodology to identify primary and secondary success factors. This could provide guidance to a project manager for the development of a risk management plan and for deciding which risks are worth taking and which are not.

Critical success factors identify what is necessary to meet the desired deliverables of the customer. We can also look at key performance indicators (KPIs), which measure the quality of the process used to achieve the end results. KPIs are internal measures or metrics that can be reviewed on a periodic basis throughout the life cycle of the project. Typical KPIs include:

- Use of the project management methodology
- Establishment of the control processes
- Use of interim metrics
- Quality of resources assigned versus planned for
- Client involvement

TABLE 2–5. SUCCESS FACTORS

Primary	Secondary
• Within time	• Follow-on work from this customer
• Within cost	• Using the customer's name as a reference on your literature
• Within quality limits	• Commercialization of a product
• Accepted by the customer	• With minimum or mutually agreed upon scope changes
	• Without disturbing the main flow of work
	• Without changing the corporate culture
	• Without violating safety requirements
	• Providing efficiency and effectiveness of operations
	• Satisfying OSHA/EPA requirements
	• Maintaining ethical conduct
	• Providing a strategic alignment
	• Maintaining a corporate reputation
	• Maintaining regulatory agency relations

Key performance indicators answer such questions as: Did we use the methodology correctly? Did we keep management informed, and how frequently? Were the proper resources assigned and were they used effectively? Were there lessons learned that could necessitate updating the methodology or its use? Companies excellent in project management measure success both internally and externally using CSFs and KPIs.

2.11 THE MANY FACES OF FAILURE[5]

Previously we stated that success might be a cube rather than a point. If we stay within the cube but miss the point, is that a failure? Probably not! The true definition of failure is when the final results are not what were expected, even though the original expectations may or may not have been reasonable. Sometimes customers and even internal executives set performance targets that are totally unrealistic in hopes of achieving 80–90 percent. For simplicity's sake, let us define failure as unmet expectations.

With unmeetable expectations, failure is virtually assured since we have defined failure as unmet expectations. This is called a *planning failure* and is the difference between what was planned and what was, in fact, achieved. The second component of failure is poor performance or *actual failure*. This is the difference between what was achievable and what was actually accomplished.

Perceived failure is the net sum of *actual failure* and *planning failure*. Figures 2–14 and 2–15 illustrate the components of perceived failure. In Figure 2–14, *project management* has planned a level of accomplishment (C) lower than what is achievable given project circumstances and resources (D). This is a classic underplanning situation. Actual accomplishment (B), however, was less than planned.

A slightly different case is illustrated in Figure 2–15. Here, we have planned to accomplish more than is achievable. Planning failure is again assured even if no actual failure occurs. In both of these situations (overplanning and underplanning), the actual failure is the same, but the perceived failure can vary considerably.

Today, most project management practitioners focus on the *planning failure* term. If this term can be compressed or even eliminated, then the magnitude of the actual failure, should it occur, would be diminished. A good project management methodology helps to reduce this term. We now believe that the existence of this term is largely due to the project manager's inability to perform effective risk management. In the 1980s, we believed that the failure of a project was largely a quantitative failure due to:

- Ineffective planning
- Ineffective scheduling
- Ineffective estimating
- Ineffective cost control
- Project objectives being "moving targets"

---

5. Adapted from Robert D. Gilbreath, *Winning at Project Management.* New York: Wiley, 1986, pp. 2–6.

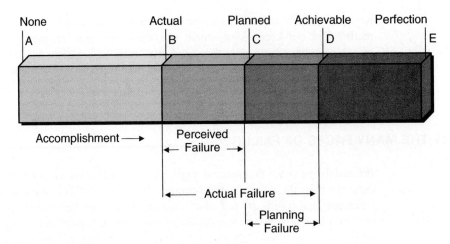

FIGURE 2–14.   Components of failure (pessimistic planning).

During the 1990s, we changed our view of failure from being quantitatively oriented to qualitatively oriented. A failure in the 1990s was largely attributed to:

- Poor morale
- Poor motivation
- Poor human relations
- Poor productivity
- No employee commitment

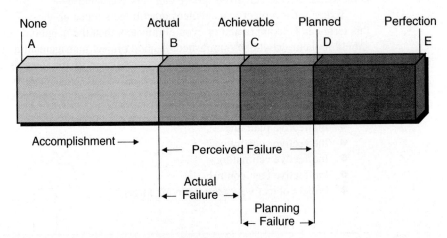

FIGURE 2–15.   Components of failure (optimistic planning).

- No functional commitment
- Delays in problem solving
- Too many unresolved policy issues
- Conflicting priorities between executives, line managers, and project managers

Although these quantitative and qualitative approaches still hold true to some degree, today we believe that the major component of planning failure is inappropriate or inadequate risk management, or having a project management methodology that does not provide any guidance for risk management.

Sometimes, the risk management component of failure is not readily identified. For example, look at Figure 2–16. The actual performance delivered by the contractor was significantly less than the customer's expectations. Is the difference due to poor technical ability or a combination of technical inability and poor risk management? Today we believe that it is a combination.

When a project is completed, companies perform a lessons-learned review. Sometimes lessons learned are inappropriately labeled and the true reason for the risk event is not known. Figure 2–17 illustrates the relationship between the marketing personnel and technical personnel when undertaking a project to develop a new product. If the project is completed with actual performance being less than customer expectations, is it because of poor risk management by the technical assessment and forecasting personnel or poor marketing risk assessment? The relationship between marketing and technical risk management is not always clear.

Figure 2–17 also shows that opportunities for trade-offs diminish as we get further downstream on the project. There are numerous opportunities for trade-offs prior to establishing the final objectives for the project. In other words, if the project fails, it may be because of the timing when the risks were analyzed.

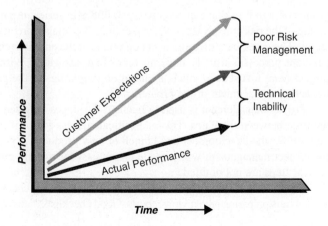

FIGURE 2–16. Risk planning.

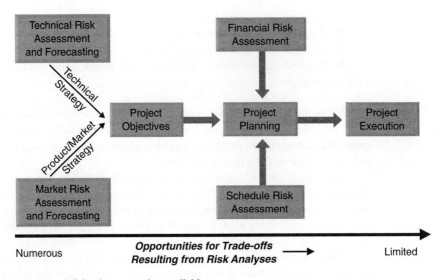

FIGURE 2–17. Mitigation strategies available.

2.12 THE STAGE-GATE PROCESS

PMBOK® Guide, 4th Edition

Chapter 2 Project Life Cycle and Organization

2.1.1 Characteristics of Project Phases

When companies recognize the need to begin developing processes for project management, the starting point is normally the stage-gate process. The stage-gate process was created because the traditional organizational structure was designed primarily for top-down, centralized management, control, and communications, all of which were no longer practical for organizations that use project management and horizontal work flow. The stage-gate process eventually evolved into life-cycle phases.

Just as the words imply, the process is composed of stages and gates. Stages are groups of activities that can be performed either in series or parallel based upon the magnitude of the risks the project team can endure. The stages are managed by cross-functional teams. The gates are structured decision points at the end of each stage. Good project management processes usually have no more than six gates. With more than six gates, the project team focuses too much attention on preparing for the gate reviews rather than on the actual management of the project.

Project management is used to manage the stages between the gates, and can shorten the time between the gates. This is a critical success factor if the stage-gate process is to be used for the development and launch of new products. A good corporate methodology for project management will provide checklists, forms, and guidelines to make sure that critical steps are not omitted.

Checklists for gate reviews are critical. Without these checklists, project managers can waste hours preparing gate review reports. Good checklists focus on answering these questions:

● Where are we today (i.e., time and cost)?
● Where will we end up (i.e., time and cost)?

- What are the present and future risks?
- What assistance is needed from management?

Project managers are never allowed to function as their own gatekeepers. The gatekeepers are either individuals (i.e., sponsors) or groups of individuals designated by senior management and empowered to enforce the structured decision-making process. The gatekeepers are authorized to evaluate the performance to date against predetermined criteria and to provide the project team with additional business and technical information.

Gatekeepers must be willing to make decisions. The four most common decisions are:

- Proceed to the next gate based upon the original objectives
- Proceed to the next gate based upon revised objectives
- Delay making a gate decision until further information is obtained
- Cancel the project

Sponsors must also have the courage to terminate a project. The purpose of the gates is not only to obtain authorization to proceed, but to identify failure early enough so that resources will not be wasted but will be assigned to more promising activities.

We can now identify the three major benefits of the stage-gate process:

- Providing structure to project management
- Providing possible standardization in planning, scheduling, and control (i.e., forms, checklists, and guidelines)
- Allowing for a structured decision-making process

Companies embark upon the stage-gate process with good intentions, but there are pitfalls that may disrupt the process. These include:

- Assigning gatekeepers and not empowering them to make decisions
- Assigning gatekeepers who are afraid to terminate a project
- Denying the project team access to critical information
- Allowing the project team to focus more on the gates than on the stages

It should be recognized that the stage-gate process is neither an end result nor a self-sufficient methodology. Instead, it is just one of several processes that provide structure to the overall project management methodology.

Today, the stage-gate process appears to have been replaced by life-cycle phases. Although there is some truth in this, the stage-gate process is making a comeback. Since the stage-gate process focuses on decision-making more than life-cycle phases, the stage-gate process is being used as an internal, decision-making tool within each of the life-cycle phases. The advantage is that, while life-cycle phases are the same for every project, the stage-gate process can be custom-designed for each project to facilitate decision-making and risk management. The stage-gate process is now an integral part of project management, whereas previously it was used primarily for new product development efforts.

2.13 PROJECT LIFE CYCLES

PMBOK® Guide, 4th Edition
Chapter 2
2.1.2 and 2.1.3

Every program, project, or product has certain phases of development known as life-cycle phases. A clear understanding of these phases permits managers and executives to better control resources to achieve goals.

During the past few years, there has been at least partial agreement about the life-cycle phases of a product. They include:

- Research and development
- Market introduction
- Growth
- Maturity
- Deterioration
- Death

Today, there is no agreement among industries, or even companies within the same industry, about the life-cycle phases of a project. This is understandable because of the complex nature and diversity of projects.

The theoretical definitions of the life-cycle phases of a system can be applied to a project. These phases include:

- Conceptual
- Planning
- Testing
- Implementation
- Closure

The first phase, the conceptual phase, includes the preliminary evaluation of an idea. Most important in this phase is a preliminary analysis of risk and the resulting impact on the time, cost, and performance requirements, together with the potential impact on company resources. The conceptual phase also includes a "first cut" at the feasibility of the effort.

The second phase is the planning phase. It is mainly a refinement of the elements in the conceptual phase and requires a firm identification of the resources required and the establishment of realistic time, cost, and performance parameters. This phase also includes the initial preparation of documentation necessary to support the system. For a project based on competitive bidding, the conceptual phase would include the decision of whether to bid, and the planning phase would include the development of the total bid package (i.e., time, schedule, cost, and performance).

Because of the amount of estimating involved, analyzing system costs during the conceptual and planning phases is not an easy task. As shown in Figure 2–18, most project or system costs can be broken down into operating (recurring) and implementation (nonrecurring) categories. Implementation costs include one-time expenses such as construction of a new facility, purchasing computer hardware, or detailed planning. Operating costs include recurring expenses such as manpower. The operating costs may be reduced as

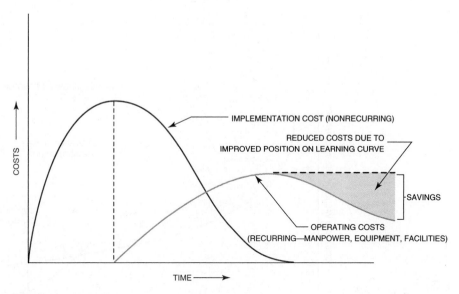

FIGURE 2–18. System costs.

shown in Figure 2–18 if personnel perform at a higher position on the learning curve. The identification of a learning curve position is vitally important during the planning phase when firm cost positions must be established. Of course, it is not always possible to know what individuals will be available or how soon they will perform at a higher learning curve position.

Once the approximate total cost of the project is determined, a cost-benefit analysis should be conducted (see Figure 2–19) to determine if the estimated value of the information obtained from the system exceeds the cost of obtaining the information. This analysis is often included as part of a feasibility study. There are several situations, such as in competitive bidding, where the feasibility study is actually the conceptual and definition phases. Because of the costs that can be incurred during these two phases, top-management approval is almost always necessary before the initiation of such a feasibility study.

The third phase—testing—is predominantly a testing and final standardization effort so that operations can begin. Almost all documentation must be completed in this phase.

The fourth phase is the implementation phase, which integrates the project's product or services into the existing organization. If the project was developed for establishment of a marketable product, then this phase could include the product life-cycle phases of market introduction, growth, maturity, and a portion of deterioration.

The final phase is closure and includes the reallocation of resources. Consider a company that sells products to consumers. As one product begins the deterioration and death phases of its life cycle (i.e., the divestment phase of a system), new products or projects must be established. Such a company would, therefore, require a continuous stream of projects to survive, as shown in Figure 2–20. As projects A and B begin their decline, new

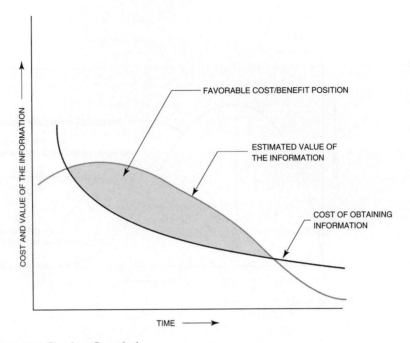

FIGURE 2–19. Cost-benefit analysis.

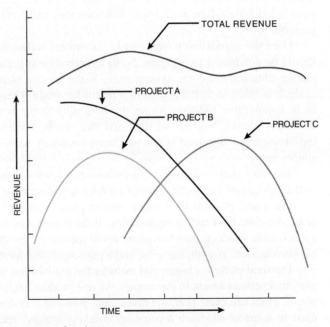

FIGURE 2–20. A stream of projects.

efforts (project C) must be developed for resource reallocation. In the ideal situation these new projects will be established at such a rate that total revenue will increase and company growth will be clearly visible.

The closure phase evaluates the efforts of the total system and serves as input to the conceptual phases for new projects and systems. This final phase also has an impact on other ongoing projects with regard to identifying priorities.

Thus far no attempt has been made to identify the size of a project or system. Large projects generally require full-time staffs, whereas small projects, although they undergo the same system life-cycle phases, may require only part-time people. This implies that an individual can be responsible for multiple projects, possibly with each project existing in a different life-cycle phase. The following questions must be considered in multiproject management:

- Are the project objectives the same?
    - For the good of the project?
    - For the good of the company?
- Is there a distinction between large and small projects?
- How do we handle conflicting priorities?
    - Critical versus critical projects
    - Critical versus noncritical projects
    - Noncritical versus noncritical projects

Later chapters discuss methods of resolving conflicts and establishing priorities.

The phases of a project and those of a product are compared in Figure 2–21. Notice that the life-cycle phases of a product generally do not overlap, whereas the phases of a project can and often do overlap.

Table 2–6 identifies the various life-cycle phases that are commonly used. Even in mature project management industries such as construction, one could survey ten different construction companies and find ten different definitions for the life-cycle phases.

The life-cycle phases for computer programming, as listed in Table 2–6, are also shown in Figure 2–22, which illustrates how manpower resources can build up and decline during a project. In Figure 2–22, PMO stands for the present method of operations, and PMO′ will be the "new" present method of operations after conversion. This life cycle would probably be representative of a twelve-month activity. Most executives prefer short data processing life cycles because computer technology changes rapidly. An executive of a major utility commented that his company was having trouble determining how to terminate a computer programming project to improve customer service because, by the time a package is ready for full implementation, an updated version appears on the scene. Should the original project be canceled and a new project begun? The solution appears to lie in establishing short data processing project life-cycle phases, perhaps through segmented implementation.

Top management is responsible for the periodic review of major projects. This should be accomplished, at a minimum, at the completion of each life-cycle phase.

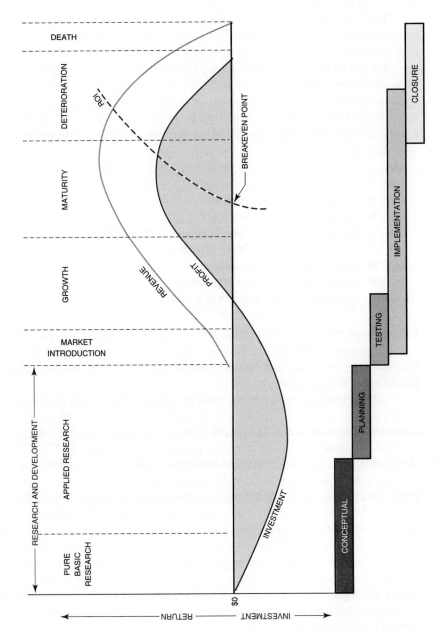

FIGURE 2–21. System/product life cycles.

TABLE 2–6.   LIFE-CYCLE PHASE DEFINITIONS

Engineering	Manufacturing	Computer Programming	Construction
• Start-up • Definition • Main • Termination	• Formation • Buildup • Production • Phase-out • Final audit	• Conceptual • Planning • Definition and design • Implementation • Conversion	• Planning, data gathering, and procedures • Studies and basic engineering • Major review • Detail engineering • Detail engineering/ construction overlap • Construction • Testing and commissioning

More companies are preparing procedural manuals for project management and for structuring work using life-cycle phases. There are several reasons for this trend:

- Clear delineation of the work to be accomplished in each phase may be possible.
- Pricing and estimating may be easier if well-structured work definitions exist.
- Key decision points exist at the end of each life-cycle phase so that incremental funding is possible.

As a final note, the reader should be aware that not all projects can be simply transposed into life-cycle phases (e.g., R&D). It might be possible (even in the same company) for different definitions of life-cycle phases to exist because of schedule length, complexity, or just the difficulty of managing the phases.

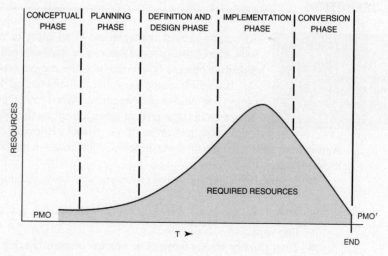

FIGURE 2–22.   Definition of a project life cycle.

2.14 GATE REVIEW MEETINGS (PROJECT CLOSURE)

Gate review meetings are a form of project closure. Gate review meetings could result in the closure of a life-cycle phase or the closure of the entire project. Gate review meetings must be planned for, and this includes the gathering, analysis, and dissemination of pertinent information. This can be done effectively with the use of forms, templates, and checklists.

There are two forms of closure pertinent to gate review meetings: contractual closure and administrative closure. Contractual closure precedes administrative closure. Contractual closure is the verification and signoff that all deliverables required for this phase have been completed and all action items have been fulfilled. Contractual closure is the responsibility of both the project manager and the contract administrator.

Administrative closure is the updating of all pertinent records required for both the customer and the contractor. Customers are particularly interested in documentation on any as-built or as-installed changes or deviations from the specifications. Also required is an archived trail of all scope changes agreed to during the life of the project. Contractors are interested in archived data that include project records, minutes, memos, newsletters, change management documentation, project acceptance documentation, and the history of audits for lessons learned and continuous improvement.

A subset of administrative closure is financial closure, which is the closing out of all charge numbers for the work completed. Even though contractual closure may have taken place, there may still exist open charge numbers for the repair of defects or to complete archived paperwork. Closure must be planned for, and this includes setting up a timetable and budget.

2.15 PROJECT MANAGEMENT METHODOLOGIES: A DEFINITION

> **PMBOK® Guide, 4th Edition**
> Chapter 4 Integration Management
> 4.2.1.4 Project Management Methodology
> 2.4.3 Organizational Process Assets

Achieving project management excellence, or maturity, is more likely with a repetitive process that can be used on each and every project. This repetitive process is referred to as the project management methodology.

If possible, companies should maintain and support a single methodology for project management. Good methodologies integrate other processes into the project management methodology, as shown in Figure 2–23. Companies such as Nortel, Ericsson, and Johnson Controls Automotive have all five of these processes integrated into their project management methodology.

During the 1990s, the following processes were integrated into a single methodology:

- *Project Management:* The basic principles of planning, scheduling, and controlling work
- *Total Quality Management:* The process of ensuring that the end result will meet the quality expectations of the customer

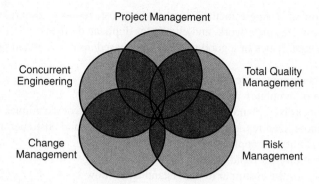

FIGURE 2–23. Integrated processes for the twenty-first century.

- *Concurrent Engineering:* The process of performing work in parallel rather than series in order to compress the schedule without incurring serious risks
- *Scope Change Control:* The process of controlling the configuration of the end result such that value added is provided to the customer
- *Risk Management:* The process of identifying, quantifying, and responding to the risks of the project without any material impact on the project's objectives

In the coming years, companies can be expected to integrate more of their business processes in the project management methodology. This is shown in Figure 2–24.

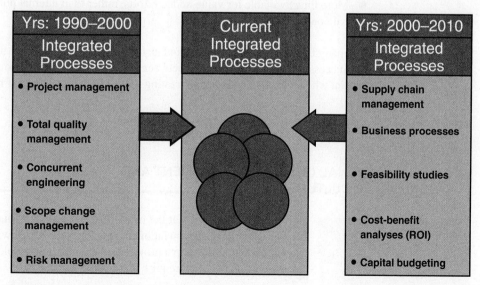

FIGURE 2–24. Integrated processes (past, present, and future).

Managing off of a single methodology lowers cost, reduces resource requirements for support, minimizes paperwork, and eliminates duplicated efforts.

The characteristics of a good methodology based upon integrated processes include:

- A recommended level of detail
- Use of templates
- Standardized planning, scheduling, and cost control techniques
- Standardized reporting format for both in-house and customer use
- Flexibility for application to all projects
- Flexibility for rapid improvements
- Easy for the customer to understand and follow
- Readily accepted and used throughout the entire company
- Use of standardized life-cycle phases (which can overlap) and end of phase reviews (Section 2.13)
- Based upon guidelines rather than policies and procedures (Section 2.9)
- Based upon a good work ethic

Methodologies do not manage projects; people do. It is the corporate culture that executes the methodology. Senior management must create a corporate culture that supports project management and demonstrates faith in the methodology. If this is done successfully, then the following benefits can be expected:

- Faster "time to market" through better control of the project's scope
- Lower overall project risk
- Better decision-making process
- Greater customer satisfaction, which leads to increased business
- More time available for value-added efforts, rather than internal politics and internal competition

One company found that its customers liked its methodology so much and that the projects were so successful, that the relationship between the contractor and the customer improved to the point where the customers began treating the contractor as a partner rather than as a supplier.

2.16 ORGANIZATIONAL CHANGE MANAGEMENT AND CORPORATE CULTURES

PMBOK® Guide, 4th Edition
Chapter 4 Integration Management
4.5 Integrated Change Control
2.4.1 Organizational Culture

It has often been said that the most difficult projects to manage are those that involve the management of change. Figure 2–25 shows the four basic inputs needed to develop a project management methodology. Each has a "human" side that may require that people change.

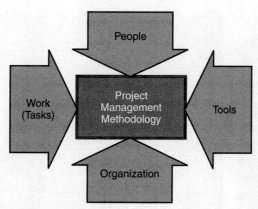

FIGURE 2–25. Methodology inputs.

Successful development and implementation of a project management methodology requires:

- Identification of the most common reasons for change in project management
- Identification of the ways to overcome the resistance to change
- Application of the principles of organizational change management to ensure that the desired project management environment will be created and sustained

For simplicity's sake, resistance can be classified as professional resistance and personal resistance to change. Professional resistance occurs when each functional unit as a whole feels threatened by project management. This is shown in Figure 2–26. Examples include:

- *Sales:* The sales staff's resistance to change arises from fear that project management will take credit for corporate profits, thus reducing the year-end bonuses for the sales force. Sales personnel fear that project managers may become involved in the sales effort, thus diminishing the power of the sales force.
- *Marketing:* Marketing people fear that project managers will end up working so closely with customers that project managers may eventually be given some of the marketing and sales functions. This fear is not without merit because customers often want to communicate with the personnel managing the project rather than those who may disappear after the sale is closed.
- *Finance (and Accounting):* These departments fear that project management will require the development of a project accounting system (such as earned value measurement) that will increase the workload in accounting and finance, and that they will have to perform accounting both horizontally (i.e., in projects) and vertically (i.e., in line groups).

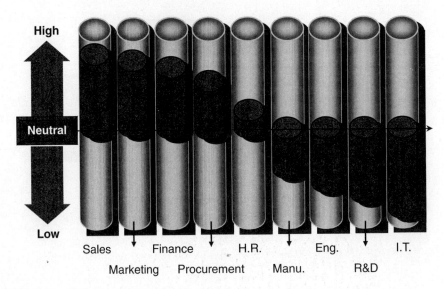

High

Neutral

Low

Sales   Finance   H.R.   Eng.   I.T.

Marketing   Procurement   Manu.   R&D

FIGURE 2–26. Resistance to change.

- *Procurement:* The fear in this group is that a project procurement system will be implemented in parallel with the corporate procurement system, and that the project managers will perform their own procurement, thus bypassing the procurement department.
- *Human Resources Management:* The HR department may fear that a project management career path ladder will be created, requiring new training programs. This will increase their workloads.
- *Manufacturing:* Little resistance is found here because, although the manufacturing segment is not project-driven, there are numerous capital installation and maintenance projects which will have required the use of project management.
- *Engineering, R&D, and Information Technology:* These departments are almost entirely project-driven with very little resistance to project management.

Getting the support of and partnership with functional management can usually overcome the functional resistance. However, the individual resistance is usually more complex and more difficult to overcome. Individual resistance can stem from:

- Potential changes in work habits
- Potential changes in the social groups
- Embedded fears
- Potential changes in the wage and salary administration program

Tables 2–7 through 2–10 show the causes of resistance and possible solutions. Workers tend to seek constancy and often fear that new initiatives will push them outside their comfort zones. Most workers are already pressed for time in their current jobs and fear that new programs will require more time and energy.

TABLE 2–7. RESISTANCE: WORK HABITS

Cause of Resistance	Ways to Overcome
• New guidelines/processes • Need to share "power" information • Creation of a fragmented work environment • Need to give up established work patterns (learn new skills) • Change in comfort zones	• Dictate mandatory conformance from above • Create new comfort zones at an acceptable pace • Identify tangible/intangible individual benefits

Some companies feel compelled to continually undertake new initiatives, and people may become skeptical of these programs, especially if previous initiatives have not been successful. The worst case scenario is when employees are asked to undertake new initiatives, procedures, and processes that they do not understand.

It is imperative that we understand resistance to change. If individuals are happy with their current environment, there will be resistance to change. But what if people are unhappy? There will still be resistance to change unless (1) people believe that the change is possible, and (2) people believe that they will somehow benefit from the change.

Management is the architect of the change process and must develop the appropriate strategies so the organization can change. This is done best by developing a shared understanding with employees by doing the following:

- Explaining the reasons for the change and soliciting feedback
- Explaining the desired outcomes and rationale
- Championing the change process
- Empowering the appropriate individuals to institutionalize the changes
- Investing in training necessary to support the changes

For most companies, the change management process will follow the pattern shown in Figure 2–27. Employees initially refuse to admit the need for change. As management begins pursuing the change, the support for the change diminishes and pockets of resistance crop up. Continuous support for the change by management encourages employees to explore the potential opportunities that will result from the change about to take place. Unfortunately, this exploration often causes additional negative information to surface, thus reinforcing the resistance to change. As pressure by management increases, and employees begin to recognize the benefits of the proposed change, support begins to grow.

TABLE 2–8. RESISTANCE: SOCIAL GROUPS

Cause of Resistance	Ways to Overcome
• Unknown new relationships • Multiple bosses • Multiple, temporary assignments • Severing of established ties	• Maintain existing relationships • Avoid cultural shock • Find an acceptable pace for rate of change

TABLE 2–9. RESISTANCE: EMBEDDED FEARS

Cause of Resistance	Ways to Overcome
• Fear of failure • Fear of termination • Fear of added workload • Fear or dislike of uncertainty/unknowns • Fear of embarrassment • Fear of a "we/they" organization	• Educate workforce on benefits of changes to the individual/corporation • Show willingness to admit/accept mistakes • Show willingness to pitch in • Transform unknowns into opportunities • Share information

TABLE 2–10. RESISTANCE: WAGE AND SALARY ADMINISTRATION

Causes of Resistance	Ways to Overcome
• Shifts in authority and power • Lack of recognition after the changes • Unknown rewards and punishment • Improper evaluation of personal performance • Multiple bosses	• Link incentives to change • Identify future advancement opportunities/career path

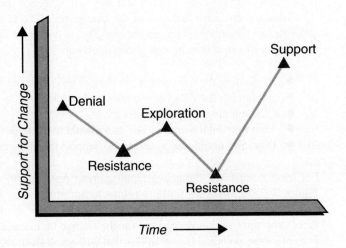

FIGURE 2–27. Change process.

The ideal purpose of change management is to create a superior culture. There are different types of project management cultures based upon the nature of the business, the amount of trust and cooperation, and the competitive environment. Typical types of cultures include:

● *Cooperative cultures:*   These are based upon trust and effective communications, internally and externally.
● *Noncooperative cultures:*   In these cultures, mistrust prevails. Employees worry more about themselves and their personal interests than what's best for the team, company, or customer.

- *Competitive cultures:* These cultures force project teams to compete with one another for valuable corporate resources. In these cultures, project managers often demand that the employees demonstrate more loyalty to the project than to their line managers. This can be disastrous when employees are working on many projects at the same time.
- *Isolated cultures:* These occur when a large organization allows functional units to develop their own project management cultures and can result in a culture-within-a-culture environment.
- *Fragmented cultures:* These occur when part of the team is geographically separated from the rest of the team. Fragmented cultures also occur on multinational projects, where the home office or corporate team may have a strong culture for project management but the foreign team has no sustainable project management culture.

Cooperative cultures thrive on effective communication, trust, and cooperation. Decisions are based upon the best interest of all of the stakeholders. Executive sponsorship is passive, and very few problems go to the executive levels for resolution. Projects are managed informally and with minimal documentation and few meetings. This culture takes years to achieve and functions well during favorable and unfavorable economic conditions.

Noncooperative cultures are reflections of senior management's inability to cooperate among themselves and with the workforce. Respect is nonexistent. These cultures are not as successful as a cooperative culture.

Competitive cultures can be healthy in the short term, especially if there is abundant work. Long-term effects are usually not favorable. In one instance, an electronics firm regularly bid on projects that required the cooperation of three departments. Management then implemented the unhealthy decision of allowing each of the three departments to bid on every job. The two "losing" departments would be treated as subcontractors.

Management believed that this competitiveness was healthy. Unfortunately, the long-term results were disastrous. The three departments refused to talk to one another and stopped sharing information. In order to get the job done for the price quoted, the departments began outsourcing small amounts of work rather than using the other departments that were more expensive. As more work was outsourced, layoffs occurred. Management then realized the disadvantages of the competitive culture it had fostered.

2.17 PROJECT MANAGEMENT INTELLECTUAL PROPERTY

We believe today that we are managing our business by projects. As such, project managers are expected to make business decisions as well as project decisions. This also implies that we must capture not only project-related best practices, but business best practices as well.

For the past decade, whenever we would capture project management best practices, they would be placed in a project management best practices library. But as we capture business best practices, we begin replacing the project management best practices library with a knowledge repository that includes both project management and business-related best practices. This is shown in Figure 2–28.

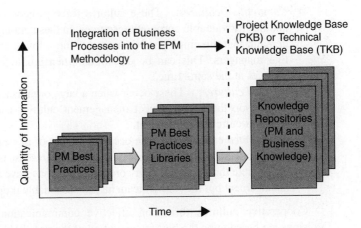

Figure 2–28. Growth of knowledge management.

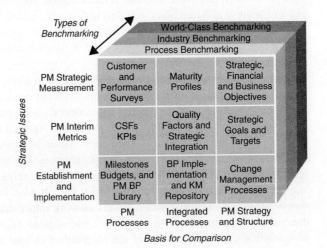

Figure 2–29. PM benchmarking and knowledge management (KM).

Another reason for the growth in intellectual property is because of the benchmarking activities that companies are performing, most likely using the project management office. Figure 2–29 shows typical benchmarking activities and the types of information being sought.

2.18 SYSTEMS THINKING

Ultimately, all decisions and policies are made on the basis of judgments; there is no other way, and there never will be. In the end, analysis is but an aid to the judgment and intuition of the decision maker. These principles hold true for project management as well as for systems management.

The systems approach may be defined as a logical and disciplined process of problem-solving. The word *process* indicates an active ongoing system that is fed by input from its parts. The systems approach:

- Forces review of the relationship of the various subsystems
- Is a dynamic process that integrates all activities into a meaningful total system
- Systematically assembles and matches the parts of the system into a unified whole
- Seeks an optimal solution or strategy in solving a problem

The systems approach to problem-solving has phases of development similar to the life-cycle phases shown in Figure 2–21. These phases are defined as follows:

- *Translation:* Terminology, problem objective, and criteria and constraints are defined and accepted by all participants.
- *Analysis:* All possible approaches to or alternatives to the solution of the problem are stated.
- *Trade-off:* Selection criteria and constraints are applied to the alternatives to meet the objective.
- *Synthesis:* The best solution in reaching the objective of the system is the result of the combination of analysis and trade-off phases.

Other terms essential to the systems approach are:

- *Objective:* The function of the system or the strategy that must be achieved.
- *Requirement:* A partial need to satisfy the objective.
- *Alternative:* One of the selected ways to implement and satisfy a requirement.
- *Selection criteria:* Performance factors used in evaluating the alternatives to select a preferable alternative.
- *Constraint:* An absolute factor that describes conditions that the alternatives *must* meet.

A common error by potential decision makers (those dissatisfied individuals with authority to act) who base their thinking solely on subjective experience, judgment, and intuition is that they fail to recognize the existence of alternatives. Subjective thinking is inhibited or affected by personal bias.

Objective thinking, on the other hand, is a fundamental characteristic of the systems approach and is exhibited or characterized by emphasis on the tendency to view events, phenomena, and ideas as external and apart from self-consciousness. Objective thinking is unprejudiced.

The systems analysis process, as shown in Figure 2–30, begins with systematic examination and comparison of those alternative actions that are related to the accomplishment of the desired objective. The alternatives are then compared on the basis of the resource costs and the associated benefits. The loop is then completed using feedback to determine how compatible each alternative is with the objectives of the organization.

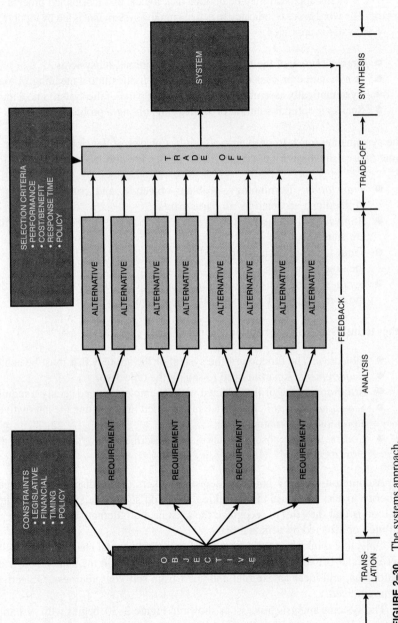

FIGURE 2–30. The systems approach.

The above analysis can be arranged in steps:

- Input data to mental process
- Analyze data
- Predict outcomes
- Evaluate outcomes and compare alternatives
- Choose the best alternative
- Take action
- Measure results and compare them with predictions

The systems approach is most effective if individuals can be trained to be ready with alternative actions that directly tie in with the prediction of outcomes. The basic tool is the outcome array, which represents the matrix of all possible circumstances. This outcome array can be developed only if the decision maker thinks in terms of the wide scope of possible outcomes. Outcome descriptions force the decision maker to spell out clearly just what he is trying to achieve (i.e., his objectives).

Systems thinking is vital for the success of a project. Project management systems urgently need new ways of strategically viewing, questioning, and analyzing project needs for alternative nontechnical and technical solutions. The ability to analyze the total project, rather than the individual parts, is essential for successful project management.

2.19 STUDYING TIPS FOR THE PMI® PROJECT MANAGEMENT CERTIFICATION EXAM

This section is applicable as a review of the principles to support the knowledge areas and domain groups in the PMBOK® Guide. This chapter addresses:

- Integration Management
- Scope Management
- Closure

Understanding the following principles is beneficial if the reader is using this text to study for the PMP® Certification Exam:

- Brief historical background of project management
- That, early on, project managers were assigned from engineering
- Benefits of project management
- Barriers to project management implementation and how to overcome them
- Differences between a program and a project
- What is meant by informal project management
- How to identify success and failure in project management
- Project life-cycle phases
- What is meant by closure to a life-cycle phase or to the entire project

- What is meant by a project management methodology
- What is meant by critical success factors (CSFs) and key performance indicators (KPIs)

In Appendix C, the following Dorale Products mini–case studies are applicable:

- Dorale Products (A) [Integration and Scope Management]
- Dorale Products (B) [Integration and Scope Management]
- Dorale Products (C) [Integration and Scope Management]
- Dorale Products (D) [Integration and Scope Management]
- Dorale Products (E) [Integration and Scope Management]
- Dorale Products (F) [Integration and Scope Management]

The following multiple-choice questions will be helpful in reviewing the principles of this chapter:

1. A structured process for managing a multitude of projects is most commonly referred to as:
   A. Project management policies
   B. Project management guidelines
   C. Industrywide templates
   D. A project management methodology

2. The most common terminology for a reusable project management methodology is:
   A. Template
   B. Concurrent scheduling technique
   C. Concurrent planning technique
   D. Skeleton framework document

3. The major behavioral issue in getting an organization to accept and use a project management methodology effectively is:
   A. Lack of executive sponsorship
   B. Multiple boss reporting
   C. Inadequate policies and procedures
   D. Limited project management applications

4. The major difference between a project and a program is usually:
   A. The role of the sponsor
   B. The role of the line manager
   C. The timeframe
   D. The specifications

5. Projects that remain almost entirely within one functional area are best managed by the:
   A. Project manager
   B. Project sponsor
   C. Functional manager
   D. Assigned functional employees

6. Large projects are managed by:
   A. The executive sponsor
   B. The project or program office for that project

    C. The manager of project managers

    D. The director of marketing

7. The most common threshold limits on when to use the project management methodology are:

    A. The importance of the customer and potential profitability

    B. The size of the project (i.e., $) and duration

    C. The reporting requirements and position of the sponsor

    D. The desires of management and functional boundaries crossed

8. A grouping of projects is called a:

    A. Program

    B. Project template

    C. Business template

    D. Business plan

9. Project management methodologies often work best if they are structured around:

    A. Rigid policies

    B. Rigid procedures

    C. Minimal forms and checklists

    D. Life-cycle phases

10. One way to validate the successful implementation of project management is by looking at the number and magnitude of the conflicts requiring:

    A. Executive involvement

    B. Customer involvement

    C. Line management involvement

    D. Project manager involvement

11. Standardization and control are benefits usually attributed to:

    A. Laissez-faire management

    B. Project management on R&D efforts

    C. Use of life cycle-phases

    D. An organization with weak executive sponsorship

12. The most difficult decision for an executive sponsor to make at the end-of-phase review meeting is to:

    A. Allow the project to proceed to the next phase based upon the original objective

    B. Allow the project to proceed to the next phase based upon a revised objective

    C. Postpone making a decision until more information is processed

    D. Cancel the project

13. Having too many life-cycle phases may be detrimental because:

    A. Executive sponsors will micromanage.

    B. Executive sponsors will become "invisible."

    C. The project manager will spend too much time planning for gate review meetings rather than managing the phases.

    D. The project manager will need to develop many different plans for each phase.

14. A project is terminated early because the technology cannot be developed, and the resources are applied to another project that ends up being successful. Which of the following is true concerning the first project?

    A. The first project is regarded as a failure.

    B. The first project is a success if the termination is done early enough before additional resources are squandered.

C. The first project is a success if the project manager gets promoted.

D. The first project is a failure if the project manager gets reassigned to a less important project.

15. Which of the following would *not* be regarded as a secondary definition of project success?

A. The customer is unhappy with the deliverable, but follow-on business is awarded based on effective customer relations.

B. The deliverables are met but OSHA and EPA laws are violated.

C. The customer is displeased with the performance, but you have developed a new technology that could generate many new products.

D. The project's costs were overrun by 40 percent, but the customer funds an enhancement project.

ANSWERS

1. D
2. A
3. B
4. C
5. C
6. B
7. B
8. A
9. D
10. A
11. C
12. D
13. C
14. B
15. B

PROBLEMS

2–1   Can the organizational chart of a company be considered as a systems model? If so, what kind of systems model?

2–2   Do you think that someone could be a good systems manager but a poor project manager? What about the reverse situation? State any assumptions that you may have to make.

2–3   Can we consider R&D as a system? If so, under what circumstances?

2–4   For each of the following projects, state whether we are discussing an open, closed, or extended system:

a. A high-technology project
b. New product R&D

    c. An on-line computer system for a bank

    d. Construction of a chemical plant

    e. Developing an in-house cost accounting reporting system

2–5   Can an entire organization be considered as a model? If so, what type?

2–6   Systems can be defined as a combination or interrelationship of subsystems. Does a project have subsystems?

2–7   If a system can, in fact, be broken down into subsystems, what problems can occur during integration?

2–8   How could suboptimization occur during systems thinking and analysis?

2–9   Would a cost-benefit analysis be easier or harder to perform in a traditional or project management organizational structure?

2–10 What impact could the product life cycle have on the selection of the project organizational structure?

2–11 In the development of a system, what criteria should be used to determine where one phase begins and another ends and where overlap can occur?

2–12 Consider the following expression: "Damn the torpedoes: full-speed ahead." Is it possible that this military philosophy can be applied to project management and lead to project success?

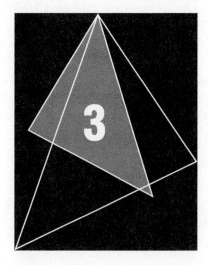

Organizational Structures

Related Case Studies (from Kerzner/*Project Management Case Studies*, 3rd Edition)	Related Workbook Exercises (from Kerzner/*Project Management Workbook and PMP®/CAPM® Exam Study Guide*, 10th Edition)	PMBOK® Guide, 4th Edition, Reference Section for the PMP® Certification Exam
• Quasar Communications, Inc. • Jones and Shephard Accountants, Inc.* • Fargo Foods • Mohawk National Bank	• The Struggle with Implementation • Multiple Choice Exam	• Human Resource Management

3.0 INTRODUCTION

PMBOK® Guide, 4th Edition
2.4.2 Organizational Structure
Chapter 9 Human Resource
  Management

During the past thirty years there has been a so-called hidden revolution in the introduction and development of new organizational structures. Management has come to realize that organizations must be dynamic in nature; that is, they must be capable of rapid restructuring should environmental conditions so dictate. These environmental factors evolved

---

* Case Study also appears at end of chapter.

from the increasing competitiveness of the market, changes in technology, and a requirement for better control of resources for multiproduct firms. More than forty years ago, Wallace identified four major factors that caused the onset of the organizational revolution[1]:

- The technology revolution (complexity and variety of products, new materials and processes, and the effects of massive research)
- Competition and the profit squeeze (saturated markets, inflation of wage and material costs, and production efficiency)
- The high cost of marketing
- The unpredictability of consumer demands (due to high income, wide range of choices available, and shifting tastes)

Much has been written about how to identify and interpret those signs that indicate that a new organizational form may be necessary. According to Grinnell and Apple, there are five general indications that the traditional structure may not be adequate for managing projects[2]:

- Management is satisfied with its technical skills, but projects are not meeting time, cost, and other project requirements.
- There is a high commitment to getting project work done, but great fluctuations in how well performance specifications are met.
- Highly talented specialists involved in the project feel exploited and misused.
- Particular technical groups or individuals constantly blame each other for failure to meet specifications or delivery dates.
- Projects are on time and to specifications, but groups and individuals aren't satisfied with the achievement.

Unfortunately, many companies do not realize the necessity for organizational change until it is too late. Management looks externally (i.e., to the environment) rather than internally for solutions to problems. A typical example would be that new product costs are rising while the product life cycle may be decreasing. Should emphasis be placed on lowering costs or developing new products?

If we assume that an organizational system is composed of both human and nonhuman resources, then we must analyze the sociotechnical subsystem whenever organizational changes are being considered. The social system is represented by the organization's personnel and their group behavior. The technical system includes the technology, materials, and machines necessary to perform the required tasks.

Behavioralists contend that there is no one best structure to meet the challenges of tomorrow's organizations. The structure used, however, must be one that optimizes company performance by achieving a balance between the social and the technical requirements. According to Sadler[3]:

> Since the relative influence of these (sociotechnical) factors change from situation to situation, there can be no such thing as an ideal structure making for effectiveness in organizations of all kinds, or even appropriate to a single type of organization at different stages in its development.

---

1. W. L. Wallace, "The Winchester-Western Division Concept of Product Planning" (New Haven: Olin Mathieson Corporation, January 1963), pp. 2–3.

2. S. K. Grinnell and H. P. Apple, "When Two Bosses Are Better Than One," *Machine Design,* January 1975, pp. 84–87.

3. Philip Sadler, "Designing an Organizational Structure," *Management International Review,* Vol. 11, No. 6, 1971, pp. 19–33.

There are often real and important conflicts between the type of organizational structure called for if the tasks are to be achieved with minimum cost, and the structure that will be required if human beings are to have their needs satisfied. Considerable management judgment is called for when decisions are made as to the allocation of work activities to individuals and groups. High standardization of performance, high manpower utilization and other economic advantages associated with a high level of specialization and routinization of work have to be balanced against the possible effects of extreme specialization in lowering employee attitudes and motivation.

Organizations can be defined as groups of people who must coordinate their activities in order to meet organizational objectives. The coordination function requires strong communications and a clear understanding of the relationships and interdependencies among people. Organizational structures are dictated by such factors as technology and its rate of change, complexity, resource availability, products and/or services, competition, and decision-making requirements. The reader must keep in mind that *there is no such thing as a good or bad organizational structure; there are only appropriate or inappropriate ones.*

Even the simplest type of organizational change can induce major conflicts. The creation of a new position, the need for better planning, the lengthening or shortening of the span of control, the need for additional technology (knowledge), and centralization or decentralization can result in major changes in the sociotechnical subsystem. Argyris has defined five conditions that form the basis for organizational change requirements[4]:

These requirements . . . depend upon (1) continuous and open access between individuals and groups, (2) free, reliable communication, where (3) independence is the foundation for individual and departmental cohesiveness and (4) trust, risk-taking and helping each other is prevalent so that (5) conflict is identified and managed in such a way that the destructive win-lose stances with their accompanying polarization of views are minimized. . . . Unfortunately these conditions are difficult to create. . . . There is a tendency toward conformity, mistrust and lack of risk-taking among the peers that results in focusing upon individual survival, requiring the seeking out of the scarce rewards, identifying one's self with a successful venture (be a hero) and being careful to avoid being blamed for or identified with a failure, thereby becoming a "bum." All these adaptive behaviors tend to induce low interpersonal competence and can lead the organization, over the long-run, to become rigid, sticky, and less innovative, resulting in less than effective decisions with even less internal commitment to the decision on the part of those involved.

Organizational restructuring is a compromise between the traditional (classical) and the behavioral schools of thought; management must consider the needs of individuals as well as the needs of the company. Is the organization structured to manage people or to manage work?

There is a wide variety of organizational forms for restructuring management. The exact method depends on the people in the organization, the company's product lines, and management's philosophy. A poorly restructured organization can sever communication channels that may have taken months or years to cultivate; cause a restructuring of the informal organization, thus creating new power, status, and political positions; and eliminate job satisfaction and motivational factors to such a degree that complete discontent results.

Sadler defines three tasks that must be considered because of the varied nature of organizations: control, integration, and external relationships.[5] If the company's position is very sensitive to the environment, then management may be most concerned with the control task. For an organization with multiple products, each requiring a high degree of engineering and technology, the integration task can become primary. Finally, for

---

4. Chris Argyris, "Today's Problems with Tomorrow's Organizations," *The Journal of Management Studies,* February 1967, pp. 31–55.

5. See note 3.

situations with strong labor unions and repetitive tasks, external relations can predominate, especially in strong technological and scientific environments where strict government regulations must be adhered to.

In the sections that follow, a variety of organizational forms will be presented. Obviously, it is an impossible task to describe all possible organizational structures. Each form describes how the project management organization evolved from the classical theories of management. Advantages and disadvantages are listed for technology and social systems. Sadler has prepared a six-question checklist that explores a company's tasks, social climate, and relationship to the environment.[6]

- To what extent does the task of organization call for close control if it is to be performed efficiently?
- What are the needs and attitudes of the people performing the tasks? What are the likely effects of control mechanisms on their motivation and performance?
- What are the natural social groupings with which people identify themselves? To what extent are satisfying social relationships important in relation to motivation and performance?
- What aspect of the organization's activities needs to be closely integrated if the overall task is to be achieved?
- What organizational measures can be developed that will provide an appropriate measure of control and integration of work activities, while at the same time meeting the needs of people and providing adequate motivation?
- What environmental changes are likely to affect the future trend of company operations? What organizational measures can be taken to insure that the enterprise responds to these effectively?

The answers to these questions are not easy. For the most part, they are a matter of the judgment exercised by organizational and behavioral managers.

3.1 ORGANIZATIONAL WORK FLOW

Organizations are continually restructured to meet the demands imposed by the environment. Restructuring can change the role of individuals in the formal and the informal organization. Many researchers believe that the greatest usefulness of behavioralists lies in their ability to help the informal organization adapt to changes and resolve the resulting conflicts. Unfortunately, behavioralists cannot be totally effective unless they have input into the formal organization as well. Whatever organizational form is finally selected, formal channels must be developed so that each individual has a clear description of the authority, responsibility, and accountability necessary for the work to proceed.

In the discussion of organizational structures, the following definitions will be used:

- *Authority* is the power granted to individuals (possibly by their position) so that they can make final decisions.
- *Responsibility* is the obligation incurred by individuals in their roles in the formal organization to effectively perform assignments.

---

6. See note 3.

- *Accountability* is being answerable for the satisfactory completion of a specific assignment. (Accountability = authority + responsibility.)

Authority and responsibility can be delegated to lower levels in the organization, whereas accountability usually rests with the individual. Yet many executives refuse to delegate and argue that an individual can have total accountability just through responsibility.

Even with these clearly definable divisions of authority, responsibility, and accountability, establishing good relationships between project and functional managers can take a great deal of time, especially during the conversion from a traditional to a project organizational form. Trust is the key to success here. The normal progression in the growth of trust is as follows:

- Even though a problem exists, both the project and functional managers deny that any problem exists.
- When the problem finally surfaces, each manager blames the other.
- As trust develops, both managers readily admit responsibility for the problems.
- The project and functional managers meet face-to-face to work out the problem.
- The project and functional managers begin to formally and informally anticipate problems.

For each of the organizational structures described in the following sections, advantages and disadvantages are listed. Many of the disadvantages stem from possible conflicts arising from problems in authority, responsibility, and accountability.

3.2 TRADITIONAL (CLASSICAL) ORGANIZATION

The traditional management structure has survived for more than two centuries. However, recent business developments, such as the rapid rate of change in technology and increased stockholder demands, have created strains on existing organizational forms. Fifty years ago companies could survive with only one or two product lines. The classical management organization, as shown in Figure 3–1, was satisfactory for control, and conflicts were minimal.[7]

However, with the passing of time, companies found that survival depended on multiple product lines (i.e., diversification) and vigorous integration of technology into the existing organization. As organizations grew and matured, managers found that company activities were not being integrated effectively, and that new conflicts were arising in the well-established formal and informal channels. Managers began searching for more innovative organizational forms that would alleviate these problems.

---

7. Many authors refer to classical organizations as pure functional organizations. This can be seen from Figure 3–1. Also note that the department level is below the division level. In some organizations these titles are reversed.

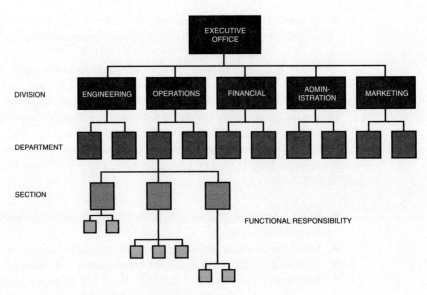

FIGURE 3–1. The traditional management structure.

Before a valid comparison can be made with the newer forms, the advantages and disadvantages of the traditional structure must be shown. Table 3–1 lists the advantages of the traditional organization. As seen in Figure 3–1, the general manager has all of the functional entities necessary to perform R&D or develop and manufacture a product. All activities are performed within the functional groups and are headed by a department (or, in some cases, a division) head. Each department maintains a strong concentration of technical expertise. Since all projects must flow through the functional departments, each project can benefit from the most advanced technology, thus making this organizational form well suited to mass production. Functional managers can hire a wide variety of specialists and provide them with easily definable paths for career progression.

TABLE 3–1.   ADVANTAGES OF THE TRADITIONAL (CLASSICAL) ORGANIZATION

- Easier budgeting and cost control are possible.
- Better technical control is possible.
  - Specialists can be grouped to share knowledge and responsibility.
  - Personnel can be used on many different projects.
  - All projects will benefit from the most advanced technology (better utilization of scarce personnel).
- Flexibility in the use of manpower.
- A broad manpower base to work with.
- Continuity in the functional disciplines; policies, procedures, and lines of responsibility are easily defined and understandable.
- Admits mass production activities within established specifications.
- Good control over personnel, since each employee has one and only one person to report to.
- Communication channels are vertical and well established.
- Quick reaction capability exists, but may be dependent upon the priorities of the functional managers.

TABLE 3–2. DISADVANTAGES OF THE TRADITIONAL (CLASSICAL) ORGANIZATION

- No one individual is directly responsible for the total project (i.e., no formal authority; committee solutions).
- Does not provide the project-oriented emphasis necessary to accomplish the project tasks.
- Coordination becomes complex, and additional lead time is required for approval of decisions.
- Decisions normally favor the strongest functional groups.
- No customer focal point.
- Response to customer needs is slow.
- Difficulty in pinpointing responsibility; this is the result of little or no direct project reporting, very little project-oriented planning, and no project authority.
- Motivation and innovation are decreased.
- Ideas tend to be functionally oriented with little regard for ongoing projects.

The functional managers maintain absolute control over the budget. They establish their own budgets, on approval from above, and specify requirements for additional personnel. Because the functional manager has manpower flexibility and a broad base from which to work, most projects are normally completed within cost.

Both the formal and informal organizations are well established, and levels of authority and responsibility are clearly defined. Because each person reports to only one individual, communication channels are well structured. If a structure has this many advantages, then why are we looking for other structures?

For each advantage, there is almost always a corresponding disadvantage (see Table 3–2). The majority of these disadvantages are related to the absence of a strong central authority or individual responsible for the total project. As a result, integration of activities that cross functional lines becomes difficult, and top-level executives must get involved with the daily routine. Conflicts occur as each functional group struggles for power. Ideas may remain functionally oriented with very little regard for ongoing projects, and the decision-making process will be slow and tedious.

Because there is no customer focal point, all communications must be channeled through upper-level management. Upper-level managers then act in a customer-relations capacity and refer all complex problems down through the vertical chain of command to the functional managers. The response to the customer's needs therefore becomes a slow and aggravating process.

Projects have a tendency to fall behind schedule in the classical organizational structure. Incredibly large lead times are required. Functional managers attend to those tasks that provide better benefits to themselves and their subordinates first.

With the growth of project management in the late 1960s, executives began to realize that many of the problems were the result of weaknesses in the traditional structure. William Goggin identified the problems that faced Dow Corning[8]:

> Although Dow Corning was a healthy corporation in 1967, it showed difficulties that troubled many of us in top management. These symptoms were, and still are, common ones in

---

8. Reprinted by permission of *Harvard Business Review.* From William C. Goggin, "How the Multidimensional Structure Works at Dow Corning," *Harvard Business Review,* January–February 1974, p. 54. Copyright © 1973 by the Harvard Business School Publishing Corporation; all rights reserved.

U.S. business and have been described countless times in reports, audits, articles and speeches. Our symptoms took such form as:

- Executives did not have adequate financial information and control of their operations. Marketing managers, for example, did not know how much it cost to produce a product. Prices and margins were set by division managers.
- Cumbersome communications channels existed between key functions, especially manufacturing and marketing.
- In the face of stiffening competition, the corporation remained too internalized in its thinking and organizational structure. It was insufficiently oriented to the outside world.
- Lack of communications between divisions not only created the antithesis of a corporate team effort but also was wasteful of a precious resource—people.
- Long-range corporate planning was sporadic and superficial; this was leading to overstaffing, duplicated effort and inefficiency.

3.3 DEVELOPING WORK INTEGRATION POSITIONS

As companies grew in size, more emphasis was placed on multiple ongoing programs with high-technology requirements. Organizational pitfalls soon appeared, especially in the integration of the flow of work. As management discovered that the critical point in any program is the interface between functional units, the new theories of "interface management" developed.

Because of the interfacing problems, management began searching for innovative methods to coordinate the flow of work between functional units without modification to the existing organizational structure. This coordination was achieved through several integrating mechanisms[9]:

- Rules and procedures
- Planning processes
- Hierarchical referral
- Direct contact

By specifying and documenting management policies and procedures, management attempted to eliminate conflicts between functional departments. Management felt that, even though many of the projects were different, the actions required by the functional personnel were repetitive and predictable. The behavior of the individuals should therefore be easily integrated into the flow of work with minimum communication between individuals or functional groups.

---

9. Jay R. Galbraith, "Matrix Organization Designs." Reprinted with permission from *Business Horizons,* February 1971, pp. 29–40. Copyright © 1971 by the Board of Trustees at Indiana University. Galbraith defines a fifth mechanism, liaison departments, that will be discussed later in this section.

Another means of reducing conflicts and minimizing the need for communication was detailed planning. Functional representation would be present at all planning, scheduling, and budget meetings. This method worked best for nonrepetitive tasks and projects.

In the traditional organization, one of the most important responsibilities of upper-level management was the resolution of conflicts through "hierarchical referral." The continuous conflicts and struggle for power between the functional units consistently required that upper-level personnel resolve those problems resulting from situations that were either nonroutine or unpredictable and for which no policies or procedures existed.

The fourth method is direct contact and interactions by the functional managers. The rules and procedures, as well as the planning process method, were designed to minimize ongoing communications between functional groups. The quantity of conflicts that executives had to resolve forced key personnel to spend a great percentage of their time as arbitrators, rather than as managers. To alleviate problems of hierarchical referral, upper-level management requested that all conflicts be resolved at the lowest possible levels. This required that functional managers meet face-to-face to resolve conflicts.

In many organizations, these new methods proved ineffective, primarily because there still existed a need for a focal point for the project to ensure that all activities would be properly integrated.

When the need for project managers was acknowledged, the next logical question was where in the organization to place them. Executives preferred to keep project managers low in the organization. After all, if they reported to someone high up, they would have to be paid more and would pose a continuous threat to management.

The first attempt to resolve this problem was to develop project leaders or coordinators within each functional department, as shown in Figure 3–2. Section-level personnel were temporarily assigned as project leaders and would return to their former positions at project termination. This is why the term "project leader" is used rather than "project manager," as the word "manager" implies a permanent relationship. This arrangement proved effective for coordinating and integrating work within one department, provided that the correct project leader was selected. Some employees considered this position an increase in power and status, and conflicts occurred about whether assignments should be based on experience, seniority, or capability. Furthermore, the project leaders had almost no authority, and section-level managers refused to take directions from them, fearing that the project leaders might be next in line for the department manager's position.

When the activities required efforts that crossed more than one functional boundary, conflicts arose. The project leader in one department did not have the authority to coordinate activities in any other department. Furthermore, the creation of this new position caused internal conflicts within each department. As a result, many employees refused to become dedicated to project management and were anxious to return to their "secure" jobs. Quite often, especially when cross-functional integration was required, the division manager was forced to act as the project manager. If the employee enjoyed the assignment of project leader, he would try to "stretch out" the project as long as possible.

Even though we have criticized this organizational form, it does not mean that it cannot work. Any organizational form will work if the employees want it to work. As an example, a computer manufacturer has a midwestern division with three departments, as in Figure 3–2, and approximately fourteen people per department. When a project comes

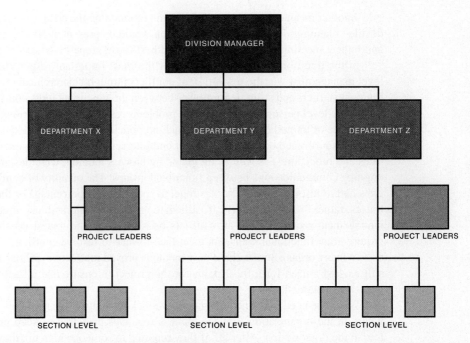

FIGURE 3–2. Departmental project management.

in, the division manager determines which department will handle most of the work. Let us say that the work load is 60 percent department X, 30 percent department Y, and 10 percent department Z. Since most of the effort is in department X, the project leader is selected from that department. When the project leader goes into the other two departments to get resources, he will almost always get the resources he wants. This organizational form works in this case because:

- The other department managers know that they may have to supply the project leader on the next activity.
- There are only three functional boundaries or departments involved (i.e., a small organization).

The next step in the evolution of project management was the task force concept. The rationale behind the task force concept was that integration could be achieved if each functional unit placed a representative on the task force. The group could then jointly solve problems as they occurred, provided that budget limitations were still adhered to. Theoretically, decisions could now be made at the lowest possible levels, thus expediting information and reducing, or even eliminating, delay time.

The task force was composed of both part-time and full-time personnel from each department involved. Daily meetings were held to review activities and discuss potential problems. Functional managers soon found that their task force employees were spending more time in unproductive meetings than in performing functional activities. In addition, the nature of the task force position caused many individuals to shift membership within the informal organization. Many functional managers then placed nonqualified and inexperienced individuals on task forces. The result was that the group soon became ineffective because they either did not have the information necessary to make the decisions, or lacked the authority (delegated by the functional managers) to allocate resources and assign work.

Development of the task force concept was a giant step toward conflict resolution: Work was being accomplished on time, schedules were being maintained, and costs were usually within budget. But integration and coordination were still problems because there were no specified authority relationships or individuals to oversee the entire project through completion. Attempts were made to overcome this by placing various people in charge of the task force: Functional managers, division heads, and even upper-level management had opportunities to direct task forces. However, without formal project authority relationships, task force members remained loyal to their functional organizations, and when conflicts came about between the project and functional organization, the project always suffered.

Although the task force concept was a step in the right direction, the disadvantages strongly outweighed the advantages. A strength of the approach was that it could be established very rapidly and with very little paperwork. Integration, however, was complicated; work flow was difficult to control; and functional support was difficult to obtain because it was almost always strictly controlled by the functional manager. In addition, task forces were found to be grossly ineffective on long-range projects.

The next step in the evolution of work integration was the establishment of liaison departments, particularly in engineering divisions that perform multiple projects involving a high level of technology (see Figure 3–3). The purpose of the liaison department was to

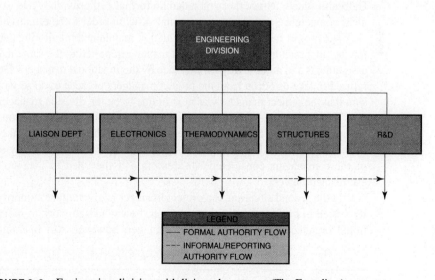

FIGURE 3–3. Engineering division with liaison department (The Expeditor).

handle transactions between functional units within the (engineering) division. The liaison personnel received their authority through the division head. The liaison department did not actually resolve conflicts. Their prime function was to assure that all departments worked toward the same requirements and goals. Liaison departments are still in existence in many large companies and typically handle engineering changes and design problems.

Unfortunately, the liaison department is simply a scaleup of the project coordinator within the department. The authority given to the liaison department extends only to the outer boundaries of the division. If a conflict arose between the manufacturing and engineering divisions, for example, it would still be referred to upper management for resolution. Today, liaison departments are synonymous with project engineering and systems engineering departments, and the individuals in these departments have the authority to span the entire organization.

3.4 LINE–STAFF ORGANIZATION (PROJECT COORDINATOR)

It soon became obvious that control of a project must be given to personnel whose first loyalty is directed toward the completion of the project. Thus the project management position must not be controlled by the functional managers. Figure 3–4 shows a typical line–staff organization.

Two possible situations can exist with this form of line–staff project control. In the first, the project manager serves only as the focal point for activity control, that is, a center for information. The prime responsibility of the project manager is to keep the division manager informed of the status of the project and to "harass" or attempt to "influence" managers into completing activities on time. Referring to such early project managers, Galbraith stated, "Since these men had no formal authority, they had to resort to their technical competence and their interpersonal skills in order to be effective."[10]

The project manager in the first situation maintained monitoring authority only, despite the fact that both he and the functional manager reported to the same individual. Both work assignments and merit reviews were made by the functional managers. Department managers refused to take direction from the project managers because to do so would seem an admission that the project manager was next in line to be the division manager.

The amount of authority given to the project manager posed serious problems. Almost all upper-level and division managers were from the classical management schools and therefore maintained serious reservations about how much authority to relinquish. Many of these managers considered it a demotion if they had to give up any of their long-established powers.

In the second situation, the project manager is given more authority; using the authority vested in him by the division manager, he can assign work to individuals in the functional organizations. The functional manager, however, still maintains the authority to

10. Jay R. Galbraith, "Matrix Organization Designs." *Business Horizons,* February 1971, pp. 29–40.

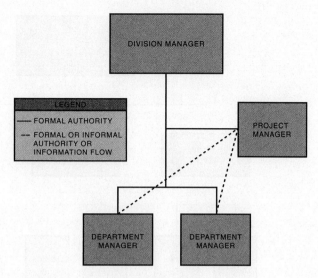

FIGURE 3–4. Line–staff organization (Project Coordinator).

perform merit reviews, but cannot enforce both professional and organizational standards in the completion of an activity. The individual performing the work is now caught in a web of authority relationships, and additional conflicts develop because functional managers are forced to share their authority with the project manager.

Although this second situation did occur during the early stages of matrix project management, it did not last because:

- Upper-level management was not ready to cope with the problems arising from shared authority.
- Upper-level management was reluctant to relinquish any of its power and authority to project managers.
- Line–staff project managers who reported to a division head did not have any authority or control over those portions of a project in other divisions; that is, the project manager in the engineering division could not direct activities in the manufacturing division.

3.5 PURE PRODUCT (PROJECTIZED) ORGANIZATION

The pure product organization, as shown in Figure 3–5, develops as a division within a division. As long as there exists a continuous flow of projects, work is stable and conflicts are at a minimum. The major advantage of this organizational flow is that one individual, the program manager, maintains complete line authority over the entire project. Not only does he assign work, but he also conducts merit reviews. Because each individual reports

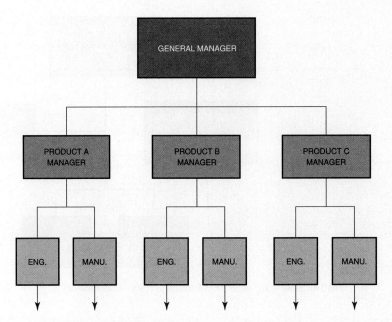

FIGURE 3–5.   Pure product or projectized structure.

to only one person, strong communication channels develop that result in a very rapid reaction time.

In pure product organizations, long lead times became a thing of the past. Trade-off studies could be conducted as fast as time would permit without the need to look at the impact on other projects (unless, of course, identical facilities or equipment were required). Functional managers were able to maintain qualified staffs for new product development without sharing personnel with other programs and projects.

The responsibilities attributed to the project manager were entirely new. First, his authority was now granted by the vice president and general manager. The program manager handled all conflicts, both those within his organization and those involving other projects. Interface management was conducted at the program manager level. Upper-level management was now able to spend more time on executive decision-making than on conflict arbitration.

The major disadvantage with the pure project form is the cost of maintaining the organization. There is no chance for sharing an individual with another project in order to reduce costs. Personnel are usually attached to these projects long after they are needed because once an employee is given up, the project manager might not be able to get him back. Motivating personnel becomes a problem. At project completion, functional personnel do not "have a home" to return to. Many organizations place these individuals into an overhead labor pool from which selection can be made during new project development. People remaining in the labor pool may be laid off. As each project comes to a close,

---

TABLE 3–3. ADVANTAGES OF THE PRODUCT ORGANIZATIONAL FORM

- Provides complete line authority over the project (i.e., strong control through a single project authority).
- Participants work directly for the project manager. Unprofitable product lines are easily identified and can be eliminated.
- Strong communications channels.
- Staffs can maintain expertise on a given project without sharing key personnel.
- Very rapid reaction time is provided.
- Personnel demonstrate loyalty to the project; better morale with product identification.
- A focal point develops for out-of-company customer relations.
- Flexibility in determining time (schedule), cost, and performance trade-offs.
- Interface management becomes easier as unit size is decreased.
- Upper-level management maintains more free time for executive decision-making.

---

people become uneasy and often strive to prove their worth to the company by over-achieving, a condition that is only temporary. It is very difficult for management to convince key functional personnel that they do, in fact, have career opportunities in this type of organization.

In pure functional (traditional) structures, technologies are well developed, but project schedules often fall behind. In the pure project structure, the fast reaction time keeps activities on schedule, but technology suffers because without strong functional groups, which maintain interactive technical communication, the company's outlook for meeting the competition may be severely hampered. The engineering department for one project might not communicate with its counterpart on other projects, resulting in duplication of efforts.

The last major disadvantage of this organizational form lies in the control of facilities and equipment. The most frequent conflict occurs when two projects require use of the same piece of equipment or facilities at the same time. Upper-level management must then assign priorities to these projects. This is normally accomplished by defining certain projects as strategic, tactical, or operational—the same definitions usually given to plans.

Tables 3–3 and 3–4 summarize the advantages and disadvantages of this organizational form.

TABLE 3–4. DISADVANTAGES OF THE PRODUCT ORGANIZATIONAL FORM

- Cost of maintaining this form in a multiproduct company would be prohibitive due to duplication of effort, facilities, and personnel; inefficient usage.
- A tendency to retain personnel on a project long after they are needed. Upper-level management must balance workloads as projects start up and are phased out.
- Technology suffers because, without strong functional groups, outlook of the future to improve company's capabilities for new programs would be hampered (i.e., no perpetuation of technology).
- Control of functional (i.e., organizational) specialists requires top-level coordination.
- Lack of opportunities for technical interchange between projects.
- Lack of career continuity and opportunities for project personnel.

3.6 MATRIX ORGANIZATIONAL FORM

PMBOK® Guide, 4th Edition
Matrix Organizational Structures
Figures 2–8, 2–9, 2–10

The matrix organizational form is an attempt to combine the advantages of the pure functional structure and the product organizational structure. This form is ideally suited for companies, such as construction, that are "project-driven." Figure 3–6 shows a typical matrix structure. Each project manager reports directly to the vice president and general manager. Since each project represents a potential profit center, the power and authority used by the project manager come directly from the general manager. The project manager has total responsibility and accountability for project success. The functional departments, on the other hand, have functional responsibility to maintain technical excellence on the project. Each functional unit is headed by a department manager whose prime responsibility is to ensure that a unified technical base is maintained and that all available information can be exchanged for each project. Department managers must also keep their people aware of the latest technical accomplishments in the industry.

Project management is a "coordinative" function, whereas matrix management is a collaborative function division of project management. In the coordinative or project organization, work is generally assigned to specific people or units who "do their own thing." In the collaborative or matrix organization, information sharing may be mandatory, and several people may be required for the same piece of work. In a project organization, authority for decision-making and direction rests with the project leader, whereas in a matrix it rests with the team.

Certain ground rules exist for matrix development:

● Participants must spend full time on the project; this ensures a degree of loyalty.
● Horizontal as well as vertical channels must exist for making commitments.
● There must be quick and effective methods for conflict resolution.

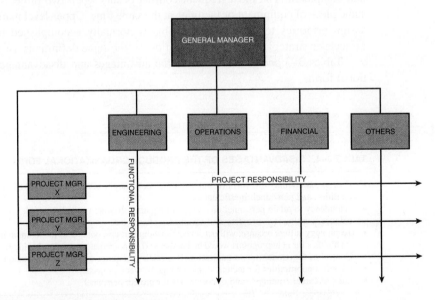

FIGURE 3–6.   Typical matrix structure.

- There must be good communication channels and free access between managers.
- All managers must have input into the planning process.
- Both horizontally and vertically oriented managers must be willing to negotiate for resources.
- The horizontal line must be permitted to operate as a separate entity except for administrative purposes.

Before describing the advantages and disadvantages of this structure, the organization concepts must be introduced. The basis for the matrix approach is an attempt to create synergism through shared responsibility between project and functional management. Yet this is easier said than done. *No two working environments are the same, and, therefore, no two companies will have the same matrix design.* The following questions must be answered before a matrix structure can be successful:

- If each functional unit is responsible for one aspect of a project, and other parts are conducted elsewhere (possibly subcontracted to other companies), how can a synergistic environment be created?
- Who decides which element of a project is most important?
- How can a functional unit (operating in a vertical structure) answer questions and achieve project goals and objectives that are compatible with other projects?

The answers to these questions depend on mutual understanding between the project and functional managers. Since both individuals maintain some degree of authority, responsibility, and accountability on each project, they must continuously negotiate. Unfortunately, the program manager might only consider what is best for his project (disregarding all others), whereas the functional manager might consider his organization more important than each project.

In order to get the job done, project managers need organizational status and authority. A corporate executive contends that the organization chart shown in Figure 3–6 can be modified to show that the project managers have adequate organizational authority by placing the department manager boxes at the tip of the functional responsibility arrowheads. With this approach, the project managers appear to be higher in the organization than their departmental counterparts but are actually equal in status. Executives who prefer this method must exercise caution because the line and project managers may not feel that there is still a balance of power.

Problem-solving in this environment is fragmented and diffused. The project manager acts as a unifying agent for project control of resources and technology. He must maintain open channels of communication to prevent suboptimization of individual projects.

In many situations, functional managers have the power to make a project manager look good, if they can be motivated to think about what is best for the project. Unfortunately, this is not always accomplished. As stated by Mantell[11]:

> There exists an inevitable tendency for hierarchically arrayed units to seek solutions and to identify problems in terms of scope of duties of particular units rather than looking

---

11. Leroy H. Mantell, "The Systems Approach and Good Management." Reprinted with permission from *Business Horizons,* October 1972 (p. 50). Copyright © 1972 by the Board of Trustees at Indiana University.

beyond them. This phenomenon exists without regard for the competence of the executive concerned. It comes about because of authority delegation and functionalism.

The project environment and functional environment cannot be separated; they must interact. The location of the project and functional unit interface is the focal point for all activities.

The functional manager controls departmental resources (i.e., people). This poses a problem because, although the project manager maintains the maximum control (through the line managers) over all resources including cost and personnel, the functional manager must provide staff for the project's requirements. It is therefore inevitable that conflicts occur between functional and project managers[12]:

> These conflicts revolve about items such as project priority, manpower costs, and the assignment of functional personnel to the project manager. Each project manager will, of course, want the best functional operators assigned to his program. In addition to these problems, the accountability for profit and loss is much more difficult in a matrix organization than in a project organization. Project managers have a tendency to blame overruns on functional managers, stating that the cost of the function was excessive. Whereas functional managers have a tendency to blame excessive costs on project managers with the argument that there were too many changes, more work required than defined initially and other such arguments.

The individual placed at the interface position has two bosses: He must take direction from both the project manager and the functional manager. The merit review and hiring and firing responsibilities still rest with the department manager. Merit reviews are normally made by the functional manager after discussions with the program manager. The functional manager may not have the time to measure the progress of this individual continuously. He must rely on the word of the program manager for merit review and promotion. The interface members generally give loyalty to the person signing their merit review. This poses a problem, especially if conflicting orders are given by the functional and project managers. The simplest solution is for the individual at the interface to ask the functional and project managers to communicate with each other to resolve the problem. This type of situation poses a problem for project managers:

- How does a project manager motivate an individual working on a project (either part-time or full-time) so that his loyalties are with the project?
- How does a project manager convince an individual to perform work according to project direction and specifications when these requests may be in conflict with department policy, especially if the individual feels that his functional boss may not regard him favorably?

There are many advantages to matrix structures, as shown in Table 3–5. Functional units exist primarily to support a project. Because of this, key people can be shared and

12. William P. Killian, "Project Management—Future Organizational Concepts," *Marquette Business Review,* Vol. 2, 1971, pp. 90–107.

TABLE 3–5. ADVANTAGES OF A PURE MATRIX ORGANIZATIONAL FORM

- The project manager maintains maximum project control (through the line managers) over all resources, including cost and personnel.
- Policies and procedures can be set up independently for each project, provided that they do not contradict company policies and procedures.
- The project manager has the authority to commit company resources, provided that scheduling does not cause conflicts with other projects.
- Rapid responses are possible to changes, conflict resolution, and project needs (as technology or schedule).
- The functional organizations exist primarily as support for the project.
- Each person has a "home" after project completion. People are susceptible to motivation and end-item identification. Each person can be shown a career path.
- Because key people can be shared, the program cost is minimized. People can work on a variety of problems; that is, better people control is possible.
- A strong technical base can be developed, and much more time can be devoted to complex problem-solving. Knowledge is available for all projects on an equal basis.
- Conflicts are minimal, and those requiring hierarchical referrals are more easily resolved.
- There is a better balance among time, cost, and performance.
- Rapid development of specialists and generalists occurs.
- Authority and responsibility are shared.
- Stress is distributed among the team (and the functional managers).

costs can be minimized. People can be assigned to a variety of challenging problems. Each person, therefore, has a "home" after project completion and a career path. People in these organizations are especially responsive to motivation and end-item identification. Functional managers find it easy to develop and maintain a strong technical base and can, therefore, spend more time on complex problem-solving. Knowledge can be shared for all projects.

The matrix structure can provide a rapid response to changes, conflicts, and other project needs. Conflicts are normally minimal, but those requiring resolution are easily resolved using hierarchical referral.

This rapid response is a result of the project manager's authority to commit company resources, provided that scheduling conflicts with other projects can be eliminated. Furthermore, the project manager has the authority independently to establish his own project policies and procedures, provided that they do not conflict with company policies. This can do away with red tape and permit a better balance among time, cost, and performance.

The matrix structure provides us with the best of two worlds: the traditional structure and the matrix structure. The advantages of the matrix structure eliminate almost all of the disadvantages of the traditional structure. The word "matrix" often brings fear to the hearts of executives because it implies radical change, or at least they think that it does. If we take a close look at Figure 3–6, we can see that the traditional structure is still there. The matrix is simply horizontal lines superimposed over the traditional structure. The horizontal lines will come and go as projects start up and terminate, but the traditional structure will remain.

Matrix structures are not without their disadvantages, as shown in Table 3–6. The first three elements are due to the horizontal and vertical work flow requirements of a matrix. Actually the flow may even be multidimensional if the project manager has to report to

TABLE 3–6. DISADVANTAGES OF A PURE MATRIX ORGANIZATIONAL FORM

- Multidimensional information flow.
- Multidimensional work flow.
- Dual reporting.
- Continuously changing priorities.
- Management goals different from project goals.
- Potential for continuous conflict and conflict resolution.
- Difficulty in monitoring and control.
- Company-wide, the organizational structure is not cost-effective because more people than necessary are required, primarily administrative.
- Each project organization operates independently. Care must be taken that duplication of efforts does not occur.
- More effort and time are needed initially to define policies and procedures, compared to traditional form.
- Functional managers may be biased according to their own set of priorities.
- Balance of power between functional and project organizations must be watched.
- Balance of time, cost, and performance must be monitored.
- Although rapid response time is possible for individual problem resolution, the reaction time can become quite slow.
- Employees and managers are more susceptible to role ambiguity than in traditional form.
- Conflicts and their resolution may be a continuous process (possibly requiring support of an organizational development specialist).
- People do not feel that they have any control over their own destiny when continuously reporting to multiple managers.

customers or corporate or other personnel in addition to his superior and the functional line managers.

Most companies believe that if they have enough resources to staff all of the projects that come along, then the company is "overstaffed." As a result of this philosophy, priorities may change continuously, perhaps even daily. Management's goals for a project may be drastically different from the project's goals, especially if executive involvement is lacking during the definition of a project's requirements in the planning phase. In a matrix, conflicts and their resolution may be a continuous process, especially if priorities change continuously. Regardless of how mature an organization becomes, there will always exist difficulty in monitoring and control because of the complex, multidirectional work flow. Another disadvantage of the matrix organization is that more administrative personnel are needed to develop policies and procedures, and therefore both direct and indirect administrative costs will increase. In addition, it is impossible to manage projects with a matrix if there are steep horizontal or vertical pyramids for supervision and reporting, because each manager in the pyramid will want to reduce the authority of the managers operating within the matrix. Each project organization operates independently. Duplication of effort can easily occur; for example, two projects might be developing the same cost accounting procedure, or functional personnel may be doing similar R&D efforts on different projects. Both vertical and horizontal communication is a must in a project matrix organization.

One of the advantages of the matrix is a rapid response time for problem resolution. This rapid response generally applies to slow-moving projects where problems occur within each functional unit. On fast-moving projects, the reaction time can become quite slow, especially if the problem spans more than one functional unit. This slow reaction time exists because the functional employees assigned to the project do not have the authority to make decisions, allocate functional resources, or change schedules. Only the line managers have this authority. Therefore, in times of crisis, functional managers must be actively brought into the "big picture" and invited to team meetings.

Middleton has listed four additional undesirable results of matrix organizations, results that can affect company capabilities[13]:

- Project priorities and competition for talent may interrupt the stability of the organization and interfere with its long-range interests by upsetting the traditional business of functional organizations.
- Long-range plans may suffer as the company gets more involved in meeting schedules and fulfilling the requirements of temporary projects.
- Shifting people from project to project may disrupt the training of employees and specialists, thereby hindering the growth and development within their fields of specialization.
- Lessons learned on one project may not be communicated to other projects.

Davis and Lawrence have identified nine additional matrix pathologies[14]:

- Power struggles: The horizontal versus vertical hierarchy.
- Anarchy: Formation of organizational islands during periods of stress.
- Groupitis: Confusing the matrix as being synonymous with group decision making.
- Collapse during economic crunch: Flourishing during periods of growth and collapsing during lean times.
- Excessive overhead: How much matrix supervision is actually necessary?
- Decision strangulation: Too many people involved in decision-making.
- Sinking: Pushing the matrix down into the depths of the organization.
- Layering: A matrix within a matrix.
- Navel gazing: Becoming overly involved in the internal relationships of the organization.

The matrix structure therefore becomes a compromise in an attempt to obtain the best of two worlds. In pure product management, technology suffered because there wasn't a single group for planning and integration. In the pure functional organization, time and schedule were sacrificed. Matrix project management is an attempt to obtain maximum technology and performance in a cost-effective manner and within time and schedule constraints.

We should note that with proper executive-level planning and control, all of the disadvantages can be eliminated. This is the only organizational form where such control is possible. But companies must resist creating more positions in executive management than are

13. Reprinted by permission of *Harvard Business Review.* From C. J. Middleton, "How to Set Up a Project Organization," *Harvard Business Review,* March–April 1967. Copyright © 1967 by the Harvard Business School Publishing Corporation; all rights reserved.

14. Stanley M. Davis and Paul R. Lawrence, *Matrix* (adapted from pp. 129–144), © 1977. Adapted by permission of Pearson Education, Inc., Upper Saddle River, NJ.

actually necessary as this will drive up overhead rates. However, there is a point where the matrix will become mature and fewer people will be required at the top levels of management.

Previously we identified the necessity for the project manager to be able to establish his own policies, procedures, rules, and guidelines. Obviously, with personnel reporting in two directions and to multiple managers, conflicts over administration can easily occur.

Most practitioners consider the matrix to be a two-dimensional system where each project represents a potential profit center and each functional department represents a cost center. (This interpretation can also create conflict because functional departments may feel that they no longer have an input into corporate profits.) For large corporations with multiple divisions, the matrix is no longer two-dimensional, but multidimensional.

William C. Goggin has described geographical area and space and time as the third and fourth dimensions of the Dow Corning matrix[15]:

> Geographical areas . . . business development varied widely from area to area, and the profit-center and cost-center dimensions could not be carried out everywhere in the same manner. . . . Dow Corning area organizations are patterned after our major U.S. organizations. Although somewhat autonomous in their operation, they subscribe to the overall corporate objectives, operating guidelines, and planning criteria. During the annual planning cycle, for example, there is a mutual exchange of sales, expense, and profit projections between the functional and business managers headquartered in the United States and the area managers around the world.
>
> Space and time. . . . A fourth dimension of the organization denotes fluidity and movement through time. . . . The multidimensional organization is far from rigid; it is constantly changing. Unlike centralized or decentralized systems that are too often rooted deep in the past, the multidimensional organization is geared toward the future: Long-term planning is an inherent part of its operation.

Goggin then went on to describe the advantages that Dow Corning expected to gain from the multidimensional organization:

- Higher profit generation even in an industry (silicones) price-squeezed by competition. (Much of our favorable profit picture seems due to a better overall understanding and practice of expense controls through the company.)
- Increased competitive ability based on technological innovation and product quality without a sacrifice in profitability.
- Sound, fast decision-making at all levels in the organization, facilitated by stratified but open channels of communications, and by a totally participative working environment.
- A healthy and effective balance of authority among the businesses, functions, and areas.
- Progress in developing short- and long-range planning with the support of all employees.

---

15. Reprinted by permission of *Harvard Business Review.* From William C. Goggin, "How the Multidimensional Structure Works at Dow Corning," *Harvard Business Review,* January–February 1974, pp. 56–57. Copyright © 1973 by the Harvard Business School Publishing Corporation; all rights reserved.

- Resource allocations that are proportional to expected results.
- More stimulating and effective on-the-job training.
- Accountability that is more closely related to responsibility and authority.
- Results that are visible and measurable.
- More top-management time for long-range planning and less need to become involved in day-to-day operations.

Obviously, the matrix structure is the most complex of all organizational forms. Grinnell and Apple define four situations where it is most practical to consider a matrix[16]:

- When complex, short-run products are the organization's primary output.
- When a complicated design calls for both innovation and timely completion.
- When several kinds of sophisticated skills are needed in designing, building, and testing the products—skills then need constant updating and development.
- When a rapidly changing marketplace calls for significant changes in products, perhaps between the time they are conceived and delivered.

Matrix implementation requires:

- Training in matrix operations
- Training in how to maintain open communications
- Training in problem solving
- Compatible reward systems
- Role definitions

3.7 MODIFICATION OF MATRIX STRUCTURES

The matrix can take many forms, but there are basically three common varieties. Each type represents a different degree of authority attributed to the program manager and indirectly identifies the relative size of the company. As an example, in the matrix of Figure 3–6, all program managers report directly to the general manager. This type of arrangement works best for small companies that have few projects and assumes that the general manager has sufficient time to coordinate activities between his project managers. In this type of arrangement, all conflicts between projects are referred to the general manager for resolution.

As companies grow in size and the number of projects, the general manager will find it increasingly difficult to act as the focal point for all projects. A new position must be created, that of director of programs, or manager of programs or projects, who is responsible for all program management. See Figure 3–7.

Executives contend that an effective span of control is five to seven people. Does this apply to the director of project management as well? Consider a company that has fifteen

---

16. S. K. Grinnell and H. P. Apple, "When Two Bosses Are Better Than One," *Machine Design,* January 1975, pp. 84–87.

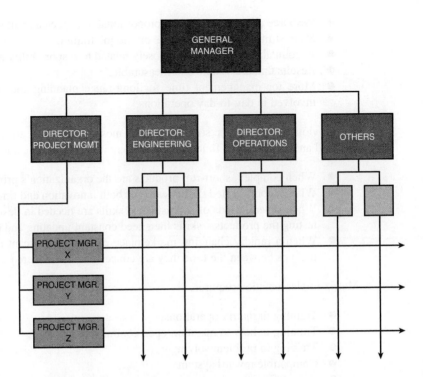

FIGURE 3–7. Development of a director of project management.

projects going on at once. There are three projects over $5 million, seven are between $1 and $3 million, and five projects are under $700,000. Each project has a full-time project manager. Can all fifteen project managers report to the same person? The company solved this problem by creating a deputy director of project management. All projects over $1 million reported to the director, and all projects under $1 million went to the deputy director. The director's rationale soon fell by the wayside when he found that the more severe problems that were occupying his time were occurring on projects with a smaller dollar volume, where flexibility in time, cost, and performance was nonexistent and trade-offs were almost impossible. If the project manager is actually a general manager, then the director of project management should be able to supervise effectively more than seven project managers. The desired span of control, of course, will vary from company to company and must take into account:

● The demands imposed on the organization by task complexity
● Available technology
● The external environment
● The needs of the organizational membership
● The types of customers and/or products

As companies expand, it is inevitable that new and more complex conflicts arise. The control of the engineering functions poses such a problem:

Should the project manager have ultimate responsibility for the engineering functions of a project, or should there be a deputy project manager who reports to the director of engineering and controls all technical activity?

Although there are pros and cons for both arrangements, the problem resolved itself in the company mentioned above when projects grew so large that the project manager became unable to handle both the project management and project engineering functions. Then, as shown in Figure 3–8, a chief project engineer was assigned to each project as deputy project manager, but remained functionally assigned to the director of engineering. The project manager was now responsible for time and cost considerations, whereas the project engineer was concerned with technical performance. The project engineer can be either "solid" vertically and "dotted" horizontally, or vice versa. There are also situations where the project engineer may be "solid" in both directions. The decision usually rests with the director of engineering. Of course, in a project where the project engineer would be needed on a part-time basis only, he would be solid vertically and dotted horizontally.

Engineering directors usually demand that the project engineer be solid vertically in order to give technical direction. As one director of engineering stated, "Only engineers that report to me will have the authority to give technical direction to other engineers. After all, how else can I be responsible for the technical integrity of the product when direction comes from outside my organization?"

This subdivision of functions is necessary in order to control large projects adequately. However, for small projects, say $100,000 or less, it is quite common on R&D projects for

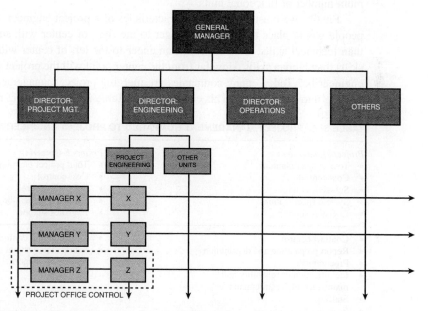

FIGURE 3–8. Placing project engineering in the project office.

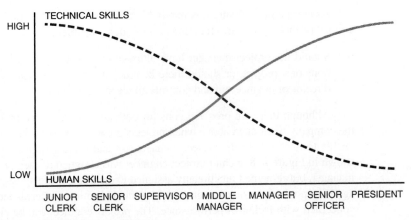

FIGURE 3–9. Philosophy of management.

an engineer to serve as the project manager as well as the project engineer. Here, the project manager must have technical expertise, not merely understanding. Furthermore, this individual can still be attached to a functional engineering support unit other than project engineering. As an example, a mechanical engineering department receives a government contract for $75,000 to perform tests on a new material. The proposal is written by an engineer attached to the department. When the contract is awarded, this individual, although not in the project engineering department, can fulfill the role of project manager and project engineer while still reporting to the manager of the mechanical engineering department. This arrangement works best (and is cost-effective) for short-duration projects that cross a minimum number of functional units.

Finally, we must discuss the characteristics of a project engineer. In Figure 3–9, most people would place the project manager to the right of center with stronger human skills than technical skills, and the project engineer to the left of center with stronger technical skills than human skills. How far from the center point will the project manager and project engineer be? Today, many companies are merging project management and project engineering into one position. This can be seen in Table 3–7. The project manager and project

TABLE 3–7.   PROJECT MANAGEMENT COMPARED TO PROJECT ENGINEERING

Project Management	*Project Engineering*
• Total project planning	• Total project planning
• Cost control	• Cost control
• Schedule control	• Schedule control
• System specifications	• System specifications
• Logistics support	• Logistics support
• Contract control	• Configuration control
• Report preparation and distribution	• Fabrication, testing, and production technical
• Procurement	leadership support
• Identification of reliability and	
maintainability requirements	
• Staffing	
• Priority scheduling	
• Management information systems	

engineer have similar functions above the line but different ones below the line.[17] The main reason for separating project management from project engineering is so that the project engineer will remain "solid" to the director of engineering in order to have the full authority to give technical direction to engineering.

3.8 THE STRONG, WEAK, OR BALANCED MATRIX

> **PMBOK® Guide, 4th Edition**
> Matrix Organizational Structures
> Figures 2–8, 2–9, 2–10

Matrix structures can be strong, weak, or balanced. The strength of the matrix is based upon who has more influence over the daily performance of the workers: project manager or line managers. If the project manager has more influence over the worker, then the matrix structure functions as a strong matrix as seen through the eyes of the project manager. If the line manager has more influence than does the project manager, then the organization functions as a weak matrix as seen by the project manager.

The most common differentiator between a strong and weak matrix is where the command of technology resides: project manager or line managers. If the project manager has a command of technology and is recognized by the line managers and the workers as being a technical expert, then the line managers will allow the workers to take technical direction from the project manager. This will result in a strong matrix structure. Workers will seek solutions to their problems from the project manager first and the line managers second. The reverse is true for a weak matrix. Project managers in a strong matrix generally possess more authority than in a weak matrix.

When a company desires a strong matrix, the project manager is generally promoted from within the organization and may have had assignments in several line functions throughout the organization. In a weak matrix, the company may hire from outside the organization but should at least require that the person selected understand the technology and the industry.

3.9 CENTER FOR PROJECT MANAGEMENT EXPERTISE

> **PMBOK® Guide, 4th Edition**
> 1.4.4 Project Management Office

In project-driven companies, the creation of a project management division is readily accepted as a necessity to conduct business. Organizational restructuring can quite often occur based on environmental changes and customer needs. In non–project-driven organizations, employees are less tolerant of organizational change. Power, authority, and turf become important. The implementation of a separate division for project management is extremely difficult. Resistance can become so strong that the entire project management process can suffer.

Recently, non–project-driven companies have created centers for project management expertise. These centers are not necessarily formal line organizations, but more informal committees whose membership may come from each functional unit of the company. The assignment to the center for expertise can be part-time or full-time; it may be only for six

---

17. Procurement, reliability, and maintainability may fall under the responsibility of the project engineer in some companies.

months to a year; and it may or may not require the individual to manage projects. Usually, the center for expertise has as its charter:

- To develop and update a methodology for project management. The methodology usually advocates informal project management.
- To act as a facilitator or trainer in conducting project management training programs.
- To provide project management assistance to any employee who is currently managing projects and requires support in planning, scheduling, and controlling projects.
- To develop or maintain files on "lessons learned" and to see that this information is made available to all project managers.

Since these centers pose no threat to the power and authority of line managers, support is usually easy to obtain.

3.10 MATRIX LAYERING

Matrix layering can be defined as the creation of one matrix within a second matrix. For example, a company can have a total company matrix, and each division or department (i.e., project engineering) can have its own internalized matrix. In the situation of a matrix within a matrix, all matrices are formal operations.

Matrix layering can also be a mix of formal and informal organizations. The formal matrix exists for work flow, but there can also exist an informal matrix for information flow. There are also authority matrices, leadership matrices, reporting matrices, and informal technical direction matrices.

An example of layering would be the multidimensional matrix, shown in Figure 3–10, where each slice represents either time, distance, or geographic area. For example, a New York

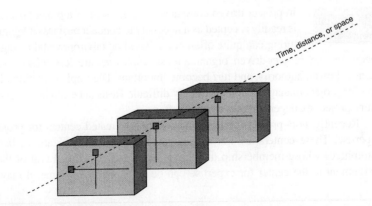

FIGURE 3–10. The multidimensional matrix.

bank utilizes a multinational matrix to control operations in foreign countries. In this case, each foreign country would represent a different slice of the total matrix.

3.11 SELECTING THE ORGANIZATIONAL FORM

PMBOK® Guide, 4th Edition
2.4 Organizational Influences

Project management has matured as an outgrowth of the need to develop and produce complex and/or large projects in the shortest possible time, within anticipated cost, with required reliability and performance, and (when applicable) to realize a profit. Granted that organizations have become so complex that traditional organizational structures and relationships no longer allow for effective management, how can executives determine which organizational form is best, especially since some projects last for only a few weeks or months while others may take years?

To answer this question, we must first determine whether the necessary characteristics exist to warrant a project management organizational form. Generally speaking, the project management approach can be effectively applied to a onetime undertaking that is[18]:

- Definable in terms of a specific goal
- Infrequent, unique, or unfamiliar to the present organization
- Complex with respect to interdependence of detailed tasks
- Critical to the company

Once a group of tasks is selected and considered to be a project, the next step is to define the kinds of projects, described in Section 2.5. These include individual, staff, special, and matrix or aggregate projects.

Unfortunately, many companies do not have a clear definition of what a project is. As a result, large project teams are often constructed for small projects when they could be handled more quickly and effectively by some other structural form. All structural forms have their advantages and disadvantages, but the project management approach appears to be the best possible alternative.

The basic factors that influence the selection of a project organizational form are:

- Project size
- Project length
- Experience with project management organization
- Philosophy and visibility of upper-level management
- Project location
- Available resources
- Unique aspects of the project

18. John M. Stewart, "Making Project Management Work." Reprinted with permission from *Business Horizons,* Fall 1965 (p. 54). Copyright © 1964 by the Board of Trustees at Indiana University.

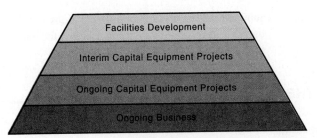

FIGURE 3–11. Matrix development in manufacturing.

This last item requires further comment. Project management (especially with a matrix) usually works best for the control of human resources and thus may be more applicable to labor-intensive projects rather than capital-intensive projects. Labor-intensive organizations have formal project management, whereas capital-intensive organizations may use informal project management. Figure 3–11 shows how matrix management was implemented by an electric equipment manufacturer. The company decided to use fragmented matrix management for facility development projects. After observing the success of the fragmented matrix, the executives expanded matrix operations to include interim and ongoing capital equipment projects. The first three levels were easy to implement. The fourth level, ongoing business, was more difficult to convert to matrix because of functional management resistance and the fear of losing authority.

Four fundamental parameters must be analyzed when considering implementation of a project organizational form:

- Integrating devices
- Authority structure
- Influence distribution
- Information system

Project management is a means of integrating all company efforts, especially research and development, by selecting an appropriate organizational form. Two questions arise when we think of designing the organization to facilitate the work of the integrators[19]:

- Is it better to establish a formal integration department, or simply to set up integrating positions independent of one another?
- If individual integrating positions are set up, how should they be related to the larger structure?

---

19. William P. Killian, "Project Management—Future Organizational Concepts," *Marquette Business Review,* Vol. 2, 1971, pp. 90–107.

Informal integration works best if, and only if, effective collaboration can be achieved between conflicting units. Without any clearly defined authority, the role of the integrator is simply to act as an exchange medium across the interface of two functional units. As the size of the organization increases, formal integration positions must exist, especially in situations where intense conflict can occur (e.g., research and development).

Not all organizations need a pure matrix structure to achieve this integration. Many problems can be solved simply through the chain of command, depending on the size of the organization and the nature of the project. The organization needed to achieve project control can vary in size from one person to several thousand people. The organizational structure needed for effective project control is governed by the desires of top management and project circumstances.

Top management must decide on the authority structure that will control the integration mechanism. The authority structure can range from pure functional authority (traditional management), to product authority (product management), and finally to dual authority (matrix management). This range is shown in Figure 3–12. From a management point of view, organizational forms are often selected based on how much authority top management wishes to delegate or surrender.

Integration of activities across functional boundaries can also be accomplished by influence. Influence includes such factors as participation in budget planning and approval, design changes, location and size of offices, salaries, and so on. Influence can also cut administrative red tape and develop a much more unified informal organization.

Matrix structures are characterized as strong or weak based on the relative influence that the project manager possesses over the assigned functional resources. When the

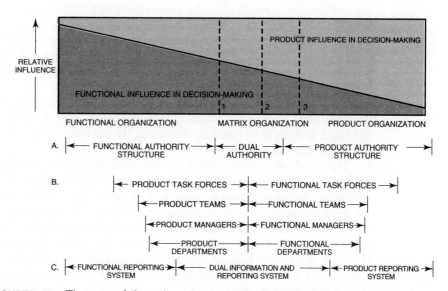

FIGURE 3–12. The range of alternatives. *Source:* Jay R. Galbraith, "Matrix Organization Designs." Reprinted with permission from *Business Horizons,* February 1971 (p. 37). Copyright © 1971 by the Board of Trustees at Indiana University.

project manager has more "relative influence" over the performance of the assigned resources than does the line manager, the matrix structure is a strong matrix. In this case, the project manager usually has the knowledge to provide technical direction, assign responsibilities, and may even have a strong input into the performance evaluation of the assigned personnel. If the balance of influence tilts in favor of the line manager, then the matrix is referred to as a weak matrix.

Information systems also play an important role. Previously we stated that one of the advantages of several project management structures is the ability to make both rapid and timely decisions with almost immediate response to environmental changes. Information systems are designed to get the right information to the right person at the right time in a cost-effective manner. Organizational functions must facilitate the flow of information through the management network.

Galbraith has described additional factors that can influence organizational selection. These factors are[20]:

- Diversity of product lines
- Rate of change of the product lines
- Interdependencies among subunits
- Level of technology
- Presence of economies of scale
- Organizational size

A diversity of project lines requires both top-level and functional managers to maintain knowledge in all areas. Diversity makes it more difficult for managers to make realistic estimates concerning resource allocations and the control of time, cost, schedules, and technology. The systems approach to management requires sufficient information and alternatives to be available so that effective trade-offs can be established. For diversity in a high-technology environment, the organizational choice might, in fact, be a trade-off between the flow of work and the flow of information. Diversity tends toward strong product authority and control.

Many functional organizations consider themselves companies within a company and pride themselves on their independence. This attitude poses a severe problem in trying to develop a synergistic atmosphere. Successful project management requires that functional units recognize the interdependence that must exist in order for technology to be shared and schedule dates to be met. Interdependency is also required in order to develop strong communication channels and coordination.

The use of new technologies poses a serious problem in that technical expertise must be established in all specialties, including engineering, production, material control, and safety. Maintaining technical expertise works best in strong functional disciplines, provided the information is not purchased outside the organization. The main problem, however, is how to communicate this expertise across functional lines. Independent R&D units can be established, as opposed to integrating R&D into each functional department's routine efforts. Organizational control requirements are much more difficult in high-technology industries with ongoing research and development than with pure production groups.

---

20. Jay R. Galbraith, "Matrix Organization Designs." Reprinted with permission from *Business Horizons,* February 1971, pp. 29–40. Copyright © 1971 by the Board of Trustees at Indiana University.

Economies of scale and size can also affect organizational selection. The economies of scale are most often controlled by the amount of physical resources that a company has available. For example, a company with limited facilities and resources might find it impossible to compete with other companies on production or competitive bidding for larger dollar-volume products. Such a company must rely heavily on maintaining multiple projects (or products), each of low cost or volume, whereas a larger organization may need only three or four projects large enough to sustain the organization. The larger the economies of scale, the more the organization tends to favor pure functional management.

The size of the organization is important in that it can limit the amount of technical expertise in the economies of scale. While size may have little effect on the organizational structure, it does have a severe impact on the economies of scale. Small companies, for example, cannot maintain large specialist staffs and, therefore, incur a larger cost for lost specialization and lost economies of scale.

The four factors described above for organizational form selections together with the six alternatives of Galbraith can be regarded as universal. Beyond these universal factors, we must look at the company in terms of its product, business base, and personnel. Goodman has defined a set of subfactors related to R&D groups[21]:

- Clear location of responsibility
- Ease and accuracy of communication
- Effective cost control
- Ability to provide good technical supervision
- Flexibility of staffing
- Importance to the company
- Quick reaction capability to sudden changes in the project
- Complexity of the project
- Size of the project with relation to other work in-house
- Form desired by customer
- Ability to provide a clear path for individual promotion

Goodman asked general managers and project managers to select from the above list and rank the factors from most important to least important in designing an organization. With one exception—flexibility of staffing—the response from both groups correlated to a coefficient of 0.811. Clear location of responsibility was seen as the most important factor, and a path for promotion the least important.

Middleton conducted a mail survey of aerospace firms in an attempt to determine how well the companies using project management met their objectives. Forty-seven responses were received. Tables 3–8 and 3–9 identify the results. Middleton stated, "In evaluating the results of the survey, it appears that a company taking the project organization approach can be reasonably certain that it will improve controls and customer (out-of-company) relations, but internal operations will be more complex."[22]

---

21. Richard A. Goodman, "Organizational Preference in Research and Development," *Human Relations,* Vol. 3, No. 4, 1970, pp. 279–298.

22. Reprinted with permission of *Harvard Business Review.* From C. J. Middleton, "How to Set Up a Project Organization," *Harvard Business Review,* March–April 1967, pp. 73–82. Copyright © 1967 by the Harvard Business School Publishing Corporation; all rights reserved.

TABLE 3–8. MAJOR COMPANY ADVANTAGES OF PROJECT MANAGEMENT

Advantages	Percent of Respondents
• Better control of projects	92%
• Better customer relations	80%
• Shorter product development time	40%
• Lower program costs	30%
• Improved quality and reliability	26%
• Higher profit margins	24%
• Better control over program security	13%

Other Benefits

- Better project visibility and focus on results
- Improved coordination among company divisions doing work on the project
- Higher morale and better mission orientation for employees working on the project
- Accelerated development of managers due to breadth of project responsibilities

Source: Reprinted by permission of *Harvard Business Review.* An exhibit from "How to Set Up a Project Organization," by C. J. Middleton, March–April, 1967 (pp. 73–82). Copyright © 1967 by the Harvard Business School Publishing Corporation; all rights reserved.

The way in which companies operate their project organization is bound to affect the organization, both during the operation of the project and after the project has been completed and personnel have been disbanded. The overall effects on the company must be looked at from a personnel and cost control standpoint. This will be accomplished, in depth, in later chapters. Although project management is growing, the creation of a project organization does not necessarily ensure that an assigned objective will be accomplished successfully. Furthermore, weaknesses can develop in the areas of maintaining capability and structural changes.

TABLE 3–9. MAJOR COMPANY DISADVANTAGES OF PROJECT MANAGEMENT

Disadvantages	Percent of Respondents
• More complex internal operations	51%
• Inconsistency in application of company policy	32%
• Lower utilization of personnel	13%
• Higher program costs	13%
• More difficult to manage	13%
• Lower profit margins	2%

Other Disadvantages

- Tendency for functional groups to neglect their job and let the project organization do everything
- Too much shifting of personnel from project to project
- Duplication of functional skills in project organization

Source: Reprinted by permission of *Harvard Business Review.* An exhibit from "How to Set Up a Project Organization," by C. J. Middleton, March–April, 1967 (pp. 73–82). Copyright © 1967 by the Harvard Business School Publishing Corporation; all rights reserved.

An almost predictable result of using the project management approach is the increase in management positions. Killian describes the results of two surveys[23]:

> One company compared its organization and management structure as it existed before it began forming project units with the structure that existed afterward. The number of departments had increased from 65 to 106, while total employment remained practically the same. The number of employees for every supervisor had dropped from 13.4 to 12.8. The company concluded that a major cause of this change was the project groups [see footnote 26 for reference article].
>
> Another company uncovered proof of its conclusion when it counted the number of second-level and higher management positions. It found that it had 11 more vice presidents and directors, 35 more managers, and 56 more second-level supervisors. Although the company attributed part of this growth to an upgrading of titles, the effect of the project organization was the creation of 60 more management positions.

Although the project organization is a specialized, task-oriented entity, it seldom, if ever, exists apart from the traditional structure of the organization.[24] All project management structures overlap the traditional structure. Furthermore, companies can have more than one project organizational form in existence at one time. A major steel product, for example, has a matrix structure for R&D and a product structure elsewhere.

Accepting a project management structure is a giant step from which there may be no return. The company may have to create more management positions without changing the total employment levels. In addition, incorporation of a project organization is almost always accompanied by the upgrading of jobs. In any event, management must realize that whichever project management structure is selected, a dynamic state of equilibrium will be necessary.

3.12 STRUCTURING THE SMALL COMPANY

Small and medium companies generally prefer to have the project manager report fairly high up in the chain of command, even though the project manager may be working on a relatively low-priority project. Project managers are usually viewed as less of a threat in small organizations than in the larger ones, thus creating less of a problem if they report high up.

Organizing the small company for projects involves two major questions:

- Where should the project manager be placed within the organization?
- Are the majority of the projects internal or external to the organization?

23. William P. Killian, "Project Management—Future Organizational Concepts," *Marquette Business Review,* Vol. 2, 1971, pp. 90–107.

24. Allen R. Janger, "Anatomy of the Project Organization," *Business Management Record,* November 1963, pp. 12–18.

These two questions are implicitly related. For either large, complex projects or those involving outside customers, project managers generally report to a high level in the organization. For small or internal projects, the project manager reports to a middle- or lower-level manager.

Small and medium companies have been very successful in managing internal projects using departmental project management (see Figure 3–2), especially when only a few functional groups must interface with one another. Quite often, line managers are permitted to wear multiple hats and also act as project managers, thereby reducing the need for hiring additional project managers.

Customers external to the organization are usually favorably impressed if a small company identifies a project manager who is dedicated and committed to their project, even if only on a part-time basis. Thus outside customers, particularly through a competitive bidding environment, respond favorably to a matrix structure, even if the matrix structure is simply eyewash for the customer. For example, consider the matrix structure shown in Figure 3–13. Both large and small companies that operate on a matrix usually develop a separate organizational chart for each customer. Figure 3–13 represents the organizational chart that would be presented to Alpha Company. The Alpha Company project would be identified with bold lines and would be placed immediately below the vice president, regardless of the priority of the project. After all, if you were the Alpha Company customer, would you want your project to appear at the bottom of the list?

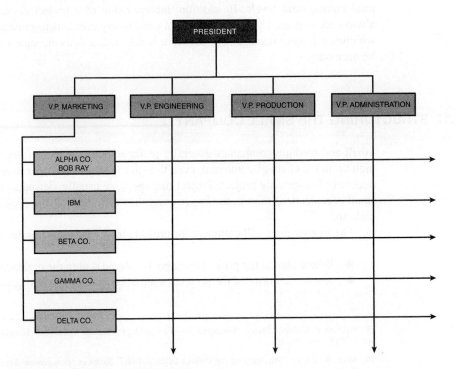

FIGURE 3–13. Matrix for a small company.

Figure 3–13 also identifies two other key points that are important to small companies. First, only the name of the Alpha Company project manager, Bob Ray, need be identified. The reason for this is that Bob Ray may also be the project manager for one or more of the other projects, and it is usually not a good practice to let the customer know that Bob Ray will have loyalties split among several projects. Actually, the organization chart shown in Figure 3–13 is for a machine tool company employing 280 people, with five major and thirty minor projects. The company has only two full-time project managers. Bob Ray manages the projects for Alpha, Gamma, and Delta Companies; the Beta Company project has the second full-time project manager; and the IBM project is being managed personally by the vice president of engineering, who happens to be wearing two hats.

The second key point is that small companies generally should not identify the names of functional employees because:

- The functional employees are probably part-time.
- It is usually best in small companies for all communications to be transmitted through the project manager.

Another example of how a simple matrix structure can be used to impress customers is shown in Figure 3–14. The company identified here actually employs only thirty-eight people. Very small companies normally assign the estimating department to report directly to the president, as shown in Figure 3–14. In addition, the senior engineers, who appear to

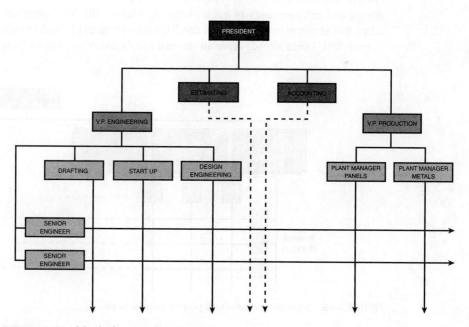

FIGURE 3–14. Matrix for a small company.

be acting in the role of project managers, may simply be the department managers for drafting, startup, and/or design engineering. Yet, from an outside customer's perspective, the company has a dedicated and committed project manager for the project.

3.13 STRATEGIC BUSINESS UNIT (SBU) PROJECT MANAGEMENT

During the past ten years, large companies have restructured into strategic business units (SBUs). An SBU is a grouping of functional units that have the responsibility for profit (or loss) of part of the organization's core businesses. Figure 3–15 shows how one of the automotive suppliers restructured into three SBUs; one each for Ford, Chrysler, and General Motors. Each strategic business unit is large enough to maintain its own project and program managers. The executive in charge of the strategic business unit may act as the sponsor for all of the program and project managers within the SBU. The major benefit of these types of project management SBUs is that it allows the SBU to work more closely with the customer. It is a customer-focused organizational structure.

It is possible for some resources to be shared across several SBUs. Manufacturing plants can end up supporting more than one SBU. Also, corporate may provide the resources for cost accounting, human resource management, and training.

A more recent organizational structure, and a more complex one, is shown in Figure 3–16. In this structure, each SBU may end up using the same platform (i.e., powertrain, chassis, and other underneath components). The platform managers are responsible for the design and enhancements of each platform, whereas the SBU program managers must adapt this platform to a new model car. This type of matrix is multidimensional inasmuch as each SBU could already have an internal matrix. Also, each manufacturing plant could

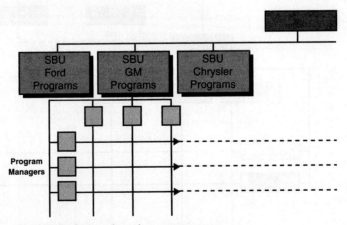

FIGURE 3–15. Strategic business unit project management.

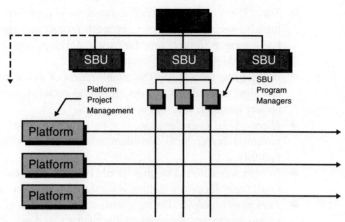

FIGURE 3–16. SBU project management using platform management.

be located outside of the continental United States, making this structure a multinational, multidimensional matrix.

3.14 TRANSITIONAL MANAGEMENT

Organizational redesign is occurring at a rapid rate because of shorter product life cycles, rapidly changing environments, accelerated development of sophisticated information systems, and increased marketplace competitiveness. Because of these factors, more companies are considering project management organizations as a solution.

Why have some companies been able to implement this change in a short period of time while other companies require years? The answer is that successful implementation requires good transitional management.

Transitional management is the art and science of managing the conversion period from one organizational design to another. Transitional management necessitates an understanding of the new goals, objectives, roles, expectations, and employees' fears.

A survey was conducted of executives, managers, and employees in thirty-eight companies that had implemented matrix management. Almost all executives felt that the greatest success could be achieved through proper training and education, both during and after transition. In addition to training, executives stated that the following fifteen challenges must be accounted for during transition:

- *Transfer of power.* Some line managers will find it extremely difficult to accept someone else managing their projects, whereas some project managers will find it difficult to give orders to workers who belong to someone else.

- *Trust.* The secret to a successful transition without formal executive authority will be trust between line managers, between project managers, and between project and line managers. It takes time for trust to develop. Senior management should encourage it throughout the transition life cycle.
- *Policies and procedures.* The establishment of well-accepted policies and procedures is a slow and tedious process. Trying to establish rigid policies and procedures at project initiation will lead to difficulties.
- *Hierarchical consideration.* During transition, every attempt should be made to minimize hierarchical considerations that could affect successful organizational maturity.
- *Priority scheduling.* Priorities should be established only when needed, not on a continual basis. If priority shifting is continual, confusion and disenchantment will occur.
- *Personnel problems.* During transition there will be personnel problems brought on by moving to new locations, status changes, and new informal organizations. These problems should be addressed on a continual basis.
- *Communications.* During transition, new channels of communications should be built but not at the expense of old ones. Transition phases should show employees that communication can be multidirectional, for example, a project manager talking directly to functional employees.
- *Project manager acceptance.* Resistance to the project manager position can be controlled through proper training. People tend to resist what they do not understand.
- *Competition.* Although some competition is healthy within an organization, it can be detrimental during transition. Competition should not be encouraged at the expense of the total organization.
- *Tools.* It is common practice for each line organization to establish its own tools and techniques. During transition, no attempt should be made to force the line organizations to depart from their current practices. Rather, it is better for the project managers to develop tools and techniques that can be integrated with those in the functional groups.
- *Contradicting demands.* During transition and after maturity, contradicting demands will be a way of life. When they first occur during transition, they should be handled in a "working atmosphere" rather than a crisis mode.
- *Reporting.* If any type of standardization is to be developed, it should be for project status reporting, regardless of the size of the project.
- *Teamwork.* Systematic planning with strong functional input will produce teamwork. Using planning groups during transition will not obtain the necessary functional and project commitments.
- *Theory X–Theory Y.* During transition, functional employees may soon find themselves managed under either Theory X or Theory Y approaches. People must realize (through training) that this is a way of life in project management, especially during crises.
- *Overmanagement costs.* A mistake often made by executives is thinking that projects can be managed with fewer resources. This usually leads to disaster because undermanagement costs may be an order of magnitude greater than overmanagement costs.

Transition to a project-driven matrix organization is not easy. Managers and professionals contemplating such a move should know:

- Proper planning and organization of the transition on a life-cycle basis will facilitate a successful change.
- Training of the executives, line managers, and employees in project management knowledge, skills, and attitudes is critical to a successful transition and probably will shorten the transition time.
- Employee involvement and acceptance may be the single most important function during transition.
- The strongest driving force of success during transition is a demonstration of commitment to and involvement in project management by senior executives.
- Organizational behavior becomes important during transition.
- Commitments made by senior executives prior to transition must be preserved during and following transition.
- Major concessions by senior management will come slowly.
- Schedule or performance compromises are not acceptable during transition; cost overruns may be acceptable.
- Conflict among participants increases during transition.
- If project managers are willing to manage with only implied authority during transition, then the total transition time may be drastically reduced.
- It is not clear how long transition will take.

Transition from a classical or product organization to a project-driven organization is not easy. With proper understanding, training, demonstrated commitment, and patience, transition will have a good chance for success.

3.15 STUDYING TIPS FOR THE PMI® PROJECT MANAGEMENT CERTIFICATION EXAM

This section is applicable as a review of the principles to support the knowledge areas and domain groups in the PMBOK® Guide. This chapter addresses:

- Human Resources Management
- Planning

Understanding the following principles is beneficial if the reader is using this text to study for the PMP® Certification Exam:

- Different types of organizational structures
- Advantages and disadvantages of each structure

- In which structure the project manager possesses the greatest amount of authority
- In which structure the project manager possesses the least amount of authority
- Three types of matrix structures

In Appendix C, the following Dorale Products mini–case studies are applicable:

- Dorale Products (G) [Human Resources Management]
- Dorale Products (H) [Human Resources Management]
- Dorale Products (J) [Human Resources Management]
- Dorale Products (K) [Human Resources Management]

The following multiple-choice questions will be helpful in reviewing the principles of this chapter:

1. In which organizational form is it most difficult to integrate project activities?
   A. Classical/traditional
   B. Projectized
   C. Strong matrix
   D. Weak matrix

2. In which organization form would the project manager possess the greatest amount of authority?
   A. Classical/traditional
   B. Projectized
   C. Strong matrix
   D. Weak matrix

3. In which organizational form does the project manager often have the least amount of authority?
   A. Classical/traditional
   B. Projectized
   C. Strong matrix
   D. Weak matrix

4. In which organizational form is the project manager least likely to share resources with other projects?
   A. Classical/traditional
   B. Projectized
   C. Strong matrix
   D. Weak matrix

5. In which organizational form do project managers have the greatest likelihood of possessing reward power and have a wage-and-salary administration function? (The project and line manager are the same person.)
   A. Classical/traditional
   B. Projectized
   C. Strong matrix
   D. Weak matrix

6. In which organizational form is the worker in the greatest jeopardy of losing his or her job if the project gets canceled?
A. Classical/traditional
B. Projectized
C. Strong matrix
D. Weak matrix

7. In which type of matrix structure would a project manager most likely have a command of technology?
A. Strong matrix
B. Balanced matrix
C. Weak matrix
D. Cross-cultural matrix

ANSWERS

1. A
2. B
3. D
4. B
5. A
6. B
7. A

PROBLEMS

3–1 Much has been written about how to identify and interpret signs that indicate that a new organizational form is needed. Grinnell and Apple have identified five signs in addition to those previously described in Section 3.6[25]:

- Management is satisfied with its technical skills, but projects are not meeting time, cost, and other project requirements.
- There is a high commitment to getting project work done, but great fluctuation in how well performance specifications are met.
- Highly talented specialists involved in the project feel exploited and misused.
- Particular technical groups or individuals constantly blame each other for failure to meet specifications or delivery dates.
- Projects are on time and to specification, but groups and individuals aren't satisfied with the achievement.

Grinnell and Apple state that there is a good chance that a matrix structure will eliminate or alleviate these problems. Do you agree or disagree? Does your answer depend on the type of project? Give examples or counterexamples to defend your answers.

25. See note 16.

3–2 One of the most difficult problems facing management is that of how to minimize the transition time between changeover from a purely traditional organizational form to a project organizational form. Managing the changeover is difficult in that management must consistently "provide individual training on teamwork and group problem solving; also, provide the project and functional groups with assignments to help build teamwork."

3–3 Do you think that personnel working in a project organizational structure should undergo "therapy" sessions or seminars on a regular basis so as to better understand their working environment? If yes, how frequently? Does the frequency depend upon the project organizational form selected, or should they all be treated equally?

3–4 Which organizational form would be best for the following corporate strategies?

    a. Developing, manufacturing, and marketing many diverse but interrelated technological products and materials
    b. Having market interests that span virtually every major industry
    c. Becoming multinational with a rapidly expanding global business
    d. Working in a business environment of rapid and drastic change, together with strong competition

3–5 Do you think that documenting relationships is necessary in order to operate effectively in any project organizational structure? How would you relate your answer to a statement made in the previous chapter that each project can set up its own policies, procedures, rules, and directives as long as they conform to company guidelines?

3–6 In general, how could each of the following parameters influence your choice for an organizational structure? Explain your answers in as much depth as possible.

    a. The project cost
    b. The project schedule
    c. The project duration
    d. The technology requirements
    e. The geographical locations
    f. The required working relationships with the customer

3–7 In general, what are the overall advantages and disadvantages of superimposing one organizational form over another?

3–8 In deciding to go to a new organizational form, what impact should the capabilities of the following groups have on your decision?

    a. Top management
    b. Middle management
    c. Lower-level management

3–9 Should a company be willing to accept a project that requires immediate organizational restructuring? If so, what factors should it consider?

3–10 Table 2–6 identifies the different life cycles of programs, projects, systems, and products. For each of the life cycles' phases, select a project organizational form that you feel would work best. Defend your answer with examples, advantages, and disadvantages.

3–11 A major steel producer in the United States uses a matrix structure for R&D. Once the product is developed, the product organizational structure is used. Are there any advantages to this setup?

3–12 A major American manufacturer of automobile parts has a division that has successfully existed for the past ten years with multiple products, a highly sophisticated R&D section, and a pure traditional structure. The growth rate for the past five years has been 12 percent. Almost all middle and upper-level managers who have worked in this division have received promotions and transfers to either another division or corporate headquarters. According to "the book," this division has all the prerequisites signifying that they should have a project organizational form of some sort, and yet they are extremely successful without it. Just from the amount of information presented, how can you account for their continued success? What do you think would be the major obstacles in convincing the personnel that a new organizational form would be better? Do you think that continued success can be achieved under the present structure?

3–13 Several authors contend that technology suffers in a pure product organizational form because there is no one group responsible for long-range planning, whereas the pure functional organization tends to sacrifice time and schedule. Do you agree or disagree with this statement? Defend your choice with examples.

3–14 Below are three statements that are often used to describe the environment of a matrix. Do you agree or disagree? Defend your answer.

  a. Project management in a matrix allows for fuller utilization of personnel.
  b. The project manager and functional manager must agree on priorities.
  c. Decision-making in a matrix requires continual trade-offs on time, cost, technical risk, and uncertainty.

3–15 Assume that you have to select a project organizational form for a small company. For each form described in this chapter, discuss the applicability and state the advantages and disadvantages as they apply to this small company. (You may find it necessary to first determine the business base of the small company.)

3–16 How would each person identified below respond to the question, "How many bosses do you have?"

  a. Project manager
  b. Functional team member
  c. Functional manager

(Repeat for each organizational form discussed in this chapter.)

3–17 If a project were large enough to contain its own resources, would a matrix organizational form be acceptable?

3–18 One of the most common reasons for not wanting to adopt a matrix is the excessive administrative costs and accompanying overhead rates. Would you expect the overhead rates to decrease as the matrix matures? (Disregard other factors that can influence the overhead rates, such as business base, growth rate, etc.)

3–19 Which type of organizational structure is best for R&D personnel to keep in touch with other researchers?

3–20 Which type of organizational form fosters teamwork in the best manner?

3–21 Canadian bankers have been using the matrix organizational structure to create "banking general managers" for all levels of a bank. Does the matrix structure readily admit itself to a banking environment in order to create future managers? Can we consider a branch manager as a matrix project manager?

3–22 A major utility company in Cleveland has what is commonly called "fragmented" project management, where each department maintains project managers through staff positions. The project managers occasionally have to integrate activities that involve departments other than their own. Each project normally requires involvement of several people. The company also has product managers operating out of a rather crude project (product) organizational structure. Recently, the product managers and project managers were competing for resources within the same departments.

To complicate matters further, management has put a freeze on hiring. Last week top management identified 120 different projects that could be undertaken. Unfortunately, under the current structure there are not enough staff project managers available to handle these projects. Also, management would like to make better use of the scarce functional resources.

Staff personnel contend that the solution to the above problems is the establishment of a project management division under which there will be a project management department and a product management department. The staff people feel that under this arrangement better utilization of line personnel will be made, and that each project can be run with fewer staff people, thus providing the opportunity for more projects. Do you agree or disagree, and what problems do you foresee?

3–23 Some organizational structures are considered to be "project-driven." Define what is meant by "project-driven." Which organizational forms described in this chapter would fall under your definition?

3–24 Are there any advantages to having a single project engineer as opposed to having a committee of key functional employees who report to the director of engineering?

3–25 The major difficulty in the selection of a project organizational form involves placement of the project manager. In the evolutionary process, the project manager started out reporting to a department head and ultimately ended up reporting to a senior executive. In general, what were the major reasons for having the project manager report higher and higher in the organizational structure?

3–26 Ralph is a department manager who is quite concerned about the performance of the people beneath him. After several months of analysis, Ralph has won the acceptance of his superiors for setting up a project management structure in his department. Out of the twenty-three departments in the company, his will be the only one with formalized project management. Can this situation be successful even though several projects require interfacing with other departments?

3–27 A large electronics corporation has a multimillion dollar project in which 90 percent of the work stays within one division. The division manager wants to be the project manager. Should this be allowed even though there exists a project management division?

3–28 The internal functioning of an organization must consider:

- The demands imposed on the organization by task complexity
- Available technology
- The external environment
- The needs of the organizational membership

Considering these facts, should an organization search for the one best way to organize under all conditions? Should managers examine the functioning of an organization relative to its needs, or vice versa?

3–29 Project managers, in order to get the job accomplished, need adequate organizational status and authority. One corporate executive contends that an organizational chart such as that in Figure 3–6 can be modified to show that the project managers have adequate authority by placing the department managers in boxes at the top of the functional responsibility arrowheads. The executive further contends that, with this approach, the project managers appear to be higher in the organization than their departmental counterparts but are actually equal in status. Do you agree or disagree with the executive's idea? Will there be a proper balance of power between project and department managers with this organizational structure?

3–30 Defend or attack the following two statements concerning the operation of a matrix:

- There should be no disruption due to dual accountability.
- A difference in judgment should not delay work in progress.

3–31 A company has fifteen projects going on at once. Three projects are over $5 million, seven projects are between $1 million and $3 million, and five projects are between $500,000 and $700,000. Each project has a full-time project manager. Just based upon this information, which organizational form would be best? Can all the project managers report to the same person?

3–32 A major insurance company is considering the implementation of project management. The majority of the projects in the company are two weeks in duration, with very few existing beyond one month. Can project management work here?

3–33 The definition of project management in Section 1.9 identifies project teams and task forces. How would you distinguish between a project team and a task force, and what industries and/or projects would be applicable to each?

3–34 Can informal project management work in a structured environment at the same time as formal project management and share the same resources?

3–35 Several people believe that the matrix structure can be multidimensional (as shown in Figure 3–12). Explain the usefulness of such a structure.

3–36 Many companies have informal project management where work flows horizontally, but in an informal manner. What are the characteristics of informal project management? Which types of companies can operate effectively with informal project management?

3–37 Some companies have tried to develop a matrix within a matrix. Is it possible to have a matrix for formal project control and an internal authority matrix, communication matrix, responsibility matrix, or a combination of several of these?

3–38 Is it possible for a matrix to get out of control because of too many small projects, each competing for the same shared resources? If so, how many projects are too many? How can management control the number of projects? Does your answer depend on whether the organization is project-driven or non–project-driven?

3–39 A government subcontractor operates with a pure specialized product management organizational structure and has four product lines. All employees are required to have a top secret security clearance. The subcontractor's plant is structured such that each of the four product lines occupies a secured area in the building. Employees wear security badges that give them access to the different areas. Most of the employees are authorized to have access only to their area. Only the executives have access to all four areas. For security reasons, functional employees are not permitted to discuss the product lines with each other.

Many of the projects performed in each of the product lines are identical, and severe duplication of efforts exist. Management is interested in converting over to a matrix structure to minimize the duplication of effort. What problems must be overcome before and during matrix implementation?

3–40 A company has decided to go to full project management utilizing a matrix structure. Can the implementation be done in stages? Can the matrix be partially implemented, say, in one portion of the organization, and then gradually expanded across the rest of the company?

3–41 A company has two major divisions, both housed under the same roof. One division is the aerospace group, where all activities are performed within a formal matrix. The second division is the industrial group, which operates with pure product management, except for the MIS department, which has an informal matrix. If both divisions have to share common corporate resources, what problems can occur?

3–42 Several Fortune 100 corporations have a corporate engineering group that assumes the responsibility of the project management–project engineering function for all major capital projects in all divisions worldwide. Explain how the corporate engineering function should work, as well as its advantages and disadvantages.

CASE STUDY

JONES AND SHEPHARD ACCOUNTANTS, INC.*

By 1990, Jones and Shephard Accountants, Inc. (J&S) was a midsized company and ranked 38th in size by the American Association of Accountants. In order to compete with the larger firms, J&S formed an Information Services Division designed primarily for studies and analyses. By 1995, the Information Services Division (ISD) had fifteen employees.

In 1997, the ISD purchased three largecomputers. With this increased capacity, J&S expanded its services to help satisfy the needs of outside customers. By September 1998, the internal and external workloads had increased to a point where the ISD now employed over fifty people.

The director of the division was very disappointed in the way that activities were being handled. There was no single person assigned to push through a project, and outside customers did not know whom to call to get answers regarding project status. The director found that most of his time was being spent on day-to-day activities such as conflict resolution instead of strategic planning and policy formulation.

The biggest problems facing the director were the two continuous internal projects (called Project X and Project Y, for simplicity) that required month-end data collation and reporting. The director felt that these two projects were important enough to require a full-time project manager on each effort.

* Revised 2007.

In October 1998, corporate management announced that the ISD director would be reassigned on February 1, 1999, and that the announcement of his replacement would not be made until the middle of January. The same week that the announcement was made, two individuals were hired from outside the company to take charge of Project X and Project Y. Exhibit 3–1 shows the organizational structure of the ISD.

Within the next thirty days, rumors spread throughout the organization about who would become the new director. Most people felt that the position would be filled from within the division and that the most likely candidates would be the two new project managers. In addition, the associate director was due to retire in December, thus creating two openings.

On January 3, 1999, a confidential meeting was held between the ISD director and the systems manager.

ISD Director: "Corporate has approved my request to promote you to division director. Unfortunately, your job will not be an easy one. You're going to have to restructure the organization somehow so that our employees will not have as many conflicts as they are now faced with. My secretary is typing up a confidential memo for you explaining my observations on the problems within our division.

"Remember, your promotion should be held in the strictest confidence until the final announcement later this month. I'm telling you this now so that you can begin planning the restructuring. My memo should help you." (See Exhibit 3–2 for the memo.)

The systems manager read the memo and, after due consideration, decided that some form of matrix would be best. To help him structure the organization properly, an outside consultant

Exhibit 3–1. ISD organizational chart

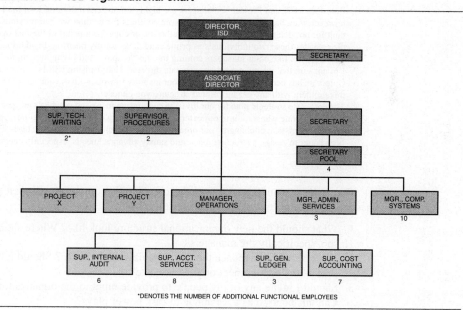

*DENOTES THE NUMBER OF ADDITIONAL FUNCTIONAL EMPLOYEES

was hired to help identify the potential problems with changing over to a matrix. The following problem areas were identified by the consultant:

1. The operations manager controls more than 50 percent of the people resources. You might want to break up his empire. This will have to be done very carefully.
2. The secretary pool is placed too high in the organization.
3. The supervisors who now report to the associate director will have to be reassigned lower in the organization if the associate director's position is abolished.
4. One of the major problem areas will be trying to convince corporate management that their change will be beneficial. You'll have to convince them that this change can be accomplished without having to increase division manpower.
5. You might wish to set up a separate department or a separate project for customer relations.
6. Introducing your employees to the matrix will be a problem. Each employee will look at the change differently. Most people have the tendency of looking first at the shift in the balance of power—have I gained or have I lost power and status?

Exhibit 3–2. Confidential memo

From: **ISD Director**
To: **Systems Manager**
Date: **January 3, 1999**

Congratulations on your promotion to division director. I sincerely hope that your tenure will be productive both personally and for corporate. I have prepared a short list of the major obstacles that you will have to consider when you take over the controls.

1. Both Project X and Project Y managers are highly competent individuals. In the last four or five days, however, they have appeared to create more conflicts for us than we had previously. This could be my fault for not delegating them sufficient authority, or could be a result of the fact that several of our people consider these two individuals as prime candidates for my position. In addition, the operations manager does not like other managers coming into his "empire" and giving direction.
2. I'm not sure that we even need an associate director. That decision will be up to you.
3. Corporate has been very displeased with our inability to work with outside customers. You must consider this problem with any organizational structure you choose.
4. The corporate strategic plan for our division contains an increased emphasis on special, internal MIS projects. Corporate wants to limit our external activities for a while until we get our internal affairs in order.
5. I made the mistake of changing our organizational structure on a day-to-day basis. Perhaps it would have been better to design a structure that could satisfy advanced needs, especially one that we can grow into.

The systems manager evaluated the consultant's comments and then prepared a list of questions to ask the consultant at their next meeting:

1. What should the new organizational structure look like? Where should I put each person, specifically the managers?
2. When should I announce the new organizational change? Should it be at the same time as my appointment or at a later date?
3. Should I invite any of my people to provide input to the organizational restructuring? Can this be used as a technique to ease power plays?
4. Should I provide inside or outside seminars to train my people for the new organizational structure? How soon should they be held?

Organizing and Staffing the Project Office and Team

Related Case Studies (from Kerzner/*Project Management Case Studies,* 3rd Edition)	Related Workbook Exercises (from Kerzner/*Project Management Workbook and PMP®/CAPM® Exam Study Guide,* 10th Edition)	PMBOK® Guide, 4th Edition, Reference Section for the PMP® Certification Exam
• Government Project Management • Falls Engineering • White Manufacturing • Martig Construction Company • Ducor Chemical • The Carlson Project	• The Bad Apple • Multiple Choice Exam	• Human Resource Management

4.0 INTRODUCTION

PMBOK® Guide, 4th Edition
Chapter 9 Human Resource Management

Successful project management, regardless of the organizational structure, is only as good as the individuals and leaders who are managing the key functions. Project management is not a one-person operation; it requires a group of individuals dedicated to the achievement of a specific goal. Project management includes:

- A project manager
- An assistant project manager

- A project (home) office
- A project team

Generally, project office personnel are assigned full-time to the project and work out of the project office, whereas the project team members work out of the functional units and may spend only a small percentage of their time on the project. Normally, project office personnel report directly to the project manager, but they may still be solid to their line function just for administrative control. A project office usually is not required on small projects, and sometimes the project can be accomplished by just one person who may fill all of the project office positions.

Before the staffing function begins, five basic questions are usually considered:

- What are the requirements for an individual to become a successful project manager?
- Who should be a member of the project team?
- Who should be a member of the project office?
- What problems can occur during recruiting activities?
- What can happen downstream to cause the loss of key team members?

On the surface, these questions may not seem especially complex. But when we apply them to a project environment (which is by definition a "temporary" situation) where a constant stream of projects is necessary for corporate growth, the staffing problems become complex, especially if the organization is understaffed.

4.1 THE STAFFING ENVIRONMENT

PMBOK® Guide, 4th Edition
9.1 Human Resource Planning

To understand the problems that occur during staffing, we must first investigate the characteristics of project management, including the project environment, the project management process, and the project manager.

Two major kinds of problems are related to the project environment: personnel performance problems and personnel policy problems. Performance is difficult for many individuals in the project environment because it represents a change in the way of doing business. Individuals, regardless of how competent they are, find it difficult to adapt continually to a changing situation in which they report to multiple managers.

On the other hand, many individuals thrive on temporary assignments because it gives them a "chance for glory." Unfortunately, some employees might consider the chance for glory more important than the project. For example, an employee may pay no attention to the instructions of the project manager and instead perform the task his own way. In this situation, the employee wants only to be recognized as an achiever and really does not care if the project is a success or failure, as long as he still has a functional home to return to where he will be identified as an achiever with good ideas.

The second major performance problem lies in the project–functional interface, where an individual suddenly finds himself reporting to two bosses, the functional manager and the project manager. If the functional manager and the project manager are in agreement about the work to be accomplished, then performance may not be hampered. But if conflicting directions are received, then the individual may let his performance suffer because of his compromising position. In this case, the employee will "bend" in the direction of the manager who controls his purse strings.

Personnel policy problems can create havoc in an organization, especially if the "grass is greener" in a project environment than in the functional environment. Functional organizations normally specify grades and salaries for employees. Project offices, on the other hand, have no such requirements and can promote and pay according to achievement. The difficulty here is that one can distinguish between employees in grades 7, 8, 9, 10, and 11 in a line organization, whereas for a project manager the distinction might appear only in the size of the project or the amount of responsibility. Bonuses are also easier to obtain in the project office but may create conflict and jealousy between the horizontal and vertical elements.

Because each project is different, the project management process allows each project to have its own policies, procedures, rules, and standards, provided they fall within broad company guidelines. Each project must be recognized as a project by top management so that the project manager has the delegated authority necessary to enforce the policies, procedures, rules, and standards.

Project management is successful only if the project manager and his team are totally dedicated to the successful completion of the project. This requires each team member of the project team and office to have a good understanding of the fundamental project requirements, which include:

- Customer liaison
- Project direction
- Project planning
- Project control
- Project evaluation
- Project reporting

Ultimately, the person with the greatest influence during the staffing phase is the project manager. The personal attributes and abilities of project managers will either attract or deter highly desirable individuals. Basic characteristics include:

- Honesty and integrity
- Understanding of personnel problems
- Understanding of project technology
- Business management competence
  - Management principles
  - Communications
- Alertness and quickness
- Versatility
- Energy and toughness
- Decision-making ability
- Ability to evaluate risk and uncertainty

Project managers must exhibit honesty and integrity to foster an atmosphere of trust. They should not make impossible promises, such as immediate promotions for everyone if a follow-on contract is received. Also, on temporarily assigned activities, such as a project, managers cannot wait for personnel to iron out their own problems because time, cost, and performance requirements will not be satisfied.

Project managers should have both business management and technical expertise. They must understand the fundamental principles of management, especially those involving the rapid development of temporary communication channels. Project managers must understand the technical implications of a problem, since they are ultimately responsible for all decision-making. However, many good technically oriented managers have failed because they have become too involved with the technical side of the project rather than the management side. There are strong arguments for having a project manager who has more than just an understanding of the necessary technology.

Because a project has a relatively short time duration, decision-making must be rapid and effective. Managers must be alert and quick in their ability to perceive "red flags" that can eventually lead to serious problems. They must demonstrate their versatility and toughness in order to keep subordinates dedicated to goal accomplishment. Executives must realize that the project manager's objectives during staffing are to:

- Acquire the best available assets and try to improve them
- Provide a good working environment for all personnel
- Make sure that all resources are applied effectively and efficiently so that all constraints are met, if possible

4.2 SELECTING THE PROJECT MANAGER: AN EXECUTIVE DECISION

PMBOK® Guide, 4th Edition
9.2.1 Acquire Project Team
9.3.1 General Management Skills

Probably the most difficult decision facing upper-level management is the selection of project managers. Some managers work best on long-duration projects where decision-making can be slow; others may thrive on short-duration projects that can result in a constant-pressure environment. A director was asked whom he would choose for a key project manager position—an individual who had been a project manager on previous programs in which there were severe problems and cost overruns, or a new aggressive individual who might have the capability to be a good project manager but had never had the opportunity. The director responded that he would go with the seasoned veteran assuming that the previous mistakes would not be made again. The argument here is that the project manager must learn from his own mistakes so they will not be made again. The new individual is apt to make the same mistakes the veteran made. However, this may limit career path opportunities for younger personnel. Stewart has commented on the importance of experience[1]:

> Though the project manager's previous experience is apt to have been confined to a single functional area of business, he must be able to function on the project as a kind of general

1. John M. Stewart, "Making Project Management Work." Reprinted with permission from *Business Horizons*, Fall 1965, p. 63. Copyright © 1965 by the Board of Trustees at Indiana University.

manager in miniature. He must not only keep track of what is happening but also play the crucial role of advocate for the project. Even for a seasoned manager, this task is not likely to be easy. Hence, it is important to assign an individual whose administrative abilities and skills in personal relations have been convincingly demonstrated under fire.

The selection process for project managers is not easy. Five basic questions must be considered:

- What are the internal and external sources?
- How do we select?
- How do we provide career development in project management?
- How can we develop project management skills?
- How do we evaluate project management performance?

Project management cannot succeed unless a good project manager is at the controls. It is far more likely that project managers will succeed if it is obvious to the subordinates that the general manager has appointed them. Usually, a brief memo to the line managers will suffice. The major responsibilities of the project manager include:

- To produce the end-item with the available resources and within the constraints of time, cost, and performance/technology
- To meet contractual profit objectives
- To make all required decisions whether they be for alternatives or termination
- To act as the customer (external) and upper-level and functional management (internal) communications focal point
- To "negotiate" with all functional disciplines for accomplishment of the necessary work packages within the constraints of time, cost, and performance/technology
- To resolve all conflicts

If these responsibilities were applied to the total organization, they might reflect the job description of the general manager. This analogy between project and general managers is one of the reasons why future general managers are asked to perform functions that are implied, rather than spelled out, in the job description. As an example, you are the project manager on a high-technology project. As the project winds down, an executive asks you to write a paper so that he can present it at a technical meeting in Tokyo. His name will appear first on the paper. Should this be a part of your job? As this author sees it, you really don't have much of a choice.

In order for project managers to fulfill their responsibilities successfully, they are constantly required to demonstrate their skills in interface, resource, and planning and control management. These implicit responsibilities are shown below:

- Interface Management
  - Product interfaces
    - Performance of parts or subsections
    - Physical connection of parts or subsections
  - Project interfaces
  - Customer
  - Management (functional and upper-level)

- Change of responsibilities
- Information flow
- Material interfaces (inventory control)
- Resource Management
  - Time (schedule)
  - Manpower
  - Money
  - Facilities
  - Equipment
  - Material
  - Information/technology
- Planning and Control Management
  - Increased equipment utilization
  - Increased performance efficiency
  - Reduced risks
  - Identification of alternatives to problems
  - Identification of alternative resolutions to conflicts

Consider the following advertisement for a facilities planning and development project manager (adapted from *The New York Times*, January 2, 1972):

> Personable, well-educated, literate individual with college degree in Engineering to work for a small firm. Long hours, no fringe benefits, no security, little chance for advancement are among the inducements offered. Job requires wide knowledge and experience in manufacturing, materials, construction techniques, economics, management and mathematics. Competence in the use of the spoken and written English is required. Must be willing to suffer personal indignities from clients, professional derision from peers in the more conventional jobs, and slanderous insults from colleagues.
>
> Job involves frequent extended trips to inaccessible locations throughout the world, manual labor and extreme frustration from the lack of data on which to base decisions.
>
> Applicant must be willing to risk personal and professional future on decisions based upon inadequate information and complete lack of control over acceptance of recommendations by clients. Responsibilities for the work are unclear and little or no guidance is offered. Authority commensurate with responsibility is not provided either by the firm or its clients.
>
> Applicant should send resume, list of publications, references and other supporting documentation to. . . .

Fortunately, these types of job descriptions are very rare today.

Finding the person with the right qualifications is not an easy task because the selection of project managers is based more on personal characteristics than on the job description. In Section 4.1 a brief outline of desired characteristics was presented. Russell Archibald defines a broader range of desired personal characteristics[2]:

- Flexibility and adaptability
- Preference for significant initiative and leadership

---

2. Russell D. Archibald, *Managing High-Technology Programs and Projects* (New York: Wiley, 1976), p. 55. Copyright © 1976 by John Wiley & Sons, Inc. Reprinted by permission of the publisher.

PMBOK® Guide, 4th Edition
9.3 Develop Project Team

- Aggressiveness, confidence, persuasiveness, verbal fluency
- Ambition, activity, forcefulness
- Effectiveness as a communicator and integrator
- Broad scope of personal interests
- Poise, enthusiasm, imagination, spontaneity
- Able to balance technical solutions with time, cost, and human factors
- Well organized and disciplined
- A generalist rather than a specialist
- Able and willing to devote most of his time to planning and controlling
- Able to identify problems
- Willing to make decisions
- Able to maintain proper balance in the use of time

This ideal project manager would probably have doctorates in engineering, business, and psychology, and experience with ten different companies in a variety of project office positions, and would be about twenty-five years old. Good project managers in industry today would probably be lucky to have 70 to 80 percent of these characteristics. The best project managers are willing and able to identify their own shortcomings and know when to ask for help.

The difficulty in staffing, especially for project managers or assistant project managers, is in determining what questions to ask during an interview to see if an individual has the necessary or desired characteristics. Individuals may be qualified to be promoted vertically but not horizontally. An individual with poor communication skills and interpersonal skills can be promoted to a line management slot because of his technical expertise, but this same individual is not qualified for project management promotion.

One of the best ways to interview is to read each element of the job description to the potential candidate. Many individuals want a career path in project management but are totally unaware of what the project manager's duties are.

So far we have discussed the personal characteristics of the project manager. There are also job-related questions to consider, such as:

- Are feasibility and economic analyses necessary?
- Is complex technical expertise required? If so, is it within the individual's capabilities?
- If the individual is lacking expertise, will there be sufficient backup strength in the line organizations?
- Is this the company's or the individual's first exposure to this type of project and/or client? If so, what are the risks to be considered?
- What is the priority for this project, and what are the risks?
- With whom must the project manager interface, both inside and outside the organization?

Most good project managers know how to perform feasibility studies and cost-benefit analyses. Sometimes these studies create organizational conflict. A major utility company begins each computer project with a feasibility study in which a cost-benefit analysis is performed. The project managers, all of whom report to a project management division, perform the study themselves without any direct functional support. The functional managers argue that the results are grossly inaccurate because the functional experts are not involved. The project managers, on the other hand, argue that they never have sufficient time or money to perform a complete analysis. Some companies resolve this by having a special group perform these studies.

Most companies would prefer to find project managers from within. Unfortunately, this is easier said than done.

There are also good reasons for recruiting from outside the company. A new project manager hired from the outside would be less likely to have strong informal ties to any one line organization and thus could be impartial. Some companies further require that the individual spend an apprenticeship period of twelve to eighteen months in a line organization to find out how the company functions, to become acquainted with the people, and to understand the company's policies and procedures.

One of the most important but often least understood characteristics of good project managers is the ability to know their own strengths and weaknesses and those of their employees. Managers must understand that in order for employees to perform efficiently:

- They must know what they are supposed to do.
- They must have a clear understanding of authority and its limits.
- They must know what their relationship with other people is.
- They should know what constitutes a job well done in terms of specific results.
- They should know where and when they are falling short.
- They must be made aware of what can and should be done to correct unsatisfactory results.
- They must feel that their superior has an interest in them as individuals.
- They must feel that their superior believes in them and wants them to succeed.

4.3 SKILL REQUIREMENTS FOR PROJECT AND PROGRAM MANAGERS

PMBOK® Guide, 4th Edition
Chapter 9 Human Resources
   Management
9.3.2.1 Interpersonal Skills
1.4.2 Program Management

Managing complex programs represents a challenge requiring skills in team building, leadership, conflict resolution, technical expertise, planning, organization, entrepreneurship, administration, management support, and the allocation of resources. This section examines these skills relative to program management effectiveness. A key factor to good program performance is the program manager's ability to integrate personnel from many disciplines into an effective work team.

To get results, the program manager must relate to (1) the people to be managed, (2) the task to be done, (3) the tools available, (4) the organizational structure, and (5) the organizational environment, including the customer community.

With an understanding of the interaction of corporate organization and behavior elements, the manager can build an environment conducive to the working team's needs. The internal and external forces that impinge on the organization of the project must be reconciled to mutual goals. Thus the program manager must be both socially and technically aware to understand how the organization functions and how these functions will affect the program organization of the particular job to be done. In addition, the program manager must understand the culture and value system of the organization he is working with. Effective program management is directly related to proficiency in these ten skills:

- Team building
- Leadership
- Conflict resolution
- Technical expertise
- Planning
- Organization
- Entrepreneurship
- Administration
- Management support
- Resource allocation

It is important that the personal management style underlying these skills facilitate the integration of multidisciplinary program resources for synergistic operation. The days of the manager who gets by with technical expertise alone or pure administrative skills are gone.

Team-Building Skills
Building the program team is one of the prime responsibilities of the program manager. Team building involves a whole spectrum of management skills required to identify, commit, and integrate the various task groups from the traditional functional organization into a single program management system.

To be effective, the program manager must provide an atmosphere conducive to teamwork. He must nurture a climate with the following characteristics:

- Team members committed to the program
- Good interpersonal relations and team spirit
- The necessary expertise and resources
- Clearly defined goals and program objectives
- Involved and supportive top management
- Good program leadership
- Open communication among team members and support organizations
- A low degree of detrimental interpersonal and intergroup conflict

Three major considerations are involved in all of the above factors: (1) effective communications, (2) sincere interest in the professional growth of team members, and (3) commitment to the project.

Leadership Skills
A prerequisite for program success is the program manager's ability to lead the team within a relatively unstructured environment. It involves

dealing effectively with managers and supporting personnel across functional lines and the ability to collect and filter relevant data for decision-making in a dynamic environment. It involves the ability to integrate individual demands, requirements, and limitations into decisions and to resolve intergroup conflicts.

As with a general manager, quality leadership depends heavily on the program manager's personal experience and credibility within the organization. An effective management style might be characterized this way:

- Clear project leadership and direction
- Assistance in problem-solving
- Facilitating the integration of new members into the team
- Ability to handle interpersonal conflict
- Facilitating group decisions
- Capability to plan and elicit commitments
- Ability to communicate clearly
- Presentation of the team to higher management
- Ability to balance technical solutions against economic and human factors

The personal traits desirable and supportive of the above skills are:

- Project management experience
- Flexibility and change orientation
- Innovative thinking
- Initiative and enthusiasm
- Charisma and persuasiveness
- Organization and discipline

Conflict Resolution Skills   Conflict is fundamental to complex task management. Understanding the determinants of conflicts is important to the program manager's ability to deal with conflicts effectively. When conflict becomes dysfunctional, it often results in poor program decision-making, lengthy delays over issues, and a disruption of the team's efforts, all negative influences to program performance. However, conflict can be beneficial when it produces involvement and new information and enhances the competitive spirit.

To successfully resolve conflict and improve overall program performance, program managers must:

- Understand interaction of the organizational and behavioral elements in order to build an environment conducive to their team's motivational needs. This will enhance active participation and minimize unproductive conflict.
- Communicate effectively with all organizational levels regarding both project objectives and decisions. Regularly scheduled status review meetings can be an important communication vehicle.
- Recognize the determinants of conflict and their timing in the project life cycle. Effective project planning, contingency planning, securing of commitments, and

involving top management can help to avoid or minimize many conflicts before they impede project performance.

The accomplished manager needs a "sixth sense" to indicate when conflict is desirable, what kind of conflict will be useful, and how much conflict is optimal for a given situation. In the final analysis, he has the sole responsibility for his program and how conflict will contribute to its success or failure.

Technical Skills

The program manager rarely has all the technical, administrative, and marketing expertise needed to direct the program single-handedly. It is essential, however, for the program manager to understand the technology, the markets, and the environment of the business. Without this understanding, the consequences of local decisions on the total program, the potential growth ramifications, and relationships to other business opportunities cannot be foreseen by the manager. Further technical expertise is necessary to evaluate technical concepts and solutions, to communicate effectively in technical terms with the project team, and to assess risks and make trade-offs between cost, schedule, and technical issues. This is why in complex problem-solving situations so many project managers must have an engineering background.

Technical expertise is composed of an understanding of the:

- Technology involved
- Engineering tools and techniques employed
- Specific markets, their customers, and requirements
- Product applications
- Technological trends and evolutions
- Relationship among supporting technologies
- People who are part of the technical community

The technical expertise required for effective management of engineering programs is normally developed through progressive growth in engineering or supportive project assignments in a specific technology area. Frequently, the project begins with an exploratory phase leading into a proposal. This is normally an excellent testing ground for the future program manager. It also allows top management to judge the new candidate's capacity for managing the technological innovations and integration of solutions.

Planning Skills

Planning skills are helpful for any undertaking; they are absolutely essential for the successful management of large complex programs. The project plan is the road map that defines how to get from the start to the final results.

Program planning is an ongoing activity at all organizational levels. However, the preparation of a project summary plan, prior to project start, is the responsibility of the program manager. Effective project planning requires particular skills far beyond writing a document with schedules and budgets. It requires communication and information processing skills to define the actual resource requirements and administrative support necessary. It requires the ability to negotiate the necessary resources and commitments from key personnel in various support organizations with little or no formal authority.

Effective planning requires skills in the areas of:

- Information processing
- Communication
- Resource negotiations
- Securing commitments
- Incremental and modular planning
- Assuring measurable milestones
- Facilitating top management involvement

In addition, the program manager must assure that the plan remains a viable document. Changes in project scope and depth are inevitable. The plan should reflect necessary changes through formal revisions and should be the guiding document throughout the life cycle of the program. An obsolete or irrelevant plan is useless.

Finally, program managers need to be aware that planning can be overdone. If not controlled, planning can become an end in itself and a poor substitute for innovative work. It is the responsibility of the program manager to build flexibility into the plan and police it against misuse.

Organizational Skills

The program manager must be a social architect; that is, he must understand how the organization works and how to work with the organization. Organizational skills are particularly important during project formation and startup when the program manager is integrating people from many different disciplines into an effective work team. It requires defining the reporting relationships, responsibilities, lines of control, and information needs. A good program plan and a task matrix are useful organizational tools. In addition, the organizational effort is facilitated by clearly defined program objectives, open communication channels, good program leadership, and senior management support.

Entrepreneurial Skills

The program manager also needs a general management perspective. For example, economic considerations affect the organization's financial performance, but objectives often are much broader than profits. Customer satisfaction, future growth, cultivation of related market activities, and minimum organizational disruptions of other programs might be equally important goals. The effective program manager is concerned with all these issues.

Entrepreneurial skills are developed through actual experience. However, formal MBA-type training, special seminars, and cross-functional training programs can help to develop the entrepreneurial skills needed by program managers.

Administrative Skills

Administrative skills are essential. The program manager must be experienced in planning, staffing, budgeting, scheduling, and other control techniques. In dealing with technical personnel, the problem is seldom to make people understand administrative techniques such as budgeting and scheduling, but to impress on them that costs and schedules are just as important as elegant technical solutions.

Particularly on larger programs, managers rarely have all the administrative skills required. While it is important that program managers understand the company's operating procedures and available tools, it is often necessary for the program manager to free himself from administrative details regardless of his ability to handle them. He has to delegate considerable administrative tasks to support groups or hire a project administrator.

Some helpful tools for the manager in the administration of his program include: (1) the meeting, (2) the report, (3) the review, and (4) budget and schedule controls. Program managers must be thoroughly familiar with these available tools and know how to use them effectively.

Management Support Building Skills

The program manager is surrounded by a myriad of organizations that either support him or control his activities. An understanding of these interfaces is important to program managers as it enhances their ability to build favorable relationships with senior management. Project organizations are shared-power systems with personnel of many diverse interests and "ways of doing things." Only a strong leader backed by senior management can prevent the development of unfavorable biases.

Four key variables influence the project manager's ability to create favorable relationships with senior management: (1) his ongoing credibility, (2) the visibility of his program, (3) the priority of his program relative to other organizational undertakings, and (4) his own accessibility.

Resource Allocation Skills

A program organization has many bosses. Functional lines often shield support organizations from direct financial control by the project office. Once a task has been authorized, it is often impossible to control the personnel assignments, priorities, and indirect manpower costs. In addition, profit accountability is difficult owing to the interdependencies of various support departments and the often changing work scope and contents.

Effective and detailed program planning may facilitate commitment and reinforce control. Part of the plan is the "Statement of Work," which establishes a basis for resource allocation. It is also important to work out specific agreements with all key contributors and their superiors on the tasks to be performed and the associated budgets and schedules. Measurable milestones are not only important for hardware components, but also for the "invisible" program components such as systems and software tasks. Ideally, these commitments on specs, schedules, and budgets should be established through involvement by key personnel in the early phases of project formation, such as the proposal phase. This is the time when requirements are still flexible, and trade-offs among performance, schedule, and budget parameters are possible. Further, this is normally the time when the competitive spirit among potential contributors is highest, often leading to a more cohesive and challenging work plan.

4.4 SPECIAL CASES IN PROJECT MANAGER SELECTION

Thus far we have assumed that the project is large enough for a full-time project manager to be appointed. This is not always the case. There are four major problem areas in staffing projects:

- Part-time versus full-time assignments
- Several projects assigned to one project manager
- Projects assigned to functional managers
- The project manager role retained by the general manager

The first problem is generally related to the size of the project. If the project is small (in time duration or cost), a part-time project manager may be selected. Many executives have fallen into the trap of letting line personnel act as part-time project managers while still performing line functions. If the employee has a conflict between what is best for the project and what is best for his line organization, the project will suffer. It is only natural that the employee will favor the place the salary increases come from.

It is a common practice for one project manager to control several projects, especially if they are either related or similar. Problems come about when the projects have drastically different priorities. The low-priority efforts will be neglected.

If the project is a high-technology effort that requires specialization and can be performed by one department, then it is not unusual for the line manager to take on a dual role and act as project manager as well. This can be difficult to do, especially if the project manager is required to establish the priorities for the work under his supervision. The line manager may keep the best resources for the project, regardless of the priority. Then that project will be a success at the expense of every other project he must supply resources to.

Probably the worst situation is that in which an executive fills the role of project manager for a particular effort. The executive may not have the time necessary for total dedication to the achievement of the project. He cannot make effective decisions as a project manager while still discharging normal duties. Additionally, the executive may hoard the best resources for his project.

4.5 SELECTING THE WRONG PROJECT MANAGER

Even though executives know the personal characteristics and traits that project managers should possess, and even though job descriptions are often clearly defined, management may still select the wrong person because they base their decision on the following criteria.

Maturity   Some executives consider gray hair to be a sure indication of maturity, but this is not the type of maturity needed for project management. Maturity in project management generally comes from exposure to several types of projects in a variety of project office positions. In aerospace and defense, it is possible for a project manager to manage the same type of project for ten years or more. When placed on a new project, the individual may try to force personnel and project requirements to adhere to the same policies and procedures that existed on the ten-year project. The project manager may know only one way of managing projects.

Hard-Nosed Tactics   Applying hard-nosed tactics to subordinates can be very demoralizing. Project managers must give people sufficient freedom to get the job done, without providing continuous supervision and direction. A line employee who is given "freedom" by his line manager but suddenly finds himself closely supervised by the project manager will be very unhappy.

Line managers, because of their ability to control an employee's salary, need only one leadership style and can force the employees to adapt. The project manager, on the other hand, cannot control salaries and must have a wide variety of leadership styles. The project manager must adapt a leadership style to the project employees, whereas the reverse is true in the line organization.

Availability   Executives should not assign individuals as project managers simply because of availability. People have a tendency to cringe when you suggest that project managers be switched halfway through a project. For example, manager X is halfway through his project. Manager Y is waiting for an assignment. A new project comes up, and the executive switches managers X and Y. There are several reasons for this. The most important phase of a project is planning, and, if it is accomplished correctly, the project could conceivably run itself. Therefore, manager Y should be able to handle manager X's project.

There are several other reasons why this switch may be necessary. The new project may have a higher priority and require a more experienced manager. Second, not all project managers are equal, especially when it comes to planning. When an executive finds a project manager who demonstrates extraordinary talents at planning, there is a natural tendency for the executive to want this project manager to plan all projects.

Technical Expertise   Executives quite often promote technical line managers without realizing the consequences. Technical specialists may not be able to divorce themselves from the technical side of the house and become project managers rather than project doers. There are also strong reasons to promote technical specialists to project managers. These people often:

- Have better relationships with fellow researchers
- Can prevent duplication of effort
- Can foster teamwork
- Have progressed up through the technical ranks
- Are knowledgeable in many technical fields
- Understand the meaning of profitability and general management philosophy

- Are interested in training and teaching
- Understand how to work with perfectionists

Promoting an employee to project management because of his technical expertise may be acceptable if, and only if, the project requires this expertise and technical direction, as in R&D efforts. For projects in which a "generalist" is acceptable as a project manager, there may be a great danger in assigning highly technical personnel. According to Wilemon and Cicero[3]:

- The greater the project manager's technical expertise, the higher the propensity that he will overly involve himself in the technical details of the project.
- The greater the project manager's difficulty in delegating technical task responsibilities, the more likely it is that he will overinvolve himself in the technical details of the project. (Depending upon his expertise to do so.)
- The greater the project manager's interest in the technical details of the project, the more likely it is that he will defend the project manager's role as one of a technical specialist.
- The lower the project manager's technical expertise, the more likely it is that he will overstress the nontechnical project functions (administrative functions).

Customer Orientation   Executives quite often place individuals as project managers simply to satisfy a customer request. Being able to communicate with the customer does not guarantee project success, however. If the choice of project manager is simply a concession to the customer, then the executive must insist on providing a strong supporting team.

New Exposure   Executives run the risk of project failure if an individual is appointed project manager simply to gain exposure to project management. An executive of a utility company wanted to rotate his line personnel into project management for twelve to eighteen months and then return them to the line organization where they would be more well-rounded individuals and better understand the working relationship between project management and line management. There are two major problems with this. First, the individual may become technically obsolete after eighteen months in project management. Second, and more important, individuals who get a taste of project management will generally not want to return to the line organization.

Company Exposure   The mere fact that individuals have worked in a variety of divisions does not guarantee that they will make good project managers. Their working in a variety of divisions may indicate that they couldn't hold any one job. In that case, they have reached their true level of incompetency, and putting them into project

---

3. D. L. Wilemon and J. P. Cicero, "The Project Manager—Anomalies and Ambiguities," *Academy of Management Journal,* Vol. 13, 1970, pp. 269–282.

TABLE 4–1. METHODS AND TECHNIQUES FOR DEVELOPING PROJECT MANAGERS

I. Experiential training/on-the-job
    Working with experienced professional leader
    Working with project team member
    Assigning a variety of project management responsibilities, consecutively
    Job rotation
    Formal on-the-job training
    Supporting multifunctional activities
    Customer liaison activities
II. Conceptual training/schooling
    Courses, seminars, workshops
    Simulations, games, cases
    Group exercises
    Hands-on exercises in using project management techniques
    Professional meetings
    Conventions, symposia
    Readings, books, trade journals, professional magazines
III. Organizational development
    Formally established and recognized project management function
    Proper project organization
    Project support systems
    Project charter
    Project management directives, policies, and procedures

management will only maximize the damage they can do to the company. Some executives contend that the best way to train a project manager is by rotation through the various functional disciplines for two weeks to a month in each organization. Other executives maintain that this is useless because the individual cannot learn anything in so short a period of time.

Tables 4–1 and 4–2 identify current thinking on methods for training project managers.

Finally, there are three special points to consider:

- Individuals should not be promoted to project management simply because they are at the top of their pay grade.
- Project managers should be promoted and paid based on performance, not on the number of people supervised.
- It is not necessary for the project manager to be the highest ranking or salaried individual on the project team with the rationale that sufficient "clout" is needed.

TABLE 4–2. HOW TO TRAIN PROJECT MANAGERS

Company Management Say Project Managers Can Be Trained in a Combination of Ways:	
Experiential learning, on-the-job	60%
Formal education and special courses	20%
Professional activities, seminars	10%
Readings	10%

4.6 NEXT GENERATION PROJECT MANAGERS

The skills needed to be an effective, twenty-first century project manager have changed from those needed during the 1980s. Historically, only engineers were given the opportunity to become project managers. The belief was that the project manager had to have a command of technology in order to make all of the technical decisions. As projects became larger and more complex, it became obvious that project managers might need simply an understanding rather than a command of technology. The true technical expertise would reside with the line managers, except for special situations such as R&D project management.

As project management began to grow and mature, the project manager was converted from a technical manager to a business manager. The primary skills needed to be an effective project manager in the twenty-first century are:

- Knowledge of the business
- Risk management
- Integration skills

The critical skill is risk management. However, to perform risk management effectively, a sound knowledge of the business is required. Figure 4–1 shows the changes in project management skills needed between 1985 and 2008.

As projects become larger, the complexities of integration management become more pronounced. Figure 4–2 illustrates the importance of integration management. In 1985, project managers spent most of their time planning and replanning with their team. This was necessary because the project manager was the technical expert. Today, line managers are the technical experts and perform the majority of the planning and replanning within their line. The project manager's efforts are now heavily oriented toward integration of the

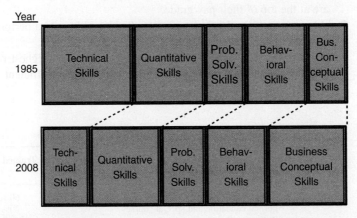

FIGURE 4–1. Project management skills.

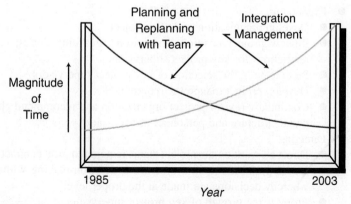

FIGURE 4–2. How do project managers spend their time?

function plans into a total project plan. Some people contend that, with the increased risks and complexities of integration management, the project manager of the future will become an expert in damage control.

4.7 DUTIES AND JOB DESCRIPTIONS

Since projects, environments, and organizations differ from company to company as well as project to project, it is not unusual for companies to struggle to provide reasonable job descriptions of the project manager and associated personnel. Below is a simple list identifying the duties of a project manager in the construction industry[4]:

- Planning
  - Become completely familiar with all contract documents
  - Develop the basic plan for executing and controlling the project
  - Direct the preparation of project procedures
  - Direct the preparation of the project budget
  - Direct the preparation of the project schedule
  - Direct the preparation of basic project design criteria and general specifications
  - Direct the preparation of the plan for organizing, executing, and controlling field construction activities
  - Review plans and procedures periodically and institute changes if necessary

4. Source unknown.

- Organizing
  - Develop organization chart for project
  - Review project position descriptions, outlining duties, responsibilities, and restrictions for key project supervisors
  - Participate in the selection of key project supervisors
  - Develop project manpower requirements
  - Continually review project organization and recommend changes in organizational structure and personnel, if necessary
- Directing
  - Direct all work on the project that is required to meet contract obligations
  - Develop and maintain a system for decision-making within the project team whereby decisions are made at the proper level
  - Promote the growth of key project supervisors
  - Establish objectives for project manager and performance goals for key project supervisors
  - Foster and develop a spirit of project team effort
  - Assist in resolution of differences or problems between departments or groups on assigned projects
  - Anticipate and avoid or minimize potential problems by maintaining current knowledge of overall project status
  - Develop clear written strategy guidelines for all major problems with clear definitions of responsibilities and restraints
- Controlling
  - Monitor project activities for compliance with company purpose and philosophy and general corporate policies
  - Interpret, communicate, and require compliance with the contract, the approved plan, project procedures, and directives of the client
  - Maintain personal control of adherence to contract warranty and guarantee provisions
  - Closely monitor project activities for conformity to contract scope provisions. Establish change notice procedure to evaluate and communicate scope changes
  - See that the plans for controlling and reporting on costs, schedule, and quality are effectively utilized
  - Maintain effective communications with the client and all groups performing project work

A more detailed job description of a construction project manager (for a utility company) appears below:

Duties

Under minimum supervision establishes the priorities for and directs the efforts of personnel (including their consultants or contractors) involved or to be involved on project controlled tasks to provide required achievement of an integrated approved set of technical, manpower, cost, and schedule requirements.

1. Directs the development of initial and revised detailed task descriptions and forecasts of their associated technical, manpower, cost, and schedule requirements for tasks assigned to the Division.
2. Directs the regular integration of initial and revised task forecasts into Divisional technical, manpower, cost, and schedule reports and initiates the approval cycle for the reports.
3. Reviews conflicting inter- and extra-divisional task recommendations or actions that may occur from initial task description and forecast development until final task completion and directs uniform methods for their resolution.
4. Evaluates available and planned additions to Division manpower resources, including their tasks applications, against integrated technical and manpower reports and initiates actions to assure that Division manpower resources needs are met by the most economical mix of available qualified consultant and contractor personnel.
5. Evaluates Divisional cost and schedule reports in light of new tasks and changes in existing tasks and initiates actions to assure that increases or decreases in task cost and schedule are acceptable and are appropriately approved.
6. Prioritizes, adjusts, and directs the efforts of Division personnel (including their consultants and contractors) resource allocations as necessary to both assure the scheduled achievement of state and federal regulatory commitments and maintain Divisional adherence to integrated manpower, cost, and schedule reports.
7. Regularly reports the results of Divisional manpower, cost, and schedule evaluations to higher management.
8. Regularly directs the development and issue of individual task and integrated Project programs reports.
9. Recommends new or revised Division strategies, goals, and objectives in light of anticipated long-term manpower and budget needs.
10. Directly supervises project personnel in the regular preparation and issue of individual task descriptions and their associated forecasts, integrated Division manpower, cost, and schedule reports, and both task and Project progress reports.
11. Establishes basic organizational and personnel qualification requirements for Division (including their consultants or contractors) performance on tasks.
12. Establishes the requirements for, directs the development of, and approves control programs to standardize methods used for controlling similar types of activities in the Project and in other Division Departments.
13. Establishes the requirements for, directs the development of, and approves administrative and technical training programs for Divisional personnel.
14. Approves recommendations for the placement of services or material purchase orders by Division personnel and assures that the cost and schedule data associated with such orders is consistent with approved integrated cost and schedule reports.
15. Promotes harmonious relations among Division organizations involved with Project tasks.
16. Exercises other duties related to Divisional project controls as assigned by the project manager.

PMBOK® Guide, 4th Edition
Chapter 9 Human Resources
Management

TABLE 4–3.   PROJECT MANAGEMENT POSITIONS AND RESPONSIBILITIES

Project Management Position	Typical Responsibility	Skill Requirements
• Project Administrator • Project Coordinator • Technical Assistant	Coordinating and integrating of subsystem tasks. Assisting in determining technical and manpower requirements, schedules, and budgets. Measuring and analyzing project performance regarding technical progress, schedules, and budgets.	• Planning • Coordinating • Analyzing • Understanding the organization
• Task Manager • Project Engineer • Assistant Project Manager	Same as above, but stronger role in establishing and maintaining project requirements. Conducting trade-offs. Directing the technical implementation according to established schedules and budgets.	• Technical expertise • Assessing trade-offs • Managing task implementation • Leading task specialists
• Project Manager • Program Manager	Same as above, but stronger role in project planning and controlling. Coordinating and negotiating requirements between sponsor and performing organizations. Bid proposal development and pricing. Establishing project organization and staffing. Overall leadership toward implementing project plan. Project profit. New business development.	• Overall program leadership • Team building • Resolving conflict • Managing multidisciplinary tasks • Planning and allocating resources • Interfacing with customers/sponsors
• Executive Program Manager	Title reserved for very large programs relative to host organization. Responsibilities same as above. Focus is on directing overall program toward desired business results. Customer liaison. Profit performance. New business development. Organizational development.	• Business leadership • Managing overall program businesses • Building program organizations • Developing personnel • Developing new business
• Director of Programs • V.P. Program Development	Responsible for managing multiprogram businesses via various project organizations, each led by a project manager. Focus is on business planning and development, profit performance, technology development, establishing policies and procedures, program management guidelines, personnel development, organizational development.	• Leadership • Strategic planning • Directing and managing program businesses • Building organizations • Selecting and developing key personnel • Identifying and developing new business

Qualifications

1. A Bachelor of Science Degree in Engineering or a Business Degree with a minor in Engineering or Science from an accredited four (4) year college or university.
2. a) (For Engineering Graduate) Ten (10) or more years of Engineering and Construction experience including a minimum of five (5) years of supervisory experience and two (2) years of management and electric utility experience.
   b) (For Business Graduate) Ten (10) or more years of management experience including a minimum of five (5) years of supervisory experience in an engineering and construction related management area and two (2) years of experience as the manager or assistant manager of major engineering and construction related projects and two (2) recent years of electric utility experience.
3. Working knowledge of state and federal regulations and requirements that apply to major design and construction projects such as fossil and nuclear power stations.
4. Demonstrated ability to develop high level management control programs.
5. Experience related to computer processing of cost and schedule information.
6. Registered Professional Engineer and membership in appropriate management and technical societies is desirable (but not necessary).
7.[5] At least four (4) years of experience as a staff management member in an operating nuclear power station or in an engineering support on- or off-site capacity.
8.[5] Detailed knowledge of federal licensing requirement for nuclear power stations.
9.[5] Reasonably effective public speaker.

Because of the potential overlapping nature of job descriptions in a project management environment, some companies try to define responsibilities for each project management position, as shown in Table 4–3.

4.8 THE ORGANIZATIONAL STAFFING PROCESS

PMBOK® Guide, 4th Edition
Chapter 9 Human Resources Management
9.2 Acquire Project Team

Staffing the project organization can become a long and tedious effort, especially on large and complex engineering projects. Three major questions must be answered:

- What people resources are required?
- Where will the people come from?
- What type of project organizational structure will be best?

To determine the people resources required, the types of individuals (possibly job descriptions) must be decided on, as well as how many individuals from each job category are necessary and when these individuals will be needed.

---

5. Qualifications 7 through 9 apply only for Nuclear Project Directors.

Consider the following situation: As a project manager, you have an activity that requires three separate tasks, all performed within the same line organization. The line manager promises you the best available resources right now for the first task but cannot make any commitments beyond that. The line manager may have only below-average workers available for the second and third tasks. However, the line manager is willing to make a deal with you. He can give you an employee who can do the work but will only give an average performance. If you accept the average employee, the line manager will guarantee that the employee will be available to you for all three tasks. How important is continuity to you? There is no clearly definable answer to this question. Some people will always want the best resources and are willing to fight for them, whereas others prefer continuity and dislike seeing new people coming and going. The author prefers continuity, provided that the assigned employee has the ability to do the up-front planning needed during the first task. The danger in selecting the best employee is that a higher-priority project may come along, and you will lose the employee; or if the employee is an exceptional worker, he may simply be promoted off your project.

Sometimes, a project manager may have to make concessions to get the right people. For example, during the seventh, eighth, and ninth months of your project you need two individuals with special qualifications. The functional manager says that they will be available two months earlier, and that if you don't pick them up then, there will be no guarantee of their availability during the seventh month. Obviously, the line manager is pressuring you, and you may have to give in. There is also the situation in which the line manager says that he'll have to borrow people from another department in order to fulfill his commitments for your project. You may have to live with this situation, but be very careful—these employees will be working at a low level on the learning curve, and overtime will not necessarily resolve the problem. You must expect mistakes here.

Line managers often place new employees on projects so they can be upgraded. Project managers often resent this and immediately go to top management for help. If a line manager says that he can do the work with lower-level people, then the project manager must believe the line manager. After all, the line manager, not the assigned employees, makes the commitment to do the work, and it is the line manager's neck that is stuck out.

Mutual trust between project and line managers is crucial, especially during staffing sessions. Once a project manager has developed a good working relationship with employees, the project manager would like to keep those individuals assigned to his activities. There is nothing wrong with a project manager requesting the same administrative and/or technical staff as before. Line managers realize this and usually agree to it.

There must also be mutual trust between the project managers themselves. Project managers must work as a team, recognize each other's needs, and be willing to make decisions that are in the best interest of the company.

Once the resources are defined, the next question must be whether staffing will be from within the existing organization or from outside sources, such as new hires or consultants. Outside consultants are advisable if, and only if, internal manpower resources are being fully utilized on other programs, or if the company does not possess the required project skills. The answer to this question will indicate which organizational form is best for achievement of the objectives. The form might be a matrix, product, or staff project management structure.

Not all companies permit a variety of project organizational forms to exist within the main company structure. Those that do, however, consider the basic questions of classical management before making a decision. These include:

- How is labor specialized?
- What should the span of management be?
  - How much planning is required?
  - Are authority relationships delegated and understood?
  - Are there established performance standards?
  - What is the rate of change of the job requirements?
- Should we have a horizontal or vertical organization?
  - What are the economics?
  - What are the morale implications?
- Do we need a unity-of-command position?

As in any organization, the subordinates can make the superior look good in the performance of his duties. Unfortunately, the project environment is symbolized by temporary assignments in which the main effort put forth by the project manager is to motivate his (temporary) subordinates toward project dedication and to make them fully understand that:

- Teamwork is vital for success.
- Esprit de corps contributes to success.
- Conflicts can occur between project and functional tiers.
- Communication is essential for success.
- Conflicting orders may be given by the:
  - Project manager
  - Functional manager
  - Upper-level manager
- Unsuccessful performance may result in transfer or dismissal from the project as well as disciplinary action.

Earlier we stated that a project operates as a separate entity but remains attached to the company through company administration policies and procedures. Although project managers can establish their own policies, procedures, and rules, the criteria for promotion must be based on company standards. Project managers should be careful about making commitments they can't keep. After unkept promises on previous projects, a project manager will find it very difficult to get top-quality personnel to volunteer for another project. Even if top management orders key individuals to be assigned to his project, they will always be skeptical about any promises that he may make.

Selecting the project manager is only one-third of the staffing problem. The next step, selecting the project office personnel and team members, can be a time-consuming chore. The project office consists of personnel who are usually assigned as full-time members of the project. The evaluation process should include active project team members, functional team members available for promotion or transfer, and outside applicants.

Upon completion of the evaluation process, the project manager meets with upper-level management. This coordination is required to assure that:

- All assignments fall within current policies on rank, salary, and promotion.
- The individuals selected can work well with both the project manager (formal reporting) and upper-level management (informal reporting).
- The individuals selected have good working relationships with the functional personnel.

Good project office personnel usually have experience with several types of projects and are self-disciplined.

The third and final step in the staffing of the project office is a meeting between the project manager, upper-level management, and the project manager on whose project the requested individuals are currently assigned. Project managers are very reluctant to give up qualified personnel to other projects, but unfortunately, this procedure is a way of life in a project environment. Upper-level management attends these meetings to show all negotiating parties that top management is concerned with maintaining the best possible mix of individuals from available resources and to help resolve staffing conflicts. Staffing from within is a negotiation process in which upper-level management establishes the ground rules and priorities.

The selected individuals are then notified of the anticipated change and asked their opinions. If individuals have strong resentment to being transferred or reassigned, alternate personnel may be selected to avoid potential problems.

Figure 4–3 shows the typical staffing pattern as a function of time. There is a manpower buildup in the early phases and a manpower decline in the later stages. This means

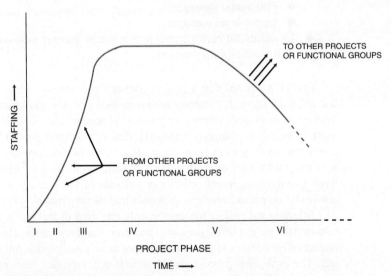

FIGURE 4–3. Staffing pattern versus time.

that the project manager should bring people on board as *needed* and release them as *early* as possible.

There are several psychological approaches that the project manager can use during the recruitment and staffing process. Consider the following:

- Line managers often receive no visibility or credit for a job well done. Be willing to introduce line managers to the customer.
- Be sure to show people how they can benefit by working for you or on your project.
- Any promises made during recruitment should be documented. The functional organization will remember them long after your project terminates.
- As strange as it may seem, the project manager should encourage conflicts to take place during recruiting and staffing. These conflicts should be brought to the surface and resolved. It is better for conflicts to be resolved during the initial planning stages than to have major confrontations later.

It is unfortunate that recruiting and retaining good personnel are more difficult in a project organizational structure than in a purely traditional one. Clayton Reeser identifies nine potential problems that can exist in project organizations[6]:

- Personnel connected with project forms of organization suffer more anxieties about possible loss of employment than members of functional organizations.
- Individuals temporarily assigned to matrix organizations are more frustrated by authority ambiguity than permanent members of functional organizations.
- Personnel connected with project forms of organization that are nearing their phase-out are more frustrated by what they perceive to be "make work" assignments than members of functional organizations.
- Personnel connected with project forms of organization feel more frustrated because of lack of formal procedures and role definitions than members of functional organizations.
- Personnel connected with project forms of organization worry more about being set back in their careers than members of functional organizations.
- Personnel connected with project forms of organization feel less loyal to their organization than members of functional organizations.
- Personnel connected with project forms of organization have more anxieties in feeling that there is no one concerned about their personal development than members of functional organizations.
- Permanent members of project forms of organization are more frustrated by multiple levels of management than members of functional organizations.
- Frustrations caused by conflict are perceived more seriously by personnel connected with project forms of organization than members of functional organizations.

---

6. Clayton Reeser, "Some Potential Human Problems of the Project Form of Organization," *Academy of Management Journal,* Vol. XII, 1969, pp. 462–466.

Employees are more likely to be motivated to working on a project if the employee had been given the right to accept or refuse the assignment. Although employees usually do not refuse assignments, there is still the question of how much permissiveness should be given to the worker. The following would be a listing or possible degrees of permissiveness:

- The line manager (or project manager) explains the project to the worker and the worker has the right to refuse the assignment. The worker does not need to explain the reason for refusing the assignment and the refusal does not limit the worker's opportunity for advancement or assignment to other project teams.
- With this degree of permissiveness, the worker has the right to refuse the assignment but must provide a reason for the refusal. The reason could be due to personal or career preference considerations such as having to travel, relocation, health reasons, possibly too much overtime involved, simply not an assignment that is viewed as enhancing the individual's career, or the employee wants an assignment on some other project.
- With this degree of permissiveness, the worker has no choice but to accept the assignment. Only an emergency would be considered as a valid reason for refusing the assignment. In this case, refusing the assignment might be damaging to the employee's career.

Grinnell and Apple have identified four additional major problems associated with staffing[7]:

- People trained in single line-of-command organizations find it hard to serve more than one boss.
- People may give lip service to teamwork, but not really know how to develop and maintain a good working team.
- Project and functional managers sometimes tend to compete rather than cooperate with each other.
- Individuals must learn to do more "managing" of themselves.

Thus far we have discussed staffing the project. Unfortunately, there are also situations in which employees must be terminated from the project because of:

- Nonacceptance of rules, policies, and procedures
- Nonacceptance of established formal authority
- Professionalism being more important to them than company loyalty
- Focusing on technical aspects at the expense of the budget and schedule
- Incompetence

There are three possible solutions for working with incompetent personnel. First, the project manager can provide an on-the-spot appraisal of the employee. This includes identification of weaknesses, corrective action to be taken, and threat of punishment if the situation continues. A second solution is reassignment of the employee to less critical activities. This solution is usually not preferred by project managers. The third and most frequent solution is the removal of the employee.

7. S. K. Grinnell and H. P. Apple, "When Two Bosses Are Better Than One," *Machine Design,* January 1975, pp. 84–87.

Although project managers can get project office people (who report to the project manager) removed directly, the removal of a line employee is an indirect process and must be accomplished through the line manager. The removal of the line employee should be made to look like a transfer; otherwise the project manager will be branded as an individual who fires people.

Executives must be ready to cope with the staffing problems that can occur in a project environment. C. Ray Gullett has summarized these major problems[8]:

- Staffing levels are more variable in a project environment.
- Performance evaluation is more complex and more subject to error in a matrix form of organization.
- Wage and salary grades are more difficult to maintain under a matrix form of organization. Job descriptions are often of less value.
- Training and development are more complex and at the same time more necessary under a project form of organization.
- Morale problems are potentially greater in a matrix organization.

4.9 THE PROJECT OFFICE

PMBOK® Guide, 4th Edition
1.4.4 Project Management Office

The project team is a combination of the project office and functional employees as shown in Figure 4–4. Although the figure identifies the project office personnel as assistant project managers, some employees may not have any such title. The advantage of such a title is that it entitles the employee to speak directly to the customer. For example, the project engineer might also be called the assistant project manager for engineering. The title is important because when the assistant project manager speaks to the customer, he represents the company, whereas the functional employee represents himself.

The project office is an organization developed to support the project manager in carrying out his duties. Project office personnel must have the same dedication toward the project as the project manager and must have good working relationships with both the project and functional managers. The responsibilities of the project office include:

- Acting as the focal point of information for both in-house control and customer reporting
- Controlling time, cost, and performance to adhere to contractual requirements
- Ensuring that all work required is documented and distributed to all key personnel
- Ensuring that all work performed is both authorized and funded by contractual documentation

The major responsibility of the project manager and the project office personnel is the integration of work across the functional lines of the organization. Functional units, such as engineering, R&D, and manufacturing, together with extra-company subcontractors, must work toward the same specifications, designs, and even objectives. The lack of proper

---

8. C. Ray Gullett, "Personnel Management in the Project Environment," *Personnel Administration/Public Personnel Review,* November–December 1972, pp. 17–22.

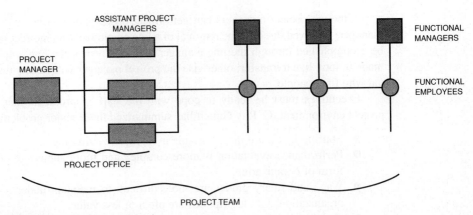

FIGURE 4–4. Project organization.

integration of these functional units is the most common cause of project failure. The team members must be dedicated to all activities required for project success, not just their own functional responsibilities. The problems resulting from lack of integration can best be solved by full-time membership and participation of project office personnel. Not all team members are part of the project office. Functional representatives, performing at the interface position, also act as integrators but at a closer position to where the work is finally accomplished (i.e., the line organization).

One of the biggest challenges facing project managers is determining the size of the project office. The optimal size is determined by a trade-off between the maximum number of members necessary to assure compliance with requirements and the maximum number for keeping the total administrative costs under control. Membership is determined by factors such as project size, internal support requirements, type of project (i.e., R&D, qualification, production), level of technical competency required, and customer support requirements. Membership size is also influenced by how strategic management views the project to be. There is a tendency to enlarge project offices if the project is considered strategic, especially if follow-on work is possible.

On large projects, and even on some smaller efforts, it is often impossible to achieve project success without permanently assigned personnel. The four major activities of the project office, shown below, indicate the need for using full-time people:

- Integration of activities
- In-house and out-of-house communication
- Scheduling with risk and uncertainty
- Effective control

These four activities require continuous monitoring by trained project personnel. The training of good project office members may take weeks or even months, and can extend beyond the time allocated for a project. Because key personnel are always in demand, project managers should ask themselves and upper-level management one pivotal question when attempting to staff the project office:

Are there any projects downstream that could cause me to lose key members of my team?

If the answer to this question is yes, then it might benefit the project to have the second- or third-choice person selected for the position or even to staff the position on a part-time basis. Another alternative, of course, would be to assign the key members to activities that are not so important and that can be readily performed by replacement personnel. This, however, is impractical because such personnel will not be employed efficiently.

Program managers would like nothing better than to have all of their key personnel assigned full-time for the duration of the program. Unfortunately, this is undesirable, if not impossible, for many projects because[9]:

- Skills required by the project vary considerably as the project matures through each of its life-cycle phases.
- Building up large permanently assigned project offices for each project inevitably causes duplication of certain skills (often those in short supply), carrying of people who are not needed on a full-time basis or for a long period, and personnel difficulties in reassignment.
- The project manager may be diverted from his primary task and become the project engineer, for example, in addition to his duties of supervision, administration, and dealing with the personnel problems of a large office rather than concentrating on managing all aspects of the project itself.
- Professionally trained people often prefer to work within a group devoted to their professional area, with permanent management having qualifications in the same field, rather than becoming isolated from their specialty peers by being assigned to a project staff.
- Projects are subject to sudden shifts in priority or even to cancellation, and full-time members of a project office are thus exposed to potentially serious threats to their job security; this often causes a reluctance on the part of some people to accept a project assignment.

All of these factors favor keeping the full-time project office as small as possible and dependent on established functional departments and specialized staffs. The approach places great emphasis on the planning and control procedures used on the project. On the other hand, there are valid reasons for assigning particular people of various specialties to the project office. These specialties usually include:

- Systems analysis and engineering (or equivalent technical discipline) and product quality and configuration control, if the product requires such an effort
- Project planning, scheduling, control, and administrative support

Many times a project office is staffed by promotion of functional specialists. This situation is quite common to engineering firms with a high percentage of technical employees, but is not without problems.

---

9. Russell D. Archibald, *Managing High-Technology Programs and Projects* (New York: Wiley, 1976), p. 82. Copyright © 1976 by John Wiley & Sons, Inc. Reprinted by permission of the publisher.

In professional firms, personnel are generally promoted to management on the basis of their professional or technical competence rather than their managerial ability. While this practice may be unavoidable, it does tend to promote men with insufficient knowledge of management techniques and creates a frustrating environment for the professional down the line.[10]

There is an unfortunate tendency for executives to create an environment where line employees feel that the "grass is greener" in project management and project engineering than in the line organization. How should an executive handle a situation where line specialists continually apply for transfer to project management? One solution is the development of a dual ladder system, with a pay scale called "consultant." This particular company created the consultant position because:

- There were several technical specialists who were worth more money to the company but who refused to accept a management position to get it.
- Technical specialists could not be paid more money than line managers.

Promoting technical specialists to a management slot simply to give them more money can:

- Create a poor line manager
- Turn a specialist into a generalist
- Leave a large technical gap in the line organization

Line managers often argue that they cannot perform their managerial duties and control these "prima donnas" who earn more money and have a higher pay grade than the line managers. That is faulty reasoning. Every time the consultants do something well, it reflects on the entire line organization, not merely on themselves.

The concept of having functional employees with a higher pay grade than the line manager can also be applied to the horizontal project. It is possible for a junior project manager suddenly to find that the line managers have a higher pay grade than the project manager. It is also possible for assistant project managers (as project engineers) to have a higher pay grade than the project manager. Project management is designed to put together the best mix of people to achieve the objective. If this best mix requires that a grade 7 report to a grade 9 (on a "temporary" project), then so be it. Executives should not let salaries, and pay grades, stand in the way of constructing a good project organization.

Another major concern is the relationship that exists between project office personnel and functional managers. In many organizations, membership in the project office is considered to be more important than in the functional department. Functional members have a tendency to resent an individual who has just been promoted out of a functional department and into project management. Killian has described ways of resolving potential conflicts[11]:

It must be kept in mind that veteran functional managers cannot be expected to accept direction readily from some lesser executive who is suddenly labelled a Project Manager. Management can avoid this problem by:

- Selecting a man who already has a high position of responsibility or placing him high enough in the organization.

10. William P. Killian, "Project Management—Future Organizational Concept," *Marquette Business Review,* 1971, pp. 90–107.
11. William P. Killian, "Project Management—Future Organizational Concept," *Marquette Business Review,* 1971, pp. 90–107.

- Assigning him a title as important-sounding as those of functional managers.
- Supporting him in his dealings with functional managers.

If the Project Manager is expected to exercise project control over the functional departments, then he must report to the same level as the departments, or higher.

Executives can severely hinder project managers by limiting their authority to select and organize (when necessary) a project office and team. According to Cleland[12]:

> His [project manager's] staff should be qualified to provide personal administrative and technical support. He should have sufficient authority to increase or decrease his staff as necessary throughout the life of the project. The authorization should include selective augmentation for varying periods of time from the supporting functional areas.

Many executives have a misconception concerning the makeup and usefulness of the project office. People who work in the project office should be individuals whose first concern is project management, not the enhancement of their technical expertise. It is almost impossible for individuals to perform for any extended period of time in the project office without becoming cross-trained in a second or third project office function. For example, the project manager for cost could acquire enough expertise eventually to act as the assistant to the assistant project manager for procurement. This technique of project office cross-training is an excellent mechanism for creating good project managers.

We have mentioned two important facts concerning the project management staffing process:

- The individual who aspires to become a project manager must be willing to give up technical expertise and become a generalist.
- Individuals can be qualified to be promoted vertically but not horizontally.

Once an employee has demonstrated the necessary attributes to be a good project manager, there are three ways the individual can become a project manager or part of the project office. The executive can:

- Promote the individual in salary and grade and transfer him into project management.
- Laterally transfer the individual into project management without any salary or grade increase. If, after three to six months, the employee demonstrates that he can perform, he will receive an appropriate salary and grade increase.
- Give the employee a small salary increase without any grade increase or a grade increase without any salary increase, with the stipulation that additional awards will be forthcoming after the observation period, assuming that the employee can handle the position.

---

12. David I. Cleland, "Why Project Management?" Reprinted with permission from *Business Horizons,* Winter 1964, p. 85. Copyright © 1964 by the Board of Trustees at Indiana University.

Many executives believe in the philosophy that once an individual enters the world of project management, there are only two places to go: up in the organization or out the door. If an individual is given a promotion and pay increase and is placed in project management and fails, his salary may not be compatible with that of his previous line organization, and now there is no place for him to go. Most executives, and employees, prefer the second method because it actually provides some protection for the employee.

Many companies don't realize until it is too late that promotions to project management may be based on a different set of criteria from promotions to line management. Promotions on the horizontal line are strongly based on communicative skills, whereas line management promotions are based on technical skills.

4.10 THE FUNCTIONAL TEAM

> **PMBOK® Guide, 4th Edition**
> Chapter 9 Human Resources
>   Management
> 2.3 Project Team Definition

The project team consists of the project manager, the project office (whose members may or may not report directly to the project manager), and the functional or interface members (who must report horizontally as well as vertically for information flow). Functional team members are often shown on organizational charts as project office team members. This is normally done to satisfy customer requirements.

Upper-level management can have an input into the selection process for functional team members but should not take an active role unless the project and functional managers cannot agree. Functional management must be represented at all staffing meetings because functional staffing is directly dependent on project requirements and because:

- Functional managers generally have more expertise and can identify high-risk areas.
- Functional managers must develop a positive attitude toward project success. This is best achieved by inviting their participation in the early activities of the planning phase.

Functional team members are not always full-time. They can be full-time or part-time for either the duration of the project or only specific phases.

The selection process for both the functional team member and the project office must include evaluation of any special requirements. The most common special requirements develop from:

- Changes in technical specifications
- Special customer requests
- Organizational restructuring because of deviations from existing policies
- Compatibility with the customer's project office

A typical project office may include between ten and thirty members, whereas the total project team may be in excess of a hundred people, causing information to be shared slowly. For large projects, it is desirable to have a full-time functional representative from each major division or department assigned permanently to the project, and perhaps even to the project office. Such representation might include:

- Program management
- Project engineering
- Engineering operations
- Manufacturing operations
- Procurement
- Quality control
- Cost accounting
- Publications
- Marketing
- Sales

Both the project manager and team members must understand fully the responsibilities and functions of each other team member so that total integration can be achieved rapidly and effectively. On high-technology programs the chief project engineer assumes the role of deputy project manager. Project managers must understand the problems that the line managers have when selecting and assigning the project staff. Line managers try to staff with people who understand the need for teamwork.

When employees are attached to a project, the project manager must identify the "star" employees. These are the employees who are vital for the success of the project and who can either make or break the project manager. Most of the time, star employees are found in the line organization, not the project office.

As a final point, project managers can assign line employees added responsibilities within the scope of the project. If the added responsibilities can result in upgrading, then the project manager should consult with the line manager before such situations are initiated. Quite often, line managers (or even personnel representatives) send "check" people into the projects to verify that employees are performing at their proper pay grade. This is very important when working with blue-collar workers who, by union contractual agreements, must be paid at the grade level at which they are performing.

Also, project managers must be willing to surrender resources when they are no longer required. If the project manager constantly cries wolf in a situation where a problem really does not exist, the line manager will simply pull away the resources (this is the line manager's right), and a deteriorating working relationship will result.

4.11 THE PROJECT ORGANIZATIONAL CHART

One of the first requirements of the project startup phase is to develop the organizational chart for the project and determine its relationship to the parent organizational structure. Figure 4–5 shows, in abbreviated form, the six major programs at Dalton Corporation. Our concern is with the Midas Program. Although the Midas Program may have the lowest priority of the six programs, it is placed at the top, and in boldface, to give the impression

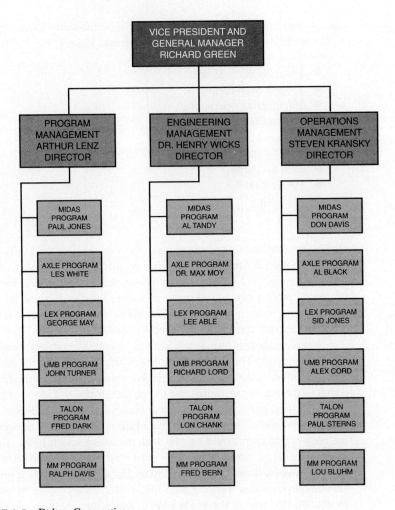

FIGURE 4–5.   Dalton Corporation.

that it is the top priority. This type of representation usually makes the client or customer feel that his program is important to the contractor.

The employees shown in Figure 4–5 may be part-time or full-time, depending upon the project's requirements. Perturbations on Figure 4–5 might include one employee's name identified on two or more vertical positions (i.e., the project engineer on two projects) or the same name in two horizontal boxes (i.e., for a small project, the same person could be the project manager and project engineer). Remember, this type of chart is for the customer's benefit and may not show the true "dotted/solid" reporting relationships in the company.

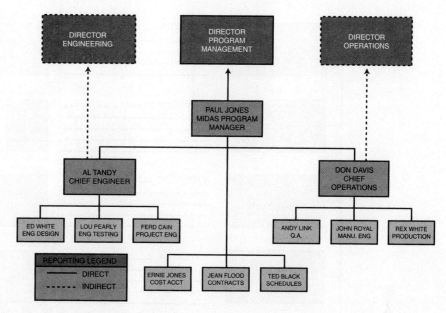

FIGURE 4–6. Midas Program office.

The next step is to show the program office structure, as illustrated in Figure 4–6. Note that the chief of operations and the chief engineer have dual reporting responsibility; they report directly to the program manager and indirectly to the directors. Again, this may be just for the customer's benefit with the real reporting structure being reversed. Beneath the chief engineer, there are three positions. Although these positions appear as solid lines, they might actually be dotted lines. For example, Ed White might be working only part-time on the Midas Program but is still shown on the chart as a permanent program office member. Jean Flood, under contracts, might be spending only ten hours per week on the Midas Program.

If the function of two positions on the organizational chart takes place at different times, then both positions may be shown as manned by the same person. For example, Ed White may have his name under both engineering design and engineering testing if the two activities are far enough apart that he can perform them independently.

The people shown in the project office organizational chart, whether full-time or part-time, may not be physically sitting in the project office. For full-time, long-term assignments, as in construction projects, the employees may be physically sitting side by side, whereas for part-time assignments, it may be imperative for them to sit in their functional group. Remember, these types of charts may simply be eyewash for the customer.

Most customers realize that the top-quality personnel may be shared with other programs and projects. Project manning charts, such as the one shown in Figure 4–7, can be used for this purpose. These manning charts are also helpful in preparing the management volume of proposals to show the customer that key personnel will be readily available on his project.

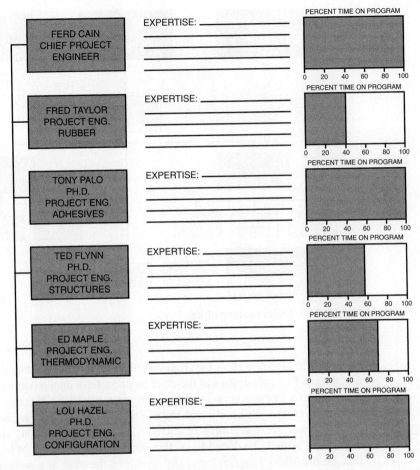

FIGURE 4–7.   Project engineering department manning for the Midas Program.

4.12 SPECIAL PROBLEMS

There are always special problems that influence the organizational staffing process. For example, the department shown in Figure 4–8 has a departmental matrix. All activities stay within the department. Project X and project Y are managed by line employees who have been temporarily assigned to the projects, whereas project Z is headed by supervisor B. The department's activities involve high-technology engineering as well as R&D.

The biggest problem facing the department managers is that of training their new employees. The training process requires nine to twelve months. The employees become familiar with the functioning of all three sections, and only after training is an employee assigned to one of the sections. Line managers claim that they do not have sufficient time to supervise training. As a result, the department manager in the example found staff person C to be the most competent person to supervise training. A special department training project was set up, as shown in Figure 4–8.

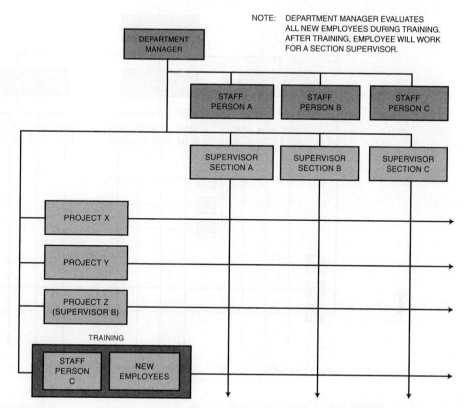

FIGURE 4–8. The training problem.

NOTE: DEPARTMENT MANAGER EVALUATES ALL NEW EMPLOYEES DURING TRAINING. AFTER TRAINING, EMPLOYEE WILL WORK FOR A SECTION SUPERVISOR.

Figure 4–9 shows a utility company that has three full-time project managers controlling three projects, all of which cut across the central division. Unfortunately, the three full-time project managers cannot get sufficient resources from the central division because the line managers are also acting as divisional project managers and saving the best resources for their own projects.

The obvious solution to the problem is that the central division line managers not be permitted to wear two hats. Instead, one full-time project manager can be added to the left division to manage all three central division projects. It is usually best for all project managers to report to the same division for priority setting and conflict resolution.

Line managers have a tendency to feel demoted when they are suddenly told that they can no longer wear two hats. For example, Mr. Adams was a department manager with thirty years of experience in a company. For the last several years, he had worn two hats and acted as both project manager and functional manager on a variety of projects. He was regarded as an expert in his field. The company decided to incorporate formal project management and established a project management department. Mr. Bell, a thirty-year-old employee with three years of experience with the company, was assigned as the project manager. In order to staff his project, Bell asked Adams for Mr. Cane (Bell's friend) to be assigned to the project as the functional representative. Cane had been with the company for two years. Adams agreed to the request and informed Cane of his new assignment,

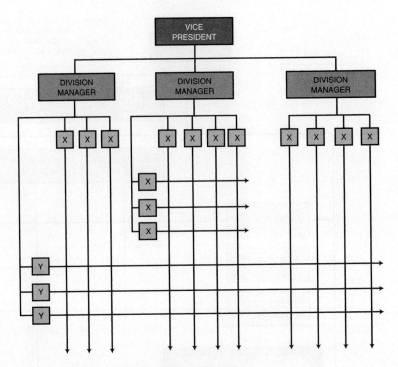

NOTE: X INDICATES FULL–TIME FUNCTIONAL MANAGERS
Y INDICATES FULL–TIME PROJECT MANAGERS

FIGURE 4–9. Utility service organization.

closing with the remarks, "This project is yours all the way. I don't want to have anything to do with it. I'll be busy with paperwork as a result of the new organizational structure. Just send me a memo once in a while telling me what's happening."

During the project kickoff meeting, it became obvious to everyone that the only person with the necessary expertise was Adams. Without his support, the duration of the project could be expected to double.

The real problem here was that Adams wanted to feel important and needed, and was hoping that the project manager would come to him asking for his assistance. The project manager correctly analyzed the situation but refused to ask for the line manager's help. Instead, the project manager asked an executive to step in and force the line manager to help. The line manager gave his help, but with great reluctance. Today, the line manager provides poor support to the projects that come across his line organization.

4.13 SELECTING THE PROJECT MANAGEMENT IMPLEMENTATION TEAM

PMBOK® Guide, 4th Edition
Chapter 9 Human Resources
   Management
9.2 Acquire Project Team

The implementation of project management within an organization requires strong executive support and an implementation team that is dedicated to making project management work. Selecting the wrong team players can either lengthen the implementation process or reduce

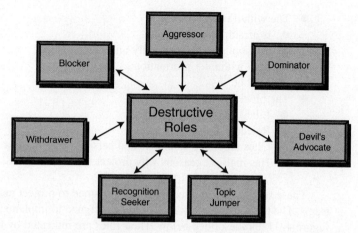

FIGURE 4–10. Roles people play that undermine project management implementation.

employee morale. Some employees may play destructive roles on a project team. These roles, which undermine project management implementation, are shown in Figure 4–10 and described below:

- The aggressor
  - Criticizes everybody and everything on project management
  - Deflates the status and ego of other team members
  - Always acts aggressively
- The dominator
  - Always tries to take over
  - Professes to know everything about project management
  - Tries to manipulate people
  - Will challenge those in charge for leadership role
- The devil's advocate
  - Finds fault in all areas of project management
  - Refuses to support project management unless threatened
  - Acts more of a devil than an advocate
- The topic jumper
  - Must be the first one with a new idea/approach to project management
  - Constantly changes topics
  - Cannot focus on ideas for a long time unless it is his/her idea
  - Tries to keep project management implementation as an action item forever
- The recognition seeker
  - Always argues in favor of his/her own ideas
  - Always demonstrates status consciousness
  - Volunteers to become the project manager if status is recognized
  - Likes to hear himself/herself talk
  - Likes to boast rather than provide meaningful information

- The withdrawer
  - Is afraid to be criticized
  - Will not participate openly unless threatened
  - May withhold information
  - May be shy
- The blocker
  - Likes to criticize
  - Rejects the views of others
  - Cites unrelated examples and personal experiences
  - Has multiple reasons why project management will not work

These types of people should not be assigned to project management implementation teams. The types of people who should be assigned to implementation teams are shown in Figure 4–11 and described below. Their roles are indicated by their words:

- The initiators
  - "Is there a chance that this might work?"
  - "Let's try this."
- The information seekers
  - "Have we tried anything like this before?"
  - "Do we know other companies where this has worked?"
  - "Can we get this information?"
- The information givers
  - "Other companies found that . . ."
  - "The literature says that . . ."
  - "Benchmarking studies indicate that . . ."
- The encouragers
  - "Your idea has a lot of merit."
  - "The idea is workable, but we may have to make small changes."
  - "What you said will really help us."

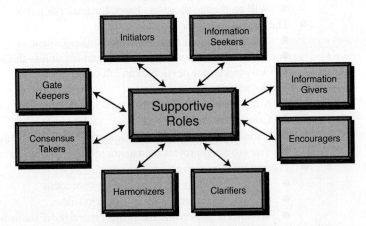

FIGURE 4–11. Roles people play that support project management implementation.

- The clarifiers
  - "Are we saying that . . . ?"
  - "Let me state in my own words what I'm hearing from the team."
  - "Let's see if we can put this into perspective."
- The harmonizers
  - "We sort of agree, don't we?"
  - "Your ideas and mine are close together."
  - "Aren't we saying the same thing?"
- The consensus takers
  - "Let's see if the team is in agreement."
  - "Let's take a vote on this."
  - "Let's see how the rest of the group feels about this."
- The gate keepers
  - "Who has not given us their opinions on this yet?"
  - "Should we keep our options open?"
  - "Are we prepared to make a decision or recommendation, or is there additional information to be reviewed?"

4.14 STUDYING TIPS FOR THE PMI® PROJECT MANAGEMENT CERTIFICATION EXAM

This section is applicable as a review of the principles to support the knowledge areas and domain groups in the PMBOK® Guide. This chapter addresses:

- Human Resources Management
- Planning
- Project Staffing

Understanding the following principles is beneficial if the reader is using this text to study for the PMP® Certification Exam:

- What is meant by a project team
- Staffing process and environment
- Role of the line manager in staffing
- Role of the executive in staffing
- Skills needed to be a project manager
- That the project manager is responsible for helping the team members grow and learn while working on the project

In Appendix C, the following Dorale Products mini–case studies are applicable:

- Dorale Products (G) [Human Resources Management]
- Dorale Products (H) [Human Resources Management]

- Dorale Products (I) [Human Resources Management]
- Dorale Products (J) [Human Resources Management]
- Dorale Products (K) [Human Resources Management]

The following multiple-choice questions will be helpful in reviewing the principles of this chapter:

1. During project staffing, the *primary* role of senior management is in the selection of the:
   A. Project manager
   B. Assistant project managers
   C. Functional team
   D. Executives do not get involved in staffing.

2. During project staffing, the *primary* role of line management is:
   A. Approving the selection of the project manager
   B. Approving the selection of assistant project managers
   C. Assigning functional resources based upon who is available
   D. Assigning functional resources based upon availability and the skill set needed

3. A project manager is far more likely to succeed if it is obvious to everyone that:
   A. The project manager has a command of technology.
   B. The project manager is a higher pay grade than everyone else on the team.
   C. The project manager is over 45 years of age.
   D. Executive management has officially appointed the project manager.

4. Most people believe that the best way to train someone in project management is through:
   A. On-the-job training
   B. University seminars
   C. Graduate degrees in project management
   D. Professional seminars and meeting

5. In staffing negotiations with the line manager, you identify a work package that requires a skill set of a grade 7 worker. The line manager informs you that he will assign a grade 6 and a grade 8 worker. You should:
   A. Refuse to accept the grade 6 because you are not responsible for training
   B. Ask for two different people
   C. Ask the sponsor to interfere
   D. Be happy! You have two workers.

6. You priced out a project at 1000 hours assuming a grade 7 employee would be assigned. The line manager assigns a grade 9 employee. This will result in a significant cost overrun. The project manager should:
   A. Reschedule the start date of the project based upon the availability of a grade 7
   B. Ask the sponsor for a higher priority for your project
   C. Reduce the scope of the project
   D. See if the grade 9 can do the job in less time

7. As a project begins to wind down, the project manager should:
   A. Release all nonessential personnel so that they can be assigned to other projects
   B. Wait until the project is officially completed before releasing anyone
   C. Wait until the line manager officially requests that the people be released
   D. Talk to other project managers to see who wants your people

ANSWERS

1. A
2. D
3. D
4. A
5. D
6. D
7. A

PROBLEMS

4–1 From S. K. Grinnell and H. P. Apple ("When Two Bosses Are Better Than One," *Machine Design,* January 1975, pp. 84–87):

- People trained in single-line-of-command organizations find it hard to serve more than one boss.
- People may give lip service to teamwork, but not really know how to develop and maintain a good working team.
- Project and functional managers sometimes tend to compete rather than cooperate with each other.
- Individuals must learn to do more "managing" of themselves.

The authors identify the above four major problems associated with staffing. Discuss each problem and identify the type of individual most likely to be involved (i.e., engineer, contract administrator, cost accountant, etc.) and in which organizational form this problem would be most apt to occur.

4–2 David Cleland ("Why Project Management?" Reprinted from *Business Horizons,* Winter 1964, p. 85. Copyright © 1964 by the Foundation for the School of Business at Indiana University. Used with permission) made the following remarks:

His [project manager's] staff should be qualified to provide personal administrative and technical support. He should have sufficient authority to increase or decrease his staff as necessary throughout the life of the project. This authorization should include selective augmentation for varying periods of time from the supporting functional areas.

Do you agree or disagree with these statements? Should the type of project or type of organization play a dominant role in your answer?

4–3 The contractor's project office is often structured to be compatible with the customer's project office, sometimes on a one-to-one basis. Some customers view the contractor's project organization merely as an extension of their own company. Below are three statements concerning this relationship. Are these statements true or false? Defend your answers.

- There must exist mutual trust between the customer and contractor together with a close day-to-day working relationship.

- The project manager and the customer must agree on the hierarchy of decision that each must make, either independently or jointly. (Which decisions can each make independently or jointly?)
- Both the customer and contractor's project personnel must be willing to make decisions as fast as possible.

4–4 C. Ray Gullet ("Personnel Management in the Project Organization," *Personnel Administration/Public Personnel Review,* November–December 1972, pp. 17–22) has identified five personnel problems. How would you, as a project manager, cope with each problem?

- Staffing levels are more variable in a project environment.
- Performance evaluation is more complex and more subject to error in a matrix form of organization.
- Wage and salary grades are more difficult to maintain under a matrix form of organization. Job descriptions are often of less value.
- Training and development are more complex and at the same time more necessary under a project form of organization.
- Morale problems are potentially greater in a matrix organization.

4–5 Some people believe that a project manager functions, in some respects, like a physician. Is there any validity in this?

4–6 Paul is a project manager for an effort that requires twelve months. During the seventh, eighth, and ninth months he needs two individuals with special qualifications. The functional manager has promised that these individuals will be available two months before they are needed. If Paul does not assign them to his project at that time, they will be assigned elsewhere and he will have to do with whomever will be available later. What should Paul do? Do you have to make any assumptions in order to defend your answer?

4–7 Some of the strongest reasons for promoting functional engineers to project engineers are:

- Better relationships with fellow researchers
- Better prevention of duplication of effort
- Better fostering of teamwork

These reasons are usually applied to R&D situations. Could they also be applied to product life-cycle phases other than R&D?

4–8 The following have been given as qualifications for a successful advanced-technology project manager:

- Career has progressed up through the technical ranks
- Knowledgeable in many engineering fields
- Understands general management philosophy and the meaning of profitability
- Interested in training and teaching his superiors
- Understands how to work with perfectionists

Can these same qualifications be modified for non-R&D project management? If so, how?

4–9 W. J. Taylor and T. F. Watling (*Successful Project Management,* London: Business Books, 1972, p. 32) state:

It is often the case, therefore, that the Project Manager is more noted for his management technique expertise, his ability to "get things done" and his ability to "get on with people" than for his sheer technical prowess. However, it can be dangerous to minimize this latter talent when choosing Project Managers dependent upon project type and size. The

Project Manager should preferably be an expert either in the field of the project task or a subject allied to it.

How dangerous can it be if this latter talent is minimized? Will it be dangerous under all circumstances?

4–10 Frank Boone is the most knowledgeable piping engineer in the company. For five years, the company has turned down his application for transfer to project engineering and project management stating that he is too valuable to the company in his current position. If you were a project manager, would you want this individual as part of your functional team? How should an organization cope with this situation?

4–11 Tom Weeks is manager of the insulation group. During a recent group meeting, Tom commented, "The company is in trouble. As you know, we're bidding on three programs right now. If we win just one of them, we can probably maintain our current work level. If, by some slim chance, we were to win all three, you'll all be managers tomorrow." The company won all three programs, but the insulation group did not hire anyone, and there were no promotions. What would you, as a project manager on one of the new projects, expect your working relations to be with the insulation group?

4–12 You are a project engineer on a high-technology program. As the project begins to wind down, your boss asks you to write a paper so that he can present it at a technical meeting. His name goes first on the paper. Should this be part of your job? How do you feel about this situation?

4–13 Research has indicated that the matrix structure is often confusing because it requires multiple roles for people, with resulting confusion about these roles (Keith Davis, *Human Relations at Work,* New York: McGraw-Hill, 1967, pp. 296–297). Unfortunately, not all program managers, project managers, and project engineers possess the necessary skills to operate in this environment. Stuckenbruck has stated, "The path to success is strewn with the bodies of project managers who were originally functional line managers and then went into project management" (Linn Stuckenbruck, "The Effective Project Manager," *Project Management Quarterly,* Vol. VII, No. 1, March 1976, pp. 26–27). What do you feel is the major cause for this downfall of the functional manager?

4–14 For each of the organizational forms shown below, who determines what resources are needed, when they are needed, and how they will be employed? Who has the authority and responsibility to mobilize these resources?

    a. Traditional organization
    b. Matrix organization
    c. Product line organization
    d. Line/staff project organization

4–15 Do you agree or disagree that project organizational forms encourage peer-to-peer communications and dynamic problem-solving?

4–16 The XYZ Company operates on a traditional structure. The company has just received a contract to develop a new product line for a special group of customers. The company has decided to pull out selected personnel from the functional departments and set up a single product organizational structure to operate in parallel with the functional departments.

    a. Set up the organizational chart.
    b. Do you think this setup can work? Does your answer depend on how many years this situation must exist?

4–17 You are the project engineer on a program similar to one that you directed previously. Should you attempt to obtain the same administrative and/or technical staff that you had before?

4–18 A person assigned to your project is performing unsatisfactorily. What should you do? Will it make a difference if he is in the project office or a functional employee?

4–19 You have been assigned to the project office as an assistant project engineer. You are to report to the chief project engineer who reports formally to the project manager and informally to the vice president of engineering. You have never worked with this chief project engineer before. During the execution of the project, it becomes obvious to you that the chief project engineer is making decisions that do not appear to be in the best interest of the project. What should you do about this?

4–20 Should individuals be promoted to project management because they are at the top of their functional pay grade?

4–21 Should one functional department be permitted to "borrow" (on a temporary basis) people from another functional department in order to fulfill project manning requirements? Should this be permitted if overtime is involved?

4–22 Should a project manager be paid for performance or for the number of people he supervises?

4–23 Should a project manager try to upgrade his personnel?

4–24 Why should a functional manager assign his best people to you on a long-term project?

4–25 A coal company has adopted the philosophy that the project manager for new mine startup projects will be the individual who will eventually become the mine superintendent. The coal company believes that this type of "ownership" philosophy is good. Do you agree?

4–26 Can a project manager be considered as a "hired gun"?

4–27 Manufacturing organizations are using project management/project engineering strictly to give new employees exposure to total company operations. After working on one or two projects, each approximately one to two years in duration, the employee is transferred to line management for his career path and opportunities for advancement. Can a situation such as this, where there is no career path in either project management or project engineering, work successfully? Could there be any detrimental effects on the projects?

4–28 Can a project manager create dedication and a true winning spirit and still be hated by all?

4–29 Can anyone be trained to be a project manager?

4–30 A power and light company has part-time project management in which an individual acts as both a project manager and a functional employee at the same time. The utility company claims that this process prevents an employee from becoming "technically obsolete," and that when the employee returns to full-time functional duties, he is a more well-rounded individual. Do you agree or disagree? What are the arrangement's advantages and disadvantages?

4–31 Some industries consider the major criterion for promotion and advancement to be gray hair and/or baldness. Is this type of maturity advantageous?

4–32 In Figure 4–8 we showed that Al Tandy and Don Davis (as well as other project office personnel) reported directly to the project manager and indirectly to functional management. Could this situation be reversed, with the project office personnel reporting indirectly to the project manager and directly to functional management?

4–33 Most organizations have "star" people who are usually identified as those individuals who are the key to success. How does a project manager identify these people? Can they be in the project office, or must they be functional employees or managers?

4–34 Considering your own industry, what job-related or employee-related factors would you wish to know before selecting someone to be a project manager or a project engineer on an effort valued at:

    a. $30,000?
    b. $300,000?
    c. $3,000,000?
    d. $30,000,000?

4–35 One of the major controversies in project management occurs over whether the project manager needs a command of technology in order to be effective. Consider the following situation:

You are the project manager on a research and development project. Marketing informs you that they have found a customer for your product and that you must make major modifications to satisfy the customer's requirements. The engineering functional managers tell you that these modifications are impossible. Can a project manager without a command of technology make a viable decision as to whether to risk additional funds and support marketing, or should he believe the functional managers, and tell marketing that the modifications are impossible? How can a project manager, either with or without a command of technology, tell whether the functional managers are giving him an optimistic or a pessimistic opinion?

4–36 As a functional employee, you demonstrate that you have exceptionally good writing skills. You are then promoted to the position of special staff assistant to the division manager and told that you are to assume full responsibility for all proposal work that must flow through your division. How do you feel about this? Is it a promotion? Where can you go from here?

4–37 Government policymakers content that only high-ranking individuals (high GS grades) can be project managers because a good project manager needs sufficient "clout" to make the project go. In government, the project manager is generally the highest grade on the project team. How can problems of pay grade be overcome? Is the government's policy effective?

4–38 A major utility company is worried about the project manager's upgrading functional employees. On an eight-month project that employs four hundred full-time project employees, the department managers have set up "check" people whose responsibility is to see that functional employees do not have unauthorized (i.e., not approved by the functional manager) work assignments above their current grade level. Can this system work? What if the work is at a position below their grade level?

4–39 A major utility company begins each computer project with a feasibility study in which a cost-benefit analysis is performed. The project managers, all of whom report to a project management division, perform the feasibility study themselves without any functional support. The functional personnel argue that the feasibility study is inaccurate because the functional "experts" are not involved. The project managers, on the other hand, stipulate that they never have sufficient time or money to involve the functional personnel. Can this situation be resolved?

4–40 How would you go about training individuals within your company or industry to be good project managers? What assumptions are you making?

4–41 Should project teams be allowed to evolve by themselves?

4–42 At what point or phase in the life cycle of a project should a project manager be appointed?

4–43 Top management generally has two schools of thought concerning project management. One school states that the project manager should be used as a means for coordinating activities that cut across several functional departments. The second school states that the project management position should be used as a means of creating future general managers. Which school of thought is correct?

4–44 Some executives feel that personnel working in a project office should be cross-trained in several assistant project management functions. What do you think about this?

4–45 A company has a policy that employees wishing to be project managers must first spend one to one-and-a-half years in the functional employee side of the house so that they can get to know the employees and company policy. What do you think about this?

4–46 Your project has grown to a point where there now exist openings for three full-time assistant project managers. Unfortunately, there are no experienced assistant project managers available. You are told by upper-level management that you will fill these three positions by promotions from within. Where in the organization should you look? During an interview, what questions should you ask potential candidates? Is it possible that you could find candidates who are qualified to be promoted vertically but not horizontally?

4–47 A functional employee has demonstrated the necessary attributes of a potentially successful project manager. Top management can:

- Promote the individual in salary and grade and transfer him into project management.
- Laterally transfer the employee into project management without any salary or grade increase. If, after three to six months, the employee demonstrates that he can perform, he will receive an appropriate salary and grade increase.
- Give the employee either a grade increase without any salary increase, or a small salary increase without any grade increase, under the stipulation that additional awards will be given at the end of the observation period, assuming that the employee can handle the position.

If you were in top management, which method would you prefer? If you dislike the above three choices, develop your own alternative. What are the advantages and disadvantages of each choice? For each choice, discuss the ramifications if the employee cannot handle the project management position.

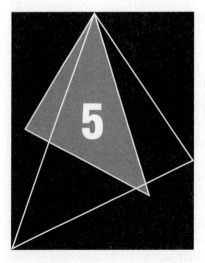

Management Functions

5

Related Case Studies (from Kerzner/*Project Management Case Studies*, 3rd Edition)	Related Workbook Exercises (from Kerzner/*Project Management Workbook and PMP®/CAPM® Exam Study Guide*, 10th Edition)	PMBOK® Guide, 4th Edition, Reference Section for the PMP® Certification Exam
• Wynn Computer Equipment (WCE) • The Trophy Project*	• The Communication Problem • Meetings, Meetings, and Meetings • The Empowerment Problem • Project Management Psychology • Multiple Choice Exam • Crossword Puzzle on Human Resource Management • Crossword Puzzle on Communications Management	• Human Resource Management • Communications Management

5.0 INTRODUCTION

PMBOK® Guide, 4th Edition
1.6 Project Management Skills
1.4.4 Role of the PMO

As we have stated, the project manager measures his success by how well he can negotiate with both upper-level and functional management for the resources necessary to achieve the project objective. Moreover, the project manager may have a great deal of delegated authority but very little

---

*Case Study also appears at end of chapter.

power. Hence, the managerial skills he requires for successful performance may be drastically different from those of his functional management counterparts.

The difficult aspect of the project management environment is that individuals at the project–functional interface must report to two bosses. Functional managers and project managers, by virtue of their different authority levels and responsibilities, treat their people in different fashions depending on their "management school" philosophies. There are generally five management schools, as described below:

- *The classical/traditional school:* Management is the process of getting things done (i.e., achieving objectives) by working both with and through people operating in organized groups. Emphasis is placed on the end-item or objective, with little regard for the people involved.
- *The empirical school:* Managerial capabilities can be developed by studying the experiences of other managers, whether or not the situations are similar.
- *The behavioral school:* Two classrooms are considered within this school. First, we have the human relations classroom in which we emphasize the interpersonal relationship between individuals and their work. The second classroom includes the social system of the individual. Management is considered to be a system of cultural relationships involving social change.
- *The decision theory school:* Management is a rational approach to decision making using a system of mathematical models and processes, such as operations research and management science.
- *The management systems school:* Management is the development of a systems model, characterized by input, processing, and output, and directly identifies the flow of resources (money, equipment, facilities, personnel, information, and material) necessary to obtain some objective by either maximizing or minimizing some objective function. The management systems school also includes contingency theory, which stresses that each situation is unique and must be optimized separately within the constraints of the system.

In a project environment, functional managers are generally practitioners of the first three schools of management, whereas project managers utilize the last two. This imposes hardships on both the project managers and functional representatives. The project manager must motivate functional representatives toward project dedication on the horizontal line using management systems theory and quantitative tools, often with little regard for the employee. After all, the employee might be assigned for a very short-term effort, whereas the end-item is the most important objective. The functional manager, however, expresses more concern for the individual needs of the employee using the traditional or behavioral schools of management.

Modern practitioners still tend to identify management responsibilities and skills in terms of the principles and functions developed in the early management schools, namely:

- Planning
- Organizing
- Staffing
- Controlling
- Directing

Although these management functions have generally been applied to traditional management structures, they have recently been redefined for temporary management positions. Their fundamental meanings remain the same, but the applications are different.

5.1 CONTROLLING

Controlling is a three-step process of measuring progress toward an objective, evaluating what remains to be done, and taking the necessary corrective action to achieve or exceed the objectives. These three steps—measuring, evaluating, and correcting—are defined as follows:

- *Measuring:* determining through formal and informal reports the degree to which progress toward objectives is being made.
- *Evaluating:* determining cause of and possible ways to act on significant deviations from planned performance.
- *Correcting:* taking control action to correct an unfavorable trend or to take advantage of an unusually favorable trend.

The project manager is responsible for ensuring the accomplishment of group and organizational goals and objectives. To effect this, he must have a thorough knowledge of standards and cost control policies and procedures so that a comparison is possible between operating results and preestablished standards. The project manager must then take the necessary corrective actions. Later chapters provide a more in-depth analysis of control, especially the cost control function.

In Chapter 1, we stated that project managers must understand organizational behavior in order to be effective and must have strong interpersonal skills. This is especially important during the controlling function. Line managers may have the luxury of time to build up relationships with each of their workers. But for a project manager time is a constraint, and it is not always easy to predict how well or how poorly an individual will interact with a group, especially if the project manager has never worked with this employee previously. Understanding the physiological and social behavior of how people perform in a group cannot happen overnight.

5.2 DIRECTING

Directing is the implementing and carrying out (through others) of those approved plans that are necessary to achieve or exceed objectives. Directing involves such steps as:

- *Staffing:* seeing that a qualified person is selected for each position.
- *Training:* teaching individuals and groups how to fulfill their duties and responsibilities.
- *Supervising:* giving others day-to-day instruction, guidance, and discipline as required so that they can fulfill their duties and responsibilities.
- *Delegating:* assigning work, responsibility, and authority so others can make maximum utilization of their abilities.
- *Motivating:* encouraging others to perform by fulfilling or appealing to their needs.

- *Counseling:* holding private discussions with another about how he might do better work, solve a personal problem, or realize his ambitions.
- *Coordinating:* seeing that activities are carried out in relation to their importance and with a minimum of conflict.

Directing subordinates is not an easy task because of both the short time duration of the project and the fact that employees might still be assigned to a functional manager while temporarily assigned to your effort. The luxury of getting to "know" one's subordinates may not be possible in a project environment.

Project managers must be decisive and move forward rapidly whenever directives are necessary. It is better to decide an issue and be 10 percent wrong than it is to wait for the last 10 percent of a problem's input and cause a schedule delay and improper use of resources. Directives are most effective when the KISS (keep it simple, stupid) rule is applied. Directives should be written with one simple and clear objective so that subordinates can work effectively and get things done right the first time. Orders must be issued in a manner that expects immediate compliance. Whether people will obey an order depends mainly on the amount of respect they have for you. Therefore, never issue an order that you cannot enforce. Oral orders and directives should be disguised as suggestions or requests. The requestor should ask the receiver to repeat the oral orders so that there is no misunderstanding.

PMBOK® Guide, 4th Edition
Chapter 9 Human Resources
   Management
9.4 Manage the Team

Project managers must understand human behavior in order to motivate people toward successful accomplishment of project objectives. Douglas McGregor advocated that most workers can be categorized according to two theories.[1] The first, often referred to as Theory X, assumes that the average worker is inherently lazy and requires supervision. Theory X further assumes that:

- The average worker dislikes work and avoids work whenever possible.
- To induce adequate effort, the supervisor must threaten punishment and exercise careful supervision.
- The average worker avoids increased responsibility and seeks to be directed.

The manager who accepts Theory X normally exercises authoritarian-type control over workers and allows little participation during decision-making. Theory X employees generally favor lack of responsibility, especially in decision-making.

According to Theory Y, employees are willing to get the job done without constant supervision. Theory Y further assumes that:

- The average worker wants to be active and finds the physical and mental effort on the job satisfying.
- Greatest results come from willing participation, which will tend to produce self-direction toward goals without coercion and control.
- The average worker seeks opportunity for personal improvement and self-respect.

---

1. Douglas McGregor, *The Human Side of Enterprise* (New York: McGraw-Hill, 1960), pp. 33–34.

The manager who accepts Theory Y normally advocates participation and a management–employee relationship. However, in working with professionals, especially engineers, special care must be exercised because these individuals often pride themselves on their ability to find a better way to achieve the end result regardless of cost. If this happens, project managers must become authoritarian leaders and treat Theory Y employees as though they are Theory X.

William Ouchi has identified a Theory Z that emphasizes the Japanese cultural values and the behavior of the Japanese workers.[2] According to Theory Z, there exist significant differences between the Japanese and American cultures and how the workers are treated. The Japanese focus on lifetime employment whereas the Americans look at short-term employment. The Japanese focus on collective decision-making such as in quality circles whereas Americans focus on individual decision-making. The Japanese emphasize informal administrative control whereas the Americans lean toward a more formal control. Japanese companies place workers on nonspecialized career paths with slow evaluation and promotion whereas Americans prefer specialized career path opportunities with rapid evaluation and promotion. Finally, Japanese managers have more of an interest in the personal life of their workers than do American managers.

> **PMBOK® Guide, 4th Edition**
> Chapter 9 Human Resources
>   Management
> 9.3.2 Develop the Team

Many psychologists have established the existence of a prioritized hierarchy of needs that motivate individuals toward satisfactory performance. Maslow was the first to identify these needs.[3] Maslow's hierarchy of needs is shown in Figure 5–1. The first level is that of the basic or physiological needs, namely, food, water, clothing, shelter, sleep, and sexual satisfaction. Simply speaking, human primal desire to satisfy these basic needs motivates him to do a good job.

After an employee has fulfilled his physiological needs, he turns to the next lower need, safety. Safety needs include economic security and protection from harm, disease, and violence. Safety can also include security. It is important that project managers realize this because these managers may find that as a project nears termination, functional employees are more interested in finding a new role for themselves than in giving their best to the current situation.

The next level contains the social needs, including love, belonging, togetherness, approval, and group membership. At this level, the informal organization plays a dominant role. Many people refuse promotions to project management (as project managers, project office personnel, or functional representatives) because they fear that they will lose their "membership" in the informal organization. This problem can occur even on short-duration projects. In a project environment, project managers generally do not belong to any informal organization and, therefore, tend to look outside the organization to fulfill this need. Project managers consider authority and funding to be very important in gaining project support. Functional personnel, however, prefer friendship and work assignments. In other words, the project manager can use the project itself as a means of helping fulfill the third level for the line employees (i.e., team spirit).

---

2. W. G. Ouchi and A. M. Jaeger, "Type Z Organization: Stability in the Midst of Mobility," *Academy of Management Review,* April 1978, pp. 305–314.

3. Abraham Maslow, *Motivation and Personality* (New York: Harper and Brothers, 1954).

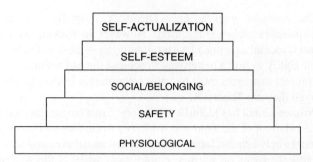

FIGURE 5–1. Maslow's hierarchy of needs.

The two lowest needs are esteem and self-actualization. The esteem need includes self-esteem (self-respect), reputation, the esteem of others, recognition, and self-confidence. Highly technical professionals are often not happy unless esteem needs are fulfilled. For example, many engineers strive to publish and invent as a means of satisfying these needs. These individuals often refuse promotions to project management because they believe that they cannot satisfy esteem needs in this position. Being called a project manager does not carry as much importance as being considered an expert in one's field by one's peers. The lowest need is self-actualization and includes doing what one can do best, desiring to utilize one's potential, full realization of one's potential, constant self-development, and a desire to be truly creative. Many good project managers find this level to be the most important and consider each new project as a challenge by which they can achieve self-actualization.

Frederick Herzberg and his associates conducted motivational research studies.[4] Herzberg concluded that Maslow's lower three levels (physiological, safety, and social needs) were hygiene factors that were either satisfied or dissatisfied. The only real motivational factors were the self-esteem and self-actualization needs. Herzberg believed that the physiological needs were hygiene factors and were extremely short-term needs. Self-esteem and self-actualization were more long-term needs and could be increased through job rotation, which includes job enrichment.

Another motivational technique can be related to the concept of expectancy theory (also referred to as the immature–mature organization), which was developed by the behaviorist Chris Argyris. Expectancy theory says that when the needs of the organization and the needs of the individual are congruent, both parties benefit and motivation increases. When there is incongruence between the needs of the individual and the needs of the organization, the individual will experience:

- Frustration
- Psychological failure
- Short-term perspectives
- Conflict

4. F. Herzberg, B. Mausner, and B. B. Snyderman, *The Motivation to Work* (New York: John Wiley & Sons, 1959).

Project managers must motivate temporarily assigned individuals by appealing to their desires to fulfill the lowest two levels, but not by making promises that cannot be met. Project managers must motivate by providing:

- A feeling of pride or satisfaction for one's ego
- Security of opportunity
- Security of approval
- Security of advancement, if possible
- Security of promotion, if possible
- Security of recognition
- A means for doing a better job, not a means to keep a job

Understanding professional needs is an important factor in helping people realize their true potential. Such needs include:

- Interesting and challenging work
- Professionally stimulating work environment
- Professional growth
- Overall leadership (ability to lead)
- Tangible rewards
- Technical expertise (within the team)
- Management assistance in problem-solving
- Clearly defined objectives
- Proper management control
- Job security
- Senior management support
- Good interpersonal relations
- Proper planning
- Clear role definition
- Open communications
- A minimum of changes

Motivating employees so that they feel secure on the job is not easy, especially since a project has a finite lifetime. Specific methods for producing security in a project environment include:

- Letting people know why they are where they are
- Making individuals feel that they belong where they are
- Placing individuals in positions for which they are properly trained
- Letting employees know how their efforts fit into the big picture

Since project managers cannot motivate by promising material gains, they must appeal to each person's pride. The guidelines for proper motivation are:

- Adopt a positive attitude
- Do not criticize management

- Do not make promises that cannot be kept
- Circulate customer reports
- Give each person the attention he requires

There are several ways of motivating project personnel. Some effective ways include:

- Giving assignments that provide challenges
- Clearly defining performance expectations
- Giving proper criticism as well as credit
- Giving honest appraisals
- Providing a good working atmosphere
- Developing a team attitude
- Providing a proper direction (even if Theory Y)

5.3 PROJECT AUTHORITY

PMBOK® Guide, 4th Edition
9.1.3 Human Resource Planning

Project management structures create a web of relationships that can cause chaos in the delegation of authority and the internal authority structure. Four questions must be considered in describing project authority:

- What is project authority?
- What is power, and how is it achieved?
- How much project authority should be granted to the project manager?
- Who settles project authority interface problems?

One form of the project manager's authority can be defined as the legal or rightful power to command, act, or direct the activities of others. Authority can be delegated from one's superiors. Power, on the other hand, is granted to an individual by his subordinates and is a measure of their respect for him. A manager's authority is a combination of his power and influence such that subordinates, peers, and associates willingly accept his judgment.

In the traditional structure, the power spectrum is realized through the hierarchy, whereas in the project structure, power comes from credibility, expertise, or being a sound decision-maker.

Authority is the key to the project management process. The project manager must manage across functional and organizational lines by bringing together activities required to accomplish the objectives of a specific project. Project authority provides the way of thinking required to unify all organizational activities toward accomplishment of the project regardless of where they are located. The project manager who fails to build and maintain his alliances will soon find opposition or indifference to his project requirements.

The amount of authority granted to the project manager varies according to project size, management philosophy, and management interpretation of potential conflicts with functional managers. There do exist, however, certain fundamental elements over which the project manager must have authority in order to maintain effective control. According to Steiner and Ryan[5]:

> The project manager should have broad authority over all elements of the project. His authority should be sufficient to permit him to engage all necessary managerial and technical actions required to complete the project successfully. He should have appropriate authority in design and in making technical decisions in development. He should be able to control funds, schedule and quality of product. If subcontractors are used, he should have maximum authority in their selection.

Generally speaking, a project manager should have more authority than his responsibility calls for, the exact amount of authority usually depending on the amount of risk that the project manager must take. The greater the risk, the greater the amount of authority. A good project manager knows where his authority ends and does not hold an employee responsible for duties that he (the project manager) does not have the authority to enforce. Some projects are directed by project managers who have only monitoring authority. These project managers are referred to as influence project managers.

Failure to establish authority relationships can result in:

● Poor communication channels
● Misleading information
● Antagonism, especially from the informal organization
● Poor working relationships with superiors, subordinates, peers, and associates
● Surprises for the customer

The following are the most common sources of power and authority problems in a project environment:

● Poorly documented or no formal authority
● Power and authority perceived incorrectly
● Dual accountability of personnel
● Two bosses (who often disagree)
● The project organization encouraging individualism
● Subordinate relations stronger than peer or superior relationships
● Shifting of personnel loyalties from vertical to horizontal lines
● Group decision-making based on the strongest group
● Ability to influence or administer rewards and punishment
● Sharing resources among several projects

The project manager does not have unilateral authority in the project effort. He frequently negotiates with the functional manager. The project manager has the authority to

---

determine the "when" and "what" of the project activities, whereas the functional manager has the authority to determine "how the support will be given." The project manager accomplishes his objectives by working with personnel who are largely professional. For professional personnel, project leadership must include explaining the rationale of the effort as well as the more obvious functions of planning, organizing, directing, and controlling.

Certain ground rules exist for authority control through negotiations:

- Negotiations should take place at the lowest level of interaction.
- Definition of the problem must be the first priority:
  - The issue
  - The impact
  - The alternative
  - The recommendations
- Higher-level authority should be used if, and only if, agreement cannot be reached.

The critical stage of any project is planning. This includes more than just planning the activities to be accomplished; it also includes the planning and establishment of the authority relationships that must exist for the duration of the project. Because the project management environment is an ever-changing one, each project establishes its own policies and procedures, a situation that can ultimately result in a variety of authority relationships. It is therefore possible for functional personnel to have different responsibilities on different projects, even if the tasks are the same.

During the planning phase the project team develops a responsibility assignment matrix (RAM) that contains such elements as:

- General management responsibility
- Operations management responsibility
- Specialized responsibility
- Who must be consulted
- Who may be consulted
- Who must be notified
- Who must approve

The responsibility matrix is often referred to as a linear responsibility chart (LRC) or responsibility assignment matrix (RAM). Linear responsibility charts identify the participants, and to what degree an activity will be performed or a decision will be made. The LRC attempts to clarify the authority relationships that can exist when functional units share common work. As described by Cleland and King[6]:

> The need for a device to clarify the authority relationships is evident from the relative unity
> of the traditional pyramidal chart, which (1) is merely a simple portrayal of the overall

---

6. From David I. Cleland and William Richard King, *Systems Analysis and Project Management* (New York: McGraw-Hill), p. 271.

functional and authority models and (2) must be combined with detailed position descriptions and organizational manuals to delineate authority relationships and work performance duties.

Figure 5–2 shows a typical linear responsibility chart. The rows, which indicate the activities, responsibilities, or functions required, can be all of the tasks in the work breakdown structure. The columns identify either positions, titles, or the people themselves. If the chart will be given to an outside customer, then only the titles should appear, or the customer will call the employees directly without going through the project manager. The symbols indicate the degrees of authority or responsibility existing between the rows and columns.

Another example of an LRC is shown in Figure 5–3. In this case, the LRC is used to describe how internal and external communications should take place. This type of chart can be used to eliminate communications conflicts. Consider a customer who is unhappy about having all of his information filtered through the project manager and requests that his line people be permitted to talk to your line people on a one-on-one basis.

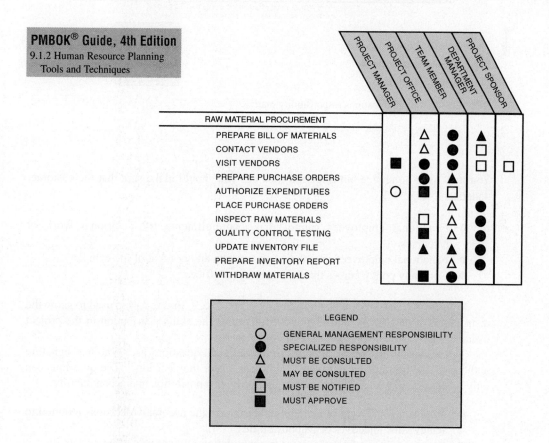

PMBOK® Guide, 4th Edition
9.1.2 Human Resource Planning
   Tools and Techniques

FIGURE 5–2. Linear responsibility chart (responsibility assignment matrix).

INITIATED FROM	Internal							External (Customer)						
	Project Manager	Project Office	Team Member	Department Members	Functional Employees	Division Manager	Executive Management	Project Manager	Project Office	Team Member	Department Members	Functional Employees	Division Manager	Executive Management
PROJECT MANAGER	▨	○	◆	◺	▲	▲	◆	○	○	■	■	■	■	◺
PROJECT OFFICE	○	▨	○	○	▲	▲	▲	○	○	◺	◺	■	■	◺
TEAM MEMBER	◆	○	▨	◆	○	■	■	■	■	▲	▲	▲	■	■
DEPARTMENT MEMBERS	▲	◺	○	▨	○	◆	■	◺	◺	◺	◺	◺	■	■
FUNCTIONAL EMPLOYEES	▲	▲	○	○	▨	■	■	▲	▲	▲	▲	▲	■	■
DIVISION MANAGERS	◺	▲	▲	▲	▲	▨	◺	■	■	■	■	■	◺	◺
EXECUTIVE MANAGEMENT	◺	▲	▲	▲	▲	▲	▨	◺	◺	▲	▲	■	◺	◺

REPORTED TO

*CAN VARY FROM TASK TO TASK AND CAN BE WRITTEN OR ORAL
** DOES NOT INCLUDE REGULARLY SCHEDULED INTERCHANGE MEETINGS

LEGEND
○ DAILY
◆ WEEKLY
○ MONTHLY
▲ AS NEEDED
◺ INFORMAL
■ NEVER

FIGURE 5–3. Communications responsibility matrix.*

You may have no choice but to permit this, but you should make sure that the customer understands that:

- Functional employees cannot make commitments for additional work or resources.
- Functional employees give their own opinion and not that of the company.
- Company policy comes through the project office.

Figures 5–4 and 5–5 are examples of modified LRCs. Figure 5–4 is used to show the distribution of data items, and Figure 5–5 identifies the skills distribution in the project office.

The responsibility matrix attempts to answer such questions as: "Who has signature authority?" "Who must be notified?" "Who can make the decision?" The questions can only be answered by clear definitions of authority, responsibility, and accountability:

- *Authority* is the right of an individual to make the necessary decisions required to achieve his objectives or responsibilities.
- *Responsibility* is the assignment for completion of a specific event or activity.
- *Accountability* is the acceptance of success or failure.

DATA ITEM DISTRIBUTION MATRIX

CUSTOMER AND CONTRACTOR'S PERSONNEL

DATA ITEM	REPORT DESCRIPTION	PROJECT MANAGER	PROJECT OFFICE	TEAM MEMBER	LINE MANAGER	EXECUTIVE MANAGEMENT
1	MONTHLY COST SUMMARIES	X	X			X
2	MILESTONE REPORTS	X	X	X	X	X
3	MANPOWER CURVES	X	X		X	
4	INVENTORY UTILIZATION	X	X			
5	PRESSURE TEST REPORT	X	X		X	
6	HUMIDITY TESTS	X	X		X	
7	HOTLINE REPORTS	X	X	X	X	X
8	SCHEDULING SUMMARIES	X	X	X	X	

FIGURE 5–4. Data distribution matrix.

The linear responsibility chart, although a valuable tool for management, does have a weakness in that it does not describe how people interact within the program. The LRC must be considered with the organization for a full understanding of how interactions between individuals and organizations take place. As described by Karger and Murdick, the LRC has merit[7]:

> Obviously the chart has weaknesses, of which one of the larger ones is that it is a mechanical aid. Just because it says that something is a fact does not make it true. It is very difficult to discover, except generally, exactly what occurs in a company—and with whom. The chart tries to express in specific terms relationships that cannot always be delineated so clearly; moreover, the degree to which it can be done depends on the specific situation. This is the difference between the formal and informal organizations mentioned. Despite this, the Linear Responsibility Chart is one of the best devices for organization analysis known to the authors.

Linear responsibility charts can result from customer-imposed requirements above and beyond normal operations. For example, the customer may require as part of its quality control that a specific engineer supervise and approve all testing of a certain item or that another individual approve all data released to the customer over and above program office

7. D. W. Karger and R. G. Murdick, *Managing Engineering and Research* (New York: Industrial Press, 1963), p. 89.

PROJECT TEAM

FUNCTIONAL AREAS OF EXPERTISE

	ABLE, J.	BAKER, P.	COOK, D.	DIRK, L.	EASLEY, P.	FRANKLIN, W.	GREEN, C.	HENRY, L.	IMHOFF, R.	JULES, C.	KLEIN, W.	LEDGER, D.	MAYER, O.	NEWTON, A.	OLIVER, G.	PRATT, L.
ADMINISTRATIVE MANAGEMENT		a				a	a	a			a	a			a	
COST CONTROL		b	b		b	b	b				b	b		b	b	
ECONOMIC ANALYSIS	c			c				c	c				c			c
ENERGY SYSTEMS		d	d		d		d			d			d		d	d
ENVIRONMENTAL IMPACT ASSESSMENT	e	e	e						e		e		e			
INDUSTRIAL ENGINEERING	f				f					f						
INSTRUMENTATION	g			g		g					g				g	
PIPING AND DESIGN LAYOUT	h		h		h	h				h			h			
PLANNING AND SCHEDULING		i		i	i		i	i				i		i		i
PROJECT MANAGEMENT	j			j		j					j				j	
PROJECT REPORTING		k	k		k			k	k			k		k		k
QUALITY CONTROL		l	l			l	l	l								
SITE EVALUATION		m				m			m	m				m		
SPECIFICATION PREPARATION			n	n			n				n		n			n
SYSTEM DESIGN		o	o		o		o	o		o		o			o	

FIGURE 5–5. Personal skills matrix.

approval. Such customer requirements necessitate LRCs and can cause disruptions and conflicts within an organization.

Several key factors affect the delegation of authority and responsibility, both from upper-level management to project management and from project management to functional management. These key factors include:

- The maturity of the project management function
- The size, nature, and business base of the company
- The size and nature of the project
- The life cycle of the project
- The capabilities of management at all levels

Once agreement has been reached as to the project manager's authority and responsibility, the results must be documented to clearly delineate his role in regard to:

- His focal position
- Conflict between the project manager and functional managers
- Influence to cut across functional and organizational lines
- Participation in major management and technical decisions
- Collaboration in staffing the project
- Control over allocation and expenditure of funds
- Selection of subcontractors
- Rights in resolving conflicts
- Voice in maintaining integrity of the project team
- Establishment of project plans
- Providing a cost-effective information system for control
- Providing leadership in preparing operational requirements
- Maintaining prime customer liaison and contact
- Promoting technological and managerial improvements
- Establishment of project organization for the duration
- Cutting red tape

Perhaps the best way to document the project manager's authority is through the project charter, which is one of the three methods, shown in Figure 5–6, by which project managers attain authority. Documenting the project manager's authority is necessary because:

- All interfacing must be kept as simple as possible.
- The project manager must have the authority to "force" functional managers to depart from existing standards and possibly incur risk.
- The project manager must gain authority over those elements of a program that are not under his control. This is normally achieved by earning the respect of the individuals concerned.
- The project manager should not attempt to fully describe the exact authority and responsibilities of his project office personnel or team members. Instead, he should encourage problem-solving rather than role definition.

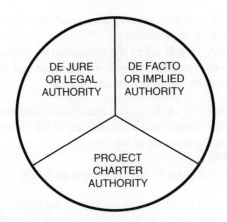

FIGURE 5–6. Types of project authority.

5.4 INTERPERSONAL INFLUENCES

PMBOK® Guide, 4th Edition
9.1.2 Human Resource Planning
Tools and Techniques

There exist a variety of relationships (although they are not always clearly definable) between power and authority. These relationships are usually measured by "relative" decision power as a function of the authority structure, and are strongly dependent on the project organizational form.

Consider the following statements made by project managers:

- "I've had good working relations with department X. They like me and I like them. I can usually push through anything ahead of schedule."
- "I know it's contrary to department policy, but the test must be conducted according to these criteria or else the results will be meaningless" (remark made to a team member by a research scientist who was temporarily promoted to project management for an advanced state-of-the-art effort).

Project managers are generally known for having a lot of delegated authority but very little formal power. They must, therefore, get jobs done through the use of interpersonal influences. There are five such interpersonal influences:

- *Legitimate power:* the ability to gain support because project personnel perceive the project manager as being officially empowered to issue orders.
- *Reward power:* the ability to gain support because project personnel perceive the project manager as capable of directly or indirectly dispensing valued organizational rewards (i.e., salary, promotion, bonus, future work assignments).
- *Penalty power:* the ability to gain support because the project personnel perceive the project manager as capable of directly or indirectly dispensing penalties that they wish to avoid. Penalty power usually derives from the same source as reward power, with one being a necessary condition for the other.

- *Expert power:* the ability to gain support because personnel perceive the project manager as possessing special knowledge or expertise (that functional personnel consider as important).
- *Referent power:* the ability to gain support because project personnel feel personally attracted to the project manager or his project.

Expert and referent power are examples of personal power that comes from the personal qualities or characteristics to which team members are attracted. Legitimate, reward,

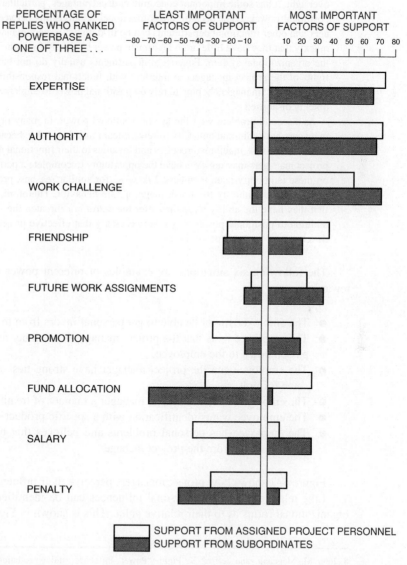

FIGURE 5–7. Significance of factors of support to project management. *Source: Seminar in Project Management Workbook,* © 1979 by Hans J. Thamhain. Reproduced by permission.

and penalty power are often referred to as examples of position power, which is directly related to one's position within the organization. Line managers generally possess a great amount of position power. But in a project environment, position power may be difficult to achieve. According to Magenau and Pinto[8]:

> Within the arena of project management, the whole issue of position power becomes more problematic. Project managers in many organizations operate outside the standard functional hierarchy. While that position allows them a certain freedom of action without direct oversight, it has some important concomitant disadvantages, particularly as they pertain to positional power. First, because cross-functional relationships between the project manager and other functional departments can be ill-defined, project managers discover rather quickly that they have little or no legitimate power to simply force their decisions through the organizational system. Functional departments usually do not have to recognize the rights of the project managers to interfere with functional responsibilities; consequently, novice project managers hoping to rely on positional power to implement their projects are quickly disabused.
>
> As a second problem with the use of positional power, in many organizations, project managers have minimal authority to reward team members who, because they are temporary subordinates, maintain direct ties and loyalties to their functional departments. In fact, project managers may not even have the opportunity to complete a performance evaluation on these temporary team members. Likewise, for similar reasons, project managers may have minimal authority to punish inappropriate behavior. Therefore, they may discover that they have the ability to neither offer the carrot nor threaten the stick. As a result, in addition to positional power, it is often necessary that effective project managers seek to develop their personal power bases.

The following six situations are examples of referent power (the first two are also reward power):

- The employee might be able to get personal favors from the project manager.
- The employee feels that the project manager is a winner and the rewards will be passed down to the employee.
- The employee and the project manager have strong ties, such as the same foursome for golf.
- The employee likes the project manager's manner of treating people.
- The employee wants identification with a specific product or product line.
- The employee has personal problems and believes that he can get empathy or understanding from the project manager.

Figure 5–7 shows how project managers perceive their influence style.

Like relative power, interpersonal influences can be identified with various project organizational forms as to their relative value. This is shown in Figure 5–8.

8. John M. Magenau, and Jeffrey K. Pinto, "Power, Influence, and Negotiation in Project Management," Peter W. G. Morris and Jeffrey Pinto, eds., *Project Organization and Project Management Competencies* (Wiley, 2007), p. 91. Reprinted by permission of John Wiley.

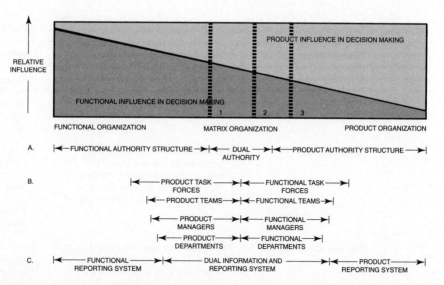

FIGURE 5–8. The range of alternatives. *Source:* Jay R. Galbraith, "Matrix Organization Designs." Reprinted with permission from *Business Horizons,* February 1971, p. 37. Copyright © 1971 by the Board of Trustees at Indiana University.

For any temporary management structure to be effective, there must exist a rational balance of power between functional and project management. Unfortunately, a balance of equal power is often impossible to obtain because each project is inherently different from others, and the project managers possess different leadership abilities.

Achievement of this balance is a never-ending challenge for management. If time and cost constraints on a project cannot be met, the project influence in decision-making increases, as can be seen in Figure 5–8. If the technology or performance constraints need reappraisal, then the functional influence in decision-making will dominate.

Regardless of how much authority and power a project manager develops over the course of the project, the ultimate factor in his ability to get the job done is usually his leadership style. Developing bonds of trust, friendship, and respect with the functional workers can promote success.

5.5 BARRIERS TO PROJECT TEAM DEVELOPMENT

PMBOK® Guide, 4th Edition
9.3 Develop Project Team

Most people within project-driven and non–project-driven organizations have differing views of project management. Table 5–1 compares the project and functional viewpoints of project management. These differing views can create severe barriers to successful project management operations.

The understanding of barriers to project team building can help in developing an environment conducive to effective teamwork. The following barriers are typical for many project environments.

TABLE 5–1. COMPARISON OF THE FUNCTIONAL AND THE PROJECT VIEWPOINTS

Phenomena	Project Viewpoint	Functional Viewpoint
Line–staff organizational dichotomy	Vestiges of the hierarchical model remain: the line functions are placed in a support position. A web of authority and responsibility exists.	Line functions have direct responsibility for accomplishing the objectives; line commands, and staff advises.
Scalar principle	Elements of the vertical chain exist, but prime emphasis is placed on horizontal and diagonal work flow. Important business is conducted as the legitimacy of the task requires.	The chain of authority relationships is from superior to subordinate throughout the organization. Central, crucial, and important business is conducted up and down the vertical hierarchy.
Superior–subordinate relationship	Peer-to-peer, manager-to-technical expert, associate-to-associate, etc., relationships are used to conduct much of the salient business.	This is the most important relationship; if kept healthy, success will follow. All important business is conducted through a pyramiding structure of superiors and subordinates
Organizational objectives	Management of a project becomes a joint venture of many relatively independent organizations. Thus, the objective becomes multilateral.	Organizational objectives are sought by the parent unit (an assembly of suborganizations) working within its environment. The objective is unilateral.
Unity of direction	The project manager manages across functional and organizational lines to accomplish a common interorganizational objective.	The general manager acts as the one head for a group of activities having the same plan.
Parity of authority and responsibility	Considerable opportunity exists for the project manager's responsibility to exceed his authority. Support people are often responsible to other managers (functional) for pay, performance reports, promotions, etc.	Consistent with functional management; the integrity of the superior–subordinate relationship is maintained through functional authority and advisory staff services.
Time duration	The project (and hence the organization) is finite in duration.	Tends to perpetuate itself to provide continuing facilitative support.

Source: David I. Cleland, "Project Management," in David I. Cleland and William R. King, eds., *Systems Organizations, Analysis, Management: A Book of Readings* (New York: McGraw-Hill, Inc., 1969), pp. 281–290. © 1969 by McGraw-Hill Inc. Reprinted with permission of the publisher.

Differing outlooks, priorities, and interests. A major barrier exists when team members have professional objectives and interests that are different from the project objectives. These problems are compounded when the team relies on support organizations that have different interests and priorities.

Role conflicts. Team development efforts are thwarted when role conflicts exist among the team members, such as ambiguity over who does what within the project team and in external support groups.

Project objectives/outcomes not clear. Unclear project objectives frequently lead to conflict, ambiguities, and power struggles. It becomes difficult, if not impossible, to define roles and responsibilities clearly.

Dynamic project environments. Many projects operate in a continual state of change. For example, senior management may keep changing the project scope, objectives, and resource base. In other situations, regulatory changes or client demands can drastically affect the internal operations of a project team.

Competition over team leadership. Project leaders frequently indicated that this barrier most likely occurs in the early phases of a project or if the project runs into severe problems. Obviously, such cases of leadership challenge can result in barriers to team building. Frequently, these challenges are covert challenges to the project leader's ability.

Lack of team definition and structure. Many senior managers complain that teamwork is severely impaired because it lacks clearly defined task responsibilities and reporting structures. We find this situation is most prevalent in dynamic, organizationally unstructured work environments such as computer systems and R&D projects. A common pattern is that a support department is charged with a task but no one leader is clearly delegated the responsibility. As a consequence, some personnel are working on the project but are not entirely clear on the extent of their responsibilities. In other cases, problems result when a project is supported by several departments without interdisciplinary coordination.

Team personnel selection. This barrier develops when personnel feel unfairly treated or threatened during the staffing of a project. In some cases, project personnel are assigned to a team by functional managers, and the project manager has little or no input into the selection process. This can impede team development efforts, especially when the project leader is given available personnel versus the best, hand-picked team members. The assignment of "available personnel" can result in several problems (e.g., low motivation levels, discontent, and uncommitted team members). We've found, as a rule, that the more power the project leader has over the selection of his team members, and the more negotiated agreement there is over the assigned task, the more likely it is that team-building efforts will be fruitful.

Credibility of project leader. Team-building efforts are hampered when the project leader suffers from poor credibility within the team or from other managers. In such cases, team members are often reluctant to make a commitment to the project or the leader. Credibility problems may come from poor managerial skills, poor technical judgments, or lack of experience relevant to the project.

Lack of team member commitment. Lack of commitment can have several sources. For example, the team members having professional interests elsewhere, the feeling of insecurity that is associated with projects, the unclear nature of the rewards that may be forthcoming upon successful completion, and intense interpersonal conflicts within the team can all lead to lack of commitment.

Lack of team member commitment may result from suspicious attitudes existing between the project leader and a functional support manager, or between two team members from two warring functional departments. Finally, low commitment levels are likely to occur when a "star" on a team "demands" too much effort from other team members or too much attention from the team leader. One team leader put it this way: "A lot of teams have their prima donnas and you learn to live and function with them. They can be critical to overall success. But some stars can be so demanding on everyone that they'll kill the team's motivation."

Communication problems. Not surprisingly, poor communication is a major enemy to effective team development. Poor communication exists on four major levels: problems of communication among team members, between the project leader and the team members, between the project team and top management, and between the project leaders and the client. Often the problem is caused by team members simply not keeping others informed on key project developments. Yet the "whys" of poor communication patterns are far more difficult to determine. The problem can result from low motivation levels, poor morale, or carelessness. It was also discovered that poor communication patterns between the team and support groups result in severe team-building problems, as does poor communication with the client. Poor communication practices often lead to unclear objectives and poor project control, coordination, and work flow.

Lack of senior management support. Project leaders often indicate that senior management support and commitment is unclear and subject to waxing and waning over the project life cycle. This behavior can result in an uneasy feeling among team members and lead to low levels of enthusiasm and project commitment. Two other common problems are that senior management often does not help set the right environment for the project team at the outset, nor do they give the team timely feedback on their performance and activities during the life of the project.

Project managers who are successfully performing their role not only recognize these barriers but also know when in the project life cycle they are most likely to occur. Moreover, these managers take preventive actions and usually foster a work environment that is conducive to effective teamwork. The effective team builder is usually a social architect who understands the interaction of organizational and behavior variables and can foster a climate of active participation and minimal conflict. This requires carefully developed skills in leadership, administration, organization, and technical expertise on the project. However, besides the delicately balanced management skills, the project manager's sensitivity to the basic issues underlying each barrier can help to increase success in developing an effective project team. Specific suggestions for team building are advanced in Table 5–2.

5.6 SUGGESTIONS FOR HANDLING THE NEWLY FORMED TEAM

A major problem faced by many project leaders is managing the anxiety that usually develops when a new team is formed. The anxiety experienced by team members is normal and predictable, but is a barrier to getting the team quickly focused on the task.

TABLE 5–2. BARRIERS TO EFFECTIVE TEAM BUILDING AND SUGGESTED HANDLING APPROACHES

Barrier	Suggestions for Effectively Managing Barriers (How to Minimize or Eliminate Barriers)
Differing outlooks, priorities, interests, and judgments of team members	Make effort early in the project life cycle to discover these conflicting differences. Fully explain the scope of the project and the rewards that may be forthcoming on successful project completion. Sell "team" concept and explain responsibilities. Try to blend individual interests with the overall project objectives.
Role conflicts	As early in a project as feasible, ask team members where they see themselves fitting into the project. Determine how the overall project can best be divided into subsystems and subtasks (e.g., the work breakdown structure). Assign/negotiate roles. Conduct regular status review meetings to keep team informed on progress and watch for unanticipated role conflicts over the project's life.
Project objectives/outcomes not clear	Assure that all parties understand the overall and interdisciplinary project objectives. Clear and frequent communication with senior management and the client becomes critically important. Status review meetings can be used for feedback. Finally, a proper team name can help to reinforce the project objectives.
Dynamic project environments	The major challenge is to stabilize external influences. First, key project personnel must work out an agreement on the principal project direction and "sell" this direction to the total team. Also educate senior management and the customer on the detrimental consequences of unwarranted change. It is critically important to forecast the "environment" within which the project will be developed. Develop contingency plans.
Competition over team leadership	Senior management must help establish the project manager's leadership role. On the other hand, the project manager needs to fulfill the leadership expectations of team members. Clear role and responsibility definition often minimizes competition over leadership.
Lack of team definition and structure	Project leaders need to sell the team concept to senior management as well as to their team members. Regular meetings with the team will reinforce the team notion as will clearly defined tasks, roles, and responsibilities. Also, visibility in memos and other forms of written media as well as senior management and client participation can unify the team.
Project personnel selection	Attempt to negotiate the project assignments with potential team members. Clearly discuss with potential team members the importance of the project, their role in it, what rewards might result on completion, and the general "rules of the road" of project management. Finally, if team members remain uninterested in the project, then replacement should be considered.
Credibility of project leader	Credibility of the project leader among team members is crucial. It grows with the image of a sound decision-maker in both general management and relevant technical expertise. Credibility can be enhanced by the project leader's relationship to other key managers who support the team's efforts.

(continues)

TABLE 5–2. BARRIERS TO EFFECTIVE TEAM BUILDING AND SUGGESTED HANDLING APPROACHES (*Continued*)

Barrier	Suggestions for Effectively Managing Barriers (How to Minimize or Eliminate Barriers)
Lack of team member commitment	Try to determine lack of team member commitment early in the life of the project and attempt to change possible negative views toward the project. Often, insecurity is a major reason for the lack of commitment; try to determine why insecurity exists, then work on reducing the team members' fears. Conflicts with other team members may be another reason for lack of commitment. It is important for the project leader to intervene and mediate the conflict quickly. Finally, if a team member's professional interests lie elsewhere, the project leader should examine ways to satisfy part of the team member's interests or consider replacement.
Communication problems	The project leader should devote considerable time communicating with individual team members about their needs and concerns. In addition, the leader should provide a vehicle for timely sessions to encourage communications among the individual team contributors. Tools for enhancing communications are status meetings, reviews, schedules, reporting system, and colocation. Similarly, the project leader should establish regular and thorough communications with the client and senior management. Emphasis is placed on written and oral communications with key issues and agreements in writing.
Lack of senior management support	Senior management support is an absolute necessity for dealing effectively with interface groups and proper resource commitment. Therefore, a major goal for project leaders is to maintain the continued interest and commitment of senior management in their projects. We suggest that senior management become an integral part of project reviews. Equally important, it is critical for senior management to provide the proper environment for the project to function effectively. Here the project leader needs to tell management at the onset of the program what resources are needed. The project manager's relationship with senior management and ability to develop senior management support is critically affected by his own credibility and the visibility and priority of his project.

This anxiety may come from several sources. For example, if the team members have never worked with the project leader, they may be concerned about his leadership style. Some team members may be concerned about the nature of the project and whether it will match their professional interests and capabilities, or help or hinder their career aspirations. Further, team members can be highly anxious about life-style/work-style disruptions. As one project manager remarked, "Moving a team member's desk from one side of the room to the other can sometimes be just about as traumatic as moving someone from Chicago to Manila."

Another common concern among newly formed teams is whether there will be an equitable distribution of the workload among team members and whether each member is capable of pulling his own weight. In some newly formed teams, members not only must do their own work, but also must train other team members. Within reason this is bearable, but when it becomes excessive, anxiety increases.

Certain steps taken early in the life of a team can minimize the above problems. First, we recommend that the project leader talk with each team member one-to-one about the following:

1. What the objectives are for the project.
2. Who will be involved and why.
3. The importance of the project to the overall organization or work unit.
4. Why the team member was selected and assigned to the project. What role he will perform.
5. What rewards might be forthcoming if the project is successfully completed.
6. What problems and constraints are likely to be encountered.
7. The rules of the road that will be followed in managing the project (e.g., regular status review meetings).
8. What suggestions the team member has for achieving success.
9. What the professional interests of the team member are.
10. What challenge the project will present to individual members and the entire team.
11. Why the team concept is so important to project management success and how it should work.

Dealing with these anxieties and helping team members feel that they are an integral part of the team can yield rich dividends. First, as noted in Figure 5–9, team members are more likely to openly share their ideas and approaches. Second, it is more likely that the team will be able to develop effective decision-making processes. Third, the team is likely

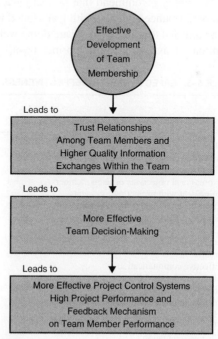

FIGURE 5–9. Team-building outcomes.

to develop more effective project control procedures, including those traditionally used to monitor project performance (PERT/CPM, networking, work breakdown structures, etc.) and those in which team members give feedback to each other regarding performance.

5.7 TEAM BUILDING AS AN ONGOING PROCESS

While proper attention to team building is critical during early phases of a project, it is a never-ending process. The project manager is continually monitoring team functioning and performance to see what corrective action may be needed to prevent or correct various team problems. Several barometers (summarized in Table 5–3) provide good clues of potential team dysfunctioning. First, noticeable changes in performance levels for the team and/or for individual team members should always be investigated. Such changes can be symptomatic of more serious problems (e.g., conflict, lack of work integration, communication problems, and unclear objectives). Second, the project leader and team members must be aware of the changing energy levels of team members. These changes, too, may signal more serious problems or that the team is tired and stressed. Sometimes changing the work pace or taking time off can reenergize team members. Third, verbal and nonverbal clues from team members may be a source of information on team functioning. It is important to hear the needs and concerns of team members (verbal clues) and to observe how they act in carrying out their responsibilities (nonverbal clues). Finally, detrimental behavior of one team member toward another can be a signal that a problem within the team warrants attention.

We highly recommend that project leaders hold regular meetings to evaluate overall team performance and deal with team functioning problems. The focus of these meetings can be directed toward "what we are doing well as a team" and "what areas need our team's attention." This approach often brings positive surprises in that the total team is informed

TABLE 5–3. EFFECTIVENESS–INEFFECTIVENESS INDICATORS

The Effective Team's Likely Characteristics	The Ineffective Team's Likely Characteristics
• High performance and task efficiency • Innovative/creative behavior • Commitment • Professional objectives of team members coincident with project requirements • Team members highly interdependent, interface effectively • Capacity for conflict resolution, but conflict encouraged when it can lead to beneficial results • Effective communication • High trust levels • Results orientation • Interest in membership • High energy levels and enthusiasm • High morale • Change orientation	• Low performance • Low commitment to project objectives • Unclear project objectives and fluid commitment levels from key participants • Unproductive gamesmanship, manipulation of others, hidden feelings, conflict avoidance at all costs • Confusion, conflict, inefficiency • Subtle sabotage, fear, disinterest, or foot-dragging • Cliques, collusion, isolation of members • Lethargy/unresponsiveness

of progress in diverse project areas (e.g., a breakthrough in technology development, a subsystem schedule met ahead of the original target, or a positive change in the client's behavior toward the project). After the positive issues have been discussed the review session should focus on actual or potential problem areas. The meeting leader should ask each team member for his observations and then open the discussion to ascertain how significant the problems really are. Assumptions should, of course, be separated from the facts of each situation. Next, assignments should be agreed on for best handling these problems. Finally, a plan for problem follow-up should be developed. The process should result in better overall performance and promote a feeling of team participation and high morale.

5.8 DYSFUNCTIONS OF A TEAM

In a pure line organization, line managers may have the luxury of "time" to build up relationships with their subordinates and provide slow guidance on how the employees should function on teams. But in a project environment, time is a constraint, and the project manager must act or react quickly to get the desired teamwork.

Understanding the dysfunctions of a team, and being able to correct the problems quickly, is essential in a project environment. Patrick Lencioni has authored a best-selling text describing the five most common dysfunctions of a team. In his text, he describes the five dysfunctions as[9]:

- Absence of trust
- Fear of conflict
- Lack of commitment
- Avoidance of accountability
- Inattention to results

In his text, he identifies the differences between the teams that have these dysfunctions and those that do not possess them:

- Members of a team with an absence of trust...
  - Conceal their weakness and mistakes from one another
  - Hesitate to ask for help or provide constructive feedback
  - Hesitate to offer help outside their own area of responsibility
  - Jump to conclusions about intentions and aptitudes of others without attempting to clarify them
  - Failing to recognize and tap into another's skills and experiences
  - Waste time and energy managing their behaviors for effect
  - Hold grudges
  - Dread meetings and find reasons to avoid spending time together

9. P. Lencioni, *The Five Dysfunctions of a Team* (New York: Jossey-Bass, 2002), pp. 197–218. Reprinted by permission of John Wiley.

- Members of trusting teams…
  - Admit weaknesses and mistakes
  - Ask for help
  - Accept questions and input about their areas of responsibility
  - Give one another the benefit of the doubt before arriving at a negative conclusion
  - Take risks in offering feedback and assistance
  - Appreciate and tap into one another's skills and experiences
  - Focus time and energy on important issues, not politics
  - Offer and accept apologies without hesitation
  - Look forward to meetings and their opportunities to work as a group

- Teams that fear conflict…
  - Have boring meetings
  - Create environments where back-channel politics and personal attacks thrive
  - Ignore controversial topics that are critical to team success
  - Fail to tap into all the opinions and perspectives of team members
  - Waste time and energy with posturing and interpersonal risk management

- Teams that engage in conflict…
  - Have lively, interesting meetings
  - Extract and exploit the ideas of all team members
  - Solve real problems quickly
  - Minimize politics
  - Put critical topics on the table for discussion

- A team that fails to commit…
  - Creates ambiguity among the team about direction and priorities
  - Watches windows and opportunities closly due to excessive analysis and unnecessary delay
  - Breeds lack of confidence and fear of failure
  - Revisits discussions and decisions again and again
  - Encourages second-guessing among team members

- A team that commits…
  - Creates clarity around direction and priorities
  - Aligns the entire team around common objectives
  - Develops an ability to learn from mistakes
  - Takes advantage of opportunities before competitors do
  - Moves forward without hesitation
  - Changes direction without hesitation or guilt

- A team that avoids accountability…
  - Creates resentment among team members who have different standards or performance
  - Encourages mediocrity
  - Misses deadlines and key deliverables
  - Places an undue burden on the team leader as the sole source of discipline

- A team that holds one another accountable…
  - Ensures that poor performers feel pressure to improve
  - Identifies potential problems quickly by questioning one another's approaches without hesitation
  - Establishes respect among team members who are held to the same high standards
  - Avoids excessive bureaucracy around performance management and corrective action

- A team that is not focused on results…
  - Stagnates/fails to grow
  - Rarely defeats competitors
  - Loses achievement-oriented employees
  - Encourages team members to focus on their own careers and individual goals
  - Is easily distracted

- A team that focuses on collective results…
  - Retains achievement-oriented employees
  - Minimizes individualistic behavior
  - Enjoys success and suffers failure acutely
  - Benefits from individuals who subjugate their own goals/interests for the good of the team
  - Avoids distractions

Another item that can lead to dysfunctional behavior among team members is when the project manager and team do not have the same shared values. According to Kouzes and Posner[10]:

> Shared values are the foundations for building productive and genuine working relationships. Although credible leaders honor the diversity of their many constituencies, they also stress their common values. Leaders build on agreement. They don't try to get everyone to be in accord on everything — this goal is unrealistic, perhaps even impossible. Moreover, to achieve it would negate the very advantages of diversity. But to take a first step, and then a second, and then a third, people must have some common core of understanding. After all, if there's no agreement about values, then what exactly is the leader — and everyone else — going to model? If disagreements over fundamental values continue, the result is intense conflict, false expectations, and diminished capacity.

Kouzes and Posner also show, through their own research, that shared values can make a difference:

- They foster strong feelings of personal effectiveness
- They promote high levels of company loyalty

---

10. James M. Kouzes and Barry Z. Posner, *The Leadership Challenge,* 4th ed. (New York: Jossey-Bass, 2007), pp. 60, 62.

- They facilitate consensus about key organizational goals and stakeholders
- They encourage ethical behavior
- They promote strong norms about working hard and caring
- They reduce levels of job stress and tension
- They foster pride in the company
- They facilitate understanding about job expectations
- They foster teamwork and esprit de corps

5.9 LEADERSHIP IN A PROJECT ENVIRONMENT

Leadership can be defined as a style of behavior designed to integrate both the organizational requirements and one's personal interests into the pursuit of some objective. All managers have some sort of leadership responsibility. If time permits, successful leadership techniques and practices can be developed.

Leadership is composed of several complex elements, the three most common being:

- The person leading
- The people being led
- The situation (i.e., the project environment)

Project managers are often selected or not selected because of their leadership styles. The most common reason for not selecting an individual is his inability to balance the technical and managerial project functions. Wilemon and Cicero have defined four characteristics of this type of situation[11]:

- The greater the project manager's technical expertise, the higher his propensity to overinvolve himself in the technical details of the project.
- The greater the project manager's difficulty in delegating technical task responsibilities, the more likely it is that he will overinvolve himself in the technical details of the project (depending on his ability to do so).
- The greater the project manager's interest in the technical details of the project, the more likely it is that he will defend the project manager's role as one of a technical specialist.
- The lower the project manager's technical expertise, the more likely it is that he will overstress the nontechnical project functions (administrative functions).

---

11. D. L. Wilemon and John P. Cicero, "The Project Manager: Anomalies and Ambiguities," *Academy of Management Journal*, Vol. 13, pp. 269–282, 1970.

There have been several surveys to determine what leadership techniques are best. The following are the results of a survey by Richard Hodgetts[12]:

- Human relations–oriented leadership techniques
  - "The project manager must make all the team members feel that their efforts are important and have a direct effect on the outcome of the program."
  - "The project manager must educate the team concerning what is to be done and how important its role is."
  - "Provide credit to project participants."
  - "Project members must be given recognition and prestige of appointment."
  - "Make the team members feel and believe that they play a vital part in the success (or failure) of the team."
  - "By working extremely closely with my team I believe that one can win a project loyalty while to a large extent minimizing the frequency of authority-gap problems."
  - "I believe that a great motivation can be created just by knowing the people in a personal sense. I know many of the line people better than their own supervisor does. In addition, I try to make them understand that they are an indispensable part of the team."
  - "I would consider the most important technique in overcoming the authority-gap to be understanding as much as possible the needs of the individuals with whom you are dealing and over whom you have no direct authority."
- Formal authority–oriented leadership techniques
  - "Point out how great the loss will be if cooperation is not forthcoming."
  - "Put all authority in functional statements."
  - "Apply pressure beginning with a tactful approach and minimum application warranted by the situation and then increasing it."
  - "Threaten to precipitate high-level intervention and do it if necessary."
  - "Convince the members that what is good for the company is good for them."
  - "Place authority on full-time assigned people in the operating division to get the necessity work done."
  - "Maintain control over expenditures."
  - "Utilize implicit threat of going to general management for resolution."
  - "It is most important that the team members recognize that the project manager has the charter to direct the project."

5.10 LIFE-CYCLE LEADERSHIP

PMBOK® Guide, 4th Edition
9.3 Develop Project Team

In the opinion of the author, Hersey and Blanchard developed the best model for analyzing leadership in a project management environment. Over the years the model has been expanded by Paul Hersey and is shown

---

12. Richard M. Hodgetts, "Leadership Techniques in Project Organizations," *Academy of Management Journal,* Vol. 11, pp. 211–219, 1968.

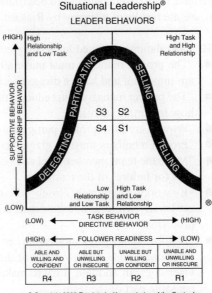

Situational Leadership®
LEADER BEHAVIORS

FIGURE 5–10. Expanded Situational Leadership Model. Adapted from Paul Hersey, *Situational Selling* (Escondido, CA: Center for Leadership Studies, 1985), p. 35. Reproduced by permission of the Center for Leadership Studies.®

in Figure 5–10 as the Situational Leadership® Model. The model contends that there are four basic leadership styles and that to use them most effectively entails matching the most appropriate leadership style to the readiness of the follower. Readiness is defined as job-related experience, willingness to accept job responsibility, and desire to achieve. It's about not only *doing* a good job but also *wanting* to do a good job. Most importantly though is the concept that this is a situational model. This is critical because it means that the same person can be more ready to perform one task than they are to do another and that the style a leader uses will have to change accordingly to be the most effective and successful at influencing the desired behaviors in that person.

Referring to Figure 5–10, suppose that a subordinate was not performing a certain task and showed through his or her behavior every indication of not wanting to (R1). According to the model, the leadership style that has the highest probability of successfully and effectively getting that person to perform is one that involves high amounts of structured task behavior that could be generally described as directive in nature (S1). This would entail telling a subordinate who, what, when, where, and how to go about performing the particular task. It would also be appropriate in this style to acknowledge steps taken in the right direction as far as performance is concerned, but this type of relationship behavior must match the magnitude of the steps being taken or the subordinate may be left with the false impression that his or her current level of performance is in fact acceptable. Some have gone so far as to equate this "telling" leadership style with the purely task-orientated behavior of an autocratic approach where the leader's main concern is the accomplishment of objective, often with very little concern for the employees or their feelings. An autocratic leader by his or her nature is very forceful and relies heavily on his or her own abilities and judgment often at the expense of other people's opinions. Note, however, in Figure 5–10 that the bell curve in the model does not go all the way to zero, indicating some relationship behavior present in this style that increases appropriately as the level of performance does.

As shown in Figure 5–10, suppose that an employee was beginning to perform the task in question but wasn't yet doing so at a sustained and acceptable level even though he or she really seemed to want to do a good job (R2). The leadership style with the highest probability of successfully and effectively influencing the desired behavior from this employee rests in quadrant S2. This employee needs everything from the leader. The employee needs structure to keep him or her on track and support not only to build the foundations of trust that help him or her continue to develop but also to give the big picture of how personal actions contribute to the success of the team. This is where the leader shares the "why" behind the behaviors in which he or she is asking the subordinate to engage.

At some point, one would hope that subordinates would begin performing the task in question at a sustained and acceptable level (R3). When this takes place, the follower is no longer in need of being told who, what, when, where and how to do the task but rather seeks autonomy and freedom as a reward for their good performance. They desire more of a collegial relationship with their superior that reflects the fact that they have arrived, but they will also be insecure about completely letting go of the involvement from the leader that made them feel so secure in the past. For a leader, the appropriate style for this readiness level would be S3. It would entail engaging in relationship behavior that gets followes to admit from their own mouths that they are indeed performing at a sustained and acceptable level and that they don't really need the leader to be so intricately involved in the process. For some this step occurs quickly, for others, they never make the leap. They must learn to have confidence in themselves and their abilities, and the leader's job at this point is to help with that process through the use of relationship behaviors that avoid making those followers feel more insecure. This means that the followers must trust the leader, and they can earn that trust by doing things like taking calculated risks that not only allow the followers small wins but also allows them to learn from their mistakes without being beaten up for them. It's a fine line the leader walks. They must be supportive without being an enabler.

Some leaders are blessed with followers who not only perform at a high and sustained level but are totally and rightfully confident about their ability to do so (R4). In such an instance, the appropriate leadership style rests in quadrant S4. It involves leadership behaviors such as monitoring and observing, which are characterized by low amounts of task and relationship-oriented behaviors. The leader is kept informed of both the good and bad in a timely manner from a person at this readiness level but is not the decision-maker. Responsibility lies with the subordinate who takes ownership of their actions and expects the leader to spend his or her energy obtaining resources for them and protecting or shielding them from other influences in the organization that could impede their performance.

Let's see where some common leadership style descriptors fall within the model.

- **Democratic or Participative Leadership:** This leadership style encourages workers to communicate with one another and get involved with decision-making either by himself or herself or with assistance of the project manager. A great deal of authority is delegated to the team members, and they are encouraged to take an active role in the management of the project. The leadership style is often found in quadrant S3 with some spill over as the manager becomes less involved in the process into quadrant S4.
- **Laissez-Faire Leadership:** With the leadership style, the project manger turns things over to the workers. This can feel like abandonment to the subordinates if they are not both performing at a high level and willing to do so. The project manager may make an occasional appearance with this style just to see how things are going, but for the main part there is no active involvement by the project manager. This leadership style is found in quadrant S4.
- **Autocratic Leadership:** With this leadership style, the project manager focuses very heavily upon the task, with little concern for the workers. With autocratic

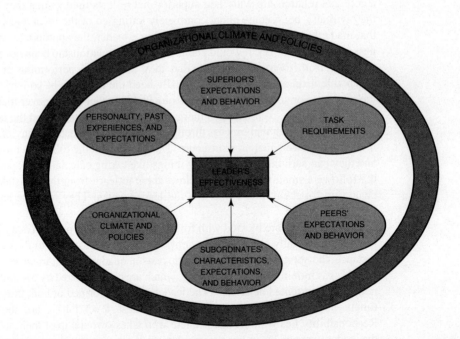

FIGURE 5–11. Personality and situational factors that influence effective leadership. *Source:* James A. F. Stoner, *Management,* 2nd ed. (Englewood Cliffs, New Jersey: Prentice-Hall Inc., 1982). Used by permission.

leadership, all authority is in the hands of the project manager and the project manager has the final say in any and all decisions. This leadership style is found in quadrant S1.

This type of situational approach to leadership is extremely important to project managers because it implies that effective leadership must be both dynamic and flexible rather than static and rigid (see Figure 5–11). Effective leaders recognize that when it comes to human behavior, there is no one best way that fits all circumstances. They need both task and relationship behavior to be able to be their most effective. Thankfully, it doesn't have to be a perfect match to work. Sometimes close is good enough, and sometimes a project manager's followers are willing to let them demonstrate a less than appropriate style because that manager has taken the time to earn their trust or perhaps even warned them of the necessity of going there when a crisis occurs. Just be wary of "living" in this mode because it may be the leader causing the crisis and the only fire needing to be put out could end up being the leader.

In pure project management, the situation is even more complex. It is not enough to have a different leadership style for each team member. Remember that any one person is more ready to do some tasks than others. For example, they may be really good at training others but detest and avoid report writing. That person's leader will have to use a different style depending on which task they are asking their follower to perform. To illustrate this graphically, the quadrants in Figure 5–10 should be three-dimensional, with the third axis being the life cycle phase of the project. In other words, the leadership is dependent not only on the situation, but also on the life-cycle phase of the project.

5.11 ORGANIZATIONAL IMPACT

In most companies, whether or not project-oriented, the impact of management emphasis on the organization is well known. In the project environment there also exists a definite impact due to leadership emphasis. The leadership emphasis is best seen by employee contributions, organizational order, employee performance, and the project manager's performance:

- Contributions from People
  - A good project manager encourages active cooperation and responsible participation. The result is that both good and bad information is contributed freely.
  - A poor project manager maintains an atmosphere of passive resistance with only responsive participation. This results in information being withheld.
- Organizational Order
  - A good project manager develops policy and encourages acceptance. A low price is paid for contributions.
  - A poor project manager goes beyond policies and attempts to develop procedures and measurements. A high price is normally paid for contributions.

- Employee Performance
  - A good project manager keeps people informed and satisfied (if possible) by aligning motives with objectives. Positive thinking and cooperation are encouraged. A good project manager is willing to give more responsibility to those willing to accept it.
  - A poor project manager keeps people uninformed, frustrated, defensive, and negative. Motives are aligned with incentives rather than objectives. The poor project manager develops a "stay out of trouble" atmosphere.
- Performance of the Project Manager
  - A good project manager assumes that employee misunderstandings can and will occur, and therefore blames himself. A good project manager constantly attempts to improve and be more communicative. He relies heavily on moral persuasion.
  - A poor project manager assumes that employees are unwilling to cooperate and therefore blames subordinates. The poor project manager demands more through authoritarian attitudes and relies heavily on material incentives.

Management emphasis also impacts the organization. The following four categories show this management emphasis resulting for both good and poor project management:

- Management Problem-Solving
  - A good project manager performs his own problem-solving at the level for which he is responsible through delegation of problem-solving responsibilities.
  - A poor project manager will do subordinate problem-solving in known areas. For areas that he does not know, he requires that his approval be given prior to idea implementation.
- Organizational Order
  - A good project manager develops, maintains, and uses a single integrated management system in which authority and responsibility are delegated to the subordinates. In addition, he knows that occasional slippages and overruns will occur, and simply tries to minimize their effect.
  - A poor project manager delegates as little authority and responsibility as possible, and runs the risk of continual slippages and overruns. A poor project manager maintains two management information systems: one informal system for himself and one formal (eyewash) system simply to impress his superiors.
- Performance of People
  - A good project manager finds that subordinates willingly accept responsibility, are decisive in attitude toward the project, and are satisfied.
  - A poor project manager finds that his subordinates are reluctant to accept responsibility, are indecisive in their actions, and seem frustrated.

- Performance of the Project Manager
  - A good project manager assumes that his key people can "run the show." He exhibits confidence in those individuals working in areas in which he has no expertise, and exhibits patience with people working in areas where he has a familiarity. A good project manager is never too busy to help his people solve personal or professional problems.
  - A poor project manager considers himself indispensable, is overcautious with work performed in unfamiliar areas, and becomes overly interested in work he knows. A poor project manager is always tied up in meetings.

5.12 EMPLOYEE–MANAGER PROBLEMS

The two major problem areas in the project environment are the "who has what authority and responsibility" question, and the resulting conflicts associated with the individual at the project–functional interface. Almost all project problems in some way or another involve these two major areas. Other problem areas found in the project environment include:

- The pyramidal structure
- Superior–subordinate relationships
- Departmentalization
- Scalar chain of command
- Organizational chain of command
- Power and authority
- Planning goals and objectives
- Decision-making
- Reward and punishment
- Span of control

The two most common employee problems involve the assignment and resulting evaluation processes. Personnel assignments were discussed in Chapter 4. In summary:

- People should be assigned to tasks commensurate with their skills.
- Whenever possible, the same person should be assigned to related tasks.
- The most critical tasks should be assigned to the most responsible people.

The evaluation process in a project environment is difficult for an employee at the functional–project interface, especially if hostilities develop between the functional and project managers. In this situation, the interfacing employee almost always suffers owing to a poor rating by either the project manager or his supervisor. Unless the employee continually keeps his superior abreast of his performance and achievements, the supervisor must rely solely on the input (often flawed) received from project office personnel.

Three additional questions must be answered with regard to employee evaluation:

● Of what value are job descriptions?
● How do we maintain wage and salary grades?
● Who provides training and development, especially under conditions where variable manloading can exist?

If each project is, in fact, different from all others, then it becomes an almost impossible task to develop accurate job descriptions. In many cases, wage and salary grades are functions of a unit manning document that specifies the number, type, and grade of all employees required on a given project. Although this might be a necessity in order to control costs, it also is difficult to achieve because variable manloading changes project priorities. Variable manloading creates several difficulties for project managers, especially if new employees are included. Project managers like to have seasoned veterans assigned to their activities because there generally does not exist sufficient time for proper and close supervision of the training and development of new employees. Functional managers, however, contend that the training has to be accomplished on someone's project, and sooner or later all project managers must come to this realization.

On the manager level, the two most common problems involve personal values and conflicts. Personal values are often attributed to the "changing of the guard." New managers have a different sense of values from that of the older, more experienced managers. Miner identifies some of these personal values attributed to new managers[13]:

● Less trust, especially of people in positions of authority.
● Increased feelings of being controlled by external forces and events, and thus belief that they cannot control their own destinies. This is a kind of change that makes for less initiation of one's own activities and a greater likelihood of responding in terms of external pressures. There is a sense of powerlessness, although not necessarily a decreased desire for power.
● Less authoritarian and more negative attitudes toward persons holding positions of power.
● More independence, often to the point of rebelliousness and defiance.
● More freedom, less control in expressing feelings, impulses, and emotions.
● Greater inclination to live in the present and to let the future take care of itself.
● More self-indulgence.
● Moral values that are relative to the situation, less absolute, and less tied to formal religion.
● A strong and increasing identification with their peer and age groups, with the youth culture.

---

13. John B. Miner, "The OD-Management Development Conflict." Reprinted with permission from *Business Horizons,* December 1973, p. 32. Copyright © 1973 by the Board of Trustees at Indiana University.

● Greater social concern and greater desire to help the less fortunate.
● More negative attitude toward business, the management role in particular. A professional position is clearly preferred to managing.
● A desire to contribute less to an employing organization and to receive more from the organization.

Previously, we defined one of the attributes of a project manager as liking risks. Unfortunately, the amount of risk that today's managers are willing to accept varies not only with their personal values but also with the impact of current economic conditions and top management philosophies. If top management views a specific project as vital for the growth of the company, then the project manager may be directed to assume virtually no risks during the execution of the project. In this case the project manager may attempt to pass all responsibility to higher or lower management claiming that "his hands are tied." Wilemon and Cicero identify problems with risk identification[14]:

● The project manager's anxiety over project risk varies in relation to his willingness to accept final responsibility for the technical success of his project. Some project managers may be willing to accept full responsibility for the success or failure of their projects. Others, by contrast, may be more willing to share responsibility and risk with their superiors.
● The greater the length of stay in project management, the greater the tendency for project managers to remain in administrative positions within an organization.
● The degree of anxiety over professional obsolescence varies with the length of time the project manager spends in project management positions.

The amount of risk that managers will accept also varies with age and experience. Older, more experienced managers tend to take few risks, whereas the younger, more aggressive managers may adopt a risk-lover policy in hopes of achieving a name for themselves.

Conflicts exist at the project–functional interface regardless of how hard we attempt to structure the work. According to Cleland and King, this interface can be defined by the following relationships[15]:

● Project Manager
  ● *What* is to be done?
  ● *When* will the task be done?
  ● *Why* will the task be done?
  ● *How much* money is available to do the task?
  ● *How well* has the total project been done?

---

14. D. L. Wilemon and John P. Cicero, "The Project Manager: Anomalies and Ambiguities," *Academy of Management Journal,* Vol. 13, 1970, pp. 269–282.

15. From David I. Cleland and William Richard King, *Systems Analysis and Project Management* (New York: McGraw-Hill), p. 237.

- Functional Manager
  - *Who* will do the task?
  - *Where* will the task be done?
  - *How* will the task be done?
  - *How well* has the functional input been integrated into the project?

The result of these differing views is inevitable conflict between the functional and project manager, as described by William Killian[16]:

> The conflicts revolve about items such as project priority, manpower costs, and the assignment of functional personnel to the project manager. Each project manager will, of course, want the best functional operators assigned to his project. In addition to these problems, the accountability for profit and loss is much more difficult in a matrix organization than in a project organization. Project managers have a tendency to blame overruns on functional managers, stating that the cost of the function was excessive. Whereas functional managers have a tendency to blame excessive costs on project managers with the argument that there were too many changes, more work required than defined initially, and other such arguments.

Major conflicts can also arise during problem resolution sessions because the time constraints imposed on the project often prevent both parties from taking a logical approach. One of the major causes of prolonged problem-solving is a lack of pertinent information. The following information should be reported by the project manager[17]:

- The problem
- The cause
- The expected impact on schedule, budget, profit, or other pertinent area
- The action taken or recommended and the results expected of that action
- What top management can do to help

5.13 MANAGEMENT PITFALLS

The project environment offers numerous opportunities for project managers and team members to get into trouble. Common types of management pitfalls are:

- Lack of self-control (knowing oneself)
- Activity traps
- Managing versus doing
- People versus task skills

---

16. William P. Killian, "Project Management—Future Organizational Concepts," *Marquette Business Review,* Vol. 2, 1971, pp. 90–107.

17. Russell D. Archibald, *Managing High-Technology Programs and Projects* (New York: Wiley, 1976), p. 230.

- Ineffective communications
- Time management
- Management bottlenecks

Knowing oneself, especially one's capabilities, strengths, and weaknesses, is the first step toward successful project management. Too often, managers will assume that they are jacks-of-all-trades, will "bite off more than they can chew," and then find that insufficient time exists for training additional personnel.

The following lines illustrate self-concept:

Four Men

It chanced upon a winter's night
Safe sheltered from the weather.
The board was spread for only one,
Yet four men dined together.
There sat the man I meant to be
In glory, spurred and booted.
And close beside him, to the right
The man I am reputed.
The man I think myself to be
His seat was occupying
Hard by the man I really am
To hold his own was trying.
And all beneath one roof we met
Yet none called his fellow brother
No sign of recognition passed
They knew not one another.

Author unknown

Activity traps result when the means become the end, rather than the means to achieve the end. The most common activity traps are team meetings, customer–technical interchange meetings, and the development of special schedules and charts that cannot be used for customer reporting but are used to inform upper-level management of project status. Sign-off documents are another activity trap and managers must evaluate whether all this paperwork is worth the effort.

We previously defined a characteristic of poor leadership as the inability to obtain a balance between management functions and technical functions. This can easily develop into an activity trap where the individual becomes a doer rather than a manager. Unfortunately, there often exists a very fine line between managing and doing. As an example, consider a project manager who was asked by one of his technical people to make a telephone call to assist him in solving a problem. Simply making the phone call is doing work that should be done by the project team members or even the functional manager. However, if the person being called requires that someone in absolute authority be included in the conversation, then this can be considered managing instead of doing.

There are several other cases where one must become a doer in order to be an effective manager and command the loyalty and respect of subordinates. Assume a special situation where you must schedule subordinates to work overtime on holidays or weekends.

By showing up at the plant during these times, just to make a brief appearance, you can create a better working atmosphere and understanding with the subordinates.

Another major pitfall is the decision to utilize either people skills or task skills. Is it better to utilize subordinates with whom you can obtain a good working relationship or to employ highly skilled people simply to get the job done? Obviously, the project manager would like nothing better than to have the best of both worlds, but this is not always possible. Consider the following situations:

- There is a task that will take three weeks to complete. John has worked for you before, but not on such a task as this. John, however, understands how to work with you. Paul is very competent but likes to work alone. He can get the job done within constraints. Should you employ people or task skills? (Would your answer change if the task were three months instead of three weeks?)
- There exist three tasks, each one requiring two months of work. Richard has the necessary people skills to handle all three tasks, but he will not be able to do so as efficiently as a technical specialist. The alternate choice is to utilize three technical specialists.

Based on the amount of information given, the author prefers task skills so as not to hinder the time or performance constraints on the project. Generally speaking, for long-duration projects that require constant communications with the customer, it might be better to have permanently assigned employees who can perform a variety of tasks. Customers dislike seeing a steady stream of new faces.

It is often said that a good project manager must be willing to work sixty to eighty hours a week to get the job done. This might be true if he is continually fighting fires or if budgeting constraints prevent employing additional staff. The major reason, however, is the result of ineffective time management. Prime examples might include the continuous flow of paperwork, unnecessary meetings, unnecessary phone calls, and acting as a tour guide for visitors.

- To be effective, the project manager must establish time management rules and then ask himself four questions:
  - What am I doing that I don't have to be doing at all?
  - What am I doing that can be done better by someone else?
  - What am I doing that could be done sufficiently well by someone else?
  - Am I establishing the right priorities for my activities?
- Rules for time management
  - Conduct a time analysis (time log)
  - Plan solid blocks for important things
  - Classify your activities
  - Establish priorities
  - Establish opportunity cost on activities
  - Train your system (boss, subordinate, peers)
  - Practice delegation
  - Practice calculated neglect
  - Practice management by exception
  - Focus on opportunities—not on problems

5.14 COMMUNICATIONS

PMBOK® Guide, 4th Edition
Chapter 10 Project
  Communications Management

Effective project communications ensure that we get the right information to the right person at the right time and in a cost-effective manner. Proper communication is vital to the success of a project. Typical definitions of effective communication include:

- An exchange of information
- An act or instance of transmitting information
- A verbal or written message
- A technique for expressing ideas effectively
- A process by which meanings are exchanged between individuals through a common system of symbols

The communications environment can be regarded as a network of channels. Most channels are two-way channels. The number of two-way channels, N, can be calculated from the formula

$$N = \frac{X(X - 1)}{2}$$

In this formula, X represents the number of people communicating with each other. For example, if four people are communicating (i.e., $X = 4$), then there are six two-way channels.

When a breakdown in communications occurs, disaster follows, as Figure 5–12 demonstrates.

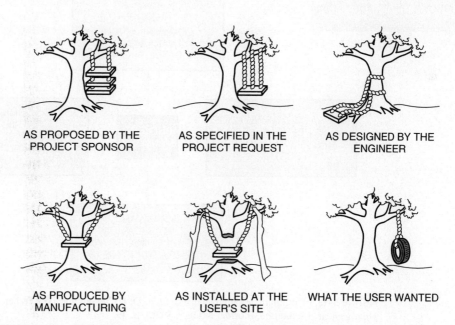

AS PROPOSED BY THE
PROJECT SPONSOR

AS SPECIFIED IN THE
PROJECT REQUEST

AS DESIGNED BY THE
ENGINEER

AS PRODUCED BY
MANUFACTURING

AS INSTALLED AT THE
USER'S SITE

WHAT THE USER WANTED

FIGURE 5–12. A breakdown in communications. (Source unknown)

Figures 5–13 and 5–14 show typical communications patterns. Some people consider Figure 5–13 "politically incorrect" because project managers should not be identified as talking "down" to people. Most project managers communicate laterally, whereas line managers communicate vertically downward to subordinates. Figure 5–15 shows the complete communication model. The screens or barriers are from one's perception, personality, attitudes, emotions, and prejudices.

PMBOK® Guide, 4th Edition
Figure 10–8 Basic Communication Model

- *Perception barriers* occur because individuals can view the same message in different ways. Factors influencing perception include the individual's level of education and region of experience. Perception problems can be minimized by using words that have precise meaning.
- *Personality and interests,* such as the likes and dislikes of individuals, affect communications. People tend to listen carefully to topics of interest but turn a deaf ear to unfamiliar or boring topics.
- *Attitudes, emotions, and prejudices* warp our sense of interpretation. Individuals who are fearful or have strong love or hate emotions will tend to protect themselves by distorting the communication process. Strong emotions rob individuals of their ability to comprehend.

Typical barriers that affect the encoding process include:

- Communication goals
- Communication skills

PMBOK® Guide, 4th Edition
10.2 Communications Planning

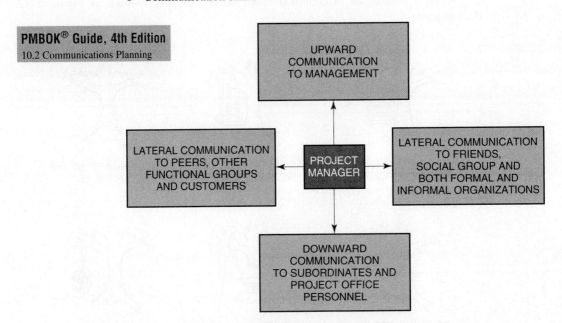

FIGURE 5–13. Communication channels. *Source:* D. I. Cleland and H. Kerzner, *Engineering Team Management* (Melbourne, Florida: Krieger, 1986), p. 39.

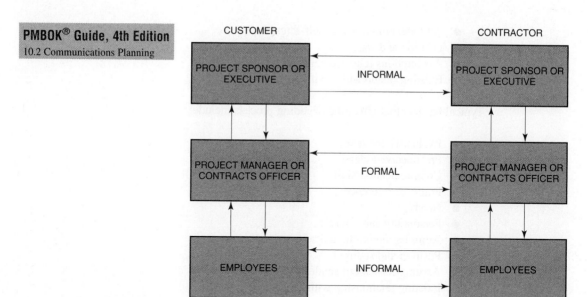

FIGURE 5–14. Customer communications. *Source:* D. I. Cleland and H. Kerzner, *Engineering Team Management* (Melbourne, Florida: Krieger, 1986), p. 64.

- Frame of reference
- Sender credibility
- Needs
- Personality and interests
- Interpersonal sensitivity

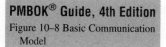

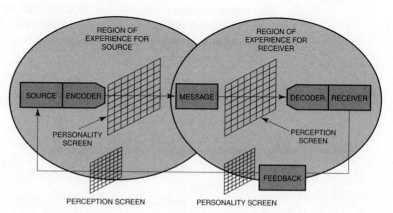

FIGURE 5–15. Total communication process. *Source:* D. I. Cleland and H. Kerzner, *Engineering Team Management* (Melbourne, Florida: Krieger, 1986), p. 46.

- Attitude, emotion, and self-interest
- Position and status
- Assumptions (about receivers)
- Existing relationships with receivers

Typical barriers that affect the decoding process include:

- Evaluative tendency
- Preconceived ideas
- Communication skills
- Frame of reference
- Needs
- Personality and interest
- Attitudes, emotion, and self-interest
- Position and status
- Assumptions about sender
- Existing relationship with sender
- Lack of responsive feedback
- Selective listening

The receiving of information can be affected by the way the information is received. The most common ways include:

- Hearing activity
- Reading skills
- Visual activity
- Tactile sensitivity
- Olfactory sensitivity
- Extrasensory perception

The communications environment is controlled by both the internal and external forces, which can act either individually or collectively. These forces can either assist or restrict the attainment of project objectives.

Typical internal factors include:

- Power games
- Withholding information
- Management by memo
- Reactive emotional behavior
- Mixed messages
- Indirect communications
- Stereotyping
- Transmitting partial information
- Blocking or selective perception

Typical external factors include:

- The business environment
- The political environment
- The economic climate
- Regulatory agencies
- The technical state-of-the-art

The communications environment is also affected by:

- Logistics/geographic separation
- Personal contact requirements
- Group meetings
- Telephone
- Correspondence (frequency and quantity)
- Electronic mail

Noise tends to distort or destroy the information within the message. Noise results from our own personality screens, which dictate the way we present the message, and perception screens, which may cause us to "perceive" what we thought was said. Noise therefore can cause ambiguity:

- Ambiguity causes us to hear what we want to hear.
- Ambiguity causes us to hear what the group wants.
- Ambiguity causes us to relate to past experiences without being discriminatory.

In a project environment, a project manager may very well spend 90 percent or more of his or her time communicating. Typical functional applications include:

- Providing project direction
  - Decision-making
  - Authorizing work
  - Directing activities
  - Negotiating
  - Reporting (including briefings)
- Attending meetings
- Overall project management
- Marketing and selling
- Public relations
- Records management
  - Minutes
  - Memos/letters/newsletters
  - Reports
  - Specifications
  - Contract documents

PMBOK® Guide, 4th Edition
10.3.3 Information Distribution
10.3.3 Outputs

Project managers are required to provide briefings for both internal and external customers. Visual aids can greatly enhance a presentation. Their advantages include:

- Enlivening a presentation, which helps to capture and hold the interest of an audience.
- Adding a visual dimension to an auditory one, which permits an audience to perceive a message through two separate senses, thereby strengthening the learning process.
- Spelling out unfamiliar words by presenting pictures, diagrams, or objects, and by portraying relations graphically, which helps in introducing material that is difficult or new.
- Remaining in view much longer than oral statements can hang in the air, which can serve the same purpose as repetition in acquainting an audience with the unfamiliar and bringing back listeners who stray from the presentation.

Meetings can be classified according to their frequency of occurrence:

- The daily meeting where people work together on the same project with a common objective and reach decisions informally by general agreement.
- The weekly or monthly meeting where members work on different but parallel projects and where there is a certain competitive element and greater likelihood that the chairman will make the final decision himself or herself.
- The irregular, occasional, or special-project meeting, composed of people whose normal work does not bring them into contact and whose work has little or no relationship to that of the others. They are united only by the project the meeting exists to promote and motivated by the desire that the project succeed. Though actual voting is uncommon, every member effectively has a veto.

There are three types of written media used in organizations:

PMBOK® Guide, 4th Edition
10.3.3 Information Distribution

- Individually oriented media: These include letters, memos, and reports.
- Legally oriented media: These include contracts, agreements, proposals, policies, directives, guidelines, and procedures.
- Organizationally oriented media: These include manuals, forms, and brochures.

Because of the time spent in a communications mode, the project manager may very well have as his or her responsibility the process of *communications management.* Communications management is the formal or informal process of conducting or supervising the exchange of information either upward, downward, laterally or diagonally. There appears to be a direct correlation between the project manager's ability to manage the communications process and project performance.

The communications process is more than simply conveying a message; it is also a source for control. Proper communications let the employees in on the act because employees need to know and understand. Communication must convey both information and motivation. The problem, therefore, is how to communicate. Below are six simple steps:

- Think through what you wish to accomplish.
- Determine the way you will communicate.
- Appeal to the interest of those affected.
- Give playback on ways others communicate to you.
- Get playback on what you communicate.
- Test effectiveness through reliance on others to carry out your instructions.

Knowing how to communicate does not guarantee that a clear message will be generated. There are techniques that can be used to improve communications. These techniques include:

- Obtaining feedback, possibly in more than one form
- Establishing multiple communications channels
- Using face-to-face communications if possible
- Determining how sensitive the receiver is to your communications
- Being aware of symbolic meaning such as expressions on people's faces
- Communicating at the proper time
- Reinforcing words with actions
- Using a simple language
- Using redundancy (i.e., saying it two different ways) whenever possible

With every effort to communicate there are always barriers. The barriers include:

- Receiver hearing what he wants to hear. This results from people doing the same job so long that they no longer listen.
- Sender and receiver having different perceptions. This is vitally important in interpreting contractual requirements, statements of work, and proposal information requests.
- Receiver evaluating the source before accepting the communications.
- Receiver ignoring conflicting information and doing as he pleases.
- Words meaning different things to different people.
- Communicators ignoring nonverbal cues.
- Receiver being emotionally upset.

The scalar chain of command can also become a barrier with regard to in-house communications. The project manager must have the authority to go to the general manager or counterpart to communicate effectively. Otherwise, filters can develop and distort the final message.

Three important conclusions can be drawn about communications techniques and barriers:

PMBOK® Guide, 4th Edition
Chapter 10 Communications
  Skills

- Don't ssume that the message you sent will be received in the form you sent it.
- The swiftest and most effective communications take place among people with common points of view. The manager who fosters good relationships with his associates will have little difficulty in communicating with them.
- Communications must be established early in the project.

In a project environment, communications are often filtered. There are several reasons for the filtering of upward communications:

- Unpleasantness for the sender
- Receiver cannot obtain information from any other source
- To embarrass a superior
- Lack of mobility or status for the sender
- Insecurity
- Mistrust

Communication is also listening. Good project managers must be willing to listen to their employees, both professionally and personally. The advantages of listening properly are that:

- Subordinates know you are sincerely interested
- You obtain feedback
- Employee acceptance is fostered.

The successful manager must be willing to listen to an individual's story from beginning to end, without interruptions, and to see the problem through the eyes of the subordinate. Finally, before making a decision, the manager should ask the subordinate for his solutions to the problem.

Project managers should ask themselves four questions:

- Do I make it easy for employees to talk to me?
- Am I sympathetic to their problems?
- Do I attempt to improve human relations?
- Do I make an extra effort to remember names and faces?

PMBOK® Guide, 4th Edition
Chapter 10 Communications
  Skills

The project manager's communication skills and personality screen often dictates the communication style. Typical communication styles include:

- Authoritarian: gives expectations and specific guidance
- Promotional: cultivates team spirit
- Facilitating: gives guidance as required, noninterfering

TABLE 5–4. COMMUNICATIONS POLICY

Program Manager	Functional Manager	Relationship
The program manager utilizes existing authorized communications media to the maximum extent rather than create new ones.		Communications up, down, and laterally are essential elements to the success of programs in a multiprogram organization, and to the morale and motivation of supporting functional organizations. In principle, communication from the program manager should be channeled through the program team member to functional managers.
Approves program plans, subdivided work description, and/or work authorizations, and schedules defining specific program requirements.	Assures his organization's compliance with all such program direction received.	Program definition must be within the scope of the contract as expressed in the program plan and work breakdown structure.
Signs correspondence that provides program direction to functional organizations. Signs correspondence addressed to the customer that pertains to the program except that which has been expressly assigned by the general manager, the function organizations, or higher management in accordance with division policy.	Assures his organization's compliance with all such program direction received. Functional manager provides the program manager with copies of all "Program" correspondence released by his organization that may affect program performance. Ensures that the program manager is aware of correspondence with unusual content, on an exception basis, through the cognizant program team member or directly if such action is warranted by the gravity of the situation.	In the program manager's absence, the signature authority is transferred upward to his reporting superior unless an acting program manager has been designated. Signature authority for correspondence will be consistent with established division policy.
Reports program results and accomplishments to the customer and to the general manager, keeping them informed of significant problems and events.	Participates in program reviews, being aware of and prepared in matters related to his functional specialty. Keeps his line or staff management and cognizant program team member informed of significant problems and events relating to any program in which his personnel are involved.	Status reporting is the responsibility of functional specialists. The program manager utilizes the specialist organizations. The specialists retain their own channels to the general manager but must keep the program manager informed.

- Conciliatory: friendly and agreeable, builds compatible team
- Judicial: uses sound judgment
- Ethical: honest, fair, by the book
- Secretive: not open or outgoing (to project detriment)
- Disruptive: breaks apart unity of group, agitator
- Intimidating: "tough guy," can lower morale
- Combative: eager to fight or be disagreeable

PMBOK® Guide, 4th Edition
Chapter 10
10.3 Performance Reporting

Team meetings are often used to exchange valuable and necessary information. The following are general guides for conducting more effective meetings:

- Start on time. If you wait for people, you reward tardy behavior.
- Develop agenda "objectives." Generate a list and proceed; avoid getting hung up on the order of topics.
- Conduct one piece of business at a time.
- Allow each member to contribute in his own way. Support, challenge, and counter; view differences as helpful; dig for reasons or views.
- Silence does not always mean agreement. Seek opinions: "What's your opinion on this, Peggy?"
- Be ready to confront the verbal member: "Okay, we've heard from Mike on this matter; now how about some other views?"
- Test for readiness to make a decision.
- Make the decision.
- Test for commitment to the decision.
- Assign roles and responsibilities (only after decision-making).
- Agree on follow-up or accountability dates.
- Indicate the next step for this group.
- Set the time and place for the next meeting.
- End on time.
- Ask yourself if the meeting was necessary.

Many times, company policies and procedures can be established for the development of communications channels. Table 5–4 illustrates such communications guidelines.

5.15 PROJECT REVIEW MEETINGS

Project review meetings are necessary to show that progress is being made on a project. There are three types of review meetings:

- Project team review meetings
- Executive management review meetings
- Customer project review meetings

Most projects have weekly, bimonthly, or monthly meetings in order to keep the project manager and his team informed about the project's status. These meetings are flexible and should be called only if they will benefit the team.

Executive management has the right to require monthly status review meetings. However, if the project manager believes that other meeting dates are better (because they occur at a point where progress can be identified), then he should request them.

Customer review meetings are often the most critical and most inflexibly scheduled. Project managers must allow time to prepare handouts and literature well in advance of the meeting.

5.16 PROJECT MANAGEMENT BOTTLENECKS

Poor communications can easily produce communications bottlenecks. The most common bottleneck occurs when all communications between the customer and the parent organization must flow through the project office. Requiring that all information pass through the project office may be necessary but slows reaction times. Regardless of the qualifications of the project office members, the client always fears that the information he receives will be "filtered" prior to disclosure.

Customers not only like firsthand information, but also prefer that their technical specialists be able to communicate directly with the parent organization's technical specialists. Many project managers dislike this arrangement, for they fear that the technical specialists may say or do something contrary to project strategy or thinking. These fears can be allayed by telling the customer that this situation will be permitted if, and only if, the customer realizes that the remarks made by the technical specialists do not, in any way, shape, or form, reflect the position of the project office or company.

For long-duration projects the customer may require that the contractor have an established customer representative office in the contractor's facilities. The idea behind this is

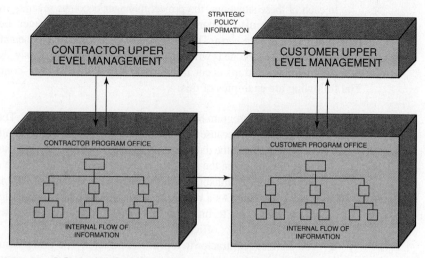

FIGURE 5–16. Information flow pattern from contractor program office.

sound in that all information to the customer must flow through the customer's project office at the contractor's facility. This creates a problem in that it attempts to sever direct communications channels between the customer and contractor project managers. The result is the establishment of a local project office to satisfy contractual requirements, while actual communications go from customer to contractor as though the local project office did not exist. This creates an antagonistic local customer project office.

Another bottleneck occurs when the customer's project manager considers himself to be in a higher position than the contractor's project manager and, therefore, seeks some higher authority with which to communicate. Project managers who seek status can often jeopardize the success of the project by creating rigid communications channels.

Figure 5–16 identifies why communications bottlenecks such as these occur. There almost always exist a minimum of two paths for communications flow to and from the customer, which can cause confusion.

5.17 COMMUNICATION TRAPS

PMBOK® Guide, 4th Edition
Chapter 10 Communications
  Management
10.2.2 Communications Planning

Projects are run by communications. The work is defined by the communications tool known as the work breakdown structure. Actually, this is the easy part of communications, where everything is well defined. Unfortunately, project managers cannot document everything they wish to say or relate to other people, regardless of the level in the company. The worst possible situation occurs when an outside customer loses faith in the contractor. When a situation of mistrust prevails, the logical sequence of events would be:

● More documentation
● More interchange meetings
● Customer representation on your site

In each of these situations, the project manager becomes severely overloaded with work. This situation can also occur in-house when a line manager begins to mistrust a project manager, or vice versa. There may suddenly appear an exponential increase in the flow of paperwork, and everyone is writing "protection" memos. Previously, everything was verbal.

Communication traps occur most frequently with customer–contractor relationships. The following are examples of this:

● Phase I of the program has just been completed successfully. The customer, however, was displeased because he had to wait three weeks to a month after all tests were completed before the data were presented. For Phase II, the customer is insisting that his people be given the raw data at the same time your people receive it.
● The customer is unhappy with the technical information that is being given by the project manager. As a result, he wants his technical people to be able to communicate with your technical people on an individual basis without having to go through the project office.
● You are a subcontractor to a prime contractor. The prime contractor is a little nervous about what information you might present during a technical interchange

meeting where the customer will be represented, and therefore wants to review all material before the meeting.

- Functional employees are supposed to be experts. In front of the customer (or even your top management) an employee makes a statement that you, the project manager, do not believe is completely true or accurate.
- On Tuesday morning, the customer's project manager calls your project manager and asks him a question. On Tuesday afternoon, the customer's project engineer calls your project engineer and asks him the same question.

Communication traps can also occur between the project office and line managers. Below are several examples:

- The project manager holds too many or too few team meetings.
- People refuse to make decisions, and ultimately the team meetings are flooded with agenda items that are irrelevant.
- Last month, Larry completed an assignment as an assistant project manager on an activity where the project manager kept him continuously informed as to project status. Now, Larry is working for a different project manager who tells him only what he needs to know to get the job done.

In a project environment, the line manager is not part of any project team; otherwise he would spend forty hours per week simply attending team meetings. Therefore, how does the line manager learn of the true project status? Written memos will not do it. The information must come firsthand from either the project manager or the assigned functional employee. Line managers would rather hear it from the project manager because line employees have the tendency to censor bad news from the respective line manager. Line managers must be provided true status by the project office.

Sometimes, project managers expect too much from their employees during problem-solving or brainstorming sessions, and communications become inhibited. There are several possible causes for having unproductive team meetings:

- Because of superior–subordinate relationships (i.e., pecking orders), creativity is inhibited.
- All seemingly crazy or unconventional ideas are ridiculed and eventually discarded. Contributors do not wish to contribute anything further.
- Meetings are dominated by upper-level management personnel.
- Many people are not given adequate notification of meeting time and subject matter.

5.18 PROVERBS AND LAWS

Below are twenty project management proverbs that show you what can go wrong[18]:

- You cannot produce a baby in one month by impregnating nine women.
- The same work under the same conditions will be estimated differently by ten different estimators or by one estimator at ten different times.

18. Source unknown.

- The most valuable and least used word in a project manager's vocabulary is "NO."
- You can con a sucker into committing to an unreasonable deadline, but you can't bully him into meeting it.
- The more ridiculous the deadline, the more it costs to try to meet it.
- The more desperate the situation, the more optimistic the situatee.
- Too few people on a project can't solve the problems—too many create more problems than they solve.
- You can freeze the user's specs but he won't stop expecting.
- Frozen specs and the abominable snowman are alike: They are both myths, and they both melt when sufficient heat is applied.
- The conditions attached to a promise are forgotten, and the promise is remembered.
- What you don't know hurts you.
- A user will tell you anything you ask about—nothing more.
- Of several possible interpretations of a communication, the least convenient one is the only correct one.
- What is not on paper has not been said.
- No major project is ever installed on time, within budget, with the same staff that started it.
- Projects progress quickly until they become 90 percent complete; then they remain at 90 percent complete forever.
- If project content is allowed to change freely, the rate of change will exceed the rate of progress.
- No major system is ever completely debugged; attempts to debug a system inevitably introduce new bugs that are even harder to find.
- Project teams detest progress reporting because it vividly demonstrates their lack of progress.
- Parkinson and Murphy are alive and well—in your project.

There are thousands of humorous laws covering all subjects, including economics, general business, engineering, management, and politics. The list below shows some of these laws that are applicable to project management:

- **Abbott's Admonitions**
  1. If you have to ask, you're not entitled to know.
  2. If you don't like the answer, you shouldn't have asked the question.
- **Acheson's Rule of the Bureaucracy:** A memorandum is written not to inform the reader but to protect the writer.
- **Anderson's Law:** I have yet to see any problem, however complicated, which, when you looked at it in the right way, did not become still more complicated.
- **Benchley's Law:** Anyone can do any amount of work provided it isn't the work he or she is supposed to be doing at that moment.
- **Bok's Law:** If you think education is expensive—try ignorance.
- **Boling's Postulate:** If you're feeling good, don't worry. You'll get over it.
- **Brook's First Law:** Adding manpower to a late software project makes it later.

- **Brook's Second Law:** Whenever a system becomes completely defined, some damn fool discovers something which either abolishes the system or expands it beyond recognition.
- **Brown's Law of Business Success:** Our customer's paperwork is profit. Our own paperwork is loss.
- **Chisholm's Second Law:** When things are going well, something will go wrong.
  Corollaries
  1. When things just can't get any worse, they will.
  2. Anytime things appear to be going better, you have overlooked something.
- **Cohn's Law:** The more time you spend reporting what you are doing, the less time you have to do anything. Stability is achieved when you spend all your time doing nothing but reporting on the nothing you are doing.
- **Connolly's Law of Cost Control:** The price of any product produced for a government agency will be not less than the square of the initial firm fixed-price contract.
- **Cooke's Law:** In any decisive situation, the amount of relevant information available is inversely proportional to the importance of the decision.
- **Mr. Cooper's Law:** If you do not understand a particular word in a piece of technical writing, ignore it. The piece will make perfect sense without it.
- **Cornuelle's Law:** Authority tends to assign jobs to those least able to do them.
- **Courtois's Rule:** If people listened to themselves more often, they'd talk less.
- **First Law of Debate:** Never argue with a fool. People might not know the difference.
- **Donsen's Law:** The specialist learns more and more about less and less until, finally, he or she knows everything about nothing; whereas the generalist learns less and less about more and more until, finally, he knows nothing about everything.
- **Douglas's Law of Practical Aeronautics:** When the weight of the paperwork equals the weight of the plane, the plane will fly.
- **Dude's Law of Duality:** Of two possible events, only the undesired one will occur.
- **Economists' Laws**
  1. What men learn from history is that men do not learn from history.
  2. If on an actuarial basis there is a 50–50 chance that something will go wrong, it will actually go wrong nine times out of ten.
- **Old Engineer's Law:** The larger the project or job, the less time there is to do it.
- **Nonreciprocal Laws of Expectations**
  1. Negative expectations yield negative results.
  2. Positive expectations yield negative results.
- **Fyffe's Axiom:** The problem-solving process will always break down at the point at which it is possible to determine who caused the problem.
- **Golub's Laws of Computerdom**
  1. Fuzzy project objectives are used to avoid the embarrassment of estimating the corresponding costs.
  2. A carelessly planned project takes three times longer to complete than expected; a carefully planned project takes only twice as long.
  3. The effort required to correct the course increases geometrically with time.
  4. Project teams detest weekly progress reporting because it so vividly manifests their lack of progress.

- **Gresham's Law:** Trivial matters are handled promptly; important matters are never resolved.
- **Hoare's Law of Large Programs:** Inside every large program is a small program struggling to get out.
- **Issawi's Law of Cynics:** Cynics are right nine times out of ten; what undoes them is their belief that they are right ten times out of ten.
- **Johnson's First Law:** When any mechanical contrivance fails, it will do so at the most inconvenient possible time.
- **Malek's Law:** Any simple idea will be worded in the most complicated way.
- **Patton's Law:** A good plan today is better than a perfect plan tomorrow.
- **Peter's Prognosis:** Spend sufficient time in confirming the need and the need will disappear.
- **Law of Political Erosion:** Once the erosion of power begins, it has a momentum all its own.
- **Pudder's Law:** Anything that begins well ends badly. Anything that begins badly ends worse.
- **Putt's Law:** Technology is dominated by two types of people—those who understand what they do not manage and those who manage what they do not understand.
- **Truman's Law:** If you cannot convince them, confuse them.
- **Von Braun's Law of Gravity:** We can lick gravity, but sometimes the paperwork is overwhelming.

5.19 HUMAN BEHAVIOR EDUCATION

If there is a weakness in some of the project management education programs, it lies in the area of human behavior education. The potential problem is that there is an abundance of courses on planning, scheduling, and cost control but not very many courses on behavioral sciences that are directly applicable to a project management environment. All too often, lectures on human behavior focus upon application of the theories and principles based upon a superior (project manager) to subordinate (team member) relationship. This approach fails because:

- Team members can be at a higher pay grade than the project manager.
- The project manager most often has little overall authority.
- The project manager most often has little formal reward power.
- Team members may be working on multiple projects at the same time.
- Team members may receive conflicting instructions from the project managers and their line manager.
- Because of the project's duration, the project manager may not have the time necessary to adequately know the people on the team on a personal basis.
- The project manager may not have any authority to have people assigned to the project team or removed.

Topics that managers and executives believe should be covered in more depth in the behavioral courses include:

- Conflict management with all levels of personnel
- Facilitation management
- Counseling skills
- Mentorship skills
- Negotiation skills
- Communication skills with all stakeholders
- Presentation skills

The problem may emanate from the limited number of textbooks on human behavior applications directly applicable to the project management environment. One of the best books in the marketplace was written by Steven Flannes and Ginger Levin.[19] The book stresses application of project management education by providing numerous examples from the authors' project management experience.

5.20 MANAGEMENT POLICIES AND PROCEDURES

Although project managers have the authority and responsibility to establish project policies and procedures, they must fall within the general guidelines established by top management. Table 5–5 identifies sample top-management guidelines. Guidelines can also be established for planning, scheduling, controlling, and communications.

5.21 STUDYING TIPS FOR THE PMI® PROJECT MANAGEMENT CERTIFICATION EXAM

This section is applicable as a review of the principles to support the knowledge areas and domain groups in the PMBOK® Guide. This chapter addresses:

- Human Resources Management
- Communications Management
- Closure

---

19. Steven W. Flannes and Ginger Levin, *People Skills for Project Managers* (Vienna, VA: Management Concepts, 2001).

TABLE 5–5. PROJECT GUIDELINES

Program Manager	Functional Manager	Relationship
The program manager is responsible for overall program direction, control, and coordination; and is the principal contact with the program management of the customer.	The functional organization managers are responsible for supporting the program manager in the performance of the contract(s) and in accordance with the terms of the contract(s) and are accountable to their cognizant managers for the total performance.	The program manager determines what will be done: he obtains, through the assigned program team members, the assistance and concurrence of the functional support organizations in determining the definitive requirements and objectives of the program.
To achieve the program objectives, the program manager utilizes the services of the functional organizations in accordance with the prescribed division policies and procedures affecting the functional organizations.		The functional organizations determine *how* the work will be done.
The program manager establishes program and technical policy as defined by management policy.	The functional support organizations perform all work within their functional areas for all programs within the cost, schedule, quality, and specifications established by contract for the program so as to assist the program manager in achieving the program objectives.	The program manager operates within prescribed division policies and procedures except where requirements of a particular program necessitate deviations or modifications as approved by the general manager. The functional support organizations provide strong, aggressive support to the program managers.
The program manager is responsible for the progress being made as well as the effectiveness of the total program.		
Integrates research, development, production, procurement, quality assurance, product support, test, and financial and contractual aspects.		
Approves detailed performance specifications, pertinent physical characteristics, and functional design criteria to meet the program's development or operational requirements.	The functional support organization management seeks out or initiates innovations, methods, improvements, or other means that will enable that function to better schedule commitments, reduce cost, improve quality, or otherwise render exemplary performance as approved by the program manager.	The program manager relies on the functional support program team members for carrying out specific program assignments.
Ensures preparation of, and approves, overall plan, budgets, and work statements essential to the integration of system elements.		Program managers and the functional support program team members are jointly responsible for ensuring that unresolved conflicts between requirements levied on functional organizations by different program managers are brought to the attention of management.
Directs the preparation and maintenance of a time, cost, and performance schedule to ensure the orderly progress of the program.		

Coordinates and approves subcontract work statement, schedules, contract type, and price for major "buy" items.

Coordinates and approves vendor evaluation and source selections in conjunction with procurement representative to the program team.

Program decision authority rests with the program manager for all matters relating to his assigned program, consistent with division policy and the responsibilities assigned by the general manager.

Program managers do not make decisions that are the responsibility of the functional support organizations as defined in division policies and procedures and/or as assigned by the general manager.

Functional organization managers do not request decisions of a program manager that are not within the program manager's delineated authority and responsibility and that do not affect the requirements of the program.

Functional organizations do not make program decisions that are the responsibility of the program manager.

Joint participation in problem solution is essential to providing satisfactory decisions that fulfill overall program and company objectives, and is accomplished by the program manager and the assigned program team members.

In arriving at program decisions, the program manager obtains the assistance and concurrence of cognizant functional support managers, through the cognizant program team member, since they are held accountable for their support of each program and for overall division functional performance.

Understanding the following principles is beneficial if the reader is using this text to study for the PMP® Certification Exam:

- How the various management theories relate to project management
- Various leadership styles
- Different types of power
- Different types of authority
- Need to document authority
- Contributions of Maslow, McGregor, Herzberg, and Ouchi
- Importance of human resources management in project management
- Need to clearly identify each team member's role and responsibility
- Various ways to motivate team members
- That both the project manager and the team are expected to solve their own problems
- That team development is an ongoing process throughout the project life cycle
- Barriers to encoding and decoding
- Need for communication feedback
- Various communication styles
- Types of meetings

In Appendix C, the following Dorale Products mini–case study is applicable:

- Dorale Products (I) [Human Resources and Communications Management]

The following multiple-choice questions will be helpful in reviewing the principles of this chapter:

1. Which of the following is not one of the sources of authority for a project manager?
   A. Project charter
   B. Job description for a project manager
   C. Delegation from senior management
   D. Delegation from subordinates

2. Which form of power do project managers that have a command of technology and are leading R&D projects most frequently use?
   A. Reward power
   B. Legitimate power
   C. Expert power
   D. Referent power

3. If a project manager possesses penalty (or coercive) power, he or she most likely also possesses:
   A. Reward power
   B. Legitimate power
   C. Expert power
   D. Referent power

4. A project manager with a history of success in meeting deliverables and in working with team members would most likely possess a great deal of:
   A. Reward power
   B. Legitimate power

C. Expert power

D. Referent power

5. Most project managers are motivated by which level of Maslow's hierarchy of human needs?

A. Safety

B. Socialization

C. Self-esteem

D. Self-actualization

6. You have been placed in charge of a project team. The majority of the team members have less than two years of experience working on project teams and most of the people have never worked with you previously. The leadership style you would most likely select would be:

A. Telling

B. Selling

C. Participating

D. Delegating

7. You have been placed in charge of a new project team and are fortunate to have been assigned the same people that worked for you on your last two projects. Both previous proj-ects were very successful and the team performed as a high-performance team. The leadership style you would most likely use on the new project would be:

A. Telling

B. Selling

C. Participating

D. Delegating

8. Five people are in attendance in a meeting and are communicating with one another. How many two-way channels of communication are present?

A. 4

B. 5

C. 10

D. 20

9. A project manager provides a verbal set of instructions to two team members on how to perform a specific test. Without agreeing or disagreeing with the project manager, the two employees leave the project manager's office. Later, the project manager discovers that the tests were not conducted according to his instructions. The most probable cause of failure would be:

A. Improper encoding

B. Improper decoding

C. Improper format for the message

D. Lack of feedback on instructions

10. A project manager that allows workers to be actively involved with the project manager in making decisions would be using which leadership style.

A. Passive

B. Participative/democratic

C. Autocratic

D. Laissez-faire

11. A project manager that dictates all decisions and does not allow for any participation by the workers would be using which leadership style.

       A. Passive
       B. Participative/democratic
       C. Autocratic
       D. Laissez-faire

12. A project manager that allows the team to make virtually all of the decisions without any involvement by the project manager would be using which leadership style.
       A. Passive
       B. Participative/democratic
       C. Autocratic
       D. Laissez-faire

ANSWERS

    1. D
    2. C
    3. A
    4. D
    5. D
    6. A
    7. D
    8. C
    9. D
  10. B
  11. C
  12. D

PROBLEMS

5–1 A project manager finds that he does not have direct reward power over salaries, bonuses, work assignments, or project funding for members of the project team with whom he interfaces. Does this mean that he is totally deficient in reward power? Explain your answer.

5–2 For each of the remarks made below, what types of interpersonal influences could exist?

    a. "I've had good working relations with department X. They like me and I like them. I can usually push through anything ahead of schedule."
    b. A research scientist was temporarily promoted to project management for an advanced state-of-the-art effort. He was overheard making the following remark to a team member: "I know it's contrary to department policy, but the test must be conducted according to these criteria or else the results will be meaningless."

5–3 Do you agree or disagree that scientists and engineers are likely to be more creative if they feel that they have sufficient freedom in their work? Can this condition backfire?

5–4 Should the amount of risk and uncertainty in the project have a direct bearing on how much authority is granted to a project manager?

5–5 Some projects are directed by project managers who have only monitoring authority. These individuals are referred to as influence project managers. What kind of projects would be under their control? What organizational structure might be best for this?

5–6 As a project nears termination, the project manager may find that the functional people are more interested in finding a new role for themselves than in giving their best to the current situation. How does this relate to Maslow's hierarchy of needs, and what should the project manager do?

5–7 Richard M. Hodgetts ("Leadership Techniques in the Project Organization," *Academy of Management Journal,* June 1968, pp. 211–219) conducted a survey on aerospace, chemical, construction, and state government workers as to whether they would rate the following leadership techniques as very important, important, or not important:

- Negotiation
- Personality and/or persuasive ability
- Competence
- Reciprocal favors

How do you think each industry answered the questionnaires?

5–8 In a project environment, time is a constraint rather than a luxury, and this creates a problem for the project manager who has previously never worked with certain team members. Some people contend that the project manager must create some sort of test to measure, early on, the ability of people to work together as a team.

Is such a test possible for people working in a project environment? Are there any project organizational forms that would be conducive for such testing?

5–9 Project managers consider authority and funding as being very important in gaining support. Functional personnel, however, prefer friendship and work assignments. How can these two outlooks be related to the theories of Maslow and McGregor?

5–10 On large projects, some people become experts at planning while others become experts at implementation. Planners never seem to put on another hat and see the problems of the people doing the implementation whereas the people responsible for implementation never seem to understand the problems of the planners. How can this problem be resolved on a continuous basis?

5–11 What kind of working relationships would result if the project manager had more reward power than the functional managers?

5–12 For each of the following remarks, state the possible situation and accompanying assumptions that you would make.

a. "A good project manager should manage by focusing on keeping people happy."
b. "A good project manager must be willing to manage tension."
c. "The responsibility for the success or failure rests with upper-level management. This is their baby."
d. Remarks by functional employee: "What if I fail on this project? What can he (the project manager) do to me?"

5–13 Can each of the following situations lead to failure?

    a. Lack of expert power
    b. Lack of referent power
    c. Lack of reward and punishment power
    d. Not having sufficient authority

5–14 One of your people comes into your office and states that he has a technical problem and would like your assistance by making a phone call.

    a. Is this managing or doing?
    b. Does your answer depend on who must be called? (That is, is it possible that authority relationships may have to be considered?)

5–15 On the LRC, can we structure the responsibility column to primary and secondary responsibilities?

5–16 Discuss the meaning of each of the two poems listed below:

> We shall have to evolve
> Problem solvers galore
> Since each problem they solve
> Creates ten problems more.
> *Author unknown*

> Jack and Jill went up the hill
>   To fetch a pail of water
> Jack fell down and broke his crown
>   And Jill came tumbling after.

> Jack could have avoided this awful lump
>   By seeking alternative choices
> Like installing some pipe and a great big pump
>   And handing Jill the invoices.[20]

5–17 What is the correct way for a project manager to invite line managers to attend team meetings?

5–18 Can a project manager sit and wait for things to happen, or should he cause things to happen?

5–19 The company has just hired a fifty-four-year-old senior engineer who holds two masters degrees in engineering disciplines. The engineer is quite competent and has worked well as a loner for the past twenty years. This same engineer has just been assigned to the R&D phase of your project. You, as project manager or project engineer, must make sure that this engineer works as a team member with other functional employees, not as a loner. How do you propose to accomplish this? If the individual persists in wanting to be a loner, should you fire him?

---

20. Stacer Holcomb, OSD (SA), as quoted in *The C/E Newsletter,* publication of the cost effectiveness section of the Operations Research Society of America, Vol. 2, No. 1, January 1967.

5–20 Suppose the linear responsibility chart is constructed with the actual names of the people involved, rather than just their titles. Should this chart be given to the customer?

5–21 How should a functional manager handle a situation where the project manager:

    a. Continually cries wolf concerning some aspect of the project when, in fact, the problem either does not exist or is not as severe as the project manager makes it out to be?

    b. Refuses to give up certain resources that are no longer needed on the project?

5–22 How do you handle a project manager or project engineer who continually tries to "bite off more than he can chew?" If he were effective at doing this, at least temporarily, would your answer change?

5–23 A functional manager says that he has fifteen people assigned to work on your project next week (according to the project plan and schedule). Unfortunately, you have just learned that the prototype is not available and that these fifteen people will have nothing to do. Now what? Who is at fault?

5–24 Manpower requirements indicate that a specific functional pool will increase sharply from eight to seventeen people over the next two weeks and then drop back to eight people. Should you question this?

5–25 Below are several sources from which legal authority can be derived. State whether each source provides the project manager with sufficient authority from which he can effectively manage the project.

    a. The project or organizational charter
    b. The project manager's position in the organization
    c. The job description and specifications for project managers
    d. Policy documents
    e. The project manager's "executive" rank
    f. Dollar value of the contract
    g. Control of funds

5–26 Is this managing or doing?[21]

MANAGING	DOING	
_____	_____	1. Making a call with one of your people to assist him in solving a technical problem.
_____	_____	2. Signing a check to approve a routine expenditure.
_____	_____	3. Conducting the initial screening interview of a job applicant.
_____	_____	4. Giving one of your experienced people your solution to a new problem without first asking for his recommendation.
_____	_____	5. Giving your solution to a recurring problem that one of your new people has just asked you about.
_____	_____	6. Conducting a meeting to explain to your people a new procedure.

---

21. From Raymond O. Leon, *Manage More by Doing Less* (New York: McGraw-Hill), p. 4. Copyright © 1971 by McGraw-Hill, Inc., New York. Used with permission of McGraw-Hill Book Company.

_____   _____ 7. Phoning a department to request help in solving a problem that one of your people is trying to solve.

_____   _____ 8. Filling out a form to give one of your people a pay increase.

_____   _____ 9. Explaining to one of your people why he is receiving a merit pay increase.

_____   _____ 10. Deciding whether to add a position.

_____   _____ 11. Asking one of your people what he thinks about an idea you have that will affect your people.

_____   _____ 12. Transferring a desirable assignment from employee A to employee B because employee A did not devote the necessary effort.

_____   _____ 13. Reviewing regular written reports to determine your people's progress toward their objectives.

_____   _____ 14. Giving a regular progress report by phone to your supervisor.

_____   _____ 15. Giving a tour to an important visitor from outside of your organization.

_____   _____ 16. Drafting an improved layout of facilities.

_____   _____ 17. Discussing with your key people the extent to which they should use staff services during the next year.

_____   _____ 18. Deciding what your expense-budget request will be for your area of responsibility.

_____   _____ 19. Attending a professional or industrial meeting to learn detailed technical developments.

_____   _____ 20. Giving a talk on your work activities to a local community group.

5–27 Below are three broad statements describing the functions of management. For each statement, are we referring to upper-level management, project management, or functional management?

    a. Acquire the best available assets and try to improve them.

    b. Provide a good working environment for all personnel.

    c. Make sure that all resources are applied effectively and efficiently such that all constraints are met, if possible.

5–28 Decide whether you agree or disagree that, in the management of people, the project manager:

- Must convert mistakes into learning experiences.
- Acts as the lubricant that eases the friction (i.e., conflicts) between the functioning parts.

5–29 Functional employees are supposed to be the experts. A functional employee makes a statement that the project manager does not believe is completely true or accurate. Should the project manager support the team member? If so, for how long? Does your answer depend on to whom the remarks are being addressed, such as upper-level management or the customer? At what point should a project manager stop supporting his team members?

5–30 Below are four statements: two statements describe a function, and two others describe a purpose. Which statements refer to project management and which refer to functional management?

- Function
  - Reduce or eliminate uncertainty
  - Minimize and assess risk
- Purpose
  - Create the environment (using transformations)
  - Perform decision-making in the transformed environment

5–31 Manager A is a department manager with thirty years of experience in the company. For the last several years, he has worn two hats and acted as both project manager and functional manager on a variety of projects. He is an expert in his field. The company has decided to incorporate formal project management and has established a project management department. Manager B, a thirty-year-old employee with three years of experience with the company, has been assigned as project manager. In order to staff his project, manager B has requested from manager A that manager C (a personal friend of manager B) be assigned to the project as the functional representative. Manager C is twenty-six years old and has been with the company for two years. Manager A agrees to the request and informs manager C of his new assignment, closing with the remarks, "This project is yours all the way. I don't want to have anything to do with it. I'll be too busy with paperwork as the result of our new organizational structure. Just send me a memo once in a while telling me what's happening."

During the project kickoff meeting it became obvious to both manager B and manager C that the only person with the necessary expertise was manager A. Without the support of manager A, the time duration for project completion could be expected to double.

This situation is ideal for role playing. Put yourself in the place of managers A, B, and C and discuss the reasons for your actions. How can this problem be overcome? How do you get manager A to support the project? Who should inform upper-level management of this situation? When should upper-level management be informed? Would any of your answers change if manager B and manager C were not close friends?

5–32 Is it possible for a product manager to have the same degree of tunnel vision that a project manager has? If so, under what circumstances?

5–33 Your company has a policy that employees can participate in an educational tuition reimbursement program, provided that the degree obtained will benefit the company and that the employee's immediate superior gives his permission. As a project manager, you authorize George, your assistant project manager who reports directly to you, to take courses leading to an MBA degree.

Midway through your project, you find that overtime is required on Monday and Wednesday evenings, the same two evenings that George has classes. George cannot change the evenings that his classes are offered. You try without success to reschedule the overtime to early mornings or other evenings. According to company policy, the project office must supervise all overtime. Since the project office consists of only you and George, you must perform the overtime if George does not. How should you handle this situation? Would your answer change if you thought that George might leave the company after receiving his degree?

5–34 Establishing good interface relationships between the project manager and functional manager can take a great deal of time, especially during the conversion from a traditional to a project organizational form. Below are five statements that represent the different stages in the development of a good interface relationship. Place these statements in the proper order and discuss the meaning of each one.

a. The project manager and functional manager meet face-to-face and try to work out the problem.

b. Both the project and functional managers deny that any problems exist between them.

c. The project and functional managers begin formally and informally to anticipate the problems that can occur.

d. Both managers readily admit responsibility for several of the problems.

e. Each manager blames the other for the problem.

5–35 John is a functional support manager with fourteen highly competent individuals beneath him. John's main concern is performance. He has a tendency to leave scheduling and cost problems up to the project managers. During the past two months, John has intermittently received phone calls and casual visits from upper-level management and senior executives asking him about his department's costs and schedules on a variety of projects. Although he can answer almost all of the performance questions, he has experienced great difficulty in responding to time and cost questions. John is a little apprehensive that if this situation continues, it may affect his evaluation and merit pay increase. What are John's alternatives?

5–36 Projects have a way of providing a "chance for glory" for many individuals. Unfortunately, they quite often give the not-so-creative individual an opportunity to demonstrate his incompetence. Examples would include the designer who always feels that he has a better way of laying out a blueprint, or the individual who intentionally closes a door when asked to open it, or vice versa. How should a project manager handle this situation? Would your answer change if the individual were quite competent but always did the opposite just to show his individuality? Should these individuals be required to have close supervision? If close supervision is required, should it be the responsibility of the functional manager, the project office, or both?

5–37 Are there situations in which a project manager can wait for long-term changes instead of an immediate response to actions?

5–38 Is it possible for functional employees to have performed a job so long or so often that they no longer listen to the instructions given by the project or functional managers?

5–39 On Tuesday morning, the customer's project manager calls the subcontractor's project manager and asks him a question. On Tuesday afternoon, the customer's project engineer calls the contractor's project engineer and asks him the same question. How do you account for this? Could this be "planned" by the customer?

5–40 Below are eight common methods that project and functional employees can use to provide communications:

a. Counseling sessions
b. Telephone conversation
c. Individual conversation
d. Formal letter

e. Project office memo
f. Project office directive
g. Project team meeting
h. Formal report

For each of the following actions, select one and only one means of communication from the above list that you would utilize in accomplishing the action:

1. Defining the project organizational structure to functional managers
2. Defining the project organizational structure to team members
3. Defining the project organizational structure to executives
4. Explaining to a functional manager the reasons for conflict between his employee and your assistant project managers
5. Requesting overtime because of schedule slippages
6. Reporting an employee's violation of company policy

7. Reporting an employee's violation of project policy
8. Trying to solve a functional employee's grievance
9. Trying to solve a project office team member's grievance
10. Directing employees to increase production
11. Directing employees to perform work in a manner that violates company policy
12. Explaining the new indirect project evaluation system to project team members
13. Asking for downstream functional commitment of resources
14. Reporting daily status to executives or the customer
15. Reporting weekly status to executives or the customer
16. Reporting monthly or quarterly status to executives or the customer
17. Explaining the reason for the cost overrun
18. Establishing project planning guidelines
19. Requesting a vice president to attend your team meeting
20. Informing functional managers of project status
21. Informing functional team members of project status
22. Asking a functional manager to perform work not originally budgeted for
23. Explaining customer grievances to your people
24. Informing employees of the results of customer interchange meetings
25. Requesting that a functional employee be removed from your project because of incompetence

5–41 Last month, Larry completed an assignment as chief project engineering on project X. It was a pleasing assignment. Larry, and all of the other project personnel, were continually kept informed (by the project manager) concerning all project activities. Larry is now working for a new project manager who tells his staff only what they have to know in order to get their job done. What can Larry do about this situation? Can this be a good situation?

5–42 Phase I of a program has just been completed successfully. The customer, however, was displeased because he always had to wait three weeks to a month after all tests were complete before data were supplied by the contractor.

For Phase II of the program, the customer is requiring that advanced quality control procedures be adhered to. This permits the customer's quality control people to observe all testing and obtain all of the raw data at the same time the contractor does. Is there anything wrong with this arrangement?

5–43 You are a subcontractor to company Z, who in turn is the prime contractor to company Q. Before any design review or technical interchange meeting, company Z requires that they review all material to be presented both in-house and with company Q prior to the meeting. Why would a situation such as this occur? Is it beneficial?

5–44 Referring to Problem 5–43, during contract negotiations between company Q and company Z, you, as project manager for the subcontractor, are sitting in your office when the phone rings. It is company Q requesting information to support its negotiation position. Should you provide the information?

5–45 How does a project manager find out if the project team members from the functional departments have the authority to make decisions?

5–46 One of your functional people has been assigned to perform a certain test and document the results. For two weeks you "hound" this individual only to find out that he is continually procrastinating on work in another program. You later find out from one of his co-workers that he hates to write. What should you do?

5–47 During a crisis, you find that all of the functional managers as well as the team members are writing letters and memos to you, whereas previously everything was verbal. How do you account for this?

5–48 Below are several problems that commonly occur in project organizations. State, if possible, the effect that each problem could have on communications and time management:

    a. People tend to resist exploration of new ideas.
    b. People tend to mistrust each other in temporary management situations.
    c. People tend to protect themselves.
    d. Functional people tend to look at day-to-day activities rather than long-range efforts.
    e. Both functional and project personnel often look for individual rather than group recognition.
    f. People tend to create win-or-lose positions.

5–49 How can executives obtain loyalty and commitments from horizontal and vertical personnel in a project organizational structure?

5–50 What is meant by polarization of communications? What are the most common causes?

5–51 Many project managers contend that project team meetings are flooded with agenda items, many of which may be irrelevant. How do you account for this?

5–52 Paul O. Gaddis ("The Project Manager," *Harvard Business Review,* May–June 1959, p. 90, copyright © 1959 by the President and Fellows of Harvard College. All rights reserved) has stated that:

> In learning to manage a group of professional employees, the usual boss–subordinate relationship must be modified. Of special importance, the how—the details or methods of work performance by a professional employee—should be established by the employee. It follows that he must be given the facts necessary to permit him to develop a rational understanding of the why of tasks assigned to him.

How would you relate this information to the employee?

5–53 The customer has asked to have a customer representative office set up in the same building as the project office. As project manager, you put the customer's office at the opposite end of the building from where you are, and on a different floor. The customer states that he wants his office next to yours. Should this be permitted, and, if so, under what conditions?

5–54 During an interchange meeting from the customer, one of the functional personnel makes a presentation stating that he personally disagrees with the company's solution to the particular problem under discussion and that the company is "all wet" in its approach. How do you, as a project manager, handle this situation?

5–55 Do you agree or disagree with the statement that documenting results "forces" people to learn?

5–56 Should a project manager encourage the flow of problems to him? If yes, should he be selective in which ones to resolve?

5–57 Is it possible for a project manager to hold too few project review meetings?

5–58 If all projects are different, should there exist a uniform company policies and procedures manual?

5–59 Of the ten items below, which are considered as part of directing and which are controlling?

    a. Supervising
    b. Communicating
    c. Delegating
    d. Evaluating
    e. Measuring
    f. Motivating
    g. Coordinating
    h. Staffing
    i. Counseling
    j. Correcting

5–60 Which of the following items is not considered to be one of the seven Ms of management?

    a. Manpower
    b. Money
    c. Machines
    d. Methods
    e. Materials
    f. Minutes
    g. Mission

5–61 Match the following leadership styles (source unknown):

1. Management by inaction _____	a. Has an executive who manages with flair, wisdom, and vision. He listens to his people, prods them, and leads them.
2. Management by detail _____	
3. Management by invisibility _____	
4. Management by consensus _____	
5. Management by manipulation _____	b. Grows out of fear and anxiety.
6. Management by rejection _____	c. Can be fair or unfair, effective or ineffective, legitimate or illegitimate. Some people are manipulators of others for power. People are not puppets.
7. Management by survival _____	
8. Management by depotism _____	
9. Management by creativity _____	
10. Management by leadership _____	
_____	d. Is the roughly negative style. Executive always has ideas; devil's advocate. Well-prepared proponents can win—so such a boss can be stimulating.
_____	e. Has an executive who needs every conceivable fact; is methodical and orderly; often is timid, inappropriate, or late.
_____	f. Is good as long as it is based on reality. The executive has a trained instinct.

_____ g. Has an executive who will do anything to survive—the jungle fighter. If it is done constructively, the executive will build instead of destroy.

_____ h. Is totalitarian. There are no clashes of ideas. The organization moves. Creative people flee. Employees always know who is boss.

_____ i. Has an executive who is not around, has good subordinates, and works in an office, offstage.

_____ j. Can be important in dealing with the unknown (R&D projects). Subordinates are independent and powerful. This style could be a substitute for decision-making. It is important for setting policy.

CASE STUDIES

THE TROPHY PROJECT

The ill-fated Trophy Project was in trouble right from the start. Reichart, who had been an assistant project manager, was involved with the project from its conception. When the Trophy Project was accepted by the company, Reichart was assigned as the project manager. The program schedules started to slip from day one, and expenditures were excessive. Reichart found that the functional managers were charging direct labor time to his project but working on their own "pet" projects. When Reichart complained of this, he was told not to meddle in the functional manager's allocation of resources and budgeted expenditures. After approximately six months, Reichart was requested to make a progress report directly to corporate and division staffs.

Reichart took this opportunity to bare his soul. The report substantiated that the project was forecasted to be one complete year behind schedule. Reichart's staff, as supplied by the line managers, was inadequate to stay at the required pace, let alone make up any time that had already been lost. The estimated cost at completion at this interval showed a cost overrun of at least 20 percent. This was Reichart's first opportunity to tell his story to people who were in a position to correct the situation. The result of Reichart's frank, candid evaluation of the Trophy Project was very predictable. Nonbelievers finally saw the light, and the line managers realized that they had a role to play in the completion of the project. Most of the problems were now out in the open and could be corrected by providing adequate staffing and resources. Corporate staff ordered immediate remedial action and staff support to provide Reichart a chance to bail out his program.

The results were not at all what Reichart had expected. He no longer reported to the project office; he now reported directly to the operations manager. Corporate staff's interest in the project became very intense, requiring a 7:00 A.M. meeting every Monday morning for complete review of the project status and plans for recovery. Reichart found himself spending more time preparing paperwork, reports, and projections for his Monday morning meetings than he did administering the Trophy Project. The main concern of corporate was to get the project back on schedule. Reichart spent many hours preparing the recovery plan and establishing manpower requirements to bring the program back onto the original schedule.

Group staff, in order to closely track the progress of the Trophy Project, assigned an assistant program manager. The assistant program manager determined that a sure cure for the Trophy Project would be to computerize the various problems and track the progress through a very complex computer program. Corporate provided Reichart with twelve additional staff members to work on the computer program. In the meantime, nothing changed. The functional managers still did not provide adequate staff for recovery, assuming that the additional manpower Reichart had received from corporate would accomplish that task.

After approximately $50,000 was spent on the computer program to track the problems, it was found that the program objectives could not be handled by the computer. Reichart discussed this problem with a computer supplier and found that $15,000 more was required for programming and additional storage capacity. It would take two months for installation of the additional storage capacity and the completion of the programming. At this point, the decision was made to abandon the computer program.

Reichart was now a year and a half into the program with no prototype units completed. The program was still nine months behind schedule with the overrun projected at 40 percent of budget. The customer had been receiving his reports on a timely basis and was well aware of the fact that the Trophy Project was behind schedule. Reichart had spent a great deal of time with the customer explaining the problems and the plan for recovery. Another problem that Reichart had to contend with was that the vendors who were supplying components for the project were also running behind schedule.

One Sunday morning, while Reichart was in his office putting together a report for the client, a corporate vice president came into his office. "Reichart," he said, "in any project I look at the top sheet of paper and the man whose name appears at the top of the sheet is the one I hold responsible. For this project your name appears at the top of the sheet. If you cannot bail this thing out, you are in serious trouble in this corporation." Reichart did not know which way to turn or what to say. He had no control over the functional managers who were creating the problems, but he was the person who was being held responsible.

After another three months the customer, becoming impatient, realized that the Trophy Project was in serious trouble and requested that the division general manager and his entire staff visit the customer's plant to give a progress and "get well" report within a week. The division general manager called Reichart into his office and said, "Reichart, go visit our customer. Take three or four functional line people with you and try to placate him with whatever you feel is necessary." Reichart and four functional line people visited the customer and gave a four-and-a-half-hour presentation defining the problems and the progress to that point. The customer was very polite and even commented that it was an excellent presentation, but the content was totally unacceptable. The program was still six to eight months late, and the customer demanded progress reports on a weekly basis. The customer made arrangements to assign a representative in Reichart's department to be "on-site" at the project on a daily basis and to interface with Reichart and his staff as required. After this turn of events, the program became very hectic.

The customer representative demanded constant updates and problem identification and then became involved in attempting to solve these problems. This involvement created many changes in the program and the product in order to eliminate some of the problems. Reichart had trouble with the customer and did not agree with the changes in the program. He expressed his disagreement vocally when, in many cases, the customer felt the changes were at no cost. This caused a deterioration of the relationship between client and producer.

One morning Reichart was called into the division general manager's office and introduced to Mr. "Red" Baron. Reichart was told to turn over the reins of the Trophy Project to Red immediately. "Reichart, you will be temporarily reassigned to some other division within the corporation. I suggest you start looking outside the company for another job." Reichart looked at Red and asked, "Who did this? Who shot me down?"

Red was program manager on the Trophy Project for approximately six months, after which, by mutual agreement, he was replaced by a third project manager. The customer reassigned his local program manager to another project. With the new team the Trophy Project was finally completed one year behind schedule and at a 40 percent cost overrun.

LEADERSHIP EFFECTIVENESS (A)

Instructions   This tabulation form on page 270 is concerned with a comparison of personal supervisory styles. Indicate your preference to the two alternatives after each item by writing appropriate figures in the blanks. Some of the alternatives may seem equally attractive or unattractive to you. Nevertheless, please attempt to choose the alternative that is relatively more characteristic of you. For each question given, you have three (3) points that you may distribute in any of the following combinations:

A. If you agree with alternative (a) and disagree with (b), write 3 in the top blank and 0 in bottom blank.
   a. $\underline{3}$
   b. $\underline{0}$

B. If you agree with (b) and disagree with (a), write:
   a. $\underline{0}$
   b. $\underline{3}$

C. If you have a slight preference for (a) over (b), write:
   a. $\underline{2}$
   b. $\underline{1}$

D. If you have a slight preference for (b) over (a), write:
   a. $\underline{1}$
   b. $\underline{2}$

Important—Use only the combinations shown above. Try to relate each item to your own personal experience. Please make a choice from every pair of alternatives.

1. On the job, a project manager should make a decision and . . .

   a. _____ tell his team to carry it out.

   b. _____ "tell" his team about the decision and then try to "sell" it.

2. After a project manager has arrived at a decision . . .

   a. _____ he should try to reduce the team's resistance to his decision by indicating what they have to gain.

   b. _____ he should provide an opportunity for his team to get a fuller explanation of his ideas.

3. When a project manager presents a problem to his subordinates . . .

   a. _____ he should get suggestions from them and then make a decision.

   b. _____ he should define it and request that the group make a decision.

4. A project manager . . .

   a. _____ is paid to make all the decisions affecting the work of his team.

   b. _____ should commit himself in advance to assist in implementing whatever decision his team selects when they are asked to solve a problem.

5. A project manager should . . .

   a. _____ permit his team an opportunity to exert some influence on decisions but reserve final decisions for himself.

   b. _____ participate with his team in group decision-making but attempt to do so with a minimum of authority.

6. In making a decision concerning the work situation, a project manager should . . .

   a. _____ present his decision and ideas and engage in a "give-and-take" session with his team to allow them to fully explore the implications of the decision.

   b. _____ present the problem to his team, get suggestions, and then make a decision.

7. A good work situation is one in which the project manager . . .

   a. _____ "tells" his team about a decision and then tries to "sell" it to them.

   b. _____ calls his team together, presents a problem, defines the problem, and requests they solve the problem with the understanding that he will support their decision(s).

8. A well-run project will include . . .

   a. _____ efforts by the project manager to reduce the team's resistance to his decisions by indicating what they have to gain from them.

   b. _____ "give-and take" sessions to enable the project manager and team to explore more fully the implications of the project manager's decisions.

9. A good way to deal with people in a work situation is . . .

   a. _____ to present problems to your team as they arise, get suggestions, and then make a decision.

   b. _____ to permit the team to make decisions, with the understanding that the project manager will assist in implementing whatever decision they make.

10. A good project manager is one who takes . . .

    a. _____ the responsibility for locating problems and arriving at solutions, then tries to persuade his team to accept them.

    b. _____ the opportunity to collect ideas from his team about problems, then he makes his decision.

11. A project manager . . .

    a. _____ should make the decisions in his organization and tell his team to carry them out.

    b. _____ should work closely with his team in solving problems, and attempt to do so with a minimum of authority.

12. To do a good job, a project manager should . . .

    a. _____ present solutions for his team's reaction.

    b. _____ present the problem and collect from the team suggested solutions, then make a decision based on the best solution offered.

13. A good method for a project manager is . . .

    a. _____ to "tell" and then try to "sell" his decision.

    b. _____ to define the problem for his team, then pass them the right to make decisions.

14. On the job, a project manager . . .

    a. _____ need not give consideration to what his team will think or feel about his decisions.

    b. _____ should present his decisions and engage in a "give-and-take" session to enable everyone concerned to explore, more fully, the implications of the decisions.

15. A project manager . . .

    a. _____ should make all decisions himself.

    b. _____ should present the problem to his team, get suggestions, and then make a decision.

16. It is good . . .

    a. _____ to permit the team an opportunity to exert some influence on decisions, but the project manager should reserve final decisions for himself.

    b. _____ for the project manager to participate with his team in group decision-making with as little authority as possible.

17. The project manager who gets the most from his team is the one who . . .

    a. _____ exercises direct authority.

    b. _____ seeks possible solutions from them and then makes a decision.

18. An effective project manager should . . .

    a. _____ make the decisions on his project and tell his team to carry them out.

    b. _____ make the decisions and then try to persuade his team to accept them.

19. A good way for a project manager to handle work problems is to . . .

    a. _____ implement decisions without giving any consideration to what his team will think or feel.

    b. _____ permit the team an opportunity to exert some influence on decisions but reserve the final decision for himself.

20. Project managers . . .

    a. _____ should seek to reduce the team's resistance to their decisions by indicating what they have to gain from them.

    b. _____ should seek possible solutions from their team when problems arise and then make a decision from the list of alternatives.

LEADERSHIP QUESTIONNAIRE
Tabulation Form

	1	2	3	4	5
1	a	b			
2		a	b		
3				a	b
4	a				b
5			a		b
6		a		b	
7		a			b
8		a	b		
9				a	b
10	a		b		
11	a				b
12			a	b	
13		a			b
14	a		b		
15	a			b	
16			a		b
17	a			b	
18	a	b			
19	a		b		
20		a		b	
TOTAL	___	___	___	___	___

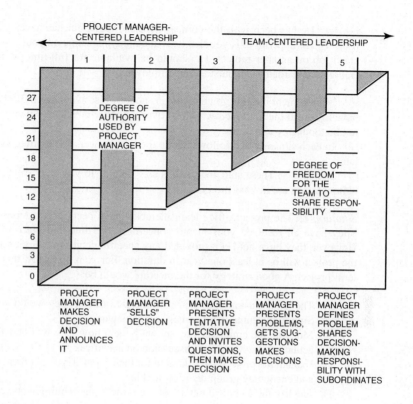

LEADERSHIP EFFECTIVENESS (B)

The Project

PMBOK® Guide, 4th Edition

Chapter 9 Human Resources
   Management
Chapter 10 Communications
   Management
Domain of Professional
   Responsibility

Your company has just won a contract for an outside customer. The contract is for one year, broken down as follows: R&D: six months; prototype testing: one month; manufacturing: five months. In addition to the risks involved in the R&D stage, both your management and the customer have stated that there will be absolutely no trade-offs on time, cost, or performance.

When you prepared the proposal six months ago, you planned and budgeted for a full-time staff of five people, in addition to the functional support personnel. Unfortunately, due to limited resources, your staff (i.e., the project office) will be as follows:

Tom: An excellent engineer, somewhat of a prima donna, but has worked very well with you on previous projects. You specifically requested Tom and were fortunate to have him assigned, although your project is not regarded as a high priority. Tom is recognized as both a technical leader and expert, and is considered as perhaps the best engineer in the company. Tom will be full-time for the duration of the project.

Bob: Started with the company a little over a year ago, and may be a little "wet behind the ears." His line manager has great expectations for him in the future but, for the time being, wants you to give him on-the-job-training as a project office team member. Bob will be full-time on your project.

Carol: She has been with the company for twenty years and does an acceptable job. She has never worked on your projects before. She is full-time on the project.

George: He has been with the company for six years, but has never worked on any of your projects. His superior tells you that he will be only half-time on your project until he finishes a crash job on another project. He should be available for full-time work in a month or two. George is regarded as an outstanding employee.

Management informs you that there is nobody else available to fill the fifth position. You'll have to spread the increased workload over the other members. Obviously, the customer may not be too happy about this.

In each situation that follows, circle the best answer. The grading system will be provided later.

Remember: These staff individuals are "dotted" to you and "solid" to their line manager, although they are in your project office.

Situation 1: The project office team members have been told to report to you this morning. They have all received your memo concerning the time and place of the kickoff meeting. However, they have not been provided any specific details concerning the project except that the project will be at least one year in duration. For your company, this is regarded as a long-term project. A good strategy for the meeting would be:

    A. The team must already be self-motivated or else they would not have been assigned. Simply welcome them and assign homework.
    B. Motivate the employees by showing them how they will benefit: esteem, pride, self-actualization. Minimize discussion on specifics.
    C. Explain the project and ask them for their input. Try to get them to identify alternatives and encourage group decision-making.
    D. Identify the technical details of the project: the requirements, performance standards, and expectations.

Situation 2: You give the team members a copy of the winning proposal and a "confidential" memo describing the assumptions and constraints you considered in developing the proposal. You tell your team to review the material and be prepared to perform detailed planning at the meeting you have scheduled for the following Monday. During Monday's planning meeting, you find that Tom (who has worked with you before) has established a take-charge role and has done some of the planning that should have been the responsibility of other team members. You should:

    A. Do nothing. This may be a beneficial situation. However, you may wish to ask if the other project office members wish to review Tom's planning.
    B. Ask each team member individually how he or she feels about Tom's role. If they complain, have a talk with Tom.
    C. Ask each team member to develop his or her own schedules and then compare results.
    D. Talk to Tom privately about the long-term effects of his behavior.

Situation 3: Your team appears to be having trouble laying out realistic schedules that will satisfy the customer's milestones. They keep asking you pertinent questions and seem to be making the right decisions, but with difficulty.

    A. Do nothing. If the team is good, they will eventually work out the problem.
    B. Encourage the team to continue but give some ideas as to possible alternatives. Let them solve the problem.

C. Become actively involved and help the team solve the problem. Supervise the planning until completion.
D. Take charge yourself and solve the problem for the team. You may have to provide continuous direction.

Situation 4: Your team has taken an optimistic approach to the schedule. The functional managers have reviewed the schedules and have sent your team strong memos stating that there is no way that they can support your schedules. Your team's morale appears to be very low. Your team expected the schedules to be returned for additional iterations and trade-offs, but not with such harsh words from the line managers. You should:

A. Take no action. This is common to these types of projects and the team must learn to cope.
B. Call a special team meeting to discuss the morale problem and ask the team for recommendations. Try to work out the problem.
C. Meet with each team member individually to reinforce his or her behavior and performance. Let members know how many other times this has occurred and been resolved through trade-offs and additional iterations. State your availability to provide advice and support.
D. Take charge and look for ways to improve morale by changing the schedules.

Situation 5: The functional departments have begun working, but are still criticizing the schedules. Your team is extremely unhappy with some of the employees assigned out of one functional department. Your team feels that these employees are not qualified to perform the required work. You should:

A. Do nothing until you are absolutely sure (with evidence) that the assigned personnel cannot perform as needed.
B. Sympathize with your team and encourage them to live with this situation until an alternative is found.
C. Assess the potential risks with the team and ask for their input and suggestions. Try to develop contingency plans if the problem is as serious as the team indicates.
D. Approach the functional manager and express your concern. Ask to have different employees assigned.

Situation 6: Bob's performance as a project office team member has begun to deteriorate. You are not sure whether he simply lacks the skills, cannot endure the pressure, or cannot assume part of the additional work that resulted from the fifth position in the project being vacant. You should:

A. Do nothing. The problem may be temporary and you cannot be sure that there is a measurable impact on the project.
B. Have a personal discussion with Bob, seek out the cause, and ask him for a solution.
C. Call a team meeting and discuss how productivity and performance are decreasing. Ask the team for recommendations and hope Bob gets the message.
D. Interview the other team members and see if they can explain Bob's actions lately. Ask the other members to assist you by talking to Bob.

Situation 7: George, who is half-time on your project, has just submitted for your approval his quarterly progress report for your project. After your signature has been attained, the report is sent to senior management and the customer. The report is marginally acceptable and not at all what you would have expected from George. George apologizes to you for the report and blames it on his other project, which is in its last two weeks. You should:

A. Sympathize with George and ask him to rewrite the report.
B. Tell George that the report is totally unacceptable and will reflect on his ability as a project office team member.
C. Ask the team to assist George in redoing the report since a bad report reflects on everyone.
D. Ask one of the other team members to rewrite the report for George.

Situation 8: You have completed the R&D stage of your project and are entering phase II: prototype testing. You are entering month seven of the twelve-month project. Unfortunately, the results of phase I R&D indicate that you were too optimistic in your estimating for phase II and a schedule slippage of at least two weeks is highly probable. The customer may not be happy. You should:

A. Do nothing. These problems occur and have a way of working themselves out. The end date of the project can still be met.
B. Call a team meeting to discuss the morale problem resulting from the slippage. If morale is improved, the slippage may be overcome.
C. Call a team meeting and seek ways of improving productivity for phase II. Hopefully, the team will come up with alternatives.
D. This is a crisis and you must exert strong leadership. You should take control and assist your team in identifying alternatives.

Situation 9: Your rescheduling efforts have been successful. The functional managers have given you adequate support and you are back on schedule. You should:

A. Do nothing. Your team has matured and is doing what they are paid to do.
B. Try to provide some sort of monetary or nonmonetary reward for your team (e.g., management-granted time off or a dinner team meeting).
C. Provide positive feedback/reinforcement for the team and search for ideas for shortening phase III.
D. Obviously, your strong leadership has been effective. Continue this role for the phase III schedule.

Situation 10: You are now at the end of the seventh month and everything is proceeding as planned. Motivation appears high. You should:

A. Leave well enough alone.
B. Look for better ways to improve the functioning of the team. Talk to them and make them feel important.
C. Call a team meeting and review the remaining schedules for the project. Look for contingency plans.
D. Make sure the team is still focusing on the goals and objectives of the project.

Situation 11: The customer unofficially informs you that his company has a problem and may have to change the design specifications before production actually begins. This would be a catastrophe for your project. The customer wants a meeting at your plant within the next seven days. This will be the customer's first visit to your plant. All previous meetings were informal and at the customer's facilities, with just you and the customer. This meeting will be formal. To prepare for the meeting, you should:

    A. Make sure the schedules are updated and assume a passive role since the customer has not officially informed you of his problem.

    B. Ask the team to improve productivity before the customer's meeting. This should please the customer.

    C. Call an immediate team meeting and ask the team to prepare an agenda and identify the items to be discussed.

    D. Assign specific responsibilities to each team member for preparation of handout material for the meeting.

Situation 12: Your team is obviously not happy with the results of the customer interface meeting because the customer has asked for a change in design specifications. The manufacturing plans and manufacturing schedules must be developed anew. You should:

    A. Do nothing. The team is already highly motivated and will take charge as before.

    B. Reemphasize the team spirit and encourage your people to proceed. Tell them that nothing is impossible for a good team.

    C. Roll up your shirt sleeves and help the team identify alternatives. Some degree of guidance is necessary.

    D. Provide strong leadership and close supervision. Your team will have to rely on you for assistance.

Situation 13: You are now in the ninth month. While your replanning is going on (as a result of changes in the specifications), the customer calls and asks for an assessment of the risks in cancelling this project right away and starting another one. You should:

    A. Wait for a formal request. Perhaps you can delay long enough for the project to finish.

    B. Tell the team that their excellent performance may result in a follow-on contract.

    C. Call a team meeting to assess the risks and look for alternatives.

    D. Accept strong leadership for this and with *minimum,* if any, team involvement.

Situation 14: One of the functional managers has asked for your evaluation of all of his functional employees currently working on your project (excluding project office personnel). Your project office personnel appear to be working more closely with the functional employees than you are. You should:

    A. Return the request to the functional manager since this is not part of your job description.

    B. Talk to each team member individually, telling them how important their input is, and ask for their evaluations.

    C. As a team, evaluate each of the functional team members, and try to come to some sort of agreement.

    D. Do not burden your team with this request. You can do it yourself.

Situation 15: You are in the tenth month of the project. Carol informs you that she has the opportunity to be the project leader for an effort starting in two weeks. She has been with the company for twenty years and this is her first opportunity as a project leader. She wants to know if she can be released from your project. You should:

A. Let Carol go. You do not want to stand in the way of her career advancement.
B. Ask the team to meet in private and conduct a vote. Tell Carol you will abide by the team vote.
C. Discuss the problem with the team since they must assume the extra workload, if necessary. Ask for their input into meeting the constraints.
D. Counsel her and explain how important it is for her to remain. You are already short-handed.

Situation 16: Your team informs you that one of the functional manufacturing managers has built up a brick wall around his department and all information requests must flow through him. The brick wall has been in existence for two years. Your team members are having trouble with status reporting, but always get the information after catering to the functional manager. You should:

A. Do nothing. This is obviously the way the line manager wants to run his department. Your team is getting the information they need.
B. Ask the team members to use their behavioral skills in obtaining the information.
C. Call a team meeting to discuss alternative ways of obtaining the information.
D. Assume strong leadership and exert your authority by calling the line manager and asking for the information.

Situation 17: The executives have given you a new man to replace Carol for the last two months of the project. Neither you nor your team have worked with this man before. You should:

A. Do nothing. Carol obviously filled him in on what he should be doing and what is involved in the project.
B. Counsel the new man individually, bring him up to speed, and assign him Carol's work.
C. Call a meeting and ask each member to explain his or her role on the project to the new man.
D. Ask each team member to talk to this man as soon as possible and help him come on board. Request that individual conversations be used.

Situation 18: One of your team members wants to take a late-afternoon course at the local college. Unfortunately, this course may conflict with his workload. You should:

A. Postpone your decision. Ask the employee to wait until the course is offered again.
B. Review the request with the team member and discuss the impact on his performance.
C. Discuss the request with the team and ask for the team's approval. The team may have to cover for this employee's workload.
D. Discuss this individually with each team member to make sure that the task requirements will still be adhered to.

Situation 19: Your functional employees have used the wrong materials in making a production run test. The cost to your project was significant, but absorbed in a small "cushion" that you saved for emergencies such as this. Your team members tell you that the test will be rerun without any slippage of the schedule. You should:

A. Do nothing. Your team seems to have the situation well under control.
B. Interview the employees that created this problem and stress the importance of productivity and following instructions.
C. Ask your team to develop contingency plans for this situation should it happen again.
D. Assume a strong leadership role for the rerun test to let people know your concern.

Situation 20: All good projects must come to an end, usually with a final report. Your project has a requirement for a final report. This final report may very well become the basis for follow-on work. You should:

A. Do nothing. Your team has things under control and knows that a final report is needed.
B. Tell your team that they have done a wonderful job and there is only one more task to do.
C. Ask your team to meet and provide an outline for the final report.
D. You must provide some degree of leadership for the final report, at least the structure. The final report could easily reflect on your ability as a manager.

Fill in the table below. The answers appear in Appendix B.

Situation	Answer	Points	Situation	Answer	Points
1			11		
2			12		
3			13		
4			14		
5			15		
6			16		
7			17		
8			18		
9			19		
10			20		
				Total	

MOTIVATIONAL QUESTIONNAIRE

On the next several pages, you will find forty statements concerning what motivates you and how you try to motivate others. Beside each statement, circle the number that corresponds to your opinion. In the example below, the choice is "Slightly Agree."

-3	Strongly Disagree
-2	Disagree
-1	Slightly Disagree
0	No Opinion
(+1)	Slightly Agree
+2	Agree
+3	Strongly Agree

Part 1

The following twenty statements involve *what motivates you.* Please rate each of the statements as honestly as possible. Circle the rating that you think is correct, *not* the one you think the instructor is looking for:

1. My company pays me a reasonable salary for the work that I do.
   $-3 \quad -2 \quad -1 \quad 0 \quad +1 \quad +2 \quad +3$

2. My company believes that every job that I do can be considered as a challenge.
   $-3 \quad -2 \quad -1 \quad 0 \quad +1 \quad +2 \quad +3$

3. The company provides me with the latest equipment (i.e., hardware, software, etc.) so I can do my job effectively.
   $-3 \quad -2 \quad -1 \quad 0 \quad +1 \quad +2 \quad +3$

4. My company provides me with recognition for work well done.
   $-3 \quad -2 \quad -1 \quad 0 \quad +1 \quad +2 \quad +3$

5. Seniority on the job, job security, and vested rights are provided by the company.
   $-3 \quad -2 \quad -1 \quad 0 \quad +1 \quad +2 \quad +3$

6. Executives provide managers with feedback of strategic or long-range information that may affect the manager's job.
   $-3 \quad -2 \quad -1 \quad 0 \quad +1 \quad +2 \quad +3$

7. My company provides off-hour clubs and organizations so that employees can socialize, as well as sponsoring social events.
   $-3 \quad -2 \quad -1 \quad 0 \quad +1 \quad +2 \quad +3$

8. Employees are allowed to either set their own work/performance standards or to at least approve/review standards set for them by management.
   $-3 \quad -2 \quad -1 \quad 0 \quad +1 \quad +2 \quad +3$

9. Employees are encouraged to maintain membership in professional societies and/or attend seminars and symposiums on work related subjects. −3 −2 −1 0 +1 +2 +3

10. The company often reminds me that the only way to have job security is to compete effectively in the marketplace.

11. Employees who develop a reputation for "excellence" are allowed to further enhance their reputation, if job related. −3 −2 −1 0 +1 +2 +3

12. Supervisors encourage a friendly, cooperative working environment for employees. −3 −2 −1 0 +1 +2 +3

13. My company provides me with a detailed job description, identifying my role and responsibilities. −3 −2 −1 0 +1 +2 +3

14. My company gives *automatic* wage wage and salary increases for the employees. −3 −2 −1 0 +1 +2 +3

15. My company gives me the opportunity to do what I do best. −3 −2 −1 0 +1 +2 +3

16. My job gives me the opportunity to be truly creative, to the point where I can solve complex problems. −3 −2 −1 0 +1 +2 +3

17. My efficiency and effectiveness is improving because the company provided me with better physical working conditions (i.e., lighting, low noise, temperature, restrooms, etc.) −3 −2 −1 0 +1 +2 +3

18. My job gives me constant self-development. −3 −2 −1 0 +1 +2 +3

19. Our supervisors have feelings for employees rather than simply treating them as "inanimate tools." −3 −2 −1 0 +1 +2 +3

20. Participation in the company's stock option/retirement plan is available to employees. −3 −2 −1 0 +1 +2 +3

Part 2

Statements 21–40 involve how project managers motivate team members. Again, it is important that your ratings honestly reflect the way you think that *you*, as project manager, try to motivate employees. Do *not* indicate the way others or the instructor might recommend motivating the employees. Your thoughts are what are important in this exercise.

21. Project managers should encourage employees to take advantage of company benefits such as stock option plans and retirement plans.

$-3 \quad -2 \quad -1 \quad 0 \quad +1 \quad +2 \quad +3$

22. Project managers should make sure that team members have a good work environment (i.e., heat, lighting, low noise, restrooms, cafeteria, etc.).

$-3 \quad -2 \quad -1 \quad 0 \quad +1 \quad +2 \quad +3$

23. Project managers should assign team members work that can enhance each team member's reputation.

$-3 \quad -2 \quad -1 \quad 0 \quad +1 \quad +2 \quad +3$

24. Project managers should create a relaxed, cooperative environment for the team members.

$-3 \quad -2 \quad -1 \quad 0 \quad +1 \quad +2 \quad +3$

25. Project managers should *continually* remind the team that job security is a function of competitiveness, staying within constraints, and good customer relations.

$-3 \quad -2 \quad -1 \quad 0 \quad +1 \quad +2 \quad +3$

26. Project managers should try to convince team members that each new assignment is a challenge.

$-3 \quad -2 \quad -1 \quad 0 \quad +1 \quad +2 \quad +3$

27. Project managers should be willing to reschedule activities, if possible, around the team's company and out-of-company social functions.

$-3 \quad -2 \quad -1 \quad 0 \quad +1 \quad +2 \quad +3$

28. Project managers should continually remind employees of how they will benefit, monetarily, by successful performance on your project.

$-3 \quad -2 \quad -1 \quad 0 \quad +1 \quad +2 \quad +3$

29. Project managers should be willing
    to "pat people on the back" and
    provide recognition where
    applicable.

    −3 −2 −1 0 +1 +2 +3

30. Project managers should encourage
    the team to maintain constant self-
    development with each assignment.

    −3 −2 −1 0 +1 +2 +3

31. Project managers should allow
    team members to set their own stan-
    dards, where applicable.

    −3 −2 −1 0 +1 +2 +3

32. Project managers should assign
    work to functional employees
    according to seniority on the job.

    −3 −2 −1 0 +1 +2 +3

33. Project managers should allow
    team members to use the informal,
    as well as formal, organization to
    get work accomplished.

    −3 −2 −1 0 +1 +2 +3

34. As a project manager, I would like
    to control the salaries of the full-
    time employees on my project.

    −3 −2 −1 0 +1 +2 +3

35. Project managers should share
    information with the team. This
    includes project information that may
    not be directly applicable to the
    team member's assignment.

    −3 −2 −1 0 +1 +2 +3

36. Project managers should encourage
    team members to be creative and
    to solve their own problems.

    −3 −2 −1 0 +1 +2 +3

37. Project managers should provide
    detailed job descriptions for team
    members, outlining the team mem-
    ber's role and responsibility.

    −3 −2 −1 0 +1 +2 +3

38. Project managers should give each
    team member the opportunity to do
    what the team member can do best.

    −3 −2 −1 0 +1 +2 +3

39. Project managers should be willing
    to interact informally with the team
    members and get to know them, as
    long as there exists sufficient time
    on the project.

    −3 −2 −1 0 +1 +2 +3

40. Most of the employees on my proj-
    ect earn a salary commensurate
    with their abilities.

    −3 −2 −1 0 +1 +2 +3

Part 1 Scoring Sheet (What Motivates You?)

Place your answers (the numerical values you circled) to questions 1–20 in the corresponding spaces in the chart below.

Basic Needs	*Safety Needs*	*Belonging Needs*
#1 _____	#5 _____	#7 _____
#3 _____	#10 _____	#9 _____
#14 _____	#13 _____	#12 _____
#17 _____	#20 _____	#19 _____
Total _____	Total _____	Total _____

Esteem/Ego Needs	*Self-Actualization Needs*
#4 _____	#2 _____
#6 _____	#15 _____
#8 _____	#16 _____
#11 _____	#18 _____
Total _____	Total _____

Transfer your total score in each category to the table on page 283 by placing an "X" in the appropriate area for motivational needs.

Part 2 Scoring Sheet (How Do You Motivate?)

Place your answers (the numerical values you circled) to questions 21–40 in the corresponding spaces in the chart below.

Basic Needs	*Safety Needs*	*Belonging Needs*
#22 _____	#21 _____	#24 _____
#28 _____	#25 _____	#27 _____
#34 _____	#32 _____	#33 _____
#40 _____	#37 _____	#39 _____
Total _____	Total _____	Total _____

Esteem/Ego Needs	*Self-Actualization Needs*
#23 _____	#26 _____
#29 _____	#30 _____
#31 _____	#36 _____
#35 _____	#38 _____
Total _____	Total _____

Transfer your total score in each category to the table on page 283 by placing an "X" in the appropriate area for motivational needs.

QUESTIONS 1–20

Points																								
Needs	−12	−11	−10	−9	−8	−7	−6	−5	−4	−3	−2	−1	0	+1	+2	+3	+4	+5	+6	+7	+8	+9	+10	+11	+12
Self-Actualization																								
Esteem/Ego																								
Belonging																								
Safety																								
Basic																								

QUESTIONS 21–40

Points																								
Needs	−12	−11	−10	−9	−8	−7	−6	−5	−4	−3	−2	−1	0	+1	+2	+3	+4	+5	+6	+7	+8	+9	+10	+11	+12
Self-Actualization																								
Esteem/Ego																								
Belonging																								
Safety																								
Basic																								

Management of Your Time and Stress

6

Related Case Studies (from Kerzner/*Project Management Case Studies,* 3rd Edition)	Related Workbook Exercises (from Kerzner/*Project Management Workbook and PMP®/CAPM® Exam Study Guide,* 10th Edition)	PMBOK® Guide, 4th Edition, Reference Section for the PMP® Certification Exam
• The Reluctant Workers* • Time Management Exercise	• Multiple Choice Exam	• Human Resource Management • Risk Management

6.0 INTRODUCTION

PMBOK® Guide, 4th Edition
Chapter 9 Human Resources Management
Chapter 6 Time Management

Managing projects within time, cost, and performance is easier said than done. The project management environment is extremely turbulent, and is composed of numerous meetings, report writing, conflict resolution, continuous planning and replanning, communications with the customer, and crisis management. Ideally, the effective project manager is a manager, not a doer, but in the "real world," project managers often compromise their time by doing both.

Disciplined time management is one of the keys to effective project management. It is often said that if the project manager cannot control his own time, then he will control nothing else on the project.

*Case Study also appears at end of chapter.

6.1 UNDERSTANDING TIME MANAGEMENT[1]

For most people, time is a resource that, when lost or misplaced, is gone forever. For a project manager, however, time is more of a constraint, and effective time management principles must be employed to make it a resource.

Most executives prefer to understaff projects, in the mistaken belief that the project manager will assume the additional workload. The project manager may already be heavily burdened with meetings, report preparation, internal and external communications, conflict resolution, and planning/replanning for crises. And yet, most project managers somehow manipulate their time to get the work done. Experienced personnel soon learn to delegate tasks and to employ effective time management principles. The following questions should help managers identify problem areas:

- Do you have trouble completing work within the allocated deadlines?
- How many interruptions are there each day?
- Do you have a procedure for handling interruptions?
- If you need a large block of uninterrupted time, is it available? With or without overtime?
- How do you handle drop-in visitors and phone calls?
- How is incoming mail handled?
- Do you have established procedures for routine work?
- Are you accomplishing more or less than you were three months ago? Six months ago?
- How difficult is it for you to say no?
- How do you approach detail work?
- Do you perform work that should be handled by your subordinates?
- Do you have sufficient time each day for personal interests?
- Do you still think about your job when away from the office?
- Do you make a list of things to do? If yes, is the list prioritized?
- Does your schedule have some degree of flexibility?

The project manager who can deal with these questions has a greater opportunity to convert time from a constraint to a resource.

6.2 TIME ROBBERS

The most challenging problem facing the project manager is his inability to say no. Consider the situation in which an employee comes into your office with a problem. The employee may be sincere when he says that he simply wants your advice but, more often

---

1. Sections 6.1, 6.2, and 6.3 are adapted from David Cleland and Harold Kerzner, *Engineering Team Management* (Melbourne, Florida: Krieger, 1986), Chapter 8.

PMBOK® Guide, 4th Edition
Chapter 6 Time Management
Chapter 11 Risk Management
1.6 General Management
   Knowledge and Skills

than not, the employee wants to take the monkey off of his back and put it onto yours. The employee's problem is now *your* problem.

To handle such situations, first screen out the problems with which you do not wish to get involved. Second, if the situation does necessitate your involvement, then you must make sure that when the employee leaves your office, he realizes that the problem is still his, not yours. Third, if you find that the problem will require your continued attention, remind the employee that all future decisions will be joint decisions and that the problem will still be on the employee's shoulders. Once employees realize that they cannot put their problems on your shoulders, they learn how to make their own decisions.

There are numerous time robbers in the project management environment. These include:

- Incomplete work
- A job poorly done that must be done over
- Telephone calls, mail, and email
- Lack of adequate responsibility and commensurate authority
- Changes without direct notification/explanation
- Waiting for people
- Failure to delegate, or unwise delegation
- Poor retrieval systems
- Lack of information in a ready-to-use format
- Day-to-day administration
- Union grievances
- Having to explain "thinking" to superiors
- Too many levels of review
- Casual office conversations
- Misplaced information
- Shifting priorities
- Indecision at any level
- Procrastination
- Setting up appointments
- Too many meetings
- Monitoring delegated work
- Unclear roles/job descriptions
- Executive meddling
- Budget adherence requirements
- Poorly educated customers
- Not enough proven managers
- Vague goals and objectives

- Lack of a job description
- Too many people involved in minor decision-making
- Lack of technical knowledge
- Lack of authorization to make decisions
- Poor functional status reporting
- Work overload
- Unreasonable time constraints
- Too much travel
- Lack of adequate project management tools
- Departmental "buck passing"
- Company politics
- Going from crisis to crisis
- Conflicting directives
- Bureaucratic roadblocks ("ego")
- Empire-building line managers
- No communication between sales and engineering
- Excessive paperwork
- Lack of clerical/administrative support
- Dealing with unreliable subcontractors
- Personnel not willing to take risks
- Demand for short-term results
- Lack of long-range planning
- Learning new company systems
- Poor lead time on projects
- Documentation (reports/red tape)
- Large number of projects
- Desire for perfection

- Lack of project organization
- Constant pressure
- Constant interruptions
- Shifting of functional personnel
- Lack of employee discipline
- Lack of qualified manpower

6.3 TIME MANAGEMENT FORMS

There are two basic forms that project managers and project engineers can use for practicing better time management. The first form is the "to do" pad as shown in Figure 6–1. The project manager or secretary prepares the list of things to do. The project manager then decides which activities he must perform himself and assigns the appropriate priorities.

The activities with the highest priorities are then transferred to the "daily calendar log," as shown in Figure 6–2. The project manager assigns these activities to the appropriate time blocks based on his own energy cycle. Unfilled time blocks are then used for unexpected crises or for lower-priority activities.

If there are more priority elements than time slots, the project manager may try to schedule well in advance. This is normally not a good practice, because it creates a backlog of high-priority activities. In addition, an activity that today is a "B" priority could easily become an "A" priority in a day or two. The moral here is do not postpone until tomorrow what you or your team can do today.

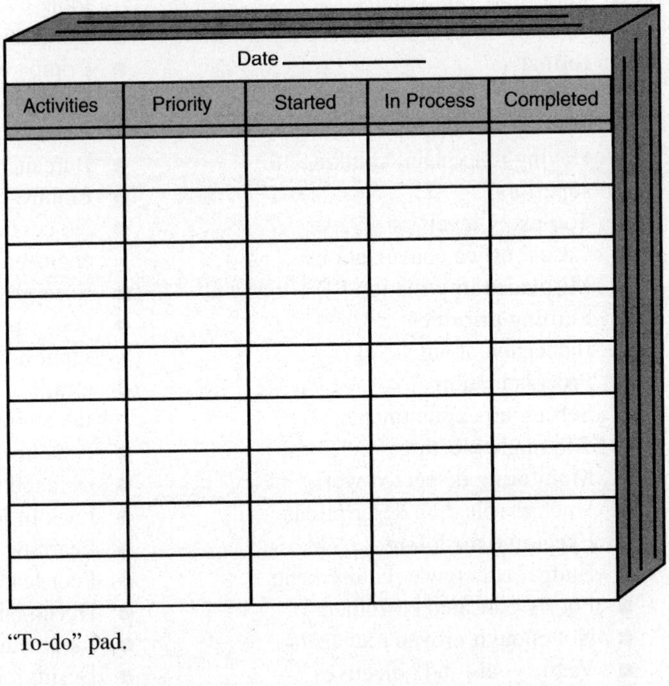

FIGURE 6–1. "To-do" pad.

Time	Activity	Priority
8:00–9:00		
9:00–10:00		
10:00–11:00		
11:00–12:00		
12:00–1:00		
1:00–2:00		
2:00–3:00		
3:00–4:00		
4:00–5:00		

Date _____

FIGURE 6–2. Daily calendar log.

6.4 EFFECTIVE TIME MANAGEMENT

There are several techniques that project managers can practice in order to make better use of their time[2]:

- Delegate.
- Follow the schedule.
- Decide fast.
- Decide who should attend.
- Learn to say no.
- Start now.
- Do the tough part first.
- Travel light.
- Work at travel stops.
- Avoid useless memos.
- Refuse to do the unimportant.
- Look ahead.
- Ask: Is this trip necessary?
- Know your energy cycle.

2. Source unknown.

- Control telephone and email time.
- Send out the meeting agenda.
- Overcome procrastination.
- Manage by exception.

As we learned in Chapter 5, the project manager, to be effective, must establish time management rules and then ask himself four questions:

- Rules for time management
  - Conduct a time analysis (time log).
  - Plan solid blocks for important things.
  - Classify your activities.
  - Establish priorities.
  - Establish opportunity cost on activities.
  - Train your system (boss, subordinate, peers).
  - Practice delegation.
  - Practice calculated neglect.
  - Practice management by exception.
  - Focus on opportunities—not on problems.
- Questions
  - What am I doing that I don't have to do at all?
  - What am I doing that can be done better by someone else?
  - What am I doing that could be done as well by someone else?
  - Am I establishing the right priorities for my activities?

6.5 STRESS AND BURNOUT

The factors that serve to make any occupation especially stressful are responsibility without the authority or ability to exert control, a necessity for perfection, the pressure of deadlines, role ambiguity, role conflict, role overload, the crossing of organizational boundaries, responsibility for the actions of subordinates, and the necessity to keep up with the information explosions or technological breakthroughs. Project managers have all of these factors in their jobs.

A project manager has his resources controlled by line management, yet the responsibilities of bringing a project to completion by a prescribed deadline are his. A project manager may be told to increase the work output, while the work force is simultaneously being cut. Project managers are expected to get work out on schedule, but are often not permitted to pay overtime. One project manager described it this way: "I have to implement plans I didn't design, but if the project fails, I'm responsible.

Project managers are subject to stress due to several different facets of their jobs. This can manifest itself in a variety of ways, such as:

1. *Being tired.* Being tired is a result of being drained of strength and energy, perhaps through physical exertion, boredom, or impatience. The definition here applies more to a short-term, rather than long-term, effect. Typical causes for feeling tired include meetings, report writing, and other forms of document preparation.

2. *Feeling depressed.* Feeling depressed is an emotional condition usually character-ized by discouragement or a feeling of inadequacy. It is usually the result of a situation that is beyond the control or capabilities of the project manager. There are several sources of depression in a project environment: Management or the client considers your report unacceptable, you are unable to get timely resources assigned, the technology is not avail-able, or the constraints of the project are unrealistic and may not be met.

3. *Being physically and emotionally exhausted.* Project managers are both managers and doers. It is quite common for project managers to perform a great deal of the work themselves, either because they consider the assigned personnel unqualified to perform the work or because they are impatient and consider themselves capable of performing the work faster. In addition, project managers often work a great deal of "self-inflicted" over-time. The most common cause of emotional exhaustion is report writing and the prepara-tion of handouts for interchange meetings.

4. *Burned out.* Being burned out is more than just a feeling; it is a condition. Being burned out implies that one is totally exhausted, both physically and emotionally, and that rest, recuperation, or vacation time may not remedy the situation. The most common cause is prolonged overtime, or the need thereof, and an inability to endure or perform under continuous pressure and stress. Burnout can occur almost overnight, often with very little warning. The solution is almost always a change in job assignment, preferably with another company.

5. *Being unhappy.* There are several factors that produce unhappiness in project management. Such factors include highly optimistic planning, unreasonable expectations by management, management cutting resources because of a "buy-in," or simply cus-tomer demands for additional data items. A major source of unhappiness is the frustra-tion caused by having limited authority that is not commensurate with the assigned responsibility.

6. *Feeling trapped.* The most common situation where project managers feel trapped is when they have no control over the assigned resources on the project and feel as though they are at the mercy of the line managers. Employees tend to favor the manager who can offer them the most rewards, and that is usually the line manager. Providing the project manager with some type of direct reward power can remedy the situation.

7. *Feeling worthless.* Feeling worthless implies that one is without worth or merit, that is, valueless. This situation occurs when project managers feel that they are manag-ing projects beneath their dignity. Most project managers look forward to the death of their project right from the onset, and expect their next project to be more important, per-haps twice the cost, and more complex. Unfortunately, there are always situations where one must take a step backwards.

8. *Feeling resentful and disillusioned about people.* This situation occurs most fre-quently in the project manager's dealings (i.e., negotiations) with the line managers.

During the planning stage of a project, line managers often make promises concerning future resource commitments, but renege on their promises during execution. Disillusionment then occurs and can easily develop into serious conflict. Another potential source of these feelings is when line managers appear to be making decisions that are not in the best interest of the project.

9. *Feeling hopeless.* The most common source of hopelessness are R&D projects where the ultimate objective is beyond the reach of the employee or even of the state-of-the-art technology. Hopelessness means showing no signs of a favorable outcome. Hopelessness is more a result of the performance constraint than of time or cost.

10. *Feeling rejected.* Feeling rejected can be the result of a poor working relationship with executives, line managers, or clients. Rejection often occurs when people with authority feel that their options or opinions are better than those of the project manager. Rejection has a demoralizing effect on the project manager because he feels that he is the "president" of the project and the true "champion" of the company.

11. *Feeling anxious.* Almost all project managers have some degree of "tunnel vision," where they look forward to the end of the project, even when the project is in its infancy. This anxious feeling is not only to see the project end, but to see it completed successfully.

Stress is not always negative, however. Without certain amounts of stress, reports would never get written or distributed, deadlines would never be met, and no one would even get to work on time. But stress can be a powerful force resulting in illness and even fatal disease, and must be understood and managed if it is to be controlled and utilized for constructive purposes.

The mind, body, and emotions are not the separate entities they were once thought to be. One affects the other, sometimes in a positive way, and sometimes in a negative way. Stress becomes detrimental when it is prolonged beyond what an individual can comfortably handle. In a project environment, with continually changing requirements, impossible deadlines, and each project being considered as a unique entity in itself, we must ask, How much prolonged stress can a project manager handle comfortably?

The stresses of project management may seem excessive for whatever rewards the position may offer. However, the project manager who is aware of the stresses inherent in the job and knows stress management techniques can face this challenge objectively and make it a rewarding experience.

6.6 STUDYING TIPS FOR THE PMI® PROJECT MANAGEMENT CERTIFICATION EXAM

This section is applicable as a review of the principles to support the knowledge areas and domain groups in the PMBOK® Guide. This chapter addresses:

- Human Resources Management
- Risk Management
- Execution

Understanding the following principles is beneficial if the reader is using this text to study for the PMP® Certification Exam:

- How stress can affect the way that the project manager works with the team
- How stress affects the performance of team members

The following multiple-choice questions will be helpful in reviewing the principles of this chapter:

1. Which of the following leadership styles most frequently creates "additional" time robbers for a project manager?
   A. Telling
   B. Selling
   C. Participating
   D. Delegating

2. Which of the following leadership styles most frequently creates "additional" time robbers for the project team?
   A. Telling
   B. Selling
   C. Participating
   D. Delegating

3. Which of the following time robbers would a project manager most likely want to handle by himself or herself rather than through delegation to equally qualified team members?
   A. Approval of procurement expenditures
   B. Status reporting to a customer
   C. Conflicting directives from the executive sponsor
   D. Earned-value status reporting

ANSWERS

1. A
2. D
3. C

PROBLEMS

6–1 Should time robbers be added to direct labor standards for pricing out work?

6–2 Is it possible for a project manager to improve his time management skills by knowing the "energy cycle" of his people? Can this energy cycle be a function of the hour of the day, day of the week, or whether overtime is required?

CASE STUDY

THE RELUCTANT WORKERS

Tim Aston had changed employers three months ago. His new position was project manager. At first he had stars in his eyes about becoming the best project manager that his company had ever seen. Now, he wasn't sure if project management was worth the effort. He made an appointment to see Phil Davies, director of project management.

Tim Aston: "Phil, I'm a little unhappy about the way things are going. I just can't seem to motivate my people. Every day, at 4:30 P.M., all of my people clean off their desks and go home. I've had people walk out of late afternoon team meetings because they were afraid that they'd miss their car pool. I have to schedule morning team meetings."

Phil Davies: "Look, Tim. You're going to have to realize that in a project environment, people think that they come first and that the project is second. This is a way of life in our organizational form."

Tim Aston: "I've continually asked my people to come to me if they have problems. I find that the people do not think that they need help and, therefore, do not want it. I just can't get my people to communicate more."

Phil Davies: "The average age of our employees is about forty-six. Most of our people have been here for twenty years. They're set in their ways. You're the first person that we've hired in the past three years. Some of our people may just resent seeing a thirty-year-old project manager."

Tim Aston: "I found one guy in the accounting department who has an excellent head on his shoulders. He's very interested in project management. I asked his boss if he'd release him for a position in project management, and his boss just laughed at me, saying something to the effect that as long as that guy is doing a good job for him, he'll never be released for an assignment elsewhere in the company. His boss seems more worried about his personal empire than he does in what's best for the company.

"We had a test scheduled for last week. The customer's top management was planning on flying in for firsthand observations. Two of my people said that they had programmed vacation days coming, and that they would not change, under any conditions. One guy was going fishing and the other guy was planning to spend a few days working with fatherless children in our community. Surely, these guys could change their plans for the test."

Phil Davies: "Many of our people have social responsibilities and outside interests. We encourage social responsibilities and only hope that the outside interests do not interfere with their jobs.

"There's one thing you should understand about our people. With an average age of forty-six, many of our people are at the top of their pay grades and have no place to go. They must look elsewhere for interests. These are the people you have to work with and motivate. Perhaps you should do some reading on human behavior."

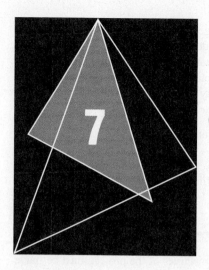

Conflicts

Related Case Studies (from Kerzner/*Project Management Case Studies*, 3rd Edition)	Related Workbook Exercises (from Kerzner/*Project Management Workbook and PMP®/CAPM® Exam Study Guide*, 10th Edition)	PMBOK® Guide, 4th Edition, Reference Section for the PMP® Certification Exam
• Facilities Scheduling at Mayer Manufacturing* • Scheduling the Safety Lab • Telestar International* • The Problem with Priorities	• Multiple Choice Exam	• Human Resource Management

7.0 INTRODUCTION

PMBOK® Guide, 4th Edition
9.4 Manage Project Team
9.4.2.3 Conflict Management

In discussing the project environment, we have purposely avoided discussion of what may be its single most important characteristic: conflicts. Opponents of project management assert that the major reason why many companies avoid changeover to a project management organizational structure is either fear or an inability to handle the resulting conflicts. Conflicts are a way of life in a project structure and can generally occur at any level in the organization, usually as a result of conflicting objectives.

*Case Study also appears at end of chapter.

The project manager has often been described as a conflict manager. In many organizations the project manager continually fights fires and crises evolving from conflicts, and delegates the day-to-day responsibility of running the project to the project team members. Although this is not the best situation, it cannot always be prevented, especially after organizational restructuring or the initiation of projects requiring new resources.

The ability to handle conflicts requires an understanding of why they occur. Asking and answering these four questions may help handle and prevent conflicts.

- What are the project objectives and are they in conflict with other projects?
- Why do conflicts occur?
- How do we resolve conflicts?
- Is there any type of analysis that could identify possible conflicts before they occur?

7.1 OBJECTIVES

Each project must have at least one objective. The objectives of the project must be made known to all project personnel and all managers, at every level of the organization. If this information is not communicated accurately, then it is entirely possible that upper-level managers, project managers, and functional managers may all have a different interpretation of the ultimate objective, a situation that invites conflicts. As an example, company X has been awarded a $100,000 government contract for surveillance of a component that appears to be fatiguing. Top management might view the objective of this project to be discovering the cause of the fatigue and eliminating it in future component production. This might give company X a "jump" on the competition. The division manager might just view it as a means of keeping people employed, with no follow-on possibilities. The department manager can consider the objective as either another job that has to be filled, or a means of establishing new surveillance technology. The department manager, therefore, can staff the necessary positions with any given degree of expertise, depending on the importance and definition of the objective.

Project objectives must be:

- Specific, not general
- Not overly complex
- Measurable, tangible, and verifiable
- Appropriate level, challenging
- Realistic and attainable
- Established within resource bounds
- Consistent with resources available or anticipated
- Consistent with organizational plans, policies, and procedures

Some practitioners use the more simplistic approach of defining an objective by saying that the project's objective must follow the SMART rule, whereby:

- S = specific
- M = measurable
- A = attainable
- R = realistic or relevant
- T = tangible or time bound

Unfortunately, the above characteristics are not always evident, especially if we consider that the project might be unique to the organization in question. As an example, research and development projects sometimes start out general, rather than specific. Research and development objectives are reestablished as time goes on because the initial objective may not be attainable. As an example, company Y believes that they can develop a high-energy rocket-motor propellant. A proposal is submitted to the government, and, after a review period, the contract is awarded. However, as is the case with all R&D projects, there always exists the question of whether the objective is attainable within time, cost, and performance constraints. It might be possible to achieve the initial objective, but at an incredibly high production cost. In this case, the specifications of the propellant (i.e., initial objectives) may be modified so as to align them closer to the available production funds.

Many projects are directed and controlled using a management-by-objective (MBO) approach. The philosophy of management by objectives:

- Is proactive rather than reactive management
- Is results oriented, emphasizing accomplishment
- Focuses on change to improve individual and organizational effectiveness

Management by objectives is a systems approach for aligning project goals with organizational goals, project goals with the goals of other subunits of the organization, and project goals with individual goals. Furthermore, management by objectives can be regarded as a:

- Systems approach to planning and obtaining project results for an organization
- Strategy of meeting individual needs at the same time that project needs are met
- Method of clarifying what each individual and organizational unit's contribution to the project should be

Whether or not MBO is utilized, project objectives must be set.

7.2 THE CONFLICT ENVIRONMENT

In the project environment, conflicts are inevitable. However, as described in Chapter 5, conflicts and their resolution can be planned for. For example, conflicts can easily develop out of a situation where members of a group have a misunderstanding of each other's roles and responsibilities. Through documentation, such as linear responsibility charts, it is possible to establish formal organizational procedures (either at the project level or company-wide). Resolution means collaboration in which people must rely on one another. Without this, mistrust will prevail.

The most common types of conflicts involve:

- Manpower resources
- Equipment and facilities
- Capital expenditures
- Costs
- Technical opinions and trade-offs

- Priorities
- Administrative procedures
- Scheduling
- Responsibilities
- Personality clashes

Each of these conflicts can vary in relative intensity over the life cycle of a project. However, project managers believe that the most frequently occurring conflicts are over schedules but the potentially damaging conflicts can occur over personality clashes. The relative intensity can vary as a function of:

- Getting closer to project constraints
- Having only two constraints instead of three (i.e., time and performance, but not cost)
- The project life cycle itself
- The person with whom the conflict occurs

Sometimes conflict is "meaningful" and produces beneficial results. These meaningful conflicts should be permitted to continue as long as project constraints are not violated and beneficial results are being received. An example of this would be two technical specialists arguing that each has a better way of solving a problem, and each trying to find additional supporting data for his hypothesis.

Conflicts can occur with anyone and over anything. Some people contend that personality conflicts are the most difficult to resolve. Below are several situations. The reader might consider what he or she would do if placed in the situations.

- Two of your functional team members appear to have personality clashes and almost always assume opposite points of view during decision-making. They are both from the same line organization.
- Manufacturing says that they cannot produce the end-item according to engineering specifications.
- R&D quality control and manufacturing operations quality control argue as to who should perform a certain test on an R&D project. R&D postulates that it is their project, and manufacturing argues that it will eventually go into production and that they wish to be involved as early as possible.
- Mr. X is the project manager of a $65 million project of which $1 million is subcontracted out to another company in which Mr. Y is the project manager. Mr. X does not consider Mr. Y as his counterpart and continually communicates with the director of engineering in Mr. Y's company.

Ideally, the project manager should report high enough so that he can get timely assistance in resolving conflicts. Unfortunately, this is easier said than done. Therefore, proj-ect managers must plan for conflict resolution. As examples of this:

- The project manager might wish to concede on a low-intensity conflict if he knows that a high-intensity conflict is expected to occur at a later point in the project.
- Jones Construction Company has recently won a $120 million effort for a local company. The effort includes three separate construction projects, each one

beginning at the same time. Two of the projects are twenty-four months in duration, and the third is thirty-six months. Each project has its own project manager. When resource conflicts occur between the projects, the customer is usually called in.

- Richard is a department manager who must supply resources to four different projects. Although each project has an established priority, the project managers continually argue that departmental resources are not being allocated effectively. Richard now holds a monthly meeting with all four of the project managers and lets them determine how the resources should be allocated.

Many executives feel that the best way of resolving conflicts is by establishing priorities. This may be true as long as priorities are not continually shifted around. As an example, Minnesota Power and Light established priorities as:

- Level 0: no completion date
- Level 1: to be completed on or before a specific date
- Level 2: to be completed in or before a given fiscal quarter
- Level 3: to be completed within a given year

This type of technique will work as long as there are not a large number of projects in any one level.

The most common factors influencing the establishment of project priorities include:

- The technical risks in development
- The risks that the company will incur, financially or competitively
- The nearness of the delivery date and the urgency
- The penalties that can accompany late delivery dates
- The expected savings, profit increase, and return on investment
- The amount of influence that the customer possesses, possibly due to the size of the project
- The impact on other projects or product lines
- The impact on affiliated organizations

The ultimate responsibility for establishing priorities rests with top-level management. Yet even with priority establishment, conflicts still develop. David Wilemon has identified several reasons why conflicts still occur[1]:

- The greater the diversity of disciplinary expertise among the participants of a project team, the greater the potential for conflict to develop among members of the team.
- The lower the project manager's degree of authority, reward, and punishment power over those individuals and organizational units supporting his project, the greater the potential for conflict to develop.

---

1. David L. Wilemon, "Managing Conflict in Temporary Management Situations," *The Journal of Management Studies,* 1973, pp. 282–296.

- The less the specific objectives of a project (cost, schedule, and technical performance) are understood by the project team members, the more likely it is that conflict will develop.
- The greater the role of ambiguity among the participants of a project team, the more likely it is that conflict will develop.
- The greater the agreement on superordinate goals by project team participants, the lower the potential for detrimental conflict.
- The more the members of functional areas perceive that the implementation of a project management system will adversely usurp their traditional roles, the greater the potential for conflict.
- The lower the percent need for interdependence among organizational units supporting a project, the greater the potential for dysfunctional conflict.
- The higher the managerial level within a project or functional area, the more likely it is that conflicts will be based upon deep-seated parochial resentments. By contrast, at the project or task level, it is more likely that cooperation will be facilitated by the task orientation and professionalism that a project requires for completion.

7.3 CONFLICT RESOLUTION

PMBOK® Guide, 4th Edition
9.4.2.3 Conflict Management

Although each project within the company may be inherently different, the company may wish to have the resulting conflicts resolved in the same manner. The four most common methods are:

1. The development of company-wide conflict resolution policies and procedures
2. The establishment of project conflict resolution procedures during the early planning activities
3. The use of hierarchical referral
4. The requirement of direct contact

Many companies have attempted to develop company-wide policies and procedures for conflict resolution, but this method is often doomed to failure because each project and conflict is different. Furthermore, project managers, by virtue of their individuality, and sometimes differing amounts of authority and responsibility, prefer to resolve conflicts in their own fashion.

A second method for resolving conflicts, and one that is often very effective, is to "plan" for conflicts during the planning activities. This can be accomplished through the use of linear responsibility charts. Planning for conflict resolution is similar to the first method except that each project manager can develop his own policies, rules, and procedures.

Hierarchial referral for conflict resolution, in theory, appears as the best method because neither the project manager nor the functional manager will dominate. Under this arrangement, the project and functional managers agree that for a proper balance to exist their common superior must resolve the conflict to protect the company's best interest. Unfortunately, this is not realistic because the common superior cannot be expected to continually resolve lower-level conflicts and it gives the impression that the functional and project managers cannot resolve their own problems.

The last method is direct contact in which conflicting parties meet face-to-face and resolve their disagreement. Unfortunately, this method does not always work and, if continually stressed, can result in conditions where individuals will either suppress the identification of problems or develop new ones during confrontation.

Many conflicts can be either reduced or eliminated by constant communication of the project objectives to the team members. This continual repetition may prevent individuals from going too far in the wrong direction.

7.4 UNDERSTANDING SUPERIOR, SUBORDINATE, AND FUNCTIONAL CONFLICTS[2]

PMBOK® Guide, 4th Edition
9.4.2.3 Conflict Management

In order for the project manager to be effective, he must understand how to work with the various employees who interface with the project. These employees include upper-level management, subordinate project team members, and functional personnel. Quite often, the project manager must demonstrate an ability for continuous adaptability by creating a different working environment with each group of employees. The need for this was shown in the previous section by the fact that the relative intensity of conflicts can vary in the life cycle of a project.

The type and intensity of conflicts can also vary with the type of employee, as shown in Figure 7–1. Both conflict causes and sources are rated according to relative conflict intensity. The data in Figure 7–1 were obtained for a 75 percent confidence level.

In the previous section we discussed the basic resolution modes for handling conflicts. The specific mode that a project manager will use might easily depend on whom the conflict is with, as shown in Figure 7–2. The data in Figure 7–2 do not necessarily show the modes that project managers would prefer, but rather identify the modes that will increase or decrease the potential conflict intensity. For example, although project managers consider, in general, that withdrawal is their least favorite mode, it can be used quite effectively with functional managers. In dealing with superiors, project managers would rather be ready for an immediate compromise than for face-to-face confrontation that could favor upper-level management.

Figure 7–3 identifies the various influence styles that project managers find effective in helping to reduce potential conflicts. Penalty power, authority, and expertise are considered as strongly unfavorable associations with respect to low conflicts. As expected, work challenge and promotions (if the project manager has the authority) are strongly favorable.

---

2. The majority of this section, including the figures, was adapted from *Seminar in Project Management Workbook*, © 1977 by Hans J. Thamhain. Reproduced by permission of Dr. Hans J. Thamhain.

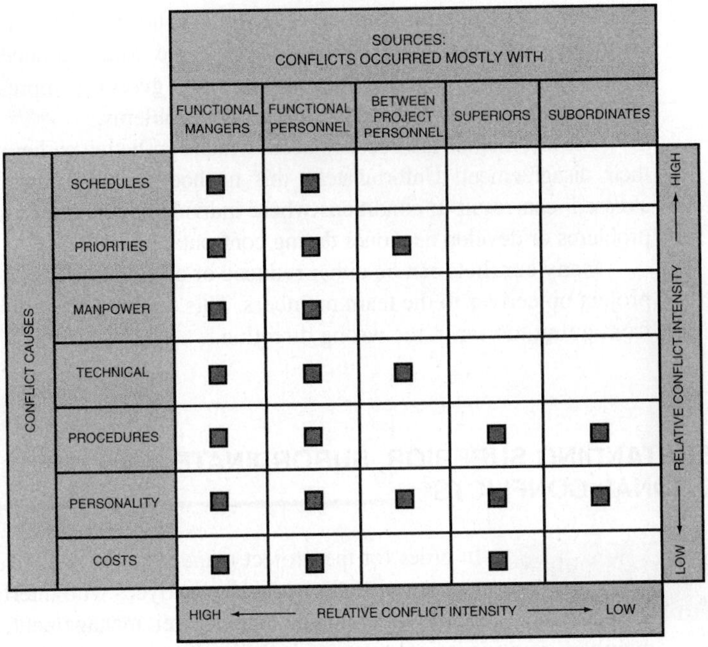

FIGURE 7–1. Relationship between conflict causes and sources.

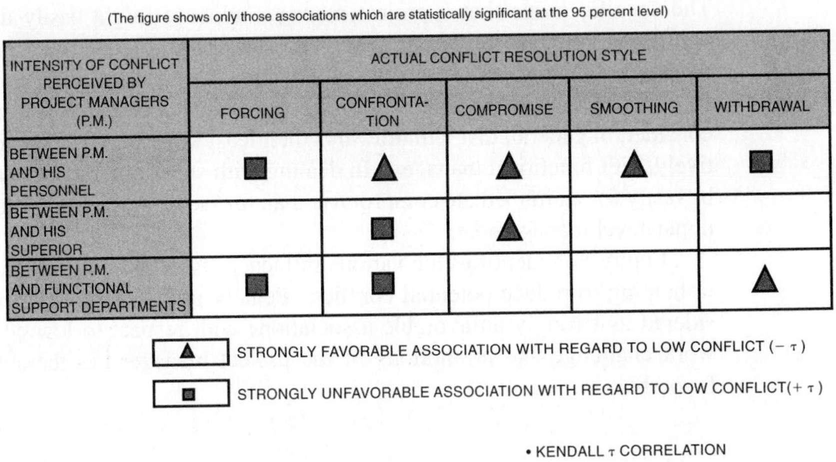

(The figure shows only those associations which are statistically significant at the 95 percent level)

▲ STRONGLY FAVORABLE ASSOCIATION WITH REGARD TO LOW CONFLICT (− τ)

■ STRONGLY UNFAVORABLE ASSOCIATION WITH REGARD TO LOW CONFLICT(+ τ)

• KENDALL τ CORRELATION

FIGURE 7–2. Association between perceived intensity of conflict and mode of conflict resolution.

(The figure shows only those associated which are statistically significant at the 95 percent level)

INTENSITY OF CONFLICT PERCEIVED BY PROJECT MANAGER (P.M.)	INFLUENCE METHODS AS PERCEIVED BY PROJECT MANAGERS						
	EXPERTISE	AUTHORITY	WORK CHALLENGE	FRIENDSHIP	PROMOTION	SALARY	PENALTY
BETWEEN P.M. AND HIS PERSONNEL	■	■	▲		▲		■
BETWEEN P.M. AND HIS SUPERIOR			▲				■
BETWEEN P.M. AND FUNCTIONAL SUPPORT DEPARTMENTS		■					■

▲ STRONGLY FAVORABLE ASSOCIATION WITH REGARD TO LOW CONFLICT ($-\tau$)

■ STRONGLY UNFAVORABLE ASSOCIATION WITH REGARD TO LOW CONFLICT($+\tau$)

• KENDALL τ CORRELATION

FIGURE 7–3. Association between influence methods of project managers and their perceived conflict intensity.

7.5 THE MANAGEMENT OF CONFLICTS[3]

PMBOK® Guide, 4th Edition
9.4.2.3 Conflict Management

Good project managers realize that conflicts are inevitable, but that good procedures or techniques can help resolve them. Once a conflict occurs, the project manager must:

- Study the problem and collect all available information
- Develop a situational approach or methodology
- Set the appropriate atmosphere or climate

If a confrontation meeting is necessary between conflicting parties, then the project manager should be aware of the logical steps and sequence of events that should be taken. These include:

- Setting the climate: establishing a willingness to participate
- Analyzing the images: how do you see yourself and others, and how do they see you?
- Collecting the information: getting feelings out in the open
- Defining the problem: defining and clarifying all positions
- Sharing the information: making the information available to all
- Setting the appropriate priorities: developing working sessions for setting priorities and timetables
- Organizing the group: forming cross-functional problem-solving groups

3. See note 2.

- Problem-solving: obtaining cross-functional involvement, securing commitments, and setting the priorities and timetable
- Developing the action plan: getting commitment
- Implementing the work: taking action on the plan
- Following up: obtaining feedback on the implementation for the action plan

The project manager or team leader should also understand conflict minimization procedures. These include:

- Pausing and thinking before reacting
- Building trust
- Trying to understand the conflict motives
- Keeping the meeting under control
- Listening to all involved parties
- Maintaining a give-and-take attitude
- Educating others tactfully on your views
- Being willing to say when you were wrong
- Not acting as a superman and leveling the discussion only once in a while

Thus, the effective manager, in conflict problem-solving situations:

- Knows the organization
- Listens with understanding rather than evaluation
- Clarifies the nature of the conflict
- Understands the feelings of others
- Suggests the procedures for resolving differences
- Maintains relationships with disputing parties
- Facilitates the communications process
- Seeks resolutions

7.6 CONFLICT RESOLUTION MODES

PMBOK® Guide, 4th Edition
9.4.2.3 Conflict Management

The management of conflicts places the project manager in the precarious situation of having to select a conflict resolution mode (previously defined in Section 7.4). Based upon the situation, the type of conflict, and whom the conflict is with, any of these modes could be justified.

Confronting (or Collaborating)

With this approach, the conflicting parties meet face-to-face and try to work through their disagreements. This approach should focus more on solving the problem and less on being combative. This approach is collaboration and integration where both parties need to win. This method should be used:

- When you and the conflicting party can both get at least what you wanted and maybe more
- To reduce cost

- To create a common power base
- To attack a common foe
- When skills are complementary
- When there is enough time
- When there is trust
- When you have confidence in the other person's ability
- When the ultimate objective is to learn

Compromising

To compromise is to bargain or to search for solutions so both parties leave with some degree of satisfaction. Compromising is often the result of confrontation. Some people argue that compromise is a "give and take" approach, which leads to a "win-win" position. Others argue that compromise is a "lose-lose" position, since neither party gets everything he/she wants or needs. Compromise should be used:

- When both parties need to be winners
- When you can't win
- When others are as strong as you are
- When you haven't time to win
- To maintain your relationship with your opponent
- When you are not sure you are right
- When you get nothing if you don't
- When stakes are moderate
- To avoid giving the impression of "fighting"

Smoothing (or Accommodating)

This approach is an attempt to reduce the emotions that exist in a conflict. This is accomplished by emphasizing areas of agreement and de-emphasizing areas of disagreement. An example of smoothing would be to tell someone, "We have agreed on three of the five points and there is no reason why we cannot agree on the last two points." Smoothing does not necessarily resolve a conflict, but tries to convince both parties to remain at the bargaining table because a solution is possible. In smoothing, one may sacrifice one's own goals in order to satisfy the needs of the other party. Smoothing should be used:

- To reach an overarching goal
- To create obligation for a trade-off at a later date
- When the stakes are low
- When liability is limited
- To maintain harmony
- When any solution will be adequate
- To create goodwill (be magnanimous)
- When you'll lose anyway
- To gain time

Forcing (or Competing, Being Uncooperative, Being Assertive) This is what happens when one party tries to impose the solution on the other party. Conflict resolution works best when resolution is achieved at the lowest possible levels. The higher up the conflict goes, the greater the tendency for the conflict to be forced, with the result being a "win-lose" situation in which one party wins at the expense of the other. Forcing should be used:

- When you are right
- When a do-or-die situation exists
- When stakes are high
- When important principles are at stake
- When you are stronger (never start a battle you can't win)
- To gain status or to gain power
- In short-term, one-shot deals
- When the relationship is unimportant
- When it's understood that a game is being played
- When a quick decision must be made

Avoiding (or Withdrawing) Avoidance is often regarded as a temporary solution to a problem. The problem and the resulting conflict can come up again and again. Some people view avoiding as cowardice and an unwillingness to be responsive to a situation. Avoiding should be used:

- When you can't win
- When the stakes are low
- When the stakes are high, but you are not ready yet
- To gain time
- To unnerve your opponent
- To preserve neutrality or reputation
- When you think the problem will go away
- When you win by delay

7.7 STUDYING TIPS FOR THE PMI® PROJECT MANAGEMENT CERTIFICATION EXAM

This section is applicable as a review of the principles to support the knowledge areas and domain groups in the PMBOK® Guide. This chapter addresses:

- Human Resources Management
- Execution

Understanding the following principles is beneficial if the reader is using this text to study for the PMP® Certification Exam:

- Components of an objective
- What is meant by a SMART criteria for an objective
- Different types of conflicts that can occur in a project environment
- Different conflict resolution modes and when each one should be used

The following multiple-choice questions will be helpful in reviewing the principles of this chapter:

1. When talking about SMART objectives, the "S" stands for:
   A. Satisfactory
   B. Static
   C. Specific
   D. Standard

2. When talking about SMART objectives, the "A" stands for:
   A. Accurate
   B. Acute
   C. Attainable
   D. Able

3. Project managers believe that the most commonly occurring conflict is:
   A. Priorities
   B. Schedules
   C. Personalities
   D. Resources

4. The conflict that generally is the most damaging to the project when it occurs is:
   A. Priorities
   B. Schedules
   C. Personalities
   D. Resources

5. The most commonly preferred conflict resolution mode for project managers is:
   A. Compromise
   B. Confrontation
   C. Smoothing
   D. Withdrawal

6. Which conflict resolution mode is equivalent to problem-solving?
   A. Compromise
   B. Confrontation
   C. Smoothing
   D. Withdrawal

7. Which conflict resolution mode avoids a conflict temporarily rather than solving it?
   A. Compromise
   B. Confrontation
   C. Smoothing
   D. Withdrawal

ANSWERS

1. C
2. C
3. B
4. C
5. B
6. B
7. D

PROBLEMS

7–1 Is it possible to establish formal organizational procedures (either at the project level or company-wide) for the resolution of conflicts? If a procedure is established, what can go wrong?

7–2 Under what conditions would a conflict result between members of a group over misunderstandings of each other's roles?

7–3 Is it possible to have a situation in which conflicts are not effectively controlled, and yet have a decision-making process that is not lengthy or cumbersome?

7–4 If conflicts develop into a situation where mistrust prevails, would you expect activity documentation to increase or decrease? Why?

7–5 If a situation occurs that can develop into meaningful conflict, should the project manager let the conflict continue as long as it produces beneficial contributions, or should he try to resolve it as soon as possible?

7–6 Consider the following remarks made by David L. Wilemon ("Managing Conflict in Temporary Management Situations," *Journal of Management Studies,* October 1973, p. 296):

> The value of the conflict produced depends upon the effectiveness of the project manager in promoting beneficial conflict while concomitantly minimizing its potential dysfunctional aspects. A good project manager needs a "sixth sense" to indicate when conflict is desirable, what kind of conflict will be useful, and how much conflict is optimal for a given situation. In the final analysis he has the sole responsibility for his project and how conflict will impact the success or failure of his project.

Based upon these remarks, would your answer to Problem 7–5 change?

7–7 Mr. X is the project manager of a $65 million project of which $1 million is subcontracted out to another company in which Mr. Y is project manager. Unfortunately, Mr. X does not consider Mr. Y as his counterpart and continually communicates with the director of engineering in Mr. Y's company. What type of conflict is that, and how should it be resolved?

7–8 Contract negotiations can easily develop into conflicts. During a disagreement, the vice president of company A ordered his director of finance, the contract negotiator, to break off contract negotiations with company B because the contract negotiator of company B did not report directly to a vice president. How can this situation be resolved?

7–9 For each part below there are two statements; one represents the traditional view and the other the project organizational view. Identify each one.

    a. Conflict should be avoided; conflict is part of change and is therefore inevitable.
    b. Conflict is the result of troublemakers and egoists; conflict is determined by the structure of the system and the relationship among components.
    c. Conflict may be beneficial; conflict is bad.

7–10 Using the modes for conflict resolution defined in Section 7.6, which would be strongly favorable and strongly unfavorable for resolving conflicts between:

    a. Project manager and his project office personnel?
    b. Project manager and the functional support departments?
    c. Project manager and his superiors?
    d. Project manager and other project managers?

7–11 Which influence methods should increase and which should decrease the opportunities for conflict between the following:

- Project manager and his project office personnel?
- Project manager and the functional support departments?
- Project manager and his superiors?
- Project manager and other project managers?

7–12 Would you agree or disagree with the statement that "Conflict resolution through collaboration needs trust; people must rely on one another."

7–13 Davis and Lawrence (*Matrix,* © 1977. Adapted by permission of Pearson Education Inc., Upper Saddle River, New Jersey) identify several situations common to the matrix that can easily develop into conflicts. For each situation, what would be the recommended cure?

    a. Compatible and incompatible personnel must work together
    b. Power struggles break the balance of power
    c. Anarchy
    d. Groupitis (people confuse matrix behavior with group decision-making)
    e. A collapse during economic crunch
    f. Decision strangulation processes
    g. Forcing the matrix organization to the lower organizational levels
    h. Navel-gazing (spending time ironing out internal disputes instead of developing better working relationships with the customer)

7–14 Determine the best conflict resolution mode for each of the following situations:

    a. Two of your functional team members appear to have personality clashes and almost always assume opposite points of view during decision-making.
    b. R&D quality control and manufacturing operations quality control continually argue as to who should perform testing on an R&D project. R&D postulates that it's their project, and manufacturing argues that it will eventually go into production and that they wish to be involved as early as possible.

c. Two functional department managers continually argue as to who should perform a certain test. You know that this situation exists, and that the department managers are trying to work it out themselves, often with great pain. However, you are not sure that they will be able to resolve the problem themselves.

7–15 Forcing a confrontation to take place assures that action will be taken. Is it possible that, by using force, a lack of trust among the participants will develop?

7–16 With regard to conflict resolution, should it matter to whom in the organization the project manager reports?

7–17 One of the most common conflicts in an organization occurs with raw materials and finished goods. Why would finance/accounting, marketing/sales, and manufacturing have disagreements?

7–18 Explain how the relative intensity of a conflict can vary as a function of:

a. Getting closer to the actual constraints
b. Having only two constraints instead of three (i.e., time and performance, but not cost)
c. The project life cycle
d. The person with whom the conflict occurs

7–19 The conflicts shown in Figure 7–1 are given relative intensities as perceived in project-driven organizations. Would this list be arranged differently for non–project-driven organizations?

7–20 Consider the responses made by the project managers in Figures 7–1 through 7–3. Which of their choices do you agree with, and which do you disagree with? Justify your answers.

7–21 As a good project manager, you try to plan for conflict avoidance. You now have a low-intensity conflict with a functional manager and, as in the past, handle the conflict with confrontation. If you knew that there would be a high-intensity conflict shortly thereafter, would you be willing to use the withdrawal mode for the low-intensity conflict in order to lay the groundwork for the high-intensity conflict?

7–22 Jones Construction Company has recently won a $120 million effort for a local company. The effort includes three separate construction projects, each one beginning at the same time. Two of the projects are eighteen months in duration and the third one is thirty months. Each project has its own project manager. How do we resolve conflicts when each project may have a different priority but they are all for the same customer?

7–23 Several years ago, Minnesota Power and Light established priorities as follows:

Level 0: no priority
Level 1: to be completed on or before a specific date
Level 2: to be completed in or before a given fiscal quarter
Level 3: to be completed within a given year

How do you feel about this system of establishing priorities?

7–24 Richard is a department manager who must supply resources to four different projects. Although each project has an established priority, the project managers continually argue that departmental resources are not being allocated effectively. Richard has decided to have a monthly group meeting with all four of the project managers and to let them determine how the resources should be allocated. Can this technique work? If so, under what conditions?

CASE STUDIES

FACILITIES SCHEDULING AT MAYER MANUFACTURING

Eddie Turner was elated with the good news that he was being promoted to section supervisor in charge of scheduling all activities in the new engineering research laboratory. The new laboratory was a necessity for Mayer Manufacturing. The engineering, manufacturing, and quality control directorates were all in desperate need of a new testing facility. Upper-level management felt that this new facility would alleviate many of the problems that previously existed.

The new organizational structure (as shown in Exhibit 7–1) required a change in policy over use of the laboratory. The new section supervisor, on approval from his department manager, would have full authority for establishing priorities for the use of the new facility. The new policy change was a necessity because upper-level management felt that there would be inevitable conflict between manufacturing, engineering, and quality control.

After one month of operations, Eddie Turner was finding his job impossible, so Eddie has a meeting with Gary Whitehead, his department manager.

Eddie: "I'm having a hell of a time trying to satisfy all of the department managers. If I give engineering prime-time use of the facility, then quality control and manufacturing say that I'm playing favorites. Imagine that! Even my own people say that I'm playing favorites with other directorates. I just can't satisfy everyone."

Gary: "Well, Eddie, you know that this problem comes with the job. You'll get the job done."

Eddie: "The problem is that I'm a section supervisor and have to work with department managers. These department managers look down on me like I'm their servant. If I were a department manager, then they'd show me some respect. What I'm really trying to say is that I would like you to send out the weekly memos to these department managers telling them of the new priorities. They wouldn't argue with you like they do with me. I can supply you with all the necessary information. All you'll have to do is to sign your name."

Exhibit 7–1. Mayer Manufacturing organizational structure

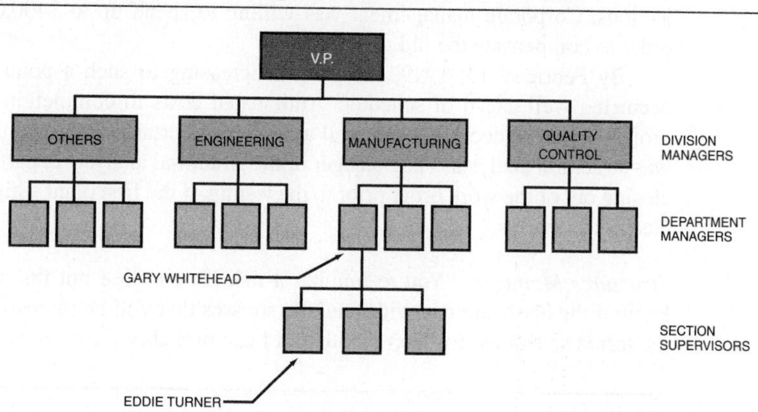

Gary: "Determining the priorities and scheduling the facilities is your job, not mine. This is a new position and I want you to handle it. I know you can because I selected you. I do not intend to interfere."

During the next two weeks, the conflicts got progressively worse. Eddie felt that he was unable to cope with the situation by himself. The department managers did not respect the authority delegated to him by his superiors. For the next two weeks, Eddie sent memos to Gary in the early part of the week asking whether Gary agreed with the priority list. There was no response to the two memos. Eddie then met with Gary to discuss the deteriorating situation.

Eddie: "Gary, I've sent you two memos to see if I'm doing anything wrong in establishing the weekly priorities and schedules. Did you get my memos?"

Gary: "Yes, I received your memos. But as I told you before, I have enough problems to worry about without doing your job for you. If you can't handle the work let me know and I'll find someone who can."

Eddie returned to his desk and contemplated his situation. Finally, he made a decision. Next week he was going to put a signature block under his for Gary to sign, with carbon copies for all division managers. "Now, let's see what happens," remarked Eddie.

TELESTAR INTERNATIONAL*

On November 15, 1998, the Department of Energy Resources awarded Telestar a $475,000 contract for the developing and testing of two waste treatment plants. Telestar had spent the better part of the last two years developing waste treatment technology under its own R&D activities. This new contract would give Telestar the opportunity to "break into a new field"—that of waste treatment.

The contract was negotiated at a firm-fixed price. Any cost overruns would have to be incurred by Telestar. The original bid was priced out at $847,000. Telestar's management, however, wanted to win this one. The decision was made that Telestar would "buy in" at $475,000 so that they could at least get their foot into the new marketplace.

The original estimate of $847,000 was very "rough" because Telestar did not have any good man-hour standards, in the area of waste treatment, on which to base their man-hour projections. Corporate management was willing to spend up to $400,000 of their own funds in order to compensate the bid of $475,000.

By February 15, 1999, costs were increasing to such a point where overrun would be occurring well ahead of schedule. Anticipated costs to completion were now $943,000. The project manager decided to stop all activities in certain functional departments, one of which was structural analysis. The manager of the structural analysis department strongly opposed the closing out of the work order prior to the testing of the first plant's high-pressure pneumatic and electrical systems.

Structures Manager: "You're running a risk if you close out this work order. How will you know if the hardware can withstand the stresses that will be imposed during the test? After all, the test is scheduled for next month and I can probably finish the analysis by then."

*Revised, 2008.

Project Manager: "I understand your concern, but I cannot risk a cost overrun. My boss expects me to do the work within cost. The plant design is similar to one that we have tested before, without any structural problems being detected. On this basis I consider your analysis unnecessary."

Structures Manager: "Just because two plants are similar does not mean that they will be identical in performance. There can be major structural deficiencies."

Project Manager: "I guess the risk is mine."

Structures Manager: "Yes, but I get concerned when a failure can reflect on the integrity of my department. You know, we're performing on schedule and within the time and money budgeted. You're setting a bad example by cutting off our budget without any real justification."

Project Manager: "I understand your concern, but we must pull out all the stops when overrun costs are inevitable."

Structures Manager: "There's no question in my mind that this analysis should be completed. However, I'm not going to complete it on my overhead budget. I'll reassign my people tomorrow. Incidentally, you had better be careful; my people are not very happy to work for a project that can be canceled immediately. I may have trouble getting volunteers next time."

Project Manager: "Well, I'm sure you'll be able to adequately handle any future work. I'll report to my boss that I have issued a work stoppage order to your department."

During the next month's test, the plant exploded. Postanalysis indicated that the failure was due to a structural deficiency.

    a. Who is at fault?
    b. Should the structures manager have been dedicated enough to continue the work on his own?
    c. Can a functional manager, who considers his organization as strictly support, still be dedicated to total project success?

HANDLING CONFLICT IN PROJECT MANAGEMENT

The next several pages contain a six-part case study in conflict management. Read the instructions carefully on how to keep score and use the boxes in the table on page 314 as the worksheet for recording your choice and the group's choice; after the case study has been completed, your instructor will provide you with the proper grading system for recording your scores.

Part 1: Facing the Conflict   As part of his first official duties, the new department manager informs you by memo that he has changed his input and output requirements for the MIS project (on which you are the project manager) because of several complaints by his departmental employees. This is contradictory to the project plan that you developed with the previous manager and are currently working toward. The department manager states that he has already discussed this with the vice president and general manager, a man to whom both of you report, and feels that the former department manager made a poor decision and did not get sufficient input from the employees who would be using the system as to the best system specifications. You telephone him and try to convince him to hold off on his request for change until a later time, but he refuses.

Changing the input–output requirements at this point in time will require a major revision and will set back total system implementation by three weeks. This will also affect other department managers who expect to see this system operational according to the original schedule. You can explain this to your superiors, but the increased project costs will be hard to absorb. The potential cost overrun might be difficult to explain at a later date.

At this point you are somewhat unhappy with yourself at having been on the search committee that found this department manager and especially at having recommended him for this position. You know that something must be done, and the following are your alternatives:

A. You can remind the department manager that you were on the search committee that recommended him and then ask him to return the favor, since he "owes you one."
B. You can tell the department manager that you will form a new search committee to replace him if he doesn't change his position.
C. You can take a tranquilizer and then ask your people to try to perform the additional work within the original time and cost constraints.
D. You can go to the vice president and general manager and request that the former requirements be adhered to, at least temporarily.
E. You can send a memo to the department manager explaining your problem and asking him to help you find a solution.
F. You can tell the department manager that your people cannot handle the request and his people will have to find alternate ways of solving their problems.
G. You can send a memo to the department manager requesting an appointment, at his earliest convenience, to help you resolve your problem.
H. You can go to the department manager's office later that afternoon and continue the discussion further.
I. You can send the department manager a memo telling him that you have decided to use the old requirements but will honor his request at a later time.

Line	Part	Personal		Group	
		Choice	Score	Choice	Score
1	1. Facing the Conflict				
2	2. Understanding Emotions	////		////	
3	3. Establishing Communications				
4	4. Conflict Resolution	////		////	
5	5. Understanding Your Choices				
6	6. Interpersonal Influences				
	TOTAL	////		////	

Although other alternatives exist, assume that these are the only ones open to you at the moment. Without discussing the answer with your group, record the letter representing your choice in the appropriate space on line 1 of the worksheet under "Personal."

As soon as all of your group have finished, discuss the problem as a group and determine that alternative that the group considers to be best. Record this answer on line 1 of the worksheet under "Group." Allow ten minutes for this part.

Part 2: Understanding Emotions

Never having worked with this department manager before, you try to predict what his reactions will be when confronted with the problem. Obviously, he can react in a variety of ways:

A. He can *accept* your solution in its entirety without asking any questions.
B. He can discuss some sort of justification in order to *defend* his position.
C. He can become extremely annoyed with having to discuss the problem again and demonstrate *hostility*.
D. He can demonstrate a willingness to *cooperate* with you in resolving the problem.
E. He can avoid making any decision at this time by *withdrawing* from the discussion.

	Your Choice					Group Choice				
	Acc.	Def.	Host.	Coop.	With.	Acc.	Def.	Host.	Coop.	With.
A. I've given my answer. See the general manager if you're not happy.										
B. I understand your problem. Let's do it your way.										
C. I understand your problem, but I'm doing what is best for my department.										
D. Let's discuss the problem. Perhaps there are alternatives.										
E. Let me explain to you why we need the new requirements.										
F. See my section supervisors. It was their recommendation.										
G. New managers are supposed to come up with new and better ways, aren't they?										

In the table above are several possible statements that could be made by the department manager when confronted with the problem. Without discussion with your group, place a check mark beside the appropriate emotion that could describe this statement. When each member of the group has completed his choice, determine the group choice. Numerical values will be assigned to your choices in the discussion that follows. Do not mark the worksheet at this time. Allow ten minutes for this part.

Part 3: Establishing Communications

Unhappy over the department manager's memo and the resulting follow-up phone conversation, you decide to walk in on the department manager. You tell him that you will have a problem trying to honor his request. He tells you that he is too busy with his own problems of restructuring his department and that your schedule and cost problems are of no concern to him at this time. You storm out of his office, leaving him with the impression that his actions and remarks are not in the best interest of either the project or the company.

The department manager's actions do not, of course, appear to be those of a dedicated manager. He should be more concerned about what's in the best interest of the company. As you contemplate the situation, you wonder if you could have received a better response from him had you approached him differently. In other words, what is your best approach to opening up communications between you and the department manager? From the list of alternatives shown below, and working alone, select the alternative that best represents how you would handle this situation. When all members of the group have selected their personal choices, repeat the process and make a group choice. Record your personal and group choices on line 3 of the worksheet. Allow ten minutes for this part.

A. Comply with the request and document all results so that you will be able to defend yourself at a later date in order to show that the department manager should be held accountable.

B. Immediately send him a memo reiterating your position and tell him that at a later time you will reconsider his new requirements. Tell him that time is of utmost importance, and you need an immediate response if he is displeased.

C. Send him a memo stating that you are holding him accountable for all cost overruns and schedule delays.

D. Send him a memo stating you are considering his request and that you plan to see him again at a later date to discuss changing the requirements.

E. See him as soon as possible. Tell him that he need not apologize for his remarks and actions, and that you have reconsidered your position and wish to discuss it with him.

F. Delay talking to him for a few days in hopes that he will cool off sufficiently and then see him in hopes that you can reopen the discussions.

G. Wait a day or so for everyone to cool off and then try to see him through an appointment; apologize for losing your temper, and ask him if he would like to help you resolve the problem.

Part 4: Conflict Resolution Modes

Having never worked with this manager before, you are unsure about which conflict resolution mode would work best. You decide to wait a few days and then set up an appointment with the department manager without stating what subject matter will be discussed. You then try to determine what conflict resolution mode appears to be dominant based on the opening remarks of the department manager. Neglecting the fact that your conversation with the department manager might already be considered as confrontation, for each statement shown below, select the conflict resolution mode that the *department manager* appears to prefer. After each member of the group has recorded his personal choices in the table below, determine the group choices. Numerical values will be attached to your answers at a later time. Allow ten minutes for this part.

    A. *Withdrawal* is retreating from a potential conflict.
    B. *Smoothing* is emphasizing areas of agreement and de-emphasizing areas of disagreement.
    C. *Compromising* is the willingness to give and take.
    D. *Forcing* is directing the resolution in one direction or another, a win-or-lose position.
    E. *Confrontation* is a face-to-face meeting to resolve the conflict.

	Personal Choice					Group Choice				
	With.	Smooth.	Comp.	Forc.	Conf.	With.	Smooth.	Comp.	Forc.	Conf.
A. The requirements are my decision, and we're doing it my way.										
B. I've thought about it and you're right. We'll do it your way.										
C. Let's discuss the problem. Perhaps there are alternatives.										
D. Let me again explain why we need the new requirements.										
E. See my section supervisors; they're handling it now.										
F. I've looked over the problem and I might be able to ease up on some of the requirements.										

Part 5: Understanding Your Choices

Assume that the department manager has refused to see you again to discuss the new requirements. Time is running out, and you would like to make a decision before the costs and schedules get out of hand. From the list below, select your personal choice and then, after each group member is finished, find a group choice.

    A.   Disregard the new requirements, since they weren't part of the original project plan.
    B.   Adhere to the new requirements, and absorb the increased costs and delays.
    C.   Ask the vice president and general manager to step in and make the final decision.
    D.   Ask the other department managers who may realize a schedule delay to try to convince this department manager to ease his request or even delay it.

Record your answer on line 5 of the worksheet. Allow five minutes for this part.

Part 6: Interpersonal Influences

Assume that upper-level management resolves the conflict in your favor. In order to complete the original work requirements you will need support from this department manager's organization. Unfortunately, you are not sure as to which type of interpersonal influence to use. Although you are considered as an expert in your field, you fear that this manager's functional employees may have a strong allegiance to the department manager and may not want to adhere to your requests. Which of the following interpersonal influence styles would be best under the given set of conditions?

A. You threaten the employees with penalty power by telling them that you will turn in a bad performance report to their department manager.

B. You can use reward power and promise the employees a good evaluation, possible promotion, and increased responsibilities on your next project.

C. You can continue your technique of trying to convince the functional personnel to do your bidding because you are the expert in the field.

D. You can try to motivate the employees to do a good job by convincing them that the work is challenging.

E. You can make sure that they understand that your authority has been delegated to you by the vice president and general manager and that they must do what you say.

F. You can try to build up friendships and off-work relationships with these people and rely on referent power.

Record your personal and group choices on line 6 of the worksheet. Allow ten minutes for completion of this part.

The solution to this exercise appears in Appendix A.

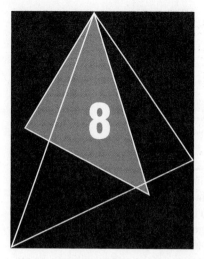

Special Topics

Related Case Studies (from Kerzner/*Project Management Case Studies,* 3rd Edition)	Related Workbook Exercises (from Kerzner/*Project Management Workbook and PMP®/CAPM® Exam Study Guide,* 10th Edition)	PMBOK® Guide, 4th Edition, Reference Section for the PMP® Certification Exam
• American Electronics International • The Tylenol Tragedies • Photolite Corporation (A) • Photolite Corporation (B) • Photolite Corporation (C) • Photolite Corporation (D) • First Security Bank of Cleveland • Jackson Industries	• The Potential Problem Audit • The Situational Audit • Multiple Choice Exam	• Integration Management • Human Resource Management • Project Management Roles and Responsibilities

8.0 INTRODUCTION

There are several situations or special topics that deserve attention. These include:

- Performance measurement
- Compensation and rewards
- Managing small projects

- Managing mega projects
- Morality, ethics and the corporate culture
- Internal partnerships
- External partnerships
- Training and education
- Integrated project teams
- Virtual teams

8.1 PERFORMANCE MEASUREMENT

PMBOK® Guide, 4th Edition
9.3.2.6 Recognition and Rewards
9.4.2 Manage Project Team

A good project manager will make it immediately clear to all new functional employees that if they perform well in the project, then he (the proj-ect manager) will inform the functional manager of their progress and achievements. This assumes that the functional manager is not providing close supervision over the functional employees and is, instead, passing on some of the responsibility to the project manager—a common situation in project management organization structures.

Many good projects as well as project management structures have failed because of the inability of the system to evaluate properly the functional employee's performance. In a project management structure, there are basically six ways that a functional employee can be evaluated on a project:

- *The project manager prepares a written, confidential evaluation and gives it to the functional manager.* The functional manager will evaluate the validity of the project manager's comments and prepare his own evaluation. Only the line manager's evaluation is shown to the employee. The use of confidential forms is not preferred because it may be contrary to government regulations and it does not provide the necessary feedback for an employee to improve.
- *The project manager prepares a nonconfidential evaluation and gives it to the functional manager.* The functional manager prepares his own evaluation form and shows both evaluations to the functional employee. This is the technique preferred by most project and functional managers. However, there are several major difficulties with this technique. If the functional employee is an average or below-average worker, and if this employee is still to be assigned to this project after his evaluation, then the project manager might rate the employee as above average simply to prevent any sabotage or bad feelings downstream. In this situation, the functional manager might want a confidential evaluation instead, knowing that the functional employee will see both evaluation forms. Functional employees tend to blame the project manager if they receive a below-average merit pay increase, but give credit to the functional manager if the increase is above average. The best bet here is for the project manager periodically to tell the functional employees how well they are doing, and to give them an honest appraisal. Several companies that use this technique allow the project manager to show the form to the line manager first (to avoid conflict later) and then show it to the employee.
- *The project manager provides the functional manager with an oral evaluation of the employee's performance.* Although this technique is commonly used, most

functional managers prefer documentation on employee progress. Again, lack of feedback may prevent the employee from improving.

- *The functional manager makes the entire evaluation without any input from the project manager.* In order for this technique to be effective, the functional manager must have sufficient time to supervise each subordinate's performance on a continual basis. Unfortunately, most functional managers do not have this luxury because of their broad span of control and must therefore rely heavily on the project manager's input.

- *The project manager makes the entire evaluation for the functional manager.* This technique can work if the functional employee spends 100 percent of his time on one project, or if he is physically located at a remote site where he cannot be observed by his functional manager.

- *All project and functional managers jointly evaluate all project functional employees at the same time.* This technique should be limited to small companies with fewer than fifty or so employees; otherwise the evaluation process might be time-consuming for key personnel. A bad evaluation will be known by everyone.

Evaluation forms can be filled out either when the employee is up for evaluation or after the project is completed. If it is to be filled out when the employee is eligible for promotion or a merit increase, then the project manager should be willing to give an *honest*

PERFORMANCE FACTORS	EXCELLENT (1 OUT OF 15)	VERY GOOD (3 OUT OF 15)	GOOD (8 OUT OF 15)	FAIR (2 OUT OF 15)	UNSATISFACTORY (1 OUT OF 15)
	FAR EXCEEDS JOB REQUIREMENTS	EXCEEDS JOB REQUIREMENTS	MEETS JOB REQUIREMENTS	NEEDS SOME IMPROVEMENT	DOES NOT MEET MINIMUM STANDARDS
QUALITY	LEAPS TALL BUILDINGS WITH A SINGLE BOUND	MUST TAKE RUNNING START TO LEAP OVER TALL BUILDING	CAN ONLY LEAP OVER A SHORT BUILDING OR MEDIUM ONE WITHOUT SPIRES	CRASHES INTO BUILDING	CANNOT RECOGNIZE BUILDINGS
TIMELINESS	IS FASTER THAN A SPEEDING BULLET	IS AS FAST AS A SPEEDING BULLET	NOT QUITE AS FAST AS A SPEEDING BULLET	WOULD YOU BELIEVE A SLOW BULLET?	WOUNDS HIMSELF WITH THE BULLET
INITIATIVE	IS STRONGER THAN A LOCOMOTIVE	IS STRONGER THAN A BULL ELEPHANT	IS STRONGER THAN A BULL	SHOOTS THE BULL	SMELLS LIKE A BULL
ADAPTABILITY	WALKS ON WATER CONSISTENTLY	WALKS ON WATER IN EMERGENCIES	WASHES WITH WATER	DRINKS WATER	PASSES WATER IN EMERGENCIES
COMMUNICATIONS	TALKS WITH GOD	TALKS WITH ANGELS	TALKS TO HIMSELF	ARGUES WITH HIMSELF	LOSES THE ARGUMENT WITH HIMSELF

FIGURE 8–1. Guide to performance appraisal.

appraisal of the employee's performance. Of course, the project manager should not fill out the evaluation form if he has not had sufficient time to observe the employee at work.

The evaluation form can be filled out at the termination of the project. This, however, may produce a problem in that the project may end the month after the employee is considered for promotion. The advantage of this technique is that the project manager may have been able to find sufficient time both to observe the employee in action and to see the output.

Figure 8–1 (see page 321) represents, in a humorous way, how project personnel perceive the evaluation form. Unfortunately, the evaluation process is very serious and can easily have a severe impact on an individual's career path with the company even though the final evaluation rests with the functional manager.

Figure 8–2 shows a simple type of evaluation form on which the project manager identifies the best description of the employee's performance. This type of form is generally used whenever the employee is up for evaluation.

Figure 8–3 shows another typical form that can be used to evaluate an employee. In each category, the employee is rated on a subjective scale. In order to minimize time and

EMPLOYEE'S NAME ______ DATE ______

PROJECT TITLE ______ JOB NUMBER ______

EMPLOYEE ASSIGNMENT ______

EMPLOYEE'S TOTAL TIME TO DATE ON PROJECT ______ EMPLOYEE'S REMAINING TIME ON PROJECT ______

TECHNICAL JUDGMENT:				
☐ Quickly reaches sound conclusions	☐ Usually makes sound conclusions	☐ Marginal decision-making ability	☐ Needs technical assistance	☐ Makes faulty conclusions
WORK PLANNING:				
☐ Good planner	☐ Plans well with help	☐ Occasionally plans well	☐ Needs detailed instructions	☐ Cannot plan at all
COMMUNICATIONS:				
☐ Always understands instructions	☐ Sometimes needs clarification	☐ Always needs clarifications	☐ Needs follow-up	☐ Needs constant instruction
ATTITUDE:				
☐ Always job interested	☐ Shows interest most of the time	☐ Shows no job interest	☐ More interested in other activities	☐ Does not care about job
COOPERATION:				
☐ Always enthusiastic	☐ Works well until job is completed	☐ Usually works well with others	☐ Works poorly with others	☐ Wants it done his/her way
WORK HABITS:				
☐ Always project oriented	☐ Most often project oriented	☐ Usually consistent with requests	☐ Works poorly with others	☐ Always works alone

ADDITIONAL COMMENTS: ______

FIGURE 8–2. Project work assignment appraisal.

EMPLOYEE'S NAME						DATE	
PROJECT TITLE						JOB NUMBER	
EMPLOYEE ASSIGNMENT							
EMPLOYEE'S TOTAL TIME TO DATE ON PROJECT						EMPLOYEE'S REMAINING TIME ON PROJECT	

	EXCELLENT	ABOVE AVERAGE	AVERAGE	BELOW AVERAGE	INADEQUATE
TECHNICAL JUDGMENT					
WORK PLANNING					
COMMUNICATIONS					
ATTITUDE					
COOPERATION					
WORK HABITS					
PROFIT CONTRIBUTION					

ADDITIONAL COMMENTS:

FIGURE 8–3. Project work assignment appraisal.

paperwork, it is also possible to have a single evaluation form at project termination for evaluation of all employees. This is shown in Figure 8–4. All employees are rated in each category on a scale of 1 to 5. Totals are obtained to provide a relative comparison of employees.

Obviously, evaluation forms such as that shown in Figure 8–4 have severe limitations, as a one-to-one comparison of all project functional personnel is of little value if the employees are from different departments. How can a project engineer be compared to a cost accountant?

Several companies are using this form by assigning coefficients of importance to each topic. For example, under a topic of technical judgment, the project engineer might have a coefficient of importance of 0.90, whereas the cost accountant's coefficient might be 0.25. These coefficients could be reversed for a topic on cost consciousness. Unfortunately, such comparisons have questionable validity, and this type of evaluation form is usually of a confidential nature.

Even though the project manager fills out an evaluation form, there is no guarantee that the functional manager will believe the project manager's evaluation. There are always situations in which the project and functional managers disagree as to either quality or direction of work.

Another problem may exist in the situation where the project manager is a "generalist," say at a grade-7 level, and requests that the functional manager assign his best

PROJECT TITLE	JOB NUMBER
EMPLOYEE ASSIGNMENT	DATE

CODE:

EXCELLENT = 5
ABOVE AVERAGE = 4
AVERAGE = 3
BELOW AVERAGE = 2
INADEQUATE = 1

NAMES	TECHNICAL JUDGMENT	WORK PLANNING	COMMUNICATIONS	ATTITUDE	COOPERATION	WORK HABITS	PROFIT CONTRIBUTION	SELF MOTIVATION	TOTAL POINTS

FIGURE 8–4. Project work assignment appraisal.

employee to the project. The functional manager agrees to the request and assigns his best employee, a grade-10 specialist. One solution to this problem is to have the project manager evaluate the expert only in certain categories such as communications, work habits, and problem-solving, but not in the area of his technical expertise.

As a final note, it is sometimes argued that functional employees should have some sort of indirect input into a project manager's evaluation. This raises rather interesting questions as to how far we can go with the indirect evaluation procedure.

From a top-management perspective, the indirect evaluation process brings with it several headaches. Wage and salary administrators readily accept the necessity for using different evaluation forms for white-collar and blue-collar workers. But now, we have a situation in which there can be more than one type of evaluation system for white-collar workers alone. Those employees who work in project-driven functional departments will be evaluated directly and indirectly, but based on formal procedures. Employees who charge their time to overhead accounts and non–project-driven departments might simply be evaluated by a single, direct evaluation procedure.

Many wage and salary administrators contend that they cannot live with a white-collar evaluation system and therefore have tried to combine the direct and indirect evaluation forms into one, as shown in Figure 8–5. Some administrators have even gone so far

I.   <u>EMPLOYEE INFORMATION:</u>

    1. NAME _____ 2. DATE OF EVALUATION _____

    3. JOB ASSIGNMENT _____ 4. DATE OF LAST EVALUATION _____

    5. PAY GRADE _____

    6. EMPLOYEE'S IMMEDIATE SUPERVISOR _____

    7. SUPERVISOR'S LEVEL: ☐ SECTION ☐ DEPT. ☐ DIVISION ☐ EXECUTIVE

II. <u>EVALUATOR'S INFORMATION:</u>

    1. EVALUATOR'S NAME _____

    2. EVALUATOR'S LEVEL: ☐ SECTION ☐ DEPT. ☐ DIVISION ☐ EXECUTIVE

    3. RATE THE EMPLOYEE ON THE FOLLOWING:

	EXCELLENT	VERY GOOD	GOOD	FAIR	POOR
ABILITY TO ASSUME RESPONSIBILITY					
WORKS WELL WITH OTHERS					
LOYAL ATTITUDE TOWARD COMPANY					
DOCUMENTS WORK WELL AND IS BOTH COST AND PROFIT CONSCIOUS					
RELIABILITY TO SEE JOB THROUGH					
ABILITY TO ACCEPT CRITCISM					
WILLINGNESS TO WORK OVERTIME					
PLANS JOB EXECUTION CAREFULLY					
TECHNICAL KNOWLEDGE					
COMMUNICATIVE SKILLS					
OVERALL RATING					

    4. RATE THE EMPLOYEE IN COMPARISON TO HIS CONTEMPORARIES:

LOWER 10%	LOWER 25%	LOWER 40%	MIDWAY	UPPER 40%	UPPER 25%	UPPER 10%

    5. RATE THE EMPLOYEE IN COMPARISON TO HIS CONTEMPORARIES:

SHOULD BE PROMOTED AT ONCE	PROMOTABLE NEXT YEAR	PROMOTABLE ALONG WITH CONTEMPORARIES	NEEDS TO MATURE IN GRADE	DEFINITELY NOT PROMOTABLE

    6. EVALUATOR'S COMMENTS: _____

_____

SIGNATURE _____

III. <u>CONCURRENCE SECTION:</u>

    1. NAME _____

    2. POSITION: ☐ DEPARTMENT ☐ DIVISION ☐ EXECUTIVE

    3. CONCURRENCE ☐ AGREE ☐ DISAGREE

    4. COMMENTS: _____

_____

SIGNATURE _____

IV. <u>PERSONNEL SECTION:</u> (to be completed by the Personnel Department only)

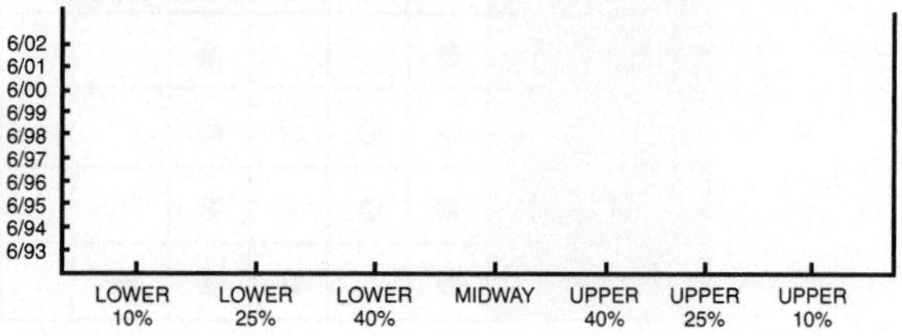

6/02 6/01 6/00 6/99 6/98 6/97 6/96 6/95 6/94 6/93

LOWER 10%   LOWER 25%   LOWER 40%   MIDWAY   UPPER 40%   UPPER 25%   UPPER 10%

V. <u>EMPLOYEE'S SIGNATURE:</u> _____ DATE: _____

FIGURE 8–5. Job evaluation.

as to adopt a single form company-wide, regardless of whether an individual is a white- or blue-collar worker.

The design of the employee's evaluation form depends on what evaluation method or procedure is being used. Generally speaking, there are nine methods available for evaluating personnel:

- Essay appraisal
- Graphic rating scale
- Field review
- Forced-choice review
- Critical incident appraisal
- Management by objectives
- Work standards approach
- Ranking methods
- Assessment center

Descriptions of these methods can be found in almost any text on wage and salary administration. Which method is best suited for a project-driven organizational structure? To answer this question, we must analyze the characteristics of the organizational form as well as those of the personnel who must perform there. An an example, project management can be described as an arena of conflict. Which of the above evaluation procedures can best be used to evaluate an employee's ability to work and progress in an atmosphere of conflict? Figure 8–6 compares the above nine evaluation procedures against the six most common project conflicts. This type of analysis must be carried out for all variables and

	Essay Appraisal	Graphic Rating Scale	Field Review	Forced-Choice Review	Critical Incident Appraisal	Management By Objectives	Work Standards Approach	Ranking Methods	Assessment Center
Conflict over schedules	●	●		●	●		●	●	
Conflict over priorities	●	●		●	●		●	●	
Conflict over technical issues	●			●			●		
Conflict over administration	●	●	●	●			●	●	●
Personality conflict	●	●		●			●		
Conflict over cost	●		●	●	●		●	●	●

Circles define areas where evaluation technique may be difficult to implement.

FIGURE 8–6. Rating evaluation techniques against types of conflict.

characteristics that describe the project management environment. Most compensation managers would agree that the management by objectives (MBO) technique offers the greatest promise for a fair and equitable evaluation of all employees. Although MBO implies that functional employees will have a say in establishing their own goals and objectives, this may not be the case. In project management, maybe the project manager or functional manager will set the objectives, and the functional employee will be told that he has to live with that. Obviously, there will be advantages and disadvantages to whatever evaluation procedures are finally selected.

8.2 FINANCIAL COMPENSATION AND REWARDS

PMBOK® Guide, 4th Edition
9.3.2.6 Recognition and Rewards

Proper financial compensation and rewards are important to the morale and motivation of people in any organization. However, there are several issues that often make it necessary to treat compensation practices of project personnel separately from the rest of the organization:

- *Job classification and job descriptions* for project personnel are usually not compatible with those existing for other professional jobs. It is often difficult to pick an existing classification and adapt it to project personnel. Without proper adjustment, the small amount of formal authority of the project and the small number of direct reports may distort the position level of project personnel in spite of their broad range of business responsibilities.
- *Dual accountability* and dual reporting relationships of project personnel raise the question of who should assess performance and control the rewards.
- *Bases for financial rewards* are often difficult to establish, quantify, and administer. The criteria for "doing a good job" are difficult to quantify.
- *Special compensations* for overtime, extensive travel, or living away from home should be considered in addition to bonus pay for preestablished results. Bonus pay is a particularly difficult and delicate issue because often many people contribute to the results of such incentives. Discretionary bonus practices can be demoralizing to the project team.

Some specific guidelines are provided here to help managers establish compensation systems for their project organizations. The foundations of these compensation practices are based on four systems: (1) job classification, (2) base pay, (3) performance appraisals, and (4) merit increases.

Job Classifications and Job Descriptions

Every effort should be made to fit the new classifications for project personnel into the existing standard classification that has already been established for the organization.

The first step is to define job titles for various project personnel and their corresponding responsibilities. Titles are noteworthy because they imply certain responsibilities, position power, organizational status, and pay level. Furthermore, titles may indicate certain functional responsibilities, as does, for example, the title of task manager.[1] Therefore, titles should be carefully selected and each of them supported by a formal job description.

The job description provides the basic charter for the job and the individual in charge of it. A good job description is brief and concise, not exceeding one page. Typically, it is broken down into three sections: (1) overall responsibilities, (2) specific duties, and (3) qualifications. A sample job description is given in Table 8–1.

Base-Pay Classifications and Incentives

After the job descriptions have been developed, one can delineate pay classes consistent with the responsibilities and accountabilities for business results. If left to the personnel specialist, these pay scales may slip toward the lower end of an equitable compensation. This is understandable because, on the surface, project positions look less senior than their functional counterparts, as formal authority over resources and direct reports are often less necessary for project positions than for traditional functional positions. The impact of such a skewed compensation system is that the project organization will attract less qualified personnel and may be seen as an inferior career path.

Many companies that have struggled with this problem have solved it by (1) working out compensation schemes as a team of senior managers and personnel specialists, and (2) applying criteria of responsibility and business/profit accountability to setting pay scales for project personnel in accord with other jobs in their organization. Managers who are hiring can choose a salary from the established range based on their judgment of actual position responsibilities, the candidate's qualifications, the available budget, and other considerations.

Performance Appraisals

Traditionally, the purpose of the performance appraisal is to:

- Assess the employee's work performance, preferably against preestablished objectives
- Provide a justification for salary actions
- Establish new goals and objectives for the next review period
- Identify and deal with work-related problems
- Serve as a basis for career discussions

In reality, however, the first two objectives are in conflict. As a result, traditional performance appraisals essentially become a salary discussion with the objective to justify subsequent

---

1. In most organizations the title of task manager indicates being responsible for managing the technical content of a project subsystem within a functional unit, having dual accountabilities to the functional superior and the project office.

TABLE 8–1. SAMPLE JOB DESCRIPTION

Job Description: Lead Project
Engineer of Processor Development

Overall Responsibility
Responsible for directing the technical development of the new Central Processor including managing the technical personnel assigned to this development. The Lead Project Engineer has dual responsibility, (1) to his/her functional superior for the technical implementation and engineering quality and (2) to the project manager for managing the development within the established budget and schedule.

Specific Duties and Responsibilities
1. Provide necessary program direction for planning, organizing, developing and integrating the engineering effort, including establishing the specific objectives, schedules, and budgets for the processor subsystem.
2. Provide technical leadership for analyzing and establishing requirements, preliminary designing, designing, prototyping, and testing of the processor subsystem.
3. Divide the work into discrete and clearly definable tasks. Assign tasks to technical personnel within the Lead Engineer's area of responsibility and other organizational units.
4. Define, negotiate, and allocate budgets and schedules according to the specific tasks and overall program requirements.
5. Measure and control cost, schedule, and technical performance against program plan.
6. Report deviations from program plan to program office.
7. Replan trade-off and redirect the development effort in case of contingencies such as to best utilize the available resources toward the overall program objectives.
8. Plan, maintain, and utilize engineering facilities to meet the long-range program requirements.

Qualifications
1. Strong technical background in state-of-the-art central processor development.
2. Prior task management experience with proven record for effective cost and schedule control of multi-disciplinary technology-based task in excess of SIM.
3. Personal skills to lead, direct, and motivate senior engineering personnel.
4. Excellent communication skills, both orally and in writing.

---

managerial actions.[2] In addition, discussions dominated by salary actions are usually not conducive for future goal setting, problem-solving, or career planning.

In order to get around this dilemma, many companies have separated the salary discussion from the other parts of the performance appraisal. Moreover, successful managers have carefully considered the complex issues involved and have built a performance appraisal system solidly based on content, measurability, and source of information.

The first challenge is in content, that is, to decide "what to review" and "how to measure performance." Modern management practices try to individualize accountability as much as possible. Furthermore, subsequent incentive or merit increases are tied to profit performance. Although most companies apply these principles to their project organizations, they do it with a great deal of skepticism. Practices are often modified to assure balance and equity for jointly performed responsibilities. A similar dilemma exists in the area

---

2. For detailed discussions, see The Conference Board, *Matrix Organizations of Complex Businesses,* 1979; plus some basic research by H. H. Meyer, E. Kay, and J. R. P. French, "Split Roles in Performance Appraisal," *Harvard Business Review,* January–February 1965.

of profit accountability. The comment of a project manager at the General Electric Company is typical of the situation faced by business managers: "Although I am responsible for business results of a large program, I really can't control more than 20 percent of its cost." Acknowledging the realities, organizations are measuring performance of their *project managers,* in at least two areas:

- *Business results* as measured by profits, contribution margin, return on investment, new business, and income; also, on-time delivery, meeting contractual requirements, and within-budget performance.
- *Managerial performance* as measured by overall project management effectiveness, organization, direction and leadership, and team performance.

The first area applies only if the project manager is indeed responsible for business results such as contractual performance or new business acquisitions. Many project managers work with company-internal sponsors, such as a company-internal new product development or a feasibility study. In these cases, producing the results within agreed-on schedule and budget constraints becomes the primary measure of performance. The second area is clearly more difficult to assess. Moreover, if handled improperly, it will lead to manipulation and game playing. Table 8–2 provides some specific measures of project management performance. Whether the sponsor is company-internal or external, project managers are usually being assessed on how long it took to organize the team, whether the project is moving along according to agreed-on schedules and budgets, and how closely they meet the global goals and objectives set by their superiors.

PMBOK® Guide, 4th Edition
9.4.2 Manage Project Team
9.4.2.2 Project Performance
    Appraisals

TABLE 8–2. PERFORMANCE MEASURES FOR PROJECT MANAGERS

Who Performs Appraisal
    Functional superior of project manager

Source of Performance Data
    Functional superior, resource managers, general managers

Primary Measures
    1. Project manager's success in leading the project toward preestablished global objectives
       • Target costs
       • Key milestones
       • Profit, net income, return on investment, contribution margin
       • Quality
       • Technical accomplishments
       • Market measures, new business, follow-on contract
    2. Project manager's effectiveness in overall project direction and leadership during all phases, including establishing:
       • Objectives and customer requirements
       • Budgets and schedules
       • Policies
       • Performance measures and controls
       • Reporting and review system

(continues)

TABLE 8–2. PERFORMANCE MEASURES FOR PROJECT MANAGERS
(*Continued*)

Secondary Measures
1. Ability to utilize organizational resources
   - Overhead cost reduction
   - Working with existing personnel
   - Cost-effective make-buy decisions
2. Ability to build effective project team
   - Project staffing
   - Interfunctional communications
   - Low team conflict complaints and hassles
   - Professionally satisfied team members
   - Work with support groups
3. Effective project planning and plan implementation
   - Plan detail and measurability
   - Commitment by key personnel and management
   - Management involvement
   - Contingency provisions
   - Reports and reviews
4. Customer/client satisfaction
   - Perception of overall project performance by sponsor
   - Communications, liaison
   - Responsiveness to changes
5. Participation in business management
   - Keeping mangement informed of new project/product/business opportunities
   - Bid proposal work
   - Business planning, policy development

Additional Considerations
1. Difficulty of tasks involved
   - Technical tasks
   - Administrative and orgnizational complexity
   - Multidisciplinary nature
   - Staffing and start-up
2. Scope of the project
   - Total project budget
   - Number of personnel involved
   - Number of organizations and subcontractors involved
3. Changing work environment
   - Nature and degree of customer changes and redirections
   - Contingencies

On the other side of the project organization, resource managers or project personnel are being assessed primarily on their ability to direct the implementation of a specific project subsystem:

- *Technical implementation* as measured against requirements, quality, schedules, and cost targets
- *Team performance* as measured by ability to staff, build an effective task group, interface with other groups, and integrate among various functions

Specific performance measures are shown in Table 8–3. In addition, the actual project performance of both project managers and their resource personnel should be assessed on the conditions under which it was achieved: the degree of task difficulty, complexity, size, changes, and general business conditions.

TABLE 8–3. PERFORMANCE MEASURES FOR PROJECT PERSONNEL

Who Performs Appraisal
  Functional superior of project person

Source of Performance Data
  Project manager and resource managers

Primary Measures
  1. Success in directing the agreed-on task toward completion
     - Technical implementation according to requirements
     - Quality
     - Key milestones/schedules
     - Target costs, design-to-cost
     - Innovation
     - Trade-offs
  2. Effectiveness as a team member or team leader
     - Building effective task team
     - Working together with others, participation, involvement
     - Interfacing with support organizations and subcontractors
     - Interfunctional coordination
     - Getting along with others
     - Change orientation
     - Making commitments

Secondary Measures
  1. Success and effectiveness in performing functional tasks in addition to project work in accordance with functional charter
     - Special assignments
     - Advancing technology
     - Developing organization
     - Resource planning
     - Functional direction and leadership
  2. Administrative support services
     - Reports and reviews
     - Special task forces and committees
     - Project planning
     - Procedure development
  3. New business development
     - Bid proposal support
     - Customer presentations
  4. Professional development
     - Keeping abreast in professional field
     - Publications
     - Liaison with society, vendors, customers, and educational institutions

Additional Considerations
  1. Difficulty of tasks involved
     - Technical challenges
     - State-of-the-art considerations
     - Changes and contingencies
  2. Managerial responsibilities
     - Task leader for number of project personnel
     - Multifunctional integration
     - Budget responsibility
     - Staffing responsibility
     - Specific accountabilities
  3. Multiproject involvement
     - Number of different projects
     - Number and magnitude of functional task and duties
     - Overall workload

Finally, one needs to decide who is to perform the performance appraisal and to make the salary adjustment. Where dual accountabilities are involved, good practices call for inputs from both bosses. Such a situation could exist for project managers who report functionally to one superior but are also accountable for specific business results to another person. While dual accountability of project managers is an exception for most organizations, it is common for project resource personnel who are responsible to their functional superior for the quality of the work and to their project manager for meeting the requirements within budget and schedule. Moreover, resource personnel may be shared among many projects. Only the functional or resource manager can judge overall performance of resource personnel.

Merit Increases and Bonuses

Professionals have come to expect merit increases as a reward for a job well done. However, under inflationary conditions, pay adjustments seldom keep up with cost-of-living increases. To deal with this salary compression and to give incentive for management performance, companies have introduced bonuses. The problem is that these standard plans for merit increases and bonuses are based on individual accountability while project personnel work in teams with shared accountabilities, responsibilities, and controls. It is usually very difficult to credit project success or failure to a single individual or a small group.

Most managers with these dilemmas have turned to the traditional remedy of the performance appraisal. If done well, the appraisal should provide particular measures of job performance that assess the level and magnitude at which the individual has contributed to the success of the project, including the managerial performance and team performance components. Therefore, a properly designed and executed performance appraisal that includes input from all accountable management elements, and the basic agreement of the employee with the conclusions, is a sound basis for future salary reviews.

8.3 CRITICAL ISSUES WITH REWARDING PROJECT TEAMS

PMBOK® Guide, 4th Edition
9.3.2.6 Recognition and Rewards

Today, most companies are using project teams. However, there still exist challenges in how to reward project teams for successful performance. The importance of how teams are rewarded is identified by Parker, McAdams, and Zielinski[3]:

> Some organizations are fond of saying, "We're all part of the team", but too often it is merely management-speak. This is especially common in conventional hierarchical organizations; they say the words but don't follow up with significant action. Their employees may read the articles and attend the conferences and come to believe that many companies have turned collaborative. Actually, though, few organizations today are genuinely team-based.

---

3. G. Parker, J. McAdams, and D. Zielinski, *Rewarding Teams* (San Francisco: Jossey-Bass, an imprint of John Wiley & Sons, 2000), p. 17; reproduced by permission of John Wiley & Sons.

Others who want to quibble point to how they reward or recognize teams with splashy bonuses or profit-sharing plans. But these do not by themselves represent a commitment to teams; they're more like a gift from a rich uncle. If top management believes that only money and a few recognition programs ("team of year" and that sort of thing) reinforce teamwork, they are wrong. These alone do not cause fundamental change in the way people and teams are managed.

But in a few organizations, teaming is a key component of the corporate strategy, involvement with teams is second nature, and collaboration happens without great thought or fanfare. There are natural work groups (teams of people who do the same or similar work in the same location), permanent cross-functional teams, ad hoc project teams, process improvement teams, and real management teams. Involvement just happens.

Why is it so difficult to reward project teams? To answer this question, we must understand what a team is and is not. According to Parker et al.[4]:

> Consider this statement: an organizational unit can act like a team, but a team is not necessarily an organizational unit, at least for describing reward plans. An organizational unit is just that, a group of employees organized into an identifiable business unit that appears on the organizational chart. They may behave in a spirit of teamwork, but for the purposes of developing reward plans they are not a "team." The organizational unit may be a whole company, a strategic business unit, a division, a department, or a work group.
>
> A "team" is a small group of people allied by a common project and sharing performance objectives. They generally have complementary skills or knowledge and an interdependence that requires that they work together to accomplish their project's objective. Team members hold themselves mutually accountable for their results. These teams are not found on an organization chart.

Incentives are difficult to apply because project teams may not appear on an organizational chart. Figure 8-7 shows the reinforcement model for employees.[5] For project teams, the emphasis is the three arrows on the right-hand side of Figure 8-7.

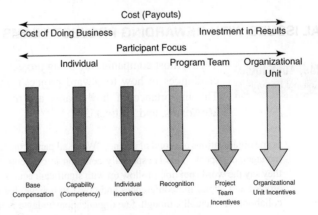

Figure 8–7. The reinforcement model.

4. See note 3, Parker et al., p. 17.
5. See note 3, Parker et al., p. 29.

Project team incentives are important because team members expect appropriate rewards and recognition for work well done. According to Parker et al.[6]:

> Project teams are usually, but not always, formed by management to tackle specific projects or challenges with a defined time frame—reviewing processes for efficiency or cost-savings recommendations, launching a new software product, or implementing enterprise resource planning systems are just a few examples. In other cases, teams self-form around specific issues or as part of continuous improvement initiatives such as team-based suggestion systems.
>
> Project teams can have cross-functional membership or simply be a subset of an existing organizational unit. The person who sponsors the team—its "champion" typically—creates an incentive plan with specific objective measures and an award schedule tied to achieving those measures. To qualify as an incentive, the plan must include pre-announced goals, with a "do this, get that" guarantee for teams. The incentive usually varies with the value added by the project.

Project team incentive plans usually have some combination of these basic measures:

- **Project Milestones:** Hit a milestone, on budget and on time, and all team members earn a defined amount. Although sound in theory, there are inherent problems in tying financial incentives to hitting milestones. Milestones often change for good reason (technological advances, market shifts, other developments) and you don't want the team and management to get into a negotiation on slipping dates to trigger the incentive. Unless milestones are set in stone and reaching them is simply a function of the team doing its normal, everyday job, it's generally best to use recognition-after-the-fact celebration of reaching milestones—rather than tying financial incentives to it.
- Rewards need not always be time-based, such that when the team hits a milestone by a certain date it earns a reward. If, for example, a product development team debugs a new piece of software on time, that's not necessarily a reason to reward it. But if it discovers and solves an unsuspected problem or writes better code before a delivery date, rewards are due.
- **Project Completion:** All team members earn a defined amount when they complete the project on budget and on time (or to the team champion's quality standards).
- **Value Added:** This award is a function of the value added by a project, and depends largely on the ability of the organization to create and track objective measures. Examples include reduced turnaround time on customer requests, improved cycle times for product development, cost savings due to new process efficiencies, or incremental profit or market share created by the product or service developed or implemented by the project team.

One warning about project incentive plans: They can be very effective in helping teams stay focused, accomplish goals, and feel like they are rewarded for their hard work, but they tend to be exclusionary. Not everyone can be on a project team. Some employees

---

6. See note 3, Parker et al., pp. 38–39.

(team members) will have an opportunity to earn an incentive that others (nonteam members) do not. There is a lack of internal equity. One way to address this is to reward core team members with incentives for reaching team goals and to recognize peripheral players who supported the team, either by offering advice, resources, or a pair of hands, or by covering for project team members back at their regular job.

Some projects are of such strategic importance that you can live with these internal equity problems and non-team members' grousing about exclusionary incentives. Bottom line, though, is this tool should be used cautiously.

Some organizations focus only on cash awards. However, Parker et al. have concluded from their research that noncash awards can work equally well, if not better, than cash awards.[7]

Many of our case organizations use noncash awards because of their staying power. Everyone loves money, but cash payments can lose their motivational impact over time. However, noncash awards carry trophy value that has great staying power because each time you took at that television set or plaque you are reminded of what you or your team did to earn it. Each of the plans encourages awards that are coveted by the recipients and, therefore, will be memorable.

If you ask employees what they want, they will invariably say cash. But providing it can be difficult if the budget is small or the targeted earnings in an incentive plan are modest. If you pay out more often than annually and take taxes out, the net amount may look pretty small, even cheap. Noncash awards tend to be more dependent on their symbolic value than their financial value.

Noncash awards come in all forms: a simple thank you, a letter of congratulations, time off with pay, a trophy, company merchandise, a plaque, gift certificates, special services, a dinner for two, a free lunch, a credit to a card issued by the company for purchases at local stores, specific items or merchandise, merchandise from an extensive catalog, travel for business or a vacation with the family, and stock options. Only the creativity and imagination of the plan creators limit the choices.

8.4 EFFECTIVE PROJECT MANAGEMENT IN THE SMALL BUSINESS ORGANIZATION

The definition of a small project could be:

- Total duration is usually three to twelve months.
- Total dollar value is $50,000 to $1.5 million (upper limit is usually capital equipment projects).
- There is continuous communication between team members, and no more than three or four cost centers are involved.
- Manual rather than computerized cost control may be acceptable.
- Project managers work closely with functional personnel and managers on a daily basis, so time-consuming detail reporting is not necessary.
- The work breakdown structure does not go beyond level three.

7. See note 3, Parker et al., pp. 190–191.

Here, we are discussing project management in both small companies and small organizations within a larger corporation. In small organizations, major differences from large companies must be accounted for:

● *In small companies, the project manager has to wear multiple hats and may have to act as a project manager and line manager at the same time.* Large companies may have the luxury of a single full-time project manager for the duration of a project. Smaller companies may not be able to afford a full-time project manager and therefore may require that functional managers wear two hats. This poses a problem in that the functional managers may be more dedicated to their own functional unit than to the project, and the project may suffer. There is also the risk that when the line manager also acts as project manager, the line manager may keep the best resources for his own project. The line manager's project may be a success at the expense of all the other projects that he must supply resources for.

In the ideal situation, the project manager works horizontally and has project dedication, whereas the line manager works vertically and has functional (or company) dedication. If the working relationship between the project and functional managers is a good one, then decisions will be made in a manner that is in the best interest of both the project and the company. Unfortunately, this may be difficult to accomplish in small companies when an individual wears multiple hats.

● *In a small company, the project manager handles multiple projects, perhaps each with a different priority.* In large companies, project managers normally handle only one project at a time. Handling multiple projects becomes a serious problem if the priorities are not close together. For this reason, many small companies avoid the establishment of priorities for fear that the lower-priority activities will never be accomplished.

● *In a small company, the project manager has limited resources.* In a large company, if the project manager is unhappy with resources that are provided, he may have the luxury of returning to the functional manager to either demand or negotiate for other resources. In a small organization, the resources assigned may be simply the only resources available.

● *In a small company, project managers must generally have a better understanding of interpersonal skills than in a larger company.* This is a necessity because a project manager in the small company has limited resources and must provide the best motivation that he can.

● *In the smaller company, the project manager generally has shorter lines of communications.* In small organizations project managers almost always report to a top-level executive, whereas in larger organizations the project managers can report to any level of management. Small companies tend to have fewer levels of management.

● *Small companies do not have a project office.* Large companies, especially in aerospace or construction, can easily support a project office of twenty to thirty people, whereas in the smaller company the project manager may have to be the entire project office. This implies that the project manager in a small company may be required to have more general and specific information about all company activities, policies, and procedures than his counterparts in the larger companies.

● *In a small company, there may be a much greater risk to the total company with the failure of as little as one project.* Large companies may be able to afford the loss of a multimillion-dollar program, whereas the smaller company may be in serious financial trouble. Thus many smaller companies avoid bidding on projects that would necessitate hiring additional resources or giving up some of its smaller accounts.

- *In a small company, there might be tighter monetary controls but with less sophisticated control techniques.* Because the smaller company incurs greater risk with the failure (or cost overrun) of as little as one project, costs are generally controlled much more tightly and more frequently than in larger companies. However, smaller companies generally rely on manual or partially computerized systems, whereas larger organizations rely heavily on sophisticated software packages.
- *In a small company, there is usually more upper-level management interference.* This is expected because in the small company there is a much greater risk with the failure of a single project. In addition, executives in smaller companies "meddle" more than executives in larger companies, and quite often delegate as little as possible to project managers.
- *Evaluation procedures for individuals are usually easier in a smaller company.* This holds true because the project manager gets to know the people better, and, as stated above, there exists a greater need for interpersonal skills on the horizontal line in a smaller company.
- *In a smaller company, project estimating is usually more precise and based on either history or standards.* This type of planning process is usually manual as opposed to computerized. In addition, functional managers in a small company usually feel obligated to live up to their commitments, whereas in larger companies, much more lip service is given.

8.5 MEGA PROJECTS

Mega projects may have a different set of rules and guidelines from those of smaller projects. For example, in large projects:

- Vast numbers of people may be required, often for short or intense periods of time.
- Continuous organizational restructuring may be necessary as each project goes through a different life-cycle phase.
- The matrix and project organizational form may be used interchangeably.
- The following elements are critical for success.
  - Training in project management
  - Rules and procedures clearly defined
  - Communications at all levels
  - Quality front-end planning

Many companies dream of winning mega project contracts only to find disaster rather than a pot of gold. The difficulty in managing mega projects stems mainly from resource restraints:

- Lack of available on-site workers (or local labor forces)
- Lack of skilled workers

- Lack of properly trained on-site supervision
- Lack of raw materials

As a result of such problems, the company immediately assigns its best employees to the mega project, thus creating severe risks for the smaller projects, many of which could lead to substantial follow-on business. Overtime is usually required, on a prolonged basis, and this results in lower efficiency and unhappy employees.

As the project schedule slips, management hires additional home-office personnel to support the project. By the time that the project is finished, the total organization is overstaffed, many smaller customers have taken their business elsewhere, and the company finds itself in the position of needing another mega project in order to survive and support the existing staff.

Mega projects are not always as glorious as people think they are. Organizational stability, accompanied by a moderate growth rate, may be more important than quantum steps to mega projects. The lesson here is that mega projects should be left to those companies that have the facilities, expertise, resources, and management know-how to handle the situation.

8.6 MORALITY, ETHICS, AND THE CORPORATE CULTURE _____

PMBOK® Guide, 4th Edition
1.1 Domain of Professional
 Responsibility and the PMP®
 Code of Conduct

Companies that promote morality and ethics in business usually have an easier time developing a cooperative culture than those that encourage unethical or immoral behavior. The adversity generated by unethical acts can be either internally or externally driven. Internally driven adversity occurs when employees or managers in your own company ask you to take action that may be in the best interest of your company but violates your own moral and ethical beliefs. Typical examples might include:

- You are asked to lie to the customer in a proposal in order to win the contract.
- You are asked to withhold bad news from your own management.
- You are asked to withhold bad news from the customer.
- You are instructed to ship a potentially defective unit to the customer in order to maintain production quotas.
- You are ordered to violate ethical accounting practices to make your numbers "look good" for senior management.
- You are asked to cover up acts of embezzlement or use the wrong charge numbers.
- You are asked to violate the confidence of a private personal decision by a team member.

External adversity occurs when your customers ask you to take action that may be in the customer's best interest (and possibly your company's best interest), but once again violates your personal moral and ethical beliefs. Typical examples might include:

- You are asked to hide or destroy information that could be damaging to the customer during legal action against your customer.
- You are asked to lie to consumers to help maintain your customer's public image.

- You are asked to release unreliable information that would be damaging to one of your customer's competitors.
- The customer's project manager asks you to lie in your proposal so that he/she will have an easier time in approving contract award.

Project managers are often placed in positions where an action must be taken for the best interest of the company and its customers, and yet the same action could be upsetting to the workers. Consider the following example as a positive way to handle this:

- A project had a delivery date where a specific number of completed units had to be on the firm's biggest customer's receiving dock by January 5. This customer represented 30% of the firm's sales and 33% of its profits. Because of product development problems and slippages, the project could not be completed early. The employees, many of whom were exempt, were informed that they would be expected to work 12-hour days, including Christmas and New Year's, to maintain the schedule. The project manager worked the same hours as his manufacturing team and was visible to all. The company allowed family members to visit the workers during the lunch and dinner hours during this period. After delivery was accomplished, the project manager arranged for all of the team members to receive two weeks of paid time off. At completion of the project, the team members were volunteering to work again for this project manager.

The project manager realized that asking his team to work these days might be viewed as immoral. Yet, because he also worked, his behavior reinforced the importance of meeting the schedule. The project manager's actions actually strengthened the cooperative nature of the culture within the firm.

Not all changes are in the best interest of both the company and the workers. Sometimes change is needed simply to survive, and this could force employees to depart from their comfort zones. The employees might even view the change as immoral. Consider the following example:

- Because of a recession, a machine tool company switched from a non–project-driven to a project-driven company. Management recognized the change and tried to convince employees that customers now wanted specialty products rather than standard products, and that the survival of the firm may be at stake. The company hired a project management consulting company to help bring in project management since the business was now project-driven. The employees vigorously resisted both the change and the training with the mistaken belief that, once the recession ended, the customers would once again want the standard, off-the-shelf products and that project management was a waste of time. The company is no longer in business and, as the employees walked out of the plant for the last time, they blamed project management for the loss of employment.

Some companies develop "Standard Practice Manuals" that describe in detail what is meant by ethical conduct in dealing with customers and suppliers. Yet, even with the existence of these manuals, well-meaning individuals may create unintended consequences that wreak havoc.

Consider the following example:

- The executive project sponsor on a government-funded R&D project decided to "massage" the raw data to make the numbers look better before presenting the data to a customer. When the customer realized what had happened, their relationship, which had been based upon trust and open communications, was now based upon mistrust and formal documentation. The entire project team suffered because of the self-serving conduct of one executive.

Sometimes, project managers find themselves in situations where the outcome most likely will be a win-lose position rather than a win-win situation. Consider the following three situations:

- An assistant project manager, Mary, had the opportunity to be promoted and manage a new large project that was about to begin. She needed her manager's permission to accept the new assignment, but if she left, her manager would have to perform her work in addition to his own for at least three months. The project manager refused to release her, and the project manager developed a reputation of preventing people from being promoted while working on his project.
- In the first month of a twelve-month project, the project manager realized that the end date was optimistic, but he purposely withheld information from the customer in hopes that a miracle would occur. Ten months later, the project manager was still withholding information waiting for the miracle. In the eleventh month, the customer was told the truth. People then labeled the project manager as an individual who would rather lie than tell the truth because it was easier.
- To maintain the customer's schedule, the project manager demanded that employees work excessive overtime, knowing that this often led to more mistakes. The company fired a tired worker who inadvertently withdrew the wrong raw materials from inventory, resulting in a $55,000 manufacturing mistake.

In all three situations, the project manager believed that his decision was in the best interest of the company at that time. Yet the final result in each case was that the project manager was labeled as unethical or immoral.

It is often said that "money is the root of all evil." Sometimes companies believe that recognizing the achievements of an individual through a financial reward system is appropriate without considering the impact on the culture. Consider the following example:

- At the end of a highly successful project, the project manager was promoted, given a $5,000 bonus and a paid vacation. The team members who were key to the project's success and who earned minimum wage, went to a fast food restaurant to celebrate their contribution to the firm and their support of each other. The project manager celebrated alone.

The company failed to recognize that project management was a team effort. The workers viewed management's reward policy as immoral and unethical because the project manager was successful due to the efforts of the entire team.

Moral and ethical conduct by project managers, project sponsors, and line managers can improve the corporate culture. Likewise, poor decisions can destroy a culture, often in much less time than it took for the culture to be developed.

8.7 PROFESSIONAL RESPONSIBILITIES

PMBOK® Guide, 4th Edition
1.1 Domain of Professional
Responsibility and the PMP®
Code of Conduct

Professional responsibilities for project managers have become increasingly important in the last few years because of the unfavorable publicity on the dealings of corporate America. These professional responsibilities have been with us for some time, especially in dealing with government agencies. Professional responsibilities for a project manager are both broad-based and encompassing. PMI® released a Project Management Professional (PMP®) Role Delineation Study in 2000 that emphasizes the professional responsibilities of the project manager. The Professional Responsibilities Domain Area in the PMBOK®Guide is based upon the Role Delineation Study and The PMI® Code of Conduct. There are five tasks emphasized under Professional Responsibilities Domain Area of the PMBOK® Guide:

- **Ensure Individual Integrity and Professionalism:** The project manager is expected to act in a professional manner at all times. This includes adhering to all legal requirements, maintaining moral and ethical standards, and protecting the community and all stakeholders even though there may be some conflicting interests among the shareholders. The project manager must be knowledgeable about legal requirements (including professional standards legal requirements), as well as multinational, ethnic, ethical, and cultural standards at both the project's location and within the team. Understanding the values set forth by the stakeholders is also necessary.

- **Contribute to the Project Management Knowledge Base:** Project managers are expected to contribute to the project management knowledge base by sharing project management knowledge on such topics as current research, best practices, lessons learned, and continuous improvement efforts. The intent of this contribution is to advance the profession, improve the quality of project management, and improve the capabilities of one's colleagues. Contributions can take the form of articles, presentations, books, and various other media.

- **Enhance Individual Competence:** Project managers are expected to enhance their own individual competencies in the same manner as they contribute to the profession. Usually, project managers that contribute to the profession enhance their own competencies at the same time.

- **Balance Stakeholder Interests:** All stakeholders may have different values and interests. These competing interests mandate that project managers not only understand stakeholder needs and objectives but also possess strong conflict resolution skills, negotiation skills, and communication skills.

- **Interact with the Team and Stakeholders in a Professional and Cooperative Manner:** Project managers are expected to understand the ethnic and cultural norms of both the team members and the stakeholders. This leads to the category of cultural diversity and socioeconomic influences such as political differences, national holidays, communication preferences, religious practices, ethical and moral beliefs, and other demographic considerations. Project managers must be willing to embrace diversity, be open-minded, exercise self-control, exhibit empathy, and exercise tolerance with a willingness to compromise.

In addition to the five tasks described under professional responsibility, PMI® has developed a Code of Professional Conduct that reinforces these five tasks. The code applies to everyone working in a project environment, not merely the project manager. As such, the code emphasizes that PMP®s must function as "role models" and exhibit characteristics such as honesty, morality, and ethical behavior.

The code has two major sections:

- Responsibilities to the Profession
- Responsibilities to the Customers and the Public

There are numerous situations that can create problems for project managers in dealing with professional responsibilities expectations. These situations include:

- Maintaining professional integrity
- Adhering to ethical standards
- Recognizing diversity
- Avoiding/reporting conflicts of interest
- Not making project decisions for personal gains
- Receiving gifts from customers and vendors
- Providing gifts to customers and vendors
- Truthfully reporting information
- Willing to identify violations
- Balancing stakeholder needs
- Succumbing to stakeholder pressure
- Managing your firm's intellectual property
- Managing your customer's intellectual property
- Adhering to security and confidentiality requirements
- Abiding by the Code of Professional Conduct

Several of these topics are explained below.

Conflict of Interest   A conflict of interest is a situation where the individual is placed in a compromising position where the individual can gain personally based upon the decisions made. This is also referred to as personal enrichment. There are numerous situations where a project manager is placed in such a position. Examples might be:

- Insider knowledge that the stock will be going up or down
- Being asked to improperly allow employees to use charge numbers on your project even though they are not working on your project
- Receiving or giving inappropriate (by dollar value) gifts
- Receiving unjustified compensation or kickbacks
- Providing the customer with false information just to keep the project alive

Project managers are expected to abide by the PMI® Code of Professional Conduct, which makes it clear that project managers should conduct themselves in an ethical manner. Unjust compensation or gains not only are frowned upon but are unacceptable. Unless these conflict-of-interest situations are understood, the legitimate interests of both the customer and the company may not be forthcoming.

Inappropriate Connections   Not all stakeholders are equal in their ability to influence the decisions made by the project manager. Some stakeholders can provide inappropriate influence/compensation, such as:

- A loan with a very low interest rate
- Ability to purchase a product/service at a price that may appear equivalent to a gift
- Ability to receive free gifts such as airline tickets, tickets to athletic events, free meals and entertainment, or even cash

Another form of inappropriate connections would be with family or friends. These individuals may provide you with information or influence by which you could gain personally in a business situation. Examples of affiliation connections might be:

- Receiving insider information
- Receiving privileged information
- Opening doors that you could not open by yourself, at least without some difficulty

Acceptance of Gifts   Today, all companies have rules concerning the acceptance of gifts and their disclosure. While it may be customary in some countries to give or accept gifts, the standard rule is usually to avoid all gifts. Some companies may stipulate limits on when gifts are permitted and the appropriateness of the gift. The gifts might be cash, free meals, or other such items.

Responsibility to Your Company (and Stakeholders)   Companies today, more than ever before, are under pressure to maintain ethical practices with customers and suppliers. This could be interpreted as a company code of ethics that stipulates the professional behavior expected from the project manager and the team members. This applies specifically to the actions of both the project manager as well as the team members. Some companies even go so far as to develop "standard practice manuals" on how to act in a professional manner. Typical sections of such manuals might be:

- Truthful representation of all information
- Full disclosure of all information
- Protection of company-proprietary information
- Responsibility to report violations
- Full compliance with groups auditing violations
- Full disclosure, and in a timely manner, of all conflicts of interest
- Ensure that all of the team members abide by the above items

8.8 INTERNAL PARTNERSHIPS

A partnership is a group of two or more individuals working together to achieve a common objective. In project management, maintaining excellent, working relations with internal partners is essential. Internally, the critical relationship is between the project and line manager.

In the early days of project management, the selection of the individual to serve as the project manager was most often dependent upon who possessed the greatest command of technology. The result, as shown in Figure 8–8, was a very poor working relationship between the project and line manager. Line managers viewed project managers as a threat, and their relationship developed into a competitive, superior-subordinate relationship. The most common form of organizational structure was a very strong matrix where the project

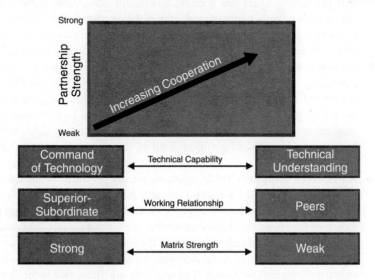

FIGURE 8–8. Partnership strength.

manager, perceived as having a command of technology, had a greater influence over the assigned employees than did their line manager.

As the magnitude and technical complexity of the projects grew, it became obvious that the project managers could not maintain a command of technology in all aspects of the project. Project managers were viewed as possessing an understanding of rather than command of technology. They became more dependent upon line managers for technical support. The project manager then found himself in the midst of a weak matrix where the employees were receiving the majority of their technical direction from the line managers.

As the partnership between the project and line managers developed, management recognized that partnerships worked best on a peer-to-peer basis. Project and line managers began to view each other as equals and share in the authority, responsibility, and accountability needed to assure project success. Good project management methodologies emphasize the cooperative working relationship that must exist between the project and line managers.

8.9 EXTERNAL PARTNERSHIPS

PMBOK® Guide, 4th Edition
2.3 Vendor Business Partners

Project management methodologies also emphasize the working relationships with external organizations such as suppliers. Outsourcing has become a major trend because it allows companies to bring their products and services to the market faster and often at a more competitive price. Therefore, external partnerships can become beneficial for both the suppliers and the customers.

There are three categories of suppliers:

- **An External Supplier:** These are suppliers that you may or may not have worked with previously. There has been no investment into a relationship with these suppliers. If they win a contract, and even if they perform well, there is no guarantee that they will receive another contract. Usually an external supplier must go through all of the requirements of the competitive bidding process for each project.
- **An Approved Supplier:** This is usually considered the lowest level of external partnering. Approved suppliers are part of an approved supplier-bidding list and are invited to bid on selected projects. If the approved supplier wins a contract, there is no guarantee that any additional contracts will be forthcoming. Some minimal relationship between the customer and supplier may exist, but the supplier may still be required to go through all of the standard protocols of competitive bidding.
- **A Preferred Supplier:** These suppliers usually get the first chance at receiving a contract but may still have to go through the entire competitive bidding process, but with a minimum amount of paperwork. Proposal information on previous history, past experience with the customer or the type of project, and other such information may not be required as part of the contractual bidding process in order to reduce time and cost. A relationship between the customer and the supplier exists. Information on lessons learned, best practices, and technological changes are often exchanged freely.

- **A Strategic Partnership Supplier:** A strong relationship exists between the customer and supplier, and they freely exchange information, especially strategic information. Each views the relationship as a long-term partnership with long-term benefits. Strategic suppliers often receive sole-source contracts without having to prepare a formal proposal, thus generating cost savings for both companies. Strategic suppliers may not be the lowest cost suppliers, but the customer's cost savings of not having to perform competitive bidding is well worth the effort.

External partnerships, if properly managed, can provide significant long-term benefits to both the customer and supplier.

The Department of Defense has been conducting research into what constitutes an effective supplier relationship.[8] Each Chrysler supplier had a Chrysler person knowledgeable about the supplier's business to contact for all supplier dealings for that commodity. These companies also interacted with key suppliers in close teaming arrangements that facilitated sharing information. Commonly called integrated product teams (IPTs), members worked together so that design, manufacturing, and cost issues were considered together. Team members were encouraged to participate as partners in meeting project goals and to interact frequently. In addition, some companies collocated suppliers with their own people or set up central working facilities with suppliers for working out issues such as how a product might be improved or be made less expensive. Motorola and Xerox saw such teams as a key vehicle for facilitating early supplier involvement in their products—one of their primary strategies. Motorola said key suppliers had building access and came in many times during a week to work with Motorola engineers.

These companies also asked suppliers to meet high standards, then differentiated the types of relationships within their pool of suppliers. Many treated key suppliers—those contributing the most to their product, such as critical parts or unique processes—differently than suppliers for noncritical or standard parts. For example, one Corning division categorized suppliers and developed relationships with them based on the extent of their impact on the customer and performance. Level 1 suppliers have a direct impact on customer satisfaction, level 2 suppliers are important to day-to-day operations, and level 3 suppliers provided commonly available products. DuPont differentiated between alliance partners—suppliers with similar goals and objectives that wish to work with DuPont for mutual benefit—and all other suppliers.

Perhaps more significantly, Chrysler's relationships with its suppliers had evolved to the point that it no longer needed to make large investments in some key technology areas. Instead, the suppliers made the technology investment themselves and had enough confidence in their relationship with Chrysler that they did not fear the long-term commitment that this entailed. For its part, Chrysler trusted the suppliers to make investments that would help keep their vehicles competitive. In this case, both supplier and product developer saw their success as that of the final product and a continuing mutually beneficial relationship.

---

8. *DoD Can Help Suppliers Contribute More to Weapon System Programs,* Best Practices Series, GAO/NSIAD-98-87, Government Accounting Office, March 1998, pp. 38, 48, 51.

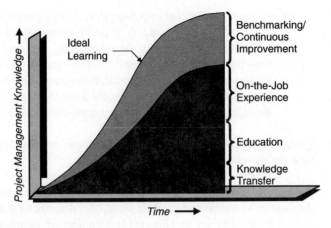

FIGURE 8–9. Project management learning curve.

8.10 TRAINING AND EDUCATION

PMBOK® Guide, 4th Edition
Task 3 of Professional
  Responsibility—Enhance
  Individual Competence

Given that most companies use the same basic tools as part of their methodology, what makes one company better than another? The answer lies in the execution of the methodology. Training and education can accelerate not only the project management maturity process but also the ability to execute the methodology.

Actual learning takes place in three areas, as shown in Figure 8–9: on-the-job experience, education, and knowledge transfer. Ideal project management knowledge would be obtained by allowing each employee to be educated on the results of the company's lessons learned studies including risk management, benchmarking, and continuous improvement efforts. Unfortunately, this is rarely done and ideal learning is hardly ever reached. To make matters worse, actual learning is less than most people believe because of lost knowledge. This lost knowledge is shown in Figure 8–10 and will occur even in companies that maintain low employee turnover ratios. These two figures also illustrate the importance of maintaining the same personnel on the project for the duration of the effort.

Companies often find themselves in a position of having to provide a key initiative for a multitude of people, or simply specialized training to a program team about to embark upon a new long-term effort. In such cases, specialized training is required, with targeted goals and results that are specifically planned for. The elements common to training on a key initiative or practice include[9]:

- A front-end analysis of the program team's needs and training requirements
- Involvement of the program teams in key decisions

---

9. Adapted from *DoD Training Can Do More to Help Weapon System Programs Implement Best Practices,* Best Practices Series, GAO/NSIAD-99-206, Government Accounting Office, August 1999, pp. 40–41, 51.

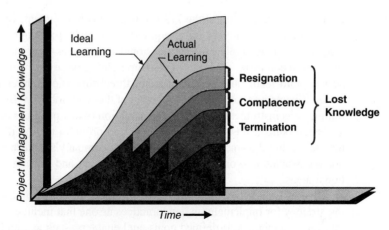

FIGURE 8–10. Project management learning curve.

- Customized training to meet program team's specific needs
- Targeted training for the implementation of specific practices
- Improved training outcomes, including better course depth, timeliness, and reach

The front-end analysis is used to determine the needs and requirements of the program office implementing the practice. The analysis is also used to identify and address barriers each program office faces when implementing new practices. According to the director of the benchmarking forum for the American Society of Training and Development, this type of analysis is crucial for an organization to be able to institute performance-improving measures. Using information from the front-end analysis, the training organizations customize the training to ensure that it directly assists program teams in implementing new practices. To ensure that the training will address the needs of the program teams, the training organizations involve the staff in making important training decisions. Program staff help decide the amount of training to be provided for certain job descriptions, course objectives, and depth of course coverage. Companies doing this believe their training approach, which includes program staff, has resulted in the right amount of course depth, timeliness, and coverage of personnel.

Officials at Boeing's Employee Training and Development organization state that their primary goal is to support their customers, the employees assigned to the Commercial Airplane Group. The training representatives develop a partnership with the staff from the beginning of the program to design and manufacture a new airplane. The training representatives form "drop teams" to collate with the program to conduct a front-end analysis and learn as much as possible about the business process and the staff's concerns. The analysis allows the drop team to determine what training is needed to support the staff implementing the new practice.

Boeing training officials said they worked side by side with the program staff to create a training program that provided team building and conflict resolution techniques and technical skills training that specifically focused on improving work competencies that would change as a result of the 777's new digital environment. To ensure all 777 staff was equally trained, employees were required to complete training before they reported to the program. For example, the professional employees—engineers and drafters—were required to complete 120 hours of start-up training on several key 777 practices, including design build teams and computer-aided three-dimensional interactive applications software.[10] Teams were often trained together at the work location. Boeing officials stated that training was instrumental to the implementation of key practices on the 777 program, such as design build teams—essentially integrated program teams (IPTs). The officials stated that design build teams were at odds with the company's culture because employees were not accustomed to working in a team environment and sharing information across functional areas.

Boeing's director of learning program development summarized the corporate training strategy for implementing new practices as one that includes a clearly stated vision or mission statement, well-defined goals, and enablers, such as training and good processes, to support the implementers. This philosophy enabled Boeing to take a year to develop the training program tailored to the 777 program—which was intended to change the corporate culture and encourage employees to rethink how they did their jobs. Both Boeing training and program officials believe that the training investment resulted in the successful implementation of the key 777 practices.

While the company officials acknowledged that training was instrumental in the implementation of the key practices, everyone also stated that training was just one of the necessary components. Creating the right environment is also key to the successful implementation of new practices, and the quality of the training was dependent on the environment. Boeing officials stressed that strong leadership is often another key force. At the inception of key programs at IBM, top leaders provide sufficient funding for training, well-defined expectations, clear direction, oversight, continued interest, and incentives to ensure that the new practices are possible to implement. The manager for the 777 program stated that Boeing's management works in teams—a key practice. He believed that it was management's ability to lead by example that helped prevent a return to the former functional way of operating. These companies believe that other factors, such as an accommodating organizational structure, good internal communication, consistent application, and supportive technology, are needed to foster the implementation of key new practices.

8.11 INTEGRATED PRODUCT/PROJECT TEAMS

PMBOK® Guide, 4th Edition
Chapter 4 Integration Management
Chapter 9 Human Resources
   Management

In recent years, there has been an effort to substantially improve the formation and makeup of teams required to develop a new product or implement a new practice. These teams have membership from across the entire organization and are called integrated product/project teams (IPTs).

---

10. This application is a computer-based design tool that allows designers the opportunity to view design drawings and the interface of millions of airplane parts as three-dimensional.

The IPT consists of a sponsor, program manager, and the core team. For the most part, members of the core team are assigned full-time to the team but may not be on the team for the duration of the entire project.

The skills needed to be a member of the core team include:

- Self-starter ability
- Work without supervision
- Good communication skills
- Cooperative
- Technical understanding
- Willing to learn backup skills
- Able to perform feasibility studies and cost/benefit analyses
- Able to perform or assist in market research studies
- Able to evaluate asset utilization
- Decision-maker
- Knowledgeable in risk management
- Understand the need for continuous validation

Each IPT is given a project charter that identifies the project's mission and identifies the assigned project manager. However, unlike traditional charters, the IPT charter can also identify the key members of the IPT by name or job responsibility.

Unlike traditional project teams, the IPT thrives on sharing information across the team and collective decision-making. IPTs eventually develop their own culture and, as such, can function in either a formal or informal capacity.

Since the concept of an IPT is well suited to large, long-term projects, it is no wonder that the Department of Defense has been researching best practices for an IPT.[11] The government looked at four projects, in both the public and private sectors, which were highly successful using the IPT approach and four government projects that had less than acceptable results. The successful IPT projects are shown in Table 8–4. The unsuccessful IPT

TABLE 8–4. EFFECTIVE IPTS

Program	Cost Status	Schedule Status	Performance Status
Daimler-Chrysler	Product cost was lowered	Decreased development cycle months by 50 percent	Improved vehicle designs
Hewlett-Packard	Lowered cost by over 60 percent	Shortened development schedule by over 60 percent	Improved system integration and product design
3M	Outperformed cost goals	Product deliveries shortened by 12 to 18 months	Improved performance by 80 percent
Advanced Amphibious Assault Vehicle	Product unit cost lower than original estimate	Ahead of original development schedule	Demonstrated fivefold increase in speed

---

11. *DoD Teaming Practices Not Achieving Potential Results,* Best Practices Series, GOA-01-501, Government Accounting Office, April 2001.

TABLE 8–5. INEFFECTIVE IPTS

Program	Cost Status	Schedule Status	Performance Status
CH-60S Helicopter	Increased cost but due to additional purchases	Schedule delayed	Software and structural difficulties
Extended Range Guided Munitions	Increases in development costs	Schedule slipped three years	Redesigning due to technical difficulties
Global Broadcast Service	Experiencing cost growth	Schedule slipped 1.5 years	Software and hardware design shortfalls
Land Warrior	Cost increase of about 50 percent	Schedule delayed four years	Overweight equipment, inadequate battery power and design

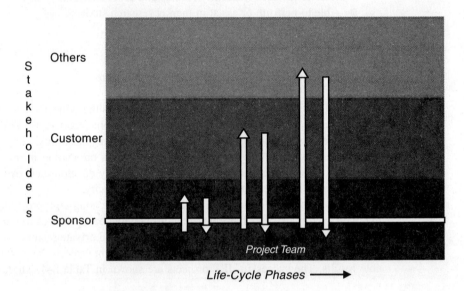

FIGURE 8–11. Knowledge and authority.

projects are shown in Table 8–5. In analyzing the data, the government came up with the results shown in Figure 8–11. Each vertical line in Figure 8–11 is a situation where the IPT must go outside of its own domain to seek information and approvals. Each time this happens, it is referred to as a "hit." The government research indicated that the greater the number of hits, the more likely it is that the time, cost, and performance constraints will not be achieved. The research confirmed that if the IPT has the knowledge necessary to make decisions, and also has the authority to make the decisions, then the desired performance would be achieved. Hits will delay decisions and cause schedule slippages.

8.12 VIRTUAL PROJECT TEAMS

Historically, project management was a face-to-face environment where team meetings involved all players convening together in one room. The team itself may even be co-located. Today, because of the size and complexity of projects, it is impossible to find all team members located under one roof. Other possible characteristics of a virtual team are shown in Table 8–6.

Duarte and Snyder define seven types of virtual teams.[12] These are shown in Table 8–7. Culture and technology can have a major impact on the performance of virtual teams. Duarte and Snyder have identified some of these relationships in Table 8–8.

TABLE 8–6. CHARACTERISTICS OF VIRTUAL TEAMS

Characteristic	Traditional Teams	Virtual Teams
Membership	Team members are all from the same company.	Team members may be multinational and all from different companies and countries.
Proximity	Team members work in close proximity with each other.	Team members may never meet face-to-face.
Methodology usage	One approach exists, perhaps an enterprise project management methodology.	Each unit can have their own methodology
Methodology structure	One approach, which is based upon either policies and procedures, or forms guidelines, templates, and checklists.	Each unit's methodology can have its own structure.
Trust	Very little trust may exist.	Trust is essential.
Authority	Leadership may focus on authority.	Leadership may focus on influence power.

TABLE 8–7. TYPES OF VIRTUAL TEAMS

Type of Team	Description
Network	Team membership is diffuse and fluid; members come and go as needed. Team lacks clear boundaries within the organization.
Parallel	Team has clear boundaries and distinct membership. Team works in the short term to develop recommendation for an improvement in a process or system.
Project or product development	Team has fluid membership, clear boundaries, and a defined customer base, technical requirement, and output. Longer-term team task is nonroutine, and the team has decision-making authority.
Work or production	Team as distinct membership and clear boundaries. Members perform regular and outgoing work, usually in one functional area.
Service	Team has distinct membership and supports ongoing customer network activity.
Management	Team has distinct membership and works on a regular basis to lead corporate activities.
Action	Team deals with immediate action, usually in an emergency situation. Membership may be fluid or distinct.

12. D. L. Duarte and N. T. Snyder, *Mastering Virtual Teams* (San Francisco: Jossey-Bass, an imprint of John Wiley & Sons, 2001), p. 10; reproduced by permission of John Wiley & Sons.

TABLE 8–8. TECHNOLOGY AND CULTURE

Cultural Factor	Technological Considerations
Power distance	Members from high-power-distance cultures may participate more freely with technologies that are asynchronous and allow anonymous input. These cultures sometimes use technology to indicate status differences between team members.
Uncertainty avoidance	People from cultures with high uncertainty avoidance may be slower adopters of technology. They may also prefer technology that is able to produce more permanent records of discussions and decisions.
Individualism–collectivism	Members from highly collectivistic cultures may prefer face-to-face interactions.
Masculinity–femininity	People from cultures with more "feminine" orientations are more prone to use technology in a nurturing way, especially during team startups.
Context	People from high-content cultures may prefer more information-rich technologies, as well as those that offer opportunities for the feeling of social presence. They may resist using technologies with low social presence to communicate with people they have never met. People from low-context cultures may prefer more asynchronous communications.

Source: D.L. Duarte and N. Tennant Snyder, *Mastering Virtual Teams.* San Francisco: Jossey-Bass, 2001, p. 60.

The importance of culture cannot be understated. Duarte and Snyder identify four important points to remember concerning the impact of culture on virtual teams. The four points are[13]:

1. There are national cultures, organizational cultures, functional cultures, and team cultures. They can be sources of competitive advantages for virtual teams that know how to use cultural differences to create synergy. Team leaders and members who understand and are sensitive to cultural differences can create more robust outcomes than can members of homogeneous teams with members who think and act alike. Cultural differences can create distinctive advantages for teams if they are understood and used in positive ways.
2. The most important aspect of understanding and working with cultural differences is to create a team culture in which problems can be surfaced and differences can he discussed in a productive, respectful manner.
3. It is essential to distinguish between problems that result from cultural differences and problems that are performance based.
4. Business practices and business ethics vary in different parts of the world. Virtual teams need to clearly articulate approaches to these that every member understands and abides by.

8.13 BREAKTHROUGH PROJECTS

Once the decision to implement project management is made, support is needed from the rest of the organization. This is best accomplished by a breakthrough project that managers

13. See note 12, Duarte and Snyder, p. 70.

and employees can follow to see project management in action. The breakthrough project should have a high probability of success, otherwise employees may erroneously blame project management for the failure when, in fact, the failure may be due to other causes.

Some people believe that the breakthrough project should be a small effort such that the employees can easily see the benefits of project management. Others contend that a large project be used as the breakthrough project because success on a small project is no guarantee that the same success can be achieved on a large project.

There are strategies and tactics that should be carefully considered during the implementation of a breakthrough project:

- **The Push for Change:** Project management implementation often requires a push for change. The push for excessive changes, such as the implementation of new and cumbersome policies and procedures, may generate more enemies than allies. Also, the recommended changes should be applicable to a broad range of projects, be easy to accept, and deemed necessary for the completion of the objectives.
- **Retention of Authority and Power:** People are more likely to accept project management if they do not feel threatened by a shift in the balance of power and authority.
- **Focus on the Deliverables:** All too often, people erroneously focus on the tools and software of project management rather than on the end result. This is a mistake. Initially, the focus should be on the deliverables and the fact that a structured project management process can improve the chances of success on each project.
- **Information Flow:** People need to see the flow of project management information in order to make an evaluation. The project information should flow according to the traditional channels.

8.14 STUDYING TIPS FOR THE PMI® PROJECT MANAGEMENT CERTIFICATION EXAM

This section is applicable as a review of the principles to support the knowledge areas and domain groups in the PMBOK® Guide. This chapter addresses:

- Human Resources Management
- Professional Responsibility
- Planning
- Execution

Understanding the following principles is beneficial if the reader is using this text to study for the PMP® Certification Exam:

- Principles and tasks included under professional responsibility
- Factors that affect professional responsibility such as conflicts of interest and gifts
- PMI® Code of Professional Conduct (this can be downloaded from the PMI® web site, pmi.org)

- That personnel performance reviews, whether formal or informal, are part of a project manager's responsibility
- Differences between project management in a large company and project management in a small company

The following multiple-choice questions will be helpful in reviewing the principles of this chapter:

1. You have been sent on a business trip to visit one of the companies bidding on a contract to be awarded by your company. You are there to determine the validity of the information in its proposal. They take you to dinner one evening at a very expensive restaurant. When the bill comes, you should:
   A. Thank them for their generosity and let them pay the bill
   B. Thank them for their generosity and tell them that you prefer to pay for your own meal
   C. Offer to pay for the meal for everyone and put it on your company's credit card
   D. Offer to pay the bill, put it on your company's credit card, and make the appropriate adjustment in the company's bid price to cover the cost of the meals

2. You are preparing a proposal in response to a Request for Proposal (RFP) from a potentially important client. The salesperson in your company working on the proposal tells you to "lie" in the proposal to improve the company's chance of winning the contract. You should:
   A. Do as you are told
   B. Refuse to work on the proposal
   C. Report the matter to your superior, the project sponsor, or the corporate legal group
   D. Resign from the company

3. You are preparing for a customer interface meeting and your project sponsor asks you to lie to the customer about certain test results. You should:
   A. Do as you are told
   B. Refuse to work on the project from this point forth
   C. Report the matter to either your superior or the corporate legal group for advice
   D. Resign from the company

4. One of the project managers in your company approaches you with a request to use some of the charge numbers from your project (which is currently running under budget) for work on their project (which is currently running over budget). Your contract is a cost-reimbursable contract for a client external to your company. You should:
   A. Do as you are requested
   B. Refuse to do this unless the project manager allows you to use his charge numbers later on
   C. Report the matter to your superior, the project sponsor, or the corporate legal group
   D. Ask the project manager to resign from the company

5. You have submitted a proposal to a client as part of a competitive bidding effort. One of the people evaluating your bid informs you that it is customary to send them some gifts in order to have a better chance of winning the contract. You should:
   A. Send them some gifts
   B. Do not send any gifts and see what happens
   C. Report the matter to your superior, the project sponsor, or the corporate legal group for advice
   D. Withdraw the proposal

6. You just discovered that the company in which your brother-in-law is employed has submitted a proposal to your company. Your brother-in-law has asked you to do everything possible to make sure that his company will win the contract because his job may be in jeopardy. You should:
   A. Do what your brother-in-law requests
   B. Refuse to look into the matter and pretend it never happened
   C. Report the conflict of interest to your superior, the project sponsor, or the corporate legal group
   D. Hire an attorney for advice

7. As part of a proposal evaluation team, you have discovered that the contract will be awarded to Alpha Company and that a formal announcement will be made in two days. The price of Alpha Company's stock may just skyrocket because of this contract award. You should:
   A. Purchase as much Alpha Company stock as you can within the next two days
   B. Tell family members to purchase the stock
   C. Tell employees in the company to purchase the stock
   D. Do nothing about stock purchases until after the formal announcement has been made

8. Your company has decided to cancel a contract with Beta Company. Only a handful of employees know about this upcoming cancellation. The announcement of the cancellation will be made in about two days. You own several shares of Beta Company stock and know full well that the stock will plunge on the bad news. You should:
   A. Sell your stock as quickly as possible
   B. Sell your stock and tell others whom you know own the stock to do the same thing
   C. Tell the executives to sell their shares if they are stockowners
   D. Do nothing until after the formal announcement is made

9. You are performing a two-day quality audit of one of your suppliers. The supplier asks you to remain a few more days so that they can take you out deep-sea fishing and gambling at the local casino. You should:
   A. Accept as long as you complete the audit within two days
   B. Accept but take vacation time for fishing and gambling
   C. Accept their invitation but at a later time so that it does not interfere with the audit
   D. Gracefully decline their invitation

10. You have been assigned as the project manager for a large project in the Pacific Rim. This is a very important project for both your company and the client. In your first meeting with the client, you are presented with a very expensive gift for yourself and another expensive gift for your husband. You were told by your company that this is considered an acceptable custom when doing work in this country. You should:
    A. Gracefully accept both gifts
    B. Gracefully accept both gifts but report only your gift to your company
    C. Gracefully accept both gifts and report both gifts to your company
    D. Gracefully refuse both gifts.

11. Your company is looking at the purchase of some property for a new plant. You are part of the committee making the final decision. You discover that the owner of a local auto dealership from whom you purchase family cars owns one of the properties. The owner of the dealership tells you in confidence that he will give you a new model car to use for free for up to three years if your company purchases his property for the new plant. You should:
    A. Say thank you and accept the offer
    B. Remove yourself from the committee for conflict of interest

    C. Report the matter to your superior, the project sponsor, or the corporate legal group for advice

    D. Accept the offer as long as the car is in your spouse's name

12. Your company has embarked upon a large project (with you as project manager) and as an output from the project there will be some toxic waste as residue from the manufacturing operations. A subsidiary plan has been developed for the containment and removal of the toxic waste and no environmental danger exists. This information on toxic waste has not been made available to the general public as yet, and the general public does not appear to know about this waste problem. During an interview with local newspaper personnel you are discussing the new project and the question of environmental concerns comes up. You should:

    A. Say there are no problems with environmental concerns

    B. Say that you have not looked at the environmental issues problems as yet

    C. Say nothing and ask for the next question

    D. Be truthful and reply as delicately as possible

13. As a project manager, you establish a project policy that you, in advance of the meeting, review all handouts presented to your external customer during project status review meetings. While reviewing the handouts, you notice that one slide contains company confidential information. Presenting this information to the customer would certainly enhance good will. You should:

    A. Present the information to the customer

    B. Remove the confidential information immediately

    C. Discuss the possible violation with senior management and the legal department before taking any action

    D. First discuss the situation with the team member that created the slide and then discuss the possible violation with senior management and the legal department before taking any action

14. You are managing a project for an external client. Your company developed a new testing procedure to validate certain properties of a product and the new testing procedure was developed entirely with internal funds. Your company owns all of the intellectual property rights associated with the new test. The workers that developed the new test used one of the components developed for your current customer as part of the experimental process. The results using the new test showed that the component would actually exceed the customer's expectations. You should:

    A. Show the results to the customer but do not discuss the fact that it came from the new test procedure

    B. Do not show the results of the new test procedure since the customer's specifications call for use of the old test procedures

    C. First change the customer's specifications and then show the customer the results

    D. Discuss the release of this information with your legal department and senior management before taking any action

15. Using the same scenario as in the previous question, assume that the new test procedure that is expected to be more accurate than the old test procedure indicates that performance will not meet customer specifications whereas the old test indicates that customer specifications will be barely met. You should:

    A. Present the old test results to the customer showing that specification requirements will be met

    B. Show both sets of test results and explain that the new procedure is unproven technology

    C. First change the customer's specifications and then show the customer the results

D. Discuss the release of this information with your legal department and senior management before taking any action

16. Your customer has demanded to see the "raw data" test results from last week's testing. Usually the test results are not released to customers until after the company reaches a conclusion on the meaning of the test results. Your customer has heard through the grapevine that the testing showed poor results. Management has left the entire decision up to you. You should:
    A. Show the results and explain that it is simply raw data and that your company's interpretation of the results will be forthcoming
    B. Withhold the information until after the results are verified
    C. Stall for time even if it means lying to the customer
    D. Explain to the customer your company's policy of not releasing raw data

17. One of your team members plays golf with your external customer's project manager. You discover that the employee has been feeding the customer company-sensitive information. You should:
    A. Inform the customer that project information from anyone other than the project manager is not official until released by the project manager
    B. Change the contractual terms and conditions and release the information
    C. Remove the employee from your project team
    D. Explain to the employee the ramifications of his actions and that he still represents the company when not at work; then report this as a violation

18. Your company has a policy that all company-sensitive material must be stored in locked filing cabinets at the end of each day. One of your employees has received several notices from the security office for violating this policy. You should:
    A. Reprimand the employee
    B. Remove the employee from your project
    C. Ask the Human Resources Group to have the employee terminated
    D. Counsel the employee as well as other team members on the importance of confidentiality and the possible consequences for violations

19. You have just received last month's earned-value information that must be shown to the customer in the monthly status review meeting. Last month's data showed unfavorable variances that exceeded the permissible threshold limits on time and cost variances. This was the result of a prolonged power outage in the manufacturing area. Your manufacturing engineer tells you that this is not a problem and next month you will be right on target on time and cost as you have been in the last five months. You should:
    A. Provide the data to the customer and be truthful in the explanation of the variances
    B. Adjust the variances so that they fall within the threshold limits since this problem will correct itself next month
    C. Do not report any variances this month
    D. Expand the threshold limits on the acceptable variances but do not tell the customer

20. You are working in a foreign country where it is customary for a customer to present gifts to the contractor's project manager throughout the project as a way of showing appreciation. Declining the gifts would be perceived by the customer as an insult. Your company has a policy on how to report gifts received. The *best* way to handle this situation would be to:
    A. Refuse all gifts
    B. Send the customer a copy of our company's policy on accepting gifts
    C. Accept the gifts and report the gifts according to policy
    D. Report all gifts even though the policy says that some gifts need not be reported

21. You are interviewing a candidate to fill a project management position in your company. On her resume, she states that she is a PMP®. One of your workers who knows the candidate informs you that she is not a PMP® yet but is planning to take the test next month and certainly expects to pass. You should:
    A. Wait until she passes the exam before interviewing her
    B. Interview her and ask her why she lied
    C. Inform PMI® of the violation
    D. Forget about it and hire her if she looks like the right person for the job

22. You are managing a multinational project from your office in Chicago. Half of your proj-ect team are from a foreign country but are living in Chicago while working on your proj-ect. These people inform you that two days during next week are national religious holidays in their country and they will be observing the holiday by not coming into work. You should:
    A. Respect their beliefs and say nothing
    B. Force them to work because they are in the United States where their holiday is not celebrated
    C. Tell them that they must work noncompensated overtime when they return to work in order to make up the lost time
    D. Remove them from the project team if possible

23. PMI® informs you that one of your team members who took the PMP® exam last week and passed may have had the answers to the questions in advance provided to him by some of your other team members who are also PMP®s and were tutoring him. PMI® is asking for your support in the investigation. You should:
    A. Assist PMI® in the investigation of the violation
    B. Call in the employee for interrogation and counseling
    C. Call in the other team members for interrogation and counseling
    D. Tell PMI® that it is their problem, not your problem

24. One of your team members has been with you for the past year since her graduation from college. The team member informs you that she is now a PMP® and shows you her certificate from PMI® acknowledging this. You wonder how she was qualified to take the exam since she had no prior work experience prior to joining your company one year ago. You should:
    A. Report this to PMI® as a possible violation
    B. Call in the employee for counseling
    C. Ask the employee to surrender her PMP® credentials
    D. Do nothing

25. Four companies have responded to your RFP. Each proposal has a different technical solution to your problem and each proposal states that the information in the proposal is company-proprietary knowledge and not to be shared with anyone. After evaluation of the proposals, you discover that the best technical approach is from the highest bidder. You are unhappy about this. You decide to show the proposal from the highest bidder to the lowest bidder to see if the lowest bidder can provide the same technical solution but at a lower cost. This situation is:
    A. Acceptable since once the proposals are submitted to your company, you have unlimited access to the intellectual property in the proposals
    B. Acceptable since all companies do this
    C. Acceptable as long as you inform the high bidder that you are showing their proposal to the lowest bidder
    D. Unacceptable and is a violation of the Code of Professional Conduct

ANSWERS

1. B
2. C
3. C
4. C
5. C
6. C
7. D
8. D
9. D
10. C
11. C
12. D
13. D
14. D
15. D
16. A
17. D
18. D
19. A
20. D
21. C
22. A
23. A
24. A
25. D

PROBLEMS

8–1 Beta Company has decided to modify its wage and salary administration program whereby line managers are evaluated for promotion and merit increases based on how well they have lived up to the commitments that they made to the project managers. What are the advantages and disadvantages of this approach?

8–2 How should a project manager handle a situation in which the functional employee (or functional manager) appears to have more loyalty to his profession, discipline, or expertise than to the project? Can a project manager also have this loyalty, say, on an R&D project?

8–3 Most wage and salary administrators contend that project management organizational structures must be "married" to the personnel evaluation process because personnel are always concerned with how they will be evaluated. Furthermore, converting from a traditional structure to a project management structure cannot be accomplished without first considering performance evaluation. What are your feelings on this?

8–4   As part of the evaluation process for functional employees, each project manager submits a written, confidential evaluation report to the employee's department manager who, in turn, makes the final judgment. The employee is permitted to see only the evaluation from his department manager. Assume that the average department merit increase is 7 percent, and that the employee could receive the merit increases shown in the following table. How would he respond in each case?

Project Manager's Evaluation	Merit Increase, %	Credit or Blame to		Reason
		P.M.	Fct. Mgr.	
Excellent	5			
Excellent	7			
Excellent	9			
Average	5			
Average	7			
Average	9			
Poor	5			
Poor	7			
Poor	9			

8–5   Should the evaluation form in Figure 8–4 be shown to the employees?

8–6   Does a functional employee have the right to challenge any items in the project manager's nonconfidential evaluation form?

8–7   Some people contend that functional employees should be able to evaluate the effectiveness of the project manager after project termination. Design an evaluation form for this purpose.

8–8   Some executives feel that evaluation forms should not include cooperation and attitude. The executives feel that a functional employee will always follow the instructions of the functional manager, and therefore attitude and cooperation are unnecessary topics. Does this kind of thinking also apply to the indirect evaluation forms that are filled out by the project managers?

8–9   Consider a situation in which the project manager (a generalist) is asked to provide an evaluation of a functional employee (a specialist). Can the project manager effectively evaluate the functional employee on technical performance? If not, then on what information can the project manager base his evaluation? Can a grade-7 generalist evaluate a grade-12 specialist?

8–10 Gary has been assigned as a part-time, assistant project manager. Gary's duties are split between assistant project management and being a functional employee. In addition, Gary reports both vertically to his functional manager and horizontally to a project manager. As part of his project responsibilities, Gary must integrate activities between his department and two other departments within his divison. His responsibilities also include writing a nonconfidential performance evaluation for all functional employees from all three departments that are assigned to his project. Can Gary effectively and honestly evaluate functional employees in his own department—people with whom he will be working side by side when the project is over? Should the project manager come to his rescue? Suppose Gary is a part-time project manager instead of a part-time assistant project manager. Can anyone come to his rescue now?

8–11 The following question was asked of executives: How do you know when to cut off research? The answers given: That's a good question, a very good question, and some people don't know when to cut it off. You have to have a feel; in some cases it depends on how much resource you have and whether you have enough resources to take a chance on sustaining research that may appear to be heading for a dead end. You don't know sometimes whether you're heading down the wrong path or not; sometimes it's pretty obvious you ought to shift directions—you've gone about as far as you can or you've taken it far enough that you can demonstrate to your own satisfaction that you just can't get there from here, or it's going to be very costly. You may discover that there are more productive ways to get around the barrier; you're always looking for faster ways. And it depends entirely on how creative the person is, whether he has tunnel vision, a very narrow vision, or whether he is fairly flexible in his conceptual thinking so that he can conceive of better ways to solve the problem. Discuss the validity of these remarks.

8–12 In a small company, can a functional manager act as director of engineering and director of project management at the same time?

8–13 In 2002, an electrical equipment manufacturer decentralized the organization, allowing each division manager to set priorities for the work in his division. The division manager of the R&D division selected as his number one priority project the development of low-cost methods for manufacturing. This project required support from the manufacturing division. The division manager for manufacturing did not assign proper resources, claiming that the results of such a project would not be realized for at least five years, and that he (the manufacturing manager) was worried only about the immediate profits. Can this problem be resolved and divisional decentralization still be maintained?

8–14 The executives of a company that produces electro-optical equipment for military use found it necessary to implement project management using a matrix. The project managers reported to corporate sales, and the engineers with the most expertise were promoted to project engineering. After the first year of operation, it became obvious to the executives that the engineering functional managers were not committed to the projects. The executives then made a critical decision. The functional employees selected by the line managers to serve on projects would report as a solid line to the project engineer and dotted to the line manager. The project engineers, who were selected for their technical expertise, were allowed to give technical direction and monetary rewards to the employees. Can this situation work? What happens if an employee has a technical question? Can he go to his line manager? Should the employees return to their former line managers at project completion? What are the authority/responsibility problems with this structure? What are the long-term implications?

8–15 Consider the four items listed on page 123 that describe what happens when a matrix goes out of control. Which of these end up creating the greatest difficulty for the company? for the project managers? for the line managers? for executives?

8–16 As a functional employee, the project manager tells you, "Sign these prints or I'll fire you from this project." How should this situation be handled?

8–17 How efficient can project management be in a unionized, immobile manpower environment?

8–18 Corporate salary structures and limited annual raise allocations often prevent proper project management performance rewards. Explain how each of the following could serve as a motivational factor:

    a. Job satisfaction
    b. Personal recognition
    c. Intellectual growth

The Variables for Success

Related Case Studies (from Kerzner/*Project Management Case Studies*, 3rd Edition)	Related Workbook Exercises (from Kerzner/*Project Management Workbook and PMP®/CAPM® Exam Study Guide*, 10th Edition)	PMBOK® Guide, 4th Edition, Reference Section for the PMP® Certification Exam
• Como Tool and Die (A) • Como Tool and Die (B)	• Multiple Choice Exam	• All PMBOK® Processes

9.0 INTRODUCTION

Project management cannot succeed unless the project manager is willing to employ the systems approach to project management by analyzing those variables that lead to success and failure. This chapter briefly discusses the dos and don'ts of project management and provides a "skeleton" checklist of the key success variables. The following four topics are included:

- ● Predicting project success
- ● Project management effectiveness
- ● Expectations

9.1 PREDICTING PROJECT SUCCESS

PMBOK® Guide, 4th Edition
3.2 Planning Process Group
Chapter 4 Project Integration
5.2 Define Scope

One of the most difficult tasks is predicting whether the project will be successful. Most goal-oriented managers look only at the time, cost, and performance parameters. If an out-of-tolerance condition exists, then additional analysis is required to identify the cause of the problem.

Looking only at time, cost, and performance might identify immediate contributions to profits, but will not identify whether the project itself was managed correctly. This takes on paramount importance if the survival of the organization is based on a steady stream of successfully managed projects. Once or twice a program manager might be able to force a project to success by continually swinging a large baseball bat. After a while, however, either the effect of the big bat will become tolerable, or people will avoid working on his projects.

Project success is often measured by the "actions" of three groups: the project manager and team, the parent organization, and the customer's organization. There are certain actions that the project manager and team can take in order to stimulate project success. These actions include:

- Insist on the right to select key project team members.
- Select key team members with proven track records in their fields.
- Develop commitment and a sense of mission from the outset.
- Seek sufficient authority and a projectized organizational form.
- Coordinate and maintain a good relationship with the client, parent, and team.
- Seek to enhance the public's image of the project.
- Have key team members assist in decision-making and problem-solving.
- Develop realistic cost, schedule, and performance estimates and goals.
- Have backup strategies in anticipation of potential problems.
- Provide a team structure that is appropriate, yet flexible and flat.
- Go beyond formal authority to maximize influence over people and key decisions.
- Employ a workable set of project planning and control tools.
- Avoid overreliance on one type of control tool.
- Stress the importance of meeting cost, schedule, and performance goals.
- Give priority to achieving the mission or function of the end-item.
- Keep changes under control.
- Seek to find ways of assuring job security for effective project team members.

In Chapter 4 we stated that a project cannot be successful unless it is recognized asa project and has the support of top-level management. Top-level management must be willing to commit company resources and provide the necessary administrative support so that the project easily adapts to the company's day-to-day routine of doing business. Furthermore, the parent organization must develop an atmosphere conducive to good working relationships between the project manager, parent organization, and client organization.

With regard to the parent organization, there exist a number of variables that can be used to evaluate parent organization support. These variables include:

- A willingness to coordinate efforts
- A willingness to maintain structural flexibility
- A willingness to adapt to change
- Effective strategic planning
- Rapport maintenance
- Proper emphasis on past experience
- External buffering
- Prompt and accurate communications
- Enthusiastic support
- Identification to all concerned parties that the project does, in fact, contribute to parent capabilities

The mere identification and existence of these variables do not guarantee project success in dealing with the parent organization. Instead, they imply that there exists a good foundation with which to work so that if the project manager and team, and the parent organization, take the appropriate actions, project success is likely. The following actions must be taken:

- Select at an early point, a project manager with a proven track record of technical skills, human skills, and administrative skills (not necessarily in that order) to lead the project team.
- Develop clear and workable guidelines for the project manager.
- Delegate sufficient authority to the project manager, and let him make important decisions in conjunction with key team members.
- Demonstrate enthusiasm for and commitment to the project and team.
- Develop and maintain short and informal lines of communication.
- Avoid excessive pressure on the project manager to win contracts.
- Avoid arbitrarily slashing or ballooning the project team's cost estimate.
- Avoid "buy-ins."
- Develop close, not meddling, working relationships with the principal client contact and project manager.

Both the parent organization and the project team must employ proper managerial techniques to ensure that judicious and adequate, but not excessive, use of planning, controlling, and communications systems can be made. These proper management techniques must also include preconditioning, such as:

- Clearly established specifications and designs
- Realistic schedules
- Realistic cost estimates
- Avoidance of "buy-ins"
- Avoidance of overoptimism

The client organization can have a great deal of influence on project success by minimizing team meetings, making rapid responses to requests for information, and simply

letting the contractor "do his thing" without any interference. The variables that exist for
the client organization include:

- A willingness to coordinate efforts
- Rapport maintenance
- Establishment of reasonable and specific goals and criteria
- Well-established procedures for changes
- Prompt and accurate communications
- Commitment of client resources
- Minimization of red tape
- Providing sufficient authority to the client contact (especially for decision-making)

With these variables as the basic foundation, it should be possible to:

- Encourage openness and honesty from the start from all participants
- Create an atmosphere that encourages healthy competition, but not cutthroat situ-
  ations or "liars'" contests
- Plan for adequate funding to complete the entire project
- Develop clear understandings of the relative importance of cost, schedule, and
  technical performance goals
- Develop short and informal lines of communication and a flat organizational
  structure
- Delegate sufficient authority to the principal client contact, and allow prompt
  approval or rejection of important project decisions
- Reject "buy-ins"
- Make prompt decisions regarding contract award or go-ahead
- Develop close, not meddling, working relationships with project participants
- Avoid arms-length relationships
- Avoid excessive reporting schemes
- Make prompt decisions regarding changes

By combining the relevant actions of the project team, parent organization, and client
organization, we can identify the fundamental lessons for management. These include:

- When starting off in project management, plan to go all the way.
  - Recognize authority conflicts—resolve.
  - Recognize change impact—be a change agent.
- Match the right people with the right jobs.
  - No system is better than the people who implement it.
- Allow adequate time and effort for laying out the project groundwork and
  defining work:
  - Work breakdown structure
  - Network planning
- Ensure that work packages are the proper size:
  - Manageable, with organizational accountability
  - Realistic in terms of effort and time

Today, there is increasing emphasis on documenting lessons learned. Boeing maintains diaries of lessons learned on each airplane project. Another company conducts a post-implementation meeting where the team is required to prepare a three- to five-page case study documenting the successes and failures on the project. The case studies are then used by the training department in preparing individuals to become future project managers. Some companies even mandate that project managers keep project notebooks document-ing all decisions as well as a project file with all project correspondence. On large projects, this may be impractical.

Most companies seem to prefer postimplementation meetings and case study docu-mentation. The problem is when to hold the postimplementation meeting. One company uses project management for new product development and production. When the first production run is complete, the company holds a postimplementation meeting to discuss what was learned. Approximately six months later, the company conducts a second postimplementation meeting to discuss customer reaction to the product. There have been situations where the reaction of the customer indicated that what the company thought they did right turned out to be a wrong decision. A follow-up case study is now prepared dur-ing the second meeting.

9.5 UNDERSTANDING BEST PRACTICES

> **PMBOK® Guide, 4th Edition**
> Chapter 9 Human Resources
>   Management
> 9.4 Manage Project Team

One of the benefits of understanding the variable of success is that it pro-vides you with a means for capturing and retaining best practices. Unfortu-nately this is easier said than done. There are multiple definitions of a best practice, such as:

> **PMBOK® Guide, 4th Edition**
> Professional Responsibility
>   Task #2—Contribute to the
>   PM Knowledge Base

- Something that works
- Something that works well
- Something that works well on a repetitive basis
- Something that leads to a competitive advantage
- Something that can be identified in a proposal to generate business

In the author's opinion, *best practices are those actions or activities undertaken by the company or individuals that lead to a sustained competitive advantage in project management.*

It has only been in recent years that the importance of best practices has been recog-nized. In the early years of project management, there were misconceptions concerning project management. Some of the misconceptions included:

- Project management is a scheduling tool such as PERT/CPM scheduling.
- Project management applies to large projects only.
- Project management is designed for government projects only.
- Project managers must be engineers and preferably with advanced degrees.
- Project managers need a "command of technology" to be successful.
- Project success is measured in technical terms only.

As project management evolved, best practices became important. Best practices can be learned from both successes and failures. In the early years of project management, private industry focused on learning best practices from successes. The government, however, focused on learning about best practices from failures. When the government finally focused on learning from successes, the knowledge on best practices came from their relationships with both their prime contractors and the subcontractors. Some of the best practices that came out of the government included:

● Use of life-cycle phases
● Standardization and consistency
● Use of templates for planning, scheduling, control, and risk
● Providing military personnel in project management positions with extended tours of duty at the same location
● Use of integrated project teams (IPTs)
● Control of contractor-generated scope changes
● Use of earned-value measurement (discussed in Chapter 15)

What to Do with a Best Practice?

With the definition that a best practice leads to a sustained competitive advantage, it is no wonder that some companies were reluctant to make their best practices known to the general public. Therefore, what should a company do with its best practices if not publicize them? The most common options available include:

● **Sharing Knowledge Internally Only:** This is accomplished using the company intranet to share information with employees. There may be a separate group within the company responsible for control of the information, perhaps even the project management officer (PMO).
● **Hidden from All But a Selected Few:** Some companies spend vast amounts of money on the preparation of forms, guidelines, templates, and checklists for project management. These documents are viewed as both company-proprietary information and best practices and are provided to only a select few on a need-to-know basis. An example of a "restricted" best practice might be specialized forms and templates for project approval wherein information contained within may be company-sensitive financial data or the company's position on profitability and market share.
● **Advertise to Your Customers:** In this approach, companies may develop a best practices brochure to market their achievements and may also maintain an extensive best practices library that is shared with their customers after contract award.

Even though companies collect best practices, not all best practices are shared outside of the company, even during benchmarking studies where all parties are expected to share information. Students often ask why textbooks do not include more information on detailed best practices such as forms and templates. One company commented to the author:

We must have spent at least $1 million over the last several years developing an extensive template on how to evaluate the risks associated with transitioning a project from engineering to manufacturing. Our company would not be happy giving this template to everyone who wants to purchase a book for $80. Some best practices templates are common knowledge and we would certainly share this information. But we view the transitioning template as proprietary knowledge not to be shared.

Critical Questions   There are several questions that must be addressed before an activity is recognized as a best practice. Three frequently asked questions are:

- Who determines that an activity is a best practice?
- How do you properly evaluate what you think is best practice to validate that in fact it is a true best practice?
- How do you get executives to recognize that best practices are true value-added activities and should be championed by executive management?

Some organizations have committees that have as their primary function the evaluation of potential best practices. Other organizations use the PMO to perform this work. These committees most often report to senior levels of management.

There is a difference between lessons learned and best practices. Lessons learned can be favorable or unfavorable, whereas best practices are usually favorable outcomes.

Evaluating whether or not something is a best practice is not time-consuming, but it is complex. Simply believing that an action is a best practice does not mean that it is a best practice. PMOs are currently developing templates and criteria for determining whether an activity may qualify as a best practice. Some items that may be included in the template are:

- Is it a measurable metric?
- Does it identify measurable efficiency?
- Does it identify measurable effectiveness?
- Does it add value to the company?
- Does it add value to the customers?
- Is it transferable to other projects?
- Does it have the potential for longevity?
- Is it applicable to multiple users?
- Does it differentiate us from our competitors?
- Will the best practice require governance?
- Will the best practice require employee training?
- Is the best practice company proprietary knowledge?

One company had two unique characteristics in its best practices template:

- Helps to avoid failure
- In a crisis, helps to resolve a critical situation

Executives must realize that these best practices are, in fact, intellectual property to benefit the entire organization. If the best practice can be quantified, then it is usually easier to convince senior management.

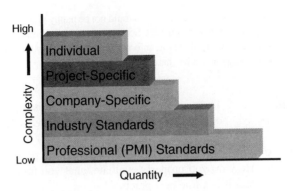

FIGURE 9–1. Levels of best practices.

Levels of Best Practices   Best practices can be discovered anywhere within or outside an organization. Figure 9–1 shows various levels of best practices. The bottom level is the professional standards level, which would include professional standards as defined by PMI®. The professional standards level contains the greatest number of best practices, but they are general rather than specific and have a low level of complexity.

The industry standards level identifies best practices related to performance within the industry. For example, the automotive industry has established standards and best practices specific to the auto industry.

As we progress to the individual best practices in Figure 9–1, the complexity of the best practices goes from general to very specific applications and, as expected, the quantity of best practices is less. An example of a best practice at each level might be (from general to specific):

● **Professional Standards:** Preparation and use of a risk management plan, including templates, guidelines, forms, and checklists for risk management.
● **Industry-Specific:** The risk management plan includes industry best practices such as the best way to transition from engineering to manufacturing.
● **Company-Specific:** The risk management plan identifies the roles and interactions of engineering, manufacturing, and quality assurance groups during transition.
● **Project-Specific:** The risk management plan identifies the roles and interactions of affected groups as they relate to a specific product/service for a customer.
● **Individual:** The risk management plan identifies the roles and interactions of affected groups based upon their personal tolerance for risk, possibly through the use of a responsibility assignment matrix prepared by the project manager.

Best practices can be extremely useful during strategic planning activities. As shown in Figure 9–2, the bottom two levels may be more useful for project strategy formulation whereas the top three levels are more appropriate for the execution of a strategy.

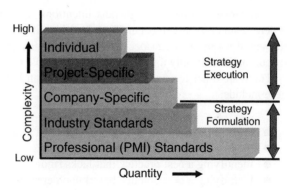

FIGURE 9–2. Usefulness of best practices.

Common Beliefs

There are several common beliefs concerning best practices. A partial list includes:

● Because best practices can be interrelated, the identification of one best practice can lead to the discovery of another best practice, especially in the same category or level of best practices.
● Because of the dependencies that can exist between best practices, it is often easier to identify categories of best practices rather than individual best practices.
● Best practices may not be transferable. What works well for one company may not work for another company.
● Even though some best practices seem simplistic and common sense in most companies, the constant reminder and use of these best practices lead to excellence and customer satisfaction.
● Best practices are not limited exclusively to companies in good financial health

Care must be taken that the implementation of a best practice does not lead to detrimental results. One company decided that the organization must recognize project management as a profession in order to maximize performance and retain qualified people. A project management career path was created and integrated into the corporate reward system.

Unfortunately the company made a severe mistake. Project managers were given significantly larger salary increases than line managers and workers. People became jealous of the project managers and applied for transfer into project management thinking that the "grass was greener." The company's technical prowess diminished and some people resigned when not given the opportunity to become project managers.

Companies can have the greatest intentions when implementing best practices and yet detrimental results can occur. Table 9–1 identifies some possible expectations and the detrimental results that can occur. The poor results could have been the result of poor expectations or not fully understanding the possible ramifications after implementation.

There are other reasons why best practices can fail or provide unsatisfactory results. These include:

● Lack of stability, clarity, or understanding of the best practice
● Failure to use best practices correctly
● Identifying a best practice that lacks rigor
● Identifying a best practice based upon erroneous judgment

Best Practices Library   With the premise that project management knowledge and best practices are intellectual property, how does a company retain this information? The solution is usually the creation of a best practices library. Figure 9–3 shows the three levels of best practices that seem most appropriate for storage in a best practices library.

Figure 9–4 shows the process of creating a best practices library. The bottom level is the discovery and understanding of what is or is not a "potential" best practice. The sources for potential best practices can originate anywhere within the organization.

The next level is the evaluation level to confirm that it is a best practice. The evaluation process can be done by the PMO or a committee but should have involvement by the senior levels of management. The evaluation process is very difficult because a one-time positive occurrence may not reflect a best practice. There must exist established criteria for the evaluation of a best practice.

Once it is agreed upon that a best practice exists, it must be classified and stored in some retrieval system such as a company intranet best practices library.

Figure 9–1 shows the levels of best practices, but the classification system for storage purposes can be significantly different. Figure 9–5 shows a typical classification system for a best practices library.

TABLE 9–1. RESULTS OF IMPLEMENTING BEST PRACTICES

Type of Best Practice	Expected Advantage	Potential Disadvantage
Use of traffic light reporting	Speed and simplicity	Poor accuracy of information
Use of a risk management template/form	Forward looking and accurate	Inability to see some potential critical risks
Highly detailed WBS	Control, accuracy, and completeness	Excessive control and cost of reporting
Using EPM on all projects	Standardization and consistency	Too expensive on certain projects
Using specialized software	Better decision-making	Too much reliance on tools

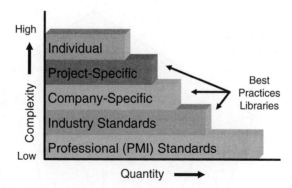

FIGURE 9–3. Levels of best practices.

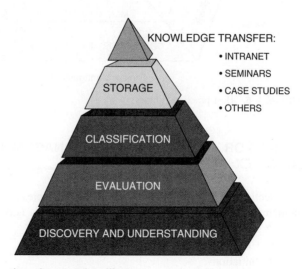

FIGURE 9–4. Creating a best practices library.

The purpose for creating a best practices library is to transfer knowledge to employees. The knowledge can be transferred through the company intranet, seminars on best practices, and case studies. Some companies require that the project team prepare case studies on lessons learned and best practices before the team is disbanded. These companies then use the case studies in company-sponsored seminars. Best practices and lessons learned must be communicated to the entire organization. The problem is determining how to do it effectively.

Another critical problem is best practices overload. One company started up a best practices library and, after a few years, had amassed hundreds of what were considered to be best practices. Nobody bothered to reevaluate whether or not all of these were still best practices. After reevaluation had taken place, it was determined that less than one-third of these were still regarded as best practices. Some were no longer best practices, others needed to be updated, and others had to be replaced with newer best practices.

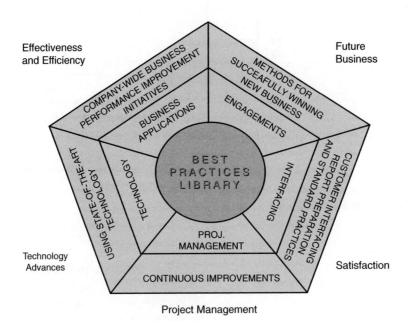

Effectiveness
and Efficiency

Future
Business

Technology
Advances

Satisfaction

Project Management

FIGURE 9–5. Best practices library.

9.6 STUDYING TIPS FOR THE PMI® PROJECT MANAGEMENT CERTIFICATION EXAM

This section is applicable as a review of the principles to support the knowledge areas and domain groups in the PMBOK® Guide. This chapter addresses:

- Communications Management
- Initiation
- Planning
- Execution
- Monitoring
- Closure

Understanding the following principles is beneficial if the reader is using this text to study for the PMP® Certification Exam:

- Importance of capturing and reporting best practices as part of all project management processes
- Variables for success

The following multiple-choice questions will be helpful in reviewing the principles of this chapter:

1. Lessons learned and best practices are captured:
   A. Only at the end of the project
   B. Only after execution is completed
   C. Only when directed to do so by the project sponsor
   D. At all times but primarily at the closure of each life-cycle phase

2. The person responsible for the identification of a best practice is the:
   A. Project manager
   B. Project sponsor
   C. Team member
   D. All of the above

3. The primary benefit of capturing lessons learned is to:
   A. Appease the customer
   B. Appease the sponsor
   C. Benefit the entire company on a continuous basis
   D. Follow the PMBOK® requirements for reporting

ANSWERS

1. D
2. D
3. C

PROBLEMS

9–1 What is an effective working relationship between project managers themselves?

9–2 Must everyone in the organization understand the "rules of the game" for project management to be effective?

9–3 Defend the statement that the first step in making project management work must be a complete definition of the boundaries across which the project manager must interact.

10 Working with Executives

Related Case Studies (from Kerzner/*Project Management Case Studies*, 3rd Edition)	Related Workbook Exercises (from Kerzner/*Project Management Workbook and PMP®/CAPM® Exam Study Guide*, 10th Edition)	PMBOK® Guide, 4th Edition, Reference Section for the PMP® Certification Exam
• Greyson Corporation • The Blue Spider Project • Corwin Corporation*	• Multiple Choice Exam	• Integration Management • Scope Management • Human Resource Management

10.0 INTRODUCTION

PMBOK® Guide, 4th Edition
Chapter 4 Integration Management
Chapter 9 Human Resources Management Chapter

In any project management environment, project managers must continually interface with executives during both the planning and execution stages. Unless the project manager understands the executive's role and thought process, a poor working relationship will develop. In order to understand the executive–project interface, two topics are discussed:

● The project sponsor
● The in-house representatives

*Case Study also appears at end of chapter.

10.1 THE PROJECT SPONSOR

PMBOK® Guide, 4th Edition
2.3 Key Stakeholders
5.1.2 Stakeholder Analysis

For more than two decades, the traditional role of senior management, as far as projects were concerned, has been to function as project sponsors. The project sponsor usually comes from the executive levels and has the primary responsibility of maintaining executive–client contact. The sponsor ensures that the correct information from the contractor's organization is reaching executives in the customer's organization, that there is no filtering of information from the contractor to the customer, and that someone at the executive levels is making sure that the customer's money is being spent wisely. The project sponsor will normally transmit cost and deliverables information to the customer, whereas schedule and performance status data come from the project manager.

In addition to executive–client contact, the sponsor also provides guidance on:

- Objective setting
- Priority setting
- Project organizational structure
- Project policies and procedures
- Project master planning
- Up-front planning
- Key staffing
- Monitoring execution
- Conflict resolution

The role of the project sponsor takes on different dimensions based on the life-cycle phase the project is in. During the planning/initiation phase of a project, the sponsor normally functions in an active role, which includes such activities as:

- Assisting the project manager in establishing the correct objectives for the project
- Providing the project manager with information on the environmental/political factors that could influence the project's execution
- Establishing the priority for the project (either individually or through consultation with other executives) and informing the project manager of the established priority and the *reason* for the priority
- Providing guidance for the establishment of policies and procedures by which to govern the project
- Functioning as the executive–client contact point

During the initiation or kickoff phase of a project, the project sponsor must be actively involved in setting objectives and priorities. It is absolutely mandatory that the executives establish the priorities in both business and technical terms.

During the execution phase of the project, the role of the executive sponsor is more passive than active. The sponsor will provide assistance to the project manager on an as-needed basis except for routine status briefings.

During the execution stage of a project, the sponsor must be *selective* in the problems that he or she wishes to help resolve. Trying to get involved in every problem will not only result in severe micromanagement, but will undermine the project manager's ability to get the job done.

The role of the sponsor is similar to that of a referee. Table 10–1 shows the working relationship between the project manager and the line managers in both mature and immature organizations. When conflicts or problems exist in the project–line interface and cannot be resolved at that level, the sponsor might find it necessary to step in and provide assistance. Table 10–2 shows the mature and immature ways that a sponsor interfaces with the project.

PMBOK® Guide, 4th Edition
1.6 Interpersonal Skills

TABLE 10–1. THE PROJECT–LINE INTERFACE

Immature Organization	Mature Organization
• Project manager is vested with power/ authority over the line managers.	• Project and line managers share authority and power.
• Project manager negotiates for best people.	• Project manager negotiates for line manager's commitment.
• Project manager works directly with functional employees.	• Project manager works through line managers.
• Project manager has no input into employee performance evaluations.	• Project manager makes recommendations to the line managers.
• Leadership is project manager-centered.	• Leadership is team-centered.

TABLE 10–2. THE EXECUTIVE INTERFACE

Immature Organization	Mature Organization
• Executive is actively involved in projects.	• Executive involvement is passive.
• Executive acts as the project champion.	• Executive acts as the project sponsor.
• Executive questions the project manager's decisions.	• Executive trusts the project manager's decisions.
• Priority shifting occurs frequently.	• Priority shifting is avoided.
• Executive views project management as a necessary evil.	• Executive views project management as beneficial.
• There is very little project management support.	• There is visible, ongoing support.
• Executive discourages bringing problems upstairs.	• Executive encourages bringing problems upstairs.
• Executive is not committed to project sponsorship.	• Executive is committed to sponsorship (and ownership).
• Executive support exists only during project start-up.	• Executive support exists on a continuous basis.
• Executive encourages project decisions to be made.	• Executive encourages business decisions to be made.
• No procedures exist for assigning project sponsors.	• Sponsorship assignment procedures are visible.
• Executives seek perfection.	• Executives seek what is possible.
• Executive discourages use of a project charter.	• Executive recognizes the importance of a charter.
• Executive is not involved in charter preparation.	• Executive takes responsibility for charter preparation.
• Executive does not understand what goes into a charter.	• Executive understands the content of a charter.
• Executives do not believe that the project team is performing.	• Executives trust that performance is taking place.

It should be understood that the sponsor exists for everyone on the project, including the line managers and their employees. Project sponsors must maintain open-door policies, even though maintaining an open-door policy can have detrimental effects. First, employees may flood the sponsor with trivial items. Second, employees may feel that they can by-pass levels of management and converse directly with the sponsor. The moral here is that employees, including the project manager, must be encouraged to be careful about how many times and under what circumstances they "go to the well."

In addition to his/her normal functional job, the sponsor must be available to provide as-needed assistance to the projects. Sponsorship can become a time-consuming effort, especially if problems occur. Therefore, executives are limited as to how many projects they can sponsor effectively at the same time.

If an executive has to function as a sponsor on several problems at once, problems can occur such as:

- Slow decision-marking resulting in problem-solving delays
- Policy issues that remain unresolved and impact decisions
- Inability to prioritize projects when necessary

As an organization matures in project management, executives begin to trust middle- and lower-level management to function as sponsors. There are several reasons for supporting this:

- Executives do not have time to function as sponsors on each and every project.
- Not all projects require sponsorship from the executive levels.
- Middle management is closer to where the work is being performed.
- Middle management is in a better position to provide advice on certain risks.
- Project personnel have easier access to middle management.

Sometimes executives in large diversified corporations are extremely busy with strategic planning activities and simply do not have the time to properly function as a sponsor. In such cases, sponsorship falls one level below senior management.

Figure 10–1 shows the major functions of a project sponsor. At the onset of a project, a senior committee meets to decide whether a given project should be deemed as priority or nonpriority. If the project is critical or strategic, then the committee may assign a senior manager as the sponsor, perhaps even a member of the committee. It is common practice for steering committee executives to function as sponsors for the projects that the steering committee oversees.

For projects that are routine, maintenance, or noncritical, a sponsor could be assigned from the middle-management levels. One organization that strongly prefers to have middle management assigned as sponsors cites the benefit of generating an atmosphere of management buy-in at the critical middle levels.

Not all projects need a project sponsor. Sponsorship is generally needed on those projects that require a multitude of resources or a large amount of integration between functional lines or that have the potential for disruptive conflicts or the need for strong customer communications. This last item requires further comment. Quite often customers wish to make sure that the contractor's project manager is spending funds prudently.

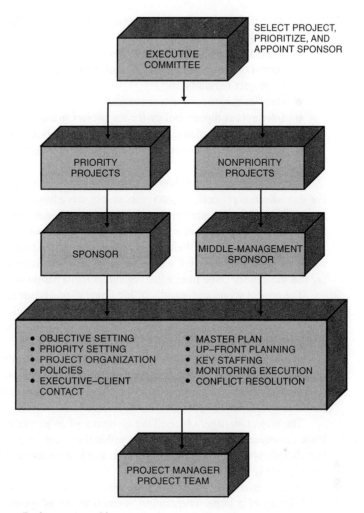

FIGURE 10–1. Project sponsorship.

Customers therefore like it when an executive sponsor supervises the project manager's funding allocation.

It is common practice for companies that are heavily involved in competitive bidding to identify in their proposal not only the resumé of the project manager, but the resumé of the executive project sponsor as well. This may give the bidder a competitive advantage, all other things being equal, because customers believe they have a direct path of communications to executive management. One such contractor identified the functions of the executive project sponsor as follows:

● Major participation in sales effort and contract negotiations
● Establishes and maintains top-level client relationships

- Assists project manager in getting the project underway (planning, procedures, staffing, etc.)
- Maintains current knowledge of major project activities (receives copies of major correspondence and reports, attends major client and project review meetings, visits project regularly, etc.)
- Handles major contractual matters
- Interprets company policy for the project manager
- Assists project manager in identifying and solving major problems
- Keeps general management and company management advised of major problems

Consider a project that is broken down into two life-cycle phases: planning and execution. For short-duration projects, say two years or less, it is advisable for the project sponsor to be the same individual for the entire project. For long-term projects of five years or so, it is possible to have a different project sponsor for each life-cycle phase, but preferably from the same level of management. The sponsor does not have to come from the same line organization as the one where the majority of the work will be taking place. Some companies even go so far as demanding that the sponsor come from a line organization that has no vested interest in the project.

The project sponsor is actually a "big brother" or advisor for the project manager. Under *no* circumstances should the project sponsor try to function as the project manager. The project sponsor should assist the project manager in solving those problems that the project manager cannot resolve by himself.

In one government organization, the project manager wanted to open up a new position on his project, and already had a woman identified to fill the position. Unfortunately, the size of the government project office was constrained by a unit-manning document that dictated the number of available positions.

The project manager obtained the assistance of an executive sponsor who, working with human resources, created a new position within thirty days. Without executive sponsorship, the bureaucratic system creating a new position would have taken months. By that time, the project would have been over.

In a second case study, the president of a medium-sized manufacturing company, a subsidiary of a larger corporation, wanted to act as sponsor on a special project. The project manager decided to make full use of this high-ranking sponsor by assigning him certain critical functions. As part of the project's schedule, four months were allocated to obtain corporate approval for tooling dollars. The project manager "assigned" this task to the project sponsor, who reluctantly agreed to fly to corporate headquarters. He returned two days later with authorization for tooling. The company actually reduced project completion time by four months, thanks to the project sponsor.

Figure 10–2 represents a situation where there were two project sponsors for one project. Alpha Company received a $25 million prime contractor project from the Air Force and subcontracted out $2 million to Beta Company. The project manager in Alpha Company earned $175,000 per year and refused to communicate directly with the project manager of Beta Company because his salary was only $90,000 per year. After all, as one executive said, "Elephants don't communicate with mice." The Alpha Company project manager instead sought out someone at Beta in his own salary range to act as the project sponsor, and the burden fell on the director of engineering.

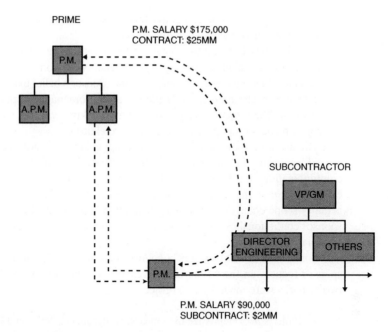

FIGURE 10–2. Multiple project sponsors.

The Alpha Company project manager reported to an Air Force colonel. The Air Force colonel considered his counterpart in Beta Company to be the vice president and general manager. Here, power and title were more important than the $100,000 differential in their salaries. Thus, there was one project sponsor for the prime contractor and a second project sponsor for the customer.

In some industries, such as construction, the project sponsor is identified in the proposal, and thus everyone knows who it is. Unfortunately, there are situations where the project sponsor is "hidden," and the project manager may not realize who it is, or know if the customer realizes who it is. This concept of invisible sponsorship occurs most frequently at the executive level and is referred to as absentee sponsorship.

There are several ways that invisible sponsorship can occur. The first is when the manager who is appointed as a sponsor refuses to act as a sponsor for fear that poor decisions or an unsuccessful project could have a negative impact on his or her career. The second type results when an executive really does not understand either sponsorship or project management and simply provides lip service to the sponsorship function. The third way involves an executive who is already overburdened and simply does not have the time to perform meaningfully as a sponsor. The fourth way occurs when the project manager refuses to keep the sponsor informed and involved. The sponsor may believe that everything is flowing smoothly and that he is not needed.

Some people contend that the best way for the project manager to work with an invisible sponsor is for the project manager to make a decision and then send a memo to the sponsor stating "This is the decision that I have made and, unless I hear from you in the next 48 hours, I will assume that you agree with my decision."

The opposite extreme is the sponsor who micromanages. One way for the project manager to handle this situation is to bury the sponsor with work in hopes that he will let go. Unfortunately this could end up reinforcing the sponsor's belief that what he is doing is correct.

The better alternative for handling a micromanaging sponsor is to ask for role clarification. The project manager should try working with the sponsor to define the roles of project manager and project sponsor more clearly.

The invisible sponsor and the overbearing sponsor are not as detrimental as the "can't-say-no" sponsor. In one company, the executive sponsor conducted executive–client communications on the golf course by playing golf with the customer's sponsor. After every golf game, the executive sponsor would return with customer requests, which were actually scope changes that were considered as no-cost changes by the customer. When a sponsor continuously says "yes" to the customer, everyone in the contractor's organization eventually suffers.

Sometimes the existence of a sponsor can do more harm than good, especially if the sponsor focuses on the wrong objectives around which to make decisions. The following two remarks were made by two project managers at an appliance manufacturer:

- Projects here emphasize time measures: deadlines! We should emphasize milestones reached and quality. We say, "We'll get you a system by a deadline." We should be saying, "We'll get you a good system."
- Upper management may not allow true project management to occur. Too many executives are "date-driven" rather than "requirements-driven." Original target dates should be for broad planning only. Specific target dates should be set utilizing the full concept of project management (i.e., available resources, separation of basic requirements from enhancements, technical and hardware constraints, unplanned activities, contingencies, etc.)

These comments illustrate the necessity of having a sponsor who understands project management rather than one who simply assists in decision-making. The goals and objectives of the sponsor must be aligned with the goals and objectives of the project, and they must be realistic. If sponsorship is to exist at the executive levels, the sponsor must be visible and constantly informed concerning the project status.

Committee Sponsorship

For years companies have assigned a single individual as the sponsor for a project. The risk was that the sponsor would show favoritism to his line group and suboptimal decision-making would occur. Recently, companies have begun looking at sponsorship by committee to correct this.

Committee sponsorship is common in those organizations committed to concurrent engineering and shortening product development time. Committees are comprised of middle managers from marketing, R&D, and operations. The idea is that the committee will be able to make decisions in the best interest of the company more easily than a single individual could.

Committee sponsorship also has its limitations. At the executive levels, it is almost impossible to find time when senior managers can convene. For a company with a large number of projects, committee sponsorship may not be a viable approach.

In time of crisis, project managers may need immediate access to their sponsors. If the sponsor is a committee, then how does the project manager get the committee to convene quickly? Also, individual project sponsors may be more dedicated than committees. Committee sponsorship has been shown to work well if one, and only one, member of the committee acts as the prime sponsor for a given project.

When to Seek Help

During status reporting, a project manager can wave either a red, yellow, or green flag. This is known as the "traffic light" reporting system, thanks in part to color printers. For each element in the status report, the project manager will illuminate one of three lights according to the following criteria:

- *Green light:* Work is progressing as planned. Sponsor involvement is not necessary.
- *Yellow light:* A potential problem may exist. The sponsor is informed but no action by the sponsor is necessary at this time.
- *Red light:* A problem exists that may affect time, cost, scope, or quality. Sponsor involvement is necessary.

Yellow flags are warnings that should be resolved at the middle levels of management or lower.

If the project manager waves a red flag, then the sponsor will probably wish to be actively involved. Red flag problems can affect the time, cost, or performance constraints of the project and an immediate decision must be made. The main function of the sponsor is to assist in making the best possible decision in a timely fashion.

Both project sponsors and project managers should not encourage employees to come to them with problems unless the employees also bring alternatives and recommendations. Usually, employees will solve most of their own problems once they prepare alternatives and recommendations.

Good corporate cultures encourage people to bring problems to the surface quickly for resolution. The quicker the potential problem is identified, the more opportunities are available for resolution.

A current problem plaguing executives is who determines the color of the light. Consider the following problem: A department manager had planned to perform 1000 hours of work in a given time frame but has completed only 500 hours at the end of the period. According to the project manager's calculation, the project is behind schedule, and he would prefer to have the traffic light colored yellow or red. The line manager, however, feels that he still has enough "wiggle room" in his schedule and that his effort will still be completed within time and cost, so he wants the traffic light colored green. Most executives seem to favor the line manager who has the responsibility for the deliverable. Although the project manager has the final say on the color of traffic light, it is most often based upon the previous working relationship between the two and the level of trust.

Some companies use more than three colors to indicate project status. One company also has an orange light for activities that are still being performed after the target milestone date.

**The New Role
of the Executive**

As project management matures, executives decentralize project sponsorship to middle- and lower-level management. Senior management then takes on new roles such as:

- Establishing a Center for Excellence in project management
- Establishing a project office or centralized project management function
- Creating a project management career path
- Creating a mentorship program for newly appointed project managers
- Creating an organization committed to benchmarking best practices in project management in other organizations
- Providing strategic information for risk management

This last bullet requires further comment. Because of the pressure placed upon the project manager for schedule compression, risk management could very well become the single most critical skill for project managers. Executives will find it necessary to provide project management with strategic business intelligence, assist in risk identification, and evaluate or prioritize risk-handling options.

**Active versus Passive
Involvement**

One of the questions facing senior management in the assigning of a project sponsor is whether or not the sponsor should have a vested interest in the project or be an impartial outsider. Table 10–3 shows the pros and cons of this. Sponsors that that do not have a vested interest in the project seem to function more as exit champions rather than project sponsors.

Managing Scope Creep

Technically oriented team members are motivated not only by meeting specifications, but also by exceeding them. Unfortunately, exceeding specifications can be quite costly. Project managers must monitor scope creep and develop plans for controlling scope changes.

But what if it is the project manager who initiates scope creep? The project sponsor must meet periodically with the project manager to review the scope baseline changes or unauthorized changes may occur and significant cost increases will result, as shown in Situation 10–1 below:

> **PMBOK® Guide, 4th Edition**
> 5.5 Scope Control
> 5.5.3.3 Change Control System

TABLE 10–3.   VESTED INTEREST OR NOT?

Vested Interest	Impartial
• Finance the fund-starved project	• Provide no funding and limited support
• Keep project alive	• Let project die
• Maximum protection from obstacles	• Limited protection from obstacles
• Fend off internal enemies	• Avoid politics and enemies
• Actively involved	• Go through motions
• Involved in personnel assignments	• Partial involvement in assignment

SITUATION 10–1: PINE LAKE AMUSEMENT PARK

After six years of debate, the board of directors of Pine Lake Amusement Park finally came to an agreement on the park's new aquarium. The aquarium would be built, at an estimated cost of $30 million and, between fundraising and bank loans, financing was possible.

After the drawings were completed and approved, the project was estimated as a two-year construction effort. Because of the project's complexity, a decision was made to have the project manager brought on board from the beginning of the design efforts, and to remain until six months after opening day. The project manager assigned was well known for his emphasis on details and his strong feelings for the aesthetic beauty of a ride or show.

The drawings were completed and a detailed construction cost estimate was undertaken. When the final cost estimate of $40 million was announced, the board of directors was faced with three alternatives: cancel the project, seek an additional $10 million in financing, or descope (i.e., reduce functionality of) the project. Additional funding was unacceptable and years of publicity on the future aquarium would be embarrassing for the board if the project were to be canceled. The only reasonable alternative was to reduce the project's scope.

After two months of intensive replanning, the project team proposed a $32 million aquarium. The board of directors agreed to the new design and the construction phase of the project began. The project manager was given specific instructions that cost overruns would not be tolerated.

At the end of the first year, more than $22 million had been spent. Not only had the project manager reinserted the scope that had been removed during the descoping efforts, but also additional scope creep had increased to the point where the final cost would now exceed $62 million. The new schedule now indicated a three-year effort. By the time that management held its review meetings with the project team, the changes had been made.

The Executive Champion

Executive champions are needed for those activities that require the implementation of change, such as a new corporate methodology for project management. Executive champions "drive" the implementation of project management down into the organization and accelerate its acceptance because their involvement implies executive-level support and interest.

10.2 HANDLING DISAGREEMENTS WITH THE SPONSOR

For years, we believed that the project sponsor had the final say on all decisions affecting the project. The sponsor usually had a vested interest in the project and was responsible for obtaining funding for the project. But what if the project manager believes that the sponsor has made the wrong decision? Should the project manager have a path for recourse action in such a situation?

There are several reasons why disagreements between the project manager and project sponsor will occur. First, the project sponsors may not have sufficient technical knowledge or information to evaluate the risks of any potential decision. Second, sponsors may be heavily burdened with other activities and unable to devote sufficient time to sponsorship. Third, some companies prefer to assign sponsors who have no vested interest in the project in hopes of getting impartial decision-making. Finally, sponsorship may be pushed down to a middle-management level where the assigned sponsor may not have all of the business knowledge necessary to make the best decisions.

Project managers are expected to challenge the project's assumptions continuously. This could lead to trade-offs. It could also lead to disagreements and conflicts between the project manager and the project sponsor. In such cases, the conflict will be brought to the executive steering committee for resolution. Sponsors must understand that their decisions as a sponsor can and should be challenged by the project manager.

Recognizing that these conflicts can exist, companies are instituting executive steering committees or executive policy board committees to quickly resolve these disputes. Few conflicts ever make it to the executive steering committee, but those that do are usually severe and may expose the company to unwanted risks.

A common conflict that may end up at the executive steering committee level is when one party wants to cancel the project and the second party wants to continue. This situation occurred at a telecommunications company where the project manager felt that the project should be canceled but the sponsor wanted the project to continue because its termination would reflect poorly upon him. Unfortunately, the steering committee sided with the sponsor and let the project continue. The company squandered precious resources for several more months before finally terminating the project.

10.3 THE COLLECTIVE BELIEF

Some projects, especially very long-term projects, often mandate that a collective belief exist. The collective belief is a fervent, and perhaps blind, desire to achieve that can permeate the entire team, the project sponsor, and even the most senior levels of management. The collective belief can make a rational organization act in an irrational manner. This is particularly true if the project sponsor spearheads the collective belief.

When a collective belief exists, people are selected based upon their support for the collective belief. Nonbelievers are pressured into supporting the collective belief and team members are not allowed to challenge the results. As the collective belief grows, both advocates and nonbelievers are trampled. The pressure of the collective belief can outweigh the reality of the results.

There are several characteristics of the collective belief, which is why some large, high-technology projects are often difficult to kill:

- Inability or refusal to recognize failure
- Refusing to see the warning signs
- Seeing only what you want to see
- Fearful of exposing mistakes
- Viewing bad news as a personal failure

- Viewing failure as a sign of weakness
- Viewing failure as damage to one's career
- Viewing failure as damage to one's reputation

10.4 THE EXIT CHAMPION

Project sponsors and project champions do everything possible to make their project successful. But what if the project champions, as well as the project team, have blind faith in the success of the project? What happens if the strongly held convictions and the collective belief disregard the early warning signs of imminent danger? What happens if the collective belief drowns out dissent?

In such cases, an exit champion must be assigned. The exit champion sometimes needs to have some direct involvement in the project in order to have credibility, but direct involvement is not always a necessity. Exit champions must be willing to put their reputation on the line and possibly face the likelihood of being cast out from the project team. According to Isabelle Royer[1]:

> Sometimes it takes an individual, rather than growing evidence, to shake the collective belief of a project team. If the problem with unbridled enthusiasm starts as an unintended consequence of the legitimate work of a project champion, then what may be needed is a countervailing force—an exit champion. These people are more than devil's advocates. Instead of simply raising questions about a project, they seek objective evidence showing that problems in fact exist. This allows them to challenge—or, given the ambiguity of existing data, conceivably even to confirm—the viability of a project. They then take action based on the data.

The larger the project and the greater the financial risk to the firm, the higher up the exit champion should reside. If the project champion just happens to be the CEO, then someone on the board of directors or even the entire board of directors should assume the role of the exit champion. Unfortunately, there are situations where the collective belief permeates the entire board of directors. In this case, the collective belief can force the board of directors to shirk their responsibility for oversight.

Large projects incur large cost overruns and schedule slippages. Making the decision to cancel such a project, once it has started, is very difficult, according to David Davis[2]:

> The difficulty of abandoning a project after several million dollars have been committed to it tends to prevent objective review and recosting. For this reason, ideally an independent management team—one not involved in the projects development—should do the recosting

1. Isabelle Royer, "Why Bad Projects are So Hard to Kill," *Harvard Business Review*, February 2003, p.11; Copyright © 2003 by the Harvard Business School Publishing Corporation. All rights reserved.
2. David Davis, "New Projects: Beware of False Economics," *Harvard Business Review*, March–April 1985, pp.100–101; Copyright © 1985 by the President and Fellows of Harvard College. All rights reserved.

and, if possible, the entire review. . . . If the numbers do not holdup in the review and recosting, the company should abandon the project. The number of bad projects that make it to the operational stage serves as proof that their supporters often balk at this decision.

. . . Senior managers need to create an environment that rewards honesty and courage and provides for more decision making on the part of project managers. Companies must have an atmosphere that encourages projects to succeed, but executives must allow them to fail.

The longer the project, the greater the necessity for the exit champions and project sponsors to make sure that the business plan has "exit ramps" such that the project can be terminated before massive resources are committed and consumed. Unfortunately, when a collective belief exists, exit ramps are purposefully omitted from the project and business plans. Another reason for having exit champions is so that the project closure process can occur as quickly as possible. As projects approach their completion, team members often have apprehension about their next assignment and try to stretch out the existing project until they are ready to leave. In this case, the role of the exit champion is to accelerate the closure process without impacting the integrity of the project.

Some organizations use members of a portfolio review board to function as exit champions. Portfolio review boards have the final say in project selection. They also have the final say as to whether or not a project should be terminated. Usually, one member of the board functions as the exit champion and makes the final presentation to the remainder of the board.

10.5 THE IN-HOUSE REPRESENTATIVES

On high-risk, high-priority projects or during periods of mistrust, customers may wish to place in-house representatives in the contractor's plant. These representatives, if treated properly, are like additional project office personnel who are not supported by your budget. They are invaluable resources for reading rough drafts of reports and making recommendations as to how their company may wish to see the report organized.

In-house representatives are normally not situated in or near the contractor's project office because of the project manager's need for some degree of privacy. The exception would be in the design phase of a construction project, where it is imperative to design what the customer wants and to obtain quick decisions and approvals.

Most in-house representatives know where their authority begins and ends. Some companies demand that in-house representatives have a project office escort when touring the plant, talking to functional employees, or simply observing the testing and manufacturing of components.

It is possible to have a disruptive in-house representative removed from the company. This usually requires strong support from the project sponsor in the contractor's shop. The important point here is that executives and project sponsors must maintain proper contact with and control over the in-house representatives, perhaps more so than the project manager.

10.6 STUDYING TIPS FOR THE PMI® PROJECT MANAGEMENT CERTIFICATION EXAM

This section is applicable as a review of the principles to support the knowledge areas and domain groups in the PMBOK® Guide. This chapter addresses:

- Integration Management
- Scope Management
- Human Resources Management
- Initiation
- Planning
- Execution
- Monitoring
- Closure

Understanding the following principles is beneficial if the reader is using this text to study for the PMP® Certification Exam:

- Role of the executive sponsor or project sponsor
- That the project sponsor need not be at the executive levels
- That some projects have committee sponsorship
- When to bring a problem to the sponsor and what information to bring with you

In Appendix C, the following Dorale Products mini–case studies are applicable:

- Dorale Products (G) [Integration and Scope Management]

The following multiple-choice questions will be helpful in reviewing the principles of this chapter:

1. The role of the project sponsor during project initiation is to assist in:
   A. Defining the project's objectives in both business and technical terms
   B. Developing the project plan
   C. Performing the project feasibility study
   D. Performing the project cost-benefit analysis

2. The role of the project sponsor during project execution is to:
   A. Validate the project's objectives
   B. Validate the execution of the plan
   C. Make all project decisions
   D. Resolve problems/conflicts that cannot be resolved elsewhere in the organization

3. The role of the project sponsor during the closure of the project or a life-cycle phase of the project is to:
   A. Validate that the profit margins are correct
   B. Sign off on the acceptance of the deliverables
   C. Administer performance reviews of the project team members
   D. All of the above

ANSWERS

    1. A

    2. D

    3. B

PROBLEMS

10–1 Should age have a bearing on how long it takes an executive to accept project management?

10–2 You have been called in by the executive management of a major utility company and asked to give a "selling" speech on why the company should go to project management. What are you going to say? What areas will you stress? What questions would you expect the executives to ask? What fears do you think the executives might have?

10–3 Some executives would prefer to have their project managers become tunnel-vision workaholics, with the project managers falling in love with their jobs and living to work instead of working to live. How do you feel about this?

10–4 Project management is designed to make effective and efficient use of resources. Most companies that adopt project management find it easier to underemploy and schedule overtime than to overemploy and either lay people off or drive up the overhead rate. A major electrical equipment manufacturer contends that with proper utilization of the project management concept, the majority of the employees who leave the company through either termination or retirement do not have to be replaced. Is this rationale reasonable?

10–5 The director of engineering services of R. P. Corporation believes that a project organizational structure of some sort would help resolve several of his problems. As part of the discussion, the director has made the following remarks: "All of our activities (or so-called projects if you wish) are loaded with up-front engineering. We have found in the past that time is the important parameter, not quality control or cost. Sometimes we rush into projects so fast that we have no choice but to cut corners, and, of course, quality must suffer."

    What questions, if any, would you like to ask before recommending a project organizational form? Which form will you recommend?

10–6 How should a project manager react when he finds inefficiency in the functional lines? Should executive management become involved?

10–7 An electrical equipment manufacturing company has just hired you to conduct a three-day seminar on project management for sixty employees. The president of the company asks you to have lunch with him on the first day of the seminar. During lunch, the executive remarks, "I inherited the matrix structure when I took over. Actually I don't think it can work here, and I'm not sure how long I'll support it." How should you continue at this point?

10–8 Should project managers be permitted to establish prerequisites for top management regarding standard company procedures?

10–9 During the implementation of project management, you find that line managers are reluctant to release any information showing utilization of resources in their line function. How should this situation be handled, and by whom?

10–10 Corporate engineering of a large corporation usually assumes control of all plant expansion projects in each of its plants for all projects over $25 million. For each case below, discuss the ramifications of this, assuming that there are several other projects going on in each plant at the same time as the plant expansion project.

    a. The project manager is supplied by corporate engineering and reports to corporate engineering, but all other resources are supplied by the plant manager.

    b. The project manager is supplied by corporate but reports to the plant manager for the duration of the project.

    c. The plant manager supplies the project manager, and the project manager reports "solid" to corporate and "dotted" to the plant manager for the duration of the project.

10–11 An aircraft company requires seven years from initial idea to full production of a military aircraft. Consider the following facts: engineering design requires a minimum of two years of R&D; manufacturing has a passive role during this time; and engineering builds its own prototype during the third year.

    a. To whom in the organization should the program manager, project manager, and project engineering report? Does your answer depend on the life-cycle phase?

    b. Can the project engineers be "solid" to the project manager and still be authorized by the engineering vice president to provide technical direction?

    c. What should be the role of marketing?

    d. Should there be a project sponsor?

10–12 Does a project sponsor have the right to have an in-house representative removed from his company?

10–13 An executive once commented that his company was having trouble managing projects, not because of a lack of tools and techniques, but because they (employees) did not know how to manage what they had. How does this relate to project management?

10–14 Ajax National is the world's largest machine tool equipment manufacturer. Its success is based on the experience of its personnel. The majority of its department managers are forty-five to fifty-five-year-old, nondegreed people who have come up from the ranks. Ajax has just hired several engineers with bachelors' and masters' degrees to control the project management and project engineering functions. Can this pose a problem? Are advanced-degreed people required because of the rapid rate of change of technology?

10–15 When does project management turn into overmanagement?

10–16 *Brainstorming at United Central Bank (Part I):* As part of the 1989 strategic policy plan for United Central Bank, the president, Joseph P. Keith, decided to embark on weekly "brainstorming meetings" in hopes of developing creative ideas that could lead to solutions to the bank's problems. The bank's executive vice president would serve as permanent chairman of the brainstorming committee. Personnel representation would be randomly selected under the constraint that 10 percent must be from division managers, 30 percent from department managers, 30 percent from section-level supervisors, and the remaining 30 percent from clerical and nonexempt personnel. President Keith further decreed that the brainstorming committee would criticize all ideas and submit only those that successfully passed the criticism test to upper-level management for review.

After six months, with only two ideas submitted to upper-level management (both ideas were made by division managers), Joseph Keith formed an inquiry committee to investigate the reasons for the lack of interest by the brainstorming committee participants. Which of the following statements might be found in the inquiry committee report? (More than one answer is possible.)

    a. Because of superior–subordinate relationships (i.e., pecking order), creativity is inhibited.

    b. Criticism and ridicule have a tendency to inhibit spontaneity.

    c. Good managers can become very conservative and unwilling to stick their necks out.

    d. Pecking orders, unless adequately controlled, can inhibit teamwork and problem solving.

    e. All seemingly crazy or unconventional ideas were ridiculed and eventually discarded.

    f. Many lower-level people, who could have had good ideas to contribute, felt inferior.

    g. Meetings were dominated by upper-level management personnel.

    h. The meetings were held at inappropriate places and times.

    i. Many people were not given adequate notification of meeting time and subject matter.

10–17 *Brainstorming at United Central Bank (Part II):* After reading the inquiry committee report, President Keith decided to reassess his thinking about brainstorming by listing the advantages and disadvantages. What are the arguments for and against brainstorming? If you were Joseph Keith, would you vote for or against the continuation of the brainstorming sessions?

10–18 *Brainstorming at United Central Bank (Part III):* President Keith evaluated all of the data and decided to give the brainstorming committee one more chance. What changes can Joseph Keith implement in order to prevent the previous problems from recurring?

10–19 Explain the meaning of the following proverb: "The first 10 percent of the work is accomplished with 90 percent of the budget. The second 90 percent of the work is accomplished with the remaining 10 percent of the budget."

10–20 You are a line manager, and two project managers (each reporting to a divisional vice president) enter your office soliciting resources. Each project manager claims that his project is top priority as assigned by his own vice president. How should you, as the line manager, handle this situation? What are the recommended solutions to keep this situation from recurring repeatedly?

10–21 Figure 10–3 shows the organizational structure for a new Environmental Protection Agency project. Alpha Company was one of three subcontractors chosen for the contract. Because this was a new effort, the project manager reported "dotted" to the board chairman, who was acting as the project sponsor. The vice president was the immediate superior to the project manager.

    Because the project manager did not believe that Alpha Company maintained the expertise to do the job, he hired an outside consultant from one of the local colleges. Both the EPA and the prime contractor approved of the consultant, and the consultant's input was excellent.

    The project manager's superior, the vice president, disapproved of the consultant, continually arguing that the company had the expertise internally. How should you, the project manager, handle this situation?

10–22 You are the customer for a twelve-month project. You have team meetings scheduled with your subcontractor on a monthly basis. The contract has a contractual requirement to prepare a twenty-five- to thirty-page handout for each team meeting. Are there any benefits for you, the customer, to see these handouts at least three to four days prior to the team meeting?

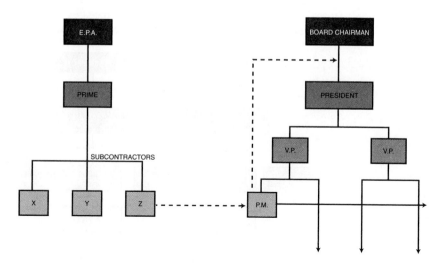

FIGURE 10–3. Organizational chart for EPA project.

10–23 You have a work breakdown structure (WBS) that is detailed to level 5. One level-5 work package requires that a technical subcontractor be selected to support one of the technical line organizations. Who should be responsible for customer–contractor communications: the project office or line manager? Does your answer depend on the life-cycle phase? The level of the WBS? Project manager's "faith" in the line manager?

10–24 Should a client have the right to communicate directly to the project staff (i.e., project office) rather than directly to the project manager, or should this be at the discretion of the project manager?

10–25 Your company has assigned one of its vice presidents to function as your project sponsor. Unfortunately, your sponsor refuses to make any critical decisions, always "passing the buck" back to you. What should you do? What are your alternatives and the pros and cons of each? Why might an executive sponsor act in this manner?

CASE STUDY

CORWIN CORPORATION*

By June 2003, Corwin Corporation had grown into a $950 million per year corporation with an international reputation for manufacturing low-cost, high-quality rubber components. Corwin maintained more than a dozen different product lines, all of which were sold as off-the-shelf

*Revised, 2007.

Exhibit 10–1. Organizational chart for Corwin Corporation

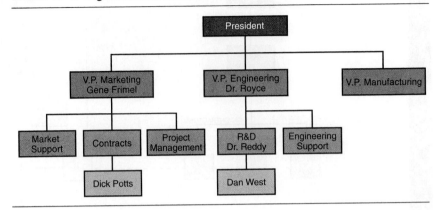

items in department stores, hardware stores, and automotive parts distributors. The name "Corwin" was now synonymous with "quality." This provided management with the luxury of having products that maintained extremely long life cycles.

Organizationally, Corwin had maintained the same structure for more than fifteen years (see Exhibit 10–1). The top management of Corwin Corporation was highly conservative and believed in a marketing approach to find new markets for existing product lines rather than to explore for new products. Under this philosophy, Corwin maintained a small R&D group whose mission was simply to evaluate state-of-the-art technology and its application to existing product lines.

Corwin's reputation was so good that they continually received inquiries about the manufacturing of specialty products. Unfortunately, the conservative nature of Corwin's management created a "do not rock the boat" atmosphere opposed to taking any type of risks. A management policy was established to evaluate all specialty-product requests. The policy required answering the following questions:

- Will the specialty product provide the same profit margin (20 percent) as existing product lines?
- What is the total projected profitability to the company in terms of follow-on contracts?
- Can the specialty product be developed into a product line?
- Can the specialty product be produced with minimum disruption to existing product lines and manufacturing operations?

These stringent requirements forced Corwin to no-bid more than 90 percent of all specialty-product inquiries.

Corwin Corporation was a marketing-driven organization, although manufacturing often had different ideas. Almost all decisions were made by marketing with the exception of product pricing and estimating, which was a joint undertaking between manufacturing and marketing. Engineering was considered as merely a support group to marketing and manufacturing.

For specialty products, the project managers would always come out of marketing even during the R&D phase of development. The company's approach was that if the specialty product should mature into a full product line, then there should be a product line manager assigned right at the onset.

The Peters Company Project

In 2000, Corwin accepted a specialty-product assignment from Peters Company because of the potential for follow-on work. In 2001 and 2002, and again in 2003, profitable follow-on contracts were received, and a good working relationship developed, despite Peter's reputation for being a difficult customer to work with.

On December 7, 2002, Gene Frimel, the vice president of marketing at Corwin, received a rather unusual phone call from Dr. Frank Delia, the marketing vice president at Peters Company.

Delia: "Gene, I have a rather strange problem on my hands. Our R&D group has $250,000 committed for research toward development of a new rubber product material, and we simply do not have the available personnel or talent to undertake the project. We have to go outside. We'd like your company to do the work. Our testing and R&D facilities are already overburdened."

Frimel: "Well, as you know, Frank, we are not a research group even though we've done this once before for you. And furthermore, I would never be able to sell our management on such an undertaking. Let some other company do the R&D work and then we'll take over on the production end."

Delia: "Let me explain our position on this. We've been burned several times in the past. Projects like this generate several patents, and the R&D company almost always requires that our contracts give them royalties or first refusal for manufacturing rights."

Frimel: "I understand your problem, but it's not within our capabilities. This project, if undertaken, could disrupt parts of our organization. We're already operating lean in engineering."

Delia: "Look, Gene! The bottom line is this: We have complete confidence in your manufacturing ability to such a point that we're willing to commit to a five-year production contract if the product can be developed. That makes it extremely profitable for you."

Frimel: "You've just gotten me interested. What additional details can you give me?"

Delia: "All I can give you is a rough set of performance specifications that we'd like to meet. Obviously, some trade-offs are possible."

Frimel: "When can you get the specification sheet to me?"

Delia: "You'll have it tomorrow morning. I'll ship it overnight express."

Frimel: "Good! I'll have my people look at it, but we won't be able to get you an answer until after the first of the year. As you know, our plant is closed down for the last two weeks in December, and most of our people have already left for extended vacations."

Delia: "That's not acceptable! My management wants a signed, sealed, and delivered contract by the end of this month. If this is not done, corporate will reduce our budget for 2003 by $250,000, thinking that we've bitten off more than we can chew. Actually, I need your answer within forty-eight hours so that I'll have some time to find another source."

Frimel: "You know, Frank, today is December 7, Pearl Harbor Day. Why do I feel as though the sky is about to fall in?"

Delia: "Don't worry, Gene! I'm not going to drop any bombs on you. Just remember, all that we have available is $250,000, and the contract must be a firm-fixed-price effort. We anticipate a six-month project with $125,000 paid on contract signing and the balance at project termination."

Frimel: "I still have that ominous feeling, but I'll talk to my people. You'll hear from us with a go or no-go decision within forty-eight hours. I'm scheduled to go on a cruise in the Caribbean, and my wife and I are leaving this evening. One of my people will get back to you on this matter."

Gene Frimel had a problem. All bid and no-bid decisions were made by a four-man committee composed of the president and the three vice presidents. The president and the vice president for manufacturing were on vacation. Frimel met with Dr. Royce, the vice president of engineering, and explained the situation.

Royce: "You know, Gene, I totally support projects like this because it would help our technical people grow intellectually. Unfortunately, my vote never appears to carry any weight."

Frimel: "The profitability potential as well as the development of good customer relations makes this attractive, but I'm not sure we want to accept such a risk. A failure could easily destroy our good working relationship with Peters Company."

Royce: "I'd have to look at the specification sheets before assessing the risks, but I would like to give it a shot."

Frimel: "I'll try to reach our president by phone."

By late afternoon, Frimel was fortunate enough to be able to contact the president and received a reluctant authorization to proceed. The problem now was how to prepare a proposal within the next two or three days and be prepared to make an oral presentation to Peters Company.

Frimel: "The Boss gave his blessing, Royce, and the ball is in your hands. I'm leaving for vacation, and you'll have total responsibility for the proposal and presentation. Delia wants the presentation this weekend. You should have his specification sheets tomorrow morning."

Royce: "Our R&D director, Dr. Reddy, left for vacation this morning. I wish he were here to help me price out the work and select the project manager. I assume that, in this case, the project manager will come out of engineering rather than marketing."

Frimel: "Yes, I agree. Marketing should not have any role in this effort. It's your baby all the way. And as for the pricing effort, you know our bid will be for $250,000. Just work backwards to justify the numbers. I'll assign one of our contracting people to assist you in the pricing. I hope I can find someone who has experience in this type of effort. I'll call Delia and tell him we'll bid it with an unsolicited proposal."

Royce selected Dan West, one of the R&D scientists, to act as the project leader. Royce had severe reservations about doing this without the R&D director, Dr. Reddy, being actively involved. With Reddy on vacation, Royce had to make an immediate decision.

On the following morning, the specification sheets arrived and Royce, West, and Dick Potts, a contracts man, began preparing the proposal. West prepared the direct labor man-hours, and Royce provided the costing data and pricing rates. Potts, being completely unfamiliar with this type of effort, simply acted as an observer and provided legal advice when necessary. Potts allowed Royce to make all decisions even though the contracts man was considered the official representative of the president.

Finally completed two days later, the proposal was actually a ten-page letter that simply contained the cost summaries (see Exhibit 10–2) and the engineering intent. West estimated that *thirty tests* would be required. The test matrix described only the test conditions for the first

Exhibit 10–2. Proposal cost summaries

Direct labor and support	$ 30,000
Testing (30 tests at $2,000 each)	60,000
Overhead at 100%	90,000
Materials	30,000
G&A (general and administrative, 10%)	21,000
Total	$231,000
Profit	19,000
Total	$250,000

five tests. The remaining twenty-five test conditions would be determined at a later date, jointly by Peters and Corwin personnel.

On Sunday morning, a meeting was held at Peters Company, and the proposal was accepted. Delia gave Royce a letter of intent authorizing Corwin Corporation to begin working on the project immediately. The final contract would not be available for signing until late January, and the letter of intent simply stated that Peters Company would assume all costs until such time that the contract was signed or the effort terminated.

West was truly excited about being selected as the project manager and being able to interface with the customer, a luxury that was usually given only to the marketing personnel. Although Corwin Corporation was closed for two weeks over Christmas, West still went into the office to prepare the project schedules and to identify the support he would need in the other areas, thinking that if he presented this information to management on the first day back to work, they would be convinced that he had everything under control.

The Work Begins . . .

On the first working day in January 2003, a meeting was held with the three vice presidents and Dr. Reddy to discuss the support needed for the project. (West was not in attendance at this meeting, although all participants had a copy of his memo.)

Reddy: "I think we're heading for trouble in accepting this project. I've worked with Peters Company previously on R&D efforts, and they're tough to get along with. West is a good man, but I would never have assigned him as the project leader. His expertise is in managing internal rather than external projects. But, no matter what happens, I'll support West the best I can."

Royce: "You're too pessimistic. You have good people in your group and I'm sure you'll be able to give him the support he needs. I'll try to look in on the project every so often. West will still be reporting to you for this project. Try not to burden him too much with other work. This project is important to the company."

West spent the first few days after vacation soliciting the support that he needed from the other line groups. Many of the other groups were upset that they had not been informed earlier and were unsure as to what support they could provide. West met with Reddy to discuss the final schedules.

Reddy: "Your schedules look pretty good, Dan. I think you have a good grasp on the problem. You won't need very much help from me. I have a lot of work to do on other activities, so I'm just going to be in the background on this project. Just drop me a note every once in a while telling me what's going on. I don't need anything formal. Just a paragraph or two will suffice."

By the end of the third week, all of the raw materials had been purchased, and initial formulations and testing were ready to begin. In addition, the contract was ready for signature. The contract contained a clause specifying that Peters Company had the right to send an in-house representative into Corwin Corporation for the duration of the project. Peters Company informed Corwin that Patrick Ray would be the in-house representative, reporting to Delia, and would assume his responsibilities on or about February 15.

By the time Pat Ray appeared at Corwin Corporation, West had completed the first three tests. The results were not what was expected, but gave promise that Corwin was heading in the right direction. Pat Ray's interpretation of the tests was completely opposite to that of West. Ray thought that Corwin was "way off base," and redirection was needed.

Ray: "Look, Dan! We have only six months to do this effort and we shouldn't waste our time on marginally acceptable data. These are the next five tests I'd like to see performed."

West: "Let me look over your request and review it with my people. That will take a couple of days, and, in the meanwhile, I'm going to run the other two tests as planned."

Ray's arrogant attitude bothered West. However, West decided that the project was too important to "knock heads" with Ray and simply decided to cater to Ray the best he could. This was not exactly the working relationship that West expected to have with the in-house representative.

West reviewed the test data and the new test matrix with engineering personnel, who felt that the test data were inconclusive as yet and preferred to withhold their opinion until the results of the fourth and fifth tests were made available. Although this displeased Ray, he agreed to wait a few more days if it meant getting Corwin Corporation on the right track.

The fourth and fifth tests appeared to be marginally acceptable just as the first three were. Corwin's engineering people analyzed the data and made their recommendations.

West: "Pat, my people feel that we're going in the right direction and that our path has greater promise than your test matrix."

Ray: "As long as we're paying the bills, we're going to have a say in what tests are conducted. Your proposal stated that we would work together in developing the other test conditions. Let's go with my test matrix. I've already reported back to my boss that the first five tests were failures and that we're changing the direction of the project."

West: "I've already purchased $30,000 worth of raw materials. Your matrix uses other materials and will require additional expenditures of $12,000."

Ray: "That's your problem. Perhaps you shouldn't have purchased all of the raw materials until we agreed on the complete test matrix."

During the month of February, West conducted fifteen tests, all under Ray's direction. The tests were scattered over such a wide range that no valid conclusions could be drawn. Ray continued sending reports back to Delia confirming that Corwin was not producing beneficial results and there was no indication that the situation would reverse itself. Delia ordered Ray to take any steps necessary to ensure a successful completion of the project.

Ray and West met again as they had done for each of the past forty-five days to discuss the status and direction of the project.

Ray: "Dan, my boss is putting tremendous pressure on me for results, and thus far I've given him nothing. I'm up for promotion in a couple of months and I can't let this project stand in my way. It's time to completely redirect the project."

West: "Your redirection of the activities is playing havoc with my scheduling. I have people in other departments who just cannot commit to this continual rescheduling. They blame me for not communicating with them when, in fact, I'm embarrassed to."

Ray: "Everybody has their problems. We'll get this problem solved. I spent this morning working with some of your lab people in designing the next fifteen tests. Here are the test conditions."

West: "I certainly would have liked to be involved with this. After all, I thought I was the project manager. Shouldn't I have been at the meeting?"

Ray: "Look, Dan! I really like you, but I'm not sure that you can handle this project. We need some good results immediately, or my neck will be stuck out for the next four months. I don't want that. Just have your lab personnel start on these tests, and we'll get along fine. Also, I'm planning on spending a great deal of time in your lab area. I want to observe the testing personally and talk to your lab personnel."

West: "We've already conducted twenty tests, and you're scheduling another fifteen tests. I priced out only thirty tests in the proposal. We're heading for a cost-overrun condition."

Ray: "Our contract is a firm-fixed-price effort. Therefore, the cost overrun is your problem."

West met with Dr. Reddy to discuss the new direction of the project and potential cost overruns. West brought along a memo projecting the costs through the end of the third month of the project (see Exhibit 10–3).

Dr. Reddy: "I'm already overburdened on other projects and won't be able to help you out. Royce picked you to be the project manager because he felt that you could do the job. Now, don't let him down. Send me a brief memo next month explaining the situation, and I'll see what I can do. Perhaps the situation will correct itself."

Exhibit 10–3. Projected cost summary at the end of the third month

	Original Proposal Cost Summary for Six-Month Project	Total Project Costs Projected at End of Third Month
Direct labor/support	$ 30,000	$ 15,000
Testing	60,000 (30 tests)	70,000 (35 tests)
Overhead	90,000 (100%)	92,000 (120%)*
Materials	30,000	50,000
G&A	21,000 (10%)	22,700 (10%)
Totals	$231,000	$249,700

*Total engineering overhead was estimated at 100%, whereas the R&D overhead was 120%.

During the month of March, the third month of the project, West received almost daily phone calls from the people in the lab stating that Pat Ray was interfering with their job. In fact, one phone call stated that Ray had changed the test conditions from what was agreed on in the latest test matrix. When West confronted Ray on his meddling, Ray asserted that Corwin personnel were very unprofessional in their attitude and that he thought this was being carried down to the testing as well. Furthermore, Ray demanded that one of the functional employees be removed immediately from the project because of incompetence. West stated that he would talk to the employee's department manager. Ray, however, felt that this would be useless and said, "Remove him or else!" The functional employee was removed from the project.

By the end of the third month, most Corwin employees were becoming disenchanted with the project and were looking for other assignments. West attributed this to Ray's harassment of the employees. To aggravate the situation even further, Ray met with Royce and Reddy, and demanded that West be removed and a new project manager be assigned.

Royce refused to remove West as project manager, and ordered Reddy to take charge and help West get the project back on track.

Reddy: "You've kept me in the dark concerning this project, West. If you want me to help you, as Royce requested, I'll need all the information tomorrow, especially the cost data. I'll expect you in my office tomorrow morning at 8:00 A.M. I'll bail you out of this mess."

West prepared the projected cost data for the remainder of the work and presented the results to Dr. Reddy (see Exhibit 10–4). Both West and Reddy agreed that the project was now out of control, and severe measures would be required to correct the situation, in addition to more than $250,000 in corporate funding.

Reddy: "Dan, I've called a meeting for 10:00 A.M. with several of our R&D people to completely construct a new test matrix. This is what we should have done right from the start."

West: "Shouldn't we invite Ray to attend this meeting? I'm sure he'd want to be involved in designing the new test matrix."

Reddy: "I'm running this show now, not Ray!! Tell Ray that I'm instituting new policies and procedures for in-house representatives. He's no longer authorized to visit the labs at his own discretion. He must be accompanied by either you or me. If he doesn't like these rules, he can get out. I'm not going to allow that guy to disrupt our organization. We're spending our money now, not his."

Exhibit 10–4. Estimate of total project completion costs

Direct labor/support	$ 47,000*
Testing (60 tests)	120,000
Overhead (120%)	200,000
Materials	103,000
G&A	47,000
	$517,000
Peters contract	250,000
Overrun	$267,000

*Includes Dr. Reddy.

West met with Ray and informed him of the new test matrix as well as the new policies and procedures for in-house representatives. Ray was furious over the new turn of events and stated that he was returning to Peters Company for a meeting with Delia.

On the following Monday, Frimel received a letter from Delia stating that Peters Company was officially canceling the contract. The reasons given by Delia were as follows:

1. Corwin had produced absolutely no data that looked promising.
2. Corwin continually changed the direction of the project and did not appear to have a systematic plan of attack.
3. Corwin did not provide a project manager capable of handling such a project.
4. Corwin did not provide sufficient support for the in-house representative.
5. Corwin's top management did not appear to be sincerely interested in the project and did not provide sufficient executive-level support.

Royce and Frimel met to decide on a course of action in order to sustain good working relations with Peters Company. Frimel wrote a strong letter refuting all of the accusations in the Peters letter, but to no avail. Even the fact that Corwin was willing to spend $250,000 of their own funds had no bearing on Delia's decision. The damage was done. Frimel was now thoroughly convinced that a contract should not be accepted on "Pearl Harbor Day."

Planning

Related Case Studies (from Kerzner/*Project Management Case Studies*, 3rd Edition)	Related Workbook Exercises (from Kerzner/*Project Management Workbook and PMP®/CAPM® Exam Study Guide*, 10th Edition)	PMBOK® Guide, 4th Edition, Reference Section for the PMP® Certification Exam
• Quantum Telecom • Concrete Masonry Corporation* • Margo Company • Project Overrun • The Two-Boss Problem • Denver International Airport (DIA)	• The Statement of Work • Technology Forecasting • The Noncompliance Project • Multiple Choice Exam • Crossword Puzzle on Scope Management	• Scope Management

11.0 INTRODUCTION

PMBOK® Guide, 4th Edition
Chapter 5 Scope Management
5.2 Define Scope

The most important responsibilities of a project manager are planning, integrating, and executing plans. Almost all projects, because of their relatively short duration and often prioritized control of resources, require formal, detailed planning. The integration of the planning activities is

* Case Study also appears in Workbook.

necessary because each functional unit may develop its own planning documentation with little regard for other functional units.

Planning, in general, can best be described as the function of selecting the enterprise objectives and establishing the policies, procedures, and programs necessary for achieving them. Planning in a project environment may be described as establishing a predetermined course of action within a forecasted environment. The project's requirements set the major milestones. If line managers cannot commit because the milestones are perceived as unrealistic, the project manager may have to develop alternatives, one of which may be to move the milestones. Upper-level management must become involved in the selection of alternatives.

The project manager is the key to successful project planning. It is desirable that the project manager be involved from project conception through execution. Project planning must be *systematic, flexible* enough to handle unique activities, *disciplined* through reviews and controls, and capable of accepting *multi-functional* inputs. Successful project managers realize that project planning is an iterative process and must be performed throughout the life of the project.

PMBOK® Guide, 4th Edition
5.2 Define Scope

One of the objectives of project planning is to completely define all work required (possibly through the development of a documented project plan) so that it will be readily identifiable to each project participant. This is a necessity in a project environment because:

- If the task is well understood prior to being performed, much of the work can be preplanned.
- If the task is not understood, then during the actual task execution more knowledge is gained that, in turn, leads to changes in resource allocations, schedules, and priorities.
- The more uncertain the task, the greater the amount of information that must be processed in order to ensure effective performance.

These considerations are important in a project environment because each project can be different from the others, requiring a variety of different resources, but having to be performed under time, cost, and performance constraints with little margin for error. Figure 11–1 identifies the type of project planning required to establish an effective monitoring and control system. The boxes at the top represent the planning activities, and the lower boxes identify the "tracking" or monitoring of the planned activities.

There are two proverbs that affect project planning:

- Failing to plan is planning to fail.
- The primary benefit of not planning is that failure will then come as a complete surprise rather than being preceded by periods of worry and depression.

Without proper planning, programs and projects can start off "behind the eight ball." Consequences of poor planning include:

- Project initiation without defined requirements
- Wild enthusiasm
- Disillusionment
- Chaos
- Search for the guilty
- Punishment of the innocent
- Promotion of the nonparticipants

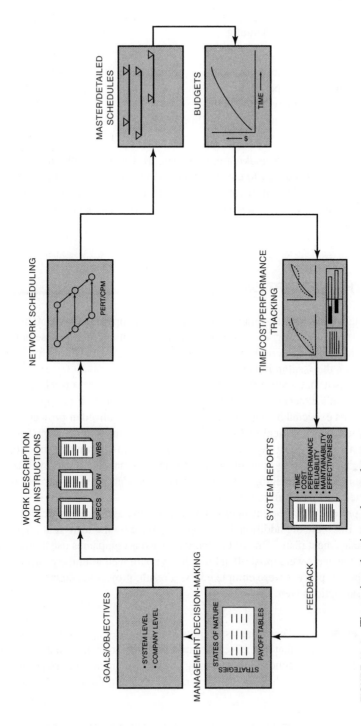

FIGURE 11–1. The project planning and control system.

There are four basic reasons for project planning:

- To eliminate or reduce uncertainty
- To improve efficiency of the operation
- To obtain a better understanding of the objectives
- To provide a basis for monitoring and controlling work

Planning is a continuous process of making entrepreneurial decisions with an eye to the future, and methodically organizing the effort needed to carry out these decisions. Furthermore, systematic planning allows an organization of set goals. The alternative to systematic planning is decision-making based on history. This generally results in reactive management leading to crisis management, conflict management, and fire fighting.

11.1 VALIDATING THE ASSUMPTIONS

Planning begins with an understanding of the assumptions. Quite often, the assumptions are made by marketing and sales personnel and then approved by senior management as part of the project selection and approval process. The expectations for the final results are based upon the assumptions made.

Why is it that, more often than not, the final results of a project do not satisfy senior management's expectations? At the beginning of a project, it is impossible to ensure that the benefits expected by senior management will be realized at project completion. While project length is a critical factor, the real culprit is changing assumptions.

Assumptions must be documented at project initiation using the project charter as a possible means. Throughout the project, the project manager must revalidate and challenge the assumptions. Changing assumptions may mandate that the project be terminated or redirected toward a different set of objectives.

A project management plan is based upon the assumptions described in the project charter. But there are additional assumptions made by the team that are inputs to the project management plan.[1] One of the primary reasons companies use a project charter is that project managers were most often brought on board well after the project selection process and approval process were completed. As a result, project managers were needed to know what assumptions were considered.

Enterprise Environmental Factors

These are assumptions about the external environmental conditions that can affect the success of the project, such as interest rates, market conditions, changing customer demands and requirements, changes in technology, and even government policies.

---

1. See *A Guide to the Project Management Body of Knowledge*®, 4th ed., 2008, Figure 4-4.

Organizational Process Assets   These are assumptions about present or future company assets that can impact the success of the project such as the capability of your enterprise project management methodology, the project management information system, forms, templates, guidelines, checklists, and the ability to capture and use lessons learned data and best practices.

11.2 GENERAL PLANNING

PMBOK® Guide, 4th Edition
Chapter 5 Scope Management
1.6 General Management
   Knowledge and Skills

Planning is determining what needs to be done, by whom, and by when, in order to fulfill one's assigned responsibility. There are nine major components of the planning phase:

- *Objective:* a goal, target, or quota to be achieved by a certain time
- *Program:* the strategy to be followed and major actions to be taken in order to achieve or exceed objectives
- *Schedule:* a plan showing when individual or group activities or accomplishments will be started and/or completed
- *Budget:* planned expenditures required to achieve or exceed objectives
- *Forecast:* a projection of what will happen by a certain time
- *Organization:* design of the number and kinds of positions, along with corresponding duties and responsibilities, required to achieve or exceed objectives
- *Policy:* a general guide for decision-making and individual actions
- *Procedure:* a detailed method for carrying out a policy
- *Standard:* a level of individual or group performance defined as adequate or acceptable

An item that has become important in recent years is documenting assumptions that go into the objectives or the project/subsidiary plans. As projects progress, even for short-term projects, assumptions can change because of the economy, technological advances, or market conditions. These changes can invalidate original assumptions or require that new assumptions be made. These changes could also mandate that projects be canceled. Companies are now validating assumptions during gate review meetings. Project charters now contain sections for documenting assumptions.

Several of these factors require additional comment. Forecasting what will happen may not be easy, especially if predictions of environmental reactions are required. For example, planning is customarily defined as either strategic, tactical, or operational. Strategic planning is generally for five years or more, tactical can be for one to five years, and operational is the here and now of six months to one year. Although most projects are operational, they can be considered as strategic, especially if spin-offs or follow-up work is promising. Forecasting also requires an understanding of strengths and weaknesses as found in:

- The competitive situation
- Marketing

- Research and development
- Production
- Financing
- Personnel
- The management structure

If project planning is strictly operational, then these factors may be clearly definable. However, if strategic or long-range planning is necessary, then the future economic outlook can vary, say, from year to year, and replanning must be done at regular intervals because the goals and objectives can change. (The procedure for this can be seen in Figure 11–1.)

The last three factors, policies, procedures, and standards, can vary from project to project because of their uniqueness. Each project manager can establish project policies, provided that they fall within the broad limits set forth by top management.

Project policies must often conform closely to company policies, and are usually similar in nature from project to project. Procedures, on the other hand, can be drastically different from project to project, even if the same activity is performed. For example, the signing off of manufacturing plans may require different signatures on two selected projects even though the same end-item is being produced.

Planning varies at each level of the organization. At the individual level, planning is required so that cognitive simulation can be established before irrevocable actions are taken. At the working group or functional level, planning must include:

- Agreement on purpose
- Assignment and acceptance of individual responsibilities
- Coordination of work activities
- Increased commitment to group goals
- Lateral communications

At the organizational or project level, planning must include:

- Recognition and resolution of group conflict on goals
- Assignment and acceptance of group responsibilities
- Increased motivation and commitment to organizational goals
- Vertical and lateral communications
- Coordination of activities between groups

The logic of planning requires answers to several questions in order for the alternatives and constraints to be fully understood. A list of questions would include:

- Prepare environmental analysis
  - Where are we?
  - How and why did we get here?

- Set objectives
  - Is this where we want to be?
  - Where would we like to be? In a year? In five years?
- List alternative strategies
  - Where will we go if we continue as before?
  - Is that where we want to go?
  - How could we get to where we want to go?
- List threats and opportunities
  - What might prevent us from getting there?
  - What might help us to get there?
- Prepare forecasts
  - Where are we capable of going?
  - What do we need to take us where we want to go?
- Select strategy portfolio
  - What is the best course for us to take?
  - What are the potential benefits?
  - What are the risks?
- Prepare action programs
  - What do we need to do?
  - When do we need to do it?
  - How will we do it?
  - Who will do it?
- Monitor and control
  - Are we on course? If not, why?
  - What do we need to do to be on course?
  - Can we do it?

One of the most difficult activities in the project environment is to keep the planning on target. These procedures can assist project managers during planning activities:

- Let functional managers do their own planning. Too often operators are operators, planners are planners, and never the twain shall meet.
- Establish goals before you plan. Otherwise short-term thinking takes over.
- Set goals for the planners. This will guard against the nonessentials and places your effort where there is payoff.
- Stay flexible. Use people-to-people contact, and stress fast response.
- Keep a balanced outlook. Don't overreact, and position yourself for an upturn.
- Welcome top-management participation. Top management has the capability to make or break a plan, and may well be the single most important variable.
- Beware of future spending plans. This may eliminate the tendency to underestimate.
- Test the assumptions behind the forecasts. This is necessary because professionals are generally too optimistic. Do not depend solely on one set of data.

- Don't focus on today's problems. Try to get away from crisis management and fire fighting.
- Reward those who dispel illusions. Avoid the Persian messenger syndrome (i.e., beheading the bearer of bad tidings). Reward the first to come forth with bad news.

11.3 LIFE-CYCLE PHASES

PMBOK® Guide, 4th Edition
Chapter 2 Project Life Cycle and
  Organization
2.1 Characteristics of Project
  Phases

Project planning takes place at two levels. The first level is the corporate cultural approach; the second method is the individual's approach. The corporate cultural approach breaks the project down into life-cycle phases, such as those shown in Table 2–6. The life-cycle phase approach is *not* an attempt to put handcuffs on the project manager but to provide a methodology for uniformity in project planning. Many companies, including government agencies, prepare checklists of activities that should be considered in each phase. These checklists are for consistency in planning. The project manager can still exercise his own planning initiatives within each phase.

A second benefit of life-cycle phases is control. At the end of each phase there is a meeting of the project manager, sponsor, senior management, and even the customer, to assess the accomplishments of this life-cycle phase and to get approval for the next phase. These meetings are often called critical design reviews, "on-off ramps," and "gates." In some companies, these meetings are used to firm up budgets and schedules for the follow-on phases. In addition to monetary considerations, life-cycle phases can be used for manpower deployment and equipment/facility utilization. Some companies go so far as to prepare project management policy and procedure manuals where all information is subdivided according to life-cycle phasing. Life-cycle phase decision points eliminate the problem where project managers do not ask for phase funding, but rather ask for funds for the whole project before the true scope of the project is known. Several companies have even gone so far as to identify the types of decisions that can be made at each end-of-phase review meeting. They include:

- Proceed with the next phase based on an approved funding level
- Proceed to the next phase but with a new or modified set of objectives
- Postpone approval to proceed based on a need for additional information
- Terminate project

Consider a company that utilizes the following life-cycle phases:

- Conceptualization
- Feasibility
- Preliminary planning
- Detail planning
- Execution
- Testing and commissioning

The conceptualization phase includes brainstorming and common sense and involves two critical factors: (1) identify and define the problem, and (2) identify and define potential solutions.

In a brainstorming session, *all* ideas are recorded and none are discarded. The brainstorming session works best if there is no formal authority present and if it lasts thirty to sixty minutes. Sessions over sixty minutes will produce ideas that may resemble science fiction.

The feasibility study phase considers the technical aspects of the conceptual alternatives and provides a firmer basis on which to decide whether to undertake the project.

The purpose of the feasibility phase is to:

● Plan the project development and implementation activities.
● Estimate the probable elapsed time, staffing, and equipment requirements.
● Identify the probable costs and consequences of investing in the new project.

If practical, the feasibility study results should evaluate the alternative conceptual solutions along with associated benefits and costs.

The objective of this step is to provide management with the predictable results of implementing a specific project and to provide generalized project requirements. This, in the form of a feasibility study report, is used as the basis on which to decide whether to proceed with the costly requirements, development, and implementation phases.

User involvement during the feasibility study is critical. The user must supply much of the required effort and information, and, in addition, must be able to judge the impact of alternative approaches. Solutions must be operationally, technically, and economically feasible. Much of the economic evaluation must be substantiated by the user. Therefore, the primary user must be highly qualified and intimately familiar with the workings of the organization and should come from the line operation.

The feasibility study also deals with the technical aspects of the proposed project and requires the development of conceptual solutions. Considerable experience and technical expertise are required to gather the proper information, analyze it, and reach practical conclusions.

Improper technical or operating decisions made during this step may go undetected or unchallenged throughout the remainder of the process. In the worst case, such an error could result in the termination of a valid project—or the continuation of a project that is not economically or technically feasible.

In the feasibility study phase, it is necessary to define the project's basic approaches and its boundaries or scope. A typical feasibility study checklist might include:

● Summary level
   ● Evaluate alternatives
   ● Evaluate market potential
   ● Evaluate cost effectiveness
   ● Evaluate producibility
   ● Evaluate technical base
● Detail level
   ● A more specific determination of the problem
   ● Analysis of the state-of-the-art technology

- Assessment of in-house technical capabilities
- Test validity of alternatives
- Quantify weaknesses and unknowns
- Conduct trade-off analysis on time, cost, and performance
- Prepare initial project goals and objectives
- Prepare preliminary cost estimates and development plan

The end result of the feasibility study is a management decision on whether to terminate the project or to approve its next phase. Although management can stop the project at several later phases, the decision is especially critical at this point, because later phases require a major commitment of resources. All too often, management review committees approve the continuation of projects merely because termination at this point might cast doubt on the group's judgment in giving earlier approval.

The decision made at the end of the feasibility study should identify those projects that are to be terminated. Once a project is deemed feasible and is approved for development, it must be prioritized with previously approved projects waiting for development (given a limited availability of capital or other resources). As development gets under way, management is given a series of checkpoints to monitor the project's actual progress as compared to the plan.

The third life-cycle phase is either preliminary planning or "defining the requirements." This is the phase where the effort is officially defined as a project. In this phase, we should consider the following:

- General scope of the work
- Objectives and related background
- Contractor's tasks
- Contractor end-item performance requirements
- Reference to related studies, documentation, and specifications
- Data items (documentation)
- Support equipment for contract end-item
- Customer-furnished property, facilities, equipment, and services
- Customer-furnished documentation
- Schedule of performance
- Exhibits, attachments, and appendices

These elements can be condensed into four core documents, as will be shown in Section 11.7. Also, it should be noted that the word "customer" can be an internal customer, such as the user group or your own executives.

The table below shows the percentage of *direct* labor hours/dollars that are spent in each phase:

Phase	Percent of Direct Labor Dollars
Conceptualization	5
Feasibility study	10
Preliminary planning	15
Detail planning	20
Execution	40
Commissioning	10

The interesting fact from this table is that as much as 50 percent of the direct labor hours and dollars can be spent before execution begins. The reason for this is simple: Quality must be planned for and designed in. Quality cannot be inspected into the project. Companies that spend less than these percentages usually find quality problems in execution.

11.4 PROPOSAL PREPARATION

There is always a question of what to do with a project manager between assignments. For companies that survive on competitive bidding, the assignment is clear: The project manager writes proposals for future work. This takes place during the feasibility study, when the company must decide whether to bid on the job. There are four ways in which proposal preparation can occur:

- *Project manager prepares entire proposal.* This occurs frequently in small companies. In large organizations, the project manager may not have access to all available data, some of which may be company proprietary, and it may not be in the best interest of the company to have the project manager spend all of his time doing this.
- *Proposal manager prepares entire proposal.* This can work as long as the project manager is allowed to review the proposal before delivery to the customer and feels committed to its direction.
- *Project manager prepares proposal but is assisted by a proposal manager.* This is common, but again places tremendous pressure on the project manager.
- *Proposal manager prepares proposal but is assisted by a project manager.* This is the preferred method. The proposal manager maintains maximum authority and control until such time as the proposal is sent to the customer, at which point the project manager takes charge. The project manager is on board right from the start, although his only effort may be preparing the technical volume of the proposal and perhaps part of the management volume.

11.5 KICKOFF MEETINGS

The typical launch of a project begins with a kickoff meeting involving the major players responsible for planning, including the project manager, assistant project managers for certain areas of knowledge, subject matter experts (SME), and functional leads. A typical sequence is shown in Figure 11–2.

There can be multiple kickoff meetings based upon the size, complexity, and time requirements for the project. The major players are usually authorized by their functional areas to make decisions concerning timing, costs, and resource requirements.

Some of the items discussed in the initial kickoff meeting include:

- Wage and salary administration, if applicable
- Letting the employees know that their boss will be informed as to how well or how poorly they perform

TYPICAL PROJECT LAUNCH

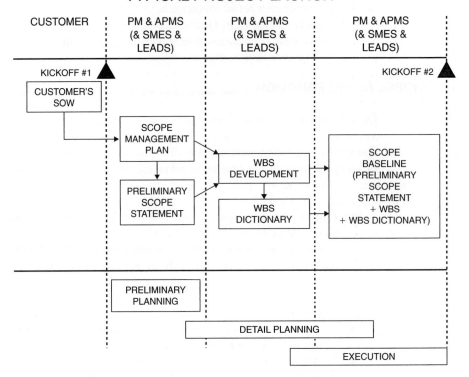

Figure 11–2.   Typical project launch.

- Initial discussion of the scope of the project including both the technical objective and the business objective
- The definition of success on this project
- The assumptions and constraints as identified in the project charter
- The project's organizational chart (if known at that time)
- The participants' roles and responsibilities

For a small or short-term project, estimates on cost and duration may be established in the kickoff meeting. In this case, there may be little need to establish a cost estimating schedule. But where the estimating cycle is expected to take several weeks, and where inputs will be required from various organizations and/or disciplines, an essential tool is an estimating schedule. In this case, there may be a need for a prekickoff meeting simply to determine the estimates. The minimum key milestones in a cost estimating schedule are (1) a "kickoff" meeting; (2) a "review of ground rules" meeting; (3) "resources input and review" meeting; and (4) summary meetings and presentations. Descriptions of these meetings and their approximate places in the estimating cycle follow.[2]

---

2. R. D. Stewart, *Cost Estimating* (New York: Wiley, 1982), pp. 56–57.

The Prekickoff Meeting

The very first formal milestone in an estimate schedule is the estimate kickoff meeting. This is a meeting of all the individuals who are expected to have an input to the cost estimate. It usually includes individuals who are proficient in technical disciplines involved in the work to be estimated; business-oriented individuals who are aware of the financial factors to be considered in making the estimate; project-oriented individuals who are familiar with the project ground rules and constraints; and, finally, the cost estimator or cost estimating team. The estimating team may not include any of the team members responsible for execution of the project.

Sufficient time should be allowed in the kickoff meeting to describe all project ground rules, constraints, and assumptions; to hand out technical specifications, drawings, schedules, and work element descriptions and resource estimating forms; and to discuss these items and answer any questions that might arise. It is also an appropriate time to clarify estimating assignments among the various disciplines represented in the event that organizational charters are not clear as to who should support which part of the estimate. This kickoff meeting may be 6 weeks to 3 months prior to the estimate completion date to allow sufficient time for the overall estimating process. If the estimate is being made in response to a request for quotation or request for bid, copies of the request for quotation document will be distributed and its salient points discussed.

The Review of Ground Rules Meeting

Several days after the estimate kickoff meeting, when the participants have had the opportunity to study the material, a review of ground rules meeting should be conducted. In this meeting the estimate manager answers questions regarding the conduct of the cost estimate, assumptions, ground rules, and estimating assignments. If the members of the estimating team are experienced in developing resource estimates for their respective disciplines, very little discussion may be needed. However, if this is the first estimating cycle for one or more of the estimating team members, it may be necessary to provide these team members with additional information, guidance, and instruction on estimating tools and methods. If the individuals who will actually perform the work are doing the estimating (which is actually the best arrangement for getting a realistic estimate), more time and support may be needed than would experienced estimators.

The Resources Input and Review Meeting

Several weeks after the kickoff and review of ground rules meetings, each team member that has a resources (man-hour and/or materials) input is asked to present his or her input before the entire estimating team. Thus starts one of the most valuable parts of the estimating process: the interaction of team members to reduce duplications, overlaps, and omissions in resource data.

The most valuable aspect of a team estimate is the synergistic effect of team interaction. In any multidisciplinary activity, it is the synthesis of information and actions that produces wise decisions rather than the mere volume of data. In this review meeting the estimator of each discipline area has the opportunity to justify and explain the rationale for his estimates in view of his peers, an activity that tends to iron out inconsistencies, overstatements, and incompatibilities in resources estimates. Occasionally, inconsistencies, overlaps, duplications, and omissions will be so significant that a second input and review meeting will be required to collect and properly synthesize all inputs for an estimate.

Summary Meetings and Presentations   Once the resources inputs have been collected, adjusted, and "priced," the cost estimate is presented to the estimating team as a "dry run" for the final presentation to the company's management or to the requesting organization. This dry run can produce visibility into further inconsistencies or errors that have crept into the estimate during the process of consolidation and reconciliation. The final review with the requesting organization or with the company's management could also bring about some changes in the estimate due to last minute changes in ground rules or budget-imposed cost ceilings.

11.6 UNDERSTANDING PARTICIPANTS' ROLES

Companies that have histories of successful plans also have employees who fully understand their roles in the planning process. Good up-front planning may not eliminate the need for changes, but may reduce the number of changes required. The responsibilities of the major players are as follows:

- Project manager will define:
    - Goals and objectives
    - Major milestones
    - Requirements
    - Ground rules and assumptions
    - Time, cost, and performance constraints
    - Operating procedures
    - Administrative policy
    - Reporting requirements
- Line manager will define:
    - Detailed task descriptions to implement objectives, requirements, and milestones
    - Detailed schedules and manpower allocations to support budget and schedule
    - Identification of areas of risk, uncertainty, and conflict
- Senior management (project sponsor) will:
    - Act as the negotiator for disagreements between project and line management
    - Provide clarification of critical issues
    - Provide communication link with customer's senior management

Successful planning requires that project, line, and senior management are in agreement with the plan.

11.7 PROJECT PLANNING

PMBOK® Guide, 4th Edition
Chapter 5 Project Scope
  Management
5.2 Define Scope

Successful project management, whether in response to an in-house project or a customer request, must utilize effective planning techniques. The first step is understanding the project objectives. These goals may be to develop expertise in a given area, to become competitive, to modify an existing facility for later use, or simply to keep key personnel employed.

The objectives are generally not independent; they are all interrelated, both implicitly and explicitly. Many times it is not possible to satisfy all objectives. At this point, management must prioritize the objectives as to which are strategic and which are not. Typical problems with developing objectives include:

- Project objectives/goals are not agreeable to all parties.
- Project objectives are too rigid to accommodate changing priorities.
- Insufficient time exists to define objectives well.
- Objectives are not adequately quantified.
- Objectives are not documented well enough.
- Efforts of client and project personnel are not coordinated.
- Personnel turnover is high.

Once the objectives are clearly defined, four questions must be considered:

- What are the major elements of the work required to satisfy the objectives, and how are these elements interrelated?
- Which functional divisions will assume responsibility for accomplishment of these objectives and the major-element work requirements?
- Are the required corporate and organizational resources available?
- What are the information flow requirements for the project?

If the project is large and complex, then careful planning and analysis must be accomplished by both the direct- and indirect-labor-charging organizational units. The project organizational structure must be designed to fit the project; work plans and schedules must be established so that maximum allocation of resources can be made; resource costing and accounting systems must be developed; and a management information and reporting system must be established.

Effective total program planning cannot be accomplished unless all of the necessary information becomes available at project initiation. These information requirements are:

- The statement of work (SOW)
- The project specifications
- The milestone schedule
- The work breakdown structure (WBS)

The statement of work (SOW) is a narrative description of the work to be accomplished. It includes the objectives of the project, a brief description of the work, the funding constraint if one exists, and the specifications and schedule. The schedule is a "gross" schedule and includes such things as the:

- Start date
- End date
- Major milestones
- Written reports (data items)

Written reports should always be identified so that if functional input is required, the functional manager will assign an individual who has writing skills.

The last major item is the work breakdown structure. The WBS is the breaking down of the statement of work into smaller elements for better visibility and control. Each of these planning items is described in the following sections.

11.8 THE STATEMENT OF WORK

PMBOK® Guide, 4th Edition
5.2.3 Scope Definition
5.2.3.1 Project Scope Statement
12.1.3.2 Contract Statement of
    Work

The PMBOK® Guide addresses four elements related to scope:

- **Scope:** Scope is the summation of all deliverables required as part of the project. This includes all products, services, and results.
- **Project Scope:** This is the work that must be completed to achieve the final scope of the project, namely the products, services, and end results. (Previously, in Section 2.7, we differentiated between project scope and product scope.)
- **Scope Statement:** This is a document that provides the basis for making future decisions such as scope changes. The intended use of the document is to make sure that all stakeholders have a common knowledge of the project scope. Included in this document are the objectives, description of the deliverables, end result or product, and justification for the project. The scope statement addresses seven questions: who, what, when, why, where, how, and how many. This document validates the project scope against the statement of work provided by the customer.
- **Statement of Work:** This is a narrative description of the end results to be provided under the contract. For the remainder of this section, we will focus our attention on the statement of work.

The statement of work (SOW) is a narrative description of the work required for the project. The complexity of the SOW is determined by the desires of top management, the customer, and/or the user groups. For projects internal to the company, the SOW is prepared by the project office with input from the user groups because the project office is usually composed of personnel with writing skills.

For projects external to the organization, as in competitive bidding, the contractor may have to prepare the SOW for the customer because the customer may not have people trained in SOW preparation. In this case, as before, the contractor would submit the SOW to the customer for approval. It is also quite common for the project manager to rewrite a customer's SOW so that the contractor's line managers can price out the effort.

In a competitive bidding environment, there are two SOWs—the SOW used in the proposal and a contract statement of work (CSOW). There might also be a proposal WBS and a contract work breakdown structure (CWBS). Special care must be taken by contract and negotiation teams to discover all discrepancies between the SOW/WBS and CSOW/CWBS, or additional costs may be incurred. A good (or winning) proposal is *no guarantee* that the customer or contractor understands the SOW. For large projects, fact-finding is usually required before final negotiations because it is *essential* that both the customer and the contractor understand and agree on the SOW, what work is required, what work is proposed, the factual basis for the costs, and other related elements. In addition, it is imperative that there be agreement between the final CSOW and CWBS.

SOW preparation is not as easy as it sounds. Consider the following:

- The SOW says that you are to conduct a *minimum* of fifteen tests to determine the material properties of a new substance. You price out twenty tests just to "play it safe." At the end of the fifteenth test, the customer says that the results are inconclusive and that you must run another fifteen tests. The cost overrun is $40,000.
- The Navy gives you a contract in which the SOW states that the prototype must be tested in "water." You drop the prototype into a swimming pool to test it. Unfortunately, the Navy's definition of "water" is the Atlantic Ocean, and it costs you $1 million to transport all of your test engineers and test equipment to the Atlantic Ocean.
- You receive a contract in which the SOW says that you must transport goods across the country using "aerated" boxcars. You select boxcars that have open tops so that air can flow in. During the trip, the train goes through an area of torrential rains, and the goods are ruined.

These three examples show that misinterpretations of the SOW can result in losses of hundreds of millions of dollars. Common causes of misinterpretation are:

- Mixing tasks, specifications, approvals, and special instructions
- Using imprecise language ("nearly," "optimum," "approximately," etc.)
- No pattern, structure, or chronological order
- Wide variation in size of tasks
- Wide variation in how to describe details of the work
- Failing to get third-party review

Misinterpretations of the statement of work can and will occur no matter how careful everyone has been. The result is creeping scope, or, as one telecommunications company calls it, "creeping elegance." The best way to control creeping scope is with a good definition of the requirements up front, if possible.

Today, both private industry and government agencies are developing manuals on SOW preparation. The following is adapted from a NASA publication on SOW preparation[3]:

- The project manager or his designees should review the documents that authorize the project and define its objectives, and also review contracts and studies leading to the present level of development. As a convenience, a bibliography of related studies should be prepared together with samples of any similar SOWs, and compliance specifications.
- A copy of the WBS should be obtained. At this point coordination between the CWBS elements and the SOW should commence. Each task element of the preliminary CWBS should be explained in the SOW, and related coding should be used.

---

3. Adapted from *Statement of Work Handbook* NHB5600.2, National Aeronautics and Space Administration, February 1975.

- The project manager should establish a SOW preparation team consisting of personnel he deems appropriate from the program or project office who are experts in the technical areas involved, and representatives from procurement, financial management, fabrication, test, logistics, configuration management, operations, safety, reliability, and quality assurance, plus any other area that may be involved in the contemplated procurement.
- Before the team actually starts preparation of the SOW, the project manager should brief program management as to the structure of the preliminary CWBS and the nature of the contemplated SOW. This briefing is used as a baseline from which to proceed further.
- The project manager may assign identified tasks to team members and identify compliance specifications, design criteria, and other requirements documentation that must be included in the SOW and assign them to responsible personnel for preparation. Assigned team members will identify and obtain copies of specifications and technical requirements documents, engineering drawings, and results of preliminary and/or related studies that may apply to various elements of the proposed procurement.
- The project manager should prepare a detailed checklist showing the mandatory items and the selected optional items as they apply to the main body or the appendixes of the SOW.
- The project manager should emphasize the use of preferred parts lists; standard subsystem designs, both existing and under development; available hardware in inventory; off-the-shelf equipment; component qualification data; design criteria handbooks; and other technical information available to design engineers to prevent deviations from the best design practices.
- Cost estimates (manning requirements, material costs, software requirements, etc.) developed by the cost estimating specialists should be reviewed by SOW contributors. Such reviews will permit early trade-off consideration on the desirability of requirements that are not directly related to essential technical objectives.
- The project manager should establish schedules for submission of coordinated SOW fragments from each task team member. He must assure that these schedules are compatible with the schedule for the request for proposal (RFP) issuance. The statement of work should be prepared sufficiently early to permit full project coordination and to ensure that all project requirements are included. It should be completed in advance of RFP preparation.

SOW preparation manuals also contain guides for editors and writers[4]:

- Every SOW that exceeds two pages in length should have a table of contents conforming to the CWBS coding structure. There should rarely be items in the SOW that are not shown on the CWBS; however, it is not absolutely necessary to restrict items to those cited in the CWBS.

4. See note 3.

- Clear and precise task descriptions are essential. The SOW writer should realize that his or her efforts will have to be read and interpreted by persons of varied background (such as lawyers, buyers, engineers, cost estimators, accountants, and specialists in production, transportation, security, audit, quality, finance, and contract management). A good SOW states precisely the product or service desired. The clarity of the SOW will affect administration of the contract, since it defines the scope of work to be performed. Any work that falls outside that scope will involve new procurement with probable increased costs.

- The most important thing to keep in mind when writing a SOW is the most likely effect the written work will have upon the reader. Therefore, every effort must be made to avoid ambiguity. All obligations of the government should be carefully spelled out. If approval actions are to be provided by the government, set a time limit. If government-furnished equipment (GFE) and/or services, etc., are to be provided, state the nature, condition, and time of delivery, if feasible.

- Remember that any provision that takes control of the work away from the contractor, even temporarily, may result in relieving the contractor of responsibility.

- In specifying requirements, use active rather than passive terminology. Say that the contractor shall conduct a test rather than that a test should be conducted. In other words, when a firm requirement is intended, use the mandatory term "shall" rather than the permissive term "should."

- Limit abbreviations to those in common usage. Provide a list of all pertinent abbreviations and acronyms at the beginning of the SOW. When using a term for the first time, spell it out and show the abbreviation or acronym in parentheses following the word or words.

- When it is important to define a division of responsibilities between the contractor, other agencies, etc., a separate section of the SOW (in an appropriate location) should be included and delineate such responsibilities.

- Include procedures. When immediate decisions cannot be made, it may be possible to include a procedure for making them (e.g., "as approved by the contracting officer," or "the contractor shall submit a report each time a failure occurs").

- Do not overspecify. Depending upon the nature of the work and the type of contract, the ideal situation may be to specify results required or end-items to be delivered and let the contractor propose his best method.

- Describe requirements in sufficient detail to assure clarity, not only for legal reasons, but for practical application. It is easy to overlook many details. It is equally easy to be repetitious. Beware of doing either. For every piece of deliverable hardware, for every report, for every immediate action, do not specify that something be done "as necessary." Rather, specify whether the judgment is to be made by the contractor or by the government. Be aware that these types of contingent actions may have an impact on price as well as schedule. Where expensive services, such as technical liaison, are to be furnished, do not say "as required." Provide a ceiling on the extent of such services, or work out a procedure (e.g., a level of effort, pool of man-hours) that will ensure adequate control.

- Avoid incorporating extraneous material and requirements. They may add unnecessary cost. Data requirements are common examples of problems in this area.

Screen out unnecessary data requirements, and specify only what is essential and when. It is recommended that data requirements be specified separately in a data requirements appendix or equivalent.

- Do not repeat detailed requirements or specifications that are already spelled out in applicable documents. Instead, incorporate them by reference. If amplification, modification, or exceptions are required, make specific reference to the applicable portions and describe the change.

Some preparation documents also contain checklists for SOW preparation.[5] A checklist is furnished below to provide considerations that SOW writers should keep in mind in preparing statements of work:

- Is the SOW (when used in conjunction with the preliminary CWBS) specific enough to permit a contractor to make a tabulation and summary of manpower and resources needed to accomplish each SOW task element?
- Are specific duties of the contractor stated so he will know what is required, and can the contracting officer's representative, who signs the acceptance report, tell whether the contractor has complied?
- Are all parts of the SOW so written that there is no question as to what the contractor is obligated to do, and when?
- When it is necessary to reference other documents, is the proper reference document described? Is it properly cited? Is all of it really pertinent to the task, or should only portions be referenced? Is it cross-referenced to the applicable SOW task element?
- Are any specifications or exhibits applicable in whole or in part? If so, are they properly cited and referenced to the appropriate SOW element?
- Are directions clearly distinguishable from general information?
- Is there a time-phased data requirement for each deliverable item? If elapsed time is used, does it specify calendar or work days?
- Are proper quantities shown?
- Have headings been checked for format and grammar? Are subheadings comparable? Is the text compatible with the title? Is a multidecimal or alphanumeric numbering system used in the SOW? Can it be cross-referenced with the CWBS?
- Have appropriate portions of procurement regulations been followed?
- Has extraneous material been eliminated?
- Can SOW task/contract line items and configuration item breakouts at lower levels be identified and defined in sufficient detail so they can be summarized to discrete third-level CWBS elements?
- Have all requirements for data been specified separately in a data requirements appendix or its equivalent? Have all extraneous data requirements been eliminated?
- Are security requirements adequately covered if required?
- Has its availability to contractors been specified?

5. See note 3.

Finally, there should be a management review of the SOW preparation interpretation[6]:

> During development of the Statement of Work, the project manager should ensure adequacy of content by holding frequent reviews with project and functional specialists to determine that technical and data requirements specified do conform to the guidelines herein and adequately support the common system objective. The CWBS/SOW matrix should be used to analyze the SOW for completeness. After all comments and inputs have been incorporated, a final team review should be held to produce a draft SOW for review by functional and project managers. Specific problems should be resolved and changes made as appropriate. A final draft should then be prepared and reviewed with the program manager, contracting officer, or with higher management if the procurement is a major acquisition. The final review should include a briefing on the total RFP package. If other program offices or other Government agencies will be involved in the procurement, obtain their concurrence also.

11.9 PROJECT SPECIFICATIONS

PMBOK® Guide, 4th Edition
5.2 Define Scope
12.1.3.2 Contract Statement
of Work

A specification list as shown in Table 11–1 is separately identified or called out as part of the statement of work. Specifications are used for man-hour, equipment, and material estimates. Small changes in a specification can cause large cost overruns.

Another reason for identifying the specifications is to make sure that there are no surprises for the customer downstream. The specifications should be the most current revision. It is not uncommon for a customer to hire outside agencies to evaluate the technical proposal and to make sure that the proper specifications are being used.

Specifications are, in fact, standards for pricing out a proposal. If specifications do not exist or are not necessary, then work standards should be included in the proposal. The work standards can also appear in the cost volume of the proposal. Labor justification backup sheets may or may not be included in the proposal, depending on RFP/RFQ (request for quotation) requirements.

Several years ago, a government agency queried contractors as to why some government programs were costing so much money. The main culprit turned out to be the specifications. Typical specifications contain twice as many pages as necessary, do not stress quality enough, are loaded with unnecessary designs and schematics, are difficult to read and update, and are obsolete before they are published. Streamlining existing specifications is a costly and time-consuming effort. The better alternative is to educate those people involved in specification preparation so that future specifications will be reasonably correct.

6. *Statement of Work Handbook* NHB5600.2, National Aeronautics and Space Administration, February 1975.

TABLE 11–1. SPECIFICATION FOR STATEMENT OF WORK

Description	Specification No.
Civil	100 (Index)
• Concrete	101
• Field equipment	102
• Piling	121
• Roofing and siding	122
• Soil testing	123
• Structural design	124
Electrical	200 (Index)
• Electrical testing	201
• Heat tracing	201
• Motors	209
• Power systems	225
• Switchgear	226
• Synchronous generators	227
HVAC	300 (Index)
• Hazardous environment	301
• Insulation	302
• Refrigeration piping	318
• Sheetmetal ductwork	319
Installation	400 (Index)
• Conveyors and chutes	401
• Fired heaters and boilers	402
• Heat exchangers	403
• Reactors	414
• Towers	415
• Vessels	416
Instruments	500 (Index)
• Alarm systems	501
• Control valves	502
• Flow instruments	503
• Level gages	536
• Pressure instruments	537
• Temperature instruments	538
Mechanical equipment	600 (Index)
• Centrifugal pumps	601
• Compressors	602
• High-speed gears	603
• Material handling equipment	640
• Mechanical agitators	641
• Steam turbines	642
Piping	700 (Index)
• Expansion joints	701
• Field pressure testing	702
• Installation of piping	703
• Pipe fabrication specs	749
• Pipe supports	750
• Steam tracing	751
Project administration	800 (Index)
• Design drawings	801
• Drafting standards	802
• General requirements	803
• Project coordination	841
• Reporting procedure	842
• Vendor data	843

(continues)

TABLE 11–1. SPECIFICATION FOR STATEMENT OF WORK
(*Continued*)

Description	Specification No.
Vessels	900 (Index)
• Fireproofing	901
• Painting	902
• Reinforced tanks	948
• Shell and tube heat exchangers	949
• Steam boilers	950
• Vessel linings	951

11.10 MILESTONE SCHEDULES

Project milestone schedules contain such information as:

- Project start date
- Project end date
- Other major milestones
- Data items (deliverables or reports)

Project start and end dates, if known, must be included. Other major milestones, such as review meetings, prototype available, procurement, testing, and so on, should also be identified. The last topic, data items, is often overlooked. There are two good reasons for preparing a separate schedule for data items. First, the separate schedule will indicate to line managers that personnel with writing skills may have to be assigned. Second, data items require direct-labor man-hours for writing, typing, editing, retyping, proofing, graphic arts, and reproduction. Many companies identify on the data item schedules the approximate number of pages per data item, and each data item is priced out at a cost per page, say $500/page. Pricing out data items separately often induces customers to require fewer reports.

The steps required to prepare a report, after the initial discovery work or collection of information, include:

- Organizing the report
- Writing
- Typing
- Editing
- Retyping
- Proofing
- Graphic arts
- Submittal for approvals
- Reproduction and distribution

Typically, 6–8 hours of work are required per page. At a burdened hourly rate of $80/hour, it is easy for the cost of documentation to become exorbitant.

11.11 WORK BREAKDOWN STRUCTURE

PMBOK® Guide, 4th Edition
5.3 Create WBS

The successful accomplishment of both contract and corporate objectives requires a plan that defines all effort to be expended, assigns responsibility to a specially identified organizational element, and establishes schedules and budgets for the accomplishment of the work. The preparation of this plan is the responsibility of the program manager, who is assisted by the program team assigned in accordance with program management system directives. The detailed planning is also established in accordance with company budgeting policy before contractual efforts are initiated.

In planning a project, the project manager must structure the work into small elements that are:

- Manageable, in that specific authority and responsibility can be assigned
- Independent, or with minimum interfacing with and dependence on other ongoing elements
- Integratable so that the total package can be seen
- Measurable in terms of progress

The first major step in the planning process after project requirements definition is the development of the work breakdown structure (WBS). A WBS is a product-oriented family tree subdivision of the hardware, services, and data required to produce the end product. The WBS is structured in accordance with the way the work will be performed and reflects the way in which project costs and data will be summarized and eventually reported. Preparation of the WBS also considers other areas that require structured data, such as scheduling, configuration management, contract funding, and technical performance parameters. The WBS is the single most important element because it provides a common framework from which:

- The total program can be described as a summation of subdivided elements.
- Planning can be performed.
- Costs and budgets can be established.
- Time, cost, and performance can be tracked.
- Objectives can be linked to company resources in a logical manner.
- Schedules and status-reporting procedures can be established.
- Network construction and control planning can be initiated.
- The responsibility assignments for each element can be established.

The work breakdown structure acts as a vehicle for breaking the work down into smaller elements, thus providing a greater probability that every major and minor activity will be accounted for. Although a variety of work breakdown structures exist, the most common is the six-level indented structure shown below:

	Level	*Description*
Managerial levels	1 2 3	Total program Project Task
Technical levels	4 5 6	Subtask Work package Level of effort

Level 1 is the total program and is composed of a set of projects. The summation of the activities and costs associated with each project must equal the total program. Each project, however, can be broken down into tasks, where the summation of all tasks equals the summation of all projects, which, in turn, comprises the total program. The reason for this subdivision of effort is simply ease of control. Program management therefore becomes synonymous with the integration of activities, and the project manager acts as the integrator, using the work breakdown structure as the common framework.

Careful consideration must be given to the design and development of the WBS. From Figure 11–3, the work breakdown structure can be used to provide the basis for:

- The responsibility matrix
- Network scheduling
- Costing
- Risk analysis
- Organizational structure
- Coordination of objectives
- Control (including contract administration)

The upper three levels of the WBS are normally specified by the customer (if part of an RFP/RFQ) as the summary levels for reporting purposes. The lower levels are generated by the contractor for in-house control. Each level serves a vital purpose: Level 1 is generally used for the authorization and release of all work, budgets are prepared at level 2, and schedules are prepared at level 3. Certain characteristics can now be generalized for these levels:

- The top three levels of the WBS reflect integrated efforts and should not be related to one specific department. Effort required by departments or sections should be defined in subtasks and work packages.
- The summation of all elements in one level must be the sum of all work in the next lower level.
- Each element of work should be assigned to one and only one level of effort. For example, the construction of the foundation of a house should be included in one project (or task), not extended over two or three. (At level 5, the work packages should be identifiable and homogeneous.)

PMBOK® Guide, 4th Edition
Figure 5–6 Sample WBS

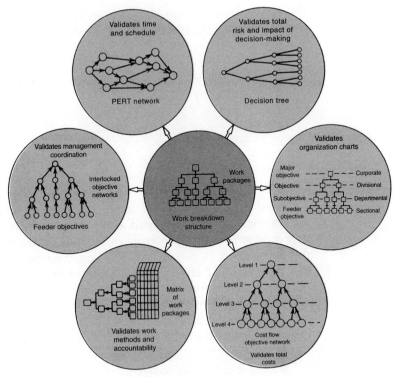

FIGURE 11–3. Work breakdown structure for objective control and evaluation. *Source:* Paul Mali, *Managing by Objectives* (New York: Wiley, 1972), p. 163. Copyright © 1972 by John Wiley & Sons. Reprinted by permission of the publisher.

- The level at which the project is managed is generally called the work package level. Actually, the work package can exist at any level below level one.
- The WBS must be accompanied by a description of the scope of effort required, or else only those individuals who issue the WBS will have a complete understanding of what work has to be accomplished. It is common practice to reproduce the customer's statement of work as the description for the WBS.
- It is often the best policy for the project manager, regardless of his technical expertise, to allow all of the line managers to assess the risks in the SOW. After all, the line managers are usually the recognized experts in the organization.

Project managers normally manage at the top three levels of the WBS and prefer to provide status reports to management at these levels also. Some companies are trying to standardize reporting to management by requiring the top three levels of the WBS to be the same for every project, the only differences being in levels 4–6. For companies with a great deal of similarity among projects, this approach has merit. For most companies, however, the differences between projects make it almost impossible to standardize the top levels of the WBS.

The work package is the critical level for managing a work breakdown structure, as shown in Figure 11–4. However, it is possible that the actual management of the work packages is supervised and performed by the line managers with status reporting provided to the project manager at higher levels of the WBS.

Work packages are natural subdivisions of cost accounts and constitute the basic building blocks used by the contractor in planning, controlling, and measuring contract performance. A work package is simply a low-level task or job assignment. It describes the work to be accomplished by a specific performing organization or a group of cost centers and serves as a vehicle for monitoring and reporting progress of work. Documents that authorize and assign work to a performing organization are designated by various names throughout industry. "Work package" is the generic term used in the criteria to identify discrete tasks that have definable end results. Ideal work packages are 80 hours and 2–4 weeks. However, this may not be possible on large projects.

It is not necessary that work package documentation contain complete, stand-alone descriptions. Supplemental documentation may augment the work package descriptions. However, the work package descriptions must permit cost account managers and work package supervisors to understand and clearly distinguish one work package effort from another. In the review of work package documentation, it may be necessary to obtain explanations from personnel routinely involved in the work, rather than requiring the work package descriptions to be completely self-explanatory.

Short-term work packages may help evaluate accomplishments. Work packages should be natural subdivisions of effort planned according to the way the work will be done. However, when work packages are relatively short, little or no assessment of work-in-process is required and the evaluation of status is possible mainly on the basis of work package completions. The longer the work packages, the more difficult and subjective the work-in-process assessment becomes unless the packages are subdivided by objective indicators such as discrete milestones with preassigned budget values or completion percentages.

In setting up the work breakdown structure, tasks should:

- Have clearly defined start and end dates
- Be usable as a communications tool in which results can be compared with expectations
- Be estimated on a "total" time duration, not when the task must start or end
- Be structured so that a minimum of project office control and documentation (i.e., forms) is necessary

For large projects, planning will be time phased at the work package level of the WBS. The work package has the following characteristics:

- Represents units of work at the level where the work is performed
- Clearly distinguishes one work package from all others assigned to a single functional group
- Contains clearly defined start and end dates that are representative of physical accomplishment (This is accomplished after scheduling has been completed.)

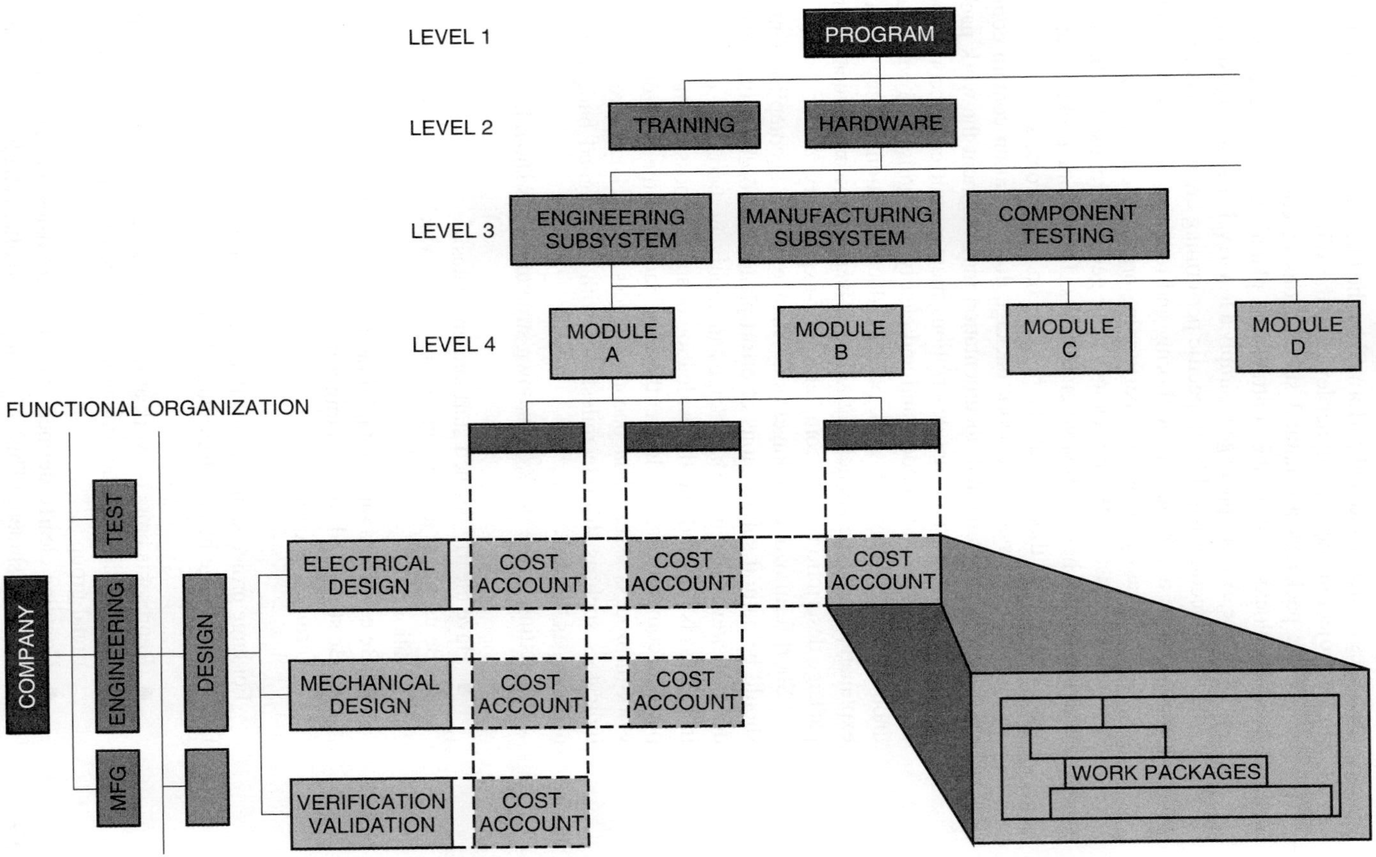

FIGURE 11–4. The cost account intersection.

- Specifies a budget in terms of dollars, man-hours, or other measurable units
- Limits the work to be performed to relatively short periods of time to minimize the work-in-process effort

Table 11–2 shows a simple work breakdown structure with the associated numbering system following the work breakdown. The first number represents the total program (in this case, it is represented by 01), the second number represents the project, and the third number identifies the task. Therefore, number 01-03-00 represents project 3 of program 01, whereas 01-03-02 represents task 2 of project 3. This type of numbering system is not standard; each company may have its own system, depending on how costs are to be controlled.

The preparation of the work breakdown structure is not easy. The WBS is a communications tool, providing detailed information to different levels of management. If it does not contain enough levels, then the integration of activities may prove difficult. If too many levels exist, then unproductive time will be made to have the same number of levels for all projects, tasks, and so on. Each major work element should be considered by itself. Remember, the WBS establishes the number of required networks for cost control.

For many programs, the work breakdown structure is established by the customer. If the contractor is required to develop a WBS, then certain guidelines must be considered including:

- The complexity and technical requirements of the program (i.e., the statement of work)
- The program cost
- The time span of the program
- The contractor's resource requirements
- The contractor's and customer's internal structure for management control and reporting
- The number of subcontracts

TABLE 11–2. WORK BREAKDOWN STRUCTURE FOR NEW PLANT CONSTRUCTION AND START-UP

Program: New Plant Construction and Start-up	01-00-00
Project 1: Analytical Study	01-01-00
Task 1: Marketing/Production Study	01-01-01
Task 2: Cost Effectiveness Analysis	01-01-02
Project 2: Design and Layout	01-02-00
Task 1: Product Processing Sketches	01-02-01
Task 2: Product Processing Blueprints	01-02-02
Project 3: Installation	01-03-00
Task 1: Fabrication	01-03-01
Task 2: Setup	01-03-02
Task 3: Testing and Run	01-03-03
Project 4: Program Support	01-04-00
Task 1: Management	01-04-01
Task 2: Purchasing Raw Materials	01-04-02

Applying these guidelines serves only to identify the complexity of the program. These data must then be subdivided and released, together with detailed information, to the different levels of the organization. The WBS should follow specified criteria because, although preparation of the WBS is performed by the program office, the actual work is performed by the doers, not the planners. Both the doers and the planners must be in agreement as to what is expected. A sample listing of criteria for developing a work breakdown structure is shown below:

- The WBS and work description should be easy to understand.
- All schedules should follow the WBS.
- No attempt should be made to subdivide work arbitrarily to the lowest possible level. The lowest level of work should not end up having a ridiculous cost in comparison to other efforts.
- Since scope of effort can change during a program, every effort should be made to maintain flexibility in the WBS.
- The WBS can act as a list of discrete and tangible milestones so that everyone will know when the milestones were achieved.
- The level of the WBS can reflect the "trust" you have in certain line groups.
- The WBS can be used to segregate recurring from nonrecurring costs.
- Most WBS elements (at the lowest control level) range from 0.5 to 2.5 percent of the total project budget.

11.12 WBS DECOMPOSITION PROBLEMS

There is a common misconception that WBS decomposition is an easy task to perform. In the development of the WBS, the top three levels or management levels are usually roll-up levels. Preparing templates at these levels is becoming common practice. However, at levels 4–6 of the WBS, templates may not be appropriate. There are reasons for this.

- Breaking the work down to extremely small and detailed work packages may require the creation of hundreds or even thousands of cost accounts and charge numbers. This could increase the management, control, and reporting costs of these small packages to a point where the costs exceed the benefits. Although a typical work package may be 200–300 hours and approximately two weeks in duration, consider the impact on a large project, which may have more than one million direct labor hours.
- Breaking the work down to small work packages can provide accurate cost control if, and only if, the line managers can determine the costs at this level of detail. Line managers must be given the right to tell project managers that costs *cannot* be determined at the requested level of detail.
- The work breakdown structure is the basis for scheduling techniques such as the Arrow Diagramming Method and the Precedence Diagramming Method. At low levels of the WBS, the interdependencies between activities can become so complex that meaningful networks cannot be constructed.

One solution to the above problems is to create "hammock" activities, which encompass several activities where exact cost identification cannot or may not be accurately determined. Some projects identify a "hammock" activity called management support (or project office), which includes overall project management, data items, management reserve, and possibly procurement. The advantage of this type of hammock activity is that the charge numbers are under the *direct* control of the project manager.

There is a common misconception that the typical dimensions of a work package are approximately 80 hours and less than two weeks to a month. Although this may be true on small projects, this would necessitate millions of work packages on large jobs and this may be impractical, even if line managers could control work packages of this size.

From a cost control point of view, cost analysis down to the fifth level is advantageous. However, it should be noted that the cost required to prepare cost analysis data to each lower level may increase exponentially, especially if the customer requires data to be presented in a specified format that is not part of the company's standard operating procedures. The level-5 work packages are normally for in-house control only. Some companies bill customers separately for each level of cost reporting below level 3.

The WBS can be subdivided into subobjectives with finer divisions of effort as we go lower into the WBS. By defining subobjectives, we add greater understanding and, it is hoped, clarity of action for those individuals who will be required to complete the objectives. Whenever work is structured, understood, easily identifiable, and within the capabilities of the individuals, there will almost always exist a high degree of confidence that the objective can be reached.

Work breakdown structures can be used to structure work for reaching such objectives as lowering cost, reducing absenteeism, improving morale, and lowering scrap factors. The lowest subdivision now becomes an end-item or subobjective, not necessarily a work package as described here. However, since we are describing project management, for the remainder of the text we will consider the lowest level as the work package.

Once the WBS is established and the program is "kicked off," it becomes a very costly procedure to either add or delete activities, or change levels of reporting because of cost control. Many companies do not give careful forethought to the importance of a properly developed WBS, and ultimately they risk cost control problems downstream. One important use of the WBS is that it serves as a cost control standard for any future activities that may follow on or may just be similar. One common mistake made by management is the combining of direct support activities with administrative activities. For example, the department manager for manufacturing engineering may be required to provide administrative support (possibly by attending team meetings) throughout the duration of the program. If the administrative support is spread out over each of the projects, a false picture is obtained as to the actual hours needed to accomplish each project in the program. If one of the projects should be canceled, then the support man-hours for the total program would be reduced when, in fact, the administrative and support functions may be constant, regardless of the number of projects and tasks.

Quite often work breakdown structures accompanying customer RFPs contain much more scope of effort, as specified by the statement of work, than the existing funding will support. This is done intentionally by the customer in hopes that a contractor may be willing to "buy in." If the contractor's price exceeds the customer's funding limitations, then the scope of effort must be reduced by eliminating activities from the WBS. By developing a separate project for administrative and indirect support activities, the customer can easily modify his costs by eliminating the direct support activities of the canceled effort.

Before we go on, there should be a brief discussion of the usefulness and applicability of the WBS system. Many companies and industries have been successful in managing programs without the use of work breakdown structures, especially on repetitive-type programs. As was the case with the SOW, there are also preparation guides for the WBS[7]:

- Develop the WBS structure by subdividing the total effort into discrete and logical subelements. Usually a program subdivides into projects, major systems, major subsystems, and various lower levels until a manageable-size element level is reached. Wide variations may occur, depending upon the type of effort (e.g., major systems development, support services, etc.). Include more than one cost center and more than one contractor if this reflects the actual situation.
- Check the proposed WBS and the contemplated efforts for completeness, compatibility, and continuity.
- Determine that the WBS satisfies both functional (engineering/manufacturing/test) and program/project (hardware, services, etc.) requirements, including recurring and nonrecurring costs.
- Check to determine if the WBS provides for logical subdivision of all project work.
- Establish assignment of responsibilities for all identified effort to specific organizations.
- Check the proposed WBS against the reporting requirements of the organizations involved.

PMBOK® Guide, 4th Edition
5.3.3.1 WBS

There are also checklists that can be used in the preparation of the WBS[8]:

- Develop a preliminary WBS to not lower than the top three levels for solicitation purposes (or lower if deemed necessary for some special reason).
- Assure that the contractor is required to extend the preliminary WBS in response to the solicitation, to identify and structure all contractor work to be compatible with his organization and management system.
- Following negotiations, the CWBS included in the contract should not normally extend lower than the third level.
- Assure that the negotiated CWBS structure is compatible with reporting requirements.

---

7. Source: *Handbook for Preparation of Work Breakdown Structures*, NHB5610.1, National Aeronautics and Space Administration, February 1975.

8. See note 7.

- Assure that the negotiated CWBS is compatible with the contractor's organization and management system.
- Review the CWBS elements to ensure correlation with:
  - The specification tree
  - Contract line items
  - End-items of the contract
  - Data items required
  - Work statement tasks
  - Configuration management requirements
- Define CWBS elements down to the level where such definitions are meaningful and necessary for management purposes (WBS dictionary).
- Specify reporting requirements for selected CWBS elements if variations from standard reporting requirements are desired.
- Assure that the CWBS covers measurable effort, level of effort, apportioned effort, and subcontracts, if applicable.
- Assure that the total costs at a particular level will equal the sum of the costs of the constituent elements at the next lower level.

On simple projects, the WBS can be constructed as a "tree diagram" (see Figure 11–5) or according to the logic flow. In Figure 11–5, the tree diagram can follow the work or even the organizational structure of the company (i.e., division, department, section, unit). The second method is to create a logic flow (see Figure 12–21) and cluster certain elements to represent tasks and projects. In the tree method, lower-level functional units may be

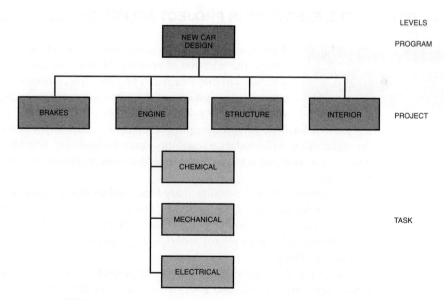

FIGURE 11–5. WBS tree diagram.

assigned to one, and only one, work element, whereas in the logic flow method the lower-level functional units may serve several WBS elements.

A tendency exists to develop guidelines, policies, and procedures for project management, but not for the development of the WBS. Some companies have been marginally successful in developing a "generic" methodology for levels 1, 2, and 3 of the WBS to use on all projects. The differences appear in levels 4, 5, and 6.

The table below shows the three most common methods for structuring the WBS:

| | Method | | |
Level	Flow	Life Cycle	Organization
Program	Program	Program	Program
Project	System	Life cycle	Division
Task	Subsystem	System	Department
Subtask	People	Subsystem	Section
Work package	People	People	People
Level of effort	People	People	People

The flow method breaks the work down into systems and major subsystems. This method is well suited for projects less than two years in length. For longer projects, we use the life-cycle method, which is similar to the flow method. The organization method is used for projects that may be repetitive or require very little integration between functional units.

11.13 ROLE OF THE EXECUTIVE IN PROJECT SELECTION

PMBOK® Guide, 4th Edition
Chapter 4 Integration
4.1.2 Develop Project Charter
Tools and Techniques

A prime responsibility of senior management (and possibly project sponsors) is the selection of projects. Most organizations have an established selection criteria, which can be subjective, objective, quantitative, qualitative, or simply a seat-of-the-pants guess. In any event, there should be a valid reason for selecting the project.

From a financial perspective, project selection is basically a two-part process. First, the organization will conduct a feasibility study to determine whether the project *can* be done. The second part is to perform a benefit-to-cost analysis to see whether the company *should* do it.

The purpose of the feasibility study is to validate that the project meets feasibility of cost, technological, safety, marketability, and ease of execution requirements. The company may use outside consultants or subject matter experts (SMEs) to assist in both feasibility studies and benefit-to-cost analyses. A project manager may not be assigned until after the feasibility study is completed.

As part of the feasibility process during project selection, senior management often solicits input from SMEs and lower-level managers through rating models. The rating models normally identify the business and/or technical criteria against which the ratings

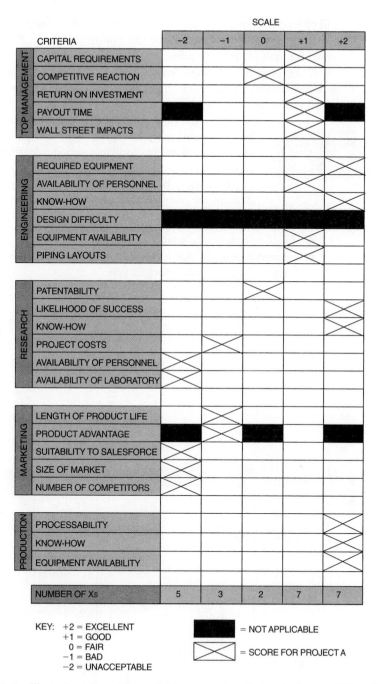

FIGURE 11–6. Illustration of a scaling model for one project, Project A. *Source:* William E. Souder, *Project Selection and Economic Appraisal,* p. 66.

will be made. Figure 11–6 shows a scaling model for a single project. Figure 11–7 shows a checklist rating system to evaluate three projects at once. Figure 11–8 shows a scoring model for multiple projects using weighted averages.

If the project is deemed feasible and a good fit with the strategic plan, then the project is prioritized for development along with other projects. Once feasibility is determined, a benefit-to-cost analysis is performed to validate that the project will, if executed correctly, provide the required financial and nonfinancial benefits. Benefit-to-cost analyses require significantly more information to be scrutinized than is usually available during a feasibility study. This can be an expensive proposition.

PMBOK® Guide, 4th Edition
5.2.2 Scope Definition
5.2.2.2 Product Analysis

Estimating benefits and costs in a timely manner is very difficult. Benefits are often defined as:

- Tangible benefits for which dollars may be reasonably quantified and measured.
- Intangible benefits that may be quantified in units other than dollars or may be identified and described subjectively.

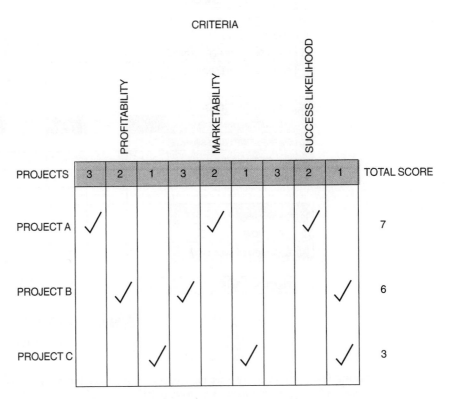

FIGURE 11–7. Illustration of a checklist for three projects. *Source:* William Souder, *Project Selection and Economic Appraisal,* p. 68.

CRITERIA	PROFITABILITY	PATENTABILITY	MARKETABILITY	PRODUCEABILITY
CRITERION WEIGHTS	4	3	2	1

PROJECTS	CRITERION SCORES*				TOTAL WEIGHTED SCORE
PROJECT D	10	6	4	3	69
PROJECT E	5	10	10	5	75
PROJECT F	3	7	10	10	63

TOTAL WEIGHTED SCORE = Σ (CRITERION SCORE × CRITERION WEIGHT)

* SCALE: 10 = EXCELLENT; 1 = UNACCEPTABLE

FIGURE 11–8. Illustration of a scoring model. *Source:* William Souder, *Project Selection and Economic Appraisal*, p. 69.

Costs are significantly more difficult to quantify. The minimum costs that must be determined are those that specifically are used for comparison to the benefits. These include:

● The current operating costs or the cost of operating in today's circumstances.
● Future period costs that are expected and can be planned for.
● Intangible costs that may be difficult to quantify. These costs are often omitted if quantification would contribute little to the decision-making process.

TABLE 11–3. FEASIBILITY STUDY AND BENEFIT-COST ANALYSIS

	Feasibility Study	Benefit-Cost Analysis
Basic Question	*Can We Do It?*	*Should We Do It?*
Life-Cycle Phase	Preconceptual	Conceptual
PM Selected	Usually not yet	Usually identified but partial involvement
Analysis	Qualitative	Quantitative
Critical Factors for Go/No-Go	• Technical	• Net present value
	• Cost	• Discounted cash flow
	• Quality	• Internal rate of return
	• Safety	• Return on investment
	• Ease of performance	• Probability of success
	• Economical	• Reality of assumptions
	• Legal	and constraints
Executive Decision Criteria	Strategic fit	Benefits exceed costs by required margin

There must be careful documentation of all known constraints and assumptions that were made in developing the costs and the benefits. Unrealistic or unrecognized assumptions are often the cause of unrealistic benefits. The go or no-go decision to continue with a project could very well rest upon the validity of the assumptions.

Table 11–3 shows the major differences between feasibility studies and benefit-to-cost analyses.

Today, the project manager may end up participating in the project selection process. In Chapter 1, we discussed the new breed of project manager, namely a person that has excellent business skills as well as project management skills. These business skills now allow us to bring the project manager on board the project at the beginning of the initiation phase rather than at the end of the initiation phase because the project manager can now make a valuable contribution to the project selection process. The project manager can be of assistance during project selection by providing business case knowledge including:

- Opportunity options (sales volume, market share, and follow-on business)
- Resource requirements (team knowledge requirements and skill set)
- Refined project costs
- Refined savings
- Benefits (financial, strategic, payback)
- Project metrics (key performance indicators and critical success factors)
- Benefits realization (consistency with the corporate business plan)
- Risks
- Exit strategies
- Organizational readiness and strengths
- Schedule/milestones
- Overall complexity
- Technology complexity and constraints, if any[9]

9. For additional factors that can influence project selection decision making, see J. R. Meredith and S. J. Mantel, Jr., *Project Management,* 3rd ed., (New York: Wiley, 1995), pp. 44–46.

11.14 ROLE OF THE EXECUTIVE IN PLANNING

Executives are responsible for selecting the project manager, and the person chosen should have planning expertise. Not all technical specialists are good planners. Likewise, some people that are excellent in execution have minimal planning skills. Executives must make sure that whomever is assigned as the project manager has both planning and execution skills. In addition, executives must take an active role during project planning activities especially if they also function as project sponsors.[10]

Executives must not arbitrarily set unrealistic milestones and then "force" line managers to fulfill them. Both project and line managers should try to adhere to unrealistic milestones, but if a line manager says he cannot, executives should comply because the line manager is supposedly the expert.

Executives should interface with project and line personnel during the planning stage in order to define the requirements and establish reasonable deadlines. Executives must realize that creating an unreasonable deadline may require the reestablishment of priorities, and, of course, changing priorities can push milestones backward.

11.15 THE PLANNING CYCLE

Previously, we stated that perhaps the most important reason for structuring projects into life-cycle phases is to provide management with control of the critical decision points in order to:

- Avoid commitment of major resources too early
- Preserve future options
- Maximize benefits of each project in relation to all other projects
- Assess risks

On long-term projects, phasing can be overdone, resulting in extra costs and delays. To prevent this, many project-driven companies resort to other types of systems, such as a management cost and control system (MCCS). No program or project can be efficiently organized and managed without some form of management cost and control system. Figure 11–9 shows the five phases of a management cost and control system. The first phase constitutes the planning cycle, and the next four phases identify the operating cycle.

Figure 11–10 shows the activities included in the planning cycle. The work breakdown structure serves as the initial control from which all planning emanates. The WBS acts as a vital artery for communications and operations in all phases. A comprehensive analysis of management cost and control systems is presented in Chapter 15.

---

10. Although this section is called "The Role of the Executive in Planning," it also applies to line management if project sponsorship is pushed down to the middle-management level or lower. This is quite common in highly mature project management organizations where senior management has sufficient faith in line management's ability to serve as project sponsors.

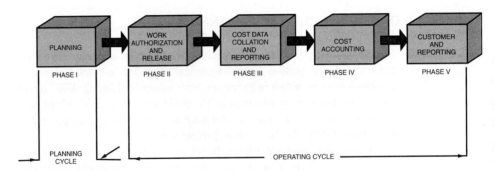

FIGURE 11–9. Phases of a management cost and control system.

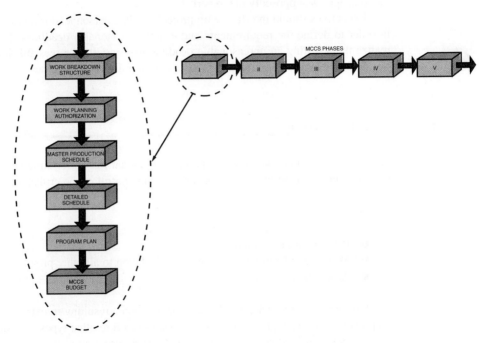

FIGURE 11–10. The planning cycle of a management cost and control system.

11.16 WORK PLANNING AUTHORIZATION

PMBOK® Guide, 4th Edition
4.3.2 Direct and Manage
Project Execution

After receipt of a contract, some form of authorization is needed before work can begin, even in the planning stage. Both work authorization and work planning authorization are used to release funds, but for different purposes. Work planning authorization releases funds (primarily for functional management) so that scheduling, costs, budgets, and all other types of plans can be prepared prior to the release of operational cycle funds, which hereafter shall be referred to simply as work authorization. Both forms of authorization require the same paperwork. In many companies this work authorization is identified as a subdivided work description (SWD), which is a narrative description of the effort to be performed by the cost center (division-level minimum). This package establishes the work to be performed, the period of performance, and possibly the maximum number of hours available. The SWD is multipurpose in that it can be used to release contract funds, authorize planning, describe activities as identified in the WBS, and, last but not least, release work.

The SWD is one of the key elements in the planning of a program as shown in Figure 11–10. Contract control and administration releases the contract funds by issuing a SWD, which sets forth general contractual requirements and authorizes program management to proceed. Program management issues the SWD to set forth the contractual guidelines and requirements for the functional units. The SWD specifies how the work will be performed, which functional organizations will be involved, and who has what specific responsibilities, and authorizes the utilization of resources within a given time period.

The SWD authorizes both the program team and functional management to begin work. As shown in Figure 11–10, the SWD provides direct input to Phase II of the MCCS. Phase I and Phase II can and do operate simultaneously because it is generally impossible for program office personnel to establish plans, procedures, and schedules without input from the functional units.

The subdivided work description package is used by the operating organizations to further subdivide the effort defined by the WBS into small segments or work packages.

Many people contend that if the data in the work authorization document are different from what was originally defined in the proposal, the project is in trouble right at the start. This may not be the case, because most projects are priced out assuming "unlimited" resources, whereas the hours and dollars in the work authorization document are based upon "limited" resources. This situation is common for companies that thrive on competitive bidding.

11.17 WHY DO PLANS FAIL?

No matter how hard we try, planning is not perfect, and sometimes plans fail. Typical reasons include:

- Corporate goals are not understood at the lower organizational levels.
- Plans encompass too much in too little time.
- Financial estimates are poor.
- Plans are based on insufficient data.
- No attempt is being made to systematize the planning process.
- Planning is performed by a planning group.
- No one knows the ultimate objective.
- No one knows the staffing requirements.
- No one knows the major milestone dates, including written reports.
- Project estimates are best guesses, and are not based on standards or history.
- Not enough time has been given for proper estimating.
- No one has bothered to see if there will be personnel available with the necessary skills.
- People are not working toward the same specifications.
- People are consistently shuffled in and out of the project with little regard for schedule.

Why do these situations occur? If corporate goals are not understood, it is because corporate executives have been negligent in providing the necessary strategic information and feedback. If a plan fails because of extreme optimism, then the responsibility lies with both the project and line managers for not assessing risk. Project managers should ask the line managers if the estimates are optimistic or pessimistic, and expect an honest answer. Erroneous financial estimates are the responsibility of the line manager. If the project fails because of a poor definition of the requirements, then the project manager is totally at fault.

Sometimes project plans fail because simple details are forgotten or overlooked. Examples of this might be:

- Neglecting to tell a line manager early enough that the prototype is not ready and that rescheduling is necessary.
- Neglecting to see if the line manager can still provide additional employees for the next two weeks because it was possible to do so six months ago.

Sometimes plans fail because the project manager "bites off more than he can chew," and then something happens, such as his becoming ill. Many projects have failed because the project manager was the only one who knew what was going on and then got sick.

11.18 STOPPING PROJECTS

PMBOK® Guide, 4th Edition
4.6 Close Projects

There are always situations in which projects have to be stopped. Nine reasons for stopping are:

- Final achievement of the objectives
- Poor initial planning and market prognosis
- A better alternative is found
- A change in the company interest and strategy
- Allocated time is exceeded
- Budgeted costs are exceeded
- Key people leave the organization
- Personal whims of management
- Problem too complex for the resources available

Today most of the reasons why projects are not completed on time and within cost are behavioral rather than quantitative. They include:

- Poor morale
- Poor human relations
- Poor labor productivity
- No commitment by those involved in the project

The last item appears to be the cause of the first three items in many situations.

Once the reasons for cancellation are defined, the next problem concerns how to stop the project. Some of the ways are:

- Orderly planned termination
- The "hatchet" (withdrawal of funds and removal of personnel)

- Reassignment of people to higher priority tasks
- Redirection of efforts toward different objectives
- Burying it or letting it die on the vine (i.e., not taking any official action)

There are three major problem areas to be considered in stopping projects:

- Worker morale
- Reassignment of personnel
- Adequate documentation and wrap-up

11.19 HANDLING PROJECT PHASEOUTS AND TRANSFERS

PMBOK® Guide, 4th Edition
4.4 Monitor and Control
Project Work

By definition, projects (and even life cycle phases) have an end point. Closing out is a very important phase in the project life cycle, which should follow particular disciplines and procedures with the objective of:

- Effectively bringing the project to closure according to agreed-on contractual requirements
- Preparing for the transition of the project into the next operational phase, such as from production to field installation, field operation, or training
- Analyzing overall project performance with regard to financial data, schedules, and technical efforts
- Closing the project office, and transferring or selling off all resources originally assigned to the project, including personnel
- Identifying and pursuing follow-on business

Although most project managers are completely cognizant of the necessity for proper planning for project start-up, many project managers neglect planning for project termination. Planning for project termination includes:

- Transferring responsibility
- Completion of project records
    - Historic reports
    - Postproject analysis
- Documenting results to reflect "as built" product or installation
- Acceptance by sponsor/user
- Satisfying contractual requirements
- Releasing resources
    - Reassignment of project office team members
    - Disposition of functional personnel
    - Disposition of materials
- Closing out work orders (financial closeout)
- Preparing for financial payments

Project success or failure often depends on management's ability to handle personnel issues properly during this final phase. If job assignments beyond the current project look undesirable or uncertain to project team members, a great deal of anxiety and conflict may develop that diverts needed energy to job hunting, foot dragging, or even sabotage. Project personnel may engage in job searches on their own and may leave the project prematurely. This creates a glaring void that is often difficult to patch.

Given business realities, it is difficult to transfer project personnel under ideal conditions. The following suggestions may increase organizational effectiveness and minimize personal stress when closing out a project:

- Carefully plan the project closeout on the part of both project and functional managers. Use a checklist to prepare the plan.
- Establish a simple project closeout procedure that identifies the major steps and responsibilities.
- Treat the closeout phase like any other project, with clearly delineated tasks, agreed-on responsibilities, schedules, budgets, and deliverable items or results.
- Understand the interaction of behavioral and organizational elements in order to build an environment conducive to teamwork during this final project phase.
- Emphasize the overall goals, applications, and utilities of the project as well as its business impact.
- Secure top-management involvement and support.
- Be aware of conflict, fatigue, shifting priorities, and technical or logistic problems. Try to identify and deal with these problems when they start to develop. Communicating progress through regularly scheduled status meetings is the key to managing these problems.
- Keep project personnel informed of upcoming job opportunities. Resource managers should discuss and negotiate new assignments with personnel and involve people already in the next project.
- Be aware of rumors. If a reorganization or layoff is inevitable, the situation should be described in a professional manner or people will assume the worst.
- Assign a contract administrator dedicated to company-oriented projects. He will protect your financial position and business interests by following through on customer sign-offs and final payment.

11.20 DETAILED SCHEDULES AND CHARTS

The scheduling of activities is the first major requirement of the program office after program go-ahead. The program office normally assumes full responsibility for activity scheduling if the activity is not too complex. For large programs, functional management input is required before scheduling can be completed. Depending on program size and contractual requirements, the program office may have a staff member whose sole responsibility is to continuously develop and update activity schedules to track program work. The resulting information is supplied to program office personnel, functional management, team members, and the customer.

Activity scheduling is probably the single most important tool for determining how company resources should be integrated. Activity schedules are invaluable for projecting time-phased resource utilization requirements, providing a basis for visually tracking performance and estimating costs. The schedules serve as master plans from which both the customer and management have an up-to-date picture of operations.

Certain guidelines should be followed in the preparation of schedules, regardless of the projected use or complexity:

- All major events and dates must be clearly identified. If a statement of work is supplied by the customer, those dates shown on the accompanying schedules must be included. If for any reason the customer's milestone dates cannot be met, the customer should be notified immediately.
- The exact sequence of work should be defined through a network in which interrelationships between events can be identified.
- Schedules should be directly relatable to the work breakdown structure. If the WBS is developed according to a specific sequence of work, then it becomes an easy task to identify work sequences in schedules using the same numbering system as in the WBS. The minimum requirement should be to show where and when all tasks start and finish.
- All schedules must identify the time constraints and, if possible, should identify those resources required for each event.

Although these four guidelines relate to schedule preparation, they do not define how complex the schedules should be. Before preparing schedules, three questions should be considered:

- How many events or activities should each network have?
- How much of a detailed technical breakdown should be included?
- Who is the intended audience for this schedule?

Most organizations develop multiple schedules: summary schedules for management and planners and detailed schedules for the doers and lower-level control. The detailed schedules may be strictly for interdepartmental activities. Program management must approve all schedules down through the first three levels of the work breakdown structure. For lower-level schedules (i.e., detailed interdepartmental), program management may or may not request a sign of approval.

One of the most difficult problems to identify in schedules is a hedge position. A hedge position is a situation in which the contractor may not be able to meet a customer's milestone date without incurring a risk, or may not be able to meet activity requirements following a milestone date because of contractual requirements. To illustrate a common hedge position, consider Example 11–1 below.

Example 11–1. Condor Corporation is currently working on a project that has three phases: design, development, and qualification of a certain component. Contractual requirements with the customer specify that no components will be fabricated for the development phase until the design review meeting is held following the design phase.

Condor has determined that if it does not begin component fabrication prior to the design review meeting, then the second and third phases will slip. Condor is willing to accept the risk that should specifications be unacceptable during the design review meeting, the costs associated with preauthorization of fabrication will be incurred. How should this be shown on a schedule? (The problems associated with performing unauthorized work are not being considered here.)

The solution is not easy. Condor must show on the master production schedule that component fabrication will begin early, at the contractor's risk. This should be followed up by a contractual letter in which both the customer and contractor understand the risks and implications.

Detailed schedules are prepared for almost every activity. It is the responsibility of the program office to marry all of the detailed schedules into one master schedule to verify that all activities can be completed as planned. The preparation sequence for schedules (and also for program plans) is shown in Figure 11–11. The program office submits a request for detailed schedules to the functional managers and the functional managers prepare summary schedules, detailed schedules, and, if time permits, interdepartmental schedules. Each functional manager then reviews his schedules with the program office. The program office, together with the functional program team members, integrates all of the plans and schedules and verifies that all contractual dates can be met.

Before the schedules are submitted to publications, rough drafts of each schedule and plan should be reviewed with the customer. This procedure accomplishes the following:

- Verifies that nothing has fallen through the cracks
- Prevents immediate revisions to a published document and can prevent embarrassing moments
- Minimizes production costs by reducing the number of early revisions
- Shows customers early in the program that you welcome their help and input into the planning phase

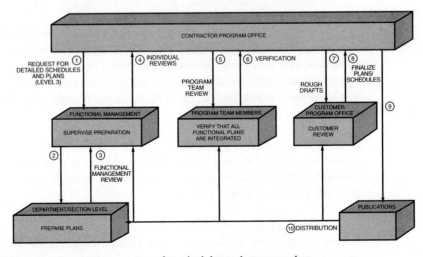

FIGURE 11–11. Preparation sequence for schedules and program plans.

After the document is published, it should be distributed to all program office personnel, functional team members, functional management, and the customer. Examples of detailed schedules are shown in Chapter 13.

In addition to the detailed schedules, the program office, with input provided by functional management, must develop organization charts. The charts show who has responsibility for each activity and display the formal (and often the informal) lines of communication. Examples were shown in Section 4.11.

The program office may also establish linear responsibility charts (LRCs). In spite of the best attempts by management, many functions in an organization may overlap between functional units. Also, management might wish to have the responsibility for a certain activity given to a functional unit that normally would not have that responsibility. This is a common occurrence on short-duration programs where management desires to cut costs and red tape.

Project personnel should keep in mind why the schedule was developed. The primary objective is usually to coordinate activities to complete the project with the:

- Best time
- Least cost
- Least risk

There are also secondary objectives of scheduling:

- Studying alternatives
- Developing an optimal schedule
- Using resources effectively
- Communicating
- Refining the estimating criteria
- Obtaining good project control
- Providing for easy revisions

Large projects, especially long-term efforts, may require a "war room." War rooms generally have only one door and no windows. All of the walls are covered with large schedules, perhaps printed on blueprint paper, and each wall could have numerous sliding panels. The schedules and charts on each wall could be updated on a daily basis. The room would be used for customer briefings, team meetings, and any other activities related specifically to this project.

11.21 MASTER PRODUCTION SCHEDULING

The release of the planning SWD, as shown in Figure 11–10, authorizes the manufacturing units to prepare a master production schedule from which detailed analysis of the utilization of company resources can be seen and tracked.

Master production scheduling is not a new concept. Earliest material control systems used a "quarterly ordering system" to produce a master production schedule (MPS) for plant production. This system uses customer order backlogs to develop a production plan over a

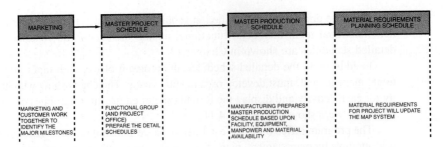

FIGURE 11–12. Material requirements planning interrelationships.

three-month period. The production plan is then exploded manually to determine what parts must be purchased or manufactured at the proper time. However, rapidly changing customer requirements and fluctuating lead times, combined with a slow response to these changes, can result in the disruption of master production scheduling.[11]

Master Production Schedule Definition

A *master production schedule* is a statement of what will be made, how many units will be made, and when they will be made. It is a production plan, not a sales plan. The MPS considers the total demand on a plant's resources, including finished product sales, spare (repair) part needs, and inter-plant needs. The MPS must also consider the capacity of the plant and the requirements imposed on vendors. Provisions are made in the overall plan for each manufacturing facility's operation. All planning for materials, manpower, plant, equipment, and financing for the facility is driven by the master production schedule.

Objectives of the MPS

Objectives of master production scheduling are:

- To provide top management with a means to authorize and control manpower levels, inventory investment, and cash flow
- To coordinate marketing, manufacturing, engineering, and finance activities by a common performance objective
- To reconcile marketing and manufacturing needs
- To provide an overall measure of performance
- To provide data for material and capacity planning

The development of a master production schedule is a very important step in a planning cycle. Master production schedules directly tie together personnel, materials, equipment, and facilities, as shown in Figure 11–12. Master production schedules also identify key dates to the customer, should he wish to visit the contractor during specific operational periods.

---

11. The master production schedule is being discussed here because of its importance in the planning cycle. The MPS cannot be fully utilized without effective inventory control procedures.

11.22 PROJECT PLAN

PMBOK® Guide, 4th Edition
Chapter 5 Project Scope
  Management
Chapter 4 Integration Management
3.2 Planning Process Group

A project plan is fundamental to the success of any project. For large and often complex projects, customers may require a project plan that documents all activities within the program. The project plan then serves as a guideline for the lifetime of the project and may be revised as often as once a month, depending on the circumstances and the type of project (i.e., research and development projects require more revisions to the project plan than manufacturing or construction projects). The project plan provides the following framework:

- Eliminates conflicts between functional managers
- Eliminates conflicts between functional management and program management
- Provides a standard communications tool throughout the lifetime of the project (It should be geared to the work breakdown structure)
- Provides verification that the contractor understands the customer's objectives and requirements
- Provides a means for identifying inconsistencies in the planning phase
- Provides a means for early identification of problem areas and risks so that no surprises occur downstream
- Contains all of the schedules defined in Section 11.18 as a basis for progress analysis and reporting

Development of a project plan can be time-consuming and costly. All levels of the organization participate. The upper levels provide summary information, and the lower levels provide the details. The project plan, like activity schedules, does not preclude departments from developing their own plans.

The project plan must identify how the company resources will be integrated. The process is similar to the sequence of events for schedule preparation, shown in Figure 11–11. Since the project plan must explain the events in Figure 11–11, additional iterations are required, which can cause changes in a project. This can be seen in Figure 11–13.

The project plan is a standard from which performance can be measured by the customer and the project and functional managers. The plan serves as a cookbook by answering these questions for all personnel identified with the project:

- What will be accomplished?
- How will it be accomplished?
- Where will it be accomplished?
- When will it be accomplished?
- Why will it be accomplished?

The answers to these questions force both the contractor and the customer to take a hard look at:

- Project requirements
- Project management

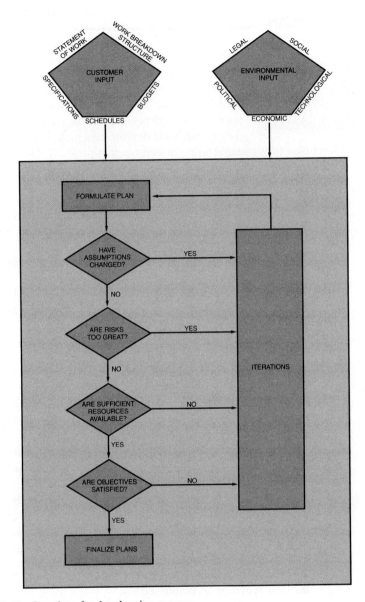

FIGURE 11–13. Iterations for the planning process.

- Project schedules
- Facility requirements
- Logistic support
- Financial support
- Manpower and organization

The project plan is more than just a set of instructions. It is an attempt to eliminate crisis by preventing anything from "falling through the cracks." The plan is documented and approved by both the customer and the contractor to determine what data, if any, are missing and the probable resulting effect. As the project matures, the project plan is revised to account for new or missing data. The most common reasons for revising a plan are:

- "Crashing" activities to meet end dates
- Trade-off decisions involving manpower, scheduling, and performance
- Adjusting and leveling manpower requests

The makeup of the project plan may vary from contractor to contractor.[12] Most project plans can be subdivided into four main sections: introduction, summary and conclusions, management, and technical. The complexity of the information is usually up to the discretion of the contractor, provided that customer requirements, as may be specified in the statement of work, are satisfied.

The introductory section contains the definition of the project and the major parts involved. If the project follows another, or is an outgrowth of similar activities, this is indicated, together with a brief summary of the background and history behind the project.

The summary and conclusion section identifies the targets and objectives of the project and includes the necessary "lip service" on how successful the project will be and how all problems can be overcome. This section must also include the project master schedule showing how all projects and activities are related. The total project master schedule should include the following:

- An appropriate scheduling system (bar charts, milestone charts, network, etc.)
- A listing of activities at the project level or lower
- The possible interrelationships between activities (can be accomplished by logic networks, critical path networks, or PERT networks)
- Activity time estimates (a natural result of the item above)

The summary and conclusion chapter is usually the second section in the project plan so that upper-level customer management can have a complete overview of the project without having to search through the technical information.

The management section of the project plan contains procedures, charts, and schedules as follows:

- The assignment of key personnel to the project is indicated. This usually refers only to the project office personnel and team members, since under normal operations these will be the only individuals interfacing with customers.

---

12. Cleland and King define fourteen subsections for a program plan. This detail appears more applicable to the technical and management volumes of a proposal. They do, however, provide a more detailed picture than presented here. See David I. Cleland and William R. King, *Systems Analysis and Project Management* (New York: McGraw-Hill, 1975), pp. 371–380.

- Manpower, planning, and training are discussed to assure customers that qualified people will be available from the functional units.
- A linear responsibility chart might also be included to identify to customers the authority relationships that will exist in the program.

Situations exist in which the management section may be omitted from the proposal. For a follow-up program, the customer may not require this section if management's positions are unchanged. Management sections are also not required if the management information was previously provided in the proposal or if the customer and contractor have continuous business dealings.

The technical section may include as much as 75 to 90 percent of the program plan, especially if the effort includes research and development, and may require constant updating as the project matures. The following items can be included as part of the technical section:

- A detailed breakdown of the charts and schedules used in the project master schedule, possibly including schedule/cost estimates.
- A listing of the testing to be accomplished for each activity. (It is best to include the exact testing matrices.)
- Procedures for accomplishment of the testing. This might also include a description of the key elements in the operations or manufacturing plans, as well as a listing of the facility and logistic requirements.
- Identification of materials and material specifications. (This might also include system specifications.)
- An attempt to identify the risks associated with specific technical requirements (not commonly included). This assessment tends to scare management personnel who are unfamiliar with the technical procedures, so it should be omitted if possible.

The project plan, as used here, contains a description of all phases of the project. For many projects, especially large ones, detailed planning is required for all major events and activities. Table 11–4 identifies the type of individual plans that may be required in place of a (total) project plan. These are often called subsidiary plans.

The project plan, once agreed on by the contractor and customer, is then used to provide project direction. This is shown in Figure 11–14. If the project plan is written clearly, then any functional manager or supervisor should be able to identify what is expected of him. The project plan should be distributed to each member of the project team, all functional managers and supervisors interfacing with the project, and all key functional personnel.

One final note need be mentioned concerning the legality of the project plan. The project plan may be specified contractually to satisfy certain requirements as identified in the customer's statement of work. The contractor retains the right to decide how to accomplish this, unless, of course, this is also identified in the SOW. If the SOW specifies that quality assurance testing will be accomplished on fifteen end-items from the production line, then fifteen is the minimum number that must be tested. The project plan may show that twenty-five items are to be tested. If the contractor develops cost overrun problems, he may wish to revert to the SOW and test only fifteen items. Contractually, he may do this without

TABLE 11–4. TYPES OF PLANS

Type of Plan	Description
Budget	How much money is allocated to each event?
Configuration management	How are technical changes made?
Facilities	What facilities resources are available?
Logistics support	How will replacements be handled?
Management	How is the program office organized?
Manufacturing	What are the time-phase manufacturing events?
Procurement	What are my sources? Should I make or buy? If vendors are not qualified, how shall I qualify them?
Quality assurance	How will I guarantee specifications will be met?
Research/development	What are the technical activities?
Scheduling	Are all critical dates accounted for?
Tooling	What are my time-phased tooling requirements?
Training	How will I maintain qualified personnel?
Transportation	How will goods and services be shipped?

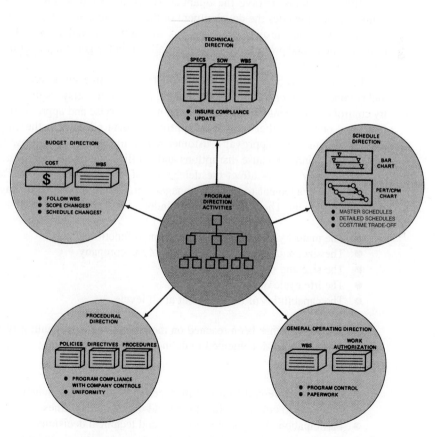

FIGURE 11–14. Project direction activities.

informing the customer. In most cases, however, the customer is notified, and the project is revised.

11.23 TOTAL PROJECT PLANNING

The difference between the good project manager and the poor project manager is often described in one word: planning. Project planning involves planning for:

- Schedule development
- Budget development
- Project administration (see Section 5.3)
- Leadership styles (interpersonal influences; see Section 5.4)
- Conflict management (see Chapter 7)

The first two items involve the quantitative aspects of planning. Planning for project administration includes the development of the linear responsibility chart.

Although each project manager has the authority and responsibility to establish project policies and procedures, they must fall within the general guidelines established by top management.

Linear responsibility charts can result from customer-imposed requirements above and beyond normal operations. For example, the customer may require as part of his quality control requirements that a specific engineer supervise and approve all testing of a certain item, or that another individual approve all data released to the customer over and above program office approval. Customer requirements similar to those identified above require LRCs and can cause disruptions and conflicts within an organization.

Several key factors affect the delegation of authority and responsibility both from upper-level management to project management, and from project management to functional management. These key factors include:

- The maturity of the project management function
- The size, nature, and business base of the company
- The size and nature of the project
- The life cycle of the project
- The capabilities of management at all levels

Once agreement has been reached on the project manager's authority and responsibility, the results may be documented to delineate that role regarding:

- Focal position
- Conflict between the project manager and functional managers
- Influence to cut across functional and organizational lines
- Participation in major management and technical decisions
- Collaboration in staffing the project
- Control over allocation and expenditure of funds
- Selection of subcontractors

- Rights in resolving conflicts
- Input in maintaining the integrity of the project team
- Establishment of project plans
- Provisions for a cost-effective information system for control
- Provisions for leadership in preparing operational requirements
- Maintenance of prime customer liaison and contact
- Promotion of technological and managerial improvements
- Establishment of project organization for the duration
- Elimination of red tape

Documenting the project manager's authority is necessary in some situations because:

- All interfacing must be kept as simple as possible.
- The project manager must have the authority to "force" functional managers to depart from existing standards and possibly incur risk.
- Gaining authority over those elements of a program that are not under the project manager's control is essential. This is normally achieved by earning the respect of the individuals concerned.
- The project manager should not attempt to fully describe the exact authority and responsibilities of the project office personnel or team members. Problem-solving rather than role definition should be encouraged.

Although documenting project authority is undesirable, it may be necessary, especially if project initiation and planning require a formal project chart. In such a case, a letter such as that shown in Table 11–5 may suffice.

Power and authority are often discussed as though they go hand in hand. Authority comes from people above you, perhaps by delegation, whereas power comes from people below you. You can have authority without power or power without authority.

In a traditional organizational structure, most individuals maintain position power. The higher up you sit, the more power you have. But in project management, the reporting level of the project might be irrelevant, especially if a project sponsor exists. In project management, the project manager's power base emanates from his

- Expertise (technical or managerial)
- Credibility with employees
- Sound decision-making ability

The last item is usually preferred. If the project manager is regarded as a sound decision-maker, then the employees normally give the project manager a great deal of power over them.

Leadership styles refer to the interpersonal influence modes that a project manager can use. Project managers may have to use several different leadership styles, depending on the makeup of the project personnel. Conflict management is important because if the project manager can predict what conflicts will occur and when they are most likely to occur, he may be able to plan for the resolution of the conflicts through project administration.

Figure 11–15 shows the complete project planning phase for the quantitative portions. The object, of course, is to develop a project plan that shows complete distribution of

TABLE 11–5.   PROJECT CHARTER

ELECTRODYNAMICS
12 Oak Avenue
Cleveland, Ohio 44114

11 June 2001

To: Distribution
From: L. White, Executive Vice President
Subject: Project Charter for the Acme Project

Mr. Robert L. James has been assigned as the Project Manager for the Acme Project.

Responsibility

Mr. James will be responsible for ensuring that all key milestones are met within the
time, cost, and performance constraints of his project, while adhering to proper quality
control standards. Furthermore, the project manager must work closely with line
managers to ensure that all assigned resources are used effectively and efficiently, and
that the project is properly staffed.

Additionally, the project manager will be responsible for:

1. All formal communications between the customer and contractor.
2. Preparation of a project plan that is realistic, and acceptable by both the customer
   and contractor.
3. Preparation of all project data items.
4. Keeping executive management informed as to project status through weekly
   (detailed) and monthly (summary) status reporting.
5. Ensuring that all functional employees and managers are kept informed as to their
   responsibilities on the project and all revisions imposed by the customer or parent
   organization.
6. Comparing actual to predicted cost and performance, and taking corrective action
   when necessary.
7. Maintaining a plan that continuously displays the project's time, cost, and performance as well as resource commitments made by the functional managers.

Authority

To ensure that the project meets its objectives, Mr. James is authorized to manage the
project and issue directives in accordance to the policies and procedures section of
the company's *Project Management Manual*. Additional directives may be issued
through the office of the executive vice-president.

The program manager's authority also includes:

1. Direct access to the customer on all matters pertaining to the Acme Project.
2. Direct access to Electrodynamics' executive management on all matters pertaining to
   the Acme Project.
3. Control and distribution of all project dollars, including procurement, such that company and project cash flow limitations are adhered to.
4. To revise the project plan as needed, and with customer approval.
5. To require periodic functional status reporting.
6. To monitor the time, cost, and performance activities in the functional departments
   and ensure that all problems are promptly identified, reported, and solved.
7. To cut across all functional lines and to interface with all levels of management as
   necessary to meet project requirements.
8. To renegotiate with functional managers for changes in personnel assignments.
9. Delegating responsibilities and authority to functional personnel, provided that the
   line manager is in approval that the employee can handle this authority/responsibility
   level.

Any questions regarding the above policies should be directed to the undersigned.

L. White
Executive Vice-President

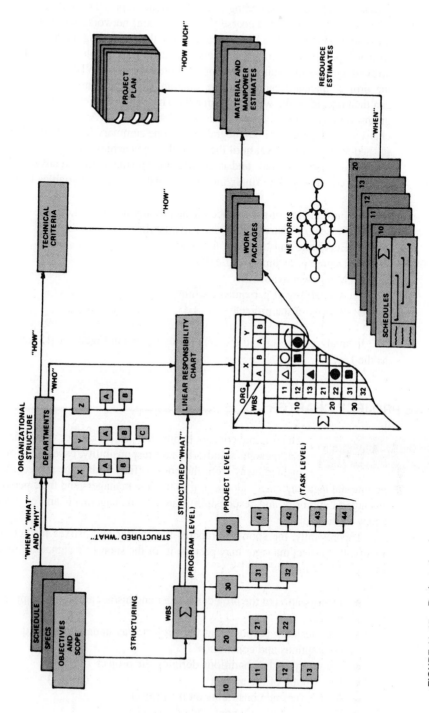

FIGURE 11–15. Project planning.

resources and the corresponding costs. The figure represents an iterative process. The project manager begins with a coarse (arrow diagram) network, and then decides on the work breakdown structure. The WBS is essential to the arrow diagram and should be constructed so that reporting elements and levels are easily identifiable. Eventually, there will be an arrow diagram and detailed chart for each element in the WBS. If there is too much detail, the project manager can refine the diagram by combining all logic into one plan and can then decide on the work assignments. There is a risk here that, by condensing the diagrams as much as possible, there may be a loss of clarity. As shown in Figure 11–15, all the charts and schedules can be integrated into one summary-level figure. This can be accomplished at each WBS level until the desired plan is achieved.

Finally, project, line, and executive management must analyze other internal and external variables before finalizing these schedules. These variables include:

- Introduction or acceptance of the product in the marketplace
- Present or planned manpower availability
- Economic constraints of the project
- Degree of technical difficulty
- Manpower availability
- Availability of personnel training
- Priority of the project

In small companies and projects, certain items in Figure 11–15 may be omitted, such as the LRCs.

11.24 THE PROJECT CHARTER

PMBOK® Guide, 4th Edition
4.1 Develop Project Charter

The original concept behind the project charter was to document the project manager's authority and responsibility, especially for projects implemented away from the home office. Today, the project charter is more of an internal legal document identifying to the line managers and their personnel the project manager's authority and responsibility and the management- and/or customer-approved scope of the project.

Theoretically, the sponsor prepares the charter and affixes his/her signature, but in reality, the project manager may prepare it for the sponsor's signature. At a minimum, the charter should include:

- Identification of the project manager and his/her authority to apply resources to the project
- The business purpose that the project was undertaken to address, including all assumptions and constraints
- Summary of the conditions defining the project
- Description of the project
- Objectives and constraints on the project
- Project scope (inclusions and exclusions)
- Key stakeholders and their roles

- Risks
- Involvement by certain stakeholders

The PMBOK® Guide provides a framework for the project charter. What is somewhat unfortunate is that every company seems to have its own idea of what should be included in a charter. The contents of a charter are often dependent upon where in the evolution and life cycle of a project the charter is prepared. (See *Advanced Project Management: Best Practices on Implementation* by Harold Kerzner, John Wiley & Sons, New York, 2004, pp. 101–102, 120, 629–630.) Some companies such as Computer Associates use both a full charter (closely aligned to the PMBOK® Guide) and an abbreviated charter based upon the size and complexity of the project.

The charter is a "legal" agreement between the project manager and the company. Some companies supplement the charter with a "contract" that functions as an agreement between the project and the line organizations.

Some companies have converted the charter into a highly detailed document containing:

- The scope baseline/scope statement
  - Scope and objectives of the project (SOW)
  - Specifications
  - WBS (template levels)
  - Timing
  - Spending plan (S-curve)
- The management plan
  - Resource requirements and manloading (if known)
  - Resumés of key personnel
  - Organizational relationships and structure
  - Responsibility assignment matrix
  - Support required from other organizations
  - Project policies and procedures
  - Change management plan
  - Management approval of above

When the project charter contains a scope baseline and management plan, the project charter may function as the project plan. This is not really an effective use of the charter, but it may be acceptable on certain types of projects for internal customers.

11.25 MANAGEMENT CONTROL

PMBOK® Guide, 4th Edition
4.5 Integrated Change Control

Because the planning phase provides the fundamental guidelines for the remainder of the project, careful management control must be established. In addition, since planning is an ongoing activity for a variety of different programs, management guidelines must be established on a company-wide basis in order to achieve unity and coherence.

All functional organizations and individuals working directly or indirectly on a program are responsible for identifying, to the project manager, scheduling and planning

TABLE 11-6. PLANNING AND REQUIREMENTS POLICIES

Program Manager	Functional Manager	Relationship
Requests the preparation of the program master schedules and provides for integration with the division composite schedules.	Develops the details of the program plans and requirements in conjunction with the program manager. Provides proposal action in support of program manager requirements and the program master schedule.	Program planning and scheduling is a functional specialty; the program manager utilizes the services of the specialist organizations. The specialists retain their own channels to the general manager but must keep the program manager informed.
Defines work to be accomplished through preparation of the subdivided work description package.		
Provides program guidance and direction for the preparation of program plans that establish program cost, schedule, and technical performance; and that define the major events and tasks to ensure the orderly progress of the program.	With guidance furnished by the program manager, participates in the preparation of program plans, schedules, and work release documents which cover cost, schedule, and technical performance; and which define major events and tasks. Provides supporting detail plans and schedules.	Program planning is also a consultative operation and is provided guidelines by the program manager. Functional organizations initiate supporting plans for program manager approval, or react to modify plans to maintain currency. Functional organizations also initiate planning studies involving trade-offs and alternative courses of action for presentation to the program manager.
Establishes priorities within the program. Obtains relative program priorities between programs managed by other programs from the director, program management, manager, marketing and product development, or the general manager as specified by the policy.	Negotiates priorities with program managers for events and tasks to be performed by his organization.	The program manager and program team members are oriented to his program, whereas the functional organizations and the functional managers are "function" and multiprogram oriented. The orientation of each director, manager, and team member

Approves program contractual data requirements.

Remains alert to new contract requirements, government regulations and directives that might affect the work, cost, or management of the program.

Provides early technical requirements definitions, and substantiates make-or-buy recommendations. Participates in the formulation of the make-or-buy plan for the program.

Approves the program bill of material for need and compliance with program need and requirements.

Directs data management including maintenance of current and historical files on programmed contractual data requirements.

Conducts analysis of contractual data requirements. Develops data plans including contractor data requirements list and obtains program manager approval.

Remains alert to new contract requirements, government regulations, and directives that might affect the work, cost, or management of his organization on any program.

Provides the necessary make-or-buy data; substantiates estimates and recommendations in the area of functional specialty.

Prepares the program bill of material.

must be mutually recognized to preclude unreasonable demands and conflicting priorities. Priority conflicts that cannot be resolved must be referred to the general manager.

Make-or-buy concurrence and approvals are obtained in accordance with current Policies and Procedures.

TABLE 11–7. SCHEDULING POLICIES

Program Manager	Functional Manager	Relationship
Provides contractual data requirements and guidance for construction of program master schedules.	The operations directorate shall construct the program master schedule. Data should include but not be limited to engineering plans, manufacturing plans, procurement plans, test plans, quality plans, and provide time spans for accomplishment of work elements defined in the work breakdown structure to the level of definition visible in the planned subdivided work description package.	The operations directorate constructs the program master schedule with data received from functional organizations and direction from the program manager. Operations shall coordinate program master schedule with functional organizations and secure program manager's approval prior to release.
Concurs with detail schedules construction by functional organizations. Provides corrective action decisions and direction as required at any time a functional organization fails to meet program master schedule requirements or when, by analysis, performance indicated by detail schedule monitoring threatens to impact the program master schedule.	Constructs detail program schedules and working schedules in consonance with program manager–approved program master schedule. Secures program manager concurrence and forwards copies to the program manager.	Program manager monitors the functional organization's detail schedules for compliance with program master schedules and reports variance items that may impact division operations to the director, program management.

problems that require corrective action during both the planning cycle and the operating cycle. The program manager bears the ultimate and final responsibility for identifying requirements for corrective actions. Management policies and directives are written specifically to assist the program manager in defining the requirements. Without clear definitions during the planning phase, many projects run off in a variety of directions.

Many companies establish planning and scheduling management policies for the project and functional managers, as well as a brief description of how they should interface. Table 11–6 identifies a typical management policy for planning and requirements, and Table 11–7 describes scheduling management policies.

11.26 THE PROJECT MANAGER–LINE MANAGER INTERFACE

PMBOK® Guide, 4th Edition
1.6 Interpersonal Skills

The utilization of management controls, such as those outlined in Section 11.25, does not necessarily guarantee successful project planning. Good project planning, as well as other project functions, requires a good working relationship between the project and line managers. At this interface:

- The project manager answers these questions:
  - What is to be done? (using the SOW, WBS)

- When will the task be done? (using the summary schedule)
- Why will the task be done? (using the SOW)
- How much money is available? (using the SOW)
- The line manager answers these questions:
  - How will the task be done? (i.e., technical criteria)
  - Where will the task be done? (i.e., technical criteria)
  - Who will do the task? (i.e., staffing)

Project managers may be able to tell line managers "how" and "where," provided that the information appears in the SOW as a requirement for the project. Even then, the line manager can take exception based on his technical expertise.

Figures 11–16 and 11–17 show what can happen when project managers overstep their bounds. In Figure 11–16, the manufacturing manager built a brick wall to keep the project managers away from his personnel because the project managers were telling his line people how to do their job. In Figure 11–17, the subproject managers (for simplicity's sake, equivalent to project engineers) would have, as their career path, promotions to assistant project managers (APMs). Unfortunately, the APMs still felt that they were technically

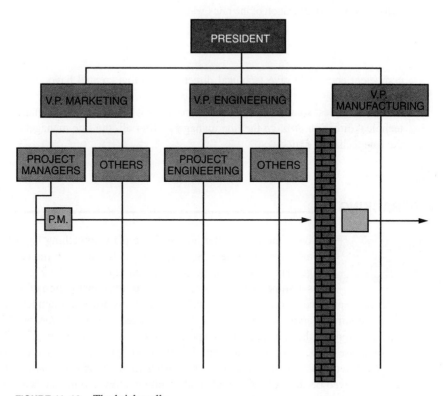

FIGURE 11–16. The brick wall.

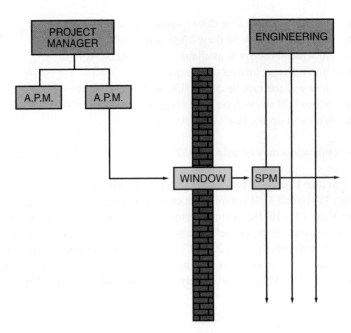

FIGURE 11–17. Modification of the brick wall.

competent enough to give technical direction, and this created havoc for the engineering managers.

The simplest solution to all of these problems is for the project manager to provide the technical direction *through* the line managers. After all, the line managers are supposedly the true technical experts.

11.27 FAST-TRACKING

PMBOK® Guide, 4th Edition
2.1 Characteristics of the Project Life Cycle

Sometimes, no matter how well we plan, something happens that causes havoc on the project. Such is the case when either the customer or management changes the project's constraints. Consider Figure 11–18 and let us assume that the execution time for the construction of the project is one year. To prepare the working drawings and specifications down through level 5 of the WBS would require an additional 35 percent of the expected execution time, and if a feasibility study is required, then an additional 40 percent will be added on. In other words, if the execution phase of the project is one year, then the entire project is almost two years.

Now, let us assume that management wishes to keep the end date fixed but the start date is delayed because of lack of adequate funding. How can this be accomplished *without* sacrificing the quality? The answer is to fast-track the project. Fast-tracking a project means that activities that are normally done in series are done in parallel. An example of

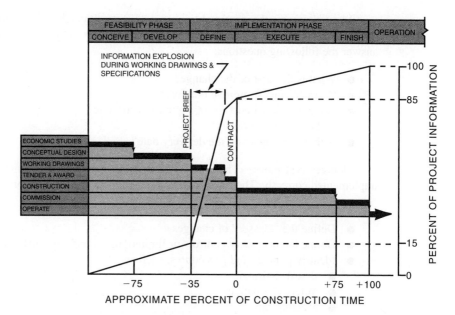

FIGURE 11–18. The information explosion. *Source:* R. M. Wideman, *Cost Control of Capital Projects* (Vancouver, B.C.: A.E.W. Services of Canada, 1983), p. 22.

this is when construction begins before detail design is completed. (See Chapter 2, Table 2–5 on life-cycle phases.)

Fast-tracking a job can accelerate the schedule but requires that additional risks be taken. If the risks materialize, then either the end date will slip or expensive rework will be needed. Almost all project-driven companies fast-track projects, but there is danger when fast-tracking becomes a way of life.

11.28 CONFIGURATION MANAGEMENT

PMBOK® Guide, 4th Edition
4.5 Integrated Change Control

A critical tool employed by a project manager is configuration management or configuration change control. As projects progress downstream through the various life-cycle phases, the cost of engineering changes can grow boundlessly. It is not uncommon for companies to bid on proposals at 40 percent below their own cost hoping to make up the difference downstream with engineering changes. It is also quite common for executives to "encourage" project managers to seek out engineering changes because of their profitability.

Configuration management is a control technique, through an orderly process, for formal review and approval of configuration changes. If properly implemented, configuration management provides

- Appropriate levels of review and approval for changes
- Focal points for those seeking to make changes
- A single point of input to contracting representatives in the customer's and contractor's office for approved changes

At a minimum, the configuration control committee should include representation from the customer, contractor, and line group initiating the change. Discussions should answer the following questions:

- What is the cost of the change?
- Do the changes improve quality?
- Is the additional cost for this quality justifiable?
- Is the change necessary?
- Is there an impact on the delivery date?

Changes cost money. Therefore, it is imperative that configuration management be implemented correctly. The following steps can enhance the implementation process:

- Define the starting point or "baseline" configuration
- Define the "classes" of changes
- Define the necessary controls or limitations on both the customer and contractor
- Identify policies and procedures, such as
  - Board chairman
  - Voters/alternatives
  - Meeting time
  - Agenda
  - Approval forums
  - Step-by-step processes
  - Expedition processes in case of emergencies

Effective configuration control pleases both customer and contractor. Overall benefits include:

- Better communication among staff
- Better communication with the customer
- Better technical intelligence
- Reduced confusion for changes
- Screening of frivolous changes
- Providing a paper trail

As a final note, it must be understood that configuration control, as used here, is not a replacement for design review meetings or customer interface meetings. These meetings are still an integral part of all projects.

11.29 ENTERPRISE PROJECT MANAGEMENT METHODOLOGIES

Enterprise project management methodologies can enhance the project planning process as well as providing some degree of standardization and consistency.

Companies have come to the realization that enterprise project management methodologies work best if the methodology is based upon templates rather than rigid policies and procedures. The International Institute for Learning has created a Unified Project Management Methodology (UPMM™) with templates categorized according to the PMBOK® Guide Areas of Knowledge[13]:

Communication
    Project Charter
    Project Procedures Document
    Project Change Requests Log
    Project Status Report
    PM Quality Assurance Report
    Procurement Management Summary
    Project Issues Log
    Project Management Plan
    Project Performance Report

Cost
    Project Schedule
    Risk Response Plan and Register
    Work Breakdown Structure (WBS)
    Work Package
    Cost Estimates Document
    Project Budget
    Project Budget Checklist

Human Resources
    Project Charter
    Work Breakdown Structure (WBS)
    Communications Management Plan
    Project Organization Chart
    Project Team Directory
    Responsibility Assignment Matrix (RAM)
    Project Management Plan
    Project Procedures Document
    Kickoff Meeting Checklist
    Project Team Performance Assessment
    Project Manager Performance Assessment

Integration
    Project Procedures Overview
    Project Proposal
    Communications Management Plan

---

13. Unified Project Management Methodology (UPMM™) is registered, copyrighted, and owned by International Institute for Learning, Inc., © 2005; reproduced by permission.

Procurement Plan
Project Budget
Project Procedures Document
Project Schedule
Responsibility Assignment Matrix (RAM)
Risk Response Plan and Register
Scope Statement
Work Breakdown Structure (WBS)
Project Management Plan
Project Change Requests Log
Project Issues Log
Project Management Plan Changes Log
Project Performance Report
Lessons Learned Document
Project Performance Feedback
Product Acceptance Document
Project Charter
Closing Process Assessment Checklist
Project Archives Report

Procurement

Project Charter
Scope Statement
Work Breakdown Structure (WBS)
Procurement Plan
Procurement Planning Checklist
Procurement Statement of Work (SOW)
Request for Proposal Document Outline
Project Change Requests Log
Contract Formation Checklist
Procurement Management Summary

Quality

Project Charter
Project Procedures Overview
Work Quality Plan
Project Management Plan
Work Breakdown Structure (WBS)
PM Quality Assurance Report
Lessons Learned Document
Project Performance Feedback
Project Team Performance Assessment
PM Process Improvement Document

Risk

Procurement Plan
Project Charter

Project Procedures Document
Work Breakdown Structure (WBS)
Risk Response Plan and Register

Scope

Project Scope Statement
Work Breakdown Structure (WBS)
Work Package
Project Charter

Time

Activity Duration Estimating Worksheet
Cost Estimates Document
Risk Response Plan and Register Medium
Work Breakdown Structure (WBS)
Work Package
Project Schedule
Project Schedule Review Checklist

11.30 PROJECT AUDITS

In recent years, the necessity for a structured independent review of various parts of a business, including projects, has taken on a more important role. Part of this can be attributed to the Sarbanes–Oxley law compliance requirements. These independent reviews are audits that focus on either discovery or decision-making. The audits can be scheduled or random and can be performed by in-house personnel or external examiners.

There are several types of audits. Some common types include:

- **Performance Audits:** These audits are used to appraise the progress and performance of a given project. The project manager, project sponsor, or an executive steering committee can conduct this audit.
- **Compliance Audits:** These audits are usually performed by the project management office (PMO) to validate that the project is using the project management methodology properly. Usually the PMO has the authority to perform the audit but may not have the authority to enforce compliance.
- **Quality Audits:** These audits ensure that the planned project quality is being met and that all laws and regulations are being followed. The quality assurance group performs this audit.
- **Exit Audits:** These audits are usually for projects that are in trouble and may need to be terminated. Personnel external to the project, such as an exit champion or an executive steering committee, conduct the audits.
- **Best Practices Audits:** These audits can be conducted at the end of each life-cycle phase or at the end of the project. Some companies have found that project managers may not be the best individuals to perform the audit. In such situations, the company may have professional facilitators trained in conducting best practices reviews.

11.31 STUDYING TIPS FOR THE PMI® PROJECT MANAGEMENT CERTIFICATION EXAM

This section is applicable as a review of the principles to support the knowledge areas and domain groups in the PMBOK® Guide. This chapter addresses:

● Scope Management
● Initiation
● Planning
● Execution
● Monitoring
● Closure

Understanding the following principles is beneficial if the reader is using this text to study for the PMP® Certification Exam:

● Need for effective planning
● Components of a project plan and subsidiary plans
● Need for and components of a statement of work (both proposal and contractual)
● How to develop a work breakdown structure and advantages and disadvantages of highly detailed levels
● Types of work breakdown structures
● Purpose of a work package
● Purpose of configuration management and role of the change control board
● Need for a project charter and components of a project charter
● Need for the project team to be involved in project-planning activities
● That changes to a plan or baseline need to be managed

In Appendix C, the following Dorale Products mini–case studies are applicable:

● Dorale Products (C) [Scope Management]
● Dorale Products (D) [Scope Management]
● Dorale Products (E) [Scope Management]

The following multiple-choice questions will be helpful in reviewing the principles of this chapter:

1. The document that officially sanctions the project is the:
   A. Project charter
   B. Project plan
   C. Feasibility study
   D. Cost-benefit analysis

2. The work breakdown structure "control points" for the management of a project are the:
   A. Milestones
   B. Work packages
   C. Activities
   D. Constraints

3. One of the most common reasons why projects undergo scope changes is:
   A. Poor work breakdown structure
   B. Poorly defined statement of work

C. Lack of resources
D. Lack of funding

4. Which of the following generally cannot be validated using a work breakdown structure?
   A. Schedule control
   B. Cost control
   C. Quality control
   D. Risk management

Answer questions 5–8 using the work breakdown structure (WBS) shown below (numbers in parentheses show the dollar value for a particular element):

1.00.00	
1.1.0	($25K)
1.1.1	
1.1.2	($12K)
1.2.0	
1.2.1	($16K)
1.2.2.0	
1.2.2.1	($20K)
1.2.2.2	($30K)

5. The cost of WBS element 1.2.2.0 is:
   A. $20K
   B. $30K
   C. $50K
   D. Cannot be determined

6. The cost of WBS element 1.1.1 is:
   A. $12K
   B. $13K
   C. $25K
   D. Cannot be determined

7. The cost of the entire program (1.00.00) is:
   A. $25K
   B. $66K
   C. $91K
   D. Cannot be determined

8. The work packages in the WBS are at WBS level(s):
   A. 2 only
   B. 3 only
   C. 4 only
   D. 3 and 4

9. One of the outputs of the PMBOK® Scope Planning Process is:
   A. A project charter
   B. A scope statement and management plan
   C. A detailed WBS
   D. None of the above

10. Which of the following is (are) the benefit(s) of developing a WBS to low levels?
    A. Better estimation of costs
    B. Better control

C. Less likely that something will "fall through the cracks"
D. All of the above

11. The PMBOK® Scope Verification Process is used to verify that:
    A. The budget is correct.
    B. The scope is correct.
    C. The schedule is correct.
    D. A life-cycle phase or the end of the project has been completed successfully.

12. Financial closeout, which is often part of following the Scope Verification Process, is used to:
    A. Close out all charge numbers
    B. Close out all charge numbers for the work performed and completed
    C. Amend the work authorization forms
    D. None of the above

13. One of your contractors has sent you an e-mail requesting that they be allowed to conduct only eight tests rather than the ten tests required by the specification. What should the project manager do first?
    A. Change the scope baseline
    B. Ask the contractor to put forth a change request
    C. Look at the penalty clauses in the contract
    D. Ask your sponsor for his or her opinion

14. One of your contractors sends you an e-mail request to use high quality raw materials in your project stating that this will be value-added and improve quality. What should the project manager do first?
    A. Change the scope baseline
    B. Ask the contractor to put forth a change request
    C. Ask your sponsor for his or her opinion
    D. Change the WBS

15. What are the maximum number of subsidiary plans a program management plan can contain?
    A. 10
    B. 15
    C. 20
    D. Unlimited number

16. The change control board, of which you are a member, approves a significant scope change. The first document that the project manager should updated would be the:
    A. Scope baseline
    B. Schedule
    C. WBS
    D. Budget

ANSWERS

1. A
2. B
3. B
4. C
5. C

6. B
7. C
8. D
9. B
10. D
11. D
12. B
13. B
14. B
15. D
16. A

PROBLEMS

11–1 Under what conditions would each of the following either not be available or not be necessary for initial planning?

    a. Work breakdown structure
    b. Statement of work
    c. Specifications
    d. Milestone schedules

11–2 What planning steps should precede total program scheduling? What steps are necessary?

11–3 How does a project manager determine how complex to make a program plan or how many schedules to include?

11–4 Can objectives always be identified and scheduled?

11–5 Can a WBS always be established for attaining an objective?

11–6 Who determines the work necessary to accomplish an objective?

11–7 What roles does a functional manager play in establishing the first three levels of the WBS?

11–8 Should the length of a program have an impact on whether to set up a separate project or task for administrative support? How about for raw materials?

11–9 Is it possible for the WBS to be designed so that resource allocation is easier to identify?

11–10 If the scope of effort of a project changes during execution of activities, what should be the role of the functional manager?

11–11 What types of conflicts can occur during the planning cycle, and what modes should be used for their resolution?

11–12 What would be the effectiveness of Figure 11–3 if the work packages were replaced by tasks?

11–13 Under what situations or projects would work planning authorization not be necessary?

11–14 On what types of projects could hedge positions be easily identified on a schedule?

11–15 Can activities 5 and 6 of Figure 11–11 be eliminated? What risks does a project manager incur if these activities are eliminated?

11–16 Where in the planning cycle should responsibility charts be prepared? Can you identify this point in Figure 11–11?

11–17 For each one of the decision points in Figure 11–13, who makes the decision? Who must input information? What is the role of the functional manager and the functional team member? Where are strategic variables identified?

11–18 Consider a project in which all project planning is performed by a group. After all planning is completed, including the program plan and schedules, a project manager is selected. Is there anything wrong with this arrangement? Can it work?

11–19 How do the customer and contractor know if each one completely understands the statement of work, the work breakdown structure, and the program plan?

11–20 Should a good project plan formulate methods for anticipating problems?

11–21 Some project managers schedule staff meetings as the primary means for planning and control. Do you agree with this philosophy?

11–22 Paul Mali (*Management by Objectives,* New York: John Wiley, 1972, p. 12) defines MBO as a five-step process:

- Finding the objective
- Setting the objective
- Validating the objective
- Implementing the objective
- Controlling and reporting status of the objective

How can the work breakdown structure be used to accomplish each of the above steps? Would you agree or disagree that the more levels the WBS contains, the greater the understanding and clarity of those steps necessary to complete the objectives?

11–23 Many textbooks on management state that you should plan like you work, by doing one thing at a time. Can this same practice be applied at the project level, or must a project manager plan all activities at once?

11–24 Is it true that project managers set the milestones and functional managers hope they can meet them?

11–25 You have been asked to develop a work breakdown structure for a project. How should you go about accomplishing this? Should the WBS be time-phased, department-phased, division-phased, or some combination?

11–26 You have just been instructed to develop a schedule for introducing a new product into the marketplace. Below are the elements that must appear in your schedule. Arrange these elements into a work breakdown structure (down through level 3), and then draw the arrow diagram. You may feel free to add additional topics as necessary.

- Production layout
- Market testing

- Review plant costs
- Select distributors

- Analyze selling cost
- Analyze customer reactions
- Storage and shipping costs
- Select salespeople
- Train salespeople
- Train distributors
- Literature to salespeople
- Literature to distributors
- Print literature
- Sales promotion
- Sales manual
- Trade advertising
- Lay out artwork
- Approve artwork
- Introduce at trade show
- Distribute to salespeople
- Establish billing procedure
- Establish credit procedure
- Revise cost of production
- Revise selling cost
- Approvals*
- Review meetings*
- Final specifications
- Material requisitions

(* Approvals and review meetings can appear several times.)

11–27 Once a project begins, a good project manager will set up checkpoints. How should this be accomplished? Will the duration of the project matter? Can checkpoints be built into a schedule? If so, how should they be identified?

11–28 Detailed schedules (through WBS levels 3, 4, 5, . . .) are prepared by the functional managers. Should these schedules be shown to the customer?

11–29 The project start-up phase is complete, and you are now ready to finalize the operational plan. Below are six steps that are often part of the finalization procedure. Place them in the appropriate order.

1. Draw diagrams for each individual WBS element.
2. Establish the work breakdown structure and identify the reporting elements and levels.
3. Create a coarse (arrow-diagram) network and decide on the WBS.
4. Refine the diagram by combining all logic into one plan. Then decide on the work assignments.
5. If necessary, try to condense the diagram as much as possible without losing clarity.
6. Integrate diagrams at each level until only one exists. Then begin integration into higher WBS levels until the desired plan is achieved.

11–30 Below are seven factors that must be considered before finalizing a schedule. Explain how a base case schedule can change as a result of each of these:

- Introduction or acceptance of the product in the marketplace
- Present or planned manpower availability
- Economic constraints of the project
- Degree of technical difficulty
- Manpower availability
- Availability of personnel training
- Priority of the project

11–31 You are the project manager of a nine-month effort. You are now in the fifth month of the project and are more than two weeks behind schedule, with very little hope of catching up. The dam breaks in a town near you, and massive flooding and mudslides take place. Fifteen of your key functional people request to take off three days from the following week to help fellow church members dig out. Their functional managers, bless their hearts, have left the entire decision up to you. Should you let them go?

11–32 Once the functional manager and project manager agree on a project schedule, who is responsible for getting the work performed? Who is accountable for getting the work performed? Why the difference, if any?

11–33 Discuss the validity of the following two statements on authority:

   a. A good project manager will have more authority than his responsibility calls for.
   b. A good project manager should not hold a subordinate responsible for duties that he (the project manager) does not have the authority to enforce.

11–34 Below are twelve instructions. Which are best described as planning, and which are best described as forecasting?

   a. Give a complete definition of the work.
   b. Lay out a proposed schedule.
   c. Establish project milestones.
   d. Determine the need for different resources.
   e. Determine the skills required for each WBS task or element.
   f. Change the scope of the effort and obtain new estimates.
   g. Estimate the total time to complete the required work.
   h. Consider changing resources.
   i. Assign appropriate personnel to each WBS element.
   j. Reschedule project resources.
   k. Begin scheduling the WBS elements.
   l. Change the project priorities.

11–35 A major utility company has a planning group that prepares budgets (with the help of functional groups) and selects the projects to be completed within a given time period. You are assigned as a project manager on one of the projects and find out that it should have been started "last month" in order to meet the completion date. What can you, the project manager, do about this? Should you delay the start of the project to replan the work?

11–36 The director of project management calls you into his office and informs you that one of your fellow project managers has had a severe heart attack midway through a project. You will be taking over his project, which is well behind schedule and overrunning costs. The director of project management then "orders" you to complete the project within time and cost. How do you propose to do it? Where do you start? Should you shut down the project to replan it?

11–37 Planning is often described as establishing, budgeting, scheduling, and resource allocation. Identify these four elements in Figure 11–1.

11–38 A company is undertaking a large development project that requires that a massive "blueprint design tree" be developed. What kind of WBS outline would be best to minimize the impact of having two systems, one for blueprints and one for WBS work?

11–39 A company allows each line organization to perform its own procurement activities (through a centralized procurement office) as long as the procurement funds have been allocated during the project planning phase. The project office does not sign off on these functional procurement requisitions and may not even know about them. Can this system work effectively? If so, under what conditions?

11–40 As part of a feasibility study, you are asked to prepare, with the assistance of functional managers, a schedule and cost summary for a project that will occur three years downstream, if the project is approved at all. Suppose that three years downstream the project is approved. How does the project manager get functional managers to accept the schedule and cost summary that they themselves prepared three years before?

11–41 "Expecting trouble." Good project managers know what type of trouble can occur at the various stages in the development of a project. The activities in the numbered list below indicate the various stages of a project. The lettered list that follows identifies major problems. For each project stage, select and list all of those problems that are applicable.

1. Request for proposal _____
2. Submittal to customer _____
3. Contract award _____
4. Design review meetings_____
5. Testing the product_____
6. Customer acceptance _____

a. Engineering does not request manufacturing input for end-item producibility.
b. The work breakdown structure is poorly defined.
c. Customer does not fully realize the impact that a technical change will have upon cost and schedule.
d. Time and cost constraints are not compatible with the state of the art.

e. The project–functional interface definition is poor.
f. Improper systems integration has created conflicts and a communications breakdown.
g. Several functional managers did not realize that they were responsible for certain risks.
h. The impact of design changes is not systematically evaluated.

11–42 Table 11–8 identifies twenty-six steps in project planning and control. Below is a description of each of the twenty-six steps. Using this information, fill in columns 1 and 2 (column 2 is a group response). After your instructor provides you with column 3, fill in the remainder of the table.

1. *Develop the linear responsibility chart.* This chart identifies the work breakdown structure and assigns specific authority/responsibility to various individuals as groups in order to be sure that all WBS elements are accounted for. The linear responsibility chart can be prepared with either the titles or names of individuals. Assume that this is prepared after you negotiate for qualified personnel, so that you know either the names or capabilities of those individuals who will be assigned.

2. *Negotiate for qualified functional personnel.* Once the work is decided on, the project manager tries to identify the qualifications for the desired personnel. This then becomes the basis for the negotiation process.

3. *Develop specifications.* This is one of the four documents needed to initially define the requirements of the project. Assume that these are either performance or material specifications, and are provided to you at the initial planning stage by either the customer or the user.

4. *Determine the means for measuring progress.* Before the project plan is finalized and project execution can begin, the project manager must identify the means for measuring progress; specifically, what is meant by an out-of-tolerance condition and what are the tolerances/variances/thresholds for each WBS base case element?

5. *Prepare the final report.* This is the final report to be prepared at the termination of the project.

TABLE 11–8. STEPS IN PROJECT PLANNING AND CONTROL

Activity	Description	Column 1: Your sequence	Column 2: Group sequence	Column 3: Expert's sequence	Column 4: Difference between 1 & 3	Column 5: Difference between 2 & 3
1.	Develop linear responsibility chart					
2.	Negotiate for qualified functional personnel					
3.	Develop specifications					
4.	Determine means for measuring progress					
5.	Prepare final report					
6.	Authorize departments to begin work					
7.	Develop work breakdown structure					
8.	Close out functional work orders					
9.	Develop scope statement and set objectives					
10.	Develop gross schedule					
11.	Develop priorities for each project element					
12.	Develop alternative courses of action					
13.	Develop PERT network					
14.	Develop detailed schedules					
15.	Establish functional personnel qualifications					
16.	Coordinate ongoing activities					
17.	Determine resource requirements					
18.	Measure progress					
19.	Decide upon a basic course of action					
20.	Establish costs for each WBS element					

(*continues*)

TABLE 11–8. STEPS IN PROJECT PLANNING AND CONTROL (*Continued*)

Activity	Description	Column 1: Your sequence	Column 2: Group sequence	Column 3: Expert's sequence	Column 4: Difference between 1 & 3	Column 5: Difference between 2 & 3
21.	Review WBS costs with each functional manager					
22.	Establish a project plan					
23.	Establish cost variances for base case elements					
24.	Price out WBS					
25.	Establish logic network with checkpoints					
26.	Review base case costs with director					

6. *Authorize departments to begin work.* This step authorizes departments to begin the actual execution of the project, *not* the planning. This step occurs generally after the project plan has been established, finalized, and perhaps even approved by the customer or user group. This is the initiation of the work orders for project implementation.

7. *Develop the work breakdown structure.* This is one of the four documents required for project definition in the early project planning stage. Assume that WBS is constructed using a bottom-up approach. In other words, the WBS is constructed from the logic network (arrow diagram) and checkpoints which will eventually become the basis for the PERT/CPM charts (see Activity 25).

8. *Close out functional work orders.* This is where the project manager tries to prevent excessive charging to his project by closing out the functional work orders (i.e., Activity 6) as work terminates. This includes canceling all work orders except those needed to administer the termination of the project and the preparation of the final report.

9. *Develop scope statement and set objectives.* This is the statement of work and is one of the four documents needed in order to identify the requirements of the project. Usually, the WBS is the structuring of the statement of work.

10. *Develop gross schedule.* This is the summary or milestone schedule needed at project initiation in order to define the four requirements documents for the project. The gross schedule includes start and end dates (if known), other major milestones, and data items.

11. *Develop priorities for each project element.* After the base case is identified and alternative courses of action are considered (i.e., contingency planning), the project team performs a sensitivity analysis for each element of the WBS. This may require

assigning priorities for each WBS element, and the highest priorities may *not* necessarily be assigned to elements on the critical path.

12. *Develop alternative courses of action.* Once the base case is known and detailed courses of action (i.e., detailed scheduling) are prepared, project managers conduct "what if" games to develop possible contingency plans.

13. *Develop PERT network.* This is the finalization of the PERT/CPM network and becomes the basis from which detailed scheduling will be performed. The logic for the PERT network can be conducted earlier in the planning cycle (see Activity 25), but the finalization of the network, together with the time durations, are usually based on who has been (or will be) assigned, and the resulting authority/responsibility of the individual. In other words, the activity time duration is a function not only of the performance standard, but also of the individual's expertise and authority/ responsibility.

14. *Develop detailed schedules.* These are the detailed project schedules, and are constructed from the PERT/CPM chart and the capabilities of the assigned individuals.

15. *Establish functional personnel qualifications.* Once senior management reviews the base case costs and approves the project, the project manager begins the task of conversion from rough to detail planning. This includes identification of the required resources, and then the respective qualifications.

16. *Coordinate ongoing activities.* These are the ongoing activities for project execution, not project planning. These are the activities that were authorized to begin in Activity 6.

17. *Determine resource requirements.* After senior management approves the estimated base case costs obtained during rough planning, detailed planning begins by determining the resource requirements, including human resources.

18. *Measure progress.* As the project team coordinates ongoing activities during project execution, the team monitors progress and prepares status reports.

19. *Decide on a basic course of action.* Once the project manager obtains the rough cost estimates for each WBS element, the project manager puts together all of the pieces and determines the basic course of action.

20. *Establish costs for each WBS element.* After deciding on the base case, the project manager establishes the base case cost for each WBS element in order to prepare for the senior management pricing review meeting. These costs are usually the same as those that were provided by the line managers.

21. *Review WBS costs with each functional manager.* Each functional manager is provided with the WBS and told to determine his role and price out his functional involvement. The project manager then reviews the WBS costs to make sure that everything was accounted for and without duplication of effort.

22. *Establish a project plan.* This is the final step in detail planning. Following this step, project execution begins. (Disregard the situation where project plan development can be run concurrently with project execution.)

23. *Establish cost variances for the base case elements.* Once the priorities are known for each base case element, the project manager establishes the allowable cost variances that will be used as a means for measuring progress. Cost reporting is minimum as long as the actual costs remain within these allowable variances.

24. *Price out the WBS.* This is where the project manager provides each functional manager with the WBS for initial activity pricing.

25. *Establish logic network with checkpoints.* This is the bottom-up approach that is often used as the basis for developing both the WBS and later the PERT/CPM network.

26. *Review base case costs with director.* Here the project manager takes the somewhat rough costs obtained during the WBS functional pricing and review and seeks management's approval to begin detail planning.

11–43 Consider the work breakdown structure shown in Figure 11–19. Can the project be managed from this one sheet of paper assuming that, at the end of each month, the project manager also receives a cost and percent-complete summary?

11–44 During 1992 and 1993, General Motors saved over $2 billion due to the cost-cutting efforts of Mr. Lopez. Rumors spread throughout the auto industry that General Motors was considering a plan to offer subcontractors ten-year contracts in exchange for a 20 percent cost reduction.

These long-term contracts provided both GM and the subcontractors the chance to develop an informal project management relationship based on trust, effective communications, and minimum documentation requirements.

   a. Is it conceivable that the cost savings of 20 percent could have been realized entirely from the decrease in formalized documentation?
   b. Philosophically, what do you think happened when Mr. Lopez departed GM in the spring of 1993 for a senior position at Volkswagen? Did his informal project management system continue without him? Explain your answer.

11–45 During the recession of 1989–1993, the auto industry began taking extreme cost-cutting measures by downsizing its organizations. The downsizing efforts created project

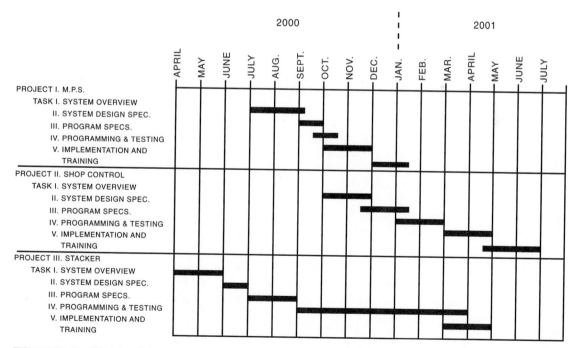

FIGURE 11–19. Work breakdown structure.

management problems for the project engineers in the manufacturing plants. With fewer resources available, more and more of the work had to be outsourced, primarily for services. The manufacturing plants had years of experience in negotiations for parts, but limited experience in negotiations for services. As a result, the service contracts were drastically overrun with engineering changes and schedule slippages. What is the real problem and your recommendation for a solution?

11–46 When to bring the project manager on board has always been a problem. For each of the following situations, identify the advantages and disadvantages.

    a. The project manager is brought on board at the beginning of the conceptual phase but acts only as an observer. The project manager neither answers questions nor provides his ideas until the brainstorming session is completed.

    b. When brainstorming is completed during the conceptual phase, senior management appoints one of the brainstorming team members to serve as the project manager.

Network Scheduling Techniques

Related Case Studies (from Kerzner/*Project Management Case Studies,* 3rd Edition)	Related Workbook Exercises (from Kerzner/*Project Management Workbook and PMP®/CAPM® Exam Study Guide,* 10th Edition)	PMBOK® Guide, 4th Edition, Reference Section for the PMP® Certification Exam
• Crosby Manufacturing Corporation*	• Crashing the Effort • Multiple Choice Exam • Crossword Puzzle on Time (Schedule) Management	• Time Management

12.0 INTRODUCTION

PMBOK® Guide, 4th Edition
Chapter 6 Project Time Management

Management is continually seeking new and better control techniques to cope with the complexities, masses of data, and tight deadlines that are characteristic of highly competitive industries. Managers also want better methods for presenting technical and cost data to customers.

Scheduling techniques help achieve these goals. The most common techniques are:

PMBOK® Guide, 4th Edition
6.1.3.3 Milestone Lists

● Gantt or bar charts

---

*Case Study also appears at end of chapter.

- Milestone charts
- Line of balance[1]
- Networks
- Program Evaluation and Review Technique (PERT)
- Arrow Diagram Method (ADM) [Sometimes called the Critical Path Method (CPM)][2]
- Precedence Diagram Method (PDM)
- Graphical Evaluation and Review Technique (GERT)

Advantages of network scheduling techniques include:

- They form the basis for all planning and predicting and help management decide how to use its resources to achieve time and cost goals.
- They provide visibility and enable management to control "one-of-a-kind" programs.
- They help management evaluate alternatives by answering such questions as how time delays will influence project completion, where slack exists between elements, and what elements are crucial to meet the completion date.
- They provide a basis for obtaining facts for decision-making.
- They utilize a so-called time network analysis as the basic method to determine manpower, material, and capital requirements, as well as to provide a means for checking progress.
- They provide the basic structure for reporting information.
- They reveal interdependencies of activities.
- They facilitate "what if" exercises.
- They identify the longest path or critical paths.
- They aid in scheduling risk analysis.

PERT was originally developed in 1958 and 1959 to meet the needs of the "age of massive engineering" where the techniques of Taylor and Gantt were inapplicable. The Special Projects Office of the U.S. Navy, concerned with performance trends on large military development programs, introduced PERT on its Polaris Weapon System in 1958, after the technique had been developed with the aid of the management consulting firm of Booz, Allen, and Hamilton. Since that time, PERT has spread rapidly throughout almost all industries. At about the same time, the DuPont Company initiated a similar technique known as the critical path method (CPM), which also has spread widely, and is particularly concentrated in the construction and process industries.

In the early 1960s, the basic requirements of PERT/time as established by the Navy were as follows:

- All of the individual tasks to complete a program must be clear enough to be put down in a network, which comprises events and activities; i.e., follow the work breakdown structure.
- Events and activities must be sequenced on the network under a highly logical set of ground rules that allow the determination of critical and subcritical paths. Networks may have more than one hundred events, but not fewer than ten.

---

1. Line of balance is more applicable to manufacturing operations for production line activities. However, it can be used for project management activities where a finite number of deliverables must be produced in a given time period. The reader need only refer to the multitude of texts on production management for more information on this technique.

2. The text uses the term CPM instead of ADM. The reader should understand that they are interchangeable.

- Time estimates must be made for each activity on a three-way basis. Optimistic, most likely, and pessimistic elapsed-time figures are estimated by the person(s) most familiar with the activity.
- Critical path and slack times are computed. The critical path is that sequence of activities and events whose accomplishment will require the greatest time.

A big advantage of PERT lies in its extensive planning. Network development and critical path analysis reveal interdependencies and problems that are not obvious with other planning methods. PERT therefore determines where the greatest effort should be made to keep a project on schedule.

The second advantage of PERT is that one can determine the probability of meeting deadlines by development of alternative plans. If the decision maker is statistically sophisticated, he can examine the standard deviations and the probability of accomplishment data. If there exists a minimum of uncertainty, one may use the single-time approach, of course, while retaining the advantage of network analysis.

A third advantage is the ability to evaluate the effect of changes in the program. For example, PERT can evaluate the effect of a contemplated shift of resources from the less critical activities to the activities identified as probable bottlenecks. PERT can also evaluate the effect of a deviation in the actual time required for an activity from what had been predicted.

Finally, PERT allows a large amount of sophisticated data to be presented in a well-organized diagram from which contractors and customers can make joint decisions.

PERT, unfortunately, is not without disadvantages. The complexity of PERT adds to implementation problems. There exist more data requirements for a PERT-organized reporting system than for most others. PERT, therefore, becomes expensive to maintain and is utilized most often on large, complex programs.

Many companies have taken a hard look at the usefulness of PERT on small projects. The result has been the development of PERT/LOB procedures, which can do the following:

- Cut project costs and time
- Coordinate and expedite planning
- Eliminate idle time
- Provide better scheduling and control of subcontractor activities
- Develop better troubleshooting procedures
- Cut the time required for routine decisions, but allow more time for decision-making

Even with these advantages, many companies should ask whether they actually need PERT because incorporating it may be difficult and costly, even with canned software packages. Criticism of PERT includes:

- Time and labor intensive
- Decision-making ability reduced
- Lacks functional ownership in estimates
- Lacks historical data for time–cost estimates
- Assumes unlimited resources
- Requires too much detail

An in-depth study of PERT would require a course or two by itself. The intent of this chapter is to familiarize the reader with the terminology, capability, and applications of networks.

12.1 NETWORK FUNDAMENTALS

PMBOK® Guide, 4th Edition
6.2 Activity Sequencing
6.2.2 Activity Sequencing Tools
    and Techniques

The major discrepancy with Gantt, milestone, or bubble charts is the inability to show the interdependencies between events and activities. These interdependencies must be identified so that a master plan can be developed that provides an up-to-date picture of operations at all times.

Interdependencies are shown through the construction of networks. Network analysis can provide valuable information for planning, integration of plans, time studies, scheduling, and resource management. The primary purpose of network planning is to eliminate the need for crisis management by providing a pictorial representation of the total program. The following management information can be obtained from such a representation:

- Interdependencies of activities
- Project completion time
- Impact of late starts
- Impact of early starts
- Trade-offs between resources and time
- "What if" exercises
- Cost of a crash program
- Slippages in planning/performance
- Evaluation of performance

Networks are composed of events and activities. The following terms are helpful in understanding networks:

- **Event:** Equivalent to a milestone indicating when an activity starts or finishes.
- **Activity:** The element of work that must be accomplished.
- **Duration:** The total time required to complete the activity.
- **Effort:** The amount of work that is actually performed within the duration. For example, the duration of an activity could be one month but the effort could be just a two-week period within the duration.
- **Critical Path:** This is the longest path through the network and determines the duration of the project. It is also the shortest amount of time necessary to accomplish the project.

Figure 12–1 shows the standard nomenclature for PERT networks. The circles represent events, and arrows represent activities. The numbers in the circles signify the specific events or accomplishments. The number over the arrow specifies the time needed (hours, days, months), to go from event 6 to event 3. The events need not be numbered in any specific order. However, event 6 must take place before event 3 can be completed (or begun). In Figure 12–2A, event 26 must take place prior to events 7, 18, and 31. In Figure 12–2B, the opposite holds true, and events 7, 18, and 31 must take place prior to event 26. Figure 12–2B is similar to "and gates" used in logic diagrams.[3]

3. PERT diagrams can, in fact, be considered as logic diagrams. Many of the symbols used in PERT have been adapted from logic flow nomenclature.

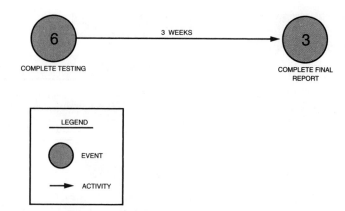

FIGURE 12–1.   Standard PERT nomenclature.

In this chapter's introduction we have summarized the advantages and disadvantages of Gantt and milestone charts. These charts, however, can be used to develop the PERT network, as shown in Figure 12–3. The bar chart in Figure 12–3A can be converted to the milestone chart in Figure 12–3B. By then defining the relationship between the events on different bars in the milestone chart, we can construct the PERT chart in Figure 12–3C.

PERT is basically a management planning and control tool. It can be considered as a road map for a particular program or project in which all of the major elements (events)

PMBOK® Guide, 4th Edition
6.2.2.2 Dependency Determination

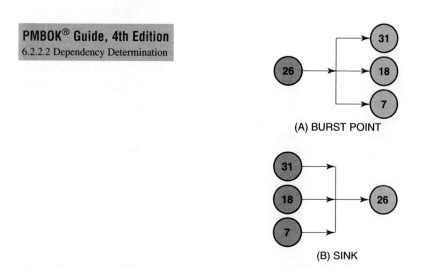

FIGURE 12–2.   PERT sources (burst points) and sinks.

PMBOK® Guide, 4th Edition
6.2 Activity Sequencing

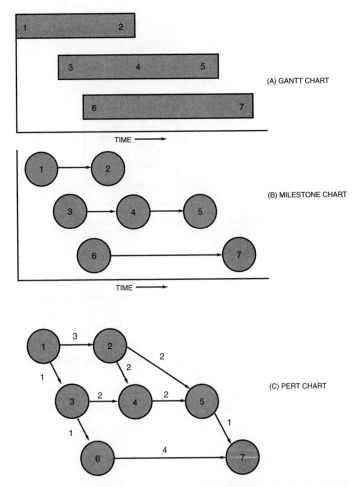

FIGURE 12–3. Conversion from bar chart to PERT chart.

have been completely identified, together with their corresponding interrelations.[4] PERT charts are often constructed from back to front because, for many projects, the end date is fixed and the contractor has front-end flexibility.

One of the purposes of constructing the PERT chart is to determine how much time is needed to complete the project. PERT, therefore, uses time as a common denominator to analyze those elements that directly influence the success of the project, namely, time, cost, and performance. The construction of the network requires two inputs. First, do events represent the start or the completion of an activity? Event completions are generally preferred. The next step is to define the sequence of events, as shown in Table 12–1,

---

4. These events in the PERT charts should be broken down to at least the same reporting levels as defined in the work breakdown structure.

TABLE 12–1. SEQUENCE OF EVENTS

Activity	Title	Immediate Predecessors	Activity Time, Weeks
1–2	A	—	1
2–3	B	A	5
2–4	C	A	2
3–5	D	B	2
3–7	E	B	2
4–5	F	C	2
4–8	G	C	3
5–6	H	D,F	2
6–7	I	H	3
7–8	J	E,I	3
8–9	K	G,J	2

which relates each event to its immediate predecessor. Large projects can easily be converted into PERT networks once the following questions are answered:

● What job immediately precedes this job?
● What job immediately follows this job?
● What jobs can be run concurrently?

Figure 12–4 shows a typical PERT network. The bold line in Figure 12–4 represents the critical path, which is established by the longest time span through the total system of events. The critical path is composed of events 1–2–3–5–6–7–8–9. The critical path is vital for successful control of the project because it tells management two things:

● Because there is no slack time in any of the events on this path, any slippage will cause a corresponding slippage in the end date of the program unless this slippage can be recovered during any of the downstream events (on the critical path).
● Because the events on this path are the most critical for the success of the project, management must take a hard look at these events in order to improve the total program.

Using PERT we can now identify the earliest possible dates on which we can expect an event to occur, or an activity to start or end. There is nothing overly mysterious about this type of calculation, but without a network analysis the information might be hard to obtain.

PERT charts can be managed from either the events or the activities. For levels 1–3 of the Work Breakdown Structure (WBS), the project manager's prime concerns are the milestones, and therefore, the events are of prime importance. For levels 4–6 of the WBS, the project manager's concerns are the activities.

The principles that we have discussed thus far also apply to CPM. The nomenclature is the same and both techniques are often referred to as arrow diagramming methods, or activity-on-arrow networks. The differences between PERT and CPM are:

● PERT uses three time estimates (optimistic, most likely, and pessimistic as shown in Section 12.7) to derive an expected time. CPM uses one time estimate that represents the normal time (i.e., better estimate accuracy with CPM).

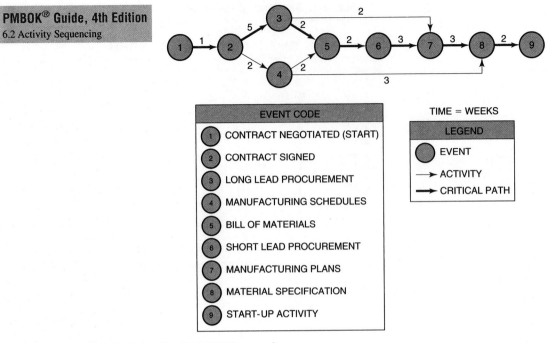

FIGURE 12–4. Simplified PERT network.

● PERT is probabilistic in nature, based on a beta distribution for each activity time and a normal distribution for expected time duration (see Section 12.7). This allows us to calculate the "risk" in completing a project. CPM is based on a single time estimate and is deterministic in nature.

● Both PERT and CPM permit the use of dummy activities in order to develop the logic.

● PERT is used for R&D projects where the risks in calculating time durations have a high variability. CPM is used for construction projects that are resource dependent and based on accurate time estimates.

● PERT is used on those projects, such as R&D, where percent complete is almost impossible to determine except at completed milestones. CPM is used for those projects, such as construction, where percent complete can be determined with reasonable accuracy and customer billing can be accomplished based on percent complete.

12.2 GRAPHICAL EVALUATION AND REVIEW TECHNIQUE (GERT)

PMBOK® Guide, 4th Edition
6.5.2 Schedule Network Analysis

Graphical evaluation and review techniques are similar to PERT but have the distinct advantages of allowing for looping, branching, and multiple project end results. With PERT one cannot easily show that if a test fails, we may have to repeat the test several times. With PERT, we cannot show that, based upon the results of a test, we can select one of several different branches to continue the project. These problems are easily overcome using GERT. [For additional information on the GERT technique, see Jack R. Meredith and Samuel J. Mantel, Jr., *Project Management*, 3rd ed. (New York: Wiley; 1995); pp. 364–367.]

12.3 DEPENDENCIES

PMBOK® Guide, 4th Edition
6.2 Activity Sequencing
6.2.2.2 Dependency Determination

There are three basic types of interrelationships or dependencies:

- *Mandatory dependencies (i.e., hard logic):* These are dependencies that cannot change, such as erecting the walls of a house before putting up the roof.
- *Discretionary dependencies (i.e., soft logic):* These are dependencies that may be at the discretion of the project manager or may simply change from project to project. As an example, one does not need to complete the entire bill of materials prior to beginning procurement.
- *External dependencies:* These are dependencies that may be beyond the control of the project manager such as having contractors sit on your critical path.

Sometimes, it is impossible to draw network dependencies without including dummy activities. Dummy activities are artificial activities, represented by a dotted line, and do not consume resources or require time. They are added into the network simply to complete the logic.

In Figure 12–5, activity C is preceded by activity B only. Now, let's assume that there exists an activity D that is preceded by both activities A and B. Without drawing a dummy activity (i.e., the dashed line), there is no way to show that activity D is preceded by both activities A and B. Using two dummy activities, one from activity A to activity D and another one from activity B to activity D, could also accomplish this representation. Software programs insert the minimum number of dummy activities, and the direction of the arrowhead is important. In Figure 12–5, the arrowhead must be pointed upward.

PMBOK® Guide, 4th Edition
6.2.2.1 Arrow Diagramming
    Method

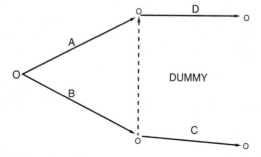

FIGURE 12–5. Dummy activity.

12.4 SLACK TIME

PMBOK® Guide, 4th Edition
6.5.2 Schedule Development
6.5.2.2 Critical Path Method

Since there exists only one path through the network that is the longest, the other paths must be either equal in length to or shorter than that path. Therefore, there must exist events and activities that can be completed before the time when they are actually needed. The time differential between the scheduled completion date and the required date to meet critical path is referred to as the slack time. In Figure 12–4, event 4 is not on the crucial path. To go from event 2 to event 5 on the critical path requires seven weeks taking the route 2–3–5. If route 2–4–5 is taken, only four weeks are required. Therefore, event 4, which requires two weeks for completion, should begin anywhere from zero to three weeks after event 2 is complete. During these three weeks, management might find another use for the resources of people, money, equipment, and facilities required to complete event 4.

The critical path is vital for resource scheduling and allocation because the project manager, with coordination from the functional manager, can reschedule those events not on the critical path for accomplishment during other time periods when maximum utilization of resources can be achieved, provided that the critical path time is not extended. This type of rescheduling through the use of slack times provides for a better balance of resources throughout the company, and may possibly reduce project costs by eliminating idle or waiting time.

Slack can be defined as the difference between the latest allowable date and the earliest expected date based on the nomenclature below:

T_E = the earliest time (date) on which an event can be expected to take place
T_L = the latest date on which an event can take place without extending the completion date of the project
Slack time = $T_L - T_E$

The calculation for slack time is performed for each event in the network, as shown in Figure 12–6, by identifying the earliest expected date and the latest starting date. For event 1, $T_L - T_E = 0$. Event 1 serves as the reference point for the network and could just as easily have been defined as a calendar date. As before, the critical path is represented as a bold line. The events on the critical path have no slack (i.e., $T_L = T_E$) and provide the boundaries for the noncritical path events.[5] Since event 2 is critical, $T_L = T_E = 3 + 7 = 10$ for event 5. Event 6 terminates the critical path with a completion time of fifteen weeks.

The earliest time for event 3, which is not on the critical path, would be two weeks ($T_E = 0 + 2 = 2$), assuming that it started as early as possible. The latest allowable date is obtained by subtracting the time required to complete the activity from events 3 to 5 from the latest starting date of event 5. Therefore, T_L (for event 3) = $10 - 5 = 5$ weeks. Event 3 can now occur anywhere between weeks 2 and 5 without interfering with the scheduled completion date of the project. This same procedure can be applied to event 4, in which case $T_E = 6$ and $T_L = 9$.

5. There are special situations where the critical path may include some slack. These cases are not considered here.

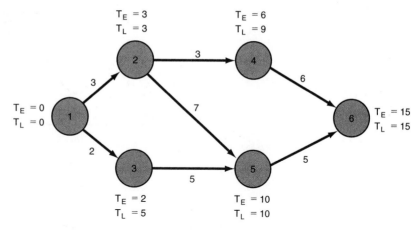

FIGURE 12–6.　Network with slack time.

Figure 12–6 contains a simple PERT network, and therefore the calculation of slack time is not too difficult. For complex networks containing multiple paths, the earliest starting dates must be found by proceeding from start to finish through the network, while the latest allowable starting date must be calculated by working backward from finish to start.

The importance of knowing exactly where the slack exists cannot be overstated. Proper use of slack time permits better technical performance. Donald Marquis has observed that those companies making proper use of slack time were 30 percent more successful than the average in completing technical requirements.[6]

Because of these slack times, PERT networks are often not plotted with a time scale. Planning requirements, however, can require that PERT charts be reconstructed with time scales, in which case a decision must be made as to whether we wish early or late time requirements for slack variables. This is shown in Figure 12–7 for comparison with total program costs and manpower planning. Early time requirements for slack variables are utilized in this figure.

The earliest times and late times can be combined to determine the probability of successfully meeting the schedule. A sample of the required information is shown in Table 12–2. The earliest and latest times are considered as random variables. The original schedule refers to the schedule for event occurrences that were established at the beginning of the project. The last column in Table 12–2 gives the probability that the earliest time will not be greater than the original schedule time for this event. The exact method for determining this probability, as well as the variances, is described in Section 12.5.

In the example shown in Figure 12–6, the earliest and latest times were calculated for each event. Some people prefer to calculate the earliest and latest times for each activity instead. Also, the earliest and latest times were identified simply as the time or date when

---

6. Donald Marquis, "Ways of Organizing Projects," *Innovation*, 1969.

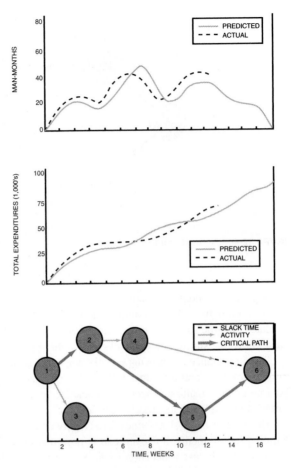

FIGURE 12–7. Comparison models for a time-phase PERT chart.

an event can be expected to take place. To make full use of the capabilities of PERT/CPM, we could identify four values:

- The earliest time when an activity can start (ES)
- The earliest time when an activity can finish (EF)
- The latest time when an activity can start (LS)
- The latest time when an activity can finish (LF)

Figure 12–8 shows the earliest and latest times identified on the activity.

To calculate the earliest starting times, we must make a forward pass through the network (i.e., left to right). The earliest starting time of a successor activity is the latest of the earliest finish dates of the predecessors. The earliest finishing time is the total of the earliest starting time and the activity duration.

TABLE 12–2. PERT CONTROL OUTPUT INFORMATION

Event Number	Earliest Time		Latest Time		Slack	Original Schedule	Probability of Meeting Schedule
	Expected	Variance	Expected	Variance			

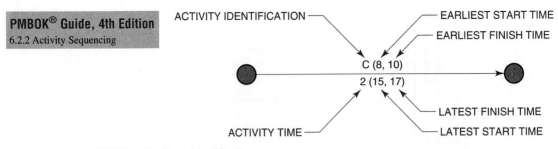

PMBOK® Guide, 4th Edition
6.2.2 Activity Sequencing

FIGURE 12–8. Slack identification.

To calculate the latest times, we must make a *backward* pass through the network by calculating the latest finish time. Since the activity time is known, the latest starting time can be calculated by subtracting the activity time from the latest finishing time. The latest finishing time for an activity entering a node is the earliest starting time of the activities exiting the node. Figure 12–9 shows the earliest and latest starting and finishing times for a typical network.

The identification of slack time can function as an early warning system for the project manager. As an example, if the total slack time available begins to decrease from one reporting period to the next, that could indicate that work is taking longer than anticipated or that more highly skilled labor is needed. A new critical path could be forming.

Looking at the earliest and latest start and finish times can identify slack. As an example, look at the two situations below:

[20, 26]	[30, 36]
[24, 30]	[25, 31]
Situation a	Situation b

PMBOK® Guide, 4th Edition
6.2.2 Activity Sequencing

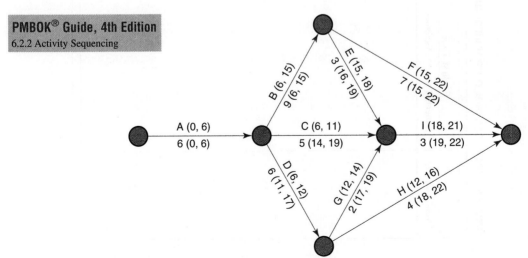

FIGURE 12–9. A typical PERT chart with slack times.

In Situation a, the slack is easily identified as four work units, where the work units can be expressed in hours, days, weeks, or even months. In Situation b, the slack is *negative* five units of work. This is referred to as negative slack or negative float.

What can cause the slack to be negative? Look at Figure 12–10. When performing a forward pass through a network, we work from left to right beginning at the customer's starting milestone (position 1). The backward pass, however, begins at the customer's end date milestone (position 2), *not* (as is often taught in the classroom) where the forward pass ends. If the forward pass ends at position 3, which is before the customer's end date, it is possible to have slack on the critical path. This slack is often called reserve time and may be added to other activities or filled with activities such as report writing so that the forward pass will extend to the customer's completion date.

Negative slack usually occurs when the forward pass extends beyond the customer's end date, as shown by position 4 in the figure. However, the backward pass is still measured from the customer's completion date, thus creating negative slack. This is most likely to result when:

- The original plan was highly optimistic, but unrealistic
- The customer's end date was unrealistic
- One or more activities slipped during project execution
- The assigned resources did not possess the correct skill levels
- The required resources would not be available until a later date

In any event, negative slack is an early warning indicator that corrective action is needed to maintain the customer's end date.

At this point, it is important to understand the physical meaning of slack. Slack measures how early or how late an event can start or finish. In Figure 12–6, the circles represented events and the slack was measured on the events. Most networks today, however,

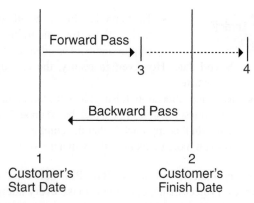

FIGURE 12–10. Slack time.

focus on the activity rather than the event, as shown in Figure 12–9. When slack is calculated on the activity, it is usually referred to as float rather than slack, but most project managers use the terms interchangeably. For activity C in Figure 12–9, the float is eight units. If the float in an activity is zero, then it is a critical path activity, such as seen in activity F. If the slack in an event is zero, then the event is a critical path event.

Another term is maximum float. The equation for maximum float is:

$$\text{Maximum float} = \text{latest finish} - \text{earliest start} - \text{duration}$$

For activity H in Figure 12–9, the maximum float is six units.

12.5 NETWORK REPLANNING

PMBOK® Guide, 4th Edition
6.5.2 Schedule Development
6.5.2.7 Schedule Compression

Once constructed, the PERT/CPM charts provide the framework from which detailed planning can be initiated and costs can be controlled and tracked. Many iterations, however, are normally made during the planning phase before the PERT/CPM chart is finished. Figure 12–11 shows this iteration process. The slack times form the basis from which additional iterations, or network replanning, can be performed. Network replanning is performed either at the conception of the program in order to reduce the length of the critical path, or during the program, should the unexpected occur. If all were to go according to schedule, then the original PERT/CPM chart would be unchanged for the duration of the project. But, how many programs or projects follow an exact schedule from start to finish?

Suppose that activities 1–2 and 1–3 in Figure 12–6 require manpower from the same functional unit. Upon inquiry by the project manager, the functional manager asserts that he can reduce activity 1–2 by one week if he shifts resources from activity 1–3 to activity 1–2. Should this happen, however, activity 1–3 will increase in length by one week. Reconstructing the PERT/CPM network as shown in Figure 12–12, the length of the critical path is reduced by one week, and the corresponding slack events are likewise changed.

There are two network replanning techniques based almost entirely upon resources: resource leveling and resource allocation.

PMBOK® Guide, 4th Edition
6.5.2.4 Resource Leveling

- Resource leveling is an attempt to eliminate the manpower peaks and valleys by smoothing out the period-to-period resource requirements. The ideal situation is to do this without changing the end date. However, in reality, the end date moves out and additional costs are incurred.
- Resource allocation (also called resource-limited planning) is an attempt to find the shortest possible critical path based upon the available or fixed resources. The problem with this approach is that the employees may not be qualified technically to perform on more than one activity in a network.

Unfortunately, not all PERT/CPM networks permit such easy rescheduling of resources. Project managers should make every attempt to reallocate resources to reduce the critical path, provided that the slack was not intentionally planned as a safety valve.

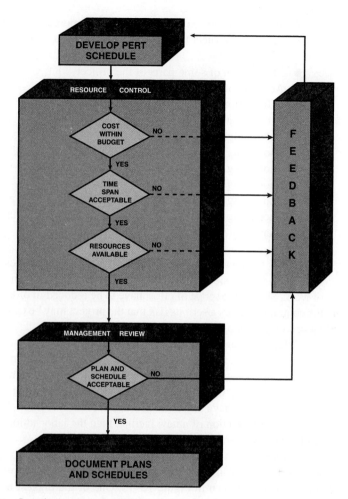

FIGURE 12–11. Iteration process for PERT schedule development.

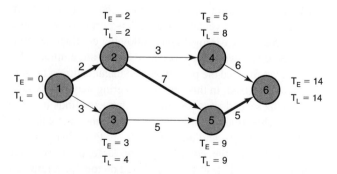

FIGURE 12–12. Network replanning of Figure 12–6.

Transferring resources from slack paths to more critical paths is only one method for reducing expected project time. Several other methods are available:

- Elimination of some parts of the project
- Addition of more resources (i.e., crashing)
- Substitution of less time-consuming components or activities
- Parallelization of activities
- Shortening critical path activities
- Shortening early activities
- Shortening longest activities
- Shortening easiest activities
- Shortening activities that are least costly to speed up
- Shortening activities for which you have more resources
- Increasing the number of work hours per day

Under the ideal situation, the project start and end dates are fixed, and performance within this time scale must be completed within the guidelines described by the statement of work. Should the scope of effort have to be reduced in order to meet other requirements, the contractor incurs a serious risk that the project may be canceled, or performance expectations may no longer be possible.

Adding resources is not always possible. If the activities requiring these added resources also call for certain expertise, then the contractor may not have qualified or experienced employees, and may avoid the risk. The contractor might still reject this idea, even if time and money were available for training new employees, because on project termination he might not have any other projects for these additional people. However, if the project is the construction of a new facility, then the labor-union pool may be able to provide additional experienced manpower.

Parallelization of activities can be regarded as accepting a risk by assuming that a certain event can begin in parallel with a second event that would normally be in sequence with it. This is shown in Figure 12–13. One of the biggest headaches at the beginning of any project is the purchasing of tooling and raw materials. As shown in Figure 12–13, four weeks can be saved by sending out purchase orders after contract negotiations are completed, but before the one-month waiting period necessary to sign the contract. Here the contractor incurs a risk. Should the effort be canceled or the statement of work change prior to the signing of the contract, the customer incurs the cost of the termination expenses from the vendors. This risk is normally overcome by the issuance of a long-lead procurement letter immediately following contract negotiations.

There are two other types of risk that are common. In the first situation, engineering has not yet finished the prototype, and manufacturing must order the tooling in order to keep the end date fixed. In this case, engineering may finally design the prototype to fit the tooling. In the second situation, the subcontractor finds it difficult to perform according to the original blueprints. In order to save time, the customer may allow the contractor to work without blueprints, and the blueprints are then changed to represent the as-built end-item.

Because of the complexities of large programs, network replanning becomes an almost impossible task when analyzed on total program activities. It is often better to have each department or division develop its own PERT/CPM networks, on approval by the

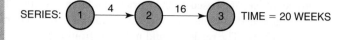

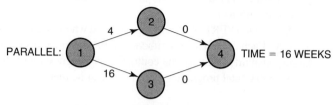

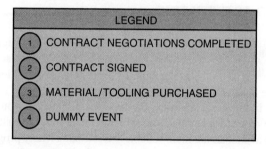

NOTE: EVENT 4 IS A DUMMY EVENT AND IS INCLUDED WITH A ZERO ACTIVITY TIME
IN ORDER TO CONSTRUCT A COMPLETE NETWORK

FIGURE 12–13. Parallelization of PERT activities.

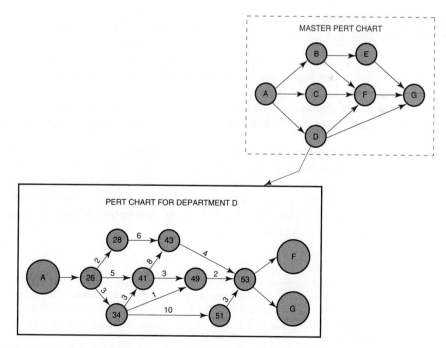

FIGURE 12–14. Master PERT chart breakdown by department.

project office, and based on the work breakdown structure. The individual PERT charts are then integrated into one master chart to identify total program critical paths, as shown in Figure 12–14. The reader should not infer from Figure 12–14 that department D does not interact with other departments or that department D is the only participant for this element of the project.

Segmented PERT charts can also be used when a number of contractors work on the same program. Each contractor (or subcontractor) develops his own PERT chart. It then becomes the responsibility of the prime contractor to integrate all of the subcontractors' PERT charts to ensure that total program requirements can be met.

12.6 ESTIMATING ACTIVITY TIME

PMBOK® Guide, 4th Edition
6.4 Activity Duration Estimating

Determining the elapsed time between events requires that responsible functional managers evaluate the situation and submit their best estimates. The calculations for critical paths and slack times in the previous sections were based on these best estimates.

In this ideal situation, the functional manager would have at his disposal a large volume of historical data from which to make his estimates. Obviously, the more historical data available, the more reliable the estimate. Many programs, however, include events and activities that are nonrepetitive. In this case, the functional managers must submit their estimates using three possible completion assumptions:

PMBOK® Guide, 4th Edition
6.4.2.4 Three-Point Estimates

- *Optimistic completion time.* This time assumes that everything will go according to plan and with minimal difficulties. This should occur approximately 1 percent of the time.
- *Pessimistic completion time.* This time assumes that everything will not go according to plan and maximum difficulties will develop. This should also occur approximately 1 percent of the time.
- *Most likely completion time.* This is the time that, in the mind of the functional manager, would most often occur should this effort be reported over and over again.[7]

Before these three times can be combined into a single expression for expected time, two assumptions must be made. The first assumption is that the standard deviation, σ, is one-sixth of the time requirement range. This assumption stems from probability theory, where the end points of a curve are three standard deviations from the mean. The second assumption requires that the probability distribution of time required for an activity be expressible as a beta distribution.[8]

---

7. It is assumed that the functional manager performs all of the estimating. The reader should be aware that there are exceptions where the program or project office would do their own estimating.

8. See F. S. Hillier and G. J. Lieberman, *Introduction to Operations Research* (San Francisco: Holden-Day, 1967), p. 229.

PMBOK® Guide, 4th Edition
6.4.2.4 Three-Point Estimates

The expected time between events can be found from the expression:

$$t_e = \frac{a + 4m + b}{6}$$

where t_e = expected time, a = most optimistic time, b = most pessimistic time, and m = most likely time.

As an example, if $a = 3$, $b = 7$, and $m = 5$ weeks, then the expected time, t_e, would be 5 weeks. This value for t_e would then be used as the activity time between two events in the construction of a PERT chart. This method for obtaining best estimates contains a large degree of uncertainty. If we change the variable times to $a = 2$, $b = 12$, and $m = 4$ weeks, then t_e will still be 5 weeks. The latter case, however, has a much higher degree of uncertainty because of the wider spread between the optimistic and pessimistic times. Care must be taken in the evaluation of risks in the expected times.

12.7 ESTIMATING TOTAL PROJECT TIME

PMBOK® Guide, 4th Edition
6.4 Activity Duration Estimates

In order to calculate the probability of completing the project on time, the standard deviations of each activity must be known. This can be found from the expression:

$$\sigma_{t_e} = \frac{b - a}{6}$$

where σ_{t_e} is the standard deviation of the expected time, t_e. Another useful expression is the variance, v, which is the square of the standard deviation. The variance is primarily useful for comparison to the expected values. However, the standard deviation can be used just as easily, except that we must identify whether it is a one, two, or three sigma limit deviation. Figure 12–15 shows the critical path of Figure 12–6, together with the corresponding values from which the expected times were calculated, as well as the standard

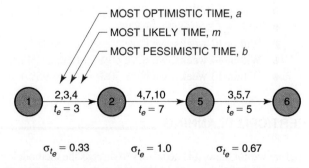

FIGURE 12–15. Expected time analysis for critical path events in Figure 12–6.

deviations. The total path standard deviation is calculated by the square root of the sum of the squares of the activity standard deviations using the following expression:

$$\sigma_{total} = \sqrt{\sigma_{1-2}^2 + \sigma_{2-5}^2 + \sigma_{5-6}^2}$$
$$= \sqrt{(0.33)^2 + (1.0)^2 + (0.67)^2}$$
$$= 1.25$$

The purpose of calculating σ is that it allows us to establish a confidence interval for each activity and the critical path. From statistics, using a normal distribution, we know that there is a 68 percent chance of completing the project within one standard deviation, a 95 percent chance within two standard deviations, and a 99.73 percent chance within three standard deviations.

This type of analysis can be used to measure the risks in the estimates, the risks in completing each activity, and the risks in completing the entire project. In other words, the standard deviation, σ, serves as a measurement of the risk. This analysis, however, assumes that normal distribution applies, which is not always the case.

As an example of measuring risk, consider a network that has only three activities on the critical path as shown below (all times in weeks):

Activity	Optimistic Time	Most Likely Time	Pessimistic Time	T_{ex}	σ	σ^2
A	3	4	5	4	$\frac{2}{6}$	$\frac{4}{36}$
B	4	4.5	8	5	$\frac{4}{6}$	$\frac{16}{36}$
C	4	6	8	6	$\frac{2}{6}$	$\frac{16}{36}$
				15		1.0

From the above table, the length of the critical path is 15 weeks. Since the variance (i.e., σ^2) is 1.0, then σ_{path}, which is the square root of the variance, must be 1 week.

We can now calculate the probability of completing the project within certain time limits:

- The probability of getting the job done within 16 weeks is
- 50% + (½)×(68%), or 84%.
- Within 17 weeks, we have 50% + (½)×(95%), or 97.5%.
- Within 14 weeks, we have 50% − (½)×(68%), or 16%.
- Within 13 weeks, we have 50% − (½)×(95%), or 2.5%.

12.8 TOTAL PERT/CPM PLANNING

Before we continue, it is necessary to discuss the methodology for preparing PERT schedules. PERT scheduling is a six-step process. Steps one and two begin with the project manager laying out a list of activities to be performed and then placing these activities in order of precedence, thus identifying the interrelationships. These charts drawn by the project manager are called either logic charts, arrow diagrams, work flow, or simply networks. The arrow diagrams will look like Figure 12–6 with two exceptions: The activity time is not identified, and neither is the critical path.

Step three is reviewing the arrow diagrams with the line managers (i.e., the true experts) in order to obtain their assurance that neither too many nor too few activities are identified, and that the interrelationships are correct.

In step four the functional manager converts the arrow diagram to a PERT chart by identifying the time duration for each activity. It should be noted here that the time estimates that the line managers provide are based on the *assumption of unlimited resources* because the calendar dates have not yet been defined.

Step five is the first iteration on the critical path. It is here that the project manager looks at the critical calendar dates in the definition of the project's requirements. If the critical path does not satisfy the calendar requirements, then the project manager must try to shorten the critical path using methods explained in Section 12.3 or by asking the line managers to take the "fat" out of their estimates.

Step six is often the most overlooked step. Here the project manager places calendar dates on each event in the PERT chart, thus converting from planning under unlimited resources to planning with *limited resources.* Even though the line manager has given you a time estimate, there is no guarantee that the correct resources will be available when needed. That is why this step is crucial. If the line manager cannot commit to the calendar dates, then replanning will be necessary. Most companies that survive on competitive bidding lay out proposal schedules based on unlimited resources. After contract award, the schedules are analyzed again because the company now has limited resources. After all, how can a company bid on three contracts simultaneously and put a detailed schedule into each proposal if it is not sure how many contracts, if any, it will win? For this reason customers require that formal project plans and schedules be provided thirty to ninety days after contract award.

Finally, PERT replanning should be an ongoing function during project execution. The best project managers continually try to assess what can go wrong and perform perturbation analysis on the schedule. (This should be obvious because the constraints and objectives of the project can change during execution.) Primary objectives on a schedule are:

- Best time
- Least cost
- Least risk

Secondary objectives include:

- Studying alternatives
- Optimum schedules
- Effective use of resources

- Communications
- Refinement of the estimating process
- Ease of project control
- Ease of time or cost revisions

Obviously, these objectives are limited by such constraints as:

- Calendar completion
- Cash or cash flow restrictions
- Limited resources
- Management approvals

12.9 CRASH TIMES

PMBOK® Guide, 4th Edition
6.5.2.7 Schedule Compression

In the preceding sections, no distinction was made between PERT and CPM. The basic difference between PERT and CPM lies in the ability to calculate percent complete. PERT is used in R&D or just development activities, where a percent-complete determination is almost impossible. Therefore, PERT is event oriented rather than activity oriented. In PERT, funding is normally provided for each milestone (i.e., event) achieved because incremental funding along the activity line has to be based on percent complete. CPM, on the other hand, is activity oriented because, in activities such as construction, percent complete along the activity line can be determined. CPM can be used as an arrow diagram network without PERT. The difference between the two methods lies in the environments in which they evolved and how they are applied. According to Archibald and Villoria[9]:

> The environmental factors which had an important role in determining the elements of the CPM techniques were:
> (a) Well-defined projects
> (b) One dominant organization
> (c) Relatively small uncertainties
> (d) One geographical location for a project

The CPM (activity-type network) has been widely used in the process industries, in construction, and in single-project industrial activities. Common problems include no place to store early arrivals of raw materials and project delays for late arrivals.

Using strictly the CPM approach, project managers can consider the cost of speeding up, or crashing, certain phases of a project. In order to accomplish this, it is necessary to

---

9. R. D. Archibald and R. L. Villoria, *Network-Based Management Systems (PERT/CPM)* (New York: Wiley, 1967), p. 14.

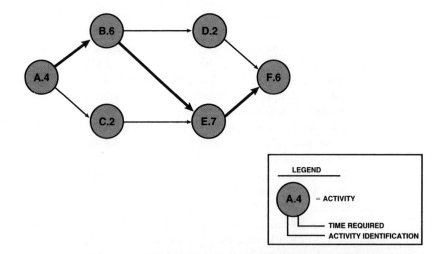

	TIME REQUIRED, WEEKS		COST $		CRASHING COST
ACTIVITY	NORMAL	CRASH	NORMAL	CRASH	PER WEEK, $
A	4	2	10,000	14,000	2,000
B	6	5	30,000	42,500	12,500
C	2	1	8,000	9,500	1,500
D	2	1	12,000	18,000	6,000
E	7	5	40,000	52,000	6,000
F	6	3	20,000	29,000	3,000

FIGURE 12–16. CPM network.

calculate a crashing cost per unit time as well as the normal expected time for each activity. CPM charts, which are closely related to PERT charts, allow visual representation of the effects of crashing. There are these requirements:

- For a CPM chart, the emphasis is on activities, not events. Therefore, the PERT chart should be redrawn with each circle representing an activity rather than an event.
- In CPM, both time and cost of each activity are considered.[10]
- Only those activities on the critical path are considered, starting with the activities for which the crashing cost per unit time is the lowest.

Figure 12–16 shows a CPM network with the corresponding crash time for all activities on and off the critical path. The activities are represented by circles and include an activity identification number and the estimated time. The costs expressed in the figure are usually direct costs only.

---

10. Although PERT considers mainly time, modifications through PERT/cost analysis can be made to consider the cost factors.

To determine crashing costs we begin with the lowest weekly crashing cost, activity A, at $2,000 per week. Although activity C has a lower crashing cost, it is not on the critical path. Only critical path activities are considered for crashing. Activity A will be the first to be crashed for a maximum of two weeks at $2,000 per week. The next activity to be considered would be F at $3,000 per week for a maximum of three weeks. These crashing costs are additional expenses above the normal estimates.

A word of caution concerning the selection and order of the activities that are to crash: There is a good possibility that as each activity is crashed, a new critical path will be developed. This new path may or may not include those elements that were bypassed because they were not on the original critical path.

Returning to Figure 12–16 (and assuming that no new critical paths are developed), activities A, F, E, and B would be crashed in that order. The crashing cost would then be an increase of $37,500 from the base of $120,000 to $157,500. The corresponding time would then be reduced from twenty-three weeks to fifteen weeks. This is shown in Figure 12–17 to illustrate how a trade-off between time and cost can be obtained. Also shown in Figure 12–17 is the increased cost of crashing elements not on the critical path. Crashing these elements would result in a cost increase of $7,500 without reducing the total project time. There is also the possibility that this figure will represent unrealistic conditions because sufficient resources are not or cannot be made available for the crashing period.

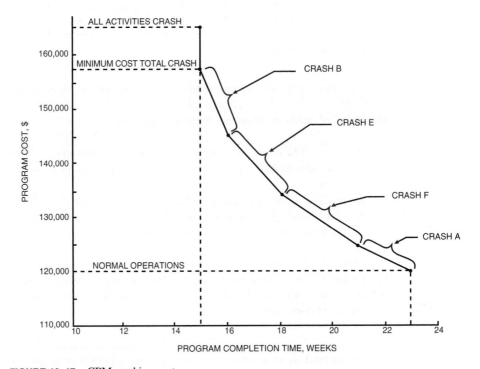

FIGURE 12–17. CPM crashing costs.

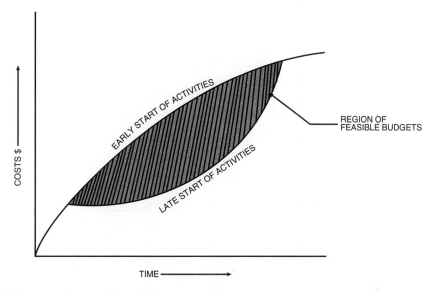

FIGURE 12–18. Region of feasible budgets.

The purpose behind balancing time and cost is to avoid wasting resources. If the direct and indirect costs can be accurately obtained, then a region of feasible budgets can be found, bounded by the early-start (crash) and late-start (or normal) activities. This is shown in Figure 12–18.

Since the direct and indirect costs are not necessarily expressible as linear functions, time–cost trade-off relationships are made by searching for the lowest possible total cost (i.e., direct and indirect) that likewise satisfies the region of feasible budgets. This method is shown in Figure 12–19.

Like PERT, CPM also contains the concept of slack time, the maximum amount of time that a job may be delayed beyond its early start without delaying the project completion time. Figure 12–20 shows a typical representation of slack time using a CPM chart. In addition, the figure shows how target activity costs can be identified. Figure 12–20 can be modified to include normal and crash times as well as normal and crash costs. In this case, the cost box in the figure would contain two numbers: The first number would be the normal cost, and the second would be the crash cost. These numbers might also appear as running totals.

12.10 PERT/CPM PROBLEM AREAS

PERT/CPM models are not without their disadvantages and problems. Even the largest organizations with years of experience in using PERT and CPM have the same ongoing problems as newer or smaller companies.

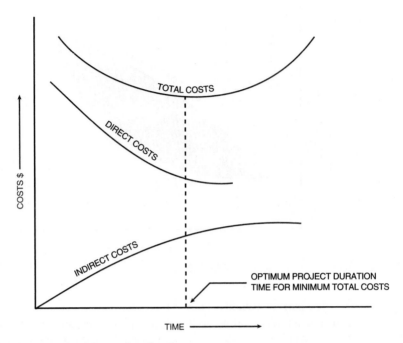

FIGURE 12–19. Determining project duration.

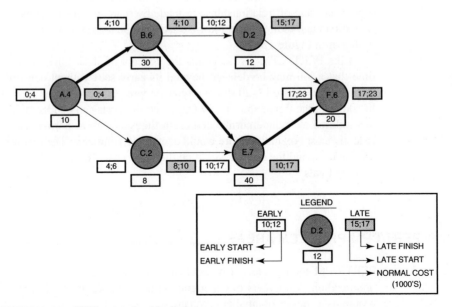

FIGURE 12–20. CPM network with slack.

Many companies have a difficult time incorporating PERT systems because PERT is end-item oriented. Many upper-level managers feel that the adoption of PERT/CPM removes a good part of their power and ability to make decisions. This is particularly evident in companies that have been forced to accept PERT/CPM as part of contractual requirements.

In PERT systems, there are planners and doers. In most organizations PERT planning is performed by the program office and functional management. Yet once the network is constructed, the planners and managers become observers and rely on the doers to accomplish the job within time and cost limitations. Management must convince the doers that they have an obligation to the successful completion of the established PERT/CPM plans.

Unless the project is repetitive, there is usually little historical information on which to base the cost estimates of most optimistic, most pessimistic, and most likely times. Problems can also involve poor predictions for overhead costs, other indirect costs, material and labor escalation factors, and crash costs. It is also possible that each major functional division of the organization has its own method for estimating costs. Engineering, for example, may use historical data, whereas manufacturing operations may prefer learning curves. PERT works best if all organizations have the same method for predicting costs and performance.

PERT networks are based on the assumption that all activities start as soon as possible. This assumes that qualified personnel and equipment are available. Regardless of how well we plan, there are almost always differences in performance times from what would normally be acceptable. For the selected model, time and cost should be well-considered estimates, not spur-of-the-moment decisions.

Cost control problems arise when the project cost and control system is not compatible with company policies. Project-oriented costs may be meshed with non-PERT-controlled jobs in order to develop the annual budget. This becomes a difficult chore for cost reporting, especially when each project may have its own method for analyzing and controlling costs.

Many people have come to expect too much of PERT-type networks. Figure 12–21 illustrates a PERT/CPM network broken down by work packages with identification of the

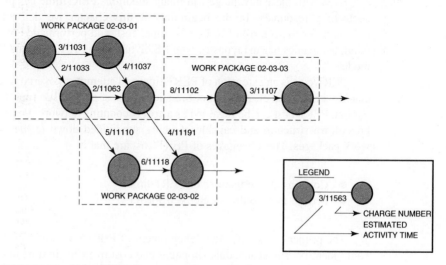

FIGURE 12–21. Using PERT for work package control.

charge numbers for each activity. Large projects may contain hundreds of charge numbers. Subdividing work packages (which are supposedly the lowest element) even further by identifying all subactivities has the advantage that direct charge numbers can be easily identified, but the time and cost for this form of detail may be prohibitive. PERT/CPM networks are tools for program control, and managers must be careful that the original game plan of using networks to identify prime and supporting objectives is still met. Additional detail may mask this all-important purpose. Remember, networks are constructed as a means for understanding program reports. Management should not be required to read reports in order to understand PERT/CPM networks.

12.11 ALTERNATIVE PERT/CPM MODELS

Because of the many advantages of PERT/time, numerous industries have found applications for this form of network. A partial list of these advantages includes capabilities for:

- Trade-off studies for resource control
- Providing contingency planning in the early stages of the project
- Visually tracking up-to-date performance
- Demonstrating integrated planning
- Providing visibility down through the lowest levels of the work breakdown structure
- Providing a regimented structure for control purposes to ensure compliance with the work breakdown structure and the statement of work
- Increasing functional members' ability to relate to the total program, thus providing participants with a sense of belonging

Even with these advantages, in many situations PERT/time has proved ineffective in controlling resources. In the beginning of this chapter we defined three parameters necessary for the control of resources: time, cost, and performance. With these factors in mind, companies began reconstructing PERT/time into PERT/cost and PERT/performance models.

PERT/cost is an extension of PERT/time and attempts to overcome the problems associated with the use of the most optimistic and most pessimistic time for estimating completion. PERT/cost can be regarded as a cost accounting network model based on the work breakdown structure and capable of being subdivided down to the lowest elements, or work packages. The advantages of PERT/cost are that it:

- Contains all the features of PERT/time
- Permits cost control at any WBS level

The primary reason for the development of PERT/cost was so that project managers could identify critical schedule slippages and cost overruns in time to correct them.

Many attempts have been made to develop effective PERT/schedule models. In almost all cases, the charts are constructed from left to right.[11] An example of such current attempts is the accomplishment/cost procedure (ACP). As described by Block[12]:

> ACP reports cost based on schedule accomplishment, rather than on the passage of time. To determine how an uncompleted task is progressing with respect to cost, ACP compares (a) cost/progress relationship budgeting with (b) the cost/progress relationship expended for the task. It utilizes data accumulated from periodic reports and from the same data base generates the following:
>
> - The relationship between cost and scheduled performance
> - The accounting relationships between cost and fiscal accounting requirements
> - The prediction of corporate cash flow needs

Unfortunately, the development of PERT/schedule techniques is still in its infancy. Although their applications have been identified, many companies feel locked in with their present method of control, whether it be PERT, CPM, or some other technique.

12.12 PRECEDENCE NETWORKS

PMBOK® Guide, 4th Edition
6.2.2.1 PDM

In recent years there has been an explosion in project management software packages. Small packages may sell for a few thousand dollars, whereas the price for larger packages may be tens of thousands of dollars. Computerized project management can provide answers to such questions as:

- How will the project be affected by limited resources?
- How will the project be affected by a change in the requirements?
- What is the cash flow for the project (and for each WBS element)?
- What is the impact of overtime?
- What additional resources are needed to meet the constraints of the project?
- How will a change in the priority of a certain WBS element affect the total project?

The more sophisticated packages can provide answers to schedule and cost based on:

- Adverse weather conditions
- Weekend activities
- Unleveled manpower requirements

---

11. See Gary E. Whitehouse, "Project Management Techniques," *Industrial Engineering,* March 1973, pp. 24–29, for a description of the technique.

12. Reprinted by permission of *Harvard Business Review.* From Ellery B. Block, "Accomplishment/Cost: Better Project Control," *Harvard Business Review,* May–June 1971, pp. 110–124. Copyright © 1971 by the Harvard Business School Publishing Corporation; all rights reserved.

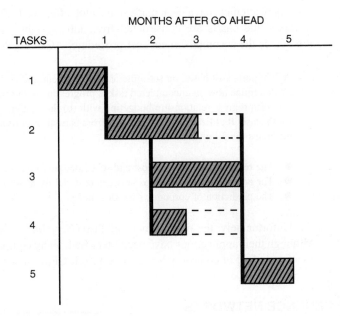

FIGURE 12–22. Precedence network.

- Variable crew size
- Splitting of activities
- Assignment of unused resources

Regardless of the sophistication of computer systems, printers and plotters prefer to draw straight lines rather than circles. Most software systems today use precedence networks, as shown in Figure 12–22, which attempt to show interrelationships on bar charts. In Figure 12–22, task 1 and task 2 are related because of the solid line between them. Task 3 and task 4 can begin when task 2 is half finished. (This cannot be shown easily on PERT without splitting activities.) The dotted lines indicate slack. The critical path can be identified by putting an asterisk (*) beside the critical elements, or by putting the critical connections in a different color or boldface.

The more sophisticated software packages display precedence networks in the format shown in Figure 12–23. In each of these figures, work is accomplished during the activity. This is sometimes referred to as the activity-on-node method. The arrow represents the relationship or constraint between activities.

Figure 12–23A illustrates a finish-to-start constraint. In this figure, activity 2 can start no earlier than the completion of activity 1. All PERT charts are finish-to-start constraints. Figure 12–23B illustrates a start-to-start constraint. Activity 2 cannot start prior to the start of activity 1. Figure 12-23C illustrates a finish-to-finish constraint. In this figure, activity 2 cannot finish until activity 1 finishes. Figure 12-23D illustrates a start-to-finish constraint.

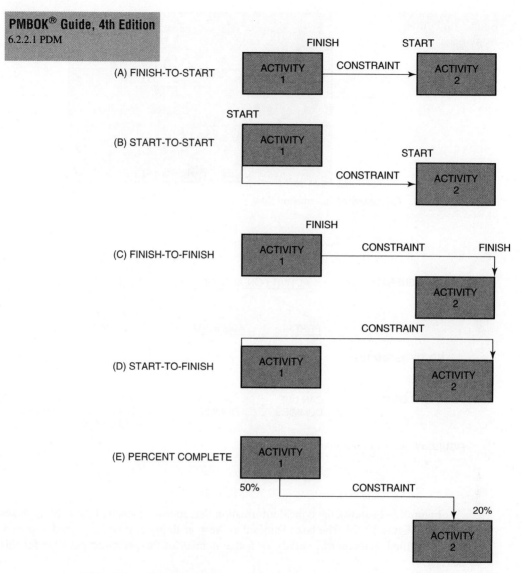

PMBOK® Guide, 4th Edition
6.2.2.1 PDM

FIGURE 12–23. Typical precedence relationships.

An example might be that you must start studying for an exam some time prior to the completion of the exam. This is the least common type of precedence chart. Figure 12-23E illustrates a percent complete constraint. In this figure, the last 20 percent of activity 2 cannot be started until 50 percent of activity 1 has been completed.[13]

---

13. Meredith and Mantel categorize precedence relationships in three broad categories; Natural Precedences, Environmental Precedences, and Preferential Precedences. For additional information on these precedence relationships, see Jack R. Meredith and Samuel J. Mantel, Jr., *Project Management*, 3rd ed. (New York: Wiley;1995), pp.385–386.

PMBOK® Guide, 4th Edition
6.1.3.2 Activity Attributes

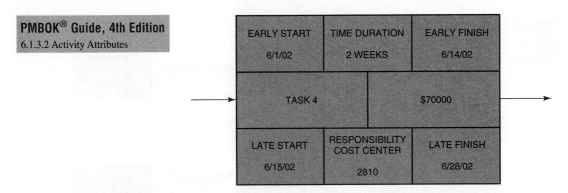

FIGURE 12–24. Computerized information flow.

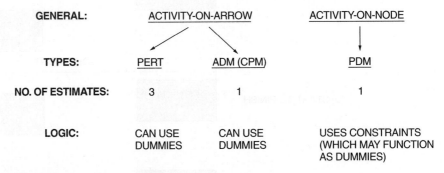

FIGURE 12–25. Comparison of networks.

Figure 12–24 shows the typical information that appears in each of the activity boxes shown in Figure 12–23. The box identified as "responsibility cost center" could also have been identified as the name, initials, or badge number of the person responsible for this activity.

Figure 12–25 shows the comparison of three of the network techniques.

12.13 LAG

PMBOK® Guide, 4th Edition
6.2.2.5 Leads and Lags
6.2.2.1 PDM

The time period between the early start or finish of one activity and the early start or finish of another activity in the sequential chain is called lag. Lag is most commonly used in conjunction with precedence networks. Figure 12–26 shows five different ways to identify lag on the constraints.

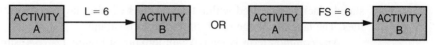

(A) FINISH-TO-START (FS) RELATIONSHIP. THE START OF B MUST LAG 6 DAYS AFTER THE FINISH OF A.

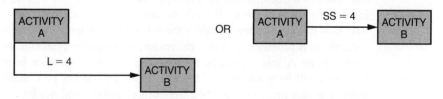

(B) START-TO-START (SS) RELATIONSHIP. THE START OF B MUST LAG 4 DAYS AFTER THE START OF A.

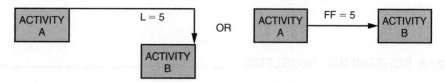

(C) FINISH-TO-FINISH (FF) RELATIONSHIP. THE FINISH OF B MUST LAG 5 DAYS AFTER THE FINISH OF A.

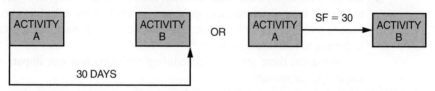

(D) START-TO-FINISH (SF) RELATIONSHIP. THE FINISH OF B MUST LAG 30 DAYS AFTER THE START OF A.

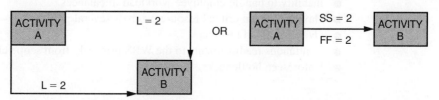

(E) COMPOSITE START-TO-START AND FINISH-TO-FINISH RELATIONSHIP. THE START OF B MUST LAG 2 DAYS AFTER THE START OF A, AND THE FINISH OF B MUST LAG 2 DAYS AFTER THE FINISH OF A.

FIGURE 12–26.   Precedence charts with lag.

Slack is measured within activities whereas lag is measured between activities. As an example, look at Figure 12–26A. Suppose that activity A ends at the end of the first week of March. Since it is a finish-to-start precedence chart, one would expect the start of activity B to be the beginning of the second week in March. But if activity B cannot start until the beginning of the third week of March, that would indicate a week of lag between activity A and activity B even though both activities can have slack within the activity. Simply stated, slack is measured within the activities whereas lag is measured between the activities. The lag may be the result of resource constraints.

Any common term is lead. Again looking at Figure 12–26A, suppose that activity A finishes on March 15 but the precedence chart shows activity B starting on March 8, seven days prior to the completion of activity A. In this case, $L = -7$, a negative value, indicating that the start of activity B leads the completion of activity A by seven days. To illustrate how this can happen, consider the following example: The line manager responsible for activity B promised you that his resources would be available on March 16, the day after activity A was scheduled to end. The line manager then informs you that these resources will be available on March 8, and if you do not pick them up on your charge number at that time, they may be assigned elsewhere and not be available on the 16th. Most project managers would take the resources on the 8th and find some work for them to do even though logic says that the work cannot begin until after activity A has finished.

12.14 SCHEDULING PROBLEMS

Every scheduling technique has advantages and disadvantages. Some scheduling problems are the result of organizational indecisiveness, such as having a project sponsor that refuses to provide the project manager guidance on whether the schedule should be based upon a least time, least cost, or least risk scheduling objective. As a result, precious time is wasted in having to redo the schedules.

However, there are some scheduling problems that can impact all scheduling techniques. These include:

- Using unrealistic estimates for effort and duration
- Inability to handle employee workload imbalances
- Having to share critical resources across several projects
- Overcommitted resources
- Continuous readjustments to the WBS primarily from scope changes
- Unforeseen bottlenecks

12.15 THE MYTHS OF SCHEDULE COMPRESSION

Simply because schedule compression techniques may exist does not mean that they will work. There is a tendency for managers to be aggressively positive in their thinking at the onset of a project, believing that compression techniques can be applied effectively. As discussed by Grey[14]:

14. Stephen Grey, *Practical Risk Assessment for Project Management* (West Sussex, England: Wiley, 1995), pp. 108–109.

There is a common tendency, especially among people who have been convinced that they must "think positive," to be unwilling to accept that an activity might take longer than planned. To the question "What is the maximum time it could take?", they respond with "It will be finished in the planned time, it will not be allowed to take longer", or words to that effect. The words "it will not be allowed to take longer" or "it must not take any longer" are so consistent that they must reflect a common feature of the way businesses manage their staff.

While most people are willing to accept that costs could exceed expectations, and might even take a perverse delight in recounting past examples, the same is not true of deadlines. This is probably due to the fact that cost over runs are resolved in-house, while schedule issues are open and visible to the customer.

There might be ways in which a schedule can be held no matter what happens. Study tasks are almost always finished on time because the scope of work is allowed to vary according to what the study turned up. This is the exception rather than the rule though. In general, you can only be sure that a task will finish on time if:

- The scope of work is flexible, at least to some extent.
- It will be possible to calibrate the task from the early part of the work to tell if the planned work rate is adequate.
- You can raise the work rate and/or reduce the scope of work to bring the task back on target in the time left after you find it is heading for an overrun.

There are five common techniques for schedule compression, and each technique has significant limitations that may make this technique more of a myth than reality. This is shown in Table 12–3.

TABLE 12–3. MYTHS AND REALITIES OF SCHEDULE COMPRESSION

Compression Technique	Myth	Reality
Use of overtime	Work will progress at the same rate on overtime.	The rate of progress is less on overtime; more mistakes may occur; and prolonged overtime may lead to burnout.
Adding more resources (i.e., crashing)	The performance rate will increase due to the added resources.	It takes time to find the resources; it takes time to get them up to speed; the resources used for the training must come from the existing resources.
Reducing scope (i.e., reducing functionality)	The customer always requests more work than actually needed.	The customer needs all of the tasks agreed to in the statement of work.
Outsourcing	Numerous qualified suppliers exist.	The quality of the suppliers' work can damage your reputation; the supplier may go out of business; and the supplier may have limited concern for your scheduled dates.
Doing series work in parallel	An activity can start before the previous activity has finished.	The risks increase and rework becomes expensive because it may involve multiple activities.

12.16 UNDERSTANDING PROJECT MANAGEMENT SOFTWARE

PMBOK® Guide, 4th Edition
6.3.2.8 Project Management
Software

Efficient project management requires more than good planning, it requires that relevant information be obtained, analyzed, and reviewed in a timely manner. This can provide early warning of pending problems and impact assessments on other activities, which can lead to alternate plans and management actions. Today, project managers have a large array of software available to help in the difficult task of tracking and controlling projects. While it is clear that even the most sophisticated software package is not a substitute for competent project leadership—and by itself does not identify or correct any task-related problems—it can be a terrific aid to the project manager in tracking the many interrelated variables and tasks that come into play with a project. Specific examples of these capabilities are:

- Project data summary: expenditure, timing, and activity
- Project management and business graphics capabilities
- Data management and reporting capabilities
- Critical path analysis
- Customized and standard reporting formats
- Multiproject tracking
- Subnetworking
- Impact analysis (what if . . .)
- Early-warning systems
- On-line analysis of recovering alternatives
- Graphical presentation of cost, time, and activity data
- Resource planning and analysis
- Cost analysis, variance analysis
- Multiple calendars
- Resource leveling

Further, many of the more sophisticated software packages are now available for personal computers and use mainly precedence networks. This offers large and small companies many advantages ranging from true user interaction, to ready access and availability, to simpler and more user-friendly interfaces, to considerably lower software cost.

12.17 SOFTWARE FEATURES OFFERED

Project management software capabilities and features vary a great deal. However, the variation is more in the depth and sophistication of the features, such as storage, display, analysis, interoperability, and user friendliness, rather than in the type of features offered, which are very similar for most software programs. Most project management software packages offer the following features:

1. *Planning, tracking, and monitoring.* These features provide for planning and tracking the projects' tasks, resources, and costs. The data format for describing the

project to the computer is usually based on standard network typologies such as the Critical Path Method (CPM), Program Evaluation and Review Technique (PERT), or Precedence Diagram Method (PDM). Task elements, with their estimated start and finish times, their assigned resources, and actual cost data, can be entered and updated as the project progresses. The software provides an analysis of the data and documents the technical and financial status of the project against its schedule and original plan. Usually, the software also provides impact assessments of plan deviations and resource and schedule projections. Many systems also provide resource leveling, a feature that averages out available resources to determine task duration and generates a leveled schedule for comparison.

2. *Reports.* Project reporting is usually achieved via a menu-driven report writer system that allows the user to request several standard reports in a standard format. The user can also modify these reports or create new ones. Depending on the sophistication of the system and its peripheral hardware, these reports are supported by a full range of Gantt charts, network diagrams, tabular summaries, and business graphics. Reporting capabilities include:

> **PMBOK® Guide, 4th Edition**
> 7.3.2 Cost Control Tools and Techniques

- Budgeted cost for work scheduled (BCWS) or planned value of work (PV)
- Budgeted cost for work performed (BCWP) or earned value of work (EV)
- Actual versus planned expenditure
- Earned value analysis
- Cost and schedule performance indices
- Cash-flow
- Critical path analysis
- Change order
- Standard government reports (DoD, DoE, NASA), formatted for the performance monitoring system (PMS)

In addition, many software packages feature a user-oriented, free-format report writer for styled project reporting.

3. *Project calendar.* This feature allows the user to establish work weeks based on actual workdays. Hence, the user can specify nonwork periods such as weekends, holidays, and vacations. The project calendar can be printed out in detail or in a summary format and is automatically the basis for all computer-assisted resource scheduling.

4. *What-if analysis.* Some software is designed to make what-if analyses easy. A separate, duplicate project database is established and the desired changes are entered. Then the software performs a comparative analysis and displays the new against the old project plan in tabular or graphical form for fast and easy management review and analysis.

5. *Multiproject analysis.* Some of the more sophisticated software packages feature a single, comprehensive database that facilitates cross-project analysis and reporting. Cost and schedule modules share common files that allow integration among projects and minimize problems of data inconsistencies and redundancies.

12.18 SOFTWARE CLASSIFICATION

For purposes of easy classification, project management software products have been divided into three categories based on the type of functions and features they provide.[15]

Level I software. Designed for single-project planning, these software packages are simple, easy to use, and their outputs are easy to understand. They do provide, however, only a limited analysis of the data. They do not provide automatic rescheduling based on specific changes. Therefore, deviations from the original project plan require complete replanning of the project and a complete new data input to the computer.

Level II software. Designed for single project management, these software packages aid project leaders in the planning, tracking, and reporting of projects. They provide a comprehensive analysis of the project, progress reports, and plan revisions, based on actual performance. This type of software is designed for managing projects beyond the planning stage, and for providing semiautomatic project control.

Level III software. These packages feature multiproject planning, monitoring, and control by utilizing a common database and sophisticated cross-project monitoring and reporting software.

Most software packages at levels II and III have the following extensive capabilities for project monitoring and control:

1. *System capacity.* The number of activities and/or number of subnetworks that may be used.
2. *Network schemes.* The network schemes are activity diagram (AD) and/or precedence relationship (PRE).
3. *Calendar dates.* An internal calendar is available to schedule the project's activities. The variations and options of the different calendar algorithms are numerous.
4. *Gantt or bar charts.* A graphic display of the output on a time scale is available if desired.
5. *Flexible report generator.* The user can specify within defined guidelines the format of the output.
6. *Updating.* The program will accept revised time estimates and completion dates and recompute the revised schedule.
7. *Cost control.* The program accepts budgeted cost figures for each activity and then the actual cost incurred, and summarizes the budgeted and actual figures on each updating run. The primary objective is to help management produce a realistic cost plan before the project is started and to assist in the control of the project expenditures as the work progresses.
8. *Scheduled dates.* A date is specified for the completion of any of the activities for purposes of planning and control. The calculations are performed with these dates as constraints.
9. *Sorting.* The program lists the activities in a sequence specified by the user.

---

15. Some standards were initially set by *PC Magazine,* "Project Management with the PC," Vol. 3, No. 24, December 11, 1984.

10. *Resource allocation.* The program attempts to allocate resources optimally using one of many heuristic algorithms.
11. *Plotter availability.* A plotter is available to plot the network diagram.
12. *Machine requirements.* This is the minimum hardware memory requirement for the program (in units of bytes).
13. *Cost.* Indicates whether the program is sold and/or leased and the purchase price and/or lease price (where available).

12.19 IMPLEMENTATION PROBLEMS

Generally speaking, mainframe software packages are more difficult to implement than smaller packages, because everyone is requested to use the same package, perhaps even the same way. The following are common difficulties during implementation:

- *Upper-level management may not like the reality of the output.* The output usually shows top management that more time and resources are needed than originally anticipated. This can also be a positive note for the project manager, who is forced to deal with severe resource constraints.
- *Upper-level management may not use the packages for planning, budgeting, and decision-making.* Upper-level personnel generally prefer the more traditional methods, or simply refuse to look at reality because of politics. As a result, the plans they submit to the board are based on an eye-pleasing approach for quick acceptance, rather than reality.
- *Day-to-day project planners may not use the packages for their own projects.* Project managers often rely on other planning methods and tools from previous assignments. They rely heavily on instinct and trial and error.
- *Upper-level management may not demonstrate support and commitment to training.* Ongoing customized training is mandatory for successful implementation, even though each project may vary.
- *Use of mainframe software requires strong internal communications lines for support.* Managers who share resources must talk to one another continually.
- *Clear, concise reports are lacking.* Large mainframe packages can generate volumes of data, even if the package has a report writer package.
- *Mainframe packages do not always provide for immediate turnabout of information.* This is often the result of not understanding how to utilize the new systems.
- *The business entity may not have any project management standards in place prior to implementation.* This relates to a lack of WBS numbering schemes, no life-cycle phases, and a poor understanding of task dependencies.
- *Implementation may highlight middle management's inexperience in project planning and organizational skills.* Fear of its use is a key factor in not obtaining proper support.
- *The business environment and organizational structure may not be appropriate to meet project management/planning needs.* If extensive sharing of resources exists, then the organizational structure should be a formal or informal matrix.

If the organization is deeply entrenched in a traditional structure, then organizational mismatch exists and the software system may not be accepted.

- *Sufficient/extensive resources (staff, equipment, etc.) are required.* Large mainframe packages consume a significant amount of resources in the implementation phase.
- *The business entity must determine the extent of, and appropriate use of, the systems within the organization.* Should it be used by all organizations? Should it be used only on high-priority projects?
- *The system may be viewed as a substitute for the extensive interpersonal skills required by the project manager.* Software systems do not replace the need for project managers with strong communications and negotiation skills.
- *Software implementation is less likely to succeed if the organization does not have sufficient training in project management principles.* This barrier is perhaps the underlying problem for all of the other barriers.

12.20 CRITICAL CHAIN[16]

PMBOK® Guide, 4th Edition
6.5.2.3 Schedule Development—
Critical Chain Method

The selection and completion of enough projects to improve an organization is often a matter of survival for executives. Witness the statistic by outplacement firm Drake, Beam, Morin stating that 57 percent of the 367 large corporations surveyed have replaced their CEOs in the past three years.[17]
Executives use projects as a primary means to meet their goals. Therefore, we can assume that many of these CEOs were unable to complete enough projects successfully in the measurement time period to keep their jobs.

In trying to meet their goals, executives often describe three major challenges in project management:

- Choosing the right projects from among a large pool
- Getting each project to completion more quickly
- Funneling more projects through the organization without adding resources

Critical Chain is a project management methodology designed to address the latter two goals. Critical Chain is based upon a general improvement methodology called the Theory of Constraints, which addresses the first executive goal—choosing the right projects. Choosing the right projects is part of strategic planning, which is discussed in depth in other books.[18]

As executives attempt to release new projects into the organization, they often hear complaints that people are overloaded. Inevitably, they face a conflict between moving

---

16. Section author Gerald I. Kendall, PMP, Principal, TOC International, www.tocinternational.com, email Gerryikendall@cs.com, 850-939-9006.

17. *USA Today,* April 8, 2002, p. B1, "Scandals, Setbacks Topple CEOs Formerly Golden Image".

18. See Gerald I. Kendall, *Viable Vision* (Boca Raton, FL: J. Ross Publishing, 2004).

resources to the new project and allowing resources to continue working on existing projects. People in the organization may also urge the executive to delay the start of the new project while the executive feels compelled to move ahead.

Most executives accept this conflict as a fact of life. They believe that their role is to push people as hard as they can to perform to high standards. As a result, the reaction of many executives to the resource conflict is to demand that existing projects be finished earlier so that their new projects can begin sooner. These demands leave project managers with their own huge conflict. In order to finish a project sooner, most project managers find that they are forced to either reduce scope or quality or add resources, which will exceed the budget. None of these alternatives is acceptable to executives.

The resulting behavior, which is now prevalent in many organizations, is the fodder for a new approach called Critical Chain Project Management. When project and resource managers fail to convince executives to delay the start of a new project, they often take three actions that lead to many other negative effects:

- Multitasking of resources
- Working toward cutting task estimates
- Managing people very closely to ensure that they meet their due dates

Since executives are a major part of the system of projects inside organizations, Critical Chain recognizes that executives are part of the problem. To solve the problem and have a major impact on project results, executives must therefore be part of the solution.

The Critical Chain solution to scheduling and managing projects was derived from a methodology called the Theory of Constraints. Dr. Eliyahu M. Goldratt is the individual most often credited with the creation and advancement of this methodology over the past twenty-five years. To derive the Critical Chain solution, Goldratt applied the five focusing steps, identified in his writings.[19] These steps are:

1. *Identify* the system's constraint.
2. Decide how to *exploit* the constraint.
3. *Subordinate* everything else to the above decision.
4. *Elevate* the system's constraint.
5. If, in a previous step, the system's constraint has been broken, *go back* to step 1.

Within any project, the Critical Chain is defined as the longest chain of dependent events where the dependency is either task or resource related. This definition assumes that the longest chain is the one that is most likely to impact negatively the overall duration of the project. The Critical Chain is not necessarily equivalent to the project duration since, sometimes, there are noncritical tasks that begin before the Critical Chain tasks begin.

The Critical Chain solution recognizes the Critical Chain as the leverage point for reducing the project's duration. The first focusing step, *identify*, recognizes that managers put practices into place that block the reduction of the Critical Chain. The *exploit* and *subordinate*

---

19. Eliyahu M. Goldratt, *Theory of Constraints* (Croton-on-Hudson, NY: North River Press, 1990).

steps implement changes to condense the Critical Chain (in other words, to shorten the amount of time it takes to complete a project).

Critical Chain implements major behavioral changes in project managers, resource managers, team members, and executives. The only way that so many people in an organization can accept such fundamental changes is through a deep understanding of the current behaviors, the new behaviors required, and the benefits. This is usually accomplished through education of executives, project managers, resource managers, and team members, followed by policy and measurement changes. These changes include:

- An end to the practice of measuring people in any way on the accuracy of their estimates
- An end to the practice of measuring people on meeting due dates for individual project tasks
- A replacement of the above two practices by "the relay runner work ethic," explained later in this chapter
- A system, agreed to by all executives and senior managers, of allowing new projects to start only when a "strategic resource" is available
- The recognition of the need to strategically protect projects from task time variations, by using properly placed buffers. This imbeds the philosophy of W. Edwards Deming, the great quality advocate, regarding the handling of "common cause" and "special cause" variation and predictability.
- The significant reduction of the practice of multitasking by moving toward dedicated work on project tasks
- The implementation of multiproject software with the data actually being used by executives, resource managers, and project managers. Critical Chain reports present a common and accurate picture of the organization's projects and a systematic and logical way to manage variances.
- The implementation of buffer management as a key management and executive process for identifying project problems during execution

The successful implementation of Critical Chain has resulted in major improvements in organizations, examples of which are documented in the case studies in this Chaper. In order to understand the magnitude of the cultural change and the problems to be overcome, this Section explains the fundamentals of the Critical Chain approach, in both individual project environments and throughout an organization.

12.21 STUDYING TIPS FOR THE PMI® PROJECT MANAGEMENT CERTIFICATION EXAM

This section is applicable as a review of the principles to support the knowledge areas and domain groups in the PMBOK® Guide. This chapter addresses:

- Time management
- Planning
- Controlling

Understanding the following principles is beneficial if the reader is using this text to study for the PMP® Certification Exam:

- How to identify the three types of scheduling techniques and their respective advantages and disadvantages
- Difference between activity-on-arrow and activity-on-node networks
- Four types of precedence networks
- Basic network terminology such as activities, events, critical path, and slack (float)
- Difference between positive and negative slack
- Schedule compression techniques and crashing and fast-tracking (concurrent engineering)
- Importance of the work breakdown structure in network development
- The steps, and their order, for the development of a network
- Three types of dependencies
- How to perform a forward and backward pass
- Resources leveling
- Resource-limited planning
- Difference between effort and duration
- Which network technique uses optimistic, most likely, and pessimistic estimates
- Use of dummy activities
- Lag
- Difference between unlimited versus limited resource planning/scheduling

The following multiple-choice questions will be helpful in reviewing the principles of this chapter:

1. The shortest time necessary to complete all of the activities in a network is called the:
   A. Activity duration length
   B. Critical path
   C. Maximum slack path
   D. Compression path

2. Which of the following *cannot* be identified after performing a forward and backward pass?
   A. Dummy activities
   B. Slack time
   C. Critical path activities
   D. How much overtime is planned

3. Which of the following is *not* a commonly used technique for schedule compression?
   A. Resource reduction
   B. Reducing scope
   C. Fast-tracking activities
   D. Use of overtime

4. A network-based schedule has four paths, namely 7, 8, 9, and 10 weeks. If the 10-week path is compressed to 8 weeks, then:
   A. We now have two critical paths.
   B. The 9-week path is now the critical path.

      C. Only the 7-week path has slack.

      D. Not enough information is provided to make a determination.

5. The major disadvantage of using bar charts to manage a project is that bar charts:

      A. Do not show dependencies between activities

      B. Are ineffective for projects under one year in length

      C. Are ineffective for projects under $1 million in size

      D. Do not identify start and end dates of a schedule

6. The first step in the development of a schedule is a:

      A. Listing of the activities

      B. Determination of dependencies

      C. Calculation of effort

      D. Calculation of durations

7. Reducing the peaks and valleys in manpower assignments in order to obtain a relatively smooth manpower curve is called:

      A. Manpower allocation

      B. Manpower leveling

      C. Resource allocation

      D. Resource commitment planning

8. Activities with no time duration are called:

      A. Reserve activities

      B. Dummy activities

      C. Zero slack activities

      D. Supervision activities

9. Optimistic, pessimistic, and most likely activity times are associated with:

      A. PERT

      B. GERT

      C. PDM

      D. ADM

10. The most common "constraint" or relationship in a precedence network is:

      A. Start-to-start

      B. Start-to-finish

      C. Finish-to-start

      D. Finish-to-finish

11. A network-based technique that allows for branching and looping is:

      A. PERT

      B. GERT

      C. PDM

      D. ADM

12. If an activity on the critical path takes longer than anticipated, then:

      A. Activities not on the critical path have additional slack.

      B. Activities not on the critical path have less slack.

      C. Additional critical path activities will appear.

      D. None of the above.

13. Which of the following is not one of the three types of dependencies?

      A. Mandatory

      B. Discretionary

C. Internal
D. External

14. You have an activity where the early start is week 6, the early finish is week 10, the latest start is week 14, and the latest finish is week 18. The slack in this activity is:
   A. 4 weeks
   B. 6 weeks
   C. 8 weeks
   D. 18 weeks

ANSWERS

1. B
2. D
3. A
4. D
5. A
6. A
7. B
8. B
9. A
10. C
11. B
12. A
13. C
14. C

PROBLEMS

12–1 Should a PERT/CPM network become a means of understanding reports and schedules, or should it be vice versa?

12–2 Before PERT diagrams are prepared, should the person performing the work have a clear definition of the requirements and objectives, both prime and supporting? Is it an absolute necessity?

12–3 Who prepares the PERT diagrams? Who is responsible for their integration?

12–4 Should PERT networks follow the work breakdown structure?

12–5 How can a PERT network be used to increase functional ability to relate to the total program?

12–6 What problems are associated with applying PERT to small programs?

12–7 Should PERT network design be dependent on the number of elements in the work breakdown structure?

12–8 Can bar charts and PERT diagrams be used to smooth out departmental manpower requirements?

12–9 Should key milestones be established at points where trade-offs are most likely to occur?

12–10 Would you agree or disagree that the cost of accelerating a project rises exponentially, especially as the project nears completion?

12–11 What are the major difficulties with PERT, and how can they be overcome?

12–12 Is PERT/cost designed to identify critical schedule slippages and cost overruns early enough that corrective action can be taken?

12–13 Draw the network and identify the critical path. Also calculate the earliest–latest starting and finishing times for each activity:

Activity	Preceding Activity	Time (Weeks)
A	—	7
B	—	8
C	—	6
D	A	6
E	B	6
F	B	8
G	C	4
H	D, E	7
I	F, G, H	3

12–14 Draw the network and identify the critical path. Also calculate the earliest–latest starting and finishing times for each activity:

Activity	Preceding Activity	Time (Weeks)
A	—	4
B	—	6
C	A, B	7
D	B	8
E	B	5
F	C	5
G	D	7
H	D, E	8
I	F, G, H	4

12–15 Consider the following network for a small maintenance project (all times are in days; network proceeds from node 1 to node 7):

    a. Draw an arrow diagram representing the project.
    b. What is the critical path and associated time?

Job Activity	Network Initial Node	Final Node	Optimistic Time	Pessimistic Time	Most Likely
A	1	2	1	3	2
B	1	4	4	6	5
C	1	3	4	6	5
D	2	6	2	4	3
E	2	4	1	3	2
F	3	4	2	4	3
G	3	5	7	15	9
H	4	6	4	6	5
I	4	7	6	14	10
J	4	5	1	3	2
K	5	7	2	4	3
L	6	7	6	14	10

    c. What is the total slack time in the network?

    d. What is the expected time for 68, 95, and 99 percent completion limits?

    e. If activity G had an estimated time of fifteen days, what impact would this have on your answer to part b?

12–16 Identify the critical path for the following network for a small MIS project (all times are in days; network proceeds from node 1 to node 10):

Job Activity	Network Initial Node	Final Node	Estimated Time
A	1	2	2
B	1	3	3
C	1	4	3
D	2	5	3
E	2	9	3
F	3	5	1
G	3	6	2
H	3	7	3
I	4	7	5
J	4	8	3
K	5	6	3
L	6	9	4
M	7	9	4
N	8	9	3
O	9	10	2

12–17 On May 1, Arnie Watson sent a memo to his boss, the director of project management, stating that the MX project would require thirteen weeks for completion according to the figure shown at the top of page 542.

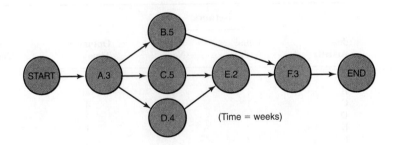

Arnie realized that the customer wanted the job completed in less time. After discussions with the functional managers, Arnie developed the table shown below:

Activity	Normal		Crash		Additional (Crash) Cost/Week
	Time	Cost	Time	Cost	
A	3	6,000	2	8,000	2,000
B	5	12,000	4	13,500	1,500
C	5	16,000	3	22,000	3,000
D	4	8,000	2	10,000	1,000
E	2	6,000	1	7,500	1,500
F	3	14,000	1	20,000	3,000
		$62,000			

a. According to the contract, there is a penalty payment of $5,000 per week for every week over six. What is the minimum amount of additional funding that Arnie should request?

b. Suppose your answer to part a gives you the same additional minimum cost for both an eight-week and a nine-week project. What factors would you consider before deciding whether to do it in eight or nine weeks?

12–18 On March 1, the project manager received three status reports indicating resource utilization to date. Shown below are the three reports as well as the PERT diagram.

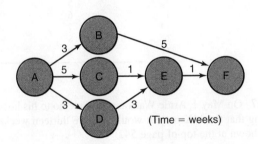

PERCENT-COMPLETION REPORT

Activity	Date Started	% Completed	Time to Complete
AB	2/1	100%	—
AC	2/1	60%	2
AD	2/1	100%	—
DE*	not started	—	3
BF	2/14	40%	3

Note: Because of priorities, resources for activity DE will not be available until 3/14. Management estimates that this activity can be crashed from 3 weeks to 2 weeks at an additional cost of $3,000

PROJECT PLANNING BUDGET: WEEKS AFTER GO-AHEAD

Activity	1	2	3	4	5	6	7	8	Total $
AB	2,000	2,000	2,000	—	—	—	—	—	6,000
AC	3,000	4,000	4,000	4,000	5,000	—	—	—	20,000
AD	2,000	3,000	2,500	—	—	—	—	—	7,500
BF	—	—	—	2,000	3,000	4,000	3,000	3,000	15,000
CE	—	—	—	—	—	2,500	—	—	2,500
DE	—	—	—	3,500	3,500	3,500	—	—	10,500
EF	—	—	—	—	—	—	3,000	—	3,000
Total	7,000	9,000	8,500	9,500	11,500	10,000	6,000	3,000	64,500

COST SUMMARY

Activity	Week Ending			Cumulative to Date		
	Budget Cost	Actual	(Over) Under	Budget Cost	Actual	(Over) Under
AB	—	—	—	6,000	6,200	(200)
AC	4,000	4,500	(500)	15,000	12,500	2,500
AD	—	2,400	(2,400)	7,500	7,400	100
BF	2,000	2,800	(800)	2,000	4,500	(2,500)
DE	3,500	—	3,500	3,500	—	3,500
Total	9,500	9,700	(200)	34,000	30,600	3,400

a. As of the end of week 4, how much time is required to complete the project (i.e., time to complete)?

b. At the end of week 4, are you over/under budget, and by how much, for the work (either partial or full) that has been completed to date? (This is *not* a cost to complete.)

c. At what point in time should the decision be made to crash activities?

d. Either construct a single table by which cost and performance data are more easily seen, or modify the above tables accordingly.

To solve this problem, you must make an assumption about the relationship between percent complete and time/cost. In the project planning budget table, assume that percent complete is *linear* with time and *nonlinear* with cost (i.e., cost must be read from table).

12–19 Can PERT charts have more depth than the WBS?

12–20 Estimating activity time is not an easy task, especially if assumptions must be made. State whether each item identified below can be accounted for in the construction of a PERT/CPM network:

    a. Consideration of weather conditions
    b. Consideration of weekend activities
    c. Unleveled manpower requirements
    d. Checking of resource allocations
    e. Variable crew size
    f. Splitting (or interrupting) of activities
    g. Assignment of unused resources
    h. Accounting for project priorities

12–21 Scheduling departmental manpower for a project is a very difficult task, even if slack time is available. Many managers would prefer to supply manpower at a constant rate rather than continually shuffle people in and out of a project.

    a. Using the information shown below, construct the PERT network, identify the critical path, and determine the slack time for each node.

Activity	Weeks	Personnel Required (Full-time)
A–B	5	3
A–C	3	3
B–D	2	4
B–E	3	5
C–E	3	5
D–F	3	5
E–F	6	3

    b. The network you have just created is a departmental PERT chart. Construct a weekly manpower plot assuming that all activities begin as early as possible. (Note: Overtime cannot be used to shorten the activity time.)
    c. The department manager wishes to assign eight people full-time for the duration of the project. However, if an employee is no longer needed on the project, he can be assigned elsewhere. Using the base of eight people, identify the standby (or idle) time and the overtime periods.
    d. Determine the standby and overtime costs, assuming that each employee is paid $300 per week and overtime is paid at time and a half. During standby time the employee draws his full salary.
    e. Repeat parts c and d and try to consider slack time in order to smooth out the manpower curve. (Hint: Some activities should begin as early as possible, while others begin as late as possible.) Identify the optimum manpower level so as to minimize the standby and overtime costs. Assume all employees must work full-time.

f. Would your answer to parts d and e change if the employees must remain for the full duration of the project, even if they are no longer required?

12–22 How does a manager decide whether the work breakdown structure should be based on a "tree" diagram or the PERT diagram?

12–23 Using Table 12–4, draw the CPM chart for the project. In this case, make all identifications on the arrows (activities) rather than the events. Show that the critical path is twenty-one weeks.

Using Table 12–5, draw the precedence chart for the project, showing interrelationships. Try to use a different color or shade for the critical path.

Calculate the *minimum* cash flow needed for the first four weeks of the project, assuming the following distribution.

Activity	Total Cost for Each Activity
A–H	16,960
I–P	5,160
Q–V	40,960
W	67,200
X	22,940

Furthermore, assume that *all* costs are linear with time, and that the activity X cost must be spent in the first two weeks. Prove that the minimum cash flow is $92,000.

TABLE 12–4. DATA FOR PROJECT CPM CHART

Activity	Preceding Activity	Normal Time (Weeks)
A	—	4
B	A	6
C	B,U,V,N	3
D	C	2
E	C	2
F	C	7
G	C	7
H	D,E	4
I	—	2
J	I,R	1
K	J	1
L	K	2
M	L	1
N	M	1
O	N	2
P	O	1
Q	—	4
R	Q	1
S	—	1
T	—	1
U	S	2
V	T	2
W*	—	*
X	—	2

*Stands for total length of project. This is management support.

TABLE 12–5. PROJECT PRECEDENCE CHART*

Weeks

Activity	1	2	3	4	5	6	7	8	9	10	11	12	13	14	15	16	17	18	19	20	21
A																					
B																					
C																					
D																					
E																					
F																					
G																					
H																					
I																					
J																					
K																					
L																					
M																					
N																					
O																					
P																					
Q																					
R																					
S																					
T																					
U																					
V																					
W																					
X																					

*Draw the appropriate bar charts into the figure, assuming that each activity starts as early as possible (identify slack). Try to show the interrelationships as in a precedence network.

12–24 For the network shown in Figure P12–24 with all times indicating weeks, answer the following questions:

    a. What is the impact on the end date of the project if activity B slips by two weeks?
    b. What is the impact on the end date of the project if activity E slips by one week?
    c. What is the impact on the end date of the project if activity D slips by two weeks?
    d. If the customer offered you a bonus for completing the project in sixteen weeks or less, which activities would you focus on first as part of compression ("crashing") analyses?

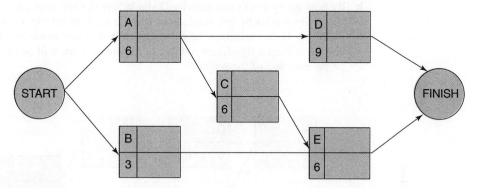

Figure P12–24

12–25 For the network shown in Figure P12–25 with all times indicating weeks, answer the following questions:

    a. What is the impact on the end date of the project if activity F slips by seven weeks?

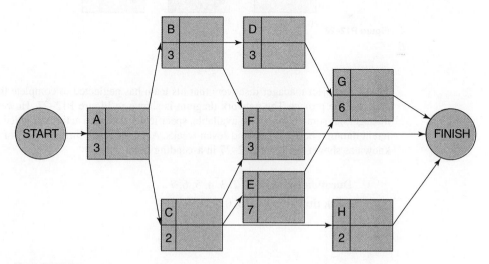

Figure P12–25

    b. What is the impact on the end date of the project if activity E slips by one week?

    c. What is the impact on activity H if activity C were to slip by two weeks?

    d. What is the impact on the end date of the project if activity B slips by two weeks?

12–26 For the network shown in Figure P12–26 with all times indicating weeks, answer the following questions:

    a. What is the impact on the end date of the project if activity I slips by three weeks?

    b. By how many weeks can activity D slip before the end date gets extended?

    c. If activity A slips by one week, how will the slack in activity G be impacted?

    d. If activity H can somehow be compressed from seven weeks to two weeks, perhaps by adding a significant number of resources, what will be the impact, if any, on the end date of the project?

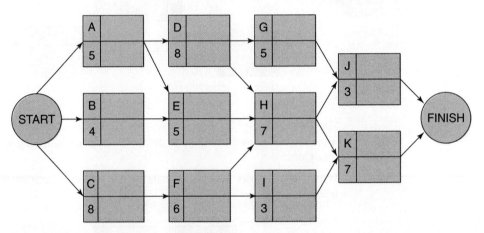

Figure P12–26

12–27 A project manager discovers that his team has neglected to complete the network diagram for the project. The network diagram is shown in Figure P12–27. However, the project manager has some information available, specifically that each activity, labeled A–G, has a different duration between one and seven weeks. Also, the slack time for each of the activities is known as shown in Figure P12–27 in ascending order.

    Duration (weeks): 1, 2, 3, 4, 5, 6, 7

    Slack time (weeks): 0, 0, 0, 2, 4, 4, 7

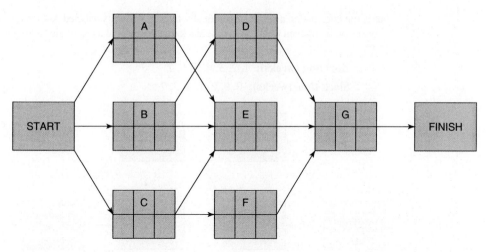

Figure P12–27

Using the clues provided below, determine the duration of each activity as well as the early start, early finish, latest start and latest finish times for each activity.

Clues

1. Activity E is on the critical path.
2. The early start (ES) time for activity F is five weeks.
3. The duration of activity B is seven weeks.
4. Activity D has four weeks of slack, but activity F has a greatest amount of slack.
5. The early finish (EF) time for activity G is seventeen weeks.
6. The latest finish (LF) time for activity E is thirteen weeks.

Activity	Duration	Early Start	Early Finish	Latest Start	Latest Finish
A	_____	_____	_____	_____	_____
B	_____	_____	_____	_____	_____
C	_____	_____	_____	_____	_____
D	_____	_____	_____	_____	_____
E	_____	_____	_____	_____	_____
F	_____	_____	_____	_____	_____
G	_____	_____	_____	_____	_____

12–28 A project manager discovers that his team has neglected to complete the network diagram for the project. The network diagram is shown in Figure P12–28. However, the project manager has some information available, specifically that each activity, labeled A–G, has a different duration between one and seven weeks. Also, the slack time for each of the activities is known as shown below.

Duration (weeks): 1, 2, 3, 4, 5, 6, 7
Slack time (weeks): 0, 0, 0, 1, 1, 3, 7

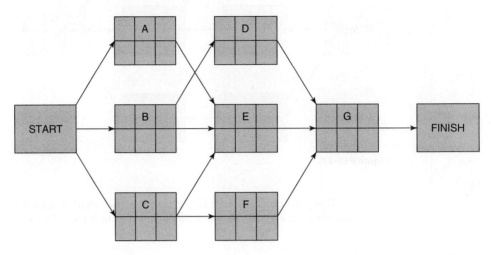

Figure P12–28

Using the clues provided below, determine the duration of each activity as well as the early start, early finish, latest start, and latest finish times for each activity.

Clues

1. Activity E is the longest duration activity and is on the critical path, which is the unlucky number 13; also, there is only one critical path.
2. The early finish (EF) time for activity F is eleven weeks.
3. The latest start (LS) time for activity D is nine weeks.
4. If activity A slips by one week, it will be on a critical path.

Activity	Duration	Early Start	Early Finish	Latest Start	Latest Finish
A					
B					
C					
D					
E					
F					
G					

12–29 A project manager discovers that his team has neglected to complete the network diagram for the project. The network diagram is shown in Figure P12–29. However, the project manager has some information available, specifically that each activity, labeled A–G, has a different duration between one and seven weeks. Also, the slack time for each of the activities is known as shown below:

Duration (weeks): 1, 2, 3, 4, 5, 6, 7

Slack time (weeks): 0, 0, 0, 3, 6, 8, 8

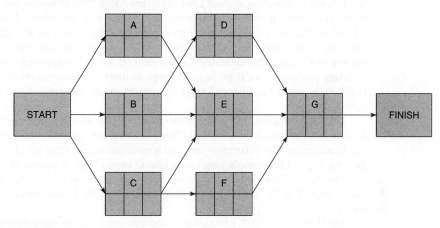

Figure P12–29

Using the clues provided below, determine the duration of each activity as well as the early start, early finish, latest start, and latest finish times for each activity.

Clues

1. There exists only one critical path, and it is the largest possible number given the possible durations shown.
2. Activity E has the smallest amount of slack that is greater than zero.
3. The early finish (EF) time for activity A is four weeks, and this does not equal the latest finish (LF) time. (Note: There is no negative slack in the network.)
4. The slack in activity C is eight weeks.
5. The duration of activity F is greater than the duration of activity C by at least two weeks.

Activity	Duration	Early Start	Early Finish	Latest Start	Latest Finish
A	_____	_____	_____	_____	_____
B	_____	_____	_____	_____	_____
C	_____	_____	_____	_____	_____
D	_____	_____	_____	_____	_____
E	_____	_____	_____	_____	_____
F	_____	_____	_____	_____	_____
G	_____	_____	_____	_____	_____

CASE STUDY

CROSBY MANUFACTURING CORPORATION

"I've called this meeting to resolve a major problem with our management cost and control system (MCCS)," remarked Wilfred Livingston, president. "We're having one hell of a time trying to meet competition with our antiquated MCCS reporting procedures. Last year we were considered nonresponsive to three large government contracts because we could not adhere to the customer's financial reporting requirements. The government has recently shown a renewed interest in Crosby Manufacturing Corporation. If we can computerize our project financial reporting procedure, we'll be in great shape to meet the competition head-on. The customer might even waive the financial reporting requirements if we show our immediate intent to convert."

Crosby Manufacturing was a $250-million-a-year electronics component manufacturing firm in 2005, at which time Wilfred "Willy" Livingston became president. His first major act was to reorganize the 700 employees into a modified matrix structure. This reorganization was the first step in Livingston's long-range plan to obtain large government contracts. The matrix provided the customer focal point policy that government agencies prefer. After three years, the matrix seemed to be working. Now they could begin the second phase, an improved MCCS policy.

On October 20, 2007, Livingston called a meeting with department managers from project management, cost accounting, MIS, data processing, and planning.

Livingston: "We have to replace our present computer with a more advanced model so as to update our MCCS reporting procedures. In order for us to grow, we'll have to develop capabilities for keeping two or even three different sets of books for our customers. Our present computer does not have this capability. We're talking about a sizable cash outlay, not necessarily to impress our customers, but to increase our business base and grow. We need weekly, or even daily, cost data so as to better control our projects."

MIS Manager: "I guess the first step in the design, development, and implementation process would be the feasibility study. I have prepared a list of the major topics which are normally included in a feasibility study of this sort" (see Exhibit 12–1).

Exhibit 12–1. Feasibility study

- Objectives of the study
- Costs
- Benefits
- Manual or computer-based solution?
- Objectives of the system
- Input requirements
- Output requirements
- Processing requirements
- Preliminary system description
- Evaluation of bids from vendors
- Financial analysis
- Conclusions

Exhibit 12–2. Typical schedule (in months)

Activity	Normal Time to Complete	Crash Time to Complete
Management go-ahead	0	0
Release of preliminary system specs	6	2
Receipt of bids on specs	2	1
Order hardware and systems software	2	1
Flowcharts completed	2	2
Applications programs completed	3	6
Receipt of hardware and systems software	3	3
Testing and debugging done	2	2
Documentation, if required	2	2
Changeover completed	22	15*

*This assumes that some of the activities can be run in parallel, instead of series.

Livingston: "What kind of costs are you considering in the feasibility study?"

MIS Manager: "The major cost items include input–output demands; processing; storage capacity; rental, purchase or lease of a system; nonrecurring expenditures; recurring expenditures; cost of supplies; facility requirements; and training requirements. We'll have to get a lot of this information from the EDP department."

EDP Manager: "You must remember that, for a short period of time, we'll end up with two computer systems in operation at the same time. This cannot be helped. However, I have prepared a typical (abbreviated) schedule of my own (see Exhibit 12–2). You'll notice from the right-hand column that I'm somewhat optimistic as to how long it should take us."

Livingston: "Have we prepared a checklist on how to evaluate a vendor?"

EDP Manager: "Besides the 'benchmark' test, I have prepared a list of topics that we must include in evaluation of any vendor (see Exhibit 12–3). We should plan to call on or visit other installations that have purchased the same equipment and see the system in action. Unfortunately, we may have to commit real early and begin developing software packages."

Exhibit 12–3. Vendor support evaluation factors

- Availability of hardware and software packages
- Hardware performance, delivery, and past track record
- Vendor proximity and service-and-support record
- Emergency backup procedure
- Availability of applications programs and their compatibility with our other systems
- Capacity for expansion
- Documentation
- Availability of consultants for systems programming and general training
- Who burdens training cost?
- Risk of obsolescence
- Ease of use

As a matter of fact, using the principle of concurrency, we should begin developing our software packages right now."

Livingston: "Because of the importance of this project, I'm going to violate our normal structure and appoint Tim Emary from our planning group as project leader. He's not as knowledgeable as you people are in regard to computers, but he does know how to lay out a schedule and get the job done. I'm sure your people will give him all the necessary support he needs. Remember, I'll be behind this project all the way. We're going to convene again one week from today, at which time I expect to see a detailed schedule with all major milestones, team meetings, design review meetings, etc., shown and identified. I'd like the project to be complete in eighteen months, if possible. If there are risks in the schedule, identify them. Any questions?"

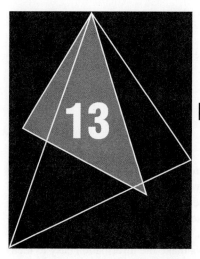

Project Graphics

Related Case Studies (from Kerzner/*Project Management Case Studies,* 3rd Edition)	Related Workbook Exercises (from Kerzner/*Project Management Workbook and PMP®/CAPM® Exam Study Guide,* 10th Edition)	PMBOK® Guide, 4th Edition, Reference Section for the PMP® Certification Exam
None	• Multiple Choice Exam	• Time Management • Communication Management

13.0 INTRODUCTION

PMBOK® Guide, 4th Ediiton
Chapter 9 Time Management
Chapter 10 Communications
  Management
6.6.2.3 Schedule Comparison
  Bar Charts

In Chapter 11, we defined the steps involved in establishing a formal program plan with detailed schedules to manage the total program. Any plan, schedule, drawing, or specification that will be read by more than one person must be expressed in a language that is understood by all recipients.

The ideal situation is to construct charts and schedules in suitable notation that can be used for both in-house control and out-of-house customer status reporting. Unfortunately, this is easier said than done. Customers and contractors are interested mainly in the three vital control parameters:

- Time
- Cost
- Performance

All schedules and charts should consider these three parameters and their relationship to corporate resources.

Information to ensure proper project evaluation is usually obtained through four methods:

- Firsthand observation
- Oral and written reports
- Review and technical interchange meetings
- Graphical displays

Firsthand observations are an excellent tool for obtaining unfiltered information, but they may not be possible on large projects. Although oral and written reports are a way of life, they often contain either too much or not enough detail, and significant information may be disguised. Review and technical interchange meetings provide face-to-face communications and can result in immediate agreement on problem definitions or solutions, such as changing a schedule. The difficulty is in the selection of attendees from the customer's and the contractor's organizations. Good graphical displays make the information easy to identify and are the prime means for tracking cost, schedule, and performance. Proper graphical displays can result in:

- Cutting project costs and reducing the time scale
- Coordinating and expediting planning
- Eliminating idle time
- Obtaining better scheduling and control of subcontractor activities
- Developing better troubleshooting procedures
- Cutting time for routine decisions, but allowing more time for decision-making

13.1 CUSTOMER REPORTING

PMBOK® Guide, 4th Edition
10.3.1.2 Performance Reporting

There are more than thirty visual methods for representing activities. The method chosen should depend on the intended audience. For example, upper-level management may be interested in costs and integration of activities, with very little detail. Summary-type charts normally suffice for this purpose. Daily practitioners, on the other hand, may require considerable detail. For customers, the presentation should include cost and performance data.

When presenting cost and performance data, figures and graphs should be easily understood and diagrams should quickly convey the intended message or objective. In many organizations, each department or division may have its own method of showing scheduling activities. Research and development organizations prefer to show the logic of activities rather than the integration of activities that would normally be representative of a manufacturing plant.

The ability to communicate is a prerequisite for successful management of a program. Program review meetings, technical interchange meetings, customer summary meetings, and in-house management control meetings all require different representative forms of current program performance status. The final form of the schedule may be bar charts, graphs, tables, bubble charts, or logic diagrams. These are described in the sections that follow.

13.2 BAR (GANTT) CHART

PMBOK® Guide, 4th Edition
6.6.2.3 Schedule Comparison
Bar Charts

The most common type of display is the bar or Gantt chart, named for Henry Gantt, who first utilized this procedure in the early 1900s. The bar chart is a means of displaying simple activities or events plotted against time or dollars. An activity represents the amount of work required to proceed from one point in time to another. Events are described as either the starting or ending point for either one or several activities.

Bar charts are most commonly used for exhibiting program progress or defining specific work required to accomplish an objective. Bar charts often include such items as listings of activities, activity duration, schedule dates, and progress-to-date. Figure 13–1 shows nine activities required to start up a production line for a new product. Each bar in the figure represents a single activity. Figure 13–1 is a typical bar chart that would be developed by the program office at program inception.

Bar charts are advantageous in that they are simple to understand and easy to change. They are the simplest and least complex means of portraying progress (or the lack of it) and can easily be expanded to identify specific elements that may be either behind or ahead of schedule.

Bar charts provide only a vague description of how the entire program or project reacts as a system, and have three major limitations. First, bar charts do not show the interdependencies of the activities, and therefore do not represent a "network" of activities. This relationship between activities is crucial for controlling program costs. Without this relationship, bar charts have little predictive value. For example, does the long-lead procurement activity in Figure 13–1 require that the contract be signed before

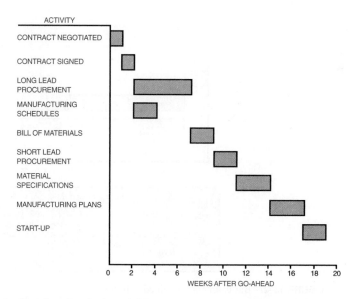

FIGURE 13–1.   Bar chart for single activities.

procurement can begin? Can the manufacturing plans be written without the material specifications activity being completed? The second major discrepancy is that the bar chart cannot show the results of either an early or a late start in activities. How will a slippage of the manufacturing schedules activity in Figure 13–1 affect the completion date of the program? Can the manufacturing schedules activity begin two weeks later than shown and still serve as an input to the bill of materials activity? What will be the result of a crash program to complete activities in sixteen weeks after go-ahead instead of the originally planned nineteen weeks? Bar charts do not reflect true project status because elements behind schedule do not mean that the program or project is behind schedule. The third limitation is that the bar chart does not show the uncertainty involved in performing the activity and, therefore, does not readily admit itself to sensitivity analysis. For instance, what is the shortest time that an activity might take? What is the longest time? What is the average or expected time to activity completion?

Even with these limitations, bar charts do, in fact, serve as useful tools for program analysis. Some of the limitations of bar charts can be overcome by combining single activities, as shown in Figure 13–2. The weakness in this method is that the numbers representing each of the activities do not indicate whether this is the beginning or the end of the activity. Therefore, the numbers should represent events rather than activities, together with proper identification. As before, no distinction is made as to whether event 2 must be completed prior to the start of event 3 or event 4. The chart also fails to define clearly the relationship between the multiple activities on a single bar. For example, must event 3 be completed prior to event 5? Often, combined activity bar charts can be converted to milestone bar charts by placing small triangles at strategic locations in the bars to indicate completion of certain milestones within each activity or grouping of activities, as shown in Figure 13–3. The exact definition of a milestone differs from company to company, but usually implies some point where major activity either begins or ends, or cost data become critical.

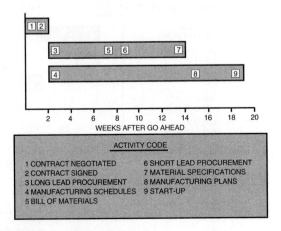

FIGURE 13–2. Bar chart for combined activities.

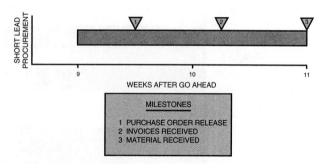

FIGURE 13–3. Bar/milestone chart.

Bar charts can be converted to partial interrelationship charts by indicating (with arrows) the order in which activities must be performed. Figure 13–4 represents the partial interrelationship of the activities in Figures 13–1 and 13–2. A full interrelationship schedule is included under the discussion of PERT networks in Chapter 12.

The most common method of presenting data to both in-house management and the customer is through the use of bar charts. Care must be taken not to make the figures overly complex so that more than one interpretation can exist. A great deal of information and color can be included in bar charts. Figure 13–5 shows a grouped bar chart for comparison of three projects performed during different years. When using different shading techniques, each area must be easily definable and no major contrast should exist between shaded areas, except for possibly the current project. When grouped bars appear on one chart, nonshaded bars should be avoided. Each bar should have some sort of shading, whether it be cross-hatched or color-coded.

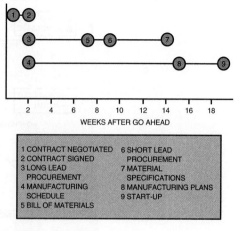

FIGURE 13–4. Partial interrelationship chart.

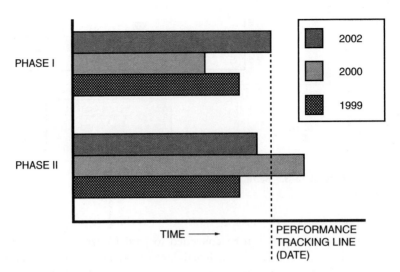

FIGURE 13–5. Grouped bar chart for performance comparison.

Contrasting shaded to nonshaded areas is normally used for comparing projected progress to actual progress, as shown in Figure 13–6. The tracking date line indicates the time when the cost data/performance data were analyzed. Project 1 is behind schedule, project 2 is ahead of schedule, and project 3 is on target. Unfortunately, the upper portion of Figure 13–6 does not indicate the costs attributed to the status of the three projects.

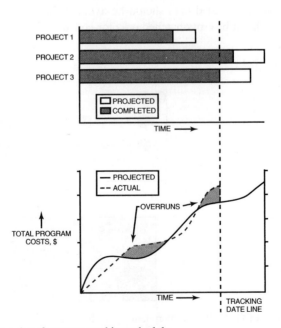

FIGURE 13–6. Cost and performance tracking schedule.

By plotting the total program costs against the same time axis (as shown in Figure 13–6), a comparison between cost and performance can be made. From the upper section of Figure 13–6 it is impossible to tell the current program cost position. From the lower section, however, it becomes evident that the program is heading for a cost overrun, possibly due to project 1. It is generally acceptable to have the same shading technique represent different situations, provided that clear separation between the shaded regions appears, as in Figure 13–6.

Another common means for comparing activities or projects is through the use of step arrangement bar charts. Figure 13–7 shows a step arrangement bar chart for a cost percentage breakdown of the five projects included within a program. Figure 13–7 can also be used for tracking, by shading certain portions of the steps that identify each project. This is not normally done, however, since this type of step arrangement tends to indicate that each step must be completed before the next step can begin.

Bar charts need not be represented horizontally. Figure 13–8 indicates the comparison between the 2000 and 2002 costs for the total program and raw materials. Three-dimensional vertical bar charts are often beautiful to behold. Figure 13–9 shows a typical three-dimensional bar chart for direct and indirect labor and material cost breakdowns.

Bar charts can be made colorful and appealing by combining them with other graphic techniques. Figure 13–10 shows a quantitative-pictorial bar chart for the distribution of total program costs. Figure 13–11 shows the same cost distribution as in Figure 13–10, but represented with the commonly used pie technique. Figure 13–12 illustrates how two quantitative bar charts can be used side by side to create a quick comparison. The right-hand side shows the labor hour percentages. Figure 13–12 works best if the scale of each axis is the same; otherwise the comparisons may appear distorted when, in fact, they are not.

The figures shown in this section do not, by any means, represent the only methods of presenting data in bar chart format. Several other methods are shown in the sections that follow.

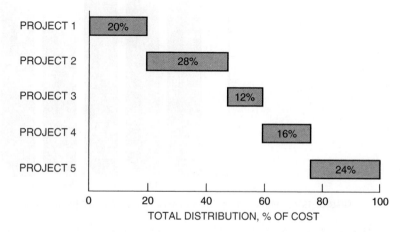

FIGURE 13–7. Step arrangement bar chart for total cost as a percentage of the five program projects.

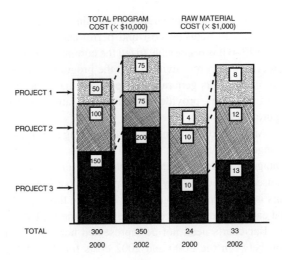

FIGURE 13–8. Cost comparison, 2000 versus 2002.

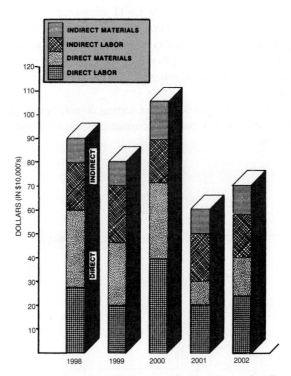

FIGURE 13–9. Direct and indirect material and labor cost breakdowns for all programs per year.

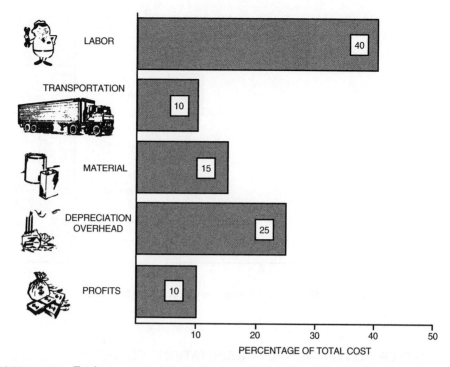

FIGURE 13–10. Total program cost distribution (quantitative-pictorial bar chart).

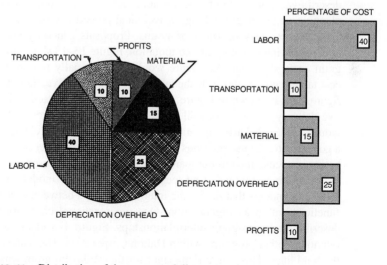

FIGURE 13–11. Distribution of the program dollar.

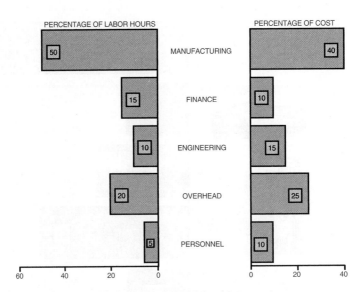

FIGURE 13–12. Divisional breakdown of costs and labor hours.

13.3 OTHER CONVENTIONAL PRESENTATION TECHNIQUES _____

Bar charts serve as a useful tool for presenting data at technical meetings. Unfortunately, programs must be won competitively or organized in-house before technical meeting presentations can be made. Competitive proposals or in-house project requests should contain descriptive figures and charts, not necessarily representing activities, but showing either planning, organizing, tracking, or technical procedures designed for the current program or used previously on other programs. Proposals generally contain figures that require either some interpolation or extrapolation. Figure 13–13 shows the breakdown of total program costs. Although this figure would also normally require interpretation, a monthly cost table accompanies it. If the table is not too extensive, then it can be included with the figure. This is shown in Figure 13–14. During proposal activities, the actual and cumulative delivery columns, as well as the dotted line in Figure 13–14, would be omitted, but would be included after updating for use in technical interchange meetings. It is normally a good practice to use previous figures and tables whenever possible because management becomes accustomed to the manner in which data are presented.

Another commonly used technique is schematic models. Organizational charts are schematic models that depict the interrelationships between individuals, organizations, or functions within an organization. One organizational chart normally cannot suffice for describing total program interrelationships. Figure 4–8 identified the Midas Program in relation to other programs within Dalton Corporation. The Midas Program is indicated by the bold lines. The program manager for the Midas Program was placed at the top of the column, even though his program may have the lowest priority. Each major unit of

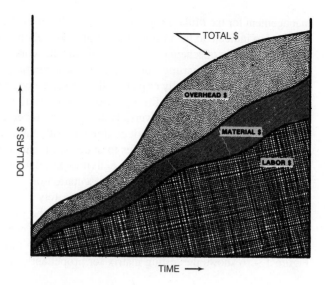

FIGURE 13–13. Total program cost breakdown.

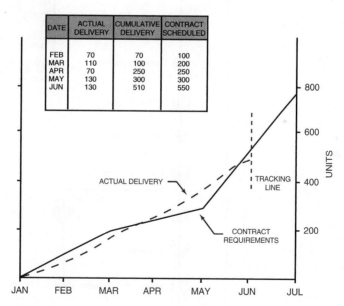

DATE	ACTUAL DELIVERY	CUMULATIVE DELIVERY	CONTRACT SCHEDULED
FEB	70	70	100
MAR	110	100	200
APR	70	250	250
MAY	130	300	300
JUN	130	510	550

FIGURE 13–14. Delivery schedule tracking (line of balance).

management for the Midas Program should be placed as close as possible to top-level management to indicate to the customer the "implied" relative importance of the program.

Another type of schematic representation is the work flowchart, synonymous with the application of flowcharting for computer programming. Flowcharts are designed to describe, either symbolically or pictorially, the sequence of events required to complete an activity. Figure 13–15 shows the logic flow for production of molding VZ-3. The symbols shown in Figure 13–15 are universally accepted by several industries.

Pictorial representation, although often a costly procedure, can add color and quality to any proposal, and they are easier to understand than a logic or bubble chart. Because customers may request tours during activities to relate to the pictorial figures, program management should avoid pictorial representation of activities that may be closed off to customer viewing, possibly due to security or safety.

Block diagrams can also be used to describe the flow of activities. Figures 4–8 and 4–9 are examples of block diagrams. Block diagrams can be used to show how information is distributed throughout an organization or how a process or activity is assembled. Figure 13–16 shows the testing matrix for propellant samples. Figures similar to this are

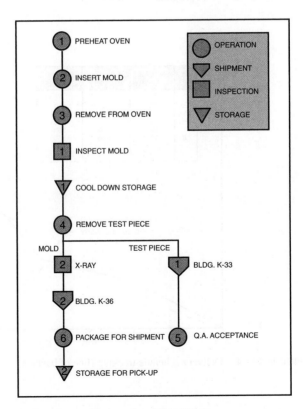

FIGURE 13–15. Logic flow for production of molding VZ-3.

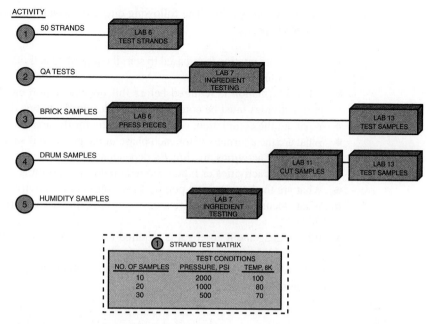

FIGURE 13–16. Propellant samples testing matrix.

developed when tours are scheduled during the production or testing phase of a program. Figure 13–16 shows the customer not only where the testing will take place, but what tests will be conducted.

Block diagrams, schematics, pictorials, and logic flows all fulfill a necessary need for describing the wide variety of activities within a company. The figures and charts are more than descriptive techniques. They can also provide management with the necessary tools for decision-making.

13.4 LOGIC DIAGRAMS/NETWORKS

Probably the most difficult figure to construct is the logic diagram. Logic diagrams are developed to illustrate the inductive and deductive reasoning necessary to achieve some objective within a given time frame. The major difficulty in developing logic diagrams is the inability to answer such key questions as: What happens if something goes wrong? Can I quantify any part of the diagram's major elements?

Logic diagrams are constructed similarly to bar charts on the supposition that nothing will go wrong and are usually accompanied by detailed questions, possibly in a checklist format, that require answering. The following questions would be representative of those asked for an R&D project:

- What documentation is released to start the described activity and possibly the elements within each activity?
- What information is required before this documentation can be released? (What prior activities must be completed, work designed, studies finalized, etc?)
- What are the completion, or success, criteria for the activity?
- What are the alternatives for each phase of the program if success is not achieved?
- What other activities are directly dependent on the result of this activity?
- What other activities or inputs are required to perform this activity?
- What are the key decision points, if any, during the activity?
- What documentation signifies completion of the activity (i.e., report, drawing, etc.)?
- What management approval is required for final documentation?

These types of questions are applicable to many other forms of data presentation, not only logic diagrams.

13.5 STUDYING TIPS FOR THE PMI® PROJECT MANAGEMENT CERTIFICATION EXAM

This section is applicable as a review of the principles to support the knowledge areas and domain groups in the PMBOK® Guide. This chapter addresses:

- Time Management
- Communication Management
- Executing
- Controlling

Understanding the following principles is beneficial if the reader is using this text to study for the PMP® Certification Exam:

- How to identify the different ways that information can be displayed for reporting purposes
- Different types of graphical reporting techniques and their advantages and disadvantages

The following multiple-choice questions will be helpful in reviewing the principles of this chapter:

1. Which of the following is a valid way of obtaining proper project performance information?
   A. First-hand observations
   B. Oral and written reports

    C. Review and technical interchange meetings
    D. All of the above

2. Proper graphical display of information can result in:
    A. Reducing paperwork costs
    B. Reducing reporting costs
    C. Reducing time for routine decisions
    D. All of the above

ANSWERS

1. D
2. D

PROBLEMS

13–1 For each type of schedule defined in this chapter answer the following questions:

    a. Who prepares the schedule?
    b. Who updates the schedule?
    c. Who should present the data to the customers?

13–2 Should the customers have the right to dictate to the contractor how the schedule should be prepared and presented? What if this request contradicts company policies and procedures?

13–3 Should a different set of schedules and charts be maintained for out-of-house as well as in-house reporting? Should separate schedules be made for each level of management? Is there a more effective way to ease these types of problems?

Pricing and Estimating

Related Case Studies (from Kerzner/*Project Management Case Studies,* 3rd Edition)	Related Workbook Exercises (from Kerzner/*Project Management Workbook and PMP®/CAPM® Exam Study Guide,* 10th Edition)	PMBOK® Guide, 4th Edition, Reference Section for the PMP® Certification Exam
• Capital Industries • Polyproducts Incorporated • Small Project Cost Estimating at Percy Company • Cory Electric • Camden Construction Corporation • Payton Corporation	• The Automobile Problem • Life-Cycle Costing • Multiple Choice Exam	• Integration Management • Scope Management • Cost Management

14.0 INTRODUCTION

PMBOK® Guide, 4th Edition
6.3.2.4 Bottom-Up Estimating
6.4.2 Activity Duration Estimating

With the complexities involved, it is not surprising that many business managers consider pricing an art. Having information on customer cost budgets and competitive pricing would certainly help. However, the reality is that whatever information is available to one bidder is generally available to the others.

571

A disciplined approach helps in developing all the input for a rational pricing recommendation. A side benefit of using a disciplined management process is that it leads to the documentation of the many factors and assumptions involved at a later time. These can be compared and analyzed, contributing to the learning experiences that make up the managerial skills needed for effective business decisions.

Estimates are *not* blind luck. They are well-thought-out decisions based on either the best available information, some type of cost estimating relationship, or some type of cost model. Cost estimating relationships (CERs) are generally the output of cost models. Typical CERs might be:

- Mathematical equations based on regression analysis
- Cost–quantity relationships such as learning curves
- Cost–cost relationships
- Cost–noncost relationships based on physical characteristics, technical parameters, or performance characteristics

14.1 GLOBAL PRICING STRATEGIES

Specific pricing strategies must be developed for each individual situation. Frequently, however, one of two situations prevails when one is pursuing project acquisitions competitively. First, the new business opportunity may be a one-of-a-kind program with little or no follow-on potential, a situation classified as type I acquisition. Second, the new business opportunity may be an entry point to a larger follow-on or repeat business, or may represent a planned penetration into a new market. This acquisition is classified as type II.

Clearly, in each case, we have specific but different business objectives. The objective for type I acquisition is to win the program and execute it profitably and satisfactorily according to contractual agreements. The type II objective is often to win the program and perform well, thereby gaining a foothold in a new market segment or a new customer community in place of making a profit. Accordingly, each acquisition type has its own, unique pricing strategy, as summarized in Table 14–1.

Comparing the two pricing strategies for the two global situations (as shown in Table 14–1) reveals a great deal of similarity for the first five points. The fundamental difference is that for a profitable new business acquisition the bid price is determined according to actual cost, whereas in a "must-win" situation the price is determined by the market forces. It should be emphasized that one of the most crucial inputs in the pricing decision is the cost estimate of the proposed baseline. The design of this baseline to the minimum requirements should be started early, in accordance with well-defined ground rules, cost models, and established cost targets. Too often the baseline design is performed in parallel with the proposal development. At the proposal stage it is too late to review and fine-tune the baseline for minimum cost. Also, such a late start does not allow much of an option for a final bid decision. Even if the price appears outside the competitive range, it makes little sense to terminate the proposal development. As all the resources have been sent anyway, one might just as well submit a bid in spite of the remote chance of winning.

Clearly, effective pricing begins a long time before proposal development. It starts with preliminary customer requirements, well-understood subtasks, and a top-down estimate with should-cost targets. This allows the functional organization to design a baseline

TABLE 14–1. TWO GLOBAL PRICING STRATEGIES

Type I Acquisition: One-of-a-Kind Program with Little or No Follow-On Business	Type II Acquisition: New Program with Potential for Large Follow-On Business or Representing a Desired Penetration into New Markets
1. Develop cost model and estimating guidelines; design proposed project/program baseline for minimum cost, to minimum customer requirements.	1. Design proposed project/program baseline compliant with customer requirements, with innovative features but minimum risks.
2. Estimate cost realistically for minimum requirements.	2. Estimate cost realistically.
3. Scrub the baseline. Squeeze out unnecessary costs.	3. Scrub baseline. Squeeze out unnecessary costs.
4. Determine realistic minimum cost. Obtain commitment from performing organizations.	4. Determine realistic minimum cost. Obtain commitment from performing organizations.
5. Adjust cost estimate for risks.	5. Determine "should-cost" including risk adjustments.
6. Add desired margins. Determine the price.	6. Compare your final cost estimate to customer budget and the "most likely" winning price.
7. Compare price to customer budget and competitive cost information.	7. Determine the gross profit margin necessary for your winning proposal. This margin could be negative!
8. Bid only if price is within competitive range.	8. Decide whether the gross margin is acceptable according to the must-win desire.
	9. Depending on the strength of your desire to win, bid the "most likely" winning price or lower.
	10. If the bid price is below cost, it is often necessary to provide a detailed explanation to the customer of where the additional funding is coming from. The source could be company profits or sharing of related activities. In any case, a clear resource picture should be given to the customer to ensure cost credibility.

to meet the customer requirements and cost targets, and gives management the time to review and redirect the design before the proposal is submitted. Furthermore, it gives management an early opportunity to assess the chances of winning during the acquisition cycle, at a point when additional resources can be allocated or the acquisition effort can be terminated before too many resources are committed to a hopeless effort.

The final pricing review session should be an integration and review of information already well known in its basic context. The process and management tools outlined here should help to provide the framework and discipline for deriving pricing decisions in an orderly and effective way.

14.2 TYPES OF ESTIMATES

PMBOK® Guide, 4th Edition
6.4.2 Activity Duration Estimating
7.1.2 Cost Estimating Tools and Techniques

Any company or corporation that wants to remain profitable must continuously improve its estimating and pricing methodologies. While it is true that some companies have been successful without good cost estimating and pricing, very few remain successful without them.

Good estimating requires that information be collected prior to the initiation of the estimating process. Typical information includes:

- Recent experience in similar work
- Professional and reference material
- Market and industry surveys
- Knowledge of the operations and processes
- Estimating software and databases if available
- Interviews with subject matter experts

Projects can range from a feasibility study, through modification of existing facilities, to complete design, procurement, and construction of a large complex. Whatever the project may be, whether large or small, the estimate and type of information desired may differ radically.

The first type of estimate is an *order-of-magnitude* analysis, which is made without any detailed engineering data. The order-of-magnitude analysis may have an accuracy of ± 35 percent within the scope of the project. This type of estimate may use past experience (not necessarily similar), scale factors, parametric curves, or capacity estimates (i.e., $/# of product or $/kW electricity).

Order-of-magnitude estimates are top-down estimates usually applied to level 1 of the WBS, and in some industries, use of parametric estimates are included. A parametric estimate is based upon statistical data. For example, assume that you live in a Chicago suburb and wish to build the home of your dreams. You contact a construction contractor who informs you that the parametric or statistical cost for a home in this suburb is $120 per square foot. In Los Angeles, the cost may be $4150 per square foot.

Next, there is the *approximate estimate* (or top-down estimate), which is also made without detailed engineering data, and may be accurate to ± 15 percent. This type of estimate is prorated from previous projects that are similar in scope and capacity, and may be titled as estimating by analogy, parametric curves, rule of thumb, and indexed cost of similar activities adjusted for capacity and technology. In such a case, the estimator may say that this activity is 50 percent more difficult than a previous (i.e., reference) activity and requires 50 percent more time, man-hours, dollars, materials, and so on.

The *definitive estimate*, or grassroots buildup estimate, is prepared from well-defined engineering data including (as a minimum) vendor quotes, fairly complete plans, specifications, unit prices, and estimate to complete. The definitive estimate, also referred to as detailed estimating, has an accuracy of ± 5 percent.

Another method for estimating is the use of *learning curves*. Learning curves are graphical representations of repetitive functions in which continuous operations will lead to a reduction in time, resources, and money. The theory behind learning curves is usually applied to manufacturing operations.

Each company may have a unique approach to estimating. However, for normal project management practices, Table 14–2 would suffice as a starting point.

Many companies try to standardize their estimating procedures by developing an *estimating manual*. The estimating manual is then used to price out the effort, perhaps as much as 90 percent. Estimating manuals usually give better estimates than industrial engineering standards because they include groups of tasks and take into consideration such items as downtime, cleanup time, lunch, and breaks. Table 14–3 shows the table of contents for a construction estimating manual.

TABLE 14–2. STANDARD PROJECT ESTIMATING

Estimating Method	Generic Type	WBS Relationship	Accuracy	Time to Prepare
Parametric	ROM*	Top down	−25% to +75%	Days
Analogy	Budget	Top down	−10% to +25%	Weeks
Engineering (grass roots)	Definitive	Bottom up	−5% to +10%	Months

*ROM = Rough order of magnitude.

TABLE 14–3. ESTIMATING MANUAL TABLE OF CONTENTS

Introduction
  Purpose and types of estimates
Major Estimating Tools
  Cataloged equipment costs
  Automated investment data system
  Automated estimate system
  Computerized methods and procedures
Classes of Estimates
  Definitive estimate
  Capital cost estimate
  Appropriation estimate
  Feasibility estimate
  Order of magnitude
  Charts—estimate specifications quantity and pricing guidelines
Data Required
  Chart—comparing data required for preparation of classes of estimates
Presentation Specifications
  Estimate procedure—general
  Estimate procedure for definitive estimate
  Estimate procedure for capital cost estimate
  Estimate procedure for appropriation estimate
  Estimate procedure for feasibility estimate

Estimating manuals, as the name implies, provide estimates. The question, of course, is "How good are the estimates?" Most estimating manuals provide accuracy limitations by defining the type of estimates (shown in Table 14–3). Using Table 14–3, we can create Tables 14–4, 14–5, and 14–6, which illustrate the use of the estimating manual.

Not all companies can use estimating manuals. Estimating manuals work best for repetitive tasks or similar tasks that can use a previous estimate adjusted by a degree-of-difficulty factor. Activities such as R&D do not lend themselves to the use of estimating manuals other than for benchmark, repetitive laboratory tests. Proposal managers must carefully consider whether the estimating manual is a viable approach. The literature

TABLE 14–4. CLASSES OF ESTIMATES

Class	Types	Accuracy
I	Definitive	±5%
II	Capital cost	±10–15%
III	Appropriation (with some capital cost)	±15–20%
IV	Appropriation	±20–25%
V	Feasibility	±25–35%
VI	Order of magnitude	> ±35%

TABLE 14–5. CHECKLIST FOR WORK NORMALLY REQUIRED FOR THE VARIOUS CLASSES (I–VI) OF ESTIMATES

Item	I	II	III	IV	V	VI
1. Inquiry	X	X	X	X	X	X
2. Legibility	X	X	X			
3. Copies	X	X				
4. Schedule	X	X	X	X		
5. Vendor inquiries	X	X	X			
6. Subcontract packages	X	X				
7. Listing	X	X	X	X	X	
8. Site visit	X	X	X	X		
9. Estimate bulks	X	X	X	X	X	
10. Labor rates	X	X	X	X	X	
11. Equipment and subcontract selection	X	X	X	X	X	
12. Taxes, insurance, and royalties	X	X	X	X	X	
13. Home office costs	X	X	X	X	X	
14. Construction indirects	X	X	X	X	X	
15. Basis of estimate	X	X	X	X	X	X
16. Equipment list	X					
17. Summary sheet	X	X	X	X	X	
18. Management review	X	X	X	X	X	X
19. Final cost	X	X	X	X	X	X
20. Management approval	X	X	X	X	X	X
21. Computer estimate	X	X	X	X		

abounds with examples of companies that have spent millions trying to develop estimating manuals for situations that just do not lend themselves to the approach.

During competitive bidding, it is important that the type of estimate be consistent with the customer's requirements. For in-house projects, the type of estimate can vary over the life cycle of a project:

- *Conceptual stage:* Venture guidance or feasibility studies for the evaluation of future work. This estimating is often based on minimum-scope information.
- *Planning stage:* Estimating for authorization of partial or full funds. These estimates are based on preliminary design and scope.
- *Main stage:* Estimating for detailed work.
- *Termination stage:* Reestimation for major scope changes or variances beyond the authorization range.

14.3 PRICING PROCESS

This activity schedules the development of the work breakdown structure and provides management with two of the three operational tools necessary for the control of a system or project. The development of these two tools is normally the responsibility of the program office with input from the functional units.

The integration of the functional unit into the project environment or system occurs through the pricing-out of the work breakdown structure. The total program costs obtained by pricing out the activities over the scheduled period of performance provide management with the third tool necessary to successfully manage the project. During the

TABLE 14–6. DATA REQUIRED FOR PREPARATION OF ESTIMATES

	Classes of Estimates					
	I	II	III	IV	V	VI
General						
Product	X	X	X	X	X	X
Process description	X	X	X	X	X	X
Capacity	X	X	X	X	X	X
Location—general					X	X
Location—specific	X	X	X	X		
Basic design criteria	X	X	X	X		
General design specifications	X	X	X	X		
Process						
Process block flow diagram						X
Process flow diagram (with equipment size and material)				X	X	
Mechanical P&Is	X	X	X			
Equipment list	X	X	X	X	X	
Catalyst/chemical specifications	X	X	X	X	X	
Site						
Soil conditions	X	X	X	X		
Site clearance	X	X	X			
Geological and meteorological data	X	X	X			
Roads, paving, and landscaping	X	X	X			
Property protection	X	X	X			
Accessibility to site	X	X	X			
Shipping and delivery conditions	X	X	X			
Major cost is factored					X	X
Major Equipment						
Preliminary sizes and materials			X	X	X	
Finalized sizes, materials, and appurtenances	X	X				
Bulk Material Quantities						
Finalized design quantity take-off		X				
Preliminary design quantity take-off	X	X	X	X		
Engineering						
Plot plan and elevations	X	X	X	X		
Routing diagrams	X	X	X			
Piping line index	X	X				
Electrical single line	X	X	X	X		
Fire protection	X	X	X			
Sewer systems	X	X	X			
Pro-services—detailed estimate	X	X				
Pro-services—ratioed estimate			X	X	X	
Catalyst/chemicals quantities	X	X	X	X	X	
Construction						
Labor wage, F/B, travel rates	X	X	X	X	X	
Labor productivity and area practices	X	X				
Detailed construction execution plan	X	X				
Field indirects—detailed estimate	X	X				
Field indirects—ratioed estimate			X	X	X	
Schedule						
Overall timing of execution				X	X	
Detailed schedule of execution	X	X	X			
Estimating preparation schedule	X	X	X			

pricing activities, the functional units have the option of consulting program management about possible changes in the activity schedules and work breakdown structure.

The work breakdown structure and activity schedules are priced out through the lowest pricing units of the company. It is the responsibility of these pricing units, whether they be sections, departments, or divisions, to provide accurate and meaningful cost data (based on historical standards, if possible). All information is priced out at the lowest level of performance required, which, from the assumption of Chapter 11, will be the task level. Costing information is rolled up to the project level and then one step further to the total program level.

Under ideal conditions, the work required (i.e., man-hours) to complete a given task can be based on historical standards. Unfortunately, for many industries, projects and programs are so diversified that realistic comparison between previous activities may not be possible. The costing information obtained from each pricing unit, whether or not it is based on historical standards, should be regarded only as an estimate. How can a company predict the salary structure three years from now? What will be the cost of raw materials two years from now? Will the business base (and therefore overhead rates) change over the duration of the program? The final response to these questions shows that costing data are explicitly related to an environment that cannot be predicted with any high degree of certainty. The systems approach to management, however, provides for a more rapid response to the environment than less structured approaches permit.

Once the cost data are assembled, they must be analyzed for their potential impact on the company resources of people, money, equipment, and facilities. It is only through a total program cost analysis that resource allocations can be analyzed. The resource allocation analysis is performed at all levels of management, ranging from the section supervisor to the vice president and general manager. For most programs, the chief executive must approve final cost data and the allocation of resources.

Proper analysis of the total program costs can provide management (both program and corporate) with a strategic planning model for integration of the current program with other programs in order to obtain a total corporate strategy. Meaningful planning and pricing models include analyses for monthly manloading schedules per department, monthly costs per department, monthly and yearly total program costs, monthly material expenditures, and total program cash-flow and man-hour requirements per month.

Previously we identified several of the problems that occur at the nodes where the horizontal hierarchy of program management interfaces with the vertical hierarchy of functional management. The pricing-out of the work breakdown structure provides the basis for effective and open communication between functional and program management where both parties have one common goal. This is shown in Figure 14–1. After the pricing effort is completed, and the program is initiated, the work breakdown structure still forms the basis of a communications tool by documenting the performance agreed on in the pricing effort, as well as establishing the criteria against which performance costs will be measured.

14.4 ORGANIZATIONAL INPUT REQUIREMENTS

Once the work breakdown structure and activity schedules are established, the program manager calls a meeting for all organizations that will submit pricing information. It is

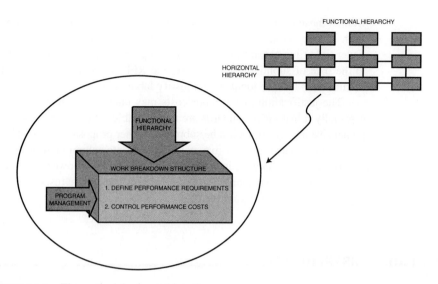

FIGURE 14–1. The vertical–horizontal interface.

imperative that all pricing or labor-costing representatives be present for the first meeting. During this "kickoff" meeting, the work breakdown structure is described in depth so that each pricing unit manager will know exactly what his responsibilities are during the program. The kickoff meeting also resolves the struggle for power among functional managers whose responsibilities may be similar. An example of this would be quality control activities. During the research and development phase of a program, research personnel may be permitted to perform their own quality control efforts, whereas during production activities the quality control department or division would have overall responsibility. Unfortunately, one meeting is not always sufficient to clarify all problems. Follow-up or status meetings are held, normally with only those parties concerned with the problems that have arisen. Some companies prefer to have all members attend the status meetings so that all personnel will be familiar with the total effort and the associated problems. The advantage of not having all program-related personnel attend is that time is of the essence when pricing out activities. Many functional divisions carry this policy one step further by having a divisional representative together with possibly key department managers or section supervisors as the only attendees at the kickoff meeting. The divisional representative then assumes all responsibility for assuring that all costing data are submitted on time. This arrangement may be beneficial in that the program office need contact only one individual in the division to learn of the activity status, but it may become a bottleneck if the representative fails to maintain proper communication between the functional units and the program office, or if the individual simply is unfamiliar with the pricing requirements of the work breakdown structure.

During proposal activities, time may be extremely important. There are many situations in which a request for proposal (RFP) requires that all responders submit their bids by a specific date. Under a proposal environment, the activities of the program office, as well as those of the functional units, are under a schedule set forth by the proposal manager. The proposal manager's schedule has very little, if any, flexibility and is normally under tight time constraints so that the proposal may be typed, edited, and published prior to the date of submittal. In this case, the RFP will indirectly define how much time the pricing units have to identify and justify labor costs.

The justification of the labor costs may take longer than the original cost estimates, especially if historical standards are not available. Many proposals often require that comprehensive labor justification be submitted. Other proposals, especially those that request an almost immediate response, may permit vendors to submit labor justification at a later date.

In the final analysis, it is the responsibility of the lowest pricing unit supervisors to maintain adequate standards, so that an almost immediate response can be given to a pricing request from a program office.

14.5 LABOR DISTRIBUTIONS

The functional units supply their input to the program office in the form of man-hours, as shown in Figure 14–2. The input may be accompanied by labor justification, if required. The man-hours are submitted for each task, assuming that the task is the lowest pricing element, and are time-phased per month. The man-hours per month per task are converted to dollars after multiplication by the appropriate labor rates. The labor rates are generally known with certainty over a twelve-month period, but from then on are only estimates. How can a company predict salary structures five years hence? If the company underestimates the salary structure, increased costs and decreased profits will occur. If the salary structure is overestimated, the company may not be competitive; if the project is government funded, then the salary structure becomes an item under contract negotiations.

The development of the labor rates to be used in the projection is based on historical costs in business base hours and dollars for the most recent month or quarter. Average hourly rates are determined for each labor unit by direct effort within the operations at the department level. The rates are only averages, and include both the highest-paid employees and lowest-paid employees, together with the department manager and the clerical support.[1] These base rates are then escalated as a percentage factor based on past experience, budget as approved by management, and the local outlook and similar industries. If the company has a predominant aerospace or defense industry business base, then these salaries are negotiated with local government agencies prior to submittal for proposals.

The labor hours submitted by the functional units are quite often overestimated for fear that management will "massage" and reduce the labor hours while attempting to maintain the same scope of effort. Many times management is forced to reduce man-hours either because of insufficient funding or just to remain competitive in the environment.

---

1. Problems can occur if the salaries of the people assigned to the program exceed the department averages. Methods to alleviate this problem are discussed later. Also, in many companies department managers are included in the overhead rate structure, not in direct labor, and therefore their salaries are not included as part of the department average.

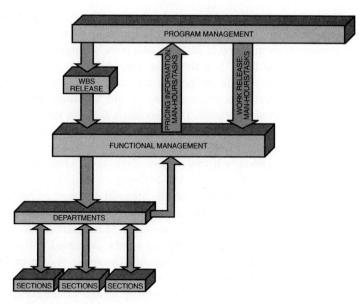

FIGURE 14–2. Functional pricing flow.

The reduction of man-hours often causes heated discussions between the functional and program managers. Program managers tend to think in terms of the best interests of the program, whereas functional managers lean toward maintaining their present staff.

The most common solution to this conflict rests with the program manager. If the program manager selects members for the program team who are knowledgeable in man-hour standards for each of the departments, then an atmosphere of trust can develop between the program office and the functional department so that man-hours can be reduced in a manner that represents the best interests of the company. This is one of the reasons why program team members are often promoted from within the functional ranks.

The man-hours submitted by the functional units provide the basis for total program cost analysis and program cost control. To illustrate this process, consider Example 14–1 below.

Example 14–1. On May 15, Apex Manufacturing decided to enter into competitive bidding for the modification and updating of an assembly line program. A work breakdown structure was developed as shown below:

PROGRAM (01-00-00): Assembly Line Modification
    PROJECT 1 (01-01-00): Initial Planning
        Task 1 (01-01-01): Engineering Control
        Task 2 (01-01-02): Engineering Development
    PROJECT 2 (01-02-00): Assembly
        Task 1 (01-02-01): Modification
        Task 2 (01-02-02): Testing

On June 1, each pricing unit was given the work breakdown structure together with the schedule shown in Figure 14–3. According to the schedule developed by the proposal manager for this project, all labor data must be submitted to the program office for review no later than June 15. It should be noted here that, in many companies, labor hours are submitted directly to the pricing department for submittal into the base case computer run. In this case, the program office would "massage" the labor hours only after the base case figures are available. This procedure assumes that sufficient time exists for analysis and modification of the base case. If the program office has sufficient personnel capable of critiquing the labor input prior to submittal to the base case, then valuable time can be saved, especially if two or three days are required to obtain computer output for the base case.

During proposal activities, the proposal manager, pricing manager, and program manager must all work together, although the program manager has the final say. The primary responsibility of the proposal manager is to integrate the proposal activities into the operational system so that the proposal will be submitted to the requestor on time. A typical schedule developed by the proposal manager is shown in Figure 14–4. The schedule includes all activities necessary to "get the proposal out of the house," with the first major step being the submittal of man-hours by the pricing organizations. Figure 14–4 also indicates the tracking of proposal costs. The proposal activity schedule is usually accompanied by a time schedule with a detailed estimates checklist if the complexity of the proposal warrants one. The checklist generally provides detailed explanations for the proposal activity schedule.

After the planning and pricing charts are approved by program team members and program managers, they are entered into an electronic data processing (EDP) system as shown in Figure 14–5. The computer then prices the hours on the planning charts using the applicable department rates for preparation of the direct budget time plan and estimate-at-completion reports. The direct budget time plan reports, once established, remain

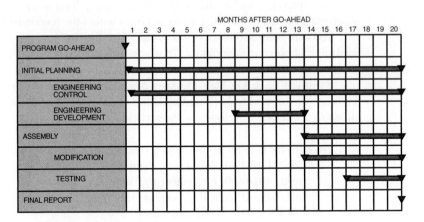

FIGURE 14–3. Activity schedule for assembly line updating.

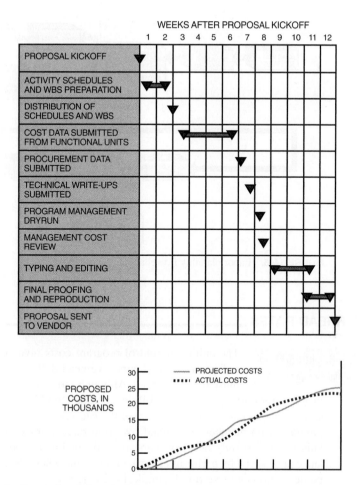

FIGURE 14–4. Proposal activity schedule.

the same for the life of the contract except for customer-directed or approved changes or when contractor management determines that a reduction in budget is advisable. However, if a budget is reduced by management, it cannot be increased without customer approval.

The time plan is normally a monthly mechanical printout of all planned effort by work package and organizational element over the life of the contract, and serves as the data bank for preparing the status completion reports.

Initially, the estimate-at-completion report is identical to the budget report, but it changes throughout the life of a program to reflect degradation or improvement in performance or any other events that will change the program cost or schedule.

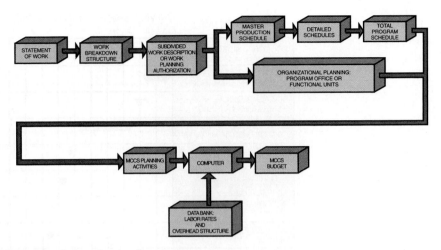

FIGURE 14–5. Labor planning flowchart.

14.6 OVERHEAD RATES

PMBOK® Guide, 4th Edition
7.1.1 Cost Estimating Inputs
The ability to control program costs involves more than tracking labor dollars and labor hours; overhead dollars, one of the biggest headaches, must also be tracked. Although most programs have an assistant program manager for cost whose responsibilities include monthly overhead rate analysis, the program manager can drastically increase the success of his program by insisting that each program team member understand overhead rates. For example, if overhead rates apply only to the first forty hours of work, then, depending on the overhead rate, program dollars can be saved by performing work on overtime where the increased salary is at a lower burden. This can be seen in Example 14–2 below.

Example 14–2. Assume that ApexManufacturing must write an interim report for task 1 of project 1 during regular shift or on overtime. The project will require 500 man-hours at $15.00 per hour. The overhead burden is 75 percent on regular shift but only 5 percent on overtime. Overtime, however, is paid at a rate of time and a half. Assuming that the report can be written on either time, which is cost-effective—regular time or overtime?

- On regular time the total cost is:

    (500 hours) $\times$ ($15.00/hour) $\times$ (100% + 75% burden) = $13,125.00

- On overtime, the total cost is:

    (500 hours) $\times$ ($15.00/hour $\times$ 1.5 overtime) $\times$ (100% + 5% burden)
    = $11,812.50

Therefore, the company can save $1,312.50 by performing the work on overtime. Scheduling overtime can produce increased profits if the overtime overhead rate burden is much less than the regular time burden. This difference can be very large in manufacturing divisions, where overhead rates between 300 and 450 percent are common.

Regardless of whether one analyzes a project or a system, all costs must have associated overhead rates. Unfortunately, many program managers and systems managers consider overhead rates as a magic number pulled out of the air. The preparation and assignment of overheads to each of the functional divisions is a science. Although the *total dollar pool* for overhead rates is relatively constant, management retains the option of deciding how to distribute the overhead among the functional divisions. A company that supports its R&D staff through competitive bidding projects may wish to keep the R&D overhead rate as low as possible. Care must be taken, however, that other divisions do not absorb additional costs so that the company no longer remains competitive on those manufactured products that may be its bread and butter.

The development of the overhead rates is a function of three separate elements: direct labor rates, direct business base projections, and projection of overhead expenses. Direct labor rates have already been discussed. The direct business base projection involves the determination of the anticipated direct labor hours and dollars along with the necessary direct materials and other direct costs required to perform and complete the program efforts included in the business base. Those items utilized in the business base projection include all contracted programs as well as the proposed or anticipated efforts. The foundation for determination of the business base required for each program can be one or more of the following:

- Actual costs to date and estimates to completion
- Proposal data
- Marketing intelligence
- Management goals
- Past performance and trends

The projection of the overhead expenses is made by an analysis of each of the elements that constitute the overhead expense. A partial listing of those items is shown in Table 14–7. Projection of expenses within the individual elements is then made based on one or more of the following:

- Historical direct/indirect labor ratios
- Regression and correlation analysis
- Manpower requirements and turnover rates
- Changes in public laws
- Anticipated changes in company benefits
- Fixed costs in relation to capital asset requirements
- Changes in business base
- Bid and proposal (B&P) tri-service agreements
- Internal research and development (IR&D) tri-service agreements

For many industries, such as aerospace and defense, the federal government funds a large percentage of the B&P and IR&D activities. This federal funding is a necessity since

TABLE 14–7. ELEMENTS OF OVERHEAD RATES

Building maintenance	New business directors
Building rent	Office supplies
Cafeteria	Payroll taxes
Clerical	Personnel recruitment
Clubs/associations	Postage
Consulting services	Professional meetings
Corporate auditing expenses	Reproduction facilities
Corporate salaries	Retirement plans
Depreciation of equipment	Sick leave
Executive salaries	Supplies/hand tools
Fringe benefits	Supervision
General ledger expenses	Telephone/telegraph facilities
Group insurance	Transportation
Holiday	Utilities
Moving/storage expenses	Vacation

many companies could not otherwise be competitive within the industry. The federal government employs this technique to stimulate research and competition. Therefore, B&P and IR&D are included in the above list.

The prime factor in the control of overhead costs is the annual budget. This budget, which is the result of goals and objectives established by the chief executive officer, is reviewed and approved at all levels of management. It is established at department level, and the department manager has direct responsibility for identifying and controlling costs against the approved plan.

The departmental budgets are summarized, in detail, for higher levels of management. This summarization permits management, at these higher organizational levels, to be aware of the authorized indirect budget in their area of responsibility.

Reports are published monthly indicating current month and year-to-date budget, actuals, and variances. These reports are published for each level of management, and an analysis is made by the budget department through coordination and review with management. Each directorate's total organization is then reviewed with the budget analyst who is assigned the overhead cost responsibility. A joint meeting is held with the directors and the vice president and general manager, at which time overhead performance is reviewed.

14.7 MATERIALS/SUPPORT COSTS

PMBOK® Guide, 4th Edition
7.1.1 Cost Estimating Inputs

The salary structure, overhead structure, and labor hours fulfill three of four major pricing input requirements. The fourth major input is the cost for materials and support. Six subtopics are included under materials/support: materials, purchased parts, subcontracts, freight, travel, and other. Freight and travel can be handled in one of two ways, both normally dependent on the size of the program. For small-dollar-volume programs, estimates are made for travel and freight. For large-dollar-volume programs, travel is normally expressed as between 3 and 5 percent of the

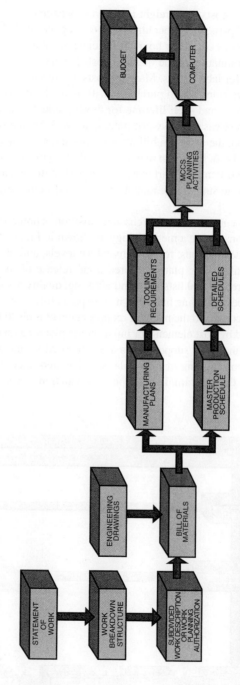

FIGURE 14–6. Material planning flowchart.

direct labor costs, and freight is likewise between 3 and 5 percent of all costs for material, purchased parts, and subcontracts. The category labeled "other support costs" may include such topics as computer hours or specialconsultants.

Determination of the material costs is very time-consuming, more so than cost determination for labor hours. Material costs are submitted via a bill of materials that includes all vendors from whom purchases will be made, projected costs throughout the program, scrap factors, and shelf lifetime for those products that may be perishable.

Upon release of the work statement, work breakdown structure, and subdivided work description, the end-item bill of materials and manufacturing plans are prepared as shown in Figure 14–6. End-item materials are those items identified as an integral part of the production end-item. Support materials consist of those materials required by engineering and operations to support the manufacture of end-items, and are identified on the manufacturing plan.

A procurement plan/purchase requisition is prepared as soon as possible after contract negotiations (using a methodology as shown in Figure 14–7). This plan is used to monitor material acquisitions, forecast inventory levels, and identify material price variances.

Manufacturing plans prepared upon release of the subdivided work descriptions are used to prepare tool lists for manufacturing, quality assurance, and engineering. From these plans a special tooling breakdown is prepared by tool engineering, which defines those tools to be procured and the material requirements of tools to be fabricated in-house. These items are priced by cost element for input on the planning charts.

The materials/support costs are submitted by month for each month of the program. If long-lead funding of materials is anticipated, then they should be assigned to the first month of the program. In addition, an escalation factor for costs of materials/support items

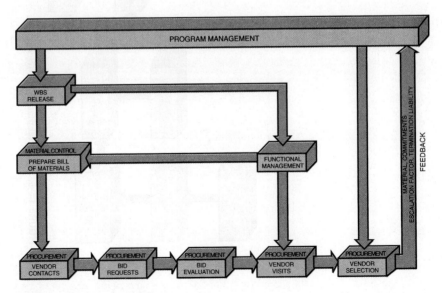

FIGURE 14–7. Procurement activity.

must be applied to all materials/support costs. Some vendors may provide fixed prices over time periods in excess of a twelve-month period. As an example, vendor Z may quote a firm-fixed price of $130.50 per unit for 650 units to be delivered over the next eighteen months if the order is placed within sixty days. There are additional factors that influence the cost of materials.

14.8 PRICING OUT THE WORK

Using logical pricing techniques will help in obtaining detailed estimates. The following thirteen steps provide a logical sequence to help a company control its limited resources. These steps may vary from company to company.

Step 1: Provide a complete definition of the work requirements.
Step 2: Establish a logic network with checkpoints.
Step 3: Develop the work breakdown structure.
Step 4: Price out the work breakdown structure.
Step 5: Review WBS costs with each functional manager.
Step 6: Decide on the basic course of action.
Step 7: Establish reasonable costs for each WBS element.
Step 8: Review the base case costs with upper-level management.
Step 9: Negotiate with functional managers for qualified personnel.
Step 10: Develop the linear responsibility chart.
Step 11: Develop the final detailed and PERT/CPM schedules.
Step 12: Establish pricing cost summary reports.
Step 13: Document the result in a program plan.

Although the pricing of a project is an iterative process, the project manager must still develop cost summary reports at each iteration point so that key project decisions can be made during the planning. Detailed pricing summaries are needed at least twice: in preparation for the pricing review meeting with management and at pricing termination. At all other times it is possible that "simple cosmetic surgery" can be performed on previous cost summaries, such as perturbations in escalation factors and procurement cost of raw materials. The list below shows the typical pricing reports:

- *A detailed cost breakdown for each WBS element.* If the work is priced out at the task level, then there should be a cost summary sheet for each task, as well as rollup sheets for each project and the total program.
- *A total program manpower curve for each department.* These manpower curves show how each department has contracted with the project office to supply functional resources. If the departmental manpower curves contain several "peaks and valleys," then the project manager may have to alter some of his schedules to obtain some degree of manpower smoothing. Functional managers always prefer manpower-smoothed resource allocations.

- *A monthly equivalent manpower cost summary.* This table normally shows the fully burdened cost for the average departmental employee carried out over the entire period of project performance. If project costs have to be reduced, the project manager performs a parametric study between this table and the manpower curve tables.
- *A yearly cost distribution table.* This table is broken down by WBS element and shows the yearly (or quarterly) costs that will be required. This table, in essence, is a project cash-flow summary per activity.
- *A functional cost and hour summary.* This table provides top management with an overall description of how many hours and dollars will be spent by each major functional unit, such as a division. Top management would use this as part of the forward planning process to make sure that there are sufficient resources available for all projects. This also includes indirect hours and dollars.
- *A monthly labor hour and dollar expenditure forecast.* This table can be combined with the yearly cost distribution, except that it is broken down by month, not activity or department. In addition, this table normally includes manpower termination liability information for premature cancellation of the project by outside customers.
- *A raw material and expenditure forecast.* This shows the cash flow for raw materials based on vendor lead times, payment schedules, commitments, and termination liability.
- *Total program termination liability per month.* This table shows the customer the monthly costs for the entire program. This is the customer's cash flow, not the contractor's. The difference is that each monthly cost contains the termination liability for man-hours and dollars, on labor and raw materials. This table is actually the monthly costs attributed to premature project termination.

These tables are used by project managers as the basis for project cost control and by upper-level executives for selecting, approving, and prioritizing projects.

14.9 SMOOTHING OUT DEPARTMENT MAN-HOURS

The dotted curve in Figure 14–8 indicates projected manpower requirements for a given department as a result of a typical program manloading schedule. Department managers, however, attempt to smooth out the manpower curve as shown by the solid line in Figure 14–8. Smoothing out the manpower requirements benefits department managers by eliminating fractional man-hours per day. The program manager must understand that if departments are permitted to eliminate peaks, valleys, and small-step functions in manpower planning, small project and task man-hour (and cost) variances can occur, but should not, in general, affect the total program cost significantly.

Two important questions to ask are whether the department has sufficient personnel available to fulfill manpower requirements and what is the rate at which the functional departments can staff the program? For example, project engineering requires approximately twenty-three people during January 2002. The functional manager, however, may

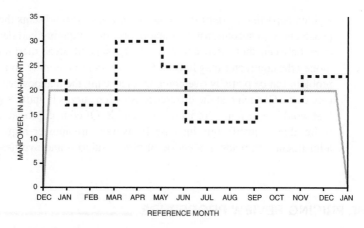

FIGURE 14–8. Typical manpower loading.

have only fifteen people available for immediate reassignment, with the remainder to be either transferred from other programs or hired from outside the company. The same situation occurs during activity termination. Will project engineering still require twenty-three people in August 2002, or can some of these people begin being phased to other programs, say, as early as June 2002? This question, specifically addressed to support and administrative tasks/projects, must be answered prior to contract negotiations. Figure 14–9 indicates the types of problems that can occur. Curve A shows the manpower requirements for

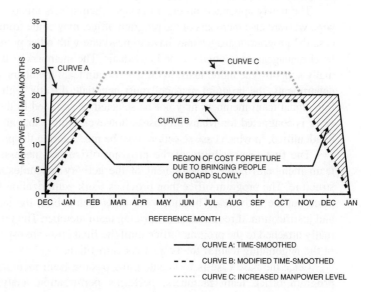

FIGURE 14–9. Linearly increased manpower loading.

a given department after time-smoothing. Curve B represents the modification to the time-phase curve to account for reasonable program manning and demanning rates. The difference between these two curves (i.e., the shaded area) therefore reflects the amount of money the contractor may have to forfeit owing to manning and demanning activities. This problem can be partially overcome by increasing the manpower levels after time-smoothing (see curve C) such that the difference between curves B and C equals the amount of money that would be forfeited from curves A and B. Of course, program management would have to be able to justify this increase in average manpower requirements, especially if the adjustments are made in a period of higher salaries and overhead rates.

14.10 THE PRICING REVIEW PROCEDURE

The ability to project, analyze, and control problem costs requires coordination of pricing information and cooperation between the functional units and upper-level management. A typical company policy for cost analysis and review is shown in Figure 14–10. Corporate management may be required to initiate or authorize activities, if corporate/company resources are or may be strained by the program, if capital expenditures are required for new facilities or equipment, or simply if corporate approval is required for all projects in excess of a certain dollar amount.

Upper-level management, upon approval by the chief executive officer of the company, approves and authorizes the initiation of the project or program. The actual performance activities, however, do not begin until the director of program management selects a program manager and authorizes either the bid and proposal budget (if the program is competitive) or project planning funds.

The newly appointed program manager then selects this program's team. Team members, who are also members of the program office, may come from other programs, in which case the program manager may have to negotiate with other program managers and upper-level management to obtain these individuals. The members of the program office are normally support-type individuals. In order to obtain team members representing the functional departments, the program manager must negotiate directly with the functional managers. Functional team members may not be selected or assigned to the program until the actual work is contracted for. Many proposals, however, require that all functional team members be identified, in which case selection must be made during the proposal stage of a program.

The first responsibility of the program office (not necessarily including functional team members) is the development of the activity schedules and the work breakdown structure. The program office then provides work authorization for the functional units to price out the activities. The functional units then submit the labor hours, material costs, and justification, if required, to the pricing team member. The pricing team member is normally attached to the program office until the final costs are established, and becomes part of the negotiating team if the project is competitive.

Once the base case is formulated, the pricing team member, together with the other program office team members, performs perturbation analyses. These analyses are designed as systems approaches to problem-solving where alternatives are developed in order to respond to management's questions during the final review.

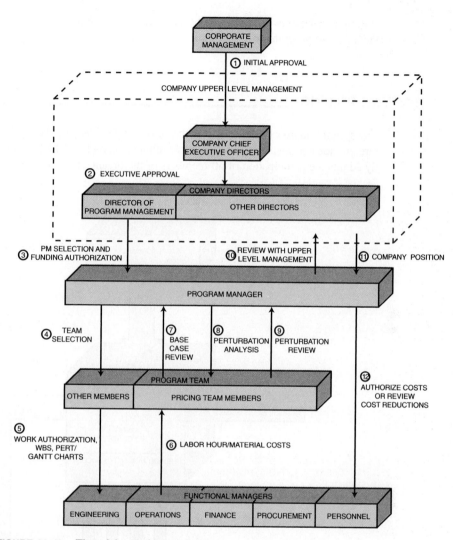

FIGURE 14–10. The pricing review procedure.

The base case, with the perturbation analysis costs, is then reviewed with upper-level management in order to formulate a company position for the program and to take a hard look at the allocation of resources required for the program. The company position may be to cut costs, authorize work, or submit a bid. Corporate approval may be required if the company's chief executive officer has a ceiling on the amount he can authorize.

If labor costs must be cut, the program manager must negotiate with the functional managers as to the size and method for the cost reductions. Otherwise, this step would simply entail authorization for the functional managers to begin the activities.

Figure 14–10 represents the system approach to determining total program costs. This procedure normally creates a synergistic environment, provides open channels of communication between all levels of management, and ensures agreement among all individuals as to program costs.

14.11 SYSTEMS PRICING

The systems approach to pricing out the activity schedules and the work breakdown structure provide a means for obtaining unity within the company. The flow of information readily admits the participation of all members of the organization in the program, even if on a

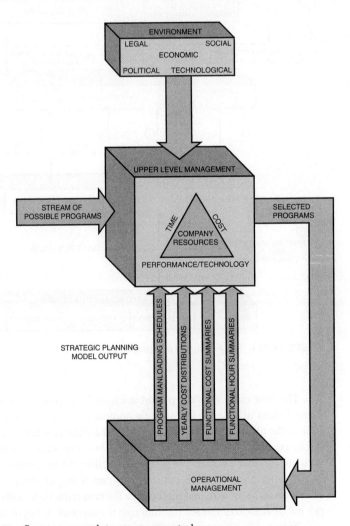

FIGURE 14–11. System approach to resource control.

part-time basis. Functional managers obtain a better understanding of how their labor fits into the total program and how their activities interface with those of other departments. For the first time, functional managers can accurately foresee how their activity can lead to corporate profits.

The project pricing model (sometimes called a strategic project planning model) acts as a management information system, forming the basis for the systems approach to resource control, as shown in Figure 14–11. The summary sheets from the computer output of the strategic pricing model help management select programs that will best utilize resources. The strategic pricing model also provides management with an invaluable tool for performing perturbation analysis on the base case costs and an opportunity for design and evaluation of contingency plans, if necessary.

14.12 DEVELOPING THE SUPPORTING/BACKUP COSTS

PMBOK® Guide, 4th Edition
7.1.2.6 Reserve Analysis

Not all cost proposals require backup support, but for those that do, the backup support should be developed along with the pricing. The itemized prices should be compatible with the supporting data. Government pricing requirements are a special case.

Most supporting data come from external (subcontractor or outside vendor) quotes. Internal data must be based on historical data, and these historical data must be updated continually as each new project is completed. The supporting data should be traceable by itemized charge numbers.

Customers may wish to audit the cost proposal. In this case, the starting point might be the supporting data. It is not uncommon on sole-source proposals to have the supporting data audited before the final cost proposal is submitted to the customer.

Not all cost proposals require supporting data; the determining factor is usually the type of contract. On a fixed-price effort, the customer may not have the right to audit your books. However, for a cost-reimbursable package, your costs are an open book, and the customer usually compares your exact costs to those of the backup support.

Most companies usually have a choice of more than one estimate to be used for backup support. In deciding which estimate to use, consideration must be given to the possibility of follow-on work:

- If your actual costs grossly exceed your backup support estimates, you may lose credibility for follow-on work.
- If your actual costs are less than the backup costs, you must use the new actual costs on follow-on efforts.

The moral here is that backup support costs provide future credibility. If you have well-documented, "livable" cost estimates, then you may wish to include them in the cost proposal even if they are not required.

Since both direct and indirect costs may be negotiated separately as part of a contract, supporting data, such as those in Tables 14–8 through 14–11 and Figure 14–12, may be necessary to justify any costs that may differ from company (or customer-approved) standards.

TABLE 14–8. OPERATIONS SKILLS MATRIX

Functional Areas of Expertise \ Technical Staff	Able, J.	Baker, P.	Cook, D.	Dirk, L.	Easley, P.	Franklin, W.	Green, C.	Henry, L.	Imhoff, R.	Jules, C.	Klein, W.	Ledger, D.	Mayer, Q.	Newton, A.	Oliver, G.	Pratt, L.
Administrative management		a		a	a	a	a	a			a	a		a	a	
Control and communications	b		b	b	b		b	b		b	b	b		b	b	b
Environmental impact assessment	c	c	c						c		c		c			
Facilities management		d					d				d		d			
Financial management	e					e			e	e	e				e	e
Human resources mangement	f							f				f				
Industrial engineering	g				g					g						
Intelligence and security								h				h		h		
Inventory control	i						i								i	i
Logistics			j		j			j				j				
OSHA	k									k			k			
Project management	l			l		l					l				l	
Quality control		m	m			m	m	m	m							
R&D			n	n							n		n			n
Wage and salary administration		o			o				o	o		o		o	o	

596

TABLE 14–9. CONTRACTOR'S MANPOWER AVAILABILITY

| | Number of Personnel | | | |
| | Total Current Staff | | Available for This Project and Other New Work 1/02 Permanent + Agency | Anticipated Growth by 1/02 Permanent + Agency |
	Permanent Employees	Agency Personnel		
Process engineers	93	—	70	4
Project managers/engineers	79	—	51	4
Cost estimating	42	—	21	2
Cost control	73	—	20	2
Scheduling/scheduling control	14	—	8	1
Procurement/purchasing	42	—	20	1
Inspection	40	—	20	2
Expediting	33	—	18	1
Home office construction management	9	—	6	0
Piping	90	13	67	6
Electrical	31	—	14	2
Instrumentation	19	—	3	1
Vessels/exchangers	24	—	19	1
Civil/structural	30	—	23	2
Other	13	—	8	0

TABLE 14–10. STAFF TURNOVER DATA

| | For Twelve-Month Period 1/1/01 to 1/1/02 | |
Hired	Number Terminated	Number
Process engineers	5	2
Project managers/engineers	1	1
Cost estimating	1	2
Cost control	12	16
Scheduling/scheduling control	2	5
Procurement/purchasing	13	7
Inspection	18	6
Expediting	4	5
Home office construction management	0	0
Design and drafting—total	37	29
Engineering specialists—total	26	45
Total	119	118

TABLE 14–11. STAFF EXPERIENCE PROFILE

	Number of Years' Employment with Contractor				
	0–1	1–2	2–3	3–5	5 or more
Process engineers	2	4	15	11	18
Project managers/engineers	1	2	5	11	8
Cost estimating	0	4	1	5	7
Cost control	5	9	4	7	12
Scheduling and scheduling control	2	2	1	3	6
Procurement/purchasing	4	12	13	2	8
Inspection	1	2	6	14	8
Expediting	6	9	4	2	3
Piping	9	6	46	31	22
Electrical	17	6	18	12	17
Instrumentation	8	8	12	13	12
Mechanical	2	5	13	27	19
Civil/structural	4	8	19	23	16
Environmental control	0	1	1	3	7
Engineering specialists	3	3	3	16	21
Total	64	81	161	180	184

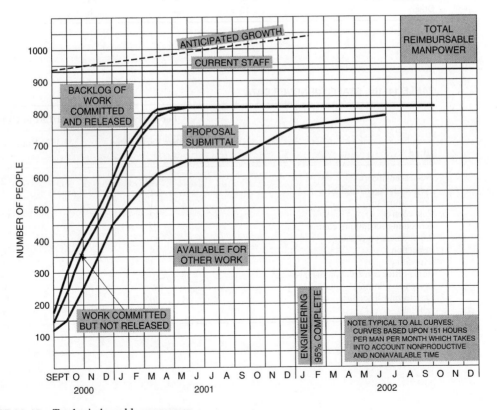

FIGURE 14–12. Total reimbursable manpower.

14.13 THE LOW-BIDDER DILEMMA

PMBOK® Guide, 4th Edition
12.3.1.3 Select Contract

There is little argument about the importance of the price tag to the proposal. The question is, what price will win the job? The decision process that leads to the final price of your proposal is highly complex with many uncertainties. Yet proposal managers, driven by the desire to win the job, may think that a very low-priced proposal will help. But winning is only the beginning. Companies have short- and long-range objectives on profit, market penetration, new product development, and so on. These objectives may be incompatible with or irrelevant to a low-price strategy. For example:

- A suspiciously low price, particularly on cost-plus type proposals, might be perceived by the customer as unrealistic, thus affecting the bidder's cost credibility or even the technical ability to perform.
- The bid price may be unnecessarily low, relative to the competition and customer budget, thus eroding profits.
- The price may be irrelevant to the bid objective, such as entering a new market. Therefore, the contractor has to sell the proposal in a credible way, e.g., using cost sharing.
- Low pricing without market information is meaningless. The price level is always relative to (1) the competitive prices, (2) the customer budget, and (3) the bidder's cost estimate.
- The bid proposal and its price may cover only part of the total program. The ability to win phase II or follow-on business depends on phase I performance and phase II price.
- The financial objectives of the customer may be more complex than just finding the lowest bidder. They may include cost objectives for total system life-cycle cost (LCC), for design to unit production cost (DTUPC), or for specific logistic support items. Presenting sound approaches for attaining these system cost–performance parameters and targets may be just as important as, if not more important than, a low bid for the system's development.

Further, it is refreshing to note that in spite of customer pressures toward low cost and fixed price, the lowest bidder is certainly not an automatic winner. Both commercial and governmental customers are increasingly concerned about cost realism and the ability to perform under contract. A compliant, sound, technical and management proposal, based on past experience with realistic, well-documented cost figures, is often chosen over the lowest bidder, who may project a risky image regarding technical performance, cost, or schedule.

14.14 SPECIAL PROBLEMS

There are always special problems that, if overlooked, can have a severe impact on the pricing effort. As an example, pricing must include an understanding of cost control—specifically, how costs are billed back to the project. There are three possible situations:

- *Work is priced out at the department average, and all work performed is charged to the project at the department average salary, regardless of who accomplished the work.* This technique is obviously the easiest, but encourages project managers to fight for the highest salary resources, since only average wages are billed to the project.

- *Work is priced out at the department average, but all work performed is billed back to the project at the actual salary of those employees who perform the work.* This method can create a severe headache for the project manager if he tries to use only the best employees on his project. If these employees are earning substantially more money than the department average, then a cost overrun will occur unless the employees can perform the work in less time. Some companies are forced to use this method by government agencies and have estimating problems when the project that has to be priced out is of a short duration where only the higher-salaried employees can be used. In such a situation it is common to "inflate" the direct labor hours to compensate for the added costs.
- *The work is priced out at the actual salary of those employees who will perform the work, and the cost is billed back the same way.* This method is the ideal situation as long as the people can be identified during the pricing effort.

Some companies use a combination of all three methods. In this case, the project office is priced out using the third method (because these people are identified early), whereas the functional employees are priced out using the first or second method.

14.15 ESTIMATING PITFALLS

PMBOK® Guide, 4th Edition
7.1.1 Cost Estimating Inputs

Several pitfalls can impede the pricing function. Probably the most serious pitfall, and the one that is usually beyond the control of the project manager, is the "buy-in" decision, which is based on the assumption that there will be "bail-out" changes or follow-on contracts later. These changes and/or contracts may be for spare parts, maintenance, maintenance manuals, equipment surveillance, optional equipment, optional services, and scrap factors. Other types of estimating pitfalls include:

- Misinterpretation of the statement of work
- Omissions or improperly defined scope
- Poorly defined or overly optimistic schedule
- Inaccurate work breakdown structure
- Applying improper skill levels to tasks
- Failure to account for risks
- Failure to understand or account for cost escalation and inflation
- Failure to use the correct estimating technique
- Failure to use forward pricing rates for overhead, general and administrative, and indirect costs

Unfortunately, many of these pitfalls do not become evident until detected by the cost control system, well into the project.

14.16 ESTIMATING HIGH-RISK PROJECTS

PMBOK® Guide, 4th Edition
6.1.2.2 Rolling Wave Planning
Chapter 11 Risk Management

Whether a project is high-risk or low-risk depends on the validity of the historical estimate. Construction companies have well-defined historical standards, which lowers their risk, whereas many R&D and MIS projects are high risk. Typical accuracies for each level of the WBS are shown in Table 14–12.

A common technique used to estimate high-risk projects is the "rolling wave" or "moving window" approach. This is shown in Figure 14–13 for a high-risk R&D project. The project lasts for twelve months. The R&D effort to be accomplished for the first six months is well defined and can be estimated to level 5 of the WBS. However, the effort for the latter six months is based on the results of the first six months and can be estimated at level 2 only, thus incurring a high risk. Now consider part B of Figure 14–13, which shows a six-month moving window. At the end of the first month, in order to maintain a six-month moving window (at level 5 of the WBS), the estimate for month seven must be improved from a level-2 to a level-5 estimate. Likewise, in parts C and D of Figure 14–13, we see the effects of completing the second and third months.

There are two key points to be considered in utilizing this technique. First, the length of the moving window can vary from project to project, and usually increases in length as you approach downstream life-cycle phases. Second, this technique works best when upper-level management understands how the technique works. All too often senior management hears only one budget and schedule number during project approval and might not realize that at least half of the project might be time/cost accurate to only 50–60 percent. Simply stated, when using this technique, the word "rough" is not synonymous with the word "detailed."

Methodologies can be developed for assessing risk. Figures 14–14, 14–15, and Table 14–13 show such methodologies.

TABLE 14–12. LOW- VERSUS HIGH-RISK ACCURACIES

WBS		Accuracy	
Level	Description	Low-Risk Projects	High-Risk Projects
1	Program	±35	±75–100
2	Project	20	50–60
3	Task	10	20–30
4	Subtask	5	10–15
5	Work package	2	5–10

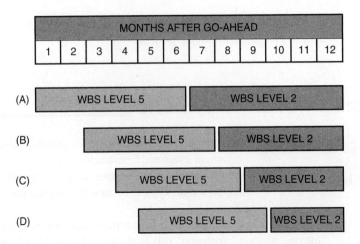

FIGURE 14–13. The moving window/rolling wave concept.

14.17 PROJECT RISKS

PMBOK® Guide, 4th Edition
11.2 Risk Identification

Project plans are "living documents" and are therefore subject to change. Changes are needed in order to prevent or rectify unfortunate situations. These unfortunate situations can be called project risks.

Risk refers to those dangerous activities or factors that, if they occur, will *increase* the probability that the project's goals of time, cost, and performance will not be met. Many risks can be anticipated and controlled. Furthermore, risk management must be an integral part of project management throughout the entire life cycle of the project.

Some common risks include:

- Poorly defined requirements
- Lack of qualified resources
- Lack of management support
- Poor estimating
- Inexperienced project manager

Risk identification is an art. It requires the project manager to probe, penetrate, and analyze all data. Tools that can be used by the project manager include:

- Decision support systems
- Expected value measures
- Trend analysis/projections
- Independent reviews and audits

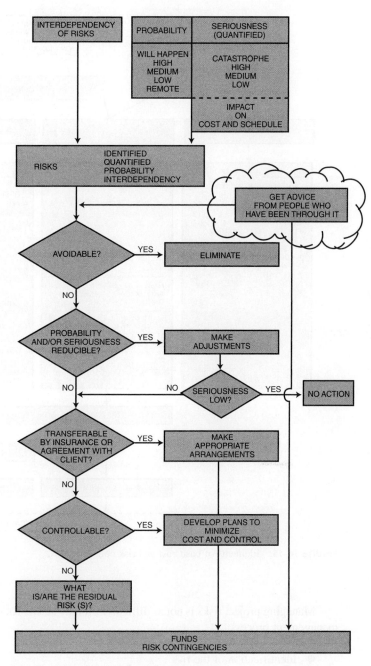

FIGURE 14–14. Decision elements for risk contingencies.

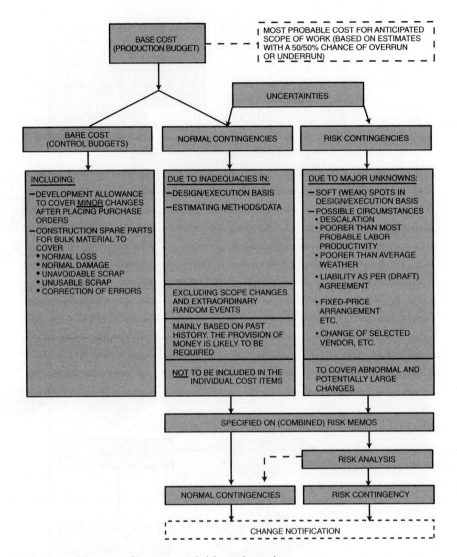

FIGURE 14–15. Elements of base cost and risk contingencies.

Managing project risks is not as difficult as it may seem. There are six steps in the risk management process:

- Identification of the risk
- Quantifying the risk
- Prioritizing the risk
- Developing a strategy for managing the risk

TABLE 14–13. STANDARD FORM FOR PROJECT RISK ANALYSIS AND RISK CONTINGENCIES

PROJECT RISK ANALYSIS & RISK CONTINGENCY   REF: PROCEDURE 0110E

Proposal/Order No. _____
Div./Dept. _____
Date _____
Issue No. _____

RISK CONTINGENCY = Σ EXPECTED VALUES

Sequence	Item	Value	Risk: Yes/No	Description of Risk of Maximum Possible Change of Item Value in %	Maximum Risk — Amount	Probability	Possible Outcome — Amount	Interdependency of Risks	Seriousness Cat./High/Med./Low	Make Adjustments	Transfer to: Insurance	Agreement	Subcontractor or Vendor	Exclusion from Scope	Inclusion in Estimate	Development Allowance	Construction Spares	Plan to Control	Accept as Residual Risk	Normal Contingency	RISK — Expected Value

605

- Project sponsor/executive review
- Taking action

Figures 14–14 and 14–15 and Table 14–13 identify the process of risk evaluation on capital projects. In all three exhibits, it is easily seen that the attempt is to quantify the risks, possibly by developing a contingency fund.

14.18 THE DISASTER OF APPLYING THE 10 PERCENT SOLUTION TO PROJECT ESTIMATES

Economic crunches can and do create chaos in all organizations. For the project manager, the worst situation is when senior management arbitrarily employs "the 10 percent solution," which is a budgetary reduction of 10 percent for each and every project, especially those that have already begun. The 10 percent solution is used to "create" funds for additional activities for which budgets are nonexistent. The 10 percent solution very rarely succeeds. For the most part, the result is simply havoc, resulting in schedule slippages, a degradation of quality and performance, and eventual budgetary increases rather than the expected decreases.

Most projects are initiated through an executive committee, governing committee, or screening committee. The two main functions of these committees are to select the projects to be undertaken and to prioritize the efforts. Budgetary considerations may also be included, as they pertain to project selection. The real budgets, however, are established from the middle-management levels and sent upstairs for approvals.

Although the role of executive committee is often ill-defined with regard to budgeting, the real problem is that the committee does not realize the impact of adopting the 10 percent solution. If the project budget is an honest one, then a reduction in budget *must* be accompanied by a trade-off in either time or performance. It is often said that 90 percent of the budget generates the first 10 percent of the desired service or quality levels, and that the remaining 10 percent of the budget will produce the remaining 90 percent of the target requirements. If this is true, then a 10 percent reduction in budget must be accompanied by a loss of performance much greater than the target reduction in cost.

It is true that some projects have "padded" estimates, and the budgetary reduction will force out the padding. Most project managers, however, provide realistic estimates and schedules with marginal padding. Likewise, a trade-off between time and cost is unlikely to help, since increasing the duration of the project will increase the cost.

Cost versus Quality   Everyone knows that reducing cost quite often results in a reduction of quality. Conversely, if the schedule is inflexible, then the only possible trade-offs available to the project manager may be cost versus quality. If the estimated budget for a project is too high, then executives often are willing to sacrifice some degree of quality to keep the budget in line. The problem, of course, is to decide how much quality degradation is acceptable.

All too often, executives believe that cost and quality are linearly related: if the budget is cut by 10 percent, then we will have an accompanying degradation of quality by 10 percent. Nothing could be further from the truth. In the table below we can see the relationship between cost, quality, and time.

Project Costs	85–90%	10–15%
Tangible Quality	10%	90%

Time ⟶

The first 85–90 percent of the budget (i.e., direct labor budget) is needed to generate the first 10 percent of the quality. The last 10–15 percent of the budget often produces the remaining 90 percent of the quality. One does not need an advanced degree in mathematics to realize that a 10 percent cost reduction could easily be accompanied by a 50 percent quality reduction, depending, of course, where the 10 percent was cut.

The following scenario shows the chain of events as they might occur in a typical organization:

- At the beginning of the fiscal year, the executive committee selects those projects to be undertaken, such that *all* available resources are consumed.
- Shortly into the fiscal year, the executive committee authorizes additional projects that must be undertaken. These projects are added to the queue.
- The executive committee recognizes that the resources available are insufficient to service the queue. Since budgets are tight, hiring additional staff is ruled out. (Even if staff could be hired, the project deadline would be at hand before the new employees were properly trained and up to speed.)
- The executive committee refuses to cancel any of the projects and takes the "easy" way out by adopting the 10 percent solution on each and every project. Furthermore, the executive committee asserts that original performance *must* be adhered to at all costs.
- Morale in the project and functional areas, which may have taken months to build, is now destroyed overnight. Functional employees lose faith in the ability of the executive committees to operate properly and make sound decisions. Employees seek transfers to other organizations.
- Functional priorities are changed on a daily basis, and resources are continuously shuffled in and out of projects, with very little regard for the schedule.
- As each project begins to suffer, project managers begin to hoard resources, refusing to surrender the people to other projects, even if the work is completed.
- As quality and performance begin to deteriorate, managers at all levels begin writing "protection" memos.

- Schedule and quality slippages become so great that several projects are extended into the next fiscal year, thus reducing the number of new projects that can be undertaken.

The 10 percent solution simply does not work. However, there are two viable alternatives. The first is to use the 10 percent solution, but only on selected projects and *after* an "impact study" has been conducted, so that the executive committee understands the impact on the time, cost, and performance constraints. The second choice, which is by far the better one, is for the executive committee to cancel or descope selected projects. Since it is impossible to reduce budget without reducing scope, canceling a project or simply delaying it until the next fiscal year is a viable choice. After all, why should all projects have to suffer?

Terminating one or two projects within the queue allows existing resources to be used more effectively, more productively, and with higher organizational morale. However, it does require strong leadership at the executive committee level for the participants to terminate a project rather than to "pass the buck" to the bottom of the organization with the 10 percent solution. Executive committees often function best if the committee is responsible for project selection, prioritization, and tracking, with the middle managers responsible for budgeting.

14.19 LIFE-CYCLE COSTING (LCC)

PMBOK® Guide, 4th Edition
7.1.1 Cost Estimating Inputs

For years, many R&D organizations have operated in a vacuum where technical decisions made during R&D were based entirely on the R&D portion of the plan, with little regard for what happens after production begins. Today, industrial firms are adopting the life-cycle costing approach that has been developed and used by military organizations. Simply stated, LCC requires that decisions made during the R&D process be evaluated against the total life-cycle cost of the system. As an example, the R&D group has two possible design configurations for a new product. Both design configurations will require the same budget for R&D and the same costs for manufacturing. However, the maintenance and support costs may be substantially greater for one of the products. If these downstream costs are not considered in the R&D phase, large unanticipated expenses may result at a point where no alternatives exist.

Life-cycle costs are the total cost to the organization for the ownership and acquisition of the product over its full life. This includes the cost of R&D, production, operation, support, and, where applicable, disposal. A typical breakdown description might include:

- *R&D costs:* The cost of feasibility studies; cost-benefit analyses; system analyses; detail design and development; fabrication, assembly, and test of engineering models; initial product evaluation; and associated documentation.
- *Production cost:* The cost of fabrication, assembly, and testing of production models; operation and maintenance of the production capability; and associated internal logistic support requirements, including test and support equipment development, spare/repair parts provisioning, technical data development, training, and entry of items into inventory.

- *Construction cost:* The cost of new manufacturing facilities or upgrading existing structures to accommodate production and operation of support requirements.
- *Operation and maintenance cost:* The cost of sustaining operational personnel and maintenance support; spare/repair parts and related inventories; test and support equipment maintenance; transportation and handling; facilities, modifications, and technical data changes; and so on.
- *Product retirement and phaseout cost (also called disposal cost):* The cost of phasing the product out of inventory due to obsolescence or wearout, and subsequent equipment item recycling and reclamation as appropriate.

Life-cycle cost analysis is the systematic analytical process of evaluating various alternative courses of action early on in a project, with the objective of choosing the best way to employ scarce resources. Life-cycle cost is employed in the evaluation of alternative design configurations, alternative manufacturing methods, alternative support schemes, and so on. This process includes:

- Defining the problem (what information is needed)
- Defining the requirements of the cost model being used
- Collecting historical data–cost relationships
- Developing estimate and test results

Successful application of LCC will:

- Provide downstream resource impact visibility
- Provide life-cycle cost management
- Influence R&D decision-making
- Support downstream strategic budgeting

There are also several limitations to life-cycle cost analyses. They include:

- The assumption that the product, as known, has a finite life-cycle
- A high cost to perform, which may not be appropriate for low-cost/low-volume production
- A high sensitivity to changing requirements

Life-cycle costing requires that early estimates be made. The estimating method selected is based on the problem context (i.e., decisions to be made, required accuracy, complexity of the product, and the development status of the product) and the operational considerations (i.e., market introduction date, time available for analysis, and available resources).

The estimating methods available can be classified as follows:

- Informal estimating methods
  - Judgment based on experience
  - Analogy

- SWAG method
- ROM method
- Rule-of-thumb method
- Formal estimating methods
  - Detailed (from industrial engineering standards)
  - Parametric

Table 14–14 shows the advantages/disadvantages of each method.

PMBOK® Guide, 4th Edition
7.1.2 Cost Estimating Tools and
  Techniques

TABLE 14–14. ESTIMATING METHODS

Estimating Technique	Application	Advantages	Disadvantages
Engineering estimates (empirical)	Reprocurement Production Development	• Most detailed technique • Best inherent accuracy • Provides best estimating base for future program change estimates	• Requires detailed program and product definition • Time-consuming and may be expensive • Subject to engineering bias • May overlook system integration costs
Parametric estimates and scaling (statistical)	Production Development	• Application is simple and low cost • Statistical database can provide expected values and prediction intervals • Can be used for equipment or systems prior to detail design or program planning	• Requires parametric cost relationships to be established • Limited frequently to specific subsystems or functional hardware of systems • Depends on quantity and quality of the data • Limited by data and number of independent variables
Equipment/subsystem analogy estimates (comparative)	Reprocurement Production Development Program planning	• Relatively simple • Low cost • Emphasizes incremental program and product changes • Good accuracy for similar systems	• Requires analogous product and program data • Limited to stable technology • Narrow range of electronic applications • May be limited to systems and equipment built by the same firm
Expert opinion	All program phases	• Available when there are insufficient data, parametric cost relationships, or program/product definition	• Subject to bias • Increased product or program complexity can degrade estimates • Estimate substantiation is not quantifiable

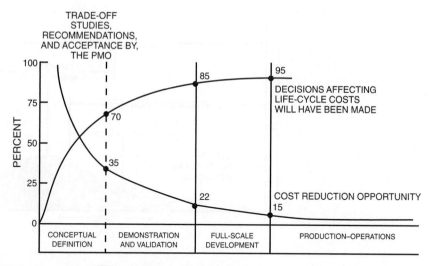

FIGURE 14–16. Department of Defense life-cycle phases.

Figure 14–16 shows the various life-cycle phases for Department of Defense projects. At the end of the demonstration and validation phase (which is the completion of R&D) 85 percent of the decisions affecting the total life-cycle cost will have been made, and the cost reduction opportunity is limited to a maximum of 22 percent (excluding the effects of learning curve experiences). Figure 14–17 shows that, at the end of the R&D phase,

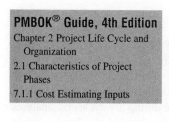

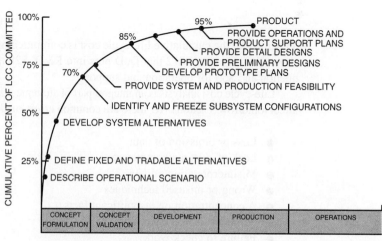

FIGURE 14–17. Actions affecting life-cycle cost (LCC).

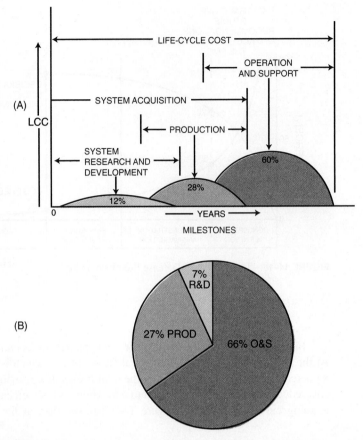

FIGURE 14–18. (A) Typical DoD system acquisition LCC profile; (B) typical communication system acquisition LCC profile.

95 percent of the cumulative life-cycle cost is committed by the government. Figure 14–18 shows that, for every $12 that DoD puts into R&D, $28 are needed downstream for production and $60 for operation and support.

Life-cycle cost analysis is an integral part of strategic planning since today's decision will affect tomorrow's actions. Yet there are common errors made during life-cycle cost analyses:

- Loss or omission of data
- Lack of systematic structure
- Misinterpretation of data
- Wrong or misused techniques
- A concentration on insignificant facts
- Failure to assess uncertainty
- Failure to check work
- Estimating the wrong items

14.20 LOGISTICS SUPPORT

There is a class of projects called "material" projects where the deliverable may require maintenance, service, and support after development. This support will continue throughout the life cycle of the deliverable. Providing service to these deliverables is referred to as logistics support.

In the previous section we showed that approximately 85 percent of the deliverable's life-cycle cost has been committed by the end of the design phase (see Figures 14–16 and 14–17). We also showed that the majority of the total life-cycle cost of a system is in operation and support, and could account for well above 60 percent of the total cost. Clearly, the decisions with the greatest chance of affecting life-cycle cost and identifying cost savings are those influencing the design of the deliverable. Simply stated, proper planning and design can save a company hundreds of millions of dollars once the deliverable is put into use.

The two key parameters used to evaluate the performance of material systems are supportability and readiness. Supportability is the ability to maintain or acquire the necessary human and nonhuman resources to support the system. Readiness is a measure of how good we are at keeping the system performing as planned and how quickly we can make repairs during a shutdown. Clearly, proper planning during the design stage of a project can reduce supportability requirements, increase operational readiness, and minimize or lower logistics support costs.

The ten elements of logistics support include:

- *Maintenance planning:* The process conducted to evolve and establish maintenance concepts and requirements for the lifetime of a materiel system.
- *Manpower and personnel:* The identification and acquisition of personnel with the skills and grades required to operate and support a material system over its lifetime.
- *Supply support:* All management actions, procedures, and techniques used to determine requirements to acquire, catalog, receive, store, transfer, issue, and dispose of secondary items. This includes provisioning for initial support as well as replenishment supply support.
- *Support equipment:* All equipment (mobile or fixed) required to support the operation and maintenance of a materiel system. This includes associated multiuse end-items; ground-handling and maintenance equipment; tools, metrology, and calibration equipment; and test and automatic test equipment. It includes the acquisition of logistics support for the support and test equipment itself.
- *Technical data:* Recorded information regardless of form or character (such as manuals and drawings) of a scientific or technical nature. Computer programs and related software are not technical data; documentation of computer programs and related software are. Also other information related to contract administration.
- *Training and training support:* The processes, procedures, techniques, training devices, and equipment used to train personnel to operate and support a materiel system. This includes individual and crew training; new equipment training; initial, formal, and on-the-job training; and logistic support planning for training equipment and training device acquisitions and installations.

- *Computer resource support:* The facilities, hardware, software, documentation, manpower, and personnel needed to operate and support embedded computer systems.
- *Facilities:* The permanent or semipermanent real property assets required to support the materiel system. Facilities management includes conducting studies to define types of facilities or facility improvement, locations, space needs, environment requirements, and equipment.
- *Packaging, handling, storage, and transportation:* The resources, processes, procedures, design considerations, and methods to ensure that all system, equipment, and support items are preserved, packaged, handled, and transported properly. This includes environmental considerations and equipment preservation requirements for short- and long-term storage and transportability.
- *Design interface:* The relationship of logistics-related design parameters to readiness and support resource requirements. These logistics-related design parameters are expressed in operational terms rather than as inherent values and specifically relate to system readiness objectives and support costs of the material system.

14.21 ECONOMIC PROJECT SELECTION CRITERIA: CAPITAL BUDGETING

PMBOK® Guide, 4th Edition
4.1.1.2 Business Case

Project managers are often called upon to be active participants during the benefit-to-cost analysis of project selection. It is highly unlikely that companies will approve a project where the costs exceed the benefits. Benefits can be measured in either financial or nonfinancial terms.

The process of identifying the financial benefits is called capital budgeting, which may be defined as the *decision-making process* by which organizations evaluate projects that include the purchase of major fixed assets such as buildings, machinery, and equipment. Sophisticated capital budgeting techniques take into consideration depreciation schedules, tax information, and cash flow. Since only the principles of capital budgeting will be discussed in this text, we will restrict ourselves to the following topics:

- Payback Period
- Discounted Cash Flow (DCF)
- Net Present Value (NPV)
- Internal Rate of Return (IRR)

14.22 PAYBACK PERIOD

PMBOK® Guide, 4th Edition
4.1.1.2 Business Case

The payback period is the exact length of time needed for a firm to recover its initial investment as calculated from cash inflows. Payback period is the *least* precise of all capital budgeting methods because the calculations are in dollars and not adjusted for the time value of money. Table 14–15 shows the cash flow stream for Project A.

TABLE 14–15. CAPITAL EXPENDITURE DATA FOR PROJECT A

Initial Investment	Expected Cash Inflows				
	Year 1	*Year 2*	*Year 3*	*Year 4*	*Year 5*
$10,000	$1,000	$2,000	$2,000	$5,000	$2,000

From Table 14–15, Project A will last for exactly five years with the cash inflows shown. The payback period will be exactly four years. If the cash inflow in Year 4 were $6,000 instead of $5,000, then the payback period would be three years and 10 months.

The problem with the payback method is that $5,000 received in Year 4 is not worth $5,000 today. This unsophisticated approach mandates that the payback method be used as a supplemental tool to accompany other methods.

14.23 THE TIME VALUE OF MONEY

PMBOK® Guide, 4th Edition
4.1.1.2 Business Case

Everyone knows that a dollar today is worth more than a dollar a year from now. The reason for this is because of the time value of money. To illustrate the time value of money, let us look at the following equation:

$$FV = PV(1 + k)^n$$

where
$$FV = \text{Future value of an investment}$$
$$PV = \text{Present value}$$
$$k = \text{Investment interest rate (or cost of capital)}$$
$$n = \text{Number of years}$$

Using this formula, we can see that an investment of $1,000 today (i.e., PV) invested at 10% (i.e., k) for one year (i.e., n) will give us a future value of $1,100. If the investment is for two years, then the future value would be worth $1,210.

Now, let us look at the formula from a different perspective. If an investment yields $1,000 a year from now, then how much is it worth *today* if the cost of money is 10%? To solve the problem, we must discount future values to the present for comparison purposes. This is referred to as "discounted cash flows."

The previous equation can be written as:

$$PV = \frac{FV}{(1 + k)^n}$$

Using the data given:

$$PV = \frac{\$1,000}{(1 + 0.1)^1} = \$909$$

Therefore, $1,000 a year from now is worth only $909 today. If the interest rate, k, is known to be 10%, then you should *not* invest more than $909 to get the $1,000 return a year from now. However, if you could purchase this investment for $875, your interest rate would be more than 10%.

Discounting cash flows to the present for comparison purposes is a viable way to assess the value of an investment. As an example, you have a choice between two investments. Investment A will generate $100,000 two years from now and investment B will generate $110,000 three years from now. If the cost of capital is 15%, which investment is better?

Using the formula for discounted cash flow, we find that:

$$PV_A = \$75,614$$
$$PV_B = \$72,327$$

This implies that a return of $100,000 in two years is worth more to the firm than a $110,000 return three years from now.

14.24 NET PRESENT VALUE (NPV)

PMBOK® Guide, 4th Edition
4.1.1.2 Business Case

The net present value (NPV) method is a sophisticated capital budgeting technique that equates the discounted cash flows against the initial investment. Mathematically,

$$NPV = \sum_{t=1}^{n} \left[\frac{FV_t}{(1 + k)^t} \right] - II$$

where FV is the future value of the cash inflows, II represents the initial investment, and k is the discount rate equal to the firm's cost of capital.

Table 14–16 calculates the NPV for the data provided previously in Table 14–15 using a discount rate of 10%.

TABLE 14–16. NPV CALCULATION FOR PROJECT A

Year	Cash Inflows	Present Value
1	$1,000	$ 909
2	2,000	1,653
3	2,000	1,503
4	5,000	3,415
5	2,000	1,242
	Present value of cash inflows	$ 8,722
	Less investment	10,000
	Net Present Value	<1,278>

This indicates that the cash inflows discounted to the present will *not* recover the initial investment. This, in fact, is a bad investment to consider. Previously, we stated that the cash flow stream yielded a payback period of four years. However, using discounted cash flow, the actual payback is greater than five years, assuming that there will be cash inflow in years 6 and 7.

If in Table 14–16 the initial investment was $5,000, then the net present value would be $3,722. The decision-making criteria using NPV are as follows:

● If the NPV is greater than or equal to zero dollars, accept the project.
● If the NPV is less than zero dollars, reject the project.

A positive value of NPV indicates that the firm will earn a return equal to or greater than its cost of capital.

14.25 INTERNAL RATE OF RETURN (IRR)

PMBOK® Guide, 4th Edition
4.1.1.2 Business Case

The internal rate of return (IRR) is perhaps the most sophisticated capital budgeting technique and also more difficult to calculate than NPV. The internal rate of return is the discount rate where the present value of the cash inflows exactly equals the initial investment. In other words, IRR is the discount rate when NPV = 0. Mathematically

$$\sum_{t=1}^{n} \left[\frac{FV_t}{(1 + IRR)^t} \right] - II = 0$$

The solution to problems involving IRR is basically a trial-and-error solution. Table 14–17 shows that with the cash inflows provided, and with a $5,000 initial investment, an IRR of 10% yielded a value of $3,722 for NPV. Therefore, as a second guess, we should try a value greater than 10% for IRR to generate a zero value for NPV. Table 14–17 shows the final calculation.

The table implies that the cash inflows are equivalent to a 31% return on investment. Therefore, if the cost of capital were 10%, this would be an excellent investment. Also, this project is "probably" superior to other projects with a lower value for IRR.

TABLE 14–17. IRR CALCULATION FOR PROJECT A CASH INFLOWS

IRR	NPV
10%	$3,722
20%	1,593
25%	807
30%	152
31%	34
32%	<78>

TABLE 14–18. CAPITAL PROJECTS

Project	IRR	Payback Period with DCF
A	10%	1 year
B	15%	2 years
C	25%	3 years
D	35%	5 years

14.26 COMPARING IRR, NPV, AND PAYBACK

PMBOK® Guide, 4th Edition
4.1.1.2 Business Case

For most projects, both IRR and NPV will generate the same accept-reject decision. However, there are differences that can exist in the underlying assumptions that can cause the projects to be ranked differently. The major problem is the differences in the magnitude and timing of the cash inflows. NPV assumes that the cash inflows are reinvested at the cost of capital, whereas IRR assumes reinvestment at the project's IRR. NPV tends to be a more conservative approach.

The timing of the cash flows is also important. Early year cash inflows tend to be at a lower cost of capital and are more predictable than later year cash inflows. Because of the downstream uncertainty, companies prefer larger cash inflows in the early years rather than the later years.

Magnitude and timing are extremely important in the selection of capital projects. Consider Table 14–18.

If the company has sufficient funds for one and only one project, the natural assumption would be to select Project D with a 35% IRR. Unfortunately, companies shy away from long-term payback periods because of the relative uncertainties of the cash inflows after Year 1. One chemical/plastics manufacturer will not consider any capital projects unless the payback period is less than one year and has an IRR in excess of 50%!

14.27 RISK ANALYSIS

PMBOK® Guide, 4th Edition
11.4.2.2 Quantitative Risk
   Analysis and Modeling
   Techniques

Suppose you have a choice between two projects, both of which require the same initial investment, have identical net present values, and require the same yearly cash inflows to break even. If the cash inflow of the first investment has a probability of occurrence of 95% and that of the second investment is 70%, then risk analysis would indicate that the first investment is better.

Risk analysis refers to the chance that the selection of this project will prove to be unacceptable. In capital budgeting, risk analysis is almost entirely based upon how well we can predict cash inflows since the initial investment is usually known with some degree of certainty. The inflows, of course, are based upon sales projections, taxes, cost of raw materials, labor rates, and general economic conditions.

Sensitivity analysis is a simple way of assessing risk. A common approach is to estimate NPV based upon an optimistic (best case) approach, most likely (expected) approach,

TABLE 14–19. SENSITIVITY ANALYSIS

Initial Investment	Project A $10,000	Project B $10,000
	Annual Cash Inflows	
Optimistic	$ 8,000	$10,000
Most likely	5,000	5,000
Pessimistic	3,000	1,000
Range	$ 5,000	$ 9,000
	Net Present Values	
Optimistic	$20,326	$27,908
Most likely	8,954	8,954
Pessimistic	1,342	<6,209>
Range	$18,984	$34,117

and pessimistic (worst case) approach. This can be illustrated using Table 14–19. Both Projects A and B require the same initial investment of $10,000, with a cost of capital of 10%, and with expected five-year annual cash inflows of $5,000/year. The range for Project A's NPV is substantially less than that of Project B, thus implying that Project A is less risky. A risk lover might select Project B because of the potential reward of $27,908, whereas a risk avoider would select Project A, which offers perhaps no chance for loss.

14.28 CAPITAL RATIONING

PMBOK® Guide, 4th Edition
11.4.2.2 Quantitative Risk
  Analysis and Modeling
  Techniques

Capital rationing is the process of selecting the best group of projects such that the highest overall net present value will result without exceeding the total budget available. An assumption with capital rationing is that the projects under consideration are mutually exclusive. There are two approaches often considered for capital rationing.

The internal rate of return approach plots the IRRs in descending order against the cumulative dollar investment. The resulting figure is often called an investment opportunity schedule. As an example, suppose a company has $300,000 committed for projects and must select from the projects identified in Table 14–20. Furthermore, assume that the cost of capital is 10%.

TABLE 14–20. PROJECTS UNDER CONSIDERATION

Project	Investment	IRR	Discounted Cash Flows at 10%
A	$ 50,000	20%	$116,000
B	120,000	18%	183,000
C	110,000	16%	147,000
D	130,000	15%	171,000
E	90,000	12%	103,000
F	180,000	11%	206,000
G	80,000	8%	66,000

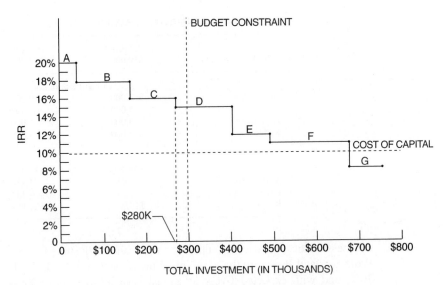

FIGURE 14–19. Investment Opportunity Schedule (IOS) for Table 14–20.

Figure 14–19 shows the investment opportunity schedule. Project G should not be considered because the IRR is less than the firm's cost of capital, but we should select Projects, A, B, and C, which will consume $280,000 out of a total budget of $300,000. This allows us to have the three largest IRRs.

The problem with the IRR approach is that it does not guarantee that the projects with the largest IRRs will maximize the total dollar returns. The reason is that not all of the funds have been consumed.

A better approach is the net present value method. In this method, the projects are again ranked according to their IRRs, but the combination of projects selected will be based upon the highest net present value. As an example, the selection of Projects A, B, and C from Table 14–20 requires an initial investment of $280,000 with resulting discounted cash flows of $446,000. The net present value of Projects A, B, and C is, therefore, $166,000. This assumes that unused portions of the original budget of $300,000 do not gain or lose money. However, if we now select Projects A, B, and D, we will invest $300,000 with a net present value of $170,000 ($470,000 less $300,000). Selection of Projects A, B, and D will, therefore, maximize net present value.

14.29 PROJECT FINANCING[2]

Project financing involves the establishment of a legally independent project company, usually for large-scale investments (LSI) and long term where the providers of funds are repaid out of cash flow and earnings, and where the assets of the unit (and only the unit)

---

2. Project financing is a relatively new topic and is now being taught in graduate programs in business. At Harvard University, it is taught as a course entitled Large Scale Investment by Professor Benjamin C. Esty. Many excellent examples appear in Professor Esty's text, *Modern Project Finance* (Hoboken, NJ: Wiley, 2004).

are used as collateral for the loans. Debt repayment would come from the project company only rather than from any other entity. A risk with project financing is that the capital assets may have a limited life. The potential limited life constraint often makes it difficult to get lenders to agree to long-term financial arrangements.

Another critical issue with project financing especially for high-technology projects is that the projects are generally long term. It may be nearly eight to ten years before service will begin, and in terms of technology, eight years can be an eternity. Project financing is often considered a "bet on the future." And if the project were to fail, the company could be worth nothing after liquidation.

There are several risks that must be considered to understand project financing. The risks commonly considered are

Financial Risks
- Use of project versus corporate financing
- Use of corporate bonds, stock, zero coupon bonds, and bank notes
- Use of secured versus unsecured debt
- The best sequence or timing for raising capital
- Bond rating changes
- Determination of the refinancing risk, if necessary

Development Risks
- Reality of the assumptions
- Reality of the technology
- Reality of development of the technology
- Risks of obsolescence

Political Risks
- Sovereignty risks
- Political instability
- Terrorism and war
- Labor availability
- Trade restrictions
- Macroeconomics such as inflation, currency conversion, and transferability of funding and technology

Organizational Risks
- Members of the board of directors
- Incentives for the officers
- Incentives for the board members
- Bonuses as a percentage of base compensation
- Process for the resolution of disputes

Execution Risks
- Timing when execution will begin
- Life expectancy of execution
- Ability to service debt during execution

14.30 STUDYING TIPS FOR THE PMI® PROJECT MANAGEMENT CERTIFICATION EXAM

This section is applicable as a review of the principles to support the knowledge areas and domain groups in the PMBOK® Guide. This chapter addresses:

- Integration Management
- Scope Management
- Time Management
- Cost Management
- Initiating
- Planning

Understanding the following principles is beneficial if the reader is using this text to study for the PMP® Certification Exam:

- What is meant by cost-estimating relationships (CER)
- Three basic types of estimates
- Relative accuracy of each type of estimate and the approximate time to prepare the estimate
- Information that is needed to prepare the estimates (i.e., labor, material, overhead rates, etc.)
- Importance of backup data for costs
- Estimating pitfalls
- Concept of rolling wave planning
- What is meant by life cycle costing
- Different ways of evaluating a project's financial feasibility or benefits (i.e., ROI, payback period, net present value, internal rate of return, depreciation, scoring models)

The following multiple-choice questions will be helpful in reviewing the principles of this chapter:

1. Which of the following is a valid way of evaluating the financial feasibility of a project?
   A. Return on investment
   B. Net present value
   C. Internal rate of return
   D. All of the above

2. The three common classification systems for estimates includes all of the following except:
   A. Parametric estimates
   B. Quick-and-dirty estimates
   C. Analogy estimates
   D. Engineering estimates

3. The most accurate estimates are:
   A. Parametric estimates
   B. Quick-and-dirty estimates

C. Analogy estimates

D. Engineering estimates

4. Which of the following is considered to be a bottom-up estimate rather than a top-down estimate?

A. Parametric estimates

B. Analogy estimates

C. Engineering estimates

D. None of the above

5. Which of the following would be considered as a cost-estimating relationship (CER)?

A. Mathematical equations based upon regression analysis

B. Learning curves

C. Cost–cost or cost–quantity relationships

D. All of the above

6. If a worker earns $30 per hour in salary but the project is charged $75 per hour for each hour the individual works, then the overhead rate is:

A. 100%

B. 150%

C. 250%

D. None of the above

7. Information supplied to a customer to support the financial data provided in a proposal is commonly called:

A. Backup data

B. Engineering support data

C. Labor justification estimates

D. Legal rights estimates

8. Estimating pitfalls can result from:

A. Poorly defined statement of work

B. Failure to account for risks in the estimates

C. Using the wrong estimating techniques

D. All of the above

9. The source of many estimating risks is:

A. Poorly defined requirements

B. An inexperienced project manager

C. Lack of management support during estimating

D. All of the above

10. A project where the scope evolves as the work takes place is called either progressive planning or:

A. Synchronous planning

B. Continuous planning

C. Rolling wave planning

D. Continuous reestimation planning

11. The calculation of the total cost of a product, from R&D to operational support and disposal, is called:

A. Birth-to-death costing

B. Life-cycle costing

C. Summary costing

D. Depreciation costing

ANSWERS

1. D
2. B
3. D
4. C
5. D
6. B
7. A
8. D
9. D
10. C
11. B

PROBLEMS

14–1 How does a project manager price out a job in which the specifications are not prepared until the job is half over?

14–2 Beta Corporation is in the process of completing a contract to produce 150 units for a given customer. The contract consisted of R&D, testing and qualification, and full production. The industrial engineering department had determined that the following number of hours were required to produce certain units:

Unit	Hours Required Per Unit
1	100
2	90
4	80
8	70
16	65
32	60
64	55
128	50

a. Plot the data points on regular graph paper with the Y-axis as hours and the X-axis as number of units produced.
b. Plot the data points on log–log paper and determine the slope of the line.
c. Compare parts a and b. What are your conclusions?
d. How much time should it take to manufacture the 150th unit?
e. How much time should it take to manufacture the 1,000th unit? Explain your answer. Is it realistic? If not, why?
f. As you are producing the 150th unit, you receive an immediate follow-on contract for another 150 units. How many manufacturing hours should you estimate for the follow-on effort (using only the learning curves)?
g. Let's assume that industrial engineering determines that the optimum number of hours (for 100 percent efficiency) of manufacturing is forty-five. At what efficiency factor are you now performing at the completion of unit number 150? After how many units in the follow-on contract will you reach the optimum level?

h. At the end of the first follow-on contract, your team and personnel are still together and performing at a 100 percent efficiency position (of part g). You have been awarded a second follow-on contract, but the work will not begin until six months from now. Assuming that you can assemble the same team, how many man-hours/unit will you estimate for the next 150-unit follow-on?

i. Would your answer to part h change if you could not assemble the same team? Explain your answer quantitatively.

j. You are now on the contract negotiation team for the second follow-on contract of 150 units (which is not scheduled to start for six months). Based on the people available and the "loss of learning" between contracts, your industrial engineering department estimates that you will be performing at a 60 percent efficiency factor. The customer says that your efficiency factor should be at least 75 percent. If your company is burdened at $40/hour, how much money is involved between the 60 and 75 percent efficiency factors?

k. What considerations should be made in deciding where to compromise in the efficiency factor?

14–3 With reference to Figure 14–10, under what conditions could *each* of the following situations occur:

a. Program manager and program office determine labor hours by pricing out the work breakdown structure without coordination with functional management.

b. Upper-level management determines the price of a bid without forming a program office or consulting functional management.

c. Perturbations on the base case are not performed.

d. The chief executive officer selects the program manager without consulting his directors.

e. Upper-level management does not wish to have a cost review meeting prior to submittal of a bid.

14–4 Can Figure 14–20 be used effectively to price out the cost of preparing reports?

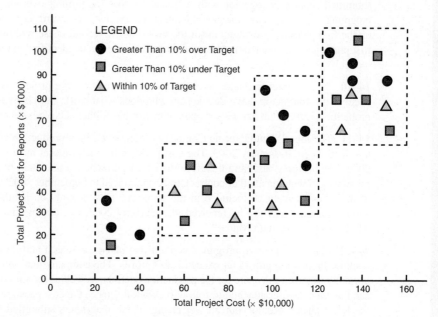

FIGURE 14–20. Project documentation costs.

14–5 Answer the following questions with reference to Figure 14–10.

    a. The base case for a program is priced out at $22 million. The company's chief executive officer is required to obtain written permission from corporate to bid on programs in excess of $20 million. During the price review meeting the chief executive states that the bid will be submitted at $19.5 million. Should you, as program manager, question this?

    b. Would your answer to part a change if this program were a follow-on to an earlier program?

    c. Proposals normally consist of management, technical, and cost volumes. Indicate in Figure 14–10 where these volumes can go to press, assuming each can be printed independently.

14–6 Under what kind of projects would each of the following parameters be selected:

    a. Salary escalation factor of 0 percent.
    b. Material termination liability of 0 percent or 100 percent.
    c. Material commitments for twenty months of a twenty-four-month program.
    d. Demanning ratio of 0 percent or 100 percent of following months' labor.

14–7 How can upper-level management use the functional cost and hour summary to determine manpower planning for the entire company? How would you expect management to react if the functional cost and hour summary indicated a shortage or an abundance of trained personnel?

14–8 Which of the figures presented in this chapter should program management make available to the functional managers? Explain your answer.

14–9 The Jennings Construction Company has decided to bid on the construction for each of the two phases of a large project. The bidding requirements are that the costs for each phase be submitted separately together with a transition cost for turning over the first phase of the program to a second contractor should Jennings not receive both awards or perform unsatisfactorily on the first phase. The evaluation for the award of the second phase will not be made until the first phase is near completion. How can the transition costs be identified in the strategic planning model?

14–10 Two contractors decide to enter into a joint venture on a project. What difficulties can occur if the contractors have decided on who does what work, but changes may take place if problems occur? What happens if one contractor has higher salary levels and overhead rates?

14–11 The Jones Manufacturing Company is competing for a production contract that requires that work begin in January 2003. The cost package for the proposal must be submitted by July 2002. The business base, and therefore the overhead rates, are uncertain because Jones has the possibility of winning another contract, to be announced in September 2002. How can the impact of the announcement be included in the proposal? How would you handle a situation where another contract may not be renewed after January 2003, i.e., assume that the announcement would not be made until March?

14–12 Many competitive programs contain two phases: research and development, and production. Production profits far exceed R&D profits. The company that wins the R&D contract normally becomes a favorite for the production contract, as well as for any follow-on work. How can the dollar figures attached to follow-on work influence the cost package that you submit for the R&D phase? Would your answer change if the man-hours submitted for the R&D phase become the basis for the production phase?

14–13 During initial pricing activities, one of the functional managers discovers that the work breakdown structure requires costing data at a level that is not normally made, and will undoubtedly incur additional costs. How should you, as a program manager, respond to this situation? What are your alternatives?

14–14 Should the project manager give the final manpower loading curves to the functional managers? If so, at what point in time?

14–15 You have been asked to price out a project for an outside customer. The project will run for eight months. Direct labor is $100,000 for each month and the overhead rate is fixed at 100 percent per month. Termination liability on the direct labor and overhead rate is 80 percent of the following month's expenses. Material expenses are as follows:

Material A: Cost is $100,000 payable 30 days net. Material is needed at the end of the fifth month. Lead time is four months with termination liability expenses as follows:

> 30 days: 25%
> 60 days: 75%
> 90 days: 100%

Material B: Cost is $200,000, payable on delivery. Material is needed at the end of the seventh month. Lead time is three months with termination liability as follows:

> 30 days: 50%
> 60 days: 100%

Complete the table below, neglecting profits.

	Month							
	1	2	3	4	5	6	7	8
Direct labor								
Overhead								
Material								
Monthly cash flow								
Cumulative cash flow								
Monthly termination liability: labor								
Cumulative termination liability: labor								
Monthly termination liability: material								
Cumulative termination liability: material								
Total project termination liability								

14–16 Should a project manager be appointed in the bidding stage of a project? If so, what authority should he have, and who is responsible for winning the contract?

14–17 Explain how useful each of the following can be during the estimating of project costs:

    a. Contingency planning and estimating
    b. Using historical databases
    c. Usefulness of computer estimating
    d. Usefulness of performance factors to account for inefficiencies and uncertainties.

Cost Control

Related Case Studies (from Kerzner/*Project Management Case Studies,* 3rd Edition)	Related Workbook Exercises (from Kerzner/*Project Management Workbook and PMP®/CAPM® Exam Study Guide,* 10th Edition)	PMBOK® Guide, 4th Edition, Reference Section for the PMP® Certification Exam
• The Bathtub Period* • Trouble in Paradise*	• Using the 50/50 Rule • Multiple Choice Exam • Crossword Puzzle on Cost Management	• Cost Management • Scope Management

15.0 INTRODUCTION

PMBOK® Guide, 4th Edition
7.3 Cost Control

Cost control is equally important to all companies, regardless of size. Small companies generally have tighter monetary controls because the failure of even one project can put the company at risk, but they have less sophisticated control techniques. Large companies may have the luxury to spread project losses over several projects, whereas the small company may have few projects.

*Case Study also appears at end of chapter.

Many people have a poor understanding of cost control. Cost control is not only "monitoring" costs and recording data, but also analyzing the data in order to take corrective action before it is too late. Cost control should be performed by all personnel who incur costs, not merely the project office.

Cost control implies good cost management, which must include:

- Cost estimating
- Cost accounting
- Project cash flow
- Company cash flow
- Direct labor costing
- Overhead rate costing
- Other tactics, such as incentives, penalties, and profit-sharing

Cost control is actually a subsystem of the management cost and control system (MCCS) rather than a complete system per se. This is shown in Figure 15–1, where the MCCS is represented as a two-cycle process: a planning cycle and an operating cycle. The operating cycle is what is commonly referred to as the cost control system. Failure of a cost control system to accurately describe the true status of a project does not necessarily imply that the cost control system is at fault. Any cost control system is only as good as the original plan against which performance will be measured. Therefore, the designing of a planning system must take into account the cost control system. For this reason, it is common for the planning cycle to be referred to as planning and control, whereas the operating cycle is referred to as cost and control.

The planning and control system must help management project the status toward objective completion. Its purpose is to establish policies, procedures, and techniques that can be used in the day-to-day management and control of projects and programs. It must, therefore, provide information that:

- Gives a picture of true work progress
- Will relate cost and schedule performance
- Identifies potential problems with respect to their sources.
- Provides information to project managers with a practical level of summarization
- Demonstrates that the milestones are valid, timely, and auditable

The planning and control system, in addition to being a tool by which objectives can be defined (i.e., hierarchy of objectives and organization accountability), exists as a tool to develop planning, measure progress, and control change. As a tool for planning, the system must be able to:

- Plan and schedule work
- Identify those indicators that will be used for measurement

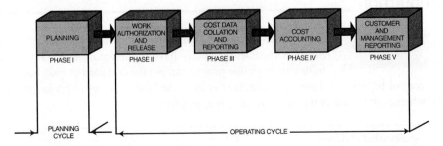

FIGURE 15–1. Phases of a management cost and control system.

- Establish direct labor budgets
- Establish overhead budgets
- Identify management reserve

The project budget that results from the planning cycle of the MCCS must be reasonable, attainable, and based on contractually negotiated costs and the statement of work. The basis for the budget is either historical cost, best estimates, or industrial engineering standards. The budget must identify planned man-power requirements, contract-allocated funds, and management reserve.

Establishing budgets requires that the planner fully understand the meaning of standards. There are two categories of standards. Performance results standards are quantitative measurements and include such items as quality of work, quantity of work, cost of work, and time-to-complete. Process standards are qualitative, including personnel, functional, and physical factors relationships. Standards are advantageous in that they provide a means for unity, a basis for effective control, and an incentive for others. The disadvantage of standards is that performance is often frozen, and employees are quite often unable to adjust to the differences.

As a tool for measuring progress and controlling change, the systems must be able to:

- Measure resources consumed
- Measure status and accomplishments
- Compare measurements to projections and standards
- Provide the basis for diagnosis and replanning

In using the MCCS, the following guidelines usually apply:

- The level of detail is specified by the project manager with approval by top management.
- Centralized authority and control over each project are the responsibility of the project management division.
- For large projects, the project manager may be supported by a project team for utilization of the MCCS.

Almost all project planning and control systems have identifiable design requirements. These include:

- A common framework from which to integrate time, cost, and technical performance
- Ability to track progress of significant parameters
- Quick response
- Capability for end-value prediction
- Accurate and appropriate data for decision-making by each level of management
- Full exception reporting with problem analysis capability
- Immediate quantitative evaluation of alternative solutions

MCCS planning activities include:

- Contract receipt (if applicable)
- Work authorization for project planning

- Work breakdown structure
- Subdivided work description
- Schedules
- Planning charts
- Budgets

MCCS planning charts are worksheets used to create the budget. These charts include planned labor in hours and material dollars.

MCCS planning is accomplished in one of these ways:

- One level below the lowest level of the WBS
- At the lowest management level
- By cost element or cost account

Even with a fully developed planning and control system, there are numerous benefits and costs. The appropriate system must consider a cost-benefit analysis, and include such items as:

- Project benefits
  - Planning and control techniques facilitate:
    —Derivation of output specifications (project objectives)
    —Delineation of required activities (work)
    —Coordination and communication between organizational units
    —Determination of type, amount, and timing of necessary resources
    —Recognition of high-risk elements and assessment of uncertainties
    —Suggestions of alternative courses of action
    —Realization of effect of resource level changes on schedule and output performance
    —Measurement and reporting of genuine progress
    —Identification of potential problems
    —Basis for problem-solving, decision-making, and corrective action
    —Assurance of coupling between planning and control
- Project cost
  - Planning and control techniques require:
    —New forms (new systems) of information from additional sources and incremental processing (managerial time, computer expense, etc.)
    —Additional personnel or smaller span of control to free managerial time for planning and control tasks (increased overhead)
    —Training in use of techniques (time and materials)

A well-disciplined MCCS will produce the following results:

- Policies and procedures that will minimize the ability to distort reporting
- Strong management emphasis on meeting commitments
- Weekly team meetings with a formalized agenda, action items, and minutes
- Top-management periodic review of the technical and financial status
- Simplified internal audit for checking compliance with procedures

For MCCS to be effective, both the scheduling and budgeting systems must be disciplined and formal in order to prevent inadvertent or arbitrary budget or schedule changes. This does *not* mean that the baseline budget and schedule, once established, is static or inflexible. Rather, it means that changes must be controlled and result only from deliberate management actions.

Disciplined use of MCCS is designed to put pressure on the project manager to perform exceptionally good project planning so that changes will be minimized. As an example, government subcontractors may not:

- Make retroactive changes to budgets or costs for work that has been completed
- Rebudget work-in-progress activities
- Transfer work or budget independently of each other
- Reopen closed work packages

In some industries, the MCCS must be used on all contracts of $2 million or more, including firm-fixed-price efforts. The fundamental test of whether to use the MCCS is to determine whether the contracts have established end-item deliverables, either hardware or computer software, that must be accomplished through measurable efforts.

Two programs are used by the government and industry in conjunction with the MCCS as an attempt to improve effectiveness in cost control. The zero-base budgeting program provides better estimating techniques for the verification portion of control. The design-to-cost program assists the decision-making part of the control process by identifying a decision-making framework from which replanning can take place.

15.1 UNDERSTANDING CONTROL

PMBOK® Guide, 4th Edition
3.2 Monitoring and Control Process Group

Effective management of a program during the operating cycle requires that a well-organized cost and control system be designed, developed, and implemented so that immediate feedback can be obtained, whereby the up-to-date usage or resources can be compared to target objectives established during the planning cycle. The requirements for an effective control system (for both cost and schedule/performance) should include[1]:

- Thorough planning of the work to be performed to complete the project
- Good estimating of time, labor, and costs
- Clear communication of the scope of required tasks
- A disciplined budget and authorization of expenditures
- Timely accounting of physical progress and cost expenditures
- Periodic reestimation of time and cost to complete remaining work
- Frequent, periodic comparison of actual progress and expenditures to schedules and budgets, both at the time of comparison and at project completion

---

1. Russell D. Archibald, *Managing High-Technology Programs and Projects* (New York: John Wiley & Sons, 1976), p. 191.

Management must compare the time, cost, and performance of the program to the budgeted time, cost, and performance, not independently but in an integrated manner. Being within one's budget at the proper time serves no useful purpose if performance is only 75 percent. Likewise, having a production line turn out exactly 200 items, as planned, loses its significance if a 50 percent cost overrun is incurred. All three resource parameters (time, cost, and performance) must be analyzed as a group, or else we might "win the battle but lose the war." The use of the expression "management cost and control system" is vague in that the implication is that only costs are controlled. This is not true—an effective control system monitors schedule and performance as well as costs by setting budgets, measuring expenditures against budgets and identifying variances, assuring that the expenditures are proper, and taking corrective action when required.

Previously we defined the work breakdown structure as the element that acts as the source from which all costs and controls must emanate. The WBS is the total project broken down into successively lower levels until the desired control levels are established. The work breakdown structure therefore serves as the tool from which performance can be subdivided into objectives and subobjectives. As work progresses, the WBS provides the framework on which costs, time, and schedule/performance can be compared against the budget for each level of the WBS.

The first purpose of control therefore becomes a verification process accomplished by the comparison of actual performance to date with the predetermined plans and standards set forth in the planning phase. The comparison serves to verify that:

● The objectives have been successfully translated into performance standards.
● The performance standards are, in fact, a reliable representation of program activities and events.
● Meaningful budgets have been established such that actual versus planned comparisons can be made.

In other words, the comparison verifies that the correct standards were selected, and that they are properly used.

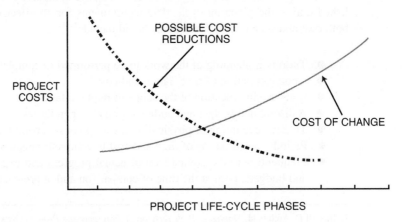

FIGURE 15–2. Cost reduction analysis.

The second purpose of control is decision-making. Three useful reports are required by management in order to make effective and timely decisions:

- The project plan, schedule, and budget prepared during the planning phase
- A detailed comparison between resources expended to date and those predetermined. This includes an estimate of the work remaining and the impact on activity completion.
- A projection of resources to be expended through program completion

These reports, supplied to the managers and the doers, provide three useful results:

- Feedback to management, the planners, and the doers
- Identification of any major deviations from the current program plan, schedule, or budget
- The opportunity to initiate contingency planning early enough that cost, performance, and time requirements can undergo corrected action without loss of resources

These reports provide management with the opportunity to minimize downstream changes by making proper corrections here and now. As shown in Figures 15–2 and 15–3, cost reductions are more available in the early project phases, but are reduced as we go further into the project life-cycle phases. Figure 15–3 identifies the people that most likely have the greatest influence on possibly initiating changes to a project. Downstream the cost of changes could easily exceed the original cost of the project. This is an example of the "iceberg" syndrome, where problems become evident too late in the project to be solved easily, resulting in a very high cost to correct them.

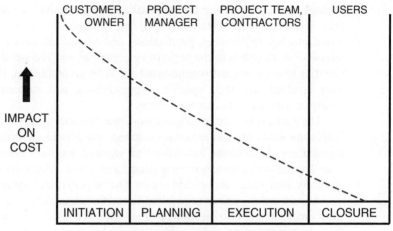

FIGURE 15–3. People with the ability to influence cost.

15.2 THE OPERATING CYCLE

The management cost and control system (MCCS) takes on paramount importance during the operating cycle of the project. The operating cycle is composed of four phases:

- Work authorization and release (phase II)
- Cost data collection and reporting (phase III)
- Cost analysis (phase IV)
- Reporting: customer and management (phase V)

These four phases, when combined with the planning cycle (phase I), constitute a closed system network that forms the basis for the management cost and control system.

Phase II is considered as work release. After planning is completed and a contract is received, work is authorized via a work description document. The work description, or project work authorization form, is a contract that contains the narrative description, organization, and time frame for *each* WBS level. This multipurpose form is used to release the contract, authorize planning, record detail description of the work outlined in the work breakdown structure, and release work to the functional departments.

Contract services may require a work description form to release the contract. The contractual work description form sets forth general contractual requirements and authorizes program management to proceed.

Program management may then issue a subdivided work description form to the functional units so that work can begin. The subdivided work description may also be issued through the combined efforts of the project team, and may be revised or amended when either the scope or the time frame changes. The subdivided work description generally is not used for efforts longer than ninety days and must be "tracked" as if a project in itself. This subdivided work description form sets forth contractual requirements and planning guidelines for the applicable performing organizations. The subdivided work description package established during the proposal and updated after negotiations by the program team is incrementally released by program management to the work control centers in manufacturing, engineering, publications, and program management as the authority for release of work orders to the performing organizations. The subdivided work description specifies how contractual requirements are to be accomplished, the functional organizations involved, and their specific responsibilities, and authorizes the expenditure of resources within a particular time frame.

The work control center assigns a work order number to the subdivided work description form, if no additional instructions are required, and releases the document to the performing organizations. If additional instructions are required, the work control center can prepare a more detailed work-release document (shop traveler, tool order, work order release), assign the applicable work order number, and release it to the performing organization.

A work order number is required for all in-house direct and indirect charging. The work order number also serves as a cross-reference number for automatic assignment of the indentured work breakdown structure number to labor and material data records in the computer.

Small companies can avoid this additional paperwork cost by going directly from an awarded contract to a single work order, which may be the only work order needed for the entire contract.

15.3 COST ACCOUNT CODES

PMBOK® Guide, 4th Edition
7.2.2.1 Cost Aggregation

Since project managers control resources through the line managers rather than directly, project managers end up controlling direct labor costs by opening and closing work orders. Work orders define the charge numbers for each cost account. By definition, a cost account is an identified level at a natural intersection point of the work breakdown structure and the organizational breakdown structure (OBS) at which functional responsibility for the work is assigned, and actual direct labor, material, and other direct costs are compared with actual work performed for management control purposes.

Cost accounts are the focal point of the MCCS and may comprise several work packages, as shown in Figure 15–4. Work packages are detailed short-span job or material items identified for the accomplishment of required work. To illustrate this, consider the cost account code breakdown shown in Figure 15–5 and the work authorization form shown in Figure 15–6. The work authorization form specifically identifies the cost centers that are "open" for this charge number, the man-hours available for each cost center, and the operational time period for the charge number. Because the exact dates of operation are completely defined, the charge number can be assigned perhaps as much as a year in advance of the work-begin date. This can be shown pictorially, as in Figure 15–7.

If the man-hours are assigned to Cost Center 2400, then any 24xx cost center can use this charge number. If the work authorization form specifies Cost Center 2610, then any 261x cost center can use the charge number. However, if Cost Center 2623 is specified, then no lower cost accounts exist, and this is the only cost center that can use this work order charge number. In other words, if a charge number is opened up at the department level, then the department manager has the right to subdivide the assigned man-hours among the various sections and subsections. Company policy usually identifies the permissible cost center levels that can be assigned in the work authorization form. These permissible levels are related to the work breakdown structure level. For example, Cost Center 5000 (i.e., divisional) can be assigned at the project level of the work breakdown structure, but only department, sectional, or subsectional cost accounts can be assigned at the task level of the work breakdown structure.

If a cost center needs additional time or additional man-hours, then a cost account change notice form must be initiated, usually by the requesting cost center, and approved by the project office. Figure 15–8 shows a typical cost account change notice form.

Large companies have computerized cost control and reporting systems. Small companies have manual or partially computerized systems. The major difficulty in using the cost account code breakdown and the work authorization form (Figures 15–5 and 15–6) is related to whether the employees fill out time cards, and frequency with which the time cards are

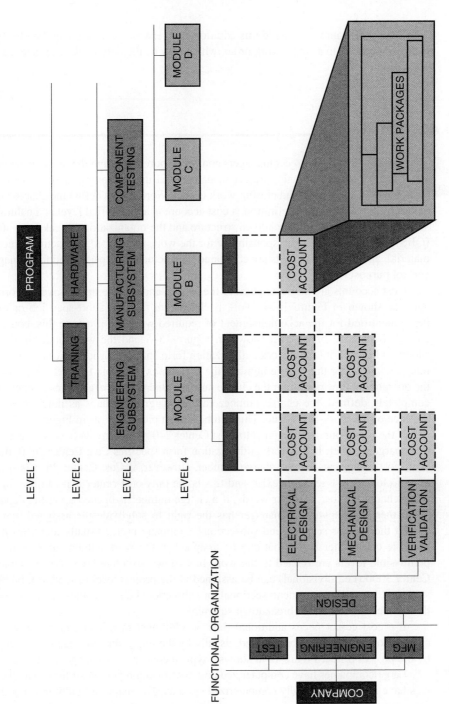

LEVEL 1

LEVEL 2

LEVEL 3

LEVEL 4

FUNCTIONAL ORGANIZATION

FIGURE 15–4. The cost account intersection.

638

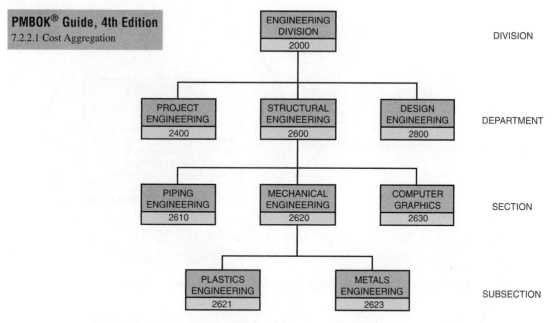

PMBOK® Guide, 4th Edition
7.2.2.1 Cost Aggregation

FIGURE 15–5. Cost account code breakdown.

PMBOK® Guide, 4th Edition
4.4 Monitor and Control
   Project Work

WORK AUTHORIZATION FORM				
WBS NO: ___31-03-02___			WORK ORDER NO: ___D1385___	
DATE OF ORIGINAL RELEASE:		___3 FEB 01___		
DATE OF REVISION:		: ___18 MAR 01___		
REVISION NUMBER:		: ___C___		
DESCRIPTION	COST CENTERS	HOURS	WORK BEGINS	WORK ENDS
TEST MATERIAL VB-2 IN ACCORDANCE WITH THE PROGRAM PLAN AND MIL STANDARD G1483-52. THIS TASK INCLUDES A WRITTEN REPORT.	2400 2610 2621 2623 5000*	150 160 140 46 600	1 AUG 01 ↓	15 SEPT 01 ↓
PROJECT OFFICE AUTHORIZATION SIGNATURE _____				

*NOTE: SOME COMPANIES DO NOT PERMIT DIVISION COST CENTERS
      TO CHARGE AT LEVEL 3 OF THE WBS

FIGURE 15–6. Work authorization form.

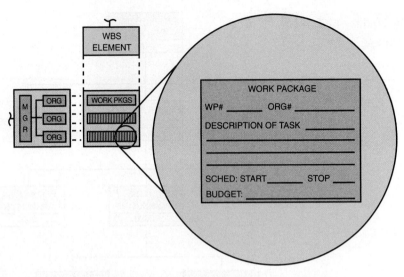

FIGURE 15–7. Planning and budgeting describe, plan, and schedule the work.

filled out. Project-driven organizations fill out time cards at least once a week, and the cards are inputted to a computerized system. Non–project-driven organizations fill out time cards on a monthly basis, with computerization depending on the size of the company.

Cost data collection and reporting constitute the second phase of the operating cycle of the MCCS. Actual cost (ACWP) and the budgeted cost for work performed (BCWP) for each contract or in-house project are accumulated in detailed cost accounts by cost center and cost element, and reported in accordance with the flow charts shown in Figure 15–9. These detailed elements, for both actual costs incurred and the budgeted cost for work performed, are usually printed out monthly for all levels of the work breakdown structure. In addition, weekly supplemental direct labor reports can be printed showing the actual labor charge incurred, and can be compared to the predicted efforts.

Table 15–1 shows a typical weekly labor report. The first column identifies the WBS number.[2] If more than one work order were assigned to this WBS element, then the work order number would appear under the WBS number. This procedure would be repeated for all work orders under the same WBS number. The second column contains the cost centers charging to this WBS element (and possibly work order numbers). Cost Center 41xx represents department 41 and is a rollup of Cost Centers 4110, 4115, and 4118. Cost Center 4xxx represents the entire division and is a rollup of all 4000-level departments. Cost Center xxxx represents the total for all divisions charging to this WBS element. The

---

2. Only three levels of cost reporting are assumed here. If work packages were used, then the WBS number would identify all five levels of control.

CACN No. ——————— Revision to Cost Account No. ——— Date ———

DESCRIPTION OF CHANGE:

_____
_____
_____
_____
_____
_____

REASON FOR CHANGE:

_____
_____
_____
_____
_____
_____

	Requested Budget	Authorized Budget	
Labor Hours	_____	_____	Period of Performance:
Material $	_____	_____	From _____
Indirect $	_____	_____	To _____

BUDGET SOURCE:

☐ Funded Contract Change
☐ Management Reserve
☐ Undistributed Budget
☐ Other _____

APPROVALS: Program Mgr. _____
INITIATED BY: _____ Prog. Control _____

FIGURE 15–8. Cost account change notice (CACN).

weekly labor reports must list all cost centers authorized to charge to this WBS element, whether or not they have incurred any costs over the last reporting period.

Most weekly labor reports provide current month subtotals and previous month totals. Although these also appear on the detailed monthly report, they are included in the weekly report for a quick-and-dirty comparison. Year-to-date totals are usually not on the weekly report unless the users request them for an immediate comparison to the estimate at completion (EAC) and the work order release.

Weekly labor output is a vital tool for members of the program office in that these reports can indicate trends in cost and performance in sufficient time for contingency plans to be established and implemented. If these reports are not available, then cost and labor overruns would not be apparent until the following month when the detailed monthly labor, cost, and materials output was obtained.

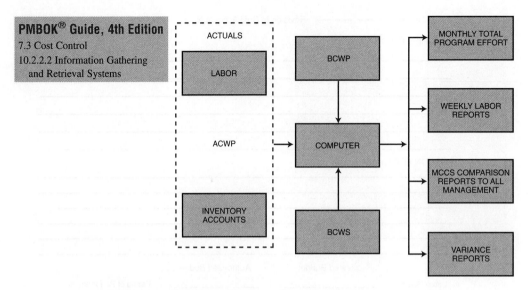

FIGURE 15–9. Cost data collection and reporting flowchart.

In Table 15–1, Cost Center 4110 has spent its entire budget. The work appears to be completed on schedule. The responsible program office team may wish to eliminate this cost center's authority to continue charging to this WBS element by issuing a new subdivided work description or work order canceling this department's efforts. Cost Center 4115 appears to be only halfway through. If time is becoming short, then Cost Center 4115 must add resources in order to meet requirements. Cost Center 4443 appears to be heading for an overrun. This could also indicate a management reserve. In this case the responsible program team member feels that the work can be accomplished in fewer hours.

Work order releases are used to authorize certain cost centers to begin charging their time to a specific cost reporting element. Work orders specify hours, not dollars. The hours indicate the "targets" that the program office would like to have the department shoot for. If the program office wished to be more specific and "compel" the departments to live within these hours, then the budgeted cost for work scheduled (BCWS) should be changed to reflect the reduced hours.

Four categories of cost data are normally accumulated:

● Labor
● Material
● Other direct charges
● Overhead

Project managers can maintain reasonable control over labor, material, and other direct charges. Overhead costs, on the other hand, are calculated yearly or monthly and applied retroactively to all applicable programs. Management reserves are often used to counterbalance the effects of adverse changes in overhead rates.

TABLE 15–1. WEEKLY LABOR REPORT

WBS No:	Cost Center	H $	Weekly Actual	Month Subtotal	Current ACWP	Current BCWP	Previous Month ACWP	Previous Month BCWP	Year to Date BCWS	Work Total EAC	Order Release
01-03-06	4110	H	200	300	300	300	1000	1000	1000	1000	1000
		$	10000	15000	15000	15000	50000	50000	50000	50000	
	4115	H	200	300	300	300	1000	1000	1000	2000	2000
		$	10000	15000	15000	15000	50000	50000	50000	100000	
	4118	H	200	300	300	300	1000	1000	1000	2000	1800
		$	10000	15000	15000	15000	50000	50000	50000	100000	
	41XX	H	600	900	900	900	900	900	900	5000	4800
		$	30000	45000	45000	45000	45000	45000	45000	250000	
	4443	H	100	200	400	360	800	700	1400	2000	1800
		$	6000	12000	24000	22600	48000	42000	84000	120000	
	4446	H	200	400	1000	1200	2000	2000	2300	3000	2500
		$	8000	16000	40000	48000	80000	80000	92000	120000	
	4448	H	300	600	1000	1200	2000	2000	2300	3000	3000
		$	15000	30000	50000	60000	100000	100000	115000	150000	
	44XX	H	600	1200	2400	2760	4800	4700	6000	8000	7300
		$	29000	58000	114000	130600	228000	222000	291000	390000	
	4XXX	H	1200	2100	3300	3660	5700	5600	6900	13000	12100
		$	59000	103000	159000	175600	273000	267000	336000	640000	
	XXXX	H	8000	18000	20000	19000	50000	48000	47000	61000	58000
		$	560000	1260000	1400000	1330000	3500000	3360000	3290000	4270000	

Note: See Table 15–4 and text for explanation of abbreviations.

643

15.4 BUDGETS

The project budget, which is the final result of the planning cycle of the MCCS, must be reasonable, attainable, and based on contractually negotiated costs and the statement of work. The basis for the budget is either historical cost, best estimates, or industrial engineering standards. The budget must identify planned manpower requirements, contract allocated funds, and management reserve.

All budgets must be traceable through the budget "log," which includes:

- Distributed budget
- Management reserve
- Undistributed budget
- Contract changes

The distributed or normal performance budget is the time-phased budget that is released through cost accounts and work packages. Management reserve is generally the dollar amount established for categories of unforeseen problems and contingencies resulting in special out-of-scope work to the performers. Sometimes, people interpret the management reserve as their own little kitty of funds for a special purpose. Below are several interpretations on how the control of the management reserve should be used.

1. The management reserve is actually excess profits and should not be used at all. It should be booked as additional profits as soon as possible. (Accounting)
2. The management reserve should be spent on any activities that add features or additional functionality to the product. Our customers will like that. It will also build up good customer relations for future work. (Marketing)
3. The management reserve should be used for those activities that add value to our company, especially our image in the community. (Senior management)
4. The management reserve should be use as part of risk management in developing mitigation strategies for risks that occur during the execution of the project. Scope changes not originally agreed to should be billed separately to the customer. (Project manager)
5. The management reserve should be used for the additional hours necessary to show that our technical community can exceed specifications rather than merely meeting them. This is our strength. The management reserve should also be used as "seed money" for exploring ideas discovered while working on this project. (Engineering and R&D)

Some people confuse the management reserve with the definition of a "reserve" or "contingency" fund. In the author's opinion, the management reserve is controlled by the project manager of the performing organization and used for escalations in salaries, raw material prices, and overhead rates. The management reserve may also be used for unforeseen problems that may occur. The management reserve should not be used to cover up bad planning estimates or budget overruns.

Also, the management reserve should not be used for scope changes. Scope changes should be paid for out of the customer's reserve or contingency fund. In other words, the management reserve generally applies to the performing organization, whereas the contingency reserve is controlled by the customer for the scope changes that may be requested by the performing organization. There is an exception. If the performing organization is requesting a small scope change, the cost of convening the change control board, paying airfares, meals, and lodgings may be prohibited. In this case, the management reserve may be used and considered as a goodwill activity for the performing organization.

The management reserve should be established based upon the project's risks. Some project may require no management reserve at all, whereas others may necessitate a reserve of 15 percent.

There is always the question of who should get to keep any unused management reserve at the end of the project. If the project is under a firm-fixed price contract, then the management reserve becomes extra profit for the performing organization. If the contract is a cost reimbursable type, all or part of the unused management reserve may have to be returned to the customer.

Although the management reserve may appear as a line item in the work breakdown structure, it is neither part of the distributed budget nor part of the cost baseline. Budgets are established on the assumption that they will be spent, whereas management reserve is money that you try not to spend. It would be inappropriate to consider the management reserve as an undistributed budget.

In addition to the "normal" performance budget and the management reserve budget, there are two other budgets:

● Undistributed budget, which is that budget associated with contract changes where time constraints prevent the necessary planning to incorporate the change into the performance budget. (This effort may be time-constrained.)
● Unallocated budget, which represents a logical grouping of contract tasks that have not yet been identified and/or authorized.

15.5 THE EARNED VALUE MEASUREMENT SYSTEM (EVMS)

In the early years of project management, it became evident that project managers were having difficulty determining project status. Some people believed that status could be determined only by a mystical approach, as shown in Figure 15–10.

The critical question was whether project managers were managing costs or just monitoring costs. The government wanted costs to be managed rather than just monitored, accounted for, or reported. This need resulted in the creation of the EVMS.

The basis for the EVMS, which some consider to be a component of the MCCS, is the determination of earned value. Earned value is a management technique that relates resource planning to schedules and technical performance requirements. Earned value management (EVM) is a systematic process that uses earned value as the primary tool for integrating cost, schedule, technical performance management, and risk management.

FIGURE 15–10. Determining the status.

Without using the EVMS, determining status can be difficult. Consider the following:

- The project
  - A total budget of $1.2 million
  - A 12-month effort
  - Produce 10 deliverables
- Reported status
  - Time elapsed: 6 months
  - Money spent to date: $700,000
  - Deliverables produced: 4 complete, 2 partial

What is the real status of the project? How far along is the project: 40, 50, 60 percent, etc.? Another problem was how to accurately relate cost to performance. If you spent 20 percent of the budget, does that imply that you are 20 percent complete? If you are 30 percent complete, then have you spent 30 percent of the budget?

The EVMS provides the following benefits:

- Accurate display of project status
- Early and accurate identification of trends
- Early and accurate identification of problems
- Basis for course corrections

The EVMS can answer the following questions:

- What is the true status of the project?
- What are the problems?
- What can be done to fix the problems?
- What is the impact of each problem?
- What are the present and future risks?

The EVMS emphasizes prevention over cures by identifying and resolving problems early. The EVMS is an early warning system allowing for early identification of trends and variances from the plan. The EVMS provides an early warning system, thus allowing the project manager sufficient time to make course corrections in small increments. It is usually easier to correct small variances as opposed to large variances. Therefore, the EVMS should be used continuously throughout the project in order to detect the variances while they are small and possibly easy to correct. Large variances are more difficult to correct and run the risk that the cost to correct the large variance may displease management to the point where the project may be canceled.

15.6 VARIANCE AND EARNED VALUE

A variance is defined as any schedule, technical performance, or cost deviation from a specific plan. Variances must be tracked and reported. They should be mitigated through corrective actions and not eliminated through a baseline change unless there is a good reason. Variances are used by all levels of management to verify the budgeting system and the scheduling system. The budgeting and scheduling system variance must be compared because:

- The cost variance compares deviations only from the budget and does not provide a measure of comparison between work scheduled and work accomplished.
- The scheduling variance provides a comparison between planned and actual performance but does not include costs.

There are two primary methods of measurement:

- *Measurable efforts:* Discrete increments of work with a definable schedule for accomplishment, whose completion produces tangible results.
- *Level of effort:* Work that does not lend itself to subdivision into discrete scheduled increments of work, such as project support and project control.

Variances are used on both types of measurement.

In order to calculate variances, we must define the three basic variances for budgeting and actual costs for work scheduled and performed. Archibald defines these variables[3]:

- Budgeted cost for work scheduled (BCWS) is the budgeted amount of cost for work scheduled to be accomplished plus the amount or level of effort or apportioned effort scheduled to be accomplished in a given time period.

PMBOK® Guide, 4th Edition
7.3.2 Cost Control Tools and Techniques
7.3.2.4 Performance Measurement Analysis

- Budget cost for work performed (BCWP) is the budgeted amount of cost for completed work, plus budgeted for level of effort or apportioned effort activity completed within a given time period. This is sometimes referred to as "earned value."
- Actual cost for work performed (ACWP) is the amount reported as actually expended in completing the work accomplished within a given time period.

3. Russell D. Archibald, *Managing High-Technology Programs and Projects* (New York: John Wiley & Sons, 1976), p. 176.

Note: The Project Management Institute has changed the nomenclature in their new version of the PMBOK® Guide whereby BCWS is now PV, BCWP is now EV, and ACWP is now AC. However, the majority of heavy users of these acronyms, specifically government contractors, still use the old acronyms. Until the PMI acronyms are accepted across all industries, we will continue to focus on the most commonly used acronyms.

BCWS represents the time-phased budget plan against which performance is measured. For the total contract, BCWS is normally the negotiated contract plus the estimated cost of authorized but unpriced work (less any management reserve). It is time-phased by the assignment of budgets to scheduled increments of work. For any given time period, BCWS is determined at the cost account level by totaling budgets for all work packages, plus the budget for the portion of in-process work (open work packages), plus the budget for level of effort and apportioned effort.

A contractor must utilize anticipated learning when developing the time-phased BCWS. Any recognized method used to apply learning is usually acceptable as long as the BCWS is established to represent as closely as possible the expected actual cost (ACWP) that will be charged to the cost account/work package.

These costs can then be applied to any level of the work breakdown structure (i.e., program, project, task, subtask, work package) for work that is completed, in-program, or anticipated. Using these definitions, the following variance definitions are obtained:

- Cost variance (CV) calculation:

$$CV = BCWP - ACWP$$

A negative variance indicates a cost-overrun condition.

- Schedule variance (SV) calculation:

$$SV = BCWP - BCWS$$

A negative variance indicates a behind-schedule condition.

In the analysis of both cost and schedule, costs are used as the lowest common denominator. In other words, the schedule variance is given as a function of cost. To alleviate this problem, the variances are usually converted to percentages:

$$\text{Cost variance \% (CVP)} = \frac{CV}{BCWP}$$

$$\text{Schedule variance \% (SVP)} = \frac{SV}{BCWS}$$

The schedule variance may be represented by hours, days, weeks, or even dollars.

As an example, consider a project that is scheduled to spend $100K for each of the first four weeks of the project. The actual expenditures at the end of week four are $325K. Therefore, BCWS = $400K and ACWP = $325K. From these two parameters alone, there are several possible explanations as to project status. However, if BCWP is now known, say $300K, then the project is behind schedule and overrunning costs.

It is important to understand the physical meaning of CV and SV. Consider the following example:

- BCWS = $1000
- BCWP = $800
- ACWP = $700

In this example, the units are dollars. The units could have just as easily been hours, days, or weeks. In this example, CV = $800 − $700 = +$100. Because CV is a positive value, it indicates that physical progress was accomplished at a lower cost than the forecasted cost. This is a favorable situation. Had CV been negative, it would have indicated that physical progress was accomplished at a greater cost than what was forecasted. If CV = 0, then the physical accomplishment was as budgeted.

Although CV is measured in hours or dollars, it is actually a measurement of the efficiency with which physical progress was accomplished compared with the plan. To correct a negative cost variance, emphasis should be placed upon the productivity rate (i.e., burn rate) at which work is being performed.

Returning to the above example, SV = $800 − $1000 = −$200. In this example, the schedule variance is a negative value, indicating that physical progress is being accomplished at a slower rate than planned. This is an unfavorable condition. If the schedule variance were positive, this would indicate physical progress being accomplished at a faster rate than planned. If SV = 0, physical progress is being accomplished as planned.

The schedule variance, SV, measures the timeliness of the physical progress compared to the plan whereas the cost variance, CV, measures the efficiency. To correct a negative schedule variance, emphasis should be placed upon improving the speed by which work is being performed.

The cost variance relates to the real cost. However, the problem with SV is how it relates to the real schedule. The schedule variance is determined from cost account or work package financial numbers and does not necessarily relate to the real schedule. The schedule variance does not distinguish between critical path and non–critical path work packages. The schedule variance by itself does not measure time. A negative schedule variance indicates a behind-schedule condition but does not mean that the critical path has slipped. On the contrary, the real schedule (i.e., precedence networks or the arrow diagramming networks) could indicate that the project will be ahead of schedule. A detailed analysis of the real schedule is still required irrespective of the value for the schedule variance.

Variances are almost always identified as critical items and are reported to all organizational levels. Critical variances are established for each level of the organization in accordance with management policies.

Not all companies have a uniform methodology for variance thresholds. Permitted variances may be dependent on such factors as:

- Life-cycle phase
- Length of life-cycle phase
- Length of project
- Type of estimate
- Accuracy of estimate

Variance controls may be different from program to program. Table 15–2 identifies sample variance criteria for program X.

For many programs and projects, variances are permitted to change over the duration of the program. For strict manufacturing programs (product management), variances may be fixed over the program time span using criteria as in Table 15–2. For programs that include research and development, larger deviations may be permitted during the earlier phases than during the later phases. Figure 15–11 shows time-phased cost variances for a program requiring research and development, qualification, and production phases. Since the risk should decrease as time goes on, the variance boundaries are reduced. Figure 15–12 shows that the variance envelope in such a case may be dependent on the type of estimate.

By using both cost and schedule variance, we can develop an integrated cost/schedule reporting system that provides the basis for variance analysis by measuring cost performance in relation to work accomplished. This system ensures that both cost budgeting and performance scheduling are constructed on the same database.

In addition to calculating the cost and schedule variances in terms of dollars or percentages, we also want to know how efficiently the work has been accomplished. The formulas used to calculate the performance efficiency as a percentage of EV are:

$$\text{Cost performance index (CPI)} = \frac{\text{BCWP}}{\text{ACWP}}$$

$$\text{Schedule performance index (SPI)} = \frac{\text{BCWP}}{\text{BCWS}}$$

TABLE 15–2. VARIANCE CONTROL FOR PROGRAM X

Organizational Level	Variance Thresholds*
Section	Variances greater than $750 that exceed 25% of costs
Section	Variances greater than $2,500 that exceed 10% of costs
Section	Variances greater than $20,000
Department	Variances greater than $2,000 that exceed 25% of costs
Department	Variances greater than $7,500 that exceed 10% of costs
Department	Variances greater than $40,000
Division	Variances greater than $10,000 that exceed 10% of costs

*Thresholds are usually tighter within company reporting system than required external to government. Thresholds for external reporting are usually adjusted during various phases of program (% lower at end).

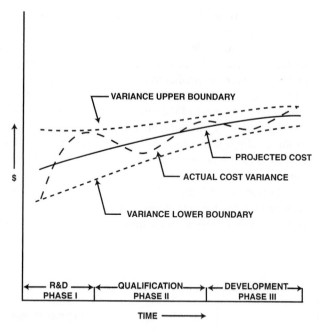

FIGURE 15–11. Project variance projection.

If CPI = 1.0, we have perfect cost performance. If CPI < 1.0, physical progress is being accomplished at a greater cost than forecasted. This is unfavorable. If CPI > 1.0, physical progress is being accomplished at less than the forecasted cost, which is favorable. Similar to CV, CPI measures the efficiency by which the physical progress was accomplished compared to the plan or baseline. For an unfavorable value of CPI, emphasis should be placed upon improving the productivity by which work was being performed.

If SPI = 1.0, we have perfect schedule performance. If SPI < 1.0, physical progress was accomplished at a slower rate than what was planned. This is unfavorable. If SPI > 1.0, physical progress was accomplished at a faster rate than what was planned, which is favorable. For an unfavorable value of SPI, emphasis should be placed upon improving the timeliness of the physical progress.

SPI and CPI are expressed as ratios compared to the performance factor of 1.0 whereas CV and SV are expressed in hours or dollars. One historic reason for this is that SPI and CPI can be used to show performance for a specified time period or trends over a

LIFE-CYCLE PHASE	MANPOWER REQUIRED	$ REQUIRED	TIME DURATION	TYPE OF ESTIMATE	ACCURACY	PERMITTED VARIANCE
MAIN	16,000 HRS.	1,285,600	6 MOS	HISTORY	±5%	±2%

FIGURE 15–12. Methodology to variance.

long time horizon without disclosing actual company sensitive numbers. This makes SPI and CPI valuable tools for customer status reporting without disclosing hard numbers.

The cost and schedule performance index is most often used for trend analysis as shown in Figure 15–13. Companies use either three-month, four-month, or six-month moving averages to predict trends. Trend analysis provides an early warning system and allows managers to take corrective action. Unfortunately, its use may be restricted to long-term projects because of the time needed to correct the situation.

Figure 15–14 shows an integrated cost/schedule system. The figure identifies a performance slippage to date. This might not be a bad situation if the costs are proportionately underrun. However, from the upper portion of Figure 15–14, we find that costs are overrun (in comparison to budget costs), thus adding to the severity of the situation.

Also shown in Figure 15–14 is the management reserve. This is identified as the difference between the contracted cost for projected performance to date and the budgeted cost. Management reserves are the contingency funds established by the program manager

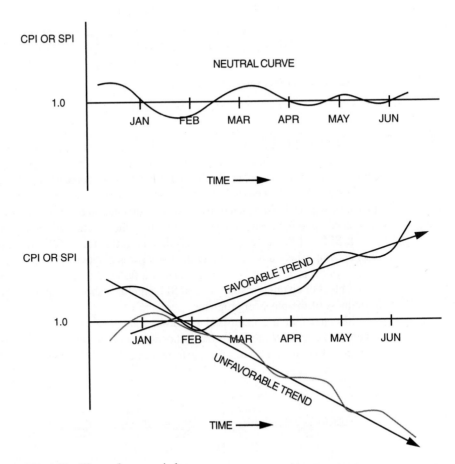

FIGURE 15–13. The performance index.

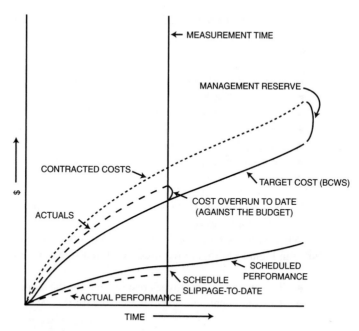

FIGURE 15–14. Integrated cost/schedule system.

to counteract unavoidable delays that can affect the project's critical path. Management reserves cover unforeseen events *within* a defined project scope, but are not used for unlikely major events or changes in scope. These changes are funded separately, perhaps through management-established contingency funds. Actually, there is a difference between management reserves (which come from project budgets) and contingency funds (which come from external sources) although most people do not differentiate. It is a natural tendency for a functional manager (and some project managers) to substantially inflate estimates to protect the particular organization and provide a certain amount of cushion. Furthermore, if the inflated budget is approved, managers will undoubtedly use all of the allocated funds, including reserves. According to Parkinson[4]:

- The work at hand expands to fill the time available.
- Expenditures rise to meet budget.

Managers must identify all such reserves for contingency plans, in time, cost, and performance (i.e., PERT slack time).

The line indicated as actual cost in Figure 15–14 shows a cost overrun compared to the budget. However, costs are still within the contractual requirement if we consider the management reserve. Therefore, things may not be as bad as they seem.

---

4. C. N. Parkinson, *Parkinson's Law* (Boston: Houghton Mifflin, 1957).

Government subcontractors are required to have a government-approved cost/schedule control system. The information requirements that must be demonstrated by such a system include:

- Budgeted cost for work scheduled (BCWS)
- Budgeted cost for work performed (BCWP)
- Actual cost for work performed (ACWP)
- Estimated cost at completion
- Budgeted cost at completion
- Cost and schedule variances/explanations
- Traceability

The last two items imply that standardized policies and procedures should exist for reporting and controlling variances.

When permitted variances are exceeded, cost account variance analysis reports, as shown in Figure 15–15, are required. Required signatures may include:

- The functional employees responsible for the work
- The functional managers responsible for the work

COST ACCOUNT NO/CAM						REPORTING LEVEL		
WBS/DESCRIPTION						AS OF		
COST PERF. DATA			VARIANCE		AT COMPLETION			
	BCWS	BCWP	ACWP	SCH	COST	BUDGET	EAC	VAR.
MONTH TO DATE ($)								
CONTRACT TO DATE ($K)								
PROBLEM CAUSE AND IMPACT								
CORRECTIVE ACTION (INCLUDE EXPECTED RECOVERY DATE)								
COST ACCOUNT MANAGER DATE		COST CENTER MGR. DATE		WBS ELEMENT MANAGER DATE				DATE

FIGURE 15–15. Cost account variance analysis report.

- The cost accountant and/or the assistant project manager for cost control
- The project manager, work breakdown structure element manager, or someone with signature authority from the project office

For variance analysis, the goal of the cost account manager (whether project officer or functional employee) is to take action that will correct the problem within the original budget or justify a new estimate.

Five questions must be addressed during variance analysis:

- What is the problem causing the variance?
- What is the impact on time, cost, and performance?
- What is the impact on other efforts, if any?
- What corrective action is planned or under way?
- What are the expected results of the corrective action?

One of the key parameters used in variance analysis is the "earned value" concept, which is the same as BCWP. Earned value is a forecasting variable used to predict whether the project will finish over or under the budget. As an example, on June 1, the budget showed that 800 hours should have been expended for a given task. However, only 600 hours appeared on the labor report. Therefore, the performance is (800/600) $\times$ 100, or 133 percent, and the task is underrunning in performance. If the actual hours were 1,000, the performance would be 80 percent, and an overrun would be occurring.

The major difficulty encountered in the determination of BCWP is the evaluation of in-process work (work packages that have been started but have not been completed at the time of cutoff for the report). The use of short-span work packages or establishment of discrete value milestones within work packages will significantly reduce the work-in-process evaluation problem, and procedures used will vary depending on work package length. For example, some contractors prefer to take no BCWP credit for a short-term work package until it is completed, while others take credit for 50 percent of the work package budget when it starts and the remaining 50 percent at completion. Some contractors use formulas that approximate the time-phasing of the effort, others use earned standards, while still others prefer to make physical assessments of the work completed to determine the applicable budget earned. For longer work packages, many contractors use discrete milestones with preestablished budget or progress values to measure work performed.

The difficulty in performing variance analysis is the calculation of BCWP because one must predict the percent complete. The simplest formula for calculating BCWP is:

$$BCWP = (\% \text{ complete}) \times BAC$$

Most people calculate "percent complete" based upon task durations. However, a more accurate representation would be to calculate "percent work complete." However, this requires a schedule that is resource loaded. To eliminate this problem, many companies use standard dollar expenditures for the project, regardless of percent complete. For example,

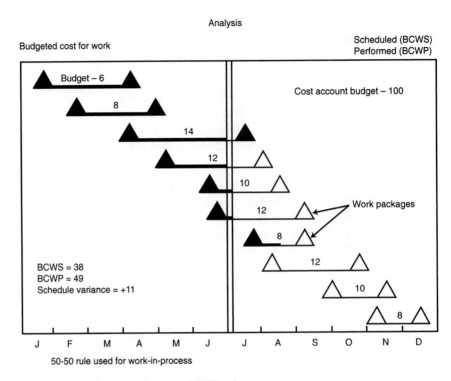

FIGURE 15–16. Analysis showing use of 50/50 rule.

we could say that 10 percent of the costs are to be "booked" for each 10 percent of the time interval. Another technique, and perhaps the most common, is the 50/50 rule:

> Half of the budget for each element is recorded at the time that the work is scheduled to begin, and the other half at the time that the work is scheduled to be completed. For a project with a large number of elements, the amount of distortion from such a procedure is minimal. (Figures 15–16 and 15–17 illustrate this technique.)

One advantage of using the 50/50 rule is that it eliminates the necessity for the continuous determination of the percent complete. However, if percent complete can be determined, then percent complete can be plotted against time expended, as shown in Figure 15–18.

There are techniques available other than the 50/50 rule[5]:

- *0/100:* Usually limited to work packages (activities) of small duration (i.e., less than one month). No value is earned until the activity is complete.
- *Milestone:* This is used for long work packages with associated interim milestones, or a functional group of activities with a milestone established at identified control points. Value is earned when the milestone is completed. In these cases, a budget is assigned to the milestone rather than the work packages.

---

5. These techniques, in addition to the 50/50 method for determining work in progress, are available in software packages.

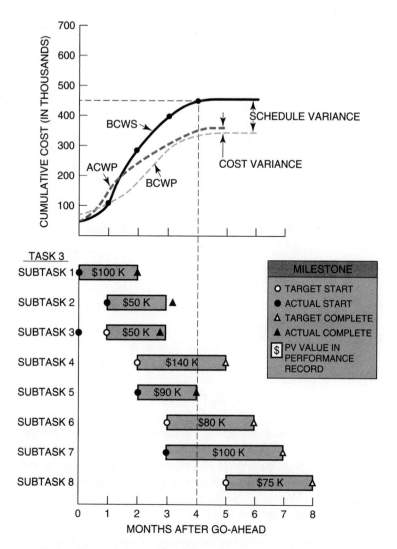

FIGURE 15–17. Project Z, task 3 cost data (contractual).

- *Percent complete:* Usually invoked for long-duration work packages (i.e., three months or more) where milestones cannot be identified. The value earned would be the reported percent of the budget.
- *Equivalent units:* Used for multiple similar-unit work packages, where earnings are on completed units, rather than labor.
- *Cost formula (80/20):* A variation of percent complete for long-duration work packages.
- *Level of effort:* This method is based on the passage of time, often used for supervision and management work packages. The value earned is based on time

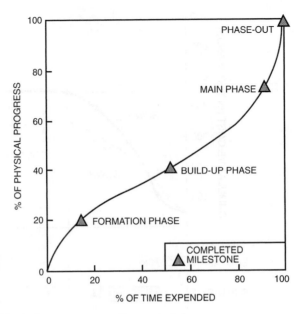

FIGURE 15–18. Physical progress versus time expended.

expended over total scheduled time. It is measured in terms of resources consumed over a given period of time and does not result in a final product.

● *Apportioned effort:* A rarely used technique, for special related work packages. As an example, a production work package might have an apportioned inspection work package of 20 percent. There are only a few applications of this technique. Many people will try to use this for supervision, which is not a valid application. This technique is used for effort that is not readily divisible into short-span work packages but that is in proportion to some other measured effort.

Generally speaking, the concept of earned value may not be an effective control tool if used in the lower levels of the WBS. Task levels and above are normally worth the effort for the calculation of earned value. As an example, consider Figure 15–17, which shows the contractual cost data for task 3 of project Z, and Table 15–3, which shows the cost data status at the end of the fourth month. The following is a brief summary of the cost data for each subtask in task 3 at the end of the fourth month:

● *Subtask 1:* All contractual funds were budgeted. Cost/performance was on time as indicated by the milestone position. Subtask is complete.
● *Subtask 2:* All contractual funds were budgeted. A cost overrun of $5,000 was incurred, and milestone was completed later than expected. Subtask is completed.
● *Subtask 3:* Subtask is completed. Costs were underrun by $10,000, probably because of early start.

TABLE 15–3. PROJECT Z, TASK 3 COST DATA STATUS AT END OF FOURTH MONTH (COST IN THOUSANDS)

Subtasks	Status	BCWS	BCWP	ACWP
1	Completed	100	100	100
2	Completed	50	50	55
3	Completed	50	50	40
4	Not started	70	0	0
5	Completed	90	90	140
6	Not started	40	0	0
7	Started	50	50	25
8	Not started	—	—	—
Total		450	340	360

Note: The data assume a 50/50 ratio for planned and earned values of budget.

- *Subtask 4:* Work is behind schedule. Actually, work has not yet begun.
- *Subtask 5:* Work is completed on schedule, but with a $50,000 cost overrun.
- *Subtask 6:* Work has not yet started. Effort is behind schedule.
- *Subtask 7:* Work has begun and appears to be 25 percent complete.
- *Subtask 8:* Work has not yet started.

To complete our analysis of the status of a project, we must determine the budget at completion (BAC) and the estimate at completion (EAC). Table 15–4 shows the parameters for variance analysis.

- The budget at completion is the sum of all budgets (BCWS) allocated to the project. This is often synonymous with the project baseline. This is what the total effort should cost.
- The estimate at completion identifies either the dollars or hours that represent a realistic appraisal of the work when performed. It is the sum of all direct and indirect costs to date plus the estimate of all authorized work remaining (EAC = cumulative actuals + the estimate-to-complete).

Using the above definitions, we can calculate the variance at completion (VAC):

$$\text{VAC} = \text{BAC} - \text{EAC}$$

TABLE 15–4. THE PARAMETERS FOR VARIANCE ANALYSIS

Question	Answer	Acronym
How much work *should* be done?	Budgeted cost of work scheduled	BCWS
How much work *is* done?	Budgeted cost of work performed	BCWP
How much *did* the "is done" work cost?	Actual cost of work performed (actuals)	ACWP
What was the total job *supposed* to cost?	Budget at completion (total budget)	BAC
What do we *now* expect the total job to cost?	Estimate at completion or latest revised estimate	EAC LRE

The estimate at completion (EAC) is the best estimate of the total cost at the completion of the project. The EAC is a periodic evaluation of the project status, usually on a monthly basis or until a significant change has been identified. It is usually the responsibility of the performing organization to prepare the EAC.

The calculation of a new EAC and subsequent revision does not imply that corrective action has been taken. Consider a three-month task that is 99 percent complete and was budgeted to spend $400K (BCWS). The actual costs to date (ACWP) are $395K. Using the 50/50 rule, BCWP is $200K. The estimated cost-to-complete (EAC) ratio is $395K/$200K, which implies that we are heading for a 100 percent cost overrun. Obviously, this is not the case.

Using the data in Table 15–5, we can calculate the estimate at completion (EAC) by the expression

$$EAC = (ACWP/BCWP) \times BAC = BAC/CPI$$
$$= (360/340) \times 579,000$$
$$= \$613,059$$

where BAC is the value of BCWS at completion.

The discussion of what value to use for BAC is argumentative. In the above calculation, we used burdened direct labor dollars. Some people prefer to use nonburdened labor with the argument that the project manager controls only direct labor hours and dollars. Also, the calculation for EAC did not include material costs or general and administrative costs.

The above calculation of EAC implies that we are overrunning labor costs by 6.38% and that the final burdened labor cost will exceed the budgeted burdened labor cost by $34,059. For a more precise calculation of EAC we would need to include material cost (assumed at $70,000) and G&A. This would give us a final cost, excluding profit, of $751,365, which is an overrun of $37,365. The resulting profit would be $86,000 less $37,365, or $48,635. The final analysis is that work is being accomplished almost on schedule except for subtask 4 and subtask 6, but costs are being overrun.

The question that remains is, "Where is the cost overrun occurring?" To answer this question, we must analyze the cost summary sheet for project Z, task 3. Table 15–5 represents a hypothetical case for the cost elements of project Z, task 3. From Table 15–5 we see that negative (overrun) variances exist for labor dollars, overhead dollars, and material costs. Because labor overhead is measured as a percentage of direct labor dollars, the problem appears to be in the direct labor dollars.

From the contractual column in Table 15–5 the project was estimated at $27.86 per hour direct labor ($241,000/8650 hours), but actuals to date are $150,000/4652 hours, or $32.24 per hour. Therefore, higher-salaried people than anticipated are being employed. This salary increase is partially offset by the fact that there exists a positive variance of 409 direct labor hours, indicating that these higher-salaried employees are performing at a more favorable position than expected on the learning curve. Since the milestones (from Figure 15–17) appear to be on target, work is progressing as planned, except for subtask 4.

TABLE 15–5. PROJECT Z, TASK 3 COST SUMMARY FOR WORK COMPLETED OR IN PROGRESS (COST IN THOUSANDS)

		Cumulative to Date			Cost Variance	Schedule Variance
	Contractual	BCWS	BCWP	ACWP		
Direct labor hours	8650	6712	5061	4652	409	
Direct labor dollars	241	187	141	150	(9)	(46)
Labor overhead (140%)	338	263	199	210	(11)	(64)
Subtotal	579	450	340	360	(20)	
Material dollars	70	66	26	30	(4)	
Subtotal	649					
G&A (10%)	65					
Subtotal	714					
Fee (12%)	86					
Total	800					

Note: This table assumes a 50/50 ratio for planned and earned values of budget.

The labor overhead rate has not changed. The contractual, BCWS, and BCWP overhead rates were estimated at 140 percent. The actuals, obtained from month-end reports, indicate that the true overhead rate is as predicted.

The following conclusions can be drawn:

● Work is being performed as planned (almost on schedule, although at a more favorable position on the learning curve), except for subtask 4, which is giving us a schedule delay.
● Direct labor costs are increasing through the use of higher-salaried employees.
● Overhead rates are as anticipated.
● Direct labor hours must be reduced even further to compensate for increased costs, or profits will be drastically reduced.

This type of analysis could have been carried out to one more level by identifying exactly which departments were using the more expensive employees. This step should probably be completed anyway to see if lower-paid employees are available and can work at the required position on the learning curve. Had the labor costs been a result of increased labor hours, this step would have definitely been necessary to identify the reason for the overrun in-house. Perhaps poor estimating was the cause.

In Table 15–5, there also appears a positive variance in materials. This likewise should undergo further analysis. The cause may be the result of improperly identified hardware, material escalation costs increasing beyond what was planned, increased scrap factors, or a change in subcontractors.

It should be obvious from the above analysis that a detailed investigation into the cause of variances appears to be the best method for identifying causes. The concept of earned value, although a crude estimate, identifies trends concerning the status of specific WBS elements. Using this concept, the budgeted cost for work scheduled (BCWS) may be called planned earned value (PEV), and the budgeted cost for work performed (BCWP) may be referred to as actual earned value (AEV). Earned values are used to determine whether costs are being incurred faster or slower than planned. However, cost overruns do not necessarily mean that there will be an eventual overrun, because the work may be getting done faster than planned.

There are several formulas that can be used to calculate EAC. Using the data shown below, we can illustrate how each of three different formulas can give a different result. Assume that your project consists of these three activities only.

Activity ACWP	% Complete	BCWS	BCWP	
A	100	1000	1000	1200
B	50	1000	500	700
C	0	1000	0	0

Formula I. $EAC = \dfrac{ACWP}{BCWP} \times BAC$

$ = \dfrac{1900}{1500}(3000) = \$3800$

Formula II. $EAC = \dfrac{ACWP}{BCWP} \times \begin{bmatrix} \text{Work completed} \\ \text{and in progress} \end{bmatrix} + \begin{bmatrix} \text{Actual (or revised) cost} \\ \text{of work packages not} \\ \text{yet begun} \end{bmatrix}$

$ = \dfrac{1900}{1500}(2000) + \$1000 = \$3533$

Formula III. $EAC = [\text{Actual to date}] + \begin{bmatrix} \text{All remaining work to be at planned} \\ \text{cost including remaining work in} \\ \text{progress} \end{bmatrix}$

$ = 1900 + [500 + 1000] = \$3400$
$$\uparrow \uparrow$$
$$B C$$

Advantages and disadvantages exist for each formula. Formula I assumes that the burn rate (i.e., ACWP/BCWP) will be the same for the remainder of the project. This is the easiest formula to use. The burn rate is updated each reporting period.

Formula II assumes that all work packages not yet opened will be completed at the planned cost. However, it is possible for planned cost to be revised based upon history from completed work packages.

Formula III assumes that all remaining work is independent of the burn rate incurred thus far. This may be unrealistic unless all remaining work can be reestimated if necessary.

TABLE 15–6. VARIANCE ANALYSIS CASE STUDIES

Case	Planned Earned Value (BCWS)	Actuals (ACWP)	Actual Earned Value (BCWP)
1	800	800	800
2	800	600	400
3	800	400	600
4	800	600	600
5	800	800	600
6	800	800	1,000
7	800	1,000	1,000
8	800	600	800
9	800	1,000	800
10	800	1,000	600
11	800	600	1,000
12	800	1,200	1,000
13	800	1,000	1,200

Other techniques are available for determining final completion costs.[6] The value of the technique selected is based upon the dollar value of the project, the risk, the quality of the cost accounting system, and the accuracy of the estimates. The estimating techniques here use only labor costs. Material costs can be added into each equation to obtain total cost.

Thirteen cases for comparing planned versus actual performance are shown in Table 15–6. Each case is described below using the relationships:

- Cost variance = actual earned value − actuals
- Schedule/performance variances = actual earned value − planned earned value

Case 1: This is the ideal planning situation where everything goes according to schedule.

Case 2: Costs are behind schedule, and the program appears to be underrunning. Work is being accomplished at less than 100 percent, since actuals exceed AEV (or BCWP). This indicates that a cost overrun can be anticipated. This situation grows even worse when we see that we are also 50 percent behind schedule. This is one of the worst possible cases.

Case 3: In this case there is good news and bad news. The good news is that we are performing the work efficiently (efficiency exceeds 100 percent). The bad news is that we are behind schedule.

Case 4: The work is not being accomplished according to schedule (i.e., is behind schedule), but the costs are being maintained for what has been accomplished.

Case 5: The costs are on target with the schedule, but the work is 25 percent behind schedule because the work is being performed at 75 percent efficiency.

Case 6: Because we are operating at 125 percent efficiency, work is ahead of schedule by 25 percent but within scheduled costs. We are performing at a more favorable position on the learning curve.

6. W. Q. Fleming and J. M. Koppelman, "Forecasting the Final Cost and Schedule Results," *PM Network,* January 1996, pp. 13–18.

Case 7: We are operating at 100 percent efficiency and work is being accomplished ahead of schedule. Costs are being maintained according to budget.

Case 8: Work is being accomplished properly, and costs are being underrun.

Case 9: Work is being accomplished properly, but costs are being overrun.

Case 10: Costs are being overrun while underaccomplishing the plan. Work is being accomplished inefficiently. This situation is very bad.

Case 11: Performance is ahead of schedule, and the costs are lower than planned. This situation results in a big Christmas bonus.

Case 12: Work is being done efficiently, and a possible cost overrun can occur. However, performance is ahead of schedule. The overall result may be either an overrun in cost or an underrun in schedule.

Case 13: Although costs are greater than those budgeted, performance is ahead of schedule, and work is being accomplished very efficiently. This is also a good situation.

In each of these cases, the concept of earned value was used to predict trends in cost and variance analysis. This method has its pros and cons.

Each of the critical variances (or earned values) identified usually requires a formal analysis to determine the cause of the variance, the corrective action to be taken, and the effect on the estimate to completion. These analyses are performed by the organization that was assigned the budget (BCWS) at the level of accumulation directed by program management.

Organization-Level Analysis

Each critical variance identified on the organizational MCCS reports may require the completion of MCCS variance analysis procedures by the supervisor of the cost center involved. Analyzing both the work breakdown and organizational structure, the supervisor systematically concentrates his efforts on cost and schedule problems appearing within his organization.

Analysis begins at the lowest organizational level by the supervisor involved. Critical variances are noted at the cost account on the MCCS report. If a schedule variance is involved and the subtask consists of a number of work packages, the supervisor may refer to a separate report that breaks down each cost account into the various work packages that are ahead or behind schedule. The supervisor can then analyze the variance on the basis of the work package involved and determine with the aid of supporting organizations the cause of the variance, the corrective action that can be taken, or the possible effect on associated or future planned effort.

Cost variances involving labor are analyzed by the supervisor on the basis of the performance of his organization in accomplishing the work assigned, within the budgeted man-hours and planned labor rate. The cause of any variance to this performance is determined, and corrective action is then implemented.

Cost variances on nonlabor efforts are analyzed by the supervisor with the aid of the program team member and other supporting organizations.

All material variance analyses are normally initiated by cost accounting as a service to the using organization. These variance analyses are completed, including cause and corrective action, to the extent that can be explained by cost accounting. They are then sent to the using organization, which reviews the analyses and completes those resulting from schedule performance or usage. If a variance is recognized as a change in the material acquisition price, this information is supplied by cost accounting to the responsible organization and a change to the estimate-to-complete is initiated by the using organization.

The supervisor should forward copies of each completed MCCS variance analysis/EAC change form to his higher-level manager and the program team member.

Program Team Analysis
The program team member may receive a team critical variance report that lists variances in his organization at the lowest level of the work breakdown structure at the division cost center level by cost element. Upon request of the program manager, analyses of variances contributing to the variances on the team critical variance report are summarized by the responsible program team member and reviewed with the program manager.

The preparation of status reports, whether they be for internal management or for the customer, should, at a minimum, answer two fundamental questions:

- Where are we today (with respect to time and cost)?
- Where will we end up (with respect to time and cost)?

The information necessary to answer these questions can be obtained from the following formulas:

- Where are we today?
  - Cost variances (in dollars/hours and percent complete)
  - Schedule variances (in dollars/hours and percent complete)
  - Percent complete
  - Percent money spent
- Where will we end up?
  - Estimate at completion (EAC)
  - The remaining critical path
  - SPI (trend analysis)
  - CPI (trend analysis)

Since SPI and CPI are used for trend analyses, we can use CPI and SPI to forecast the expected final cost and the expected end date of the project. We can express the cost at completion, EAC, as:

$$EAC = \frac{BAC}{CPI}$$

The time at completion uses SPI for the forecast and can be expressed as:

$$New\ project\ length = \frac{original\ project\ length}{SPI}$$

Care must be taken with the use of SPI to calculate the new project length because a favorable vale for SPI (i.e., >1.0) could be the result of work packages that are not on the critical path.

Once EAC and the new project length are calculated, we can calculate the variance at completion (VAC) and the estimated cost to complete (ETC) using the following two formulas:

$$\text{VAC} = \text{BAC} - \text{EAC} \qquad \text{and} \qquad \text{ETC} = \text{EAC} - \text{ACWP}$$

Percent complete and percent money spent can be obtained from the following formulas:

$$\text{Percent complete} = \frac{\text{BCWP}}{\text{BAC}}$$

$$\text{Percent money spent} = \frac{\text{ACWP}}{\text{BAC}}$$

where BAC is the budget at completion.

The program manager uses this information to review the program status with upper-level management. This review is normally on a monthly basis on large projects. In addition, the results of these analyses are used to explain variances in the contractually required reports to the customer.

After the analyses of the variances have been made, reports must be developed for both the customer and in-house (upper-level) management. Customer reporting procedures and specifications can be more detailed than in-house reporting and are often governed by the contract. Contractual requirements specify the reports required, the frequency of submission and distribution, and the customer regulation that specifies the preparation instructions for the report.

The types of reports required by the customer and management depend on the size of the program and the magnitude of the variance. Most reports contain the tracking of the vital technical parameters. These might include:

● The major milestones necessary for project success
● Comparison to specifications
● Types or conditions of testing
● Correlation of technical performance to the activity network and the work breakdown structure

One final note about reports: To save time and money, reports might be only one or two pages or fill-in-the-blank forms.

15.7 THE COST BASELINE

PMBOK® Guide, 4th Edition
7.2.3 Cost Base line

Once the project is initiated, the project team establishes the cost or financial base-line against which status will be reported and variances will be measured. Figure 15–19 represents a cost baseline. Each block represents

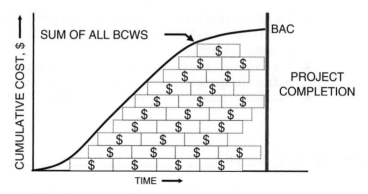

FIGURE 15–19. The cost baseline.

a cost account or work package element. The summation of all of the cost accounts or work packages would then equal the time-phased budget. Each work package would then be described through the work authorization form for that work package.

The cost baseline in Figure 15–19 is just part of the cost breakdown. An illustration of a cost breakdown appears in Figure 15–20.

There are certain distinguishing features of Figure 15–20:

- The time-phased budget, which is the released budget, is the summation of all BCWS elements.

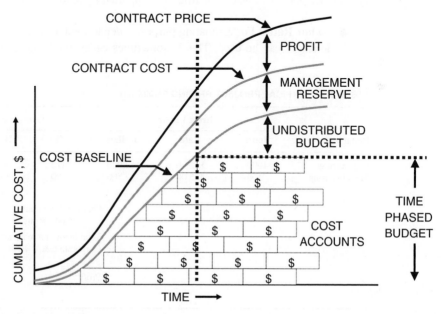

FIGURE 15–20. WBS level 1 cost breakdown.

- The cost baseline is the summation of the time-phased budget (i.e., the distributed budget) and the undistributed budget. This will equal the released, planned budget at completion (BAC).
- The contractual cost to complete the project is the summation of the cost baseline and the management reserve, assuming that a management reserve exists.
- The contract price is the contract cost plus the profit, if any.

15.8 JUSTIFYING THE COSTS

Project pricing is often based upon best guesses rather than concrete estimates. This is particularly true for companies that survive on competitive bidding and where the preparation cost of a bid may vary between $50,000 and $500,000. If the probability of winning a bid is low, then the company may spend the minimum amount of time and cost during bid preparation.

Table 15–7 shows a typical project pricing summary.

In Table 15–7, each functional area or division can have its own overhead rate. In this summary, the overhead rate for engineering is 110 percent, whereas the manufacturing overhead rate is 200 percent. If this company is a subsidiary of a larger company, then a corporate general and administrative (G&A) cost may be included. If the project is for an external customer, then a profit margin will be included.

Once the project pricing summary is completed, the costs must be justified before some executive committee. This is shown in Figure 15–21.

Every company has its own evaluation criteria cost summary approval process. Typical elements that must be justified or supported by hard data include:

- **Labor Rates:** For estimating purposes, department averages or skill set weighted averages can be used. This is sometimes called the blended rate. The best-case

TABLE 15–7. TYPICAL PROJECT PRICING SUMMARY

Department	Direct Labor			Overhead		Total
	Hours	Rate	Dollars	%	Dollars	
Engineering	1000	$42.00	42,000	110	46,200	$88,200
Manufacturing	500	$35.00	17,500	200	35,000	$52,500
					Total Labor	$140,700
				Other: Subcontracts	$10,000	
				Consultants	$2,000	$12,000
				Total labor and material		$152,700
				Corporate G&A: 10%		$ 15,270
						$167,970
				Profit: 15%		$ 25,196
						$193,166

❖Justifying the assumptions

- **–Labor rates**
- **–Use of overtime**
- **–Scrap factors**
- **–Risks**
- **–Hidden costs**

(If the assumptions are correct and justified, then the final price is most likely correct and acceptable to management.)

FIGURE 15–21.   Justifying the cost (and getting sign-off).

scenario would be estimating from the actual salary or skill set of the workers to be assigned. This may be impossible during competitive bidding because we do not know who will be available or who will be assigned assuming the contract is received. Also, if the project is a multiyear effort, we may need forward pricing rates, which are the predicted, full burdened salaries anticipated in the next few years. This is illustrated in Table 15–8.

● **Overtime:** If resources are scarce and the company has no intention of hiring additional resources, then some of the work must be accomplished on overtime. This could increase the cost of the project and an allowance must be made for possible mistakes made during this period of excessive overtime.

● **Scrap Factors:** If the project includes procurement of raw materials, then some scrap factor allowance may be necessary. This calculation may be impacted by the skill set of the resources assigned and using the materials, previous experience using these materials, and experience on these types of projects.

TABLE 15–8.   FORWARD PRICING RATES: SALARY (Departmental Pay Structure)

		Salary (per hour)		
Pay Grade	**Title**	**2009**	**2010***	**2011***
9	Engineering Consultant	$53	$56	$60
8	Senior Engineer	48	50	53
7	Engineer	39	42	45
6	Junior Engineer	34	36	39
5	Apprentice Engineer	29	31	34

*Projected rates.

Cost of Capital

Shipping/Postage

Travel

Attending Meetings

FIGURE 15–22.   Other often hidden costs.

- **Risks:** Risk analysis may be based upon the quality of the estimates and experience of those who made the estimates. Other risks considered include the company's ability to achieve the anticipated benefits or the designated profits and, if a disaster occurs, the company's exposure and liability for lawsuits.
- **Hidden Costs:** These costs, some of which are illustrated in Figure 15–22, can erode all of the profitability expected on a project. Another potentially hidden cost is the yearly or monthly workload availability. A typical calculation appears in Table 15–9. If we use Table 15–9 and all of the workers are long-term employees, then there may be less than 1840 hours available per year because senior people may have earned more than three weeks of vacation per year.

TABLE 15–9.   HOURS AVAILABLE FOR WORK

Hours available per year (52 × 40):	2080 hours
Vacation (3 weeks):	−120 hours
Sick leave (3 days):	−24 hours
Paid holidays (11 days):	−88 hours
Jury duty (1 day):	−8 hours
	1840 hours

(1840 hours/year) ÷ 12 months = 153 hours/month

15.9 THE COST OVERRUN DILEMMA

The lifeblood of most organizations is a continuous stream of new products or services. Because of the word "new," historical data may be at a minimum and cost overruns are expected. Figure 15–23 shows a typical range of overruns.

Rough order-of-magnitude (ROM) estimates are often made from "soft" data, which can result in a wide range of overruns, and are used in the initiation phase of a project. As we go from soft data to hard data and enter the planning phase of a project, the accuracy of the estimates improves and the range of the overruns narrows.

When overruns occur, the project manager looks for ways of reducing costs. The simplest way is to reduce scope. This begins with a search for items that are easy to cut. The items that are easiest to cut are those items that were poorly understood during the estimating process and were therefore underestimated. Typical items that are cut or reduced in magnitude include:

- Project management supervision
- Line management supervision
- Process controls
- Quality assurance
- Testing

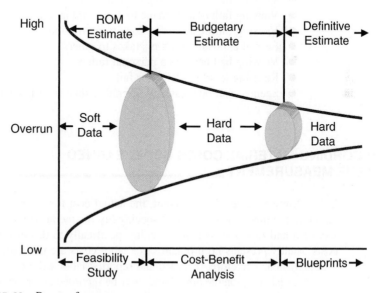

FIGURE 15–23. Range of overruns.

If the easy-to-cut items do not provide sufficient cost reductions, then a desperate search begins among the hard-to-cut items. Hard-to-cut items include:

- Direct labor hours
- Materials
- Equipment
- Facilities
- Others

If the cost reductions are unacceptable to management, then management must decide whether or not to pull the plug and cancel the project. Pulling the plug may seem like an easy decision, but it turns out to be one of the most difficult decisions for executives to make. Typical reasons for not pulling the plug include:

- Quantitative reasons
  - High exit barriers
  - Significant expenditures have been made and are unrecoverable
  - Penalty clauses
  - Breach-of-contract lawsuits
  - Payments to terminated workers
  - Low salvage value of goods and property
  - High plant closing costs
  - Moving people may end up violating seniority and labor agreements
- Qualitative reasons
  - Viewing failure as a sign of weakness
  - Viewing failure as damage to one's career
  - Viewing failure as damage to one's reputation
  - Viewing failure as a roadblock to promotion
  - Fear of exposing one's mistakes to others
  - Viewing bad news as a personal failure
  - Refusing to admit defeat or failure
  - Seeing what one wants to see rather than seeing reality

15.10 RECORDING MATERIAL COSTS USING EARNED VALUE MEASUREMENT

Using "earned value" measurement, the actual cost for work performed represents those direct and indirect costs identified specifically for the project (contract) at hand. Both the *recorded* and *reported* costs must relate specifically to this effort. Recording direct labor costs usually presents no problem since labor costs are normally recorded as the labor is accomplished. Therefore, recorded and reported labor will be the same.

Material costs, on the other hand, may be recorded at various times. Material costs can be recorded as commitments, expenditures, accruals, and applied costs. All provide useful information and are important for control purposes.

Because of the choices available for material cost analysis, material costs should be reported *separately* from the standard labor hour/labor dollar earned value report. For example, cost variances associated with the procurement of material may be determined at the time that the purchase orders are negotiated and placed with the vendors since this information provides the *earliest* visibility of potential cost variance problems. Significant variances in the anticipated and actual costs of materials can have a serious effect on the total contract cost and should be reflected promptly in the estimated cost at completion (EAC) and explained in the narrative part of the project status report.

Separating labor from material costs is essential. Consider the following example:

Example 15–1. You are budgeted to spend $1,000,000 in burdened labor and $600,000 in material. At the end of the first month of your project, the following information is made available to you:

$$
\begin{aligned}
\textit{Labor:} \quad \text{ACWP} &= \$90,000 \\
\text{BCWP} &= \$100,000 \\
\text{BAC} &= \$1,000,000 \\
\textit{Material:} \quad \text{ACWP} &= \$450,000 \\
\text{BCWP} &= \$400,000 \\
\text{BAC} &= \$600,000
\end{aligned}
$$

For simplicity's sake, let us use the following formula for EAC:

$$\text{EAC} = (\text{ACWP/BCWP}) \times \text{BAC}$$

Therefore,

EAC(labor)   = $900,000

EAC(material) = $675,000

If we add together both EACs, the estimated cost at completion will be $1,575,000, which is $25,000 *below* the planned budget. If the costs are combined before we calculate EAC, then

$$\text{EAC} = [(\$450,000 + \$90,000)/\$500,000] \times (\$1,600,000) = \$1,728,000$$

which is a $128,000 *overrun*. Therefore, it is usually best to separate material from labor in status reporting.

Another major problem is how to account for the costs of material placed on order, which does *not* reflect the cost of work completed and is not normally used in status reporting. For performance measurement purposes, it is desirable that material costs be recorded at the time that the materials are received, paid for, or used rather than as of the time that they are ordered. Therefore, the actual costs reported for materials should be derived in accordance with established procedures, and normally will be recorded for earned value measurement purposes at or after time of material receipt. In addition, costs should always be recorded on the same basis as budgets are prepared in order to make comparisons between budgeted and actual costs meaningful. For example, material should not be budgeted on the basis of when it is used and then have its costs collected/reported on the basis of when it is received. Consider the following situations:

Situation I: An equipment manufacturer receives a contract to build five machines for the *same* customer, but each machine is slightly different. The manufacturer purchases and receives five of the same electric motors, one for each machine. What is the earliest time that the manufacturer should take credit for the electric motors?

  a. When ordered
  b. When received
  c. When paid for
  d. When withdrawn from inventory
  e. When installed

Situation II: The same manufacturer has purchased large quantities of steel plate for the five machines as well as for machines for other customers. By ordering in large quantities, the manufacturer received a substantial price break. What is the earliest time the manufacturer should take credit for the steel plate?

  a. When ordered
  b. When received
  c. When paid for
  d. When withdrawn from inventory
  e. When installed

Situation III: Assume that the manufacturer in Situation II purchases the steel plate for a single customer rather than for multiple customers. What is the earliest time the manufacturer should take credit for the steel plate?

  a. When ordered
  b. When paid for
  c. When received
  d. When applied

In Situations I and III, the recommended answer is "when received." In Situation II, any answer can be argued, but the preferred answer is "when installed."

15.11 THE MATERIAL ACCOUNTING CRITERION[7]

At a minimum, the contractor's material accounting system must provide for the following:

a. Accurate cost accumulation and assignment of costs to cost accounts in a manner consistent with budgets using recognized, acceptable costing techniques.

b. Determination of material price variances by comparing planned versus actual commitments.

c. Cost performance measurement at the point in time most suitable for the category of material involved, but no earlier than the time of actual receipt of material.

d. Determination of material cost variances attributable to the excess usage of material.

e. Determination of unit or lot costs when applicable.

f. Full accountability for all material purchased for the project, including residual inventory.

In order to satisfy these six system requirements, the following accounting practices should be adhered to:

a. The material cost actuals (ACWP) must equate to its material plans (BCWS), and be carried down to the cost account level of the WBS.

b. The material price variances must be determinable by comparing planned commitments (estimated material value) to actual commitments (actual cost of the material).

c. Physical work progress or earned value (BCWP) must be determinable, but not before the materials have been received.

d. Usage cost variances (to be discussed in the next section) must be determinable from excess material usage.

e. Material unit costs and/or lot costs must be determinable, as applicable.

f. There must be full accountability of all materials purchased, including any residual material inventory.

Although this task appears difficult on the surface, it is easy if the organization focuses on two areas:

1. *The material plans (BCWS):* These frequently start at the point at which engineering or manufacturing or others have provided a definition sufficient to initiate an order for the items, regardless of when such items are actually ordered or received.

2. *The material actuals (ACWP):* This is ordinarily the point at which the costs of the parts are recorded on the firm's accounting books, that is, when the bill is paid.

Those firms that have a material commitment system in use as part of the material accounting system are usually able to establish and update the costs for their purchased goods at

---

7. Adapted from Quentin W. Fleming, *Cost/Schedule Control Systems Criteria* (Chicago: Probus Publishers, 1992), pp. 144–145.

multiple points: as an estimated liability when engineering or manufacturing defines the requirements; still as an estimated liability when someone formally initiates the request; updated to an accrued liability when an order is placed by purchasing; later updated to an actual liability when parts are received and accepted; and updated a final time when the bill is paid and the costs are recorded on the accounting books.

15.12 MATERIAL VARIANCES: PRICE AND USAGE[8]

One of the requirements of a material accounting system is that it be able to determine just why material budgets were exceeded; this is called variance analysis. When the actual material costs exceed a material budget, there are normally two causes:

1. The articles purchased cost more than was planned, called a "price variance."
2. More articles were consumed than were planned, called a "usage variance."

Price variances (PV) occur when the budgeted price value (BCWS) of the material was different than what was actually experienced (ACWP). This condition can arise for a host of reasons: poor initial estimates, inflation, different materials used than were planned, too little money available to budget, and so on.

The formula for price variance (PV) is:

$$PV = (\text{Budgeted price} - \text{Actual price}) \times (\text{Actual quantity})$$

Price variance is the difference between the budgeted cost for the bill of materials and the price paid for the bill of materials.

By contrast, usage variances (UV) occur when a greater quantity of materials is consumed than were planned. The formula for usage variance (UV) is:

$$UV = (\text{Budgeted quantity} - \text{Actual quantity}) \times (\text{Budgeted price})$$

Normally, usage variances are the resulting costs of materials used over and above the quantity called for in the bill of materials.

Consider the following example: The project manager establishes a material budget of 100 units (which includes 10 units for scrap factor) at a price of $150 per unit. Therefore, the material budget was set at $15,000. At the end of the short project, material actuals (ACWP) came in at $15,950, which was $950 over budget. What happened?

8. Adapted from Quentin W. Fleming, *Cost/Schedule Control Systems Criteria* (Chicago: Probus Publishers, 1992), pp. 151–152.

Applying the formulas defined previously,

Price variance (PV) = (BCWS price − ACWP price) × Actual quantity
= ($150 per unit − $145 per unit) × 110 units
= $550 favorable

Usage variance (UV) = (BCWP qty − ACWP qty) × BCWS price
= (100 units − 110 units) × $150 per unit
= $1,500 unfavorable

The analysis indicates that your purchase price was less than you anticipated, thus generating a cost savings. However, you used 10 units more than planned for, thus generating an unfavorable usage variance. Further investigation indicated that your line manager had increased the scrap factor from 10 to 20 units.

Good business practices indicate that such variance analyses take place to determine why actual material costs exceed the budgeted material values.

15.13 SUMMARY VARIANCES

Summary variances can be calculated for both labor and material. Consider the information shown below:

	Direct Material	Direct Labor
Planned price/unit	$ 30.00	$ 24.30
Actual units	17,853	9,000
Actual price/unit	$ 31.07	$ 26.24
Actual cost	$554,630	$236,200

We can now calculate the total price variance for direct material and the rate cost variance:

- *Total* price variance for direct material
  = Actual units × (BCWP − ACWP)
  = 17,853 × ($30.00 − $31.07)
  = $19,102.71 (unfavorable)
- Labor *rate* cost variance
  = Budgeted rate − Actual rate
  = $24.30 − $26.24
  = $1.94 (unfavorable)

15.14 STATUS REPORTING

PMBOK® Guide, 4th Edition
10.5.3.1 Performance Reporting

One of the best ways of reducing executive meddling on projects is to provide executives with frequent, meaningful status reports. Figure 15–24 shows a relatively simple status report based upon data accumulation in the form of Figures 15–25 and 15–26. These types of status reports should be short and concise, containing pertinent information only. Status can also be shown graphically as in Figure 15–27. The difference between Figure 15–27 and 15–17 is that at-completion estimates have been identified.

As the available project management software becomes more sophisticated, so does project reporting. There are four types of reports that are generally printed out from the earned value measurement system:

- **Performance Reports:** These reports indicate the physical progress to date, namely, BCWS, BCWP, and ACWP. The report might also include information on material procurement, delivery, and usage, but most companies have separate reports on materials.
- **Status Reports:** These reports identify where we are today and use the information from the performance reports to calculate SV and CV.
- **Projection Reports:** These reports calculate EAC, ETC, SPI, and CPI as well as any other forward-looking projections. These reports emphasize where we will end up.
- **Exception Reports:** These reports identify exceptions, problems, or situations that exceed the threshold limits on such items as variances, cash flow, resources assigned, and other such topics.

Reporting procedures for variance analysis should be as brief as possible. The reason for this is simple: the shorter and more concise the report, the faster that feedback can be generated and responses developed. Time is critical if rescheduling must be accomplished with limited resources. The two most common situations providing constraints on resource rescheduling are that:

- The end date is fixed
- The resources available are constant (or limited)

With a fixed end date, program rescheduling generally requires that additional resources be supplied. In the second situation, program slippage may be the only alternative unless a constant stream of resources can be redistributed so as to shorten the length of the critical path.

Once the variance analysis is completed, both project and functional management must diagnose the problem and search for corrective actions. This includes:

- Finding the cure for the problem
- Developing a plan to recover the position

1. VARIANCE ANALYSIS (cost in thousands)

Subtask	Milestone Status	Budgeted Cost Work Schedule	Budgeted Cost Work Performed	Actual Cost	Variance, % Schedule	Variance, % Cost
1	Completed	100	100	100	0	0
2	Completed	50	50	55	0	−10
3	Completed	50	50	40	0	20
4	Not started	70	0	0	−100	—
5	Completed	90	90	140	0	−55.5
6	Not started	40	0	0	−100	—
7	Started	50	50	25	0	50
8	Not started	0	0	0	—	—
Total		450	340	360	−24.4	−5.9

2. ESTIMATE AT COMPETION (EAC)

$$EAC = (360/340) \times \$579,000 = \$613,059$$
$$Overrun = 613,059 - 579,000 = \$34,059$$

3. COST SUMMARY

Costs are running approximately 5.9% over budget of higher-salaried labor.

4. SCHEDULE SUMMARY

The 24.4% behind-schedule condition is due to subtasks 4 and 6, which have not yet begun owing to lack of raw materials and the 50/50 method for booking costs. Overtime will get us back on schedule but at an additional cost of 2.5% of direct labor costs.

5. MILESTONE REPORT

Milestone/Subtask	Scheduled Completion	Projected Completion	Actual Completion
1	4/1/94		4/1/94
2	5/1/94		5/8/94
3	5/1/94		4/23/94
4	7/1/94	7/1/94	
5	6/1/94		6/1/94
6	8/1/94	8/1/94	
7	9/1/94	9/1/94	
8	10/1/94	10/1/94	

6. EVENT REPORT

Current Problem	Potential Impact	Corrective Action
(a) Lack of raw materials.	Cost overrun and behind schedule condition.	Overtime is scheduled. We will try to use lower-salaried people. Raw materials are expected to be on dock next week.
(b) Customer unhappy with test results, and wants additional work.	May need additional planning.	Customer will provide us with revised statement of work on 6/15/94.

FIGURE 15–24. Blue Spider Project, monthly project report #4.

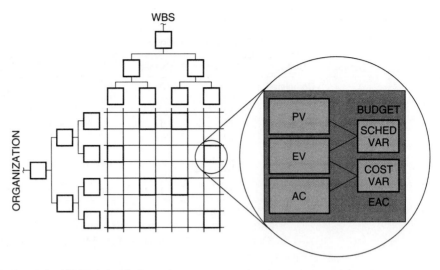

FIGURE 15–25. Data accumulation.

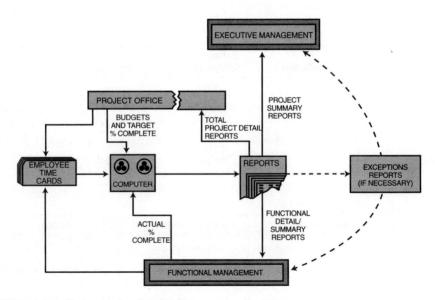

FIGURE 15–26. Cost control and report flow.

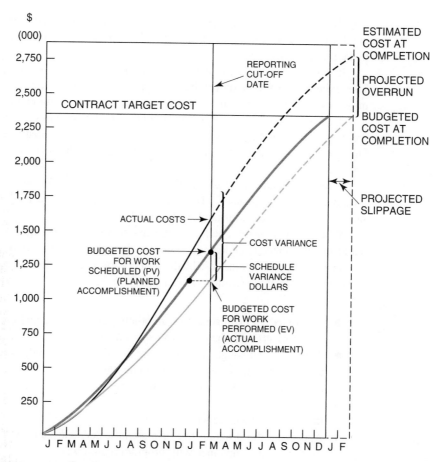

FIGURE 15–27. Graphical status reporting.

This by no means implies that all variances require corrective action. There are four major responses to a variance report:

- Ignoring it
- Functional modification
- Replanning
- System redesign

Permissible variances exist for all levels of the organization. If the variance is within these permitted deviations, then there will be no response, and the variance may be ignored. In some situations where the variance is marginal (or even within limits), corrective action may be required. This would normally occur at the functional level and might simply involve using another test procedure or possibly considering some alternative not delineated in the program plan.

If major variances occur, then either replanning or system redesign must take place. The replanning process requires the redefining and reestablishing of project goals as work progresses, but always within system specifications. This might include making trade-offs in time, cost, and performance or defining new project activities and methods of pursuing the project, such as new PERT networks. If resources are limited, then a proper redistribution or reallocation must be made. If resources are not limited, the additional personnel, financing, equipment, facilities, or information may be required.

If replanning cannot be accomplished without system redesign, then system specifications may have to be changed.[9] This is the worst possible case because performance may be sacrificed to satisfy the constraints of time and money.

Whenever companies operate on a matrix structure, information must be carefully prepared and distributed to all key individuals in the organization. To avoid dual standards and red tape, management must establish the decision-making policies associated with cost and control systems. The following is a policy guide:

- Approving all estimates, and negotiating all estimates and the definition of work requirements with the respective organizations.
- Approving the budget, and directing distribution and budgeting of available funds to all organizational levels by program element.
- Defining the work required and the schedule.
- Authorizing work release. The manager may not, however, authorize work beyond the scope of the contract.
- Approving the program bill-of-materials, detailed plans, and program schedules for need and compliance with program requirements.
- Approving the procuring work statement, the schedules, the source selection, the negotiated price, and the type of contract on major procurement.
- Monitoring the functional organization's performance against released budgets, schedules, and program requirements.
- When cost performance is unacceptable, taking appropriate action with the affected organization to modify the work requirements or to stimulate corrective action within the functional organization so as to reduce cost without changing the contracted scope of work.
- Being responsible for all communications and policy matters on contracted programs so that no communicative directives shall be issued without the signature or concurrence of the program manager.

---

9. Here we are discussing system specifications. Functional modification responses can also require specification changes, but not on the system level. Examples of functional modifications might be changes in tolerances for testing or for purchasing raw materials.

TABLE 15-10. PROGRAM INTERRELATIONSHIPS

Program Manager	Functional Manager	Relationship
Makes or approves all decisions that affect the contractually committed target time, cost, and performance requirements or objectives of the program.	Assembles and furnishes the information needed to assist the program manager in making decisions. Submits to the program manager all proposed changes that affect program cost, schedule targets, and technical requirements and objectives through the program team member.	Management controls, contract administration, budgeting, estimating, and financial controls are a functional specialty. The program manager utilizes the services of the specialist organizations. The specialists retain their own channels to the general manager but must keep the program manager informed through the program team member.
Approves all engineering change control decisions that affect the contractually committed target time, cost, and performance requirements or objectives of the program.	Implements engineering change decisions approved by the program manager. Advises him of any resulting programming impasses and negotiates adjustments through the program team member.	
Establishes program budgets in conjunction with the cognizant program team members; monitors and negotiates changes.		In all matters pertaining to budget and cost control, the program manager utilizes the services of the program team member representing the cognizant financial control organization.
Authorizes release of the budget and work authorization for the performance of approved work, and negotiates any intradirectorate reallocation above section level with the affected functional organizations through the program team members.	Within the allocated budget, provides manpower skills, facilities, and other resources pertaining to his functional specialty to the degree and level necessary to meet program schedule, cost, and technical performance requirements of the contract.	

(Continued)

TABLE 15–10. (Continued)

Program Manager	Functional Manager	Relationship
Requests the assignment of program team members to the program, and approves the release of the team member from the program.	Coordinates with the program manager in the selection and assignment of a program team member to the program or release of the program team member from the program.	Program manager does not hire or fire functional personnel. Program team members should not be removed from the program without the concurrence of the program manager.
Establishes report requirements and controls necessary for evaluation of all phases of program performance consistent with effective policies and procedures.	Works in concert with other functional organizations to ensure that he and they are proceeding satisfactorily in the completion of mutually interdependent program tasks and events.	Insofar as possible, program controls must be satisfied from existing data and controls as defined by division policies and procedures.
Measures and evaluates performance of tasks against the established plan. Identifies current and potential problems. Decides upon and authorizes corrective action.	Follows up all activities of his organization to ensure satisfactory performance to program requirements. Detects actual or potential problems. Takes timely corrective action in his organization, and when such problems involve interface with other functional organizations, notifies them and coordinates the initiation of mutually satisfactory remedial action. Keeps the program manager advised (through the program team member) of conditions affecting the program, existing, or expected problems, problems solved, and corrective action required or performed.	The program manager directs or redirects activities of functional organizations only through the cognizant program team member. Functional managers are responsible for the performance of their organizations. Functional managers do not implement decisions involving increased total program costs, changes in schedule, or changes in technical performance without prior approval of the program team members and the program manager.
Apprises the program team members and/or functional organizations of program changes affecting their function.		
Assures the establishment, coordination, and execution of support programs to the extent required or permitted by the contract.		This includes such programs as value engineering, data management, and configuration management.

Describing the responsibilities of a manager is only a portion of the management policy. Because the program manager must cross over functional boundaries to accomplish all of the above, it is also necessary to describe the responsibilities of the functional manager and the relationship between the two. Table 15–10 is an example of this. Similar tables can be developed for planning and scheduling, communications, customer relations, and contract administration.

15.15 COST CONTROL PROBLEMS

| PMBOK® Guide, 4th Edition |
| 7.3 Cost Control |

No matter how good the cost and control system is, problems can occur. Common causes of cost problem include:

- Poor estimating techniques and/or standards, resulting in unrealistic budgets
- Out-of-sequence starting and completion of activities and events
- Inadequate work breakdown structure
- No management policy on reporting and control practices
- Poor work definition at the lower levels of the organization
- Management reducing budgets or bids to be competitive or to eliminate "fat"
- Inadequent formal planning that results in unnoticed, or often uncontrolled, increases in scope of effort
- Poor comparison of actual and planned costs
- Comparison of actual and planned costs at the wrong level of management
- Unforeseen technical problems
- Schedule delays that require overtime or idle time costing
- Material escalation factors that are unrealistic

Cost overruns can occur in any phase of project development. The most common causes for cost overruns are:

- Proposal phase
  - Failure to understand customer requirements
  - Unrealistic appraisal of in-house capabilities
  - Underestimating time requirements
- Planning phase
  - Omissions
  - Inaccuracy of the work breakdown structure
  - Misinterpretation of information
  - Use of wrong estimating techniques
  - Failure to identify and concentrate on major cost elements
  - Failure to assess and provide for risks

- Negotiation phase
  - Forcing a speedy compromise
  - Procurement ceiling costs
  - Negotiation team that must "win this one"
- Contractual phase
  - Contractual discrepancies
  - SOW different from RFP requirements
  - Proposal team different from project team
- Design phase
  - Accepting customer requests without management approval
  - Problems in customer communications channels and data items
  - Problems in design review meetings
- Production phase
  - Excessive material costs
  - Specifications that are not acceptable
  - Manufacturing and engineering disagreement

15.16 STUDYING TIPS FOR THE PMI® PROJECT MANAGEMENT CERTIFICATION EXAM

This section is applicable as a review of the principles to support the knowledge areas and domain groups in the PMBOK® Guide. This chapter addresses:

- Scope Management
- Cost Management
- Initiating
- Planning
- Controlling

Understanding the following principles is beneficial if the reader is using this text to study for the PMP® Certification Exam:

- What is meant by a management cost and control system
- What is meant by earned value measurement
- The meaning of control
- Code of cost accounts
- Work authorization for and its relationship to the code of accounts
- Sources of funds for a project or changes to a project
- Four primary elements of cost monitoring and control: BCWS, BCWP, ACWP, and BAC
- How to calculate the cost and schedule variances, in hours, dollars, and percentages
- Importance of SPI and CPI in trend analysis

- Ways to forecast the time and cost to completion as well as variances at completion
- Different types of reports: performance, status, forecasting, and exception
- Use of the management reserve
- Escalation factors and how they affect a project
- What is a cost or financial baseline for a project
- Different ways to calculate either BCWP or percent complete

The following multiple-choice questions will be helpful in reviewing the principles of this chapter:

1. In earned value measurement, earned value is represented by:
   A. BCWS
   B. BCWP
   C. ACWP
   D. None of the above

2. If BCWS = 1000, BCWP = 1200, and ACWP = 1300, the project is:
   A. Ahead of schedule and under budget
   B. Ahead of schedule and over budget
   C. Behind schedule and over budget
   D. Behind schedule and under budget

3. If BAC = $20,000 and the project is 40 percent complete, then the earned value is:
   A. $5000
   B. $8000
   C. $20,000
   D. Cannot be determined

4. If BAC = $12,000 and CPI = 1.2, then the variance at completion is:
   A. −$2000
   B. +$2000
   C. −$3000
   D. +$3000

5. If BAC = $12,000 and CPI = 0.8, then the variance at completion is:
   A. −$2000
   B. +$2000
   C. −$3000
   D. +$3000

6. If BAC for a work package is $10,000 and BCWP = $4,000, then the work package is:
   A. 40 percent complete
   B. 80 percent complete
   C. 100 percent complete
   D. 120 percent complete

7. If CPI = 1.1 and SPI = 0.95, then the trend for the project is:
   A. Running over budget but ahead of schedule
   B. Running over budget but behind schedule
   C. Running under budget but ahead of schedule
   D. Running under budget but behind schedule

8. The document that describes a work package, identifies the cost centers allowed to charge against this work package, and establishes the charge number for this work package is the:
   A. Code of accounts
   B. Work breakdown structure
   C. Work authorization form
   D. None of the above

9. Unknown problems such as escalation factors are often budgeted for using the:
   A. Project manager's charge number
   B. Project sponsor's charge number
   C. Management reserve
   D. Configuration management cost account

10. EAC, ETC, SPI, and CPI most often appear in which type of report?
    A. Performance
    B. Status
    C. Forecast
    D. Exception

11. If BAC = $24,000, BCWP = 12,000, ACWP = $10,000, and CPI = 1.2, then the cost that remains to finish the project is:
    A. $10,000
    B. $12,000
    C. $14,000
    D. Cannot be determined

12. There are several purposes for the 50–50 rule, but the *primary* purpose of the 50–50 rule is to calculate:
    A. BCWS
    B. BCWP
    C. ACWP
    D. BAC

13. When a project is completed, which of the following *must* be true?
    A. BAC = ACWP
    B. ACWP = BCWP
    C. SV = 0
    D. BAC = ETC

14. In March CV = −$20,000, and in April CV = −$30,000. In order to determine whether or not the situation has really deteriorated because of a larger unfavorable cost variance, we would need to calculate:
    A. CV in percent
    B. SV in dollars
    C. SV in percent
    D. All of the above

15. If a project manager is looking for revenue for a value-added scope change, the project manager's first choice would be:
    A. Management reserve
    B. Customer-funded scope change
    C. Undistributed budget
    D. Retained profits

16. A project was originally scheduled for 20 months. If CPI is 1.25, then the new schedule date is:
A. 16 months
B. 20 months
C. 25 months
D. Cannot be determined

17. The cost or financial baseline of a project is composed of:
A. Distributed budget only
B. Distributed and undistributed budgets only
C. Distributed budget, undistributed budget, and the management reserve only
D. Distributed budget, undistributed budget, management reserve, and profit only

ANSWERS

1. B
2. B
3. B
4. B
5. C
6. A
7. D
8. C
9. C
10. C
11. A
12. B
13. C
14. A
15. B
16. D
17. B

PROBLEMS

15–1 Do cost overruns just happen, or are they caused?

15–2 Cemeteries are filled with projects that went out of control. Below are several causes that can easily develop into out-of-control conditions. In which phase of a project should each of these conditions be detected and, if possible, remedied?

a. Customer's requirements not understood
b. Project team formed after bid was prepared
c. Accepting unusual terms and conditions

    d. Permitting a grace period for changing specifications

    e. Lack of time to research specifications

    f. Overestimation of company's capabilities

15–3 Below are several factors that can result in project delays and cost overruns. Explain how these problems can be overcome.

    a. Poorly defined milestones

    b. Poor estimating techniques

    c. A missing PERT/CPM chart

    d. Functional managers not having a clear understanding of what has to be done

    e. Poor programming procedures and techniques

    f. Changes constantly being made deep in the project's life cycle

15–4 Under what circumstances would each of the figures in Chapter 13 be applicable for customer reporting? In-house reporting? Reporting to top-level management?

15–5 What impact would there be on BCWS, BCWP, ACWP, and cost and schedule variances as a result of the:

    a. Early start of an activity on a PERT chart?

    b. Late start of an activity on a PERT chart?

15–6 Alpha Company has implemented a plan whereby functional managers will be held totally responsible for all cost overruns against their (the functional managers') original estimates. Furthermore, all cost overruns must come out of the functional managers' budgets, whether they be overhead or otherwise, not the project budget. What are the advantages and disadvantages of this approach?

15–7 Karl has decided to retain a management reserve on a $400,000 project that includes a $60,000 profit. At the completion of the project, Karl finds that the management reserve fund contains $40,000. Should Karl book the management reserve as excess profits (i.e., $100,000), or should he just book the target profit of $60,000 and let the functional managers "sandbag" on the slush fund until it is depleted?

15–8 ABC Corporation has recently given out a nine-month contract to a construction subcontractor. At the end of the first month, it becomes obvious that the subcontractor is not reporting costs according to an appropriate WBS level. ABC Corporation asks the subcontractor to change its cost reporting procedures. The subcontractor states that this cannot be done without additional funding. This problem has occurred with other subcontractors as well. What can ABC Corporation do about this?

15–9 What would be the result if all project managers decided to withhold a management reserve? What criteria should be used for determining when a management reserve is necessary?

15–10 Alpha Company, a project-driven organization, pays its department managers a quarterly bonus that is dependent on two factors: the departmental overhead rate and direct labor dollars. The exact value of the bonus is proportional to how much these two factors are underrun.

    Department man-hours are priced out against the department average, which does not include the department manager's salary. His salary is included under his departmental overhead rate, but he does have the option of charging his own time as direct labor to the projects for which he must supply resources.

What do you think of this method? Is it adequate inducement for a functional manager to control resources more effectively? How would you feel, as a project manager, knowing that the functional managers got quarterly bonuses and you got none?

15–11 Many executives are reluctant to let project managers have complete control of project costs because then the project managers must know the exact salaries of almost all project personnel. Can this situation be prevented if the contract requires reporting costs as actuals?

15–12 How can a country's inflation rate influence the contractual payment policy?

15–13 Consider a situation in which several tasks may be for one to two years rather than the 200 hours normally used in the work-package level of the WBS.

    a. How will this affect cost control?
    b. Can we still use the 50/50 rule?
    c. How frequently should costs be updated?

15–14 By now you should be familiar with the various tools that can be used for planning, controlling, scheduling, and directing project activities. Table 15–11 contains a partial list of such tools and how they relate to specific project management functions. Complete the table (using the legend at the bottom) to indicate which are very useful and which are somewhat useful.

Obviously there will be some questions about what is very useful and what is somewhat useful. Be able to defend your answers.

TABLE 15–11. PROJECT PLANNING, CONTROLLING, AND DIRECTING

	Useful for			
Tool	**Planning**	**Controlling**	**Directing**	**Interface Relationships**
Project organizational charts				
Work breakdown structure				
Task descriptions				
Work packages				
Project budget				
Project plan				
Charts/schedules				
Progress reports				
Review meetings				

○ somewhat useful
● very useful

15–15 Complete the table below and plot the EAC as a function of time. What are your conclusions?

	Cumulative Cost, in Thousands			Variance $		
Week	BCWS	BCWP	ACWP	Schedule	Cost	EAC
1	50	50	25			
2	70	60	40			
3	90	80	67			
4	120	105	90			
5	130	120	115			
6	140	135	130			
7	165	150	155			
8	200	175	190			
9	250	220	230			
10	270	260	270			
11	300	295	305			
12	350	340	340			
13	380	360	370			
14	420	395	400			
15	460	460	450			

15–16 Using the information in Chapter 12, problem 12–18, complete Table 15–12.

15–17 On June 12, 2002, Delta Corporation was awarded a $160,000 contract for testing a product. The contract consisted of $143,000 for labor and materials, and the remaining $17,000 was profit. The contract had a scheduled start date of July 3. The network logic, as defined by the project manager and approved by the customer, consisted of the following:

Activity	Time (Weeks)
AB	7
AC	10
AD	8
BC	4
BE	2
CF	3
DF	5
EF	2
FG	1

TABLE 15–12. PROJECT COSTS

1	2	3	4	5	6	7
Activity	Percent Complete	Budgeted Cost for Work Scheduled	Budgeted Cost for Work Performed	Actual Cost for Work Performed	Cost Variance = 4 – 5	Schedule Variance = 4 –3
Total						

Cost variance ($) = Column 4 – Column 5 = _____

Schedule variance ($) = Column 4 – Column 3 = _____

Time-to-complete = _____

Cost-at-completion = rate of spending × Total budget = $\dfrac{\text{Column 5}}{\text{Column 4}}$ × (_____) = _____

Cost-to-complete = (Cost-at-completion) – ACWP = _____

On August 27, 2002, the executive steering committee received the following report indicating the status of the project at the end of the eighth week:

Activity	% Complete	Actual Cost	Time Remaining (Weeks)
AB	100	$23,500	0
AC	60	19,200	4
AD	87.5	37,500	1
BC	50	8,000	2
BE	50	5,500	1

The steering committee could not identify the real status of the project from this brief report. Even after comparing this brief status report with the project planning budget (see Table 15–13), the real status was not readily apparent.

Management instructed the project manager to prepare a better status report that depicted the true status of the project, as well as the amount of profit that could be expected at project completion. Your assignment is to prepare a table such as Table 15–12.

15–18 *The Alpha Machine Tool Project*

Acme Corporation has received a contractual order to build a new tooling machine for Alpha Corporation. The project started several months ago. Table 15–14 is the Monthly Cost Summary for June, 2002. Some of the entries in the table have been purposely omitted, but the following additional information is provided to help you answer the questions below:

A. Assume that the overhead of 100% is fixed over the period of performance.
B. The report you are given is at a month end, June 30, 2002.
C. The 80/20 sharing ratio says that the customer (i.e., Alpha) will pay 80 percent of the dollars above the target cost and up to the ceiling cost. Likewise, 80 percent of the cost savings below the target cost go back to Alpha.
D. The revised BCWS is revised from the released BCWS.
E. The ceiling price is based on cost (i.e., without profit).

Answer the following questions by extracting data from the Alpha Machine Tool Project's monthly summary report.

1. What is the total *negotiated* target value of the contract?
2. What is the budgeted target value for *all work authorized* under this contract?
3. What is the total budgetary amount that Acme had originally allocated/released to the Alpha Project?
4. What is the new/revised total budgetary amount that Acme has released to the Alpha Project?

TABLE 15–13. PROJECT PLANNING BUDGET

Activity	1	2	3	4	5	6	7	8	9	10	11	12	13	14	15
AB	2000	2000	3000	3000	4000	4000	3000								
AC	3000	3000	3000	4000	4000	4000	4000	2000	2000	1000					
AD	5000	5000	6000	4000	4000	4000	3000	1000							
BC								3000	4000	4000	5000				
BE								6000	6000						
CF												2000	3000	3000	
DF									3000	3000	3000	4000	4000		
EF											2000	2000			
FG															3000

Note: Table 15–13 assumes that percent is *linear* with time and *nonlinear* with cost.

TABLE 15–14. MONTHLY COST SUMMARY—JUNE 2002

Contract: Alpha Machine Tool
PM: Gary Jones
Reporting Period: June 1 - June 30, 2002
Contract Period: Feb. 1 - Oct. 30, 2002

Negotiated Cost: $2,500,000
Target Fee: 12%
Target Price: 2,800,000

Sharing Ratio: 80/20
Ceiling: 3,000,000 on cost (= $3.2 M on price)
Contract: fixed price incentive fee

Level 2 WBS Items	Current Month, $					Cumulative to Date, $					At Completion, $			
	BCWS	BCWP	ACWP	SV	CV	BCWS	BCWP	ACWP	SV	CV	Contracted BCWS	Original Released BCWS	Revised BCWS	Var.
Program mgt.	19300	19300	19300	0	0	108000	108000	108000	0	0	200000	200000	200000	
Subsystem A	23000	16600	24200	<6400>	<7600>	158000	181700	234700	23700	<53000>	250000	200000	225000	<25000>
Subsystem B	14000	15200	16800	1200	<1600>	96000	94200	93000	<1800>	1200	200000	200000	200000	
Subsystem C	0	0	0	0	0	0	0	0	0	0	300000	275000	275000	
Manu. support	11600	10400	12000	<1200>	<1600>	73000	74300	75600	1300	<1300>	200000	190000	190000	
Quality control	5900	6000	6000	100	0	5900	6000	6000	100	0	100000	100000	100000	
TOTAL DIRECT	73800	67500	78300								1250000	1165000	1190000	
OVERHEAD, 100%	73800	67500	78300								1250000	1165000	1190000	
TOTAL	147600	135000	156600								2500000	2330000	2380000	

5. How much money, if any, had Acme set aside as a management reserve based upon the original released budget? (burdened) _____

6. Has the management reserve been revised, and if so, by how much? (burdened) _____

7. Which level-2 WBS elements make up the revised management reserve? _____

8. Based upon the reviewed BCWS completion costs, how much profit can Acme expect to make on the Alpha Project? (Hint: Don't forget sharing ratio) _____

9. How much of the distributed budget that has been identified for accomplishment of work is only *indirectly* attributed to this contract? (i.e., *overhead*) _____

Answer the Following Questions for Direct Labor Only

10. Of the total direct effort budgeted for on this contract, how much work did Acme *schedule* to be performed this month? _____

11. How much of the work scheduled for accomplishment this month was actually earned (i.e., earned value)? _____

12. Did Acme do more or less work than planned for this month? How much was the schedule variance (SV)? [$ and %] _____

13. What did it actually cost Acme for the work performed this month? _____

14. What is the difference between the amount that Acme budgeted for the work performed this month and what the actual cost was? (i.e., CV) [$ and %] _____

15. Which WBS level-2 elements are the primary causes for this month's cost and schedule variances? _____

16. How much cost variance has Acme experienced to date? [$ and %] _____

17. How much schedule variance has Acme experienced to date? [$ and %] _____

18. Is the cost variance improving or getting worse?

19. Is the schedule variance improving or getting worse? _____

20. Does it appear that the scheduled end date will be met? _____

21. What is the new estimated burdened cost at completion? _____

22. How much profitability/loss can Acme expect
    from the new estimated cost at completion?   _____

23. If Acme's final burdened cost for the program
    was $3,150,000, how much profit/loss
    would it experience?   _____

15–19   Calculate the total price variance for direct labor and the labor rate cost variance
from the following data:

	Direct Material	Direct Labor
Planned price/unit	$ 10.00	$ 22.00
Actual units	9,300	12,000
Actual price/unit	$ 9.25	$ 22.50
Actual cost	$86,025,00	$270,000

15–20 One of your assistant project managers has given you an earned value report that is only
partially complete. Can you fill in the missing information?

(All numbers are in thousands of dollars)

WBS Work Packages	BCWS	BCWP	ACWP	SV	CV
A	103	115	—	12	<91>
B	0	—	40	—	—
D	42	12	33	<30>	<21>
H	66	—	94	189	161
P	87	77	116	<10>	<39>
S	175	—	184	<115>	<124>
	473	—	—	—	<144>

15–21 The following problem requires an understanding of the WBS, the cost account elements, and cost control analysis. Assume that all costs are in thousands of dollars.
   Given the partial WBS shown below, what is the total cost for the WBS element 4.0?
Assume that the costs provided are direct labors costs only and that the overhead rate is 100
percent.

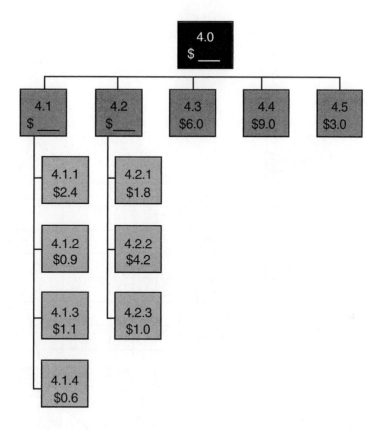

Which of the following is the value of WBS element 4.0?

    a. $60.0
    b. $30.0
    c. $24.0
    d. $54.0

Using the data in Figure 15–28, and the actual costs given below for WBS elements 5.1 through 5.4 and elements 4.1 and 4.2, answer the questions shown below:

	Actual Costs
E-1–5.1	$1.0
E-1–5.3	$1.5
E-2–5.2	$1.0
E-2–5.4	$2.0
E-3–5.1	$1.0
E-3–5.3	$2.5
E-4–5.3	$3.0
E-4–5.2	$3.5

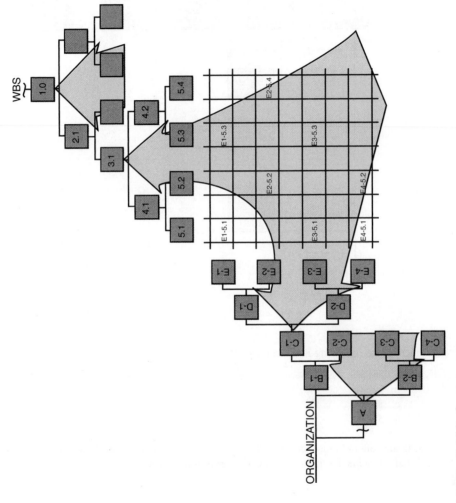

FIGURE 15–28. Exhibit of cost accounts.

WBS element 5.1	$_____
WBS element 5.2	$_____
WBS element 5.3	$_____
WBS element 5.4	$_____
WBS element 4.1	$_____
WBS element 4.2	$_____
Functional element E-1	$_____
Functional element E-2	$_____
Functional element E-3	$_____
Functional element E-4	$_____
Functional element D-1	$_____
Functional element D-2	$_____

15–22 Companies usually estimate work based upon man-months. If the work must be estimated in man-weeks, the man-month is then converted to man-weeks. The problem is in the determination of how many man-hours per month are actually available for actual direct labor work.

Your company has received a request for proposal (RFP) from one of your customers and management has decided to submit a bid. Only one department in your company will be required to perform the work and the department manager estimates that 3000 hours of *direct* labor will be required.

Your first step is to calculate the number of hours available in a typical man-month. The Human Resources Department provides you with the following *yearly* history for the average employee in the company:

- Vacation (3 weeks)
- Sick days (4 days)
- Paid holidays (10 days)
- Jury duty (1 day)

a. How many direct labor hours are available per month per person?
b. If only one employee can be assigned to the project, what will be the duration of the effort, in months?
c. If the customer wants the job completed within one year, how many employees should be assigned?

15–23 In a status report, executives want to know not only where we are today, but also where we will end up. Calculating where we will end up financially is not as easy as it sounds. Selecting the wrong formula can leave the executives and customers with a faulty impression.

There are several formulas available for the calculation of the estimated cost at completion (EAC). For simplicity, consider the following three formulas:

I. EAC = (ACWP/BCWP) × (budget at completion)
II. EAC = [(ACWP/BCWP) × (BCWS for work completed and in progress)] + (planned or revised planned costs of work packages not yet begun)
III. EAC = (actual to date) + (all remaining work, including work in progress, to be completed at the planned or budgeted costs)
   a. Using the table below, determine the value of EAC for each of the three formulas. Assume that A, B, and C are the only work packages in the project, and

BCWS(Total) is the total value for PV for each work package rather than PV for the reporting period. Use the following formula for calculating EV:

$$EV = [\% \text{ Complete}] \times BCWS(Total)$$

Activity	% Complete	BCWS(Total)	ACWP
A	100	1000	1100
B	50	1000	800
C	0	1000	0

  b. Considering only activity B, if the reason for the cost overrun is attributed to a one-time occurrence, which of the three formulas would be best to use?
  c. If the reason for the overrun in activity B is because of the higher than expected salaries of the assigned employees and these same employees will be assigned to activity C as well, which of the three formulas would be best to use?
  d. Considering only activity B, if the reason for the overrun is attributed to overtime and the overtime will continue but only through the completion of activity B, which of the three formulas would be best to use?
  e. Considering your answers to the above four parts, should a company be willing to change the formula for calculating EAC during the execution of the project as well as at each reporting period or gate review meeting?

15–24 Project managers must not only calculate variances but also determine the root cause of the variance. Some variances may be allowable while others must be explained together with a corrective plan for recovery.

In the table below, you must demonstrate your ability to calculate the cost and schedule variances as well as determine the root cause of the variances, if possible. Consider the following table, which shows a partial status report for a project composed of five work packages (i.e., activities):

Activity	BCWS	BCWP	ACWP	SV	CV
A	$800	$1000	$1100		
B	1200	1000	900		
C	800	1000	700		
D	1200	1000	1100		
E	1000	1000	800		

  a. Calculate the cost and schedule variances, in dollars, for each activity.
  b. For each activity, which of the following could be the root cause reason for the variances? Select as many as you think may apply.

  ● Accelerated schedule due to higher salaried personnel
  ● Accelerated schedule due to overlapping of activities
  ● Accelerated schedule due to overtime
  ● Accelerated schedule due to additional resources
  ● Slippage due to lack of resources
  ● Slippage due to people working on other projects
  ● Slippage due to mistakes
  ● People are working but progress is poor

- On schedule
- On budget
- Over budget

c. Management wants to know the status of the total project at level 1 of the WBS. Add up all of the activity cost and schedule variances to determine the level 1 cost and schedule variances. What are your conclusions?

d. Would your conclusions from part c above change if activities B and D were the only two activities on the critical path?

15–25 Sometimes, the root cause of a variance requires that variance analysis be performed in both hours and dollars. It is possible that calculating the variances in both hours and dollars is the only way to determine the root cause of a problem.

Problem: In Table 15–15, the cost and schedule variances are measured in fully burdened dollars for Cost Center 2834 only.

From Table 15-15, you are ahead of schedule and over budget. There could be several possible causes for this, including schedule compression, using higher salaried labor, overtime, additional resources, or other causes. How can we determine which of these causes is the real reason for the variances?

Table 15–16 shows the variance data in both hours and dollars.

a. Calculate the cost variances in both hours and dollars. Compare the results. What are your conclusions?

b. Calculate the planned fully burdened labor rate using BCWS (or BCWP).

c. Calculate the actual fully burdened labor rate using ACWP.

d. Explain the possible reasons for the differences in labor rates and how this affects your answer to part a.

e. Table 15–17 shows the departmental pay structure for Cost Center 2834. Determine the departmental overhead rate, in percent.

f. How does Table 15–17 affect your answer to part d?

TABLE 15–15. COST CENTER 2834, JUNE

BCWS	Dollars	29,750	
	Hours		SV = +$4250
BCWP	Dollars	34,000	
	Hours		
ACWP	Dollars	38,400	CV = −$4400
	Hours		

TABLE 15–16. COST CENTER 2834, JUNE

BCWS	Dollars	29,750
	Hours	350
BCWP	Dollars	34,000
	Hours	400
ACWP	Dollars	38,400
	Hours	320

CV ($$$) = negative   CV(hrs) = positive

TABLE 15–17. DEPARTMENTAL PAY STRUCTURE

Pay Grade	Title	Unburdened Salary	Burdened Salary
9	Engineering Consultant	$53/hr	$132.50
8	Senior Engineer	48	120.00
7	Engineer	39	97.50
6	Junior Engineer	34	85.00
5	Apprentice Engineer	29	72.50

15–26 Projects that span more than one year or cut across the date of corporate salary increases may require the use of forward-pricing rates. Forward-pricing rates are determined from economic data, industry surveys, and best-guess predictions.

As an example, consider Table 15–17 in the previous problem, which shows the salary structure for an engineering department. For simplicity, we shall make the following assumptions:

- Promotions and salary increases, including cost-of-living adjustments, are effective January 1 and are then held constant for the entire year.
- The overhead rate is 150 percent and fixed for the entire year.
- All projects are priced out using the salary of a pay grade 7.
- Most of the departmental workers are pay grade 7 employees.

Situation: Your company has just won a one-year contract. The contract was planned to start on January 1, 2006, and be completed by December 31, 2006. The work that was to be performed by this department was estimated at 1000 hours per month for the duration of the twelve-month project using pay grade 7 employees. The customer informs you that they wish to start the project on July 1 rather than January 1, and they assume that there is no financial impact on the total cost of the project.

The Finance Department provides you with the forward pricing rate data in Table 15–18 and tells you that the overhead rate for 2007 is expected to increase to 155 percent. Is there a financial impact on the total cost of the project, and if so, how much of an impact?

TABLE 15–18. DEPARTMENTAL PAY STRUCTURE (dollars/hour)

Pay Grade	Title	Salary		
		2006	2007*	2008*
9	Engineering Consultant	53	56	60
8	Senior Engineer	48	50	53
7	Engineer	39	42	45
6	Junior Engineer	34	36	39
5	Apprentice Engineer	29	31	34

*Projected rates.

TABLE 15–19. PROJECT PRICING SUMMARY

Department	Direct Labor			Overhead		Total
	Hours	Rate	Dollars	%	Dollars	
Engineering	1000	$42.00	42,000	110	46,200	$88,200
Manufacturing	500	$35.00	17,500	200	35,000	$52,500
					Total Labor	$140,700
				Other: Subcontracts	$10,000	
				Consultants	$2,000	$12,000
				Total labor and material		$152,700
				Corporate G&A: 10%		$ 15,270
						$167,970
				Profit: 15%		$ 25,196
						$193,166

15–27 Pricing out a customer's request for proposal is a trade-off between time, cost, and accuracy. If time and money are not an issue, then we could determine a very accurate bid. But if the company is reluctant to invest heavily in the preparation of the bid, care must be taken that there are no hidden costs.

Situation: You have been asked to price out a project for a customer, and this pricing is an activity with which you have very little previous experience. Table 15–19 shows the numbers that you arrived at in determining that a bid of $193,166 should be submitted.

Before a bid is submitted to a potential customer, the bid must be reviewed by a committee of senior managers that can question the validity of the numbers as well as look for "hidden" costs that may have been omitted. For each of the situations below, which line item in the pricing summary would most likely be impacted assuming that these hidden costs were not already included?

    a. Management tells you that, during the execution of the project, the customer will want three interface meetings with the customer held at the customer's location. The Travel Group within your company informs you that airfare, ground travel, meals, and lodging are expected to be approximately $2000 per meeting.

    b. One of the executives comments, "The shipping costs for the deliverables, including insurance, packaging, and handling, will be about $1000. I do not see this included in your summary."

    c. The RPF for the project stated that the contract would be firm-fixed-price with a lump-sum payment at the end of the project after approval/acceptance of the final deliverables. The cost of capital is expected to be approximately $6000.

    d. Engineering believes that the engineering hours in the summary could be low by about 10 percent if the risks in the estimates provided actually occur. The executives believe that a management reserve of 10 percent should be included in the summary costs.

    e. Using the information in parts a through d above, what final price should be submitted to the customer?

15–28 A homeowner hires a contractor to build a four-sided square fence around his home. The homeowner provides the materials and the contractor supplies the labor. The contractor estimates each side will cost $2000 and require one week in duration.

At the end of week 1, the first side is completed at a cost of $2000. During week 2, the second side is completed but at a cost of $2400 because the contractor damaged a water line that the contractor had to repair. During week 3, the contractor completed only half of the fence for $1000 because it rained for the remaining half of the week.

At the end of week 3, the contractor must prepare an earned value measurement status report. Calculate the following:

BCWS: _____

BCWP: _____

ACWP: _____

BAC: _____

SV($): _____

CV($): _____

EAC = ([ACWP/BCWP] × BAC): _____

ETC: _____

VAC: _____

% COMPLETE: _____

% $$ SPENT: _____

CPI: _____

SPI: _____

15–29 The data identified below was listed in a project's latest status report:

- BCWS = $36,000
- BCWP = $30,000
- ACWP = $33,000
- BAC = $120,000
- Original length of the project 10 months

Using these data, calculate the following:

a. What are the values for CPI and SPI?

b. What is the expected cost at completion (EAC)?

c. How much money will be needed from the time of the report to complete the project?

d. What is the cost variance at completion (VAC)?

e. Using SPI, what is the new expected length of the project?

15–30 In the problem in Figure P15–30, the 50% / 50% rule is being used to determine the project's status. Assume that all amounts are in dollars.

Assuming that ACWP = $40,000, determine BCWS, BCWP, BAC, SV, and CV.

15–31 In Figures P15–31A to P15–31C identify the status of part of a project using the graphical technique (i.e. S curves) rather that tables. For each of the three figures, select from one of the following fives choices as to what each figure illustrates:

a. Over budget

b. Under budget

c. Ahead of schedule

d. Behind schedule

e. Status cannot be determined

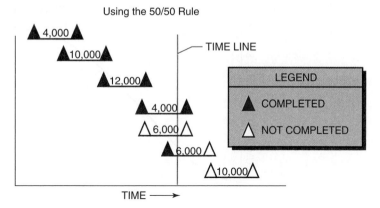

Figure P15-30

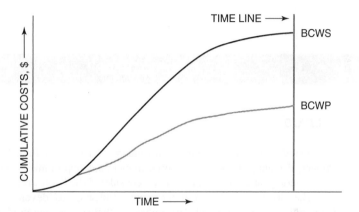

Figure P15-31A

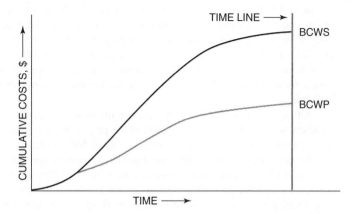

Figure P15-31B

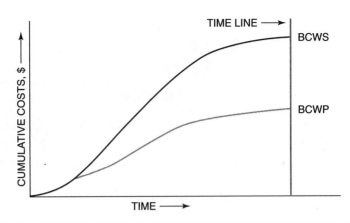

Figure P15-31C

CASE STUDIES

THE BATHTUB PERIOD

The award of the Scott contract on January 3, 1987, left Park Industries elated. The Scott Project, if managed correctly, offered tremendous opportunities for follow-on work over the next several years. Park's management considered the Scott Project as strategic in nature.

The Scott Project was a ten-month endeavor to develop a new product for Scott Corporation. Scott informed Park Industries that sole-source production contracts would follow, for at least five years, assuming that the initial R&D effort proved satisfactory. All follow-on contracts were to be negotiated on a year-to-year basis.

Jerry Dunlap was selected as project manager. Although he was young and eager, he understood the importance of the effort for future growth of the company. Dunlap was given some of the best employees to fill out his project office as part of Park's matrix organization. The Scott Project maintained a project office of seven full-time people, including Dunlap, throughout the duration of the project. In addition, eight people from the functional department were selected for representation as functional project team members, four full-time and four half-time.

Although the workload fluctuated, the manpower level for the project office and team members was constant for the duration of the project at 2,080 hours per month. The company assumed that each hour worked incurred a cost of $60.00 per person, fully burdened.

At the end of June, with four months remaining on the project, Scott Corporation informed Park Industries that, owing to a projected cash flow problem, follow-on work would not be awarded until the first week in March (1988). This posed a tremendous problem for Jerry Dunlap because he did not wish to break up the project office. If he permitted his key people to be assigned to other projects, there would be no guarantee that he could get them back at the beginning of the follow-on work. Good project office personnel are always in demand.

Jerry estimated that he needed $40,000 per month during the "bathtub" period to support and maintain his key people. Fortunately, the bathtub period fell over Christmas and New Year's, a time when the plant would be shut down for seventeen days. Between the vacation days that his key employees would be taking, and the small special projects that his people could be temporarily assigned to on other programs, Jerry revised his estimate to $125,000 for the entire bathtub period.

At the weekly team meeting, Jerry told the program team members that they would have to "tighten their belts" in order to establish a management reserve of $125,000. The project team understood the necessity for this action and began rescheduling and replanning until a management reserve of this size could be realized. Because the contract was firm-fixed-price, all schedules for administrative support (i.e., project office and project team members) were extended through February 28 on the supposition that this additional time was needed for final cost data accountability and program report documentation.

Jerry informed his boss, Frank Howard, the division head for project management, as to the problems with the bathtub period. Frank was the intermediary between Jerry and the general manager. Frank agreed with Jerry's approach to the problem and requested to be kept informed.

On September 15, Frank told Jerry that he wanted to "book" the management reserve of $125,000 as excess profit since it would influence his (Frank's) Christmas bonus. Frank and Jerry argued for a while, with Frank constantly saying, "Don't worry! You'll get your key people back. I'll see to that. But I want those uncommitted funds recorded as profit and the program closed out by November 1."

Jerry was furious with Frank's lack of interest in maintaining the current organizational membership.

    a. Should Jerry go to the general manager?

    b. Should the key people be supported on overhead?

    c. If this were a cost-plus program, would you consider approaching the customer with your problem in hopes of relief?

    d. If you were the customer of this cost-plus program, what would your response be for additional funds for the bathtub period, assuming cost overrun?

    e. Would your previous answer change if the program had the money available as a result of an underrun?

    f. How do you prevent this situation from recurring on all yearly follow-on contracts?

FRANKLIN ELECTRONICS

In October 2003 Franklin Electronics won an 18-month labor-intensive product development contract awarded by Spokane Industries. The award was a cost reimbursable contract with a cost target of $2.66 million and a fixed fee of 6.75 percent of the target. This contract would be Franklin's first attempt at using formal project management, including a newly developed project management methodology.

Franklin had won several previous contracts from Spokane Industries, but they were all fixed-price contracts with no requirement to use formal project management with earned value reporting. The terms and conditions of this contract included the following key points:

- Project management (formalized) was to be used.
- Earned value cost schedule reporting was a requirement.
- The first earned value report was due at the end of the second month's effort and monthly thereafter.
- There would be two technical interchange meetings, one at the end of the sixth month and another at the end of the twelfth month.

Earned value reporting was new to Franklin Electronics. In order to respond to the original request for proposal (RFP), a consultant was hired to conduct a four-hour seminar on earned value management. In attendance were the project manager who was assigned to the Spokane RFP and would manage the contract after contract award, the entire cost accounting department, and two line managers. The cost accounting group was not happy about having to learn earned value management techniques, but they reluctantly agreed in order to bid on the Spokane RFP. On previous projects with Spokane Industries, monthly interchange meetings were held. On this contract, it seemed that Spokane Industries believed that fewer interchange meeting would be necessary because the information necessary could just as easily be obtained through the earned value status reports. Spokane appeared to have tremendous faith in the ability of the earned value measurement system to provide meaningful information. In the past, Spokane had never mentioned that it was considering the possible implementation of an earned value measurement system as a requirement on all future contracts.

Franklin Electronics won the contact by being the lowest bidder. During the planning phase, a work breakdown structure was developed containing 45 work packages of which only 4 work packages would be occurring during the first four months of the project.

Franklin Electronics designed a very simple status report for the project. The table below contains the financial data provided to Spokane at the end of the third month.

Work Packages	Totals at End of Month 2					Totals at End of Month 3				
	PV	EV	AC	CV	SV	PV	EV	AC	CV	SV
A	38K	30K	36K	<6K>	<8K>	86K	74K	81K	<7K>	<12K>
B	17K	16K	18K	<2K>	<1K>	55K	52K	55K	<3K>	<3K>
C	26K	24K	27K	<3K>	<2K>	72K	68K	73K	<5K>	<4K>
D	40K	20K	23K	<3K>	<20K>	86K	60K	70K	<10K>	<26K>

Note: BCWS = PV, BCWP = EV, and ACWP = AC.

A week after sending the status report to Spokane Industries, Franklin's project manager was asked to attend an emergency meeting requested by Spokane's vice president for engineering, who was functioning as the project sponsor. The vice president was threatening to cancel the project because of poor performance. At the meeting, the vice president commented, "Over the past month the cost variance overrun has increased by 78 percent from $14,000 to $25,000, and the schedule variance slippage has increased by 45 percent from $31,000 to $45,000. At these rates, we are easily looking at a 500 percent cost overrun and a schedule slippage of at least one year. We cannot afford to let this project continue at this lackluster performance rate. If we cannot develop a plan to control time and cost any better than we have in the past three months, then I will just cancel the contract now, and we will find another contractor who can perform."

QUESTIONS

1. Are the vice president's comments about cost and schedule variance correct?
2 What information did the vice president fail to analyze?
3. What additional information should have been included in the status report?
4. Does Franklin Electronics understand earned value measurement? If not, then what went wrong?
5. Does Spokane Industries understand project management?
6. Does proper earned value measurement serve as a replacement for interchange meetings?
7. What should the project manager from Franklin say in his defense?

TROUBLE IN PARADISE

As a reward for becoming Acme Corporation's first PMP, Acme assigned the new PMP, Wiley Coyote, the leadership role of an important project in which the timing of the deliverables was critical to the success of the project. A delay in the schedule could cost Acme a loss of at least $100,000 per month. Wiley Coyote's first responsibility as project manager was the preparation of a solicitation package for the selection of an engineering contractor.

Eight companies prepared bids based on the solicitation package. Wiley Coyote decided to negotiate only with the low bidder, who happened to be at a significantly lower final cost than the other bidders. The contractor's project manager, Ima Roadrunner, would be handling the negotiations for the contractor. This was a contractor that Wiley Coyote had never worked with previously. Wiley Coyote reviewed the critical information in the proposal from the contractor:

- All work would be accomplished by engineering.
- Total burdened labor was 2000 hours at $120/hour.
- The duration of the project would be approximately 6 months and would be completed in 2006 (labor rates might be different in 2007).
- The contractor's overhead rate applied was 150 percent for engineering.
- All of the assigned workers would be at the same pay grade and would be assigned full time for the duration of the project.
- Profit requested was 12.5 percent, but subject to negotiations.
- Ima Roadrunner's salary would be included in the overhead structure.
- No materials were required.

During negotiations, Ima Roadrunner provided Wiley Coyote with the salary structure for engineering, shown in Exhibit 15–1.

Exhibit 15–1. Departmental pay structure

Pay Grade	Title	Unburdened Salary	Burdened Salary
9	Eng. Consultant	$53/hr.	$132.50
8	Senior Engineer	48	120.00
7	Engineer	39	97.50
6	Junior Engineer	34	85.00
5	Apprentice Eng.	29	72.50

Wiley Coyote asked Ima Roadrunner for the timing (i.e., manpower curve) of when the resources would be assigned. The result, provided by Ima Roadrunner, is shown in Exhibit 15–2 in the format of an *S*-curve that also shows the payment plan from the customer to the contractor.

The solicitation package identified the contract as a fixed-price contract with penalties for late delivery. Ima Roadrunner argued that the penalty clauses were unfair in their current wording and that a higher profit margin would be required to compensate for the risks. At this point, Wiley Coyote's superior intellect became apparent; he agreed to eliminate the penalty clauses if Ima Roadrunner agreed to lower the profit margin from 12.5 to 10 percent. Ima Roadrunner countered that the contract should then be a fixed-price-incentive-fee contract so that Ima Roadrunner could make up the lost 2.5 percent of profit by completing the project under budget. Both parties agreed to this and the deal was done.

The contract did not call for any type of earned value reporting, but Ima Roadrunner, who had very limited knowledge of earned value measurement and had never used it before, agreed to provide a monthly report that would show simply planned value (BCWS), earned value (BCWP), and actual costs (ACWP). Wiley Coyote agreed to this. After all, now that Wiley Coyote was a PMP, he knew how to extract all of the remaining information, such as variance analyses, cost-at-completion, and cost-to-completion, from these three values in the monthly report.

Exhibit 15–2. S-curve or spending curve

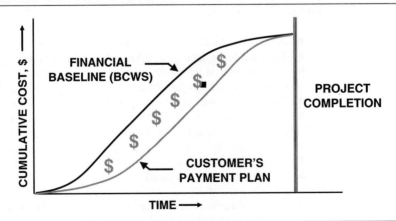

Exhibit 15–3. End of month 1

Eng. Cost Center	BCWS	Dollars	$42,000
		Hours	350
	BCWP	Dollars	$48,000
		Hours	400
	ACWP	Dollars	$34,000
		Hours	400

At the end of the first month, Wiley Coyote received the highly simplified earned value status report shown in Exhibit 15–3. Wiley Coyote was delighted with the results thus far. Now it looked like the project would be completed at least one month ahead of schedule and significantly under budget. Wiley Coyote provided Acme with significant cost savings by going with the lowest cost supplier, got the profit margin reduced by 20 percent, and would most likely come in ahead of schedule and at additional cost savings. Wiley Coyote was about to become a "superstar" in the eyes of Acme's executives. *Everyone would realize that Wiley Coyote had finally outsmarted the Roadrunner.* Wiley Coyote now began planning how he would spend the huge bonus he expected to receive at the completion of the project.

At the end of the fifth month, Ima Roadrunner informed Wiley Coyote of the good news that the project would be completed within cost, but there was also bad news that the completion date would be at the end of month 8, making the project two months late and at a significant loss to Acme Corporation. Wiley Coyote's thoughts on how to spend his bonus were now replaced with creative ideas on how to update his resume. Today, Wiley Coyote is on the lecture circuit discussing ways to identify warning signs of a potential project disaster.

Once again, the Roadrunner outsmarted Wiley Coyote. You have been hired in as a consultant to Acme Corporation to analyze what went wrong and to prepare a list of lessons learned for other project managers. Using the information in the case and all three exhibits, identify what went wrong. Also, were there any early warning signs, especially at the end of the first month, which should have warned Wiley Coyote that disaster might be imminent?

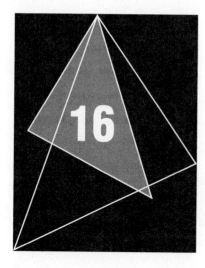

16 Trade-off Analysis in a Project Environment

"When we try to pick out anything by itself,
we find it hitched to everything else in the
universe."—MUIR'S LAW

Related Case Studies (from Kerzner/*Project Management Case Studies,* 3rd Edition)	Related Workbook Exercises (from Kerzner/*Project Management Workbook and PMP®/CAPM® Exam Study Guide,* 10th Edition)	PMBOK® Guide, 4th Edition, Reference Section for the PMP® Certification Exam
None	• Multiple Choice Exam	• Integration Management • Procurement Management • Scope Management

16.0 INTRODUCTION

PMBOK® Guide, 4th Edition
Triple-Constraint Definition

Successful project management is both an art and a science and attempts to control corporate resources within the constraints of time, cost, and performance. Most projects are unique, one-of-kind activities for which there may not have been reasonable standards for forward planning. As a result, the project manager may find it extremely difficult to stay within the time–cost–performance triangle of Figure 16–1.

The time–cost–performance triangle is the "magic combination" that is continuously pursued by the project manager throughout the life cycle of the project. If the project were to flow smoothly, according to plan, there might not be a need for trade-off analysis. Unfortunately, this rarely happens.

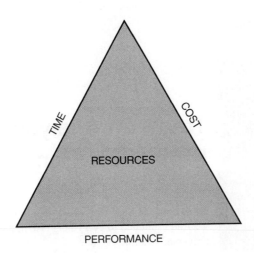

FIGURE 16–1. Overview of project management.

Trade-offs are illustrated in Figure 16–2, where the Δs represent deviations from the original estimates. The time and cost deviations are normally overruns, whereas the performance error will be an underrun. No two projects are exactly alike, and trade-off analysis will be an ongoing effort throughout the life of the project, continuously influenced by both the internal and the external environment. Experienced project managers have predetermined trade-offs in reserve, recognizing that trade-offs are part of a continuous thought process.

Trade-offs are always based on the constraints of the project. Table 16–1 illustrates the types of constraints commonly imposed. Situations A and B are the typical trade-offs encountered in project management. For example, situation A-3 portrays most research and development projects. The performance of an R&D project is usually well defined, and it is cost and time that may be allowed to go beyond budget and schedule. The determination of what to sacrifice is based on the available alternatives. If there are no alternatives to the product being developed and the potential usage is great, then cost and time are the trade-offs.

Most capital equipment projects would fall into situation A-1 or B-2, where time is of the essence. The sooner the piece of equipment gets into production, the sooner the return of investment can be realized. Often there are performance constraints that determine the profit potential of the project. If the project potential is determined to be great, cost will be the slippage factor, as in situation B-2.

Non–process-type equipment, such as air pollution control equipment, usually develops a scenario around situation B-3. Performance is fixed by the Environmental Protection Agency. The deadline for compliance can be delayed through litigation, but if the lawsuits fail, most firms then try to comply with the least expensive equipment that will meet the minimum requirements.

The professional consulting firm operates primarily under situation B-1. In situation C, the trade-off analysis will be completed based on the selection criteria and constraints. If everything is fixed (C-1), there is no room for any outcome other than total success, and if everything is variable (C-2), there are no constraints and thus no trade-off.

Many factors go into the decision to sacrifice either time, cost, or performance. It should be noted, however, that it is not always possible to sacrifice one of these items without affecting the others. For example, reducing the time could have a serious impact on performance and cost (especially if overtime is required).

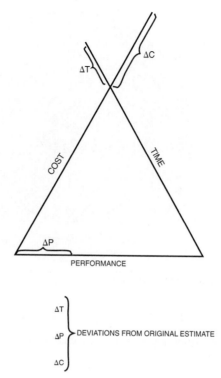

FIGURE 16–2. Project management with trade-offs.

There are several factors, such as those shown in Figure 16–3, that tend to "force" trade-offs. Poorly written documents (e.g., statements of work, contracts, and specifications) are almost always inward forces for conflict in which the project manager tends to look for performance relief. In many projects, the initial sale and negotiation, as well as the specification writing, are done by highly technical people who are driven to create a monument rather than meet the operational needs of the customer. When the

TABLE 16–1. CATEGORIES OF CONSTRAINTS

	Time	Cost	Performance
A. One Element Fixed at a Time			
A-1	Fixed	Variable	Variable
A-2	Variable	Fixed	Variable
A-3	Variable	Variable	Fixed
B. Two Elements Fixed at a Time			
B-1	Fixed	Fixed	Variable
B-2	Fixed	Variable	Fixed
B-3	Variable	Fixed	Fixed
C. Three Elements Fixed or Variable			
C-1	Fixed	Fixed	Fixed
C-2	Variable	Variable	Variable

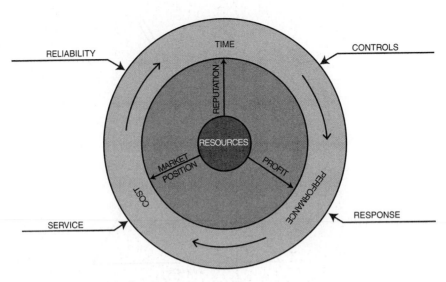

FIGURE 16–3.   Trade-off forcing factors.

operating forces dominate outward from the project to the customer, project managers may tend to seek cost relief.

16.1 METHODOLOGY FOR TRADE-OFF ANALYSIS

PMBOK® Guide, 4th Edition
3.2 Project Planning Group
Figure 3–12 Planning Process
   Group Triangle
Chapter 4 Integration Management
Chapter 5 Scope Management

Any process for managing time, cost, and performance trade-offs should emphasize the systems approach to management by recognizing that even the smallest change in a project or system could easily affect all of the organization's systems. A typical systems model is shown in Figure 16–4. Because of this, it is often better to develop a process for decision-making/trade-off analysis rather than to maintain hard-and-fast rules on trade-offs. The following six steps may help:

- Recognizing and understanding the basis for project conflicts
- Reviewing the project objectives
- Analyzing the project environment and status
- Identifying the alternative courses of action
- Analyzing and selecting the best alternative
- Revising the project plan

The first step in any decision-making process must be recognition and understanding of the conflict. Most projects have management cost and control systems that compare actual versus planned results, scrutinize the results through variance analyses, and provide status reports so that corrective action can be taken to resolve the problems. Project

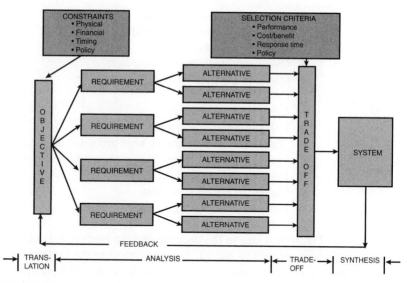

FIGURE 16–4. The systems approach.

managers must carefully evaluate information about project problems because it may not always be what it appears to be. Typical questions to ask are:

- Is the information pertinent?
- Is the information current?
- Are the data complete?
- Who has determined that this situation exists?
- How does he know this information is correct?
- If this information is true, what are the implications for the project?

The reason for this first step is to understand the cause of the conflict and the need for trade-offs. Most causes can be categorized as human errors or failures, uncertain problems, and totally unexpected problems, as shown below:

- Human errors/failures
  - Impossible schedule commitments
  - Poor control of design changes
  - Poor project cost accounting
  - Machine failures
  - Test failures
  - Failure to receive a critical input
  - Failure to receive anticipated approvals
- Uncertain problems
  - Too many concurrent projects
  - Labor contract expiration

- Change in project leadership
- Possibility of project cancellation
- Unexpected problems
    - Overcommitted company resources
    - Conflicting project priorities
    - Cash flow problems
    - Labor contract disputes
    - Delay in material shipment
    - "Fast-track" people having been promoted off the project
    - "Temporary" employees having to be returned to their home base
    - Inaccurate original forecast
    - Change in market conditions
    - New standards having been developed

The second step in the decision-making process is a complete review of the project objectives as seen by the various participants in the projects, ranging from top management to project team members. These objectives and/or priorities were originally set after considering many environmental factors, some of which may have changed over the lifetime of the project.

The nature of these objectives will usually determine the degree of rigidity that has been established between time, cost, and performance. This may require reviewing project documentation, including:

- Project objectives
- Project integration into sponsor's objectives and strategic plan
- Statement of work
- Schedule, cost, and performance specifications
- Resources consumed and projected

The third step is the analysis of the project environment and status, including a detailed measurement of the actual time, cost, and performance results with the original or revised project plan. This step should not turn into a "witch hunt" but should focus on project results, problems, and roadblocks. Factors such as financial risk, potential follow-up contracts, the status of other projects, and relative competitive positions are just a few of the environmental factors that should be reviewed. Some companies have established policies toward trade-off analysis, such as "never compromise performance." Even these policies, however, have been known to change when environmental factors add to the financial risk of the company. The following topics may be applicable under step 3:

- Discuss the project with the project management office to:
    - Determine relative priorities for time, cost, and performance
    - Determine impact on firm's profitability and strategic plan
    - Get a management assessment (even a hunch as to what the problems are)
- If the project is a contract with an outside customer, meet with the customer's project manager to assess his views relative to project status and assess the customer's priorities for time, cost, and performance.

- Meet with the functional managers to determine their views on the problem and to gain an insight regarding their commitment to a successful project. Where does this project sit in their priority list?
- Review in detail the status of each project work package. Obtain a clear and detailed appraisal by the responsible project office personnel as to:
  - Time to complete
  - Cost to complete
  - Work to complete
- Review past data to assess credibility of cost and schedule information in the previous step.

The project manager may have sufficient background to quickly assess the significance of a particular variance and the probable impact of that variance on project team performance. Knowledge of the project requirements (possibly with the assistance of the project sponsor) will usually help a project manager determine whether corrective action must be taken at all, or whether the project should simply be permitted to continue as originally conceived.

Whether or not immediate action is required, a quick analysis of why a potential problem has developed is in order. Obviously, it will not help to "cure the symptoms" if the "disease" itself is not remedied. The project manager must remain objective in such problem identification, since he himself is a key member of the project team and may be personally responsible for problems that are occurring. Suspect areas typically include:

- Inadequate planning. Either planning was not done in sufficient detail or controls were not established to determine that the project is proceeding according to the approved plan.
- Scope changes. Cost and schedule overruns are the normal result of scope changes that are permitted without formal incorporation in the project plan or increase in the resources authorized for the project.
- Poor performance. Because of the high level of interdependencies that exist within any project team structure, unacceptable performance by one individual may quickly undermine the performance of the entire team.
- Excess performance. Frequently an overzealous team member will unintentionally distort the planned balance between cost, schedule, and performance on the project.
- Environmental restraints—particularly on projects involving "third-party approvals" or dependent on outside resources. Changes, delays, or nonperformance by parties outside the project team may have an adverse impact on the team performance.

Some projects appear to be out of tolerance when, in fact, they are not. For example, some construction projects are so front-loaded with costs that there appears to be a major discrepancy when one actually does not exist. The front-end loading of cost was planned for.

The fourth step in the project trade-off process is to list alternative courses of action. This step usually means brainstorming the possible methods of completing the project by compromising some combination of time, cost, or performance. Hopefully, this step will refine these possible alternatives into the three or four most likely scenarios for project completion. At this point, some intuitive decision-making may be required to keep the list of alternatives at a manageable level.

In order fully to identify the alternatives, the project manager must have specific answers to key questions involving time, cost, and performance:

PMBOK® Guide, 4th Edition
3.2 Monitoring and Controlling
Process Group

- Time
  - Is a time delay acceptable to the customer?
  - Will the time delay change the completion date for other projects and other customers?
- What is the cause for the time delay?
- Can resources be recommitted to meet the new schedule?
- What will be the cost for the new schedule?
- Will the increased time give us added improvement?
- Will an extension of this project cause delays on other projects in the customer's house?
- What will the customer's response be?
- Will the increased time change our learning curve?
- Will this hurt our company's ability to procure future contracts?
- Cost
  - What is causing the cost overrun?
  - What can be done to reduce the remaining costs?
  - Will the customer accept an additional charge?
  - Should we absorb the extra cost?
  - Can we renegotiate the time or performance standards to stay within cost?
  - Are the budgeted costs for the remainder of the project accurate?
- Will there be any net value gains for the increased funding?
  - Is this the only way to satisfy performance?
  - Will this hurt our company's ability to procure future contracts?
  - Is this the only way to maintain the schedule?
- Performance
  - Can the original specifications be met?
  - If not, at what cost can we guarantee compliance?
  - Are the specifications negotiable?
  - What are the advantages to the company and customer for specification changes?
  - What are the disadvantages to the company and customer for performance changes?
  - Are we increasing or decreasing performance?
  - Will the customer accept a change?
  - Will there be a product or employee liability incurred?
  - Will the change in specifications cause a redistribution of project resources?
  - Will this change hurt our company's ability to procure future contracts?

Once the answers to these questions are obtained, it is often best to plot the results graphically. Graphical methods have been used during the past two decades to determine crashing costs for shortening the length of a project. To use the graphical techniques, we must decide on which of the three parameters to hold fixed.

Situation 1: Performance Is Held Constant (to Specifications)

With performance fixed, cost can be expressed as a function of time. Sample curves appear in Figures 16–5 and 16–6. In Figure 16–5, the circled X indicates the target cost and target time. Unfortunately, the cost to complete the project at the target time is higher than the budgeted cost. It may be possible to add resources and work overtime so that the time target can be met. Depending upon the way that overtime is burdened, it may be possible to find a minimum point in the curve where further delays will cause the total cost to escalate.

Curve A in Figure 16–6 shows the case where "time is money," and any additional time will increase the cost to complete. Factors such as management support time will always increase the cost to complete. There are, however, some situations where the increased costs occur in plateaus. This is shown in curve B of Figure 16–6. This could result from having to wait for temperature conditioning of a component before additional work can be completed, or simply waiting for nonscheduled resources to be available. In the latter case, the trade-off decision points may be at the end of each plateau.

With performance fixed, there are four methods available for constructing and analyzing the time–cost curves:

- Additional resources may be required. This will usually drive up the cost very fast. Assuming that the resources are available, cost control problems can occur as a result of adding resources after initial project budgeting.

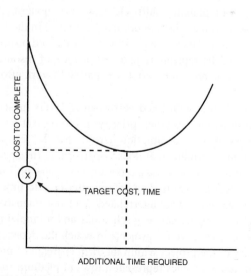

FIGURE 16–5. Trade-offs with fixed performance.

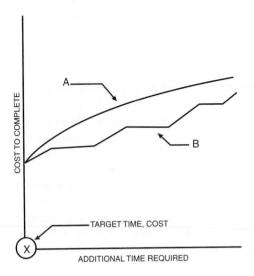

FIGURE 16–6. Trade-offs with fixed performance.

- The scope of work may be redefined and some work deleted without changing the project performance requirements. Performance standards may have been set too high, or the probability of success demanded of the project team may have been simply unrealistic. Reductions in cost and improvements in schedules would typically result from relaxing performance specifications, provided that the lower quality level will still meet the requirements of the customer.
- Available resources may be shifted in order to balance project costs or to speed up activities that are on the "critical" path work element that is trailing. This process of replanning shifts elements from noncritical to critical activities.
- Given a schedule problem, a change in the logic diagram may be needed to move from the current position to the desired position. Such a change could easily result in the replanning and reallocation of resources. An example of this would be to convert from "serial" to "parallel" work efforts. This is often risky.

Trade-offs with fixed performance levels must take into account the dependence of the firm on the customer, priority of the project within the firm, and potential for future business. A basic assumption here is that the firm may never sacrifice its reputation by delivering a product that doesn't perform to the specifications. The exception might be a change that would enhance performance and pull the project back on schedule. This is always worth investigating before entering into time–cost trade-offs.

Time and cost are interrelated in a labor-intensive project. As delivery slips, costs usually rise. Slipping delivery schedules and minimizing cost growth are usually the recommended alternative for projects in which the dependence of the firm on the customer, the priority of the project within the firm's stream of projects, and the future business potential in terms of sales represent a low- to medium-risk. Even in some high-risk situations, the contractor may have to absorb the additional cost. This decision is often based on estimating the future projects from this customer so that the loss is amortized against future business. Not all projects are financial successes.

A company's reputation for excellence is often hard to establish and can be extremely fragile. It is probably a contractor's greatest asset. This is particularly true in high-liability contracts, where the consequences of failure are extremely serious. There are companies that have been very successful in aerospace and advanced technology contracting but have seldom been the low bidder. Where the government is the contractor, performance is rated far above cost. Similarly, the consequences of a commercial aircraft crash are of such magnitude that the cost and time are relatively insignificant compared with precision manufacturing and extremely high reliability.

Sometimes projects may have fixed time and costs, leaving only the performance variable for trade-offs. However, as shown in the following scenario, the eventual outcome may be to modify the "fixed" cost constraint.

The hypothetical situation involves a government hardware subcontract, fixed-price, with delivery to the major government contractor. The major contractor had a very tight schedule, and the hardware being supplied had only a one-week "window" in which to be delivered, or the major contractor would suffer a major delay. Any delay at this point would place the general contractor in serious trouble. Both the government contracting officer and the purchasing manager of the general contractor had "emphasized" the importance of making the delivery schedule. There was no financial penalty for being late, but the contracting officer had stated in writing that any follow-on contracts, which were heavily counted on by the company's top management, would be placed with other vendors if delivery was not made on time.

Quality (performance) was critical but had never been a serious problem. In fact, performance had exceeded the contractual requirements because it had been company policy to be the "best" in the industry. This policy had, at times, caused cost problems, but it had ensured follow-on orders.

This project was in trouble at the halfway point, three months into the six-month schedule. The latest progress report indicated that the delivery would be delayed by three weeks. Costs were on target to date, but the shipping delay was expected to result in extra costs that would amount to 20 percent of the planned profit.

The project got off schedule when the flow of raw materials from a major vendor was interrupted for three weeks by a quality problem that was not discovered until the material was placed in production. Since the manufacturing time was process controlled, it was very difficult to make up lost time.

The first decision was that everything possible would be done to make delivery within one week of the original schedule. The potential lost revenue from future orders was so great that delivery must be made "at all costs," to quote the company president.

The quality system was then thoroughly investigated. It appeared that by eliminating two redundant inspection operations, one week could be saved in the total schedule. These two time-consuming inspection operations had been added when a quality problem developed on a former contract. The problem had been solved, and with present controls there was no reason to believe the inspections were still necessary. They would be eliminated with no determinable risk in performance.

Another two weeks were made up by working three production people seven days a week for the remainder of the project. This would permit delivery on the specified date of the contract, and would allow one week for other unforeseen problems so there would be a high probability of delivery within the required "window."

The cost of the seven-day-per-week work had the net effort of reducing the projected profit by 40 percent. Eliminating the two inspection operations saved 10 percent of the profit.

The plan outlined above met the time and performance specifications with increased cost that eventually reduced profit by an estimated 30 percent. The key to this situation was that only the labor, material, and overhead costs of the project were fixed, and the contractor was willing to accept a reduced profit.

Situation 2: Cost Is Fixed   With cost fixed, performance will vary as a function of time, as shown in Figure 16–7. The decision of whether to adhere to the target schedule data is usually determined by the level of performance. In curve A, performance may increase rapidly to the 90 percent level at the beginning of the project. A 10 percent increase in time may give a 20 percent increase in performance. After a certain point, a 10 percent increase in time may give only a 1 percent increase in performance. The company may not wish to risk the additional time necessary to attain the 100 percent performance level if it is possible to do so. In curve C, the additional time must be sacrificed because it is unlikely that the customer will be happy with a 30 to 40 percent performance level. Curve B is the most difficult curve to analyze unless the customer has specified exactly which level of performance will be acceptable.

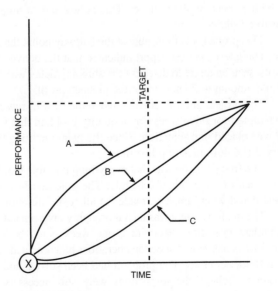

FIGURE 16–7. Trade-offs with fixed cost.

If cost is fixed, then it is imperative that the project have a carefully worded and understood contract with clear specifications as to the required level of performance and very clear statements of inclusion and exclusion. Careful attention to costs incurred because of customer changes or additional requirements can help reduce the possibility of a cost overrun. Experience in contracting ensures that costs that may be overlooked by the inexperienced project manager are included, thus minimizing the need for such trade-offs downstream. Common, overlooked items that can drive up costs include:

- Excessive detailed reporting
- Unnecessary documentation
- Excessive tracking documentation for time, cost, and performance
- Detailed specification development for equipment that could be purchased externally for less cost
- Wrong type of contract for this type of project

Often with a fixed-cost constraint, the first item that is sacrificed is performance. But such an approach can contain hidden disasters over the life of a project if the sacrificed performance turns out to have been essential to meeting some unspecified requirement such as long-term maintenance. In the long run, a degraded performance can actually increase costs rather than decrease them. Therefore, the project manager should be sure he has a good understanding of the real costs associated with trade-offs in performance.

Situation 3: Time Is Fixed　　Figure 16–8 identifies the situation in which time is fixed and cost varies with performance. Figure 16–8 is similar to Figure 16–7 in that the rate of change of performance with cost is the controlling factor. If performance is at the 90 percent level with the target cost, then the contractor may request performance relief. This is shown in curve A. However, if the actual situation reflects curve B or C, additional costs must be incurred with the same considerations of situation 1—namely, how important is the customer and what emphasis should be placed on his follow-on business?

Completing the project on schedule can be extremely important in certain cases. For example, if an aircraft pump is not delivered when the engine is ready for shipment, it can hold up the engine manufacturer, the airframe manufacturer, and ultimately the customer. All three can incur substantial losses due to the delay of a single component. Moreover, customers who are unable to perform and who incur large unanticipated costs tend to have long memories. An irate vice president in the customer's shop can kill further contracts out of all proportion to the real failure to deliver on time.

Sometimes, even though time is supposedly fixed, there may be latitude without inconvenience to the customer. This could come about because the entire program (of which your project is just one subcontract) is behind schedule, and the customer is not ready for your particular project.

Another aspect of the time factor is that "early warning" of a time overrun can often mitigate the damage to the customer and greatly increase his favorable response. Careful planning and tracking, close coordination with all functions involved, and realistic dealing with time schedules before and during the project can ensure early notification to the

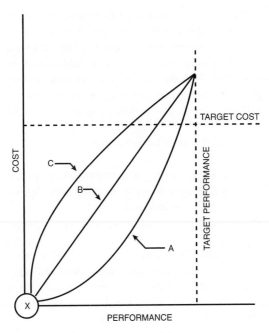

FIGURE 16–8. Trade-offs with fixed time.

customer and the possible negotiation of a trade-off of time and dollars or even technical performance. The last thing that a customer wants is to have a favorable progress report right up to the end of scheduled time and then to be surprised with a serious schedule overrun.

When time is fixed, the customer may find that he has some flexibility in determining how to arrive at the desired performance level. As shown in Figure 16–9, the contractor may be willing to accept additional costs to maximize employee safety.

Situation 4: No Constraints Are Fixed

Another common situation is that in which neither time, cost, nor performance is fixed. The best method for graphically showing the trade-off relationships is to develop parametric curves as in Figure 16–10.

Cost and time trade-offs can now be analyzed for various levels of performance. The curves can also be redrawn for various cost levels (i.e., 100, 120, 150 percent of target cost) and schedule levels.

Another method for showing a family of curves is illustrated in Figure 16–11. Here, the contractor may have several different cost paths for achieving the desired time and performance constraints. The final path selected depends on the size of the risk that the contractor wishes to take.

There have been several attempts to display the three-dimensional trade-off problem graphically. Unfortunately, such a procedure is quite complex and difficult to follow. A more common approach is to use some sort of computer model and handle the trade-off as

FIGURE 16–9. Performance versus cost.

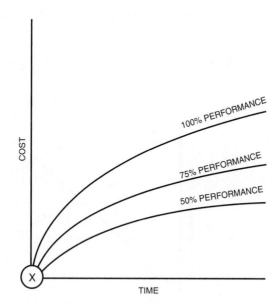

FIGURE 16–10. Trade-off analysis with family of curves.

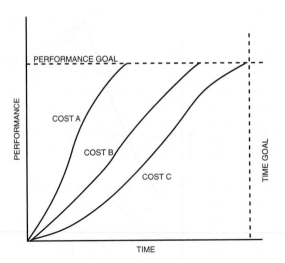

FIGURE 16–11. Cost–time–performance family of curves.

though it were a linear programming or dynamic programming problem. This too is often difficult to perform and manage.

Trade-offs can also be necessary at any point during the life cycle of a project. Figure 16–12 identifies how the relative importance of the constraints of time, cost, and performance can change over the life cycle of the project. At project initiation, costs may not have accrued to a point where they are important. On the other hand, project performance may become even more important than the schedule. At this point, additional

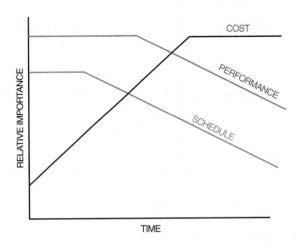

FIGURE 16–12. Life-cycle trade-offs. (Schedule not necessarily typical.)

performance can be "bought." As the project nears termination, the relative importance of the cost constraint may increase drastically, especially if project profits are the company's major source of revenue. Likewise, it is probable that the impact of performance and schedule will be lower.

Once the alternative courses of action are determined, step 5 in the methodology is employed in order to analyze and select the feasible alternatives. Analyzing the alternatives should include the preparation of the revised project objectives for cost, performance, and time, along with an analysis of the required resources, general schedules, and revised project plans necessary to support each scenario. It is then the function of top management in conjunction with the project and functional managers to choose the solution that minimizes the overall impact to the company. This impact need not be measured just in short-term financial results, but should include long-term strategic and market considerations.

The following tasks can be included in this step:

- Prepare a formal project update report including alternative work scopes, schedules, and costs to achieve.
  - Minimum cost overrun
  - Conformance to project objectives
  - Minimum schedule overrun
- Construct a decision tree including costs, work objectives, and schedules, and an estimate of the probability of success for each condition leading to the decision point.
- Present to internal and external project management the several alternatives along with an estimate of success probability.
- With management's agreement, select the appropriate completion strategy, and begin implementation. This assumes that management does not insist on an impossible task.

The last item requires further clarification. Many companies use a checklist to establish the criteria for alternative evaluation as well as for assessment of potential future problems. The following questions may be part of such a checklist:

- Will other projects be affected?
- Will rework be required in previous tasks?
- Are repair and/or maintenance made more difficult?
- Will additional tasks be required in the future?
- How will project personnel react?
- What is the effect on the project life cycle?
- Will project flexibility be reduced?
- What is the effect on key employees?
- What is the effect on the customer(s)?

The probability of occurrence and severity should be assessed for all potential future problems. If there is a high probability that the problem will recur and be severe, a plan should be developed to reduce this probability. Internal restrictions, such as manpower,

materials, machines, money, management, time, policies, quality, and changing requirements, can cause problems throughout the life cycle of a project. External restrictions of
capital, completion dates, and liability also limit project flexibility.

One of the best methods for comparing the alternatives is to list them and then rank them
in order of perceived importance relative to certain factors such as customer, potential follow-on business, cost deficit, and loss of goodwill. This is shown in Table 16–2. In the table
each of the objectives is weighted according to some method established by management.
The percentages represent the degree of satisfactory completion for each alternative. This
type of analysis, often referred to as decision-making under risk, is commonly taught in
operations research and management science coursework. Weighting factors are often used
to assist in the decision-making process. Unfortunately, this can add mass confusion to the
already confused process.

Table 16–3 shows that some companies perform trade-off analysis by equating all alternatives to a lowest common denominator—dollars. Although this conversion can be very
difficult, it does ensure that we are comparing "apples to apples." All resources such as
capital equipment can be expressed in terms of dollars. Difficulties arise in assigning dollar
values to such items as environmental pollution, safety standards, or the possible loss of life.

There are often several types of corrective action that can be utilized, including:

- Overtime
- Double shifts
- Expediting
- Additional manpower
- More money
- Change of vendors
- Change of specifications
- Shift of project resources
- Waiving equipment inspections
- Change in statement of work
- Change in work breakdown structure
- Substitution of equipment

TABLE 16–2. WEIGHING THE ALTERNATIVES

Objectives / Weights / Alternatives	Increase Future Business	Ready on Time	Meet Current Cost	Meet Current Specs	Maximize Profits
	0.4	0.25	0.10	0.20	.05
Add resources	100%	90%	30%	90%	10%
Reduce scope of work	60%	90%	90%	30%	95%
Reduce specification change	90%	80%	95%	5%	80%
Complete project late	80%	0%	20%	95%	0%
Bill customer for added cost	30%	85%	0%	60%	95%

TABLE 16–3. TRADE-OFF ANALYSIS FOR IMPROVING PERFORMANCE CAPABILITY

Assumption	Description	Capital Expenditure, $	Time to Complete, Months	Project Profit, $	Ranking in Profit, $
1	No change	0	6	100,000	5
2	Hire higher-salaried people	0	5	105,000	3
3	Refurbish equipment	10,000	7	110,000	2
4	Purchase new equipment	85,000	9	94,000	6
5	Change specifications	0	6	125,000	1
6	Subcontract	0	6	103,000	4

- Substitution of materials
- Use of outside contractors
- Providing bonus payments to contractors
- Single-sourcing
- Waiving drawing approvals

The corrective actions defined above can be used for time, cost, and performance. However, there are specific alternatives for each area. Assuming that a PERT/CPM analysis was done initially to schedule the project, then the following options are available for schedule manipulation:

- Prioritize all tasks and see the effect on the critical path of eliminating low-priority efforts.
- Use resource leveling.
- Carry the work breakdown structure to one more level, and reassess the time estimates for each task.

Performance trade-offs can be obtained as follows:

- Excessive or tight specifications that are not critical to the project may be eased. (Many times standard specifications such as mil-specs are used without regard for their necessity.)
- Requirements for testing can be altered to accommodate automation (such as accelerated life testing) to minimize costs.
- Set an absolute minimum acceptable performance requirement below which you will not pursue the project. This gives a bound at the low end of performance that can't be crossed in choosing between trade-off alternatives.
- Give up only those performance requirements that have little or no bearing on the overall project goals (including implied goals) and their achievement. This may require the project manager to itemize and prioritize major and minor objectives.

- Consider absorbing tasks with dedicated project office personnel. This is a resource trade-off that can be effective when the tasks to be performed require in-depth knowledge of the project. An example would be the use of dedicated project personnel to perform information gathering on rehabilitation-type projects. The improved performance of these people in the design and testing phases due to their strong background can save considerable time and effort.

The most promising areas for cost analysis include:

- Incremental costing (using sensitivity analysis)
- Reallocation of resources
- Material substitution where lower-cost materials are utilized without changing project specifications

Depending on the magnitude of the problem, the timeliness of its identification, and the potential impact on the project results, it may be that no actions exist that will bring the project in on time, within budget, and at an acceptable level of performance. The following viable alternatives usually remain:

- A renegotiation of project performance criteria could be attempted with the project sponsor. Such action would be based on a pragmatic view of the acceptability of the probable outcome. Personal convenience of the project manager is not a factor. Professional and legal liability for the project manager, project team, or parent organization may be very real concerns.
- If renegotiation is not considered a viable alternative, or if it is rejected, the only remaining option is to "stop loss" in completing the project. Such planning should involve both line and project management, since the parent organization is at this point seeking to defend itself. Options include:
  - Completing the project on schedule, to the minimum quality level required by the project sponsor. This results in cost overruns (financial loss) but should produce a reasonably satisfied project sponsor. (Project sponsors are not really comfortable when they know a project team is operating in a "stop-loss" mode!)
  - Controlling costs and performance, but permitting the schedule to slide. The degree of unhappiness this generates with the project sponsor will be determined by the specific situation. Risks include loss of future work or consequential damages.
  - Maintaining schedule and cost performance by allowing quality to slip. The high-risk approach has a low probability of achieving total success and a high probability of achieving total failure. Quality work done on the project will be lost if the final results are below minimum standards.
  - Seeking to achieve desired costs, schedule, and performance results in the light of impossible circumstances. This approach "hopes" that the inevitable won't happen, and offers the opportunity to fail simultaneously in all areas. Criminal liability could become an issue.

● Project cancellation, in an effort to limit exposure beyond that already encountered. This approach might terminate the career of a project manager but could enhance the career of the staff counsel!

The sixth and final step in the methodology of the management of project trade-offs is to obtain management approval and replan the project. The project manager usually identifies the alternatives and prepares his recommendation. He then submits his recommendation to top management for approval. Top-management involvement is necessary because the project manager may try to make corrective action in a vacuum. Top management normally makes decisions based on the following:

● The firm's policies on quality, integrity, and image
● The ability to develop a long-term client relationship
● Type of project (R&D, modernization, new product)
● Size and complexity of the project
● Other projects underway or planned
● Company's cash flow
● Bottom line—ROI
● Competitive risks
● Technical risks
● Impact on affiliated organizations

After choosing a new course of action from the list of alternatives, management and especially the project team must focus on achieving the revised objectives. This may require a detailed replanning of the project, including new schedules, PERT charts, work breakdown structures, and other key benchmarks. The entire management team (i.e., top management, functional managers, and project managers) must all be committed to achieving the revised project plan.

16.2 CONTRACTS: THEIR INFLUENCE ON PROJECTS

PMBOK® Guide, 4th Edition
Chapter 12 Procurement
　Management
12.3 Contract Administration

The final decision on whether to trade-off cost, time, or performance can vary depending on the type of contract. Table 16–4 identifies seven common types of contracts and the order in which trade-offs will be made.

The firm-fixed-price (FFP) contract. Time, cost, and performance are all specified within the contract, and are the contractor's responsibility. Because all constraints are equally important with respect to this type of contract, the sequence of resources sacrificed is the same as for the project-driven organization shown previously in Table 16–1.

The fixed-price-incentive-fee (FPIF) contract. Cost is measured to determine the incentive fee, and thus is the last constraint to be considered for trade-off. Because performance is usually more important than schedule for project completion, time is considered the first constraint for trade-off, and performance is the second.

TABLE 16–4. SEQUENCE OF RESOURCES SACRIFICED BASED ON TYPE OF CONTRACT

	Firm-Fixed-Price (FFP)	Fixed-Price-Incentive-Fee (FPIF)	Cost Contract	Cost Sharing	Cost-Plus-Incentive-Fee (CPIF)	Cost-Plus-Award-Fee (CPAF)	Cost-Plus-Fixed-Fee (CPFF)
Time	2	1	2	2	1	2	2
Cost	1	3	3	3	3	1	1
Performance	3	2	1	1	2	3	3

1 = first to be sacrificed.
2 = second to be sacrificed.
3 = third to be sacrificed.

The cost-plus-incentive-fee (CPIF) contract. The costs are reimbursed and measured for determination of the incentive fee. Thus cost is the last constraint to be considered for trade-off. As with the FPIF contract, performance is usually more important than schedule for project completion, and so the sequence is the same as for the FPIF contract.

The cost-plus-award-fee (CPAF) contract. The costs are reimbursed to the contractor, but the award fee is based on performance by the contractor. Thus cost would be the first constraint to be considered for trade-off, and performance would be the last constraint to be considered.

The cost-plus-fixed-fee (CPFF) contract. Costs are reimbursed to the contractor. Thus, cost would be the first constraint to be considered for trade-off. Although there are no incentives for efficiency in time or performance, there may be penalties for bad performance. Thus time is the second constraint to be considered for trade-off, and performance is the third.

16.3 INDUSTRY TRADE-OFF PREFERENCES

Table 16–5 identifies twenty-one industries that were surveyed on their preferential process for trade-offs. Obviously, there are variables that affect each decision. The data in the table reflect the interviewees' general responses, neglecting external considerations, which might have altered the order of preference.

Table 16–6 shows the relative grouping of Table 16–5 into four categories: project-driven, non–project-driven, nonprofit, and banks.

In all projects in the banking industry, whether regulated or nonregulated, cost is the first resource to be sacrificed. The major reason for this trade-off is that banks in general do not have a quantitative estimation of what actual costs they incur in providing a given service. One example of this phenomenon is that a number of commercial banks heavily emphasize the use of *Functional Cost Analysis,* a publication of the Federal Reserve, for pricing their services. This publication is a summary of data received from member banks, of which the user is one. This results in questionable output because of inaccuracies of the input.

TABLE 16–5. INDUSTRY GENERAL PREFERENCE FOR TRADE-OFFS

Industry	Time	Cost	Performance
Construction	1	3	2
Chemical	2	1	3
Electronics	2	3	1
Automotive manu.	2	1	3
Data processing	2	1	3
Government	2	1	3
Health (nonprofit)	2	3	1
Medicine (profit)	1	3	2
Nuclear	2	1	3
Manu. (plastics)	2	3	1
Manu. (metals)	1	2	3
Consulting (mgt.)	2	1	3
Consulting (eng.)	3	1	2
Office products	2	1	3
Machine tool	2	1	3
Oil	2	1	3
Primary batteries	1	3	2
Utilities	1	3	2
Aerospace	2	1	3
Retailing	3	2	1
Banking	2	1	3

Note: Numbers in table indicate the order (first, second, third) in which the three parameters are sacrificed.

In cases where federal regulations prescribe time constraints, cost is the only resource of consideration, since performance standards are also delineated by regulatory bodies.

In nonregulated banking projects, the next resource to be sacrificed depends on the competitive environment. When other competitors have developed a new service or product that a particular bank does not yet offer, then the resource of time will be less critical

TABLE 16–6. SPECIAL CASES

	Type of Organization					
	Project-Driven Organizations				Banks	
	Early Life-Cycle Phases	Late-Life-Cycle Phases	Non–Project-Driven Organizations	Nonprofit Organizations	Leader	Follower
Time	2	1	1	2	3	2
Cost	1	3	3	3	1	1
Performance	3	2	2	1	2	3

than the performance criteria. A specific case is the development of the automatic teller machine (ATM). After the initial introduction of the system by some banks (leaders), the remainder of the competitors (followers) chose to provide a more advanced ATM with little consideration for the time involved for procurement and installation. On the other hand, with the introduction of negotiable order of withdrawal (NOW) accounts, the January 1, 1981, change in federal regulations allowed banks and savings and loans to offer interest-bearing checking accounts. The ensuing scramble to offer the service by that date led to varying performance levels, especially on the part of savings and loans. In this instance the competitors sacrificed performance in order to provide a timely service.

In some banking projects, the time factor is extremely important. A number of projects depend on federal laws. The date that a specific law goes into effect sets the deadline for the project.

Generally, in a nonprofit organization, performance is the first resource that will be compromised. The United Way, free clinics, March of Dimes, American Cancer Society, and Goodwill are among the many nonprofit agencies that serve community needs. They derive their income from donations and/or federal grants, and this funding mechanism places a major constraint on their operations. Cost overruns are prohibited by the very nature of the organization. Inexperienced staff and time constraints result in poor customer service.

The non–project-driven organization is structured along the lines of the traditional vertical hierarchy. Functional managers in areas such as marketing, engineering, accounting, and sales are involved in planning, organizing, staffing, and controlling their functional areas. Many projects that materialize, specifically in a manufacturing concern, are a result of a need to improve a product or process and can be initiated by customer request, competitive climate, or internal operations. The first resource to be sacrificed in the non–project-driven organization is time, followed by performance and cost, respectively. In most manufacturing concerns, budgetary constraints outweigh performance criteria.

In a non–project-driven organization, new projects will take a back seat to the day-to-day operations of the functional departments. The organizational funds are allocated to individual departments rather than to the project itself. When functional managers are required to maintain a certain productivity level in addition to supporting projects, their main emphasis will be on operations at the expense of project development. When it becomes necessary for the firm to curtail costs, special projects will be deleted in order to maintain corporate profit margins.

Resource trade-offs in a project-driven organization depend on the life-cycle phase of a given project. During the conceptual, definition, and production phases and into the operational phase of the project, the trade-off priorities are cost first, then time, and finally performance. In these early planning phases the project is being designed to meet certain performance and time standards. At this point the cost estimates are based on the figures supplied to the project manager by the functional managers.

During the operational phase the cost factor increases in importance over time and performance, both of which begin to decrease. In this phase the organization attempts to recover its investment in the project and therefore emphasizes cost control. The performance standards may have been compromised, and the project may be behind schedule, but management will analyze the cost figures to judge the success of the project.

The project-driven organization is unique in that the resource trade-offs may vary in priority, depending on the specific project. Research and development projects may have a fixed performance level, whereas construction projects normally are constrained by a date of completion.

16.4 CONCLUSION

It is obvious from the above discussion that a project manager does have options to control a project during its execution. Project managers must be willing to control minor trade-offs as well as major ones. However, the availability of specific options is a function of the particular project environment.

Probably the greatest contribution a project manager makes to a project team organization is stability in adverse conditions. Interpersonal relationships have a great deal to do with the alternatives available and their probability of success since team performance will be required. Through a combination of management skill and sensitivity, project managers can make the trade-offs, encourage the team members, and reassure the project sponsor in order to produce a satisfactory project.

16.5 STUDYING TIPS FOR THE PMI® PROJECT MANAGEMENT CERTIFICATION EXAM

This section is applicable as a review of the principles to support the knowledge areas and domain groups in the PMBOK® Guide. This chapter addresses:

- Integration Management
- Scope Management
- Procurement Management
- Initiating
- Planning
- Execution
- Controlling

Understanding the following principles is beneficial if the reader is using this text to study for the PMP® Certification Exam:

- What is meant by a trade-off
- Who are the major players in performing trade-offs
- That assumptions and circumstances can change mandating that trade-offs take place

The following multiple-choice questions will be helpful in reviewing the principles of this chapter:

1. Trade-offs are almost always necessary because:
   A. Project managers are incapable of planning correctly.
   B. Line managers are unable to provide accurate estimates.
   C. Executives are unable to properly define project objectives.
   D. Circumstances can change, thus mandating trade-offs to take place.

2. The person who may be ultimately responsible for approving the trade-off is the:
   A. Project manager
   B. Line manager
   C. Project sponsor
   D. Customer

3. The most common trade-offs occur on:
   A. Time, cost, and quality
   B. Risk, cost, and quality
   C. Risk, time, and quality
   D. Scope, quality, and risk

4. If the start date of a project is delayed but the budget and specifications remain fixed, what would the project manager most likely trade off first?
   A. Scope
   B. Time
   C. Quality
   D. Risk

ANSWERS

1. D
2. D
3. A
4. C

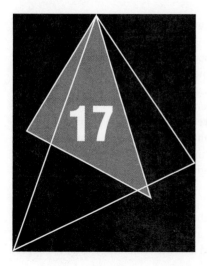

Risk Management[1]

Related Case Studies (from Kerzner/*Project Management Case Studies,* 3rd Edition)	Related Workbook Exercises (from Kerzner/*Project Management Workbook and PMP®/CAPM® Exam Study Guide,* 10th Edition)	PMBOK® Guide, 4th Edition, Reference Section for the PMP® Certification Exam
● Teloxy Engineering (A)* ● Teloxy Engineering (B)* ● The Space Shuttle ● Challenger Disaster ● Packer Telecom ● Luxor Technologies ● Altex Corporation ● Acme Corporation	● Multiple Choice Exam ● Crossword Puzzle on Risk Management	● Risk Management ● Professional ● Responsibility ● Domain

17.0 INTRODUCTION

PMBOK® Guide, 4th Edition
Chapter 11 Risk Management

In the early days of project management on many commercial programs, the majority of project decisions heavily favored cost and schedule. This favoritism occurred because we knew more about cost and scheduling

---

1. This chapter was updated by Dr. Edmund H. Conrow, CMC, CPCM, CRM, PMP. Dr. Conrow has extensive experience in developing and implementing risk management on a wide variety of projects. He is a management and technical consultant who is the author of *Effective Risk Management: Some Keys to Success,* 2nd ed. (Reston, VA: American Institute of Aeronautics and Astronautics, 2003). He can be reached at (310) 374-7975 and www.risk-services.com.

than we did about technical risks. Technology forecasting was very rarely performed other than by extrapolating past technical knowledge into the present.

Today, the state of the art of technology forecasting is being pushed to the limits on many projects. For projects with a time duration of less than one year, we normally assume that the environment is known and stable, particularly the technological environment. For projects over a year or so in length, technology forecasting must be considered. Computer technology doubles in performance about every two years. Engineering technology is said to double every three or so years. Given such rapid change, plus the inherent need to balance cost, technical performance, and schedule, how can a project manager accurately define and plan the scope of a three- or four-year project without expecting somewhat uncertain engineering changes resulting from technology improvements? With likely changing and uncertain engineering, technology, and production environments, what are the risks?

A Midwest manufacturing company embarked on an eight-year project to design the manufacturing factory of the future. How do we design the factory of the future without forecasting the technology? For example, what computer technology will exist? What types of materials will exist and what types of components will our customer require? What production rate will we need and will technology exist to support this production level?

Economists and financial institutions forecast interest rates. The forecasts appear in public newspapers and journals. Yet, every company involved in high tech does some form of technology forecasting but may be reluctant to publish the data. Technology forecasting is regarded as company proprietary information and may be part of the company's strategic planning process.

We read in the newspaper about cost overruns and schedule slips on a wide variety of medium to large-scale development projects. Several issues within the control of the buyer, seller, or major stakeholders can lead to cost growth and schedule slippage on development projects. These causes include, but are not limited to:[2]

- Starting a project with a budget and/or schedule that is inadequate for the desired level of technical performance (or proxies such as integration complexity)
- Starting a project before an adequate requirements flowdown and verification has occurred and/or before adequate resources have been committed
- Having an overall development process (or key parts of that process) that favors one or more variables over others (e.g., technical performance over cost and schedule)
- Establishing a design that is near the feasible limit of achievable technical performance at a given point in time
- Making major project design decisions before the relationship between cost, technical performance, schedule, and risk is understood

These five causes will contribute to uncertainty in forecasting technology and the associated design needed to meet technical performance requirements. And the inability to accurately forecast technology and the associated design will contribute to a project's technical risk and can also lead to cost and/or schedule risk.

Today, the competition for technical achievement has become fierce. Companies have gone through life-cycle phases of centralizing all activities, especially management functions, but are decentralizing technical expertise. By the mid-1980s, many companies recognized the need to integrate technical risks with cost and schedule risks and other activities (e.g., quality). Risk management processes were developed and implemented where risk information was made available to key decision-makers.

---

2. Edmund H. Conrow, "Some Long-Term Issues and Impediments Affecting Military Systems Acquisition Reform," *Acquisition Review Quarterly*, Defense Acquisition University, Vol. 2, No. 2, Summer 1995, pp. 199–212.

The risk management process, however, should be designed to do more than just identify potential risks. The process must also include a formal *planning* activity, *analysis* to estimate the probability and predict the impact on the project of identified risks, a *risk response* strategy for selected risks, and the ability to *monitor and control* the progress in reducing these selected risks to the desired level.

A project, by definition, is a temporary endeavor used to create something that we have not done previously and will not do again in the future. Because of this uniqueness, we have developed a "live with it" attitude on risk and attribute it as part of doing business. If risk management is set up as a continuous, disciplined process of planning, identifying, analyzing, developing risk responses, and monitoring and controlling, then the system will easily supplement other processes such as planning, budgeting, cost control, quality, and scheduling. Surprises that become problems will be diminished because the emphasis will now be on proactive rather than reactive management.

Risk management can be justified on almost all projects. The level of implementation can vary from project to project, depending on such factors as size, type of project, who the customer is, contractual requirements, relationship to the corporate strategic plan, and corporate culture. Risk management is particularly important when the overall stakes are high and/or a great deal of uncertainty exists. In the past, we treated risk as a "let's live with it." Today, risk management is a key part of overall project management. It forces us to focus on the future where uncertainty exists and develop suitable plans of action to prevent potential issues from becoming problems and adversely impacting the project.

17.1 DEFINITION OF RISK

PMBOK® Guide, 4th Edition
11.1 Plan Risk Management

Risk is a measure of the probability and consequence of not achieving a defined project goal. Most people agree that risk involves the notion of uncertainty. Can the specified aircraft range be achieved? Can the computer be produced within budgeted cost? Can the new product launch date be met? A probability measure can be used for such questions; for example, the probability of not meeting the new product introduction date is 0.15. However, when risk is considered, the consequences or damage associated with the event occurring must also be considered.

Goal A, with a probability of occurrence of only 0.05, may present a much more serious (risky) situation than goal B, with a probability of occurrence of 0.20, if the consequences of not meeting goal A are, in this case, more than four times more severe than the inability to meet goal B. Risk is not always easy to evaluate since the probability of occurrence and the consequence of occurrence are usually not directly measurable parameters and must be estimated by judgment, statistical, or other procedures.

Risk has two primary components for a given event:

- A probability of occurrence of that event
- Impact (or consequence) of the event occurring (amount at stake)

Figure 17–1 shows the components of risk.

Conceptually, the risk for each event can be defined as a function of probability and consequence (impact); that is,

$$\text{Risk} = f(\text{probability, consequence})$$

In general, as either the probability or consequence increases, so does the risk. Both the probability and consequence must be considered in risk management.

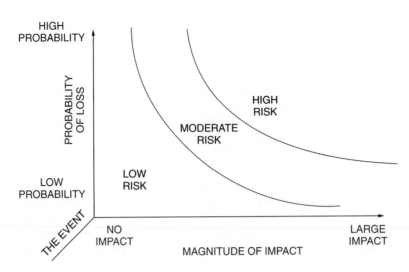

FIGURE 17–1. Overall risk is a function of its components.

Risk constitutes a lack of knowledge of future events. Typically, future events (or outcomes) that are favorable are called opportunities, whereas unfavorable events are called risks. Note, however, that risks and opportunities may be uncorrelated or only partially correlated on a given project, and risks and opportunities and their management are not simply mirror images of each other. This is because, through prospect theory (developed by Nobel Economics Laureate Daniel Kahneman and his late colleague Amos Tversky), people tend to value the same level of gains and losses differently. (See Appendix E of the Conrow book cited in footnote 1 for a discussion of risk and opportunity.)

Another element of risk is its cause. Something, or the lack of something, can induce a risky situation. We denote this source of danger as the hazard. Certain hazards can be overcome to a great extent by knowing them and taking action to overcome them. For example, a large hole in a road is a much greater danger to a driver who is unaware of it than to one who travels the road frequently and knows enough to slow down and go around the hole. This leads to the second representation of risk:

$$Risk = f(hazard, safeguard)$$

Risk increases with hazard but decreases with safeguard. The implication of this equation is that good project management should be structured to identify hazards and to allow safeguards to be developed to overcome them. If suitable safeguards are available, then the risk can be reduced to an acceptable level.

Finally, there is often confusion regarding the nature of risks, issues, and problems within the context of project management. All three items are partially related through the consequence (C) dimension but different in either the probability (P) dimension or time frame. A summary of risk, issue, problem, and opportunity with regards to probability, consequence, and time frame is given in Table 17–1.

TABLE 17–1. CONCISE DEFINITIONS OF RISK, ISSUE, AND PROBLEM

Item	Probability	Consequence	Time Frame
Risk	$0 < P < 1$	$C > 0$	Future
Issue	$P = 1$	$C > 0$	Future
Problem	$P = 1$	$C > 0$	Now
Opportunity	Unclear ($0 < P \leq 1$?)	Unclear ($C > 0$ or $C < 0$?)	Unclear (now, future?)

Source: Edmund H. Conrow, "Risk Analysis for Space Systems," Proceedings of the Space Systems Engineering and Risk Management Symposium 2008, Los Angeles, 28 February 2008. Copyright © 2008, Edmund H. Conrow. Used with permission of the author.

Both issues and problems have a probability of occurrence equal to 1—they will occur, while a risk may not occur ($P < 1$) However, an issue will occur in the future, while a problem occurs in the present. The probability dimension for an opportunity is unclear because there is no equivalent differentiation as in the probability dimension for risk, issue, and problem. Moreover, it is not possible to define the consequence dimension in a unique manner since an opportunity may represent, according to three simple definitions, a positive outcome, a less negative outcome, and an outcome that is better than expected. Finally, the time frame associated with an opportunity is also unclear as it may be now or in the future. As evident from the above discussion, while precise risk definitions have been developed, precise definitions cannot be defined for opportunity that have universal applicability. Hence, risk and opportunity are not mirror images of each other, either in terms of definitions or gains and losses.

17.2 TOLERANCE FOR RISK

There is no single textbook answer on how to manage risk—one size does not fit all projects or circumstances. The project manager must rely upon sound judgment and the use of the appropriate tools in dealing with risk. The ultimate decision on how to deal with risk is based in part upon the project manager's tolerance for risk, along with contractual requirements, stakeholder preferences, and so faith.

The three commonly used classifications of tolerance for risk appear in Figure 17–2. They include the risk averter or avoider, the neutral risk taker, and the risk taker or seeker. The *Y* axis in Figure 17–2 represents utility, which can be defined as the amount of satisfaction or pleasure that the individual receives from a payoff. The *X* axis in this case is the amount of money at stake (but can also potentially represent technical performance or schedule). Curves of this type can represent the project manager or other key decision-makers' tolerance for risk.

With the risk averter, utility rises at a *decreasing* rate. In other words, when more money is at stake, the project manager's satisfaction diminishes. With a risk neutral position, utility rises at a *constant* rate. (Note: A risk neutral position is *a* specific course of action, and not the average of risk averter and risk taker positions as is sometimes erroneously claimed.) With the risk taker, the project manager's satisfaction increases at an increasing rate when

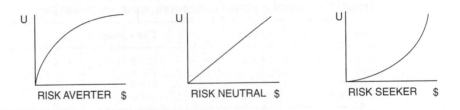

The shape of a given decision-maker's curve is derived from comparing response to alternative decision acts.

FIGURE 17–2. Risk preference and the utility function.

more money is at stake. A risk averter prefers a more certain outcome and will demand a premium to accept risk. A risk taker prefers the more uncertain outcome and may be willing to pay a penalty to take a risk. While the project manager's, or other key decision-makers', tolerance for risk may vary with time, different representations of this tolerance (e.g., risk averter and risk taker) should not exist at the same time else inconsistent decisions may be made.

17.3 DEFINITION OF RISK MANAGEMENT

PMBOK® Guide, 4th Edition
11.2 Identify Risks

Risk management is the act or practice of dealing with risk. It includes *planning* for risk, *identifying* risks, *analyzing* risks, developing *risk response* strategies, and *monitoring and controlling* risks to determine how they have changed.

Risk management is not a separate project office activity assigned to a risk management department but rather is one aspect of sound project management. Risk management should be closely coupled with key project processes, including but not limited to: overall project management, systems engineering, configuration management, cost, design/engineering, earned value, manufacturing, quality, schedule, scope, and test. (Project management and systems engineering are typically the two top-level project processes. While risk management can be linked to either of these processes, it is typically associated with project management.)

Proper risk management is proactive rather than reactive, positive rather than negative, and seeks to increase the probability of project success. As an example, an item in a network (e.g., router) requires that a new technology be developed. The schedule indicates six months for this development, but project engineers think that nine months is much more likely. If the project manager is proactive, he might develop a risk response plan right *now*. If the project manager is reactive (e.g., a "problem solver"), then he may do nothing until the problem actually occurs. At that time the project manager must react rapidly to the crisis and may have lost valuable time during which contingencies could have been developed, and at least some possible solutions may have been foreclosed. (The resulting cost, technical performance, schedule, and risk design solution space will also have likely shrunk considerably versus when the project was initiated.) Hence, proper risk management will attempt to reduce the probability of an event occurring and/or the magnitude of its impact as well as increase the probability of project success.

17.4 CERTAINTY, RISK, AND UNCERTAINTY

Decision-making falls into three categories: certainty, risk, and uncertainty. [Decision-making, including but not limited to payoff matrices, expected (monetary) value, and decision trees can be loosely linked with quantitative risk analysis discussed in Section 17.10.] Decision-making under certainty is the easiest case to work with. With certainty, we assume that all of the necessary information is available to assist us in making the right decision, and we can predict the outcome with a high level of confidence.

Decision-Making under Certainty

Decision-making under certainty implies that we know with 100 percent accuracy what the states of nature will be and what the expected payoffs will be for each state of nature. Mathematically, this can be shown with payoff matrices.

To construct a payoff matrix, we must identify (or select) the states of nature over which *we have no control*. We then select our own action to be taken for each of the states of nature. Our actions are called strategies. The elements in the payoff table are the outcomes for each strategy.

A payoff matrix based on decision-making under certainty has two controlling features.

- Regardless of which state of nature exists, there will be one strategy that will produce larger gains or smaller losses than any other strategy for all the states of nature.
- There are no probabilities assigned to each state of nature. (This could also be stated that each state of nature has an equal likelihood of occurring.)

Example 17–1. Consider a company wishing to invest $50 million to develop a new product. The company decides that the states of nature will be either a strong market demand, an even market demand, or a low market demand. The states of nature will be represented as N_1 = a strong (up) market, N_2 = an even market, and N_3 = a low market demand. The company also has narrowed their choices to one of three ways to develop the product: either A, B, or C. There also exists a strategy S_4, not to develop the product at all, in which case there would be neither profit nor loss. We shall assume that the decision is made to develop the product. The payoff matrix for this example is shown in Table 17–2. Looking for the controlling features in Table 17–2, we see that regardless of how the market reacts, strategy S_3 will always yield larger profits than the other two strategies. The project manager will therefore always select strategy S_3 in developing the new product. Strategy S_3 is the best option to take.

Table 17–2 can also be represented in subscript notation. Let $P_{i,j}$ be the elements of the matrix, where P represents profit. The subscript i is the row (strategy), and j is the column (state of nature). For example, $P_{2,3}$ = the profit from choosing strategy 2 with N_3 state of nature occurring. It should be noted that there is no restriction that the matrix be square, but at a minimum it will be a rectangle (i.e., the number of states of nature need not equal the number of possible strategies).

Decision-Making under Risk

In most cases, there usually does not exist one strategy that dominates for all states of nature. In a realistic situation, higher profits are usually accompanied by higher risks and therefore higher probable losses. When there does not exist a dominant strategy, a probability must be assigned to the occurrence of each state of nature.

Risk can be viewed as outcomes (i.e., states of nature) that can be described within established confidence limits (i.e., probability distributions). These probability distributions should ideally be either estimated or defined from experimental data.

Consider Table 17–3 in which the payoffs for strategies 1 and 3 of Table 17–2 are interchanged for the state of nature N_3.

From Table 17–3, it is obvious that there does not exist one dominant strategy. When this occurs, probabilities must be assigned to the possibility of each state of nature occurring. The best choice of strategy is, therefore, the strategy with the largest expected value, where the *expected value* is the summation of the payoff times and the probability of occurrence of the payoff for each state of nature. In mathematical formulation,

$$E_i = \sum_{j=1}^{N} P_{i,j}\, p_j$$

TABLE 17–2. PAYOFF MATRIX (PROFIT IN MILLIONS)

	STATES OF NATURE		
Strategy	N_1 = Up	N_2 = Even	N_3 = Low
S_1 = A	$50	$40	−$50
S_2 = B	$50	$50	$60
S_3 = C	$100	$80	$90

TABLE 17–3. PAYOFF TABLE (PROFIT IN MILLIONS)

	STATES OF NATURE*		
Strategy	N_1	N_2	N_3
	0.25	0.25	0.50
S_1	50	40	90
S_2	50	50	60
S_3	100	80	−50

*Numbers are assigned probabilities of occurrence for each state of nature.

where E_i is the expected payoff for strategy i, $P_{i,j}$ is the payoff element, and p_j is the probability of each state of nature occurring. The expected value for strategy S_1 is therefore

$$E_1 = (50)(0.25) + (40)(0.25) + (90)(0.50) = 67.50$$

Repeating the procedure for strategies 2 and 3, we find that $E_2 = 55$, and $E_3 = 20$. Therefore, based on the expected value, the project manager should always select strategy S_1. If two strategies of equal value occur, the decision should include other potential considerations (e.g., frequency of occurrence, resource availability, time to impact). (Note: Expected value calculations require that a risk neutral utility relationship exists. If the decision-maker is not risk neutral, such calculations may or may not be useful, and the results should be evaluated to see how they are affected by differences in risk tolerance.)

To quantify potential payoffs, we must identify the strategy we are willing to take, the expected outcome (element of the payoff table), and the probability that the outcome will occur. In the previous example, we should accept the risk associated with strategy S_1 since it gives us the greatest expected value (all else held constant). If the expected value is positive, then this strategy should be considered. If the expected value is negative, then this strategy should be proactively managed.

An important factor in decision-making under risk is the assigning of the probabilities for each of the states of nature. If the probabilities are erroneously assigned, different expected values will result, thus giving us a different perception of the best strategy to take. Suppose in Table 17–3 that the assigned probabilities of the three states of nature are 0.6, 0.2, and 0.2. The respective expected values are:

$$E_1 = 56$$
$$E_2 = 52$$
$$E_3 = 66$$

In this case, the project manager would always choose strategy S_3 (all else held constant).

Decision-Making under Uncertainty

The difference between risk and uncertainty is that under risk there are assigned specific probabilities, and under uncertainty meaningful assignments of specific probabilities are not possible. As with decision-making under risk, uncertainty also implies that there may exist no single dominant strategy. The decision-maker, however, does have at his disposal four basic criteria from which to make a management decision. The use of each criteria will depend on the type of project as well as the project manager's tolerance to risk.

The first criterion is the Hurwicz criterion, often referred to as the maximax criterion. (This criteria was developed by Nobel Economics Laureate Leonid Hurwicz.) Under the Hurwicz criterion, the decision-maker is always optimistic and attempts to maximize profits by a go-for-broke strategy. This result can be seen from the example in Table 17–3. The maximax criterion says that the decision-maker will always choose strategy S_3 because the maximum profit is 100. However, if the state of nature were N_3, then strategy S_3 would result in a maximum loss instead of a maximum gain. The use of the maximax, or Hurwicz criterion, must then be based on how big a risk can be undertaken and how much one can afford to lose. A large corporation with strong assets may use the Hurwicz criterion, whereas the small private company might be more interested in minimizing the possible losses.

A small company may be more apt to use the Wald, or maximin criterion, where the decision-maker is concerned with how much he can afford to lose. In this criteria, a pessimistic rather than optimistic position is taken with the viewpoint of minimizing the maximum loss.

In determining the Hurwicz criterion, we looked at only the maximum payoffs for each strategy in Table 17–3. For the Wald criterion, we consider only the minimum payoffs. The minimum payoffs are 40, 50, and -50 for strategies S_1, S_2, and S_3, respectively. Because the project manager wishes to minimize his maximum loss, he will always select strategy S_2 in this case. If all three minimum payoffs were negative, the project manager would select the smallest loss if these were the only options available. Depending on a company's financial position, there are situations where the project would not be undertaken if all three minimum payoffs were negative.

The third criterion is the Savage, or minimax criterion. Under this criterion, we assume that the project manager is a sore loser. To minimize the regrets of the sore loser, the project manager attempts to minimize the maximum regret; that is, the minimax criterion.

TABLE 17–4. REGRET TABLE

Strategy	States of Nature			Maximum Regrets
	N_1	N_2	N_3	
S_1	50	40	0	50
S_2	50	30	30	50
S_3	0	0	140	140

The first step in the Savage criterion is to set up a regret table by subtracting all elements in each column from the largest element. Applying this approach to Table 17–3, we obtain Table 17–4.

The regrets are obtained for each column by subtracting each element in a given column from the largest column element. The maximum regret is the largest regret for each strategy, that is, in each row. In other words, if the project manager selects strategy S_1 or S_2, he will only be sorry for a loss of 50. However, depending on the state of nature, a selection of strategy S_3 may result in a regret of 140. The Savage criterion would select either strategy S_1 or S_2 in this example.

The fourth criterion is the Laplace criterion. The Laplace criterion is an attempt to transform decision-making under uncertainty to decision-making under risk. Recall that the difference between risk and uncertainty is a knowledge of the probability of occurrence of each state of nature. The Laplace criterion makes an a priori assumption based on Bayesian statistics: If the probabilities of each state of nature are not known, then we can assume that each state of nature has an equal likelihood of occurrence. The procedure then

TABLE 17–5. LAPLACE CRITERION

Strategy	Expected Value
S_1	60
S_2	53.3
S_3	43.3

follows decision-making under risk, where the strategy with the maximum expected value is selected. Using the Laplace criterion applied to Table 17–3, and thus assuming that $P_1 = P_2 = P_3 = 1/3$, we obtain Table 17–5. The Laplace criterion would select strategy S_1 in this example.

The important conclusion to be drawn from decision-making under uncertainty is the risk that the project manager wishes to incur. For the four criteria previously mentioned, we have shown that any strategy can be chosen depending on how much money we can afford to lose and what risks we are willing to take.

The concept of expected value can also be combined with "probability" or "decision" trees to identify and quantify the potential risks. Another common term is the impact analysis diagram. Decision trees are used when a decision cannot be viewed as a single, isolated occurrence, but rather as a sequence of several interrelated decisions. In this case, the decision-maker makes an entire series of decisions simultaneously. (Again, a risk neutral utility relationship is assumed.)

Consider the following problem. A product can be manufactured using machine A or machine B. Machine A has a 40 percent chance of being used and machine B a 60 percent chance. Both machines use either process C or D. When machine A is selected, process C is selected 80 percent of the time and process D 20 percent. When machine B is selected, process C is selected 30 percent of the time and process D 70 percent of the time. What is the probability of the product being produced by the various combinations?

Figure 17–3 shows the decision tree for this problem. The probability at the end of each branch (furthest to the right) is obtained by multiplying the branch probabilities together.

For more sophisticated problems, the process of constructing a decision tree can be complicated. Decision trees contain decision points, usually represented by a box or square, where the decision-maker must select one of several available alternatives. Chance points, designated by a circle, indicate that a chance event is expected at this point. [A key assumption required for decision trees is a risk neutral position (discussed in Section 17.2). Note that the expected value computed in decision trees is not the average outcome, it is the risk neutral outcome.]

The following three steps are needed to construct a tree diagram:

- Build a logic tree, usually from left to right, including all decision points and chance points.
- Put the probabilities of the states of nature on the branches, thus forming a probability tree.
- Finally, add the conditional payoffs, thus completing the decision tree.

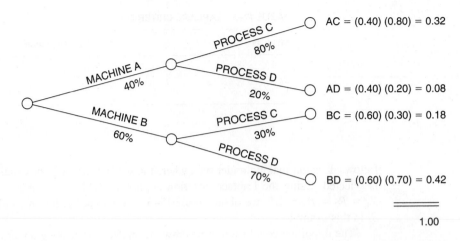

SUM OF THE PROBABILITIES
MUST EQUAL 1.00.

FIGURE 17–3. Decision tree.

Consider the following problem. You have the chance to make or buy certain widgets for resale. If you make the widgets yourself, you must purchase a new machine for $35,000. If demand is good, which is expected 70 percent of the time, an $80,000 profit will occur on the sale of the widgets. With poor market conditions, $30,000 in profits will occur, not including the cost of the machine. If we subcontract out the work, our contract administration costs will be $5000. If the market is good, profits will be $50,000; for a poor market, profits will be $15,000. Figure 17–4 shows the tree diagram for this problem. In this case, the expected value of the strategy that subcontracts out the work is $4500 greater than the expected value for the strategy that manufactures the widgets. Hence, we should select the strategy that subcontracts out the widgets.

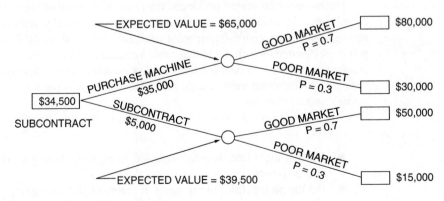

FIGURE 17–4. Expanded tree diagram.

17.5 RISK MANAGEMENT PROCESS

It is important that a risk management strategy be established early in a project and that risk be continually addressed throughout the project life cycle. Risk management includes several related actions, including risk: planning, identification, analysis, response (handling), and monitoring and control. (Numbers refer to section numbers in the 4th edition of the *PMBOK Guide.*)

- **Plan Risk Management (11.1)**: The process of developing and documenting an organized, comprehensive, and interactive strategy and methods for identifying and analyzing risks, developing risk response plans, and monitoring and controlling how risks have changed.
- **Identify Risks (11.2)**: The process of examining the program areas and each critical technical process to identify and document the associated risk.
- **Perform Risk Analysis (11.3, 11.4)**: The process of examining each identified risk to estimate the probability and predict the impact on the project. It includes both qualitative risk analysis (11.3) and quantitative risk analysis (11.4).
- **Plan Risk Response (11.5)**: The process that identifies, evaluates, selects, and implements one or more strategies in order to set risk at acceptable levels given program constraints and objectives. This includes the specifics on what should be done, when it should be accomplished, who is responsible, and associated cost and schedule. A risk or opportunity response strategy is composed of an option and implementation approach. Response options for risks include acceptance, avoidance, mitigation (also known as control), and transfer. Response options for opportunities include acceptance, enhance, exploit, and share. The most desirable response option is selected, and a specific implementation approach is then developed for this option.
- **Monitor and Control Risks (11.6)**: The process that systematically tracks and evaluates the performance of risk response actions against established metrics throughout the acquisition process and provides inputs to updating risk response strategies, as appropriate.

17.6 PLAN RISK MANAGEMENT

PMBOK® Guide, 4th Edition
11.1 Plan Risk Management

Plan for risk management (risk planning) is the detailed formulation of a program of action for the management of risk. It is the process to:

- Develop and document an organized, comprehensive, and interactive risk management strategy.
- Determine the methods to be used to execute a program's risk management strategy.
- Plan for adequate resources.

Risk planning is iterative and includes the entire risk management process, with activities to identify, analyze, respond to, and monitor and control risks. An important output of the risk planning process is the risk management plan (RMP). (Note: The RMP is an output of risk planning and *not* the risk planning process itself.)

Risk planning develops a risk management strategy, which includes both the process and implementation approach for the project. Each of these two considerations is of primary importance for achieving effective risk management. However, it is generally far easier to improve a deficient process than remedy a problematic project environment that is unsupportive or hostile toward risk management. Early efforts should establish the purpose and objective, assign responsibilities for specific areas, identify additional technical expertise needed, describe the assessment process and areas to consider, define a risk rating approach, delineate procedures for consideration of response strategies, establish monitoring and control metrics (where possible), and define the reporting, documentation, and communication needs.

The RMP is the risk-related roadmap that tells the project team how to get from where the program is today to where the program manager wants it to be in the future. The key to writing a good RMP is to provide the necessary information so the program team knows the objectives; goals; tools and techniques; reporting, documentation, and communication; organizational roles and responsibilities; and behavioral climate to achieving effective risk management. The RMP should include appropriate definitions, ground rules and assumptions associated with performing risk management on the project, candidate risk categories, suitable risk identification and analysis methodologies, a suitable risk management organizational implementation, and suitable documentation for risk management activities. The RMP should never include results (e.g., risk analysis scores) because these results may frequently change, thus necessitating updates to the RMP. Instead, risk-related results should be included in separate risk documents (e.g., risk register and its updates) to avoid unnecessary updates to the RMP.

Since the RMP is a roadmap, it may be specific in some areas, such as the assignment of responsibilities for project personnel and definitions, and general in other areas to allow users to choose the most efficient way to proceed. For example, a description of techniques that suggests several methods to perform a risk analysis is appropriate since every technique has advantages and disadvantages depending on the situation.

Another important aspect of risk planning is providing risk management training to project personnel. The vast majority of current risk management trainers and teachers have either never had long-term responsibility to make risk management work on an actual project, focus on generic process steps and bypass implementation considerations, focus on a minor subset of risk management (e.g., Monte Carlo simulations), or have a knowledge base that is far below the state of the art. It is important that risk management training be performed by individuals, whether inside or outside the project, with substantial real-world experience in making risk management work on actual projects; else the training may be nothing more than an academic exercise with little or no value. Finally risk management training should be tailored to various groups within the project as necessary, and a different emphasis may exist for decision-makers versus working-level personnel and technical versus nontechnical personnel.

17.7 RISK IDENTIFICATION

PMBOK® Guide, 4th Edition
11.2 Identify Risks
The second step in risk management is to identify risks (risk identification). This may result from a survey of the project, customer, and users for potential concerns.

Some degree of risk always exists in the project, such as in technical, test, logistics, production, engineering, and other areas. Project risks include business, contract relationship, cost, funding, management, political, and schedule risks. (Cost and schedule risks are often so fundamental to a project that they may be treated as stand-alone risk categories.) Technical risks, such as related to engineering and technology, may involve the risk of meeting a technical performance requirement, but may also involve risks in the feasibility of a design concept or the risks associated with using state-of-the-art equipment or software. Production risk includes concerns over packaging, manufacturing, lead times, and material availability. Support risks include maintainability, operability, and trainability concerns. Threat risk includes a variety of subcategories, such as security, survivability, and vulnerability—all items of increased importance in the last few years.[3] The understanding of risks in these and other areas evolves over time. Consequently, risk identification must continue through all project phases.

Project risks should be examined and dissected to a level of detail that permits an evaluator to understand the significance of the risk and its causes and to potentially examine the root cause(s). This is a practical way of addressing the large and diverse number of potential risks that often occur in moderate to large-scale programs.

The methods for identifying risks are numerous. Common practice is to classify project risk according to its source, which is typically either objective or subjective.

- Objective sources: recorded experience from past projects and the current project as it proceeds
  - Lessons learned files
  - Program documentation evaluations
  - Current performance data
- Subjective sources: experiences based upon knowledgeable experts
  - Interviews and other data from subject matter experts

Risks can also be identified according to life-cycle phases, as shown in Figure 17–5. In the early life-cycle phases, the total project risk is high in part because of the lack of information that may preclude comprehensive and accurate risk identification, and because risk response plans have yet to be developed and implemented. In the later life-cycle phases, financial risk is generally substantial both because of investments made (such as cost) and because of foreclosed options (opportunity cost).

---

3. In the broadest sense technical risk can include engineering (often termed design) and technology, production, support, and threat risks. In some cases all risks outside of project risk are termed technical risks.

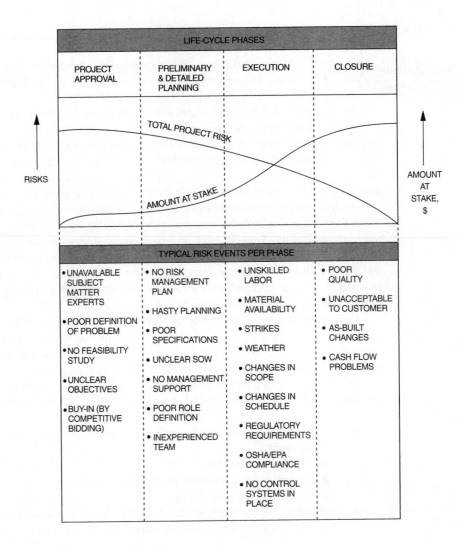

FIGURE 17–5. Life-cycle risk analysis.

Any source of information that allows recognition of a potential problem can be used for risk identification. These include, but are not limited to:

- Assumption analysis
- Baseline cost estimates
- Brainstorming
- Checklists
- Cost analysis

- Decision drivers
- Diagramming techniques (e.g., influence diagrams)
- Earned value analysis
- Expert judgment
- Lessons learned files
- Life-cycle cost analysis
- Models
- Plan/WBS decomposition
- Root cause investigations
- Schedule analysis
- Strengths, weaknesses, opportunities, and threats (SWOT)
- Systems engineering documentation
- Technical performance measurement (TPM/) planning/analysis
- Technology analysis
- Technology development/insertion projects
- Trade studies/analyses

Expert judgment techniques are applicable not only for risk identification but also for forecasting and decision-making. Two expert judgment techniques are the Delphi method and the nominal group technique. The Delphi method has the following general steps:

- Step 1: A panel of experts is selected from both inside and outside the organization. The experts do not interact on a face-to-face basis and may not even know who else sits on the panel.
- Step 2: Each expert is asked to make an anonymous prediction on a particular subject.
- Step 3: Each expert receives a composite feedback of the entire panel's answers and is asked to make new predictions based upon the feedback. The process is then repeated as necessary.

Closely related to the Delphi method is the nominal group technique, which allows for face-to-face contact and direct communication. The steps in the nominal group technique are as follows:

- Step 1: A panel is convened and asked to generate ideas in writing.
- Step 2: The ideas are listed on a board or a flip chart. Each idea is discussed among the panelists.
- Step 3: Each panelist prioritizes the ideas, which are then rank ordered. Steps 2 and 3 may be repeated as necessary.

Expert judgment techniques have the potential for bias in risk identification and risk analysis, as well as in selecting risk response strategies. These biases vary on a case-by-case basis and can affect the probability of occurrence and consequence of occurrence estimates differently.

Cognitive factors that can introduce a bias and/or noise term include, but are not limited to:

- Adjustments from an initial value
- Anchoring (biased toward the initial value)
- Availability (of past events)
- Fit ambiguous evidence into predispositions
- Insensitivity to the problem or risk
- Motivation
- Overconfidence in the reliability of the analysis
- Overconfidence in one's ability
- Proximity to project
- Relationship with other experts
- Representativeness (the degree to which A resembles B)
- Systematically omit risk components

There exist numerous ways to classify risks. In a simple business context, risk can be defined as:

- Business risk
- Insurable risk

Business risks provide us with opportunities of profit and loss. Examples of business risk would be competitor activities, bad weather, inflation, recession, customer response, and availability of resources. Insurable risks provide us with only a chance for a loss. Insurable risks include such elements as:

- *Direct property damage:* This includes insurance for assets such as fire insurance, collision insurance, and insurance for project materials, equipment, and properties.
- *Indirect consequential loss:* This includes protection for contractors for indirect losses due to third-party actions, such as equipment replacement and debris removal.
- *Legal liability:* This is protection for legal liability resulting from poor product design, design errors, product liability, and project performance failure. This does not include protection from loss of goodwill.
- *Personnel:* This provides protection resulting from employee bodily injury (worker's compensation), loss of key employees, replacement cost of key employees, and several other types of business losses due to employee actions.

On construction projects, the owner/customer usually provides "wrap-up" or "bundle" insurance, which bundles the owner, contractor, and subcontractors into one insurable package. The contractor may be given the responsibility to provide the bundled package, but it is still paid to or by the owner/customer.

The Project Management Institute has categorized risks as follows:

- *External-unpredictable:* Government regulations, natural hazards, acts of God
- *External-predictable:* Cost of money, borrowing rates, raw material availability

The external risks are outside of the project manager's control but may affect the direction of the project.

- *Internal (nontechnical):* Labor stoppages, cash flow problems, safety issues, health and benefit plans

The internal risks may be within the control of the project manager and present uncertainty that may affect the project.

- *Technical:* Changes in technology, changes in state of the art, design issues

Technical risks relate to the utilization of technology and the impact it has on the direction of the project. (Note: As previously mentioned, technical risk can also include production, support, threat, and all other nonproject risks.)

- *Legal:* Licenses, patent rights, lawsuits, subcontractor performance, contractual failure

Risk identification should be approached from the perspective of "IF" the risk occurs (e.g., probability = 1.0), "THEN" what will be the impact (consequence of occurrence) "BECAUSE" of one or more underlying causes [potentially root cause(s)]. Using this approach avoids confusion associated with the level of the probability of occurrence (and its components) and consequence of occurrence (cost, technical performance, schedule). These estimates should be performed as part of risk analysis if the candidate risk identified is approved by management.

A variety of risk identification approaches exist that are suitable for evaluating different risk categories beyond those previously mentioned. To identify risks, evaluators should break down program elements, processes, and requirements to a level where they can perform valid assessments. (The information necessary to do this varies according to the phase of the program. During the early phases, requirement and scope documents and acquisition plans may be the only program-specific data available.) One approach is based upon the project WBS and used to evaluate elements/products. A second approach is the process approach, which is used to evaluate key processes (e.g., design, manufacturing, and test). A third approach uses requirements flowdown in an attempt to isolate specifications that will likely be difficult to meet.

The WBS approach uses a WBS at an appropriate level of indenture across the project. Typically, project risks will tend to exist at two or three groupings of WBS levels. The first is WBS 1 or 2, encompassing system (or top)-level risks. (For example, the risk associated with system availability due to the use of an unproven key technology.) Many top-level risks will simply not exist at lower WBS levels and may be difficult to define. To be effective at identifying such risks, it is necessary to use the proverbial "wide angle" view of the project. The second grouping of potential risks on a moderate-to-large-scale project will typically exist at WBS levels 3–6. Here, for example, individual subsystems, boxes, software code (e.g., computer software configuration item), or assemblies may include potential risks. At this lower WBS level (vs. top-level risks) more project personnel should have specific knowledge about potential risks. While this can be very helpful, it is important not to foreclose potential risks by a myopic analysis that either intentionally or inadvertently misses certain risks. Finally, risks may exist at even lower WBS levels (e.g.,

WBS 7 and below), but such risks are often very difficult to identify unless specific information exists that points to appropriate components. (For example, an electronic component has separately been determined to have reliability issues.)

The process approach evaluates risk associated with some key processes (e.g., design, manufacturing, and test) that will exist on a project.[4] The structure is geared toward programs that are mid-to-late in the development phase but, with modifications, could be used for other programs. Templates are used for each major technical activity. Each template identifies potential areas of risk. Overlaying each template on a project allows identification of mismatched areas, which are then identified as "at risk," and thus candidate risks.

The requirements approach incorporates requirement flowdown to isolate specifications that will likely be difficult to meet. For example, a computer required to have a $10\times$ increase in throughput versus an existing state-of-the-art unit may have challenging issues associated with scalability, architecture, component design, package design, manufacturing, and the like. The following hypothetical example further demonstrates this type of analysis. Assume that an aerospace vehicle must be designed that is required to fly at very high speed at a relatively low altitude. In this case air resistance is potentially very high, and a high thrust power plant is needed to overcome this resistance. In addition, the high speed will impart considerable thermal cooling requirements to the vehicle electronics due to aerothermal heating in the atmosphere. This may well lead to design and technology constraints due to cooling requirements and the like (all else held constant). The key to performing this type of risk identification is to have an accurate, clear, and potentially comprehensive requirements flowdown coupled with experienced project personnel that can compare important aspects of requirements against potential solutions from the architecture level to the technology level.

The value in each of these approaches to risk identification lies in the methodical nature of the approach, which forces disciplined, consistent evaluation of potential risks. All three risk identification approaches mentioned above (WBS, processes, and requirements) should be considered for each project, and a mixture of approaches is likely to be superior to any single method. Note, however, using any method in a "cookbook" manner may cause unique risk aspects of the project to be overlooked, and the project manager must review the strengths and weaknesses of the approach and identify other factors that may introduce technical, schedule, cost, program, and other risks. In general, appropriate sources of information (e.g., checklists, expert judgment) are evaluated for each selected risk identification approach (e.g., WBS, key processes). The resulting candidate risks are then suitably documented in a risk register or other appropriate report. While the documentation should not be encyclopedic in nature, it should nevertheless provide sufficient information to allow decision-makers to accurately determine whether or not the candidate risk should be

4. Information on this approach is contained in the government DoD Directive 4245.7-M, "Transition from Development to Production," September 1985, which provides a standard structure for identifying technical risk areas in the transition from development to production. Note: The material in this document is somewhat dated but at least partially applicable to present-day programs. Also, the material was developed before DoD embraced integrated product teams (IPTs) in the 1990s, which may in some cases provide added focus to key program processes.

approved. For example, a risk identification form that is one half to one page in length can provide ample detail about the potential risk for decision-makers to evaluate without being burdensome to complete.

Finally, it is important that all project personnel should be involved with risk identification. Designating a small subset of people to perform risk identification almost always diminishes the results from both a technical (number of valid identified risks) and behavioral perspective (sends the "wrong message" to other project personnel) and can lead to decreased risk management effectiveness. This defective risk identification practice should be avoided whenever possible. (Note, this is different than occasionally having outside personnel brought into the project to assist in independently identifying candidate risks, which prove beneficial in challenging projects.)

17.8 RISK ANALYSIS

PMBOK® Guide, 4th Edition
11.3 Perform Qualitative Risk Analysis
11.4 Perform Quantitative Risk Analysis

Risk analysis is a systematic process to estimate the level of risk for identified and approved risks. This involves estimating the probability of occurrence and consequence of occurrence and converting the results to a corresponding risk level. The approach used depends upon the data available and requirements levied on the project. The most common form of qualitative approach is the use of probability of occurrence and consequence of occurrence scales together with a risk mapping matrix to convert the values to risk levels. Quantitative approaches include, but are not limited to, expected value [also known as expected (monetary) value for cost-based calculations], decision tree analysis (with branches specified by specific probabilities and/or distributions), payoff matrices, and modeling and simulation. Of key importance is the use of an approved, structured, repeatable methodology rather than a subjective approach that may yield uncertain and/or inaccurate results.

Risk analysis begins with a detailed evaluation of the risks that have been identified and approved by decision-makers for further evaluation. The objective is to gather enough information about the risks to estimate the probability of occurrence and consequence of occurrence if the risk occurs and convert the resulting values to a corresponding risk level. (Note: It is important that only approved risks be analyzed to prevent resources from being expended on issues that may not actually be risks.)

Risk analyses are often based on detailed information that may come from a variety of techniques, including but not limited to:

- Analysis of plans and related documents
- Comparisons with similar systems
- Data from engineering or other models
- Experience and interviewing
- Modeling and simulation
- Relevant lessons-learned studies
- Results from tests and prototype development

- Sensitivity analysis of alternatives and inputs
- Specialist and expert judgments

Each risk category (e.g., cost, technical performance, and schedule) includes a core set of evaluation tasks and is related to the other two categories. This relationship requires supportive analysis among areas to ensure the integration of the evaluation process. Some characteristics of cost, schedule, and technical evaluations follow:

Cost Evaluation
- Builds on technical and schedule evaluation results.
- Translates technical and schedule risks into cost.
- Derives cost estimate by integrating technical risk, schedule risk, and cost estimating uncertainty impacts to resources.
- Prioritizes risks for program impact.
- Documents cost basis and risks for the risk evaluation.

Schedule Evaluation
- Evaluates baseline schedule inputs.
- Reflects technical foundation, activity definition, and inputs from technical and cost areas.
- Incorporates cost and technical evaluation and schedule uncertainty inputs to program schedule model.
- Performs schedule analysis on program schedule.
- Prioritizes risks for program impact.
- Documents schedule basis and risks for the risk evaluation.

Technical Evaluation
- Provides technical foundation.
- Identifies and describes program risks (e.g., technology).
- Analyzes risks and relates them to other internal and external risks.
- Prioritizes risks for program impact.
- Analyzes inputs for cost evaluation and schedule evaluation.
- Analyzes associated program activities with both time duration and resources.
- Documents technical basis and risks for the risk evaluation.

Describing and quantifying a specific risk and the magnitude of that risk usually requires some analysis or modeling. Typical tools for use in qualitative and/or quantitative risk analysis are:

- Analysis of plans and other documents to estimate variances
- Decision analysis (including decision trees, expected value, etc.)
- Delphi techniques
- Estimating relationships
- Expert judgment
- Failure modes and effects analysis (related to reliability)

- Fault tree analysis (related to reliability)
- Graphical analysis
- Life-cycle cost analysis
- Logic analysis
- Network analysis
- Payoff matrices
- Probabilistic risk analysis (related to reliability)
- Process templates (e.g., DoD Directive 4245.7-M)
- Quick reaction rate/quantity impact analysis
- Risk mapping matrix with risk scale results
- Risk scales (typically ordinal "probability" and consequence scales)
- Schedule analysis
- Sensitivity analysis
- Simulation (e.g., Monte Carlo) for cost, technical performance, and schedule
- Technology state-of-the-art trending
- Total risk-assessing cost analysis (TRACE)
- Work breakdown structure simulation

After performing a risk analysis, it is often necessary to convert the results into risk levels. When a quantitative risk analysis methodology is used, the results can be grouped by existing cost risk, schedule risk, or technical risk boundaries that have specifically been tailored to the program, or by performing a (statistical) cluster analysis on the results.

When a qualitative risk analysis is performed, risk ratings can be used as an indication of the potential importance of risks on a program. They are typically a measure of the probability of occurrence and the consequences of occurrence, and often expressed as low, medium, and high (or possibly low, medium low, medium, medium high, and high). A representative ("strawman") set of risk rating definitions follows:

- **High risk:** Substantial impact on cost, technical performance, or schedule. Substantial action required to alleviate issue. High-priority management attention is required.
- **Medium risk:** Some impact on cost, technical performance, or schedule. Special action may be required to alleviate issue. Additional management attention may be needed.
- **Low risk:** Minimal impact on cost, technical performance, or schedule. Normal management oversight is sufficient.

It is important to use agreed-upon definitions (such as the strawman definitions above) and procedures for estimating risk levels rather than subjectively assigning them since each person could easily have a different understanding of words typically used to describe both probability and risk (e.g., low, medium, and high). Figure 17–6 shows what some probability statements mean to different people. An important point to grasp from this figure is that a nontrivial probability range (e.g., 0.3) exists for more than half of the statements evaluated. However, much larger probability response ranges than those shown in Figure 17–6 are possible. Conrow conducted a survey of 50 common subjective probability statements.

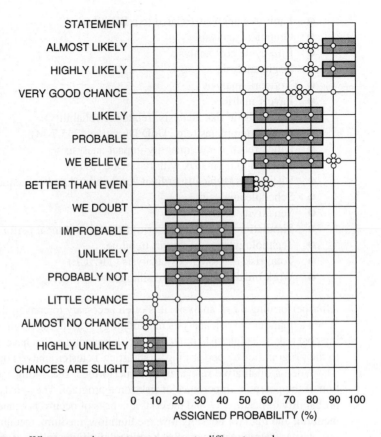

FIGURE 17–6. What uncertainty statements mean to different people.

The results of 151 completed surveys indicate that 49 of the 50 subjective probability statements had a response range (maximum–minimum) of greater than or equal to 0.70! This emphasizes the point that subjective probability statements can have substantially different meaning to different people and should never be the first choice methodology when performing a risk analysis.[5]

The prioritization of program risks should be performed after a structured risk rating approach has been applied. Here, inputs from managers and technical experts will often be necessary to separate risks analyzed to be within a rating level (e.g., to prioritize various high risks). The methodology for generating the list of prioritized risks will vary somewhat with the methodology used. For example, with risk scales, the ranking of risk levels will

---

5. Edmund H. Conrow, *Effective Risk Management: Some Keys to Success*, 2nd ed. (Reston, VA: American Institute of Aeronautics and Astronautics, 2003) Appendix J, pp. 491–513.

depend upon the probability and consequence of occurrence plus possibly frequency of occurrence, the time to impact, and interrelationships with other risks. For a cost risk analysis using a Monte Carlo simulation, the ranking may involve the percent magnitude of cost risk at the desired confidence level (e.g., 70th percentile). For decision analysis results involving the expected monetary value (EMV), the ranking may simply be a list of outcomes from the highest positive outcome to the lowest negative outcome. None of these methods of generating a list of prioritized risks is necessarily superior to any other—they are all potentially useful. The challenge in such cases is to effectively integrate the results from what may be a diverse set of results into a single risk list. There is no "best" algorithm for accomplishing this on a real-world project. Project managers should derive this list from a consensus associated with a structured evaluation of the risk analysis results and document their rationale to the extent possible.

A risk viewed as easily manageable by some managers may be considered hard to manage by less experienced or less knowledgeable managers. Consequently, the terms "high," "medium," or "low" risk are somewhat relative even when compared to the straw-man definitions. Some managers may be risk averse and choose to avoid recognized risk at all reasonable cost. Other managers may be risk seekers and actually prefer to take an approach with more risk. The terms "high," "medium," and "low" risk may change with the turnover of managers and their superiors as much as with the project events.

Program managers can use risk ratings to identify issues requiring priority management (e.g., risk response plans may be required for all medium or higher risk). Risk ratings also help to identify the areas that should be reported within and outside the program. Thus, it is important that the ratings be portrayed as accurately as possible.

High-risk areas may reflect missing capabilities in the project manager's organization or in supporting organizations. They may also reflect technical difficulties in the design or development process. In either case, "management" of risk involves using project management assets to reduce the level of risks present.

A number of different outputs are possible for both qualitative and/or quantitative risk analyses. These include, but are not limited to: (1) an overall project risk ranking, (2) a list of prioritized risks, (3) probability of exceeding project cost and/or schedule, (4) probability of not achieving project performance requirements, (5) decision analysis results, (6) failure modes and effects (reliability), (7) fault paths (reliability), and (8) probability of failure (reliability).

Another common output of risk analysis is a *watch list*. Items placed on a watch list often include indicators of the a potential risk. An example of this is the cost risk of production due to an immature technical data package. When production starts before the technical data package has been adequately engineered for producability, the first unit cost may be higher than planned. A typical watch list is structured to show the trigger event or item (e.g., long-lead items delayed) for a particular concern and the "owner" of that concern. The watch list is periodically reevaluated and items are added, modified, or deleted as appropriate. Should the trigger events occur for items on the watch list during a project, there would be immediate cause for the risk identification material to be reviewed, the risk analysis to be updated (and depending upon the estimated risk level, a risk response strategy may be developed and implemented).

17.9 QUALITATIVE RISK ANALYSIS

PMBOK® Guide, 4th Edition
11.3 Perform Qualitative Risk
Analysis

A commonly used qualitative risk analysis methodology involves risk scales (templates) for estimating probability of occurrence and consequence of occurrence, coupled with a risk mapping matrix. The risk is evaluated using expert opinion against all relevant probability of occurrence scales as well as the three consequences of occurrence scales (cost, technical performance, and schedule), and the results are then transferred onto a risk mapping matrix to convert these values to a corresponding risk level. The risk is included in a prioritized list based upon the risk level as well as other considerations (e.g., frequency of occurrence, the time to impact, and interrelationships with other risks).

Several different classes of risk scales exist.[6] The first type of scale is a nominal scale. Nominal scales have coefficients with no mathematical meaning, and the values are generally placeholders (e.g., freeway numbers). Nominal scales are not used in risk analyses.

The second type of scale is an interval scale. Interval scales, such as Fahrenheit and Celsius, are cardinal in nature. However, the scales have no meaningful zero point, and ratios between similar scales are not equivalent. Interval scales are not commonly used in risk analyses.

The third type of scale is an ordinal scale. Ordinal scales have levels that are only rank-ordered—they have no cardinal meaning *because the true scale interval values are unknown*. There is no probabilistic or mathematical justification to perform math operations (e.g., addition, multiplication, averaging) on results obtained from ordinal scale values, and any such results can have large errors. For example, relatively simple examples have been developed that contain errors of 600 percent or more when comparing actual versus assumed ordinal scale coefficients.[7] These scales may be used to represent different aspects of the probability of occurrence (e.g., technology, design, manufacturing) and consequence of occurrence (e.g., cost, schedule, and technical). Ordinal scales are commonly used in risk analyses. Such scales and a corresponding risk mapping matrix can be a helpful methodology for estimating risk. However, for the reasons discussed above great care must be taken in using this approach.

The fourth type of scale is a calibrated ordinal scale. Calibrated ordinal scales are ordinal scales whose scale-level coefficients are estimated by evaluating an additive utility function (or similar approach). These estimated cardinal coefficients replace the ordinal placeholder values. Limited mathematical operations are possible that yield valid results. However, the values are often relative rather than absolute, and the zero point may not be meaningful. Calibrated ordinal scales are not commonly used in risk analysis, in part because of the difficulty in accurately estimating the associated coefficients.

The fifth type of scale is a ratio scale. Ratio scales, such as Kelvin and Rankine, have cardinal coefficients, indicate absolute position and importance, and the zero point is

6. This section is derived from Edmund H. Conrow, *Effective Risk Management: Some Keys to Success*, 2nd ed. (Reston, VA: American Institute of Aeronautics and Astronautics, 2003), pp. 237–245. Copyright © 2003, Edmund H. Conrow. Used with permission of the author.
7. See note 5, Conrow, pp. 258–268. (American Institute of Aeronautics and Astronautics, 2003), pp. 258–268.

meaningful. In addition, intervals between scales are consistent, and ratio values between scales are meaningful. Mathematical operations can be performed on ratio scales and yield valid results. Although ratio scales are the ideal scales for use in risk analyses, they rarely exist or are used. [Claims that probability and consequence scales are true ratio scales or that results are derived from ratio scales are almost universally false. For example, in one widely distributed case the coefficients are claimed to be derived from an impact (ratio scale), yet the mathematical relationship between the coefficients is almost certainly incorrect and not representative of a meaningful ration scale.]

The sixth type of scale is based on subjective estimates assigned to different probability statements (e.g., high), termed here estimative probability. *Estimative probability* scales can either be ordinal (more common) or cardinal (less common) in nature, depending upon the source of the underlying data and the structure of the scale. In the worst case the probability estimates are point estimates or ranges developed by the scale's author with no rigorous basis to substantiate the values. In the best case the probability estimates are derived from a statistical analysis of survey data from a substantial number of respondents and include point estimates and ranges around the estimate for each probability statement. Estimative probability scales are sometimes used in risk analyses. However, this type of scale should never be the first choice when performing a risk analysis because different people may assign different probability values to the same subjective word. This can lead to nontrivial errors in both the estimated probability value and the resulting risk level.

A risk mapping matrix is typically used to convert ordinal probability of occurrence and consequence of occurrence scale values to a corresponding risk level. While there is no preset size for such a matrix, its dimensions must be less than or equal to the number of scale levels used in both the probability and consequence dimensions. With five-level probability of occurrence and consequence of occurrence scales this corresponds to a 5×5 or smaller matrix. (This is illustrated in Example 17–2.)

While risk mapping matrices are valuable to convert probability and consequence scores to risk, they have several limitations and if not carefully used can lead to errors.[8] First, such matrices are typically illustrated with the upper triangle, not the mirror image of the lower triangle as defined by a line drawn from the origin to the location of highest probability and consequence. The rationale for using asymmetric boundaries is, for example, that the lowest probability, highest consequence cell warrants being a medium risk not a low risk. Yet this approach violates utility theory, which would require that the high/medium and medium/low boundaries should be similar in shape. Second, there is no best way to map probability and consequence scale results to the matrix if there is more than one probability scale and/or consequence scale. While picking the maximum result for each category is the conservative answer, it is not necessarily the correct answer. (And remember, because these scales are ordinal, you can't take the average of the probability or consequence scales and insert that value into the matrix.)

---

8. Edmund H. Conrow, "Risk Analysis for Space Systems," Proceedings of the Space Systems Engineering and Risk Management Symposium 2008, Los Angeles, 28 February 2008. Copyright © 2008, Edmund H. Conrow. Used with permission of the author.

Third, utility preferences are different between buyers and sellers, and even different between different types of buyers and sellers. Hence, it is unclear whose utility preferences the matrix represents. Fourth, are the organizations risk averse, risk neutral, or risk takers? And more importantly, what is the risk behavior of participants in these organizations? This is important because different organizations and their individuals may hold different utility preferences, which can lead to variations in high/medium and medium/low boundaries and/or misscoring results. Fifth, ordering of risks contained within the cells of a 5 × 5 or any other risk mapping matrix is only ordinal and cannot be cardinal.

There are two additional limitations to mapping matrices when applied to opportunities. First, a different set of consequence scales are needed for opportunity vs. risk. It's unclear what these consequence scales should represent since (as mentioned in Section 17.1) there is no universal definition of opportunity, hence consequence of occurrence (or equivalent) associated with it. Thus, the very nature of an opportunity matrix is immature and problematic because of the underlying difficulty in specifying the consequence dimension. This can lead to erroneous results. Second, from prospect theory (Section 17.1) people tend to value the same level of gains and losses differently. Hence, creating opportunity and risk matrices that are identical to each other or mirror each other is wrong and can lead to erroneous decisions.

A final issue when using a risk mapping matrix is related to prioritizing the results. Clearly, low versus medium versus high risks can readily be prioritized. However, prioritizing results within a risk level (e.g., medium) is not nearly as straightforward. This is because the cells contained in the matrix almost always have ordinal, not cardinal, boundaries. This is particularly important for cells that have somewhat similar identical values (e.g., $P = 4$, $C = 3$ vs. $P = 3$, $C = 4$). Such cases require additional management attention to prioritize risks. Of course, if the results are obtained from calibrated ordinal scales, then the resulting risk factor, $RF = P * C$ directly ranks the risks without the need for a risk mapping matrix. Additional factors such as frequency of occurrence, time to impact (either when the risk response plan must be initiated or when the risk will occur), and interrelationships with other risks can also be taken into consideration by management when developing a prioritized risk list from a risk mapping matrix. (When calibrated ordinal scales are used, these considerations can become "tie breakers" for RF scores that are very close or identical to each other.)

The following simple example illustrates the use of ordinal probability of occurrence and consequence of occurrence scales, a risk mapping matrix in project risk analysis, and provides some recommendations for correctly representing the results.[9] Please note these scales should not be used on your project—they are only provided as an illustration.

Example 17–2.   A single "probability" of occurrence scale, related to technology maturity is used, and given in Table 17–6. (Note: Since ordinal probability scales almost never represent true probability but only an indicator of probability, scores derived from

9. This example is derived from Edmund H. Conrow, *Effective Risk Management: Some Keys to Success*, 2nd ed. (Reston, VA: American Institute of Aeronautics and Astronautics, 2003), Appendix I, pp. 485–489. Copyright © 2003, Edmund H. Conrow. Used with permission of the author.

TABLE 17–6. EXAMPLE ORDINAL TECHNOLOGY "PROBABILITY" MATURITY SCALE

Definition	Scale Level
Basic principles observed	E
Concept design analyzed for performance	D
Breadboard or brassboard validation in relevant environment	C
Prototype passes performance tests	B
Item deployed and operational	A

such scales are indicated as "probability" values.) In reality, technical risk will typically encompass a number of additional risk categories in addition to technology maturity, such as design, and the like. However, the use of a single risk category simplifies subsequent computations and is sufficient for illustration purposes. For the technology maturity "probability" scale, assume that low = scale levels A and B, medium = scale level C and D, and high = scale level E. (Note: This information does not correspond to low, medium, and high risk and is only an indicator of where breakpoints will occur when used in developing the risk mapping matrix later in this section. Letters are provided for scale levels instead of numbers to discourage you from attempting to perform invalid mathematical operations on the results.)

Three consequence of occurrence scales—for cost, schedule, and technical—are used and given in Table 17–7. For each of the three consequence of occurrence scales, assume that low = scale levels A and B, medium = scale level C and D, and high = scale level E. (Note: This information does not correspond to low, medium, and high risk and is only an indicator of where breakpoints will occur when used in developing the risk mapping matrix later in this section.)

Given the mapping information associated with the probability of occurrence and consequence of occurrence scales, a mapping matrix was developed and is given in Table 17–8. [Note: Setting risk boundaries is often not exact since three divisions were used for both the probability and consequence of occurrence scales versus the five possible levels (one per scale level).] A mapping matrix with different probability of occurrence and/or

TABLE 17–7. EXAMPLE ORDINAL COST, SCHEDULE, AND TECHNICAL CONSEQUENCE OF OCCURRENCE SCALE

C_C	C_S	C_T	Scale Level
≥10%	Can't achieve key team or major program milestone	Unacceptable	E
7–<10%	Major slip in key milestone or critical path impacted	Acceptable; no remaining margin	D
5–<7%	Minor slip in key milestones, not able to meet need date	Acceptable with significant reduction in margin	C
<5%	Additional resources required, able to meet need date	Acceptable with some reduction in margin	B
Minimal or no impact	Minimal or no impact	Minimal or no impact	A

TABLE 17–8. EXAMPLE RISK MAPPING MATRIX

	Consequence ⟶ Higher				
	A	B	C	D	E
E	M	M	H	H	H
D	L	M	M	H	H
C	L	L	M	M	H
B	L	L	L	M	M
A	L	L	L	L	M

(Probability axis: Higher ↑)

consequence of occurrence relationships (e.g., low = scale levels A and B, medium = scale level C, and high = scale levels D and E for both probability and consequence of occurrence scores), or five resulting risk levels (low, low medium, medium, medium high, and high), or different risk boundaries could also have been used for this example.) (As previously mentioned, I recommend using a symmetrical risk mapping matrix, such as Table 17–8, unless specific, accurate, quantitative information exists that warrants an asymmetrical matrix.)

We'll now evaluate two different items associated with a commercial high-grade digital camera using the above risk analysis methodology. Remember, these risks are hypothetical and only used to illustrate how to apply the risk analysis methodology.

In the first case, a high-performance commercial charge-coupled device (CCD) exists that is in preprototype development. The CCD will be included in a high-grade digital camera. The risk is whether or not the desired signal-to-noise ratio can be achieved to meet low-light operating requirements and avoid an increased level of image "grain" during operation. The potential cost consequence of this occurring is a 6 percent cost impact for a third design, fabrication, and test iteration (two iterations are baselined). The potential schedule consequence of this occurring is additional resources required but able to meet the need date. The potential technical consequence of this occurring is acceptable performance, but no remaining margin. In this example, the resulting probability of occurrence score from Table 17–6 is level C (preprototype maturity), and from Table 17–7 C_C = level C, C_S = level B, and C_T = level D. Given this information and the risk mapping matrix in Table 17–8, the risk level relative to cost, schedule, and technical is medium, low, and medium, respectively. Taking the maximum of the three risk scores yields an overall medium risk level for CCD low-light performance.

In the second case, a high-density digital storage card is in the concept formulation stage. This storage card will be included in the same high-grade digital camera as the CCD previously discussed. The risk is the ability to achieve the desired bit density for the card to store the desired number of very high resolution images. Here, the bit density is presumed to be a factor of 5 times greater than the existing state of the art. The potential cost consequence of not achieving the desired bit density is a 20 percent cost impact for additional technology advancement of the storage medium, plus one or more additional re-design, fabrication, and test iterations. The potential schedule consequence of this occurring is a major slip in introducing the digital camera with the desired high-density storage card. The potential technical consequence of this occurring is unacceptable performance because the desired number of high-resolution, high dynamic range images cannot be stored with existing density storage cards. (It is presumed here that multiple lower density storage cards cannot be substituted for a single high-density card.) In this example, the resulting probability of occurrence score from Table 17–6 is level D (concept design analyzed for performance), and from Table 17–7 C_C = level E, C_S = level D, and C_T = level E. Given this information and the risk mapping matrix in Table 17–8, the risk level relative to cost, schedule, and technical performance is high, high, and high, respectively. Taking the maximum of the three risk scores yields an overall high-risk level for digital storage card bit density.

Of the results for the two candidate risks, the higher-risk item is the digital storage card.

Had there been n technology risk categories instead of the single one used here (technology maturity), then there would have been $n \times 3$ total scores to report for each risk. If desired, this could be reduced to n risk scores by using a conservative mapping approach and taking the maximum of the three consequence scores per item. The $n \times 3$ total scores could also have been reduced to three risk scores per risk by using a conservative mapping approach and taking the maximum score of the n technology risk category scores per item together with each consequence score. Similarly, if desired the $n \times 3$ scores for cost consequence, technical performance consequence, and schedule consequence could be reduced to one risk score per risk by using a conservative mapping approach and taking the maximum of the n technology risk category scores per item coupled with the maximum of the cost, technical performance, and schedule consequence scores per item.

[Note: I generally *do not* recommend reporting separate risk scores associated with cost, technical performance, and schedule consequence of occurrence. This was only presented as an illustration given that a single probability scale (technology maturity) was used in this example. If you have n probability scales together with the 3 consequence scales and use the conservative mapping approach, you should generally either report the complete set of $n \times 3$ risk levels (which I don't prefer), or a single risk level (which is typically better). If you do report a single risk level and used the conservative mapping approach (selecting the maximum probability: consequence pair), it is important to identify which probability and consequence of occurrence categories led to that risk level to assist in developing a risk response plan.]

Finally, given that a medium or higher-risk level exists for both the camcorder CCD low-light performance and the digital storage card bit density, risk response plans (discussed in Section 17.12) should be developed for both risks. (Note: All risks should be analyzed before selecting risk response strategies.)

17.10 QUANTITATIVE RISK ANALYSIS

> **PMBOK® Guide, 4th Edition**
> 11.4 Perform Quantitative Risk Analysis

Several methodologies are commonly used in quantitative risk analyses. These include, but are not limited to, payoff matrices, decision analysis (typically decision trees), expected value, and a Monte Carlo process, which are discussed in Sections 17.4, 17.4, 17.4, and 17.11, respectively. If the potential probabilities of the states of nature can be represented by a point value, as in Figures 17–3 and 17–4, then the decision tree approach (which relies on expected value calculations assuming risk neutral participants) is often appropriate. On the other hand if the states of nature cannot be represented by one or more point values, then probability distributions should be used instead. A common methodology that incorporates a model structure and probability distributions is a Monte Carlo process (commonly called a Monte Carlo simulation).

Two keys to producing accurate quantitative risk analysis results include developing an accurate model structure and incorporating accurate probability information. In project risk management there is often insufficient attention paid to each of these items, and the outcome can be inaccurate results. The model structure should be carefully developed and validated before any output is used for decision-making purposes. While this is easy to do for simple decision trees (e.g., those in Figures 17–3 and 17–4), it can be much more complex when scores or hundreds of branches and potential outcomes are involved. The same is the case for Monte Carlo simulations that model complex cost estimates, technical items, or schedules. Similarly, while it may be relatively easy to estimate probabilities or probability distributions, this information may often contain both random and bias noise terms that can be very difficult to quantify and eliminate. This is particularly the case when probability information is elicited from individuals rather than when obtained from physical measurements. As mentioned in Section 17.7, a number of factors can affect probability of occurrence and/or consequence of occurrence by introducing a bias and/or noise term. These noise components can lead to errors in selecting the distribution critical values (e.g., low and high values for a triangle distribution given the deterministic most likely value).

Quantitative risk analysis outputs can be used in a variety of ways, including but not limited to developing: (1) prioritized risk lists (similar to that for calibrated ordinal scales), (2) probabilistic cost estimates at completion per project phase and probabilistic schedule estimates for key milestones to help the project manager allocate reserve accordingly, (3) probabilistic estimates of meeting desired technical performance parameters (e.g., missile accuracy) and validating technical performance of key components (e.g., real-time integrated circuit operation), and (4) estimates of the probability of meeting cost, technical performance, and schedule objectives (e.g., determining the probability of achieving the planned estimate at completion, a key schedule milestone, or top-level technical performance requirements. Trends versus time can also be developed from the above outputs by repeating the quantitative risk analyses during the course of the project phase. [Note, however, that the actual trend information will often be masked by uncertainties in the analysis that should reduce as a function of time (holding all else constant).]

17.11 PROBABILITY DISTRIBUTIONS AND THE MONTE CARLO PROCESS

PMBOK® Guide, 4th Edition
11.4.2 Quantitative Risk Analysis
Modeling Techniques

A wide variety of probability distributions can be used in performing quantitative risk analyses. These distributions broadly fall into two classes—continuous and discrete (where only finite values can exist). The obvious question some may raise is "what type of distributions should

I use in the simulation?" In reality there is no simple answer because the different phenomena being modeled will tend to be represented by different types of distributions. (And yes, when you have actual data that you're confident represents the probability distribution, then create a general distribution from this data or use the data itself as the probability distribution.) The subject of selecting probability distributions is indeed complex and you should consult one or more texts on this subject as warranted.[10] Three heuristics for selecting probability distributions follow.

First, the data should dictate the probability distribution and the probability distribution should never define the underlying data. While this sounds trite, analysts oftentimes select a particular type of probability distribution while having no idea what type of distribution should be associated with the item being modeled. If you pick a particular type of probability distribution without having a convincing argument, or if the software limits you to a particular type of distribution, then at least state your rationale and limitations. In this case reviewers will have a better understanding of the analysis limitations and how they may affect the results. Second, a strong real-world argument should exist to support the selection of a particular type of probability distribution. The contrary is also true, a distribution shouldn't be selected if it is counterintuitive to compelling real-world arguments. Third, don't blindly believe distribution fitting statistics and use that information alone as the rationale to pick a particular type of probability distribution.

For example, a paper included distribution fits on schedule risk analysis results associated with several activities. These resulting distributions were included in a cost risk analysis. However, the rationale for choosing a distribution in each case was to select the "best fitting distribution" (e.g., the distribution with the best fit statistics). This is nothing but "arm waving" and violates all three of the heuristics given above. Even worse, the authors never acknowledged any potential limitations or problems with this approach. First, there was no indication that the underlying data actually represented the "best" distribution based upon fit statistics. Second, there was no real-world argument presented that a given distribution was best let alone correct. Third, there was no probability level associated with the three estimated fit statistics for each distribution. Hence, it's unknown whether or not any of the best fitting distributions were statistically significant. And in cases where each fit statistic yielded a different distribution rank then how should the best fitting distribution be selected when no probability levels existed?

Some Common Continuous Risk Distributions

The beta, log-normal, normal, triangle, and uniform distributions are commonly used in a variety of applications. However, these distributions may not be appropriate for your use. The beta distribution is defined by two shape parameters. A variety of different beta distributions and resulting shapes are possible. In general, the distributions have a finite range, and in some cases the minimum and maximum values are specified. A constrained form of the beta distribution,

---

10. For example, Merran Evans, Nicholas Hastings, and Brian Peacock, *Statistical Distributions*, 3rd ed. (New York: Wiley-Interscience, 2000). This book provides a summary on the application of a given distribution, but more importantly it contains an excellent examination of key statistical properties for numerous distributions.

known as Program Evaluation and Review Technique (PERT) is sometimes used in performing schedule and schedule risk analyses. The distribution mean and standard deviation are constrained by the low (L), most likely (ML, which is also known as the mode), and high (H); and L and H values, respectively. Results from a PERT schedule analysis are generally not as meaningful or accurate as performed from a Monte Carlo simulation. Many software simulation packages only approximate the beta–PERT distribution (as can be verified by calculating the mean and standard deviation from the resulting sampled distribution data). In addition, the L and H estimates are subject to bias and random noise errors that can lead to errors in both specifying values for a PERT analysis, and the beta–PERT distributions and their outputs. Finally, in general, the mean of the beta–PERT distribution is weighted to the ML value (4/6) over the L (1/6) and H (1/6), while for a triangle distribution, the mean is weighted equally to the ML (1/3), L (1/3), and H (1/3) values.

The log-normal distribution may be appropriate when the data L value is greater than or equal to zero, and the upper end of the data contains relatively few extreme data points.[11] In addition, the log-normal distribution is often appropriate to model a variable that is the multiplicative product of many independent variables where no one variable dominates.[12] The user typically specifies the distribution mean (or median) and standard deviation. The distribution (probability density function) is typically right-hand skewed.

The normal distribution is applicable to modeling a large variety of natural phenomena. The normal distribution is often appropriate to model a variable that is the additive product of many independent variables where no one variable dominates.[13] The variate is specified by the mean and standard deviation and is bounded by ± infinity at the L/H ends. This may be appropriate for some project management simulation cases, but not others. For example, by definition, the residual (error) term in ordinary least squares regression, which is sometimes used to generate parametric, linear cost estimating relationships, is normally distributed.[14] The normal distribution is symmetrical and the mean = median = mode. Finally, the normal distribution can only be approximated by a three-point estimate that may include a force-fit to user specified L, ML, and H values.

The triangular distribution may be appropriate when the L and H values are known (or can be accurately estimated), the variable without risk can be modeled as the ML, and the relationship between L, ML, and H is known, and the data extends continuously from the L to the ML and the ML to the H values. (Note: the ML must be greater than or equal to the L. It must also be less than or equal to the H.) The triangle distribution is either left-hand skewed, symmetrical, or right-hand skewed depending upon the values associated with the

11. Merran Evans, Nicholas Hastings, and Brian Peacock, *Statistical Distributions*, 3rd ed. (New York: Wiley-Interscience, 2000), pp. 34–42.

12. Merran Evans, Nicholas Hastings, and Brian Peacock, Log-normal Distribution, Wikipedia Foundation, Inc., http://en.wikipedia.org/wiki/Lognormal_distribution. Last modified 26 March 2008.

13. Merran Evans, Nicholas Hastings, and Brian Peacock, Normal Distribution, Wikipedia Foundation, Inc., http://en.wikipedia.org/wiki/Normal_distribution. Last Modified 27 March 2008.

14. Note: The mean ± n standard deviation is approximately 68.3, 95.4, and 99.7% of the total distribution area for $n = 1$, 2, and 3 standard deviations, respectively. Hence, while the distribution tails extend to infinity, the chances of this occurring diminish greatly as the number of standard deviations increases. For example, the probability that a value drawn outside of the bounds of the mean ± 3 standard deviations is only three chances in 1000. This diminishes to about 4 chances in 100,000 for the mean ± 5 standard deviations.

L, ML, and H. As in the beta–PERT case, the L and H estimates are subject to bias and random noise errors that can lead to errors in both specifying values for triangular distributions and their outputs. To estimate the L value from the 10th percentile value and H value from the 90th percentile value given the 10th percentile, ML, and 90th percentile values. This is sometimes performed to reduce bias errors associated with directly estimating the endpoints. (This involves an iterative solution to a fourth-order polynomial.) Despite the possibility of reducing the bias error term associated with estimating the L and H values, it is unclear how much this error is actually reduced in real-world applications.) The triangular distribution is sometimes used as a default distribution in project risk management applications—often because additional theoretical and measurement data is missing to more accurately specify the true distribution of a variable. The obvious danger is that a triangular distribution may be specified without any exact knowledge of whether/not it even applies. In such cases the error magnitudes versus the actual underlying data distribution may be very large—several hundred percent.[15]

The uniform distribution is bounded by L and H values, and it is assumed that the values are equally distributed between the endpoints. (In effect, all values are equally likely to occur between the L and H bounds.) Uniform distributions are simple to model and visualize but may not be appropriate for well-defined elements and activities in cost and schedule estimates, respectively.

The Monte Carlo Process   The Monte Carlo process, as applied to risk management, is an attempt to create a series of probability distributions for potential risks, randomly sample these distributions, and to then transform these numbers into useful information that reflects quantification of the associated cost, technical performance, or schedule risks. While often used in technical applications (e.g., integrated circuit performance or structural response to an earthquake), Monte Carlo simulations have also been used to estimate risk in the design of service centers; time to complete key milestones in a project; the cost of developing, fabricating, and maintaining an item; inventory management; and thousands of other applications. The number of equivalent computer hours across platforms spent on noncost and nonschedule Monte Carlo simulations is probably on the order of 100,000–10,000,000 times greater than that associated with cost and schedule simulations combined. (For example, more than one million equivalent desktop computer hours were expended to model the technical performance of a single integrated circuit.) Unfortunately, this is typically not recognized nor stated in some project management literature that emphasizes cost and schedule simulations. (This is also unfortunately consistent with a near absence or mention of the technical performance dimension in some project management literature.)

The structure of cost estimating simulations is often additive—meaning that the cost sums across WBS elements or labor and material entries regardless of the estimating approach used for a particular WBS element or entry. The structure of schedule simulations

---

15. Edmund H. Conrow, "Risk Analysis for Space Systems," *Proceedings of the Space Systems Engineering and Risk Management Symposium 2008*, Los Angeles, 28 February 2008.

is generally based on a schedule network, which encompasses milestones or durations for known activities that are linked in a predefined configuration using model logic. Hence, a schedule risk analysis will generally not be represented by a simple additive model. Performance models can take on a variety of different structures that are often unique to the item being simulated and thus do not follow a simple pattern, but they are largely not represented by simple additive models.

A summary of the steps used in performing a Monte Carlo simulation for cost and schedule follows. (Technical performance simulations can have a widely varying model structure and hence may not fit into the outline below.) Although the details of implementing the Monte Carlo simulation will vary between applications, many cases use a procedure similar to this.

1. Develop and validate a suitable cost or schedule deterministic model without risk and/or uncertainty.
2. Develop the reference point estimate (e.g., cost or schedule duration) for each WBS element or activity contained within the model.
3. Check and recheck the model logic (cost and schedule) and constraints (schedule), as incorrect model logic and constraints are surprisingly common and will lead to erroneous simulation results. For example, the percentage of time spent validating the deterministic schedule logic and constraints should increase with the number of tasks present. For schedules with several thousand tasks more than half of the time should be spent validating the schedule, and less than half the time spent obtaining the probability distributions and interpreting the output.
4. Identify the lowest WBS or activity level for which probability distributions will be constructed. The level selected will depend on the program phase—often lower levels as the project matures.
5. Identify which WBS elements or activities contain estimating uncertainty and/or risk. (For example, technical risk can be present in some cost estimate WBS elements and schedule activities.)
6. Develop suitable probability distributions for each WBS element or activity with estimating uncertainty and/or risk. For cost risk analyses cost estimating uncertainty, schedule risk, and technical risk should be considered as separate distributions. For schedule risk analyses schedule estimating uncertainty, technical risk, and possibly cost risk should be considered as separate distributions. With some tools (e.g., some project scheduling software) only a single probability distribution can be used for a given WBS element or activity. In other cases multiple distributions can be used but practitioners may use only a single distribution. This is a distinct shortcoming since aggregating distributions is almost always subjective and can lead to erroneous results. (Note: Ideally for cost and schedule risk analyses the risk should be modeled rather than the elements and activities that are affected by the risk.[16] However, several available methods of implementing this approach are subjective and should be avoided.)

---

16. See note 5, Conrow, pp. 298–299.

7. Aggregate the WBS element or activity probability distributions functions using a Monte Carlo simulation. When performed for cost, the results of this step will typically be a WBS Level 1 probabilistic cost estimate at completion at the desired probability level and a cumulative distribution function (CDF) of cost versus probability. These outputs are then analyzed to determine the level of cost risk and to identify the specific cost drivers. When performed for schedule, the results of this step will be CDFs of schedule finish dates (and possibly durations) for selected tasks. These outputs are then analyzed to determine the level of schedule risk at the desired probability level and to identify the specific schedule drivers. It is also important to examine the probability density function (PDF, or histogram representation) of the selected outputs to identify whether or not they are multimodal or irregularly shaped in nature. If so, further investigation is warranted. First try running the simulation for a larger number of iterations (which holding all else constant will tend to smooth out irregularities). If the irregularities remain, investigate the inputs that feed into the output under evaluation, the associated model logic, and other considerations (e.g., the calendar used for a schedule risk analysis).

8. Sensitivity and scenario analyses should also be considered for cost and schedule risk analyses. However, they should be performed on the probabilistic (simulation) model, not the deterministic model. If the deterministic model is used, probabilistic considerations will not be taken into consideration. (For example, one schedule risk analysis software package performs a sensitivity analysis prior to performing the Monte Carlo simulation.) For cost risk analysis the sensitivity analysis identifies which variables affect the results (e.g., total program cost, cost by program phase) the most. For schedule risk analyses the percent of time a task is on the probabilistic critical path (e.g., criticality, critical index) coupled with the influence that task has on the specified output (sensitivity, usually derived from correlation) are of considerable value since neither type of information can be obtained from a deterministic analysis. Furthermore, the product of the criticality times sensitivity yields cruciality, which is a measure of the sensitivity times the percent of time the task is on the critical path.

Note: It should be recognized that the quality of Monte Carlo simulation results are only as good as the structure of the model, the quality of the reference point estimates, the selection of probability distributions used in the simulation, and how the simulation is implemented [e.g., the types of distributions, the number of distributions per element or activity, the specific critical values that define the distribution (e.g., mean and standard deviation for a normal distribution), and the number of iterations the simulation is run].[17] If this data is not carefully obtained and accurate, the results can be misleading, if not erroneous. For example, there is a tendency by some analysts performing cost and schedule simulations to default to a triangle probability distribution because the software does not support other

---

17. See note 5, Conrow, Chapter 6, for some additional considerations for Monte Carlo simulations.

probability distributions (which points to inadequate software) or that there is no specific rationale for selecting a different probability distribution. This too is a poor argument. As previously mentioned, the data should dictate the probability distribution and the probability distribution should never define the underlying data. Given these and other related considerations, decision-makers are cautioned about believing results from Monte Carlo simulations presented to several decimal places when there is often nontrivial uncertainty in the first decimal place!

Example 17–3.   The manager of a service center is contemplating the addition of a second service counter. He has observed that people are usually waiting in line. If the service center operates 12 hours per day and the cost of a checkout clerk is $60.00 (burdened) per hour, simulate the manager's problem using the Monte Carlo method assuming that the loss of goodwill is approximately $50.00 per hour.

The first step in the process is to develop procedures for defining arrival rates and service rates. The use of simulation implies that the distribution expressions are either nonexistent for this type of problem or do not apply to this case. In either event, we must construct either expressions or charts for arrival and service rates.

The arrival and service rates are obtained from sample observations over a given period of time and transformed into histograms. Let us assume that we spend some time observing and recording data at the one service counter. The data recorded will be the time between customer arrivals and the number of occurrences of these arrivals. The same procedure will be repeated for servicing. We shall record the amount of time each person spends at the checkout facility and the number of times this occurs. This data is shown in Table 17–9 and transformed to histograms in Figures 17–7 and 17–8. From Table 17–9 and Figures 17–7 and 17–8, 5 people entered the store within 5 minutes of other customers. The 5 customers may have come at the same time or different times. Likewise 18 people

TABLE 17–9.   ARRIVAL AND SERVICE RATE DATA

Arrival Time between Customers (min)	Number of Occurrences	Service Time at Checkout Counter (min)	Number of Occurrences
0	5	10	5
5	7	12	10
8	1	14	15
10	9	16	20
12	12	18	20
15	20	20	15
16	18	22	15
18	10		
20	9		
25	5		
30	4		—
Total	100		100

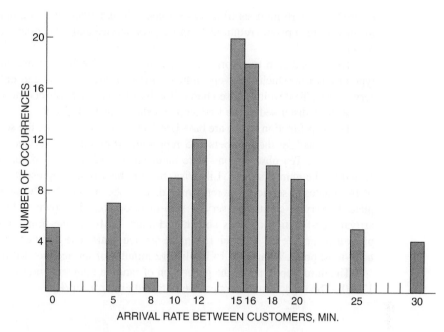

FIGURE 17–7. Arrival rate histogram.

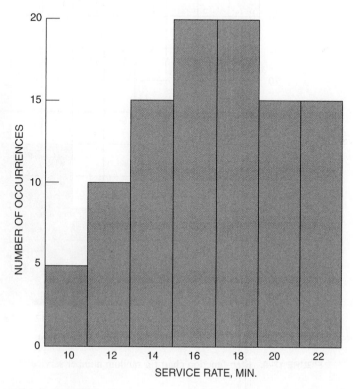

FIGURE 17–8. Service rate data.

entered within 16 minutes of other customers. The service rates are handled in the same manner. Fifteen people required 14 minutes of service and 20 people required 18 minutes of service.

The second step transforms the arrival and service histograms into a step-function type of chart in which for every number there corresponds one and only one arrival and service rate. To develop these charts, it is best to have 100 observances for both arrivals and services discussed in the first step and shown in Table 17–9.

The step-function charts are based on 100 numbers. Consider the service data in Table 17–9. We shall let the numbers 1–5 represent 10 minutes of service since there were 5 observations. Ten observations were tabulated for 12 minutes of service. This is represented by the numbers 6–15. Likewise, the numbers 16–30 represent the 15 observations of 14 minutes of service. The remaining data can be tabulated in the same manner to complete the service chart. The service step-function chart is shown in Figure 17–9 and the arrival step-function chart is shown in Figure 17–10. Some points on these charts are plateau points, as the number 15 on the service chart. The number 15 refers to the left-hand-most point. Therefore, 15 implies 12 minutes of service, not 14 minutes of service.

The third step requires the generation of random numbers and the analysis (see Table 17–10). The random numbers can be obtained either from random number tables or from

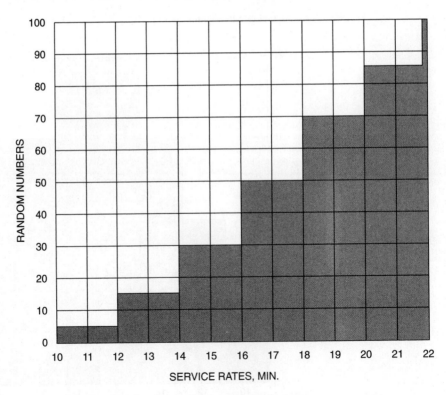

FIGURE 17–9. Step-function chart for random number service rates.

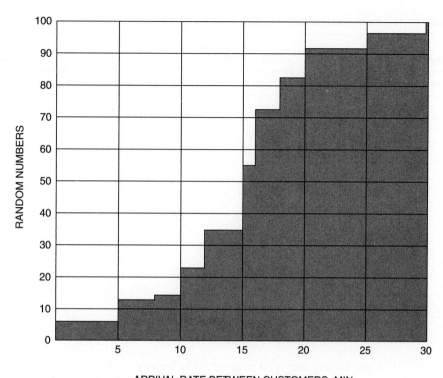

FIGURE 17–10. Step-function chart for random number arrival rates.

computer programs that contain random number generators. These random numbers are used to simulate the arrival and service rates of customers from the step-function charts in Figures 17–9 and 17–10. Random numbers are generated between 0 and 1. However, it is common practice to multiply these numbers by 100 so as to have integers between 0 and 99 or 1 and 100. As an example, consider the following 10 random numbers: 1, 8, 32, 1, 4, 15, 53, 80, 68, and 82. The numbers are read in groups of 2 with the first number representing arrivals and the second representing service. From Figure 17–10, the number 1 corresponds to a 0 arrival rate. From Figure 17–9, the number 8 corresponds to 12 minutes of service. Therefore, assuming that the store opens at 8:00 AM, the first customer arrives at the checkout facility at approximately 8:00 AM and leaves at 8:12, after requiring 12 minutes of service at the checkout counter. The second pair of points are 32 and 1. The first number, 32, indicates that the second customer arrives 12 minutes after the first customer, at 8:12. But since the first customer is through the service facility at 8:12, the second customer will not have to wait. His 10 minutes of service at the checkout counter will begin at 8:12 and he will finish at 8:22. The third customer arrives at the same time as the second customer and requires 12 minutes service. But since the second customer is in the service facility, the third customer must wait in the queue until 8:22 before entering the service facility. Therefore,

TABLE 17–10.   SINGLE-QUEUE MONTE CARLO SIMULATION MODEL

Random Number (Arrival)	Arrival Time Increment (min)	Arrival Time (Service)	Random Service Number (min)	Time Service Increment	Service Begins (min)	Waiting Ends	Time
1	0.0	8:00	8	12.00	8:00	8:12	0.0
32	12.00	8:12	1	10.00	8:12	8:22	0.0
4	0.0	8:12	15	12.00	8:22	8:34	10.00
53	15.00	8:27	80	20.00	8:34	8:54	7.00
68	16.00	8:43	82	20.00	8:54	9:14	11.00
87	20.00	9:03	83	20.00	9:14	9:34	11.00
17	10.00	9:13	47	16.00	9:34	9:50	21.00
32	12.00	9:25	64	18.00	9:50	10:08	25.00
99	30.00	9:55	10	12.00	10:08	10:20	13.00
72	16.00	10:11	39	16.00	10:20	10:36	9.00
82	18.00	10:29	41	16.00	10:36	10:52	7.00
7	5.00	10:34	65	18.00	10:52	11:10	18.00
30	12.00	10:46	92	22.00	11:10	11:32	24.00
77	18.00	11:04	32	16.00	11:32	11:48	28.00
96	25.00	11:29	82	20.00	11:48	12:08	19.00
30	12.00	11:41	41	16.00	12:08	12:24	27.00

Total waiting time = 230.00 min

his waiting time is 10 minutes and he leaves the service facility at 8:34 (8:22 + 12 minutes service). The fourth customer arrives 15 minutes after the third customer (at 8:27) and requires 20 minutes service. Since the service facility is occupied until the third customer leaves at 8:34, the fourth customer must wait 7 minutes in the queue. This process is repeated for 16 customers and the results are shown in Table 17–10.

The fourth step in the process is the final analysis of the data. The data shown in Table 17–10 consisted of 16 customers processed in the first 4 hours. The summation of the waiting time for the 4 hours is 230 minutes. Since the store is opened for 12 hours, the total waiting time is 3 × 230, or 690 minutes. At $50.00 per hour loss of goodwill, the manager loses approximately $575 per 12-hour day because of waiting-line costs. The manager can put in a second service counter. If he pays the worker $60.00 per hour burdened for a 12-hour day, the cost will be $720.00. Therefore, it is more economical for the manager to allow people to wait than to put in a second checkout facility.

17.12 PLAN RISK RESPONSE

PMBOK® Guide, 4th Edition
11.5 Plan Risk Responses

Planning risk responses (risk handling) includes specific methods and techniques to deal with known risks and opportunities, identifies who is responsible for the risk or opportunity, and provides an estimate of the resources associated with handling the risk or opportunity, if any. It involves planning and execution with the objective of reducing risks to an acceptable level and exploiting potential opportunities. There are several factors that can influence our response to a risk or opportunity, including but not limited to:

- Amount and quality of information on the actual hazards that caused the risk (descriptive uncertainty)
- Amount and quality of information on the magnitude of the damage (measurement uncertainty)
- Personal benefit to project manager for accepting the risk or opportunity (voluntary risk or opportunity)
- Risk or opportunity forced upon the project manager (involuntary risk or opportunity)
- The existence of cost-effective alternatives (equitable risks or opportunities)
- The existence of high-cost alternatives or possibly lack of options (inequitable risks or opportunities)
- Length of exposure to the risk or time available for the opportunity

Risk response planning must be compatible with the RMP and any additional guidance the program manager provides. A critical part of risk response planning involves refining and selecting the most appropriate response option(s) and specific implementation approach(es) for selected risks (often those with medium or higher risk levels) and opportunities. The selected risk response option coupled with the specific implementation approach is known as the risk response (handling) strategy, which is documented in the risk response (handling) plan. The procedure to develop a risk response strategy is straightforward. First, the most desirable risk response option [of acceptance, avoidance, control (mitigation), and transfer for risks, and acceptance, enhance, exploit, and share for opportunities] is selected; then the best implementation approach is chosen for that option. In cases where one or more backup strategies may be warranted (e.g., high risks), the above procedure is repeated. (While the selected option for a backup strategy may be the same as for the primary strategy, the implementation approach will always be different; else the primary and backup strategy would be identical.) Similarly, contingent responses can be developed for risks and opportunities where action is taken only if certain predefined conditions occur.

Personnel that evaluate candidate risk response strategies may use the following criteria as a starting point for evaluation:

- Can the strategy be feasibly implemented and still meet the user's needs?
- What is the expected effectiveness of the response strategy in reducing program risk to an acceptable level?
- Is the strategy affordable in terms of dollars and other resources (e.g., use of critical materials and test facilities)?
- Is time available to develop and implement the strategy, and what effect does that have on the overall program schedule?
- What effect does the strategy have on the system's technical performance?

A summary of risk response options for risks and opportunities is given in Table 17–11. For risks this includes acceptance, avoidance, mitigation (control), and transfer, while for opportunities this includes acceptance, enhance, exploit, and share. In addition, contingent responses are possible for both risks and opportunities.

TABLE 17–11. SUMMARY OF RESPONSE OPTIONS FOR RISKS AND OPPORTUNITIES

Type of Response	Use for Risk or Opportunity	Description
Avoidance	Risk	Eliminate risk by accepting another alternative, changing the design, or changing a requirement. Can affect the probability and/or impact.
Mitigation (control)	Risk	Reduce probability and/or impact through active measures.
Transfer	Risk	Reduce probability and/or impact by transferring ownership of all or part of the risk to another party, or by redesign across hardware/software or other interfaces, etc.
Exploit	Opportunity	Take advantage of opportunities.
Share	Opportunity	Share with another party who can increase the probability and/or impact of opportunities.
Enhance	Opportunity	Increase probability and/or impact of opportunity.
Acceptance	Risk and opportunity	Adopt a wait-and-see attitude and take action when triggers are met. Budget, schedule, and other resources must be held in reserve in case the risk occurs or opportunity is selected.

A brief discussion of the four response options for risks follows:

- *Acceptance* (i.e., retention): The project manager says, "I know the risk exists and am aware of the possible consequences. I am willing to wait and see what happens. I accept the risk should it occur."
- *Avoidance:* The project manager says, "I will not accept this option because of the potentially unfavorable results. I will either change the design to preclude the issue or requirements that lead to the issue."
- *Control* (e.g., mitigation): The project manager says, "I will take the necessary measures required to control this risk by continuously reevaluating it and developing contingency plans or fall-back positions. I will do what is expected."
- *Transfer:* The project manager says, "I will share this risk with others through insurance or a warranty or transfer the entire risk to them. I may also consider partitioning the risk across hardware and/or software interfaces or using other approaches that share the risk."

A brief discussion of the four response options for opportunities follows:

- *Acceptance* (i.e., retention): The project manager says, "I know an opportunity exists and am aware of the possible benefits. I am willing to wait and see what happens. I accept the opportunity should it occur."
- *Enhance:* The project manager says, "This is an opportunity. What can we do to increase the probability of occurrence of the opportunity, such as by using more aggressive advertising?"
- *Exploit:* The project manager says, "This is an opportunity. How can we make the most of it? Will assigning more talented resources allow us to get to the marketplace quicker?"
- *Share:* The project manager says, "This is an opportunity, but we cannot maximize the benefits alone. We should consider sharing the opportunity with a partner."

We'll now explore each of the four risk response options in somewhat greater detail.[18]

Risk assumption is an active acknowledgment of the existence of a particular risk situation and a conscious decision to accept the associated level of risk, without engaging in any special efforts to control it. However, a general cost and schedule reserve may be set aside to deal with any problems that may occur as a result of various risk assumption decisions. This risk response option recognizes that not all identified program risks warrant special handling; as such, it is most suited for those situations that have been classified as low risk. (Note: Risk assumption should not be a passive project management behavior. To the contrary, it should be a conscious decision involving active behavior to ensure that adequate resources exist to address the risk should it occur. Otherwise, project management is effectively in denial and will likely often not properly prepare for the risk to occur.)

The key to successful risk assumption is twofold:

- Identify the resources (e.g., money, people, and time) needed to overcome a risk if it occurs. This includes identifying the specific management actions (such as retesting, and additional time for further design activities) that may occur.
- Ensure that necessary administrative actions are taken to identify a management reserve to accomplish those management actions.

Risk avoidance involves a change in the concept (including design), requirements, specifications, and/or practices that reduce risk to an acceptable level. Simply stated, it eliminates the sources of high or possibly medium risk and replaces them with a lower risk solution. This method may be done in parallel with the up-front requirements analysis, supported by cost/requirement trade studies. It may also be used later in the development phase when test results indicate that some requirements cannot be met, and the potential cost and/or schedule impact would be severe.

Risk control does not attempt to eliminate the source of the risk but seeks to reduce the risk. It manages the risk in a manner that reduces the probability and/or consequence of its occurrence on the program. This option may add to the cost of a program, and the selected approach should provide an optimal mix among the candidate approaches of risk reduction, cost effectiveness, and schedule impact. A summary of some common risk control actions includes:

- **Alternative Design:** Create a backup design option that should use a lower risk approach.
- **Demonstration Events:** Demonstration events are points in the program (normally tests) that determine if risks are being successfully reduced.
- **Design of Experiments:** This engineering tool identifies critical design factors that are sensitive, therefore potentially medium or higher risk, to achieve a particular user requirement.

---

18. Material discussing the four risk response options was derived in part from "Risk Management Guide for DoD Acquisition," 5th ed., Version 2.0, Defense Acquisition University, June 2003, pp. 70–78. The cited text is an excellent summary and applicable to a wide variety of projects and industries.

- **Early Prototyping:** Build and test prototypes early in the system development.
- **Incremental Development:** Design with the intent of upgrading system parts in the future.
- **Key Parameter Control Boards:** The practice of establishing a control board for a parameter may be appropriate when a particular feature (such as system weight) is crucial to achieving the overall program requirements.
- **Manufacturing Screening:** For programs in the mid-to-late development phase, various manufacturing screens (including environmental stress screening) can be incorporated into test article production and low-rate initial production to identify deficient manufacturing processes.
- **Modeling/Simulation:** Modeling and simulation can be used to investigate various design options and system requirement levels.
- **Multiple Development Efforts:** Create systems that meet the same performance requirements. (This approach is also known as parallel development.)
- **Open Systems:** Carefully selected commercial specifications and standards whose use can result in lower risk levels.
- **Process Proofing:** Particular processes, especially manufacturing and support processes, which are critical to achieve system requirements.
- **Reviews, Walkthroughs, and Inspections:** These three actions can be used to reduce the probability and potential consequences of risks through timely assessment of actual or planned events.
- **Robust Design:** This approach uses advanced design and manufacturing techniques that promote quality and capability through design.
- **Technology Maturation Efforts:** Normally, technology maturation is used when the desired technology will replace an existing technology that is available for use in the system.
- **Test-Analyze-and-Fix (TAAF):** TAAF is the use of a period of dedicated testing to identify and correct deficiencies in a design.
- **Trade Studies:** Arrive at a balance of engineering requirements in the design of a system. Ideally, this also includes cost, schedule, and risk considerations.
- **Use of Mockups:** The use of mockups, especially human–machine interface mockups, can be utilized to conduct early exploration of design options.
- **Use of Standard Items/Software Reuse:** Use of existing and proven hardware and software, where applicable, can potentially reduce risks.

Risk transfer may reallocate risk from one part of the system to another, thereby reducing the overall system and/or lower-level risk, or redistributing risks between the buyer (e.g., government) and the seller (e.g., prime contractor), or within the buyer or seller teams. This should initially be performed as part of the requirements analysis process, but then considered as an option in developing risk response plans for approved risks. Risk transfer is a form of risk sharing and not risk abrogation on the part of the buyer or seller, and it may influence cost objectives. An example is the transfer of a function from hardware implementation to software implementation or vice versa. (Risk transfer is also not deflecting a risk because insufficient information exists about it.) The effectiveness of risk transfer depends on the use of successful system design techniques. Modularity and functional partitioning are two design techniques that support risk transfer. In some cases, risk transfer may concentrate risks in one area of the design. This allows management to focus attention and resources on that area. Other examples of risk transfer include the use of insurance, warranties, bonding (e.g., bid, performance, or payment bonds), and similar agreements. These agreements are typically between the buyer and seller such that the consequent "costs" of failure will be assumed by the seller for some agreed-to price. That price may be in terms of profit dollars, schedule changes, product performance modifications, or other considerations.

In addition to developing the risk response plan (includes the selected strategy) and identifying resources needed to implement this plan, suitable metrics should also be identified for each risk that will enable tracking progress during the risk monitoring process phase. This may include, for example, cost variance (cost), schedule variation (schedule), technical performance measurements (performance), risk variation (risk), and other metrics (e.g., process) as warranted.

Risk response options and the implemented approaches may have broad cost and other resource implications. The magnitude of these costs and resources are circumstance dependent. The approval and funding of response options and specific approaches should be done by the project manager or risk management board (or equivalent) and be part of the process that establishes the program cost, technical performance, and schedule goals. The selected response option and approach for each selected risk should be included in the project's acquisition strategy, and the detailed risk response activities should be included in the project's integrated master schedule (or equivalent).

Once the acquisition strategy includes the risk response strategy for each selected risk, the cost, schedule, and other resource impacts and can be identified and included in the program plan and schedule, respectively. The resources to implement the risk response plan should be approved and allocated by project management; else the risk management process will be viewed by participants in and outside the project as a "paper tiger."

Finally, while risks and the responses developed to address them may identify potential opportunities, pursuing opportunities will often lead to unanticipated risks. This outcome is rarely if ever discussed by opportunity proponents, yet it can lead to adverse program impacts if the resulting unexpected risks occur. One recent example involved using one class of electronic parts (opportunity) as a substitute for another class of parts that had design, manufacturing, and technology risks. A substantial amount of expensive hardware was fabricated using the substituted parts, which were claimed by the supplier to have low technical risk. Unfortunately, a complex defect associated with how the parts were fabricated that was not previously identified led to a failure rate 1000+ times higher than originally estimated. The irony here is that the original class of parts, whose risks had been gradually reduced to an acceptable level over the course of a few years, were used along with two other approaches to replace the substitute class of parts in a large-scale risk reduction activity to preclude potential early failures in billions of dollars of equipment.

17.13 MONITOR AND CONTROL RISKS _____

PMBOK® Guide, 4th Edition
11.6 Monitor & Control Risks
The monitoring and control process systematically tracks and evaluates the effectiveness of risk response actions against established metrics. Monitoring results may also provide a basis for developing additional risk response strategies, or updating existing risk response strategies, and reanalyzing known risks. In some cases monitoring results may also be used to identify new risks and revise some aspects of risk planning. The key to the risk monitoring and control process is to establish a cost, technical performance, and schedule management indicator system over the program that the program manager and other key personnel use to evaluate the status of the program. The indicator system should be designed to provide early warning of potential problems to allow management actions. Risk monitoring and control is not a problem-solving technique but, rather a proactive technique to obtain objective information on the progress to date in reducing risks to acceptable levels. Some techniques suitable for risk monitoring and control that can be used in a program-wide indicator system include:

- **Earned Value (EV):** This uses standard cost/schedule data to evaluate a program's cost performance (and provide an indicator of schedule performance) in an integrated fashion. As such, it provides a basis to determine if risk response actions are achieving their forecasted results.
- **Program Metrics:** These are formal, periodic performance assessments of the selected development processes, evaluating how well the development process is achieving its objective. This technique can be used to monitor corrective actions that emerged from an assessment of critical program processes.
- **Schedule Performance Monitoring:** This is the use of program schedule data to evaluate how well the program is progressing to completion.
- **Technical Performance Measurement (TPM):** TPM is a product design assessment that estimates, through engineering analysis and tests, the values of essential technical performance parameters of the current design as effected by risk response actions.

The indicator system and periodic reassessments of program risk should provide the program with the means to incorporate risk management into the overall program management structure. Finally, a well-defined test and evaluation program is often a key element in monitoring the performance of selected risk response strategies and updating risk analyses, and identifying candidate risks.

17.14 SOME IMPLEMENTATION CONSIDERATIONS _____

While it is important to emphasize a comprehensive, structured risk management process, it is equally important that suitable organizational and behavioral considerations exist so that the process will be properly implemented. While no single set of guidelines will suffice because implementation considerations vary on a project-by-project basis, it is important that risk management roles and responsibilities be defined in the RMP and carried out in the program. For example, you need to decide (in advance) within the project:

- Which group of managers have responsibility for risk management decision-making?
- Which group "owns" and maintains the risk management process?
- Which group or individual is responsible for risk management training and assisting others in risk management implementation?
- Who identifies candidate risks? (Everyone should!)
- How are focal points assigned for a particular approved risk?
- How are risk analyses performed and approved?
- How are risk response plans developed and approved?
- How are data for risk monitoring metrics collected?
- How are independent risk reviews performed to ensure that project risks are properly identified, analyzed, handled, and monitored?

This is but a brief list of some organizational considerations for implementing risk management, which will vary depending upon the size of the project, organizational culture, degree that effective risk management is already practiced within the organization, contractual requirements, and the like. Likewise, while behavioral considerations for effective risk management will also vary on a project-by-project basis, a few key characteristics should apply for all projects.

Risk management must be implemented in both a "top-down" and "bottom-up" manner within the project. The project manager and other decision-makers should both use risk management principles in decision-making and support and encourage all others within the project to perform risk management. The project manager should generally not be the risk manager (except on perhaps very small projects). However, top-level management must both encourage and foster a positive risk management atmosphere within the project. In addition, they must actively participate in risk management activities and use risk management principles in decision-making. Without such active support other project personnel will often view risk management as unimportant, and there may be insufficient encouragement to create or maintain a culture within the project to embrace risk management. Similarly, while it is important for key decision-makers within the project to not "shoot the messenger" for reporting risks, for example, eliminating this behavior does not in and of itself create a positive environment for performing effective risk management because, as mentioned above, a positive atmosphere that is conducive to performing risk management needs to be in place.

Working-level personnel are generally quick to decide whether or not decision-makers are committed to risk management, and if the appearance is perceived as lip service, then ineffective risk management will almost certainly exist. But working-level personnel must also be actively engaged for risk management to be effective, whereby risk management principles are assimilated as part of their job function.

A key behavioral goal associated with risk management is not to turn every person on the project into a "risk manager" but instead make them sensitive to risk management principles and to apply these principals as part of their job. Accomplishing this is often difficult but nevertheless important in order to achieve effective risk management.

17.15 THE USE OF LESSONS LEARNED

Risks that are analyzed to be medium or higher must be handled to the extent assets allow, to reduce their potential to adversely affect the program. All levels of management must be sensitive to hidden "traps" that may induce a false sense of security. If properly interpreted, these signals really indicate a developing problem in a known area of risk. Each trap is usually accompanied by several "warning signs" that show an approaching problem and the probability of failing to treat the problem at its inception.

The ability to turn traps into advantages suggests that much of the technical risk in a program can be actively handled via the risk response control or transfer option, not merely watched and resolved after a problem occurs. In some instances it may pay to watch and wait. If the probability that a certain problem will arise is low or if the cost exceeds the benefits of "fixing" the problem before it happens, risk assumption may be advisable. Effective risk management makes selection of the risk assumption option a conscious decision rather than an oversight and may trigger an appropriate addition to the risk "watch list."

"Best practices" acknowledges that all of the traps have not been identified for each risk. The traps are intended to be suggestive, and other potential issues should be examined as they arise. It is also important to recognize that sources and types of risk evolve over time. Risks may take a long time to occur on a given project. Attention must be properly focused to examine risks and lessons learned.

Lessons learned should be documented so that future project managers can learn from past mistakes. Experience is an excellent teacher in risk management. Yet, no matter how hard we try, risks will occur and projects may suffer. As an example, the project management community has considerable knowledge in going from new product development to production.[19] We plan for risk management, identify and analyze risks, and develop ways of handling and monitoring them, but some types of risks commonly occur on projects that are mid-to-late in the development phase. Some examples of these risks are now given. (Note: While these risks are closely aligned with the identify risk process approach discussed in Section 17.7, the material is provided here to illustrate how best practices may be helpful for a variety of projects.)

Risk: Design Process. The design process must reflect a sound design policy and proper engineering disciplines and practices—an integration of factors that influence the production, operation, and support of a system throughout its life cycle. Nevertheless, concepts are often selected, demonstrated, and validated with little thought given to the feasibility of producing a system employing those concepts. This omission is then carried forward into design, with voids appearing in manufacturing technology and absence of proven manufacturing methods and processes to produce the system within affordable cost. One of the most common sources of risk in the transition from development to production is failure to design for production. Some design engineers do not consider in their design the limitations in manufacturing personnel and processes. The predictable result is that an

---

19. Adapted from "Transition from Development to Production," DoD 4245.7-M, Department of Defense, September 1985. These risk areas may occur on a variety of projects, but it may not be possible to take decisive action to deal with some of them until midway in the development phase.

apparently successful design, assembled by engineers and highly skilled model shop technicians, goes to pieces in the factory environment when subjected to rate production. A design should not be produced if it cannot survive rate production without degradation.

Prevention. The potential to produce a system must be investigated carefully during the planning phase by means of appropriate producability analyses. Voids in manufacturing technology projects and manufacturing methods and processes peculiar to the design of the specific system, subsystem, and components must be addressed during engineering development.

Risk: Design Reviews. While most engineering development projects usually require formal design reviews, they often lack specific direction and discipline in the design review requirement, resulting in an unstructured review process that fails to fulfill either of the two main purposes of design review, which are (1) to bring additional knowledge to the design process to augment the basic program design and analytical activity and (2) to challenge the satisfactory accomplishment of specified design and analytical tasks needed for approval to proceed with the next step in the process.

Prevention

- The customer and their contractors recognize that design reviews represent the "front line" where readiness for transition from development to production is decided ultimately. Design review policy, schedule, budget, agenda, participants, actions, and follow-up are decided in view of this foremost need.
- Design reviews should be included in all projects in accordance with existing customer requirements. A design review plan must be developed by the contractor and approved by the customer.

Risk: Life. Life tests are intended to assess the adequacy of a particular equipment design when subjected to long-term exposure to certain operational environments. Due to the time-consuming nature of these tests, various methods have been used to accelerate test times by exposure to more stringent environments than those expected in actual operational use. These methods may give misleading results due to a lack of understanding of the acceleration factors involved.

Many projects are forced into conducting life tests after the systems are placed in use and before reliability requirements are achieved. As a result, life tests are performed after the start of production, and costly engineering change proposals (ECPs) and retrofit programs must be initiated in an attempt to "get well" with less than optimum design solutions.

Prevention

- Include life testing in the overall system integrated test plan to ensure that testing is conducted in a cost-effective manner and to meet program schedules.
- Use test data from other phases of the test program to augment the system and subsystem life testing by reducing the time required to prove that reliability requirements are met.
- Use life test data from similar equipments operating in the same environment to augment the equipment life testing, in order to gain confidence in the design.

Risk: Manufacturing Plan. Involvement of production and manufacturing engineering only *after* the design process has been completed is a fundamental error and a major transition risk. Consequences of late involvement are (1) an extended development effort required for redesign and retest of the end item for compatibility with the processes and procedures necessary to produce the item and (2) lower and inefficient rates of production due to excessive changes in the product configuration introduced on the factory floor. Increased costs and schedule delays are the result of this approach.

Prevention The following represent the key elements of a manufacturing plan:

- Master delivery schedule that identifies by each major subassembly the time spans, need dates, and who is responsible
- Durable tooling requirements to meet increased production rates as the program progresses
- Special tools
- Special test equipment
- Assembly flowcharts

Risk: Quality Manufacturing Process. The introduction of a recently developed item to the production line brings new processes and procedures to the factory floor. Changes in hardware or work flow through the manufacturing facility increase the possibility of work stoppage during rate production. Failure to qualify the manufacturing process before rate production with the same emphasis as design qualification—to confirm the adequacy of the production planning, tool design, manufacturing process, and procedures—can result in increased unit costs, schedule slippage, and degraded production performance.

Prevention

- The work breakdown structure, production statement of work, and transition and production plans do not contain any conflicting approaches. Any discrepancies among these documents are identified and resolved before production is started.
- A single-shift, 8-hour day, 5-day workweek operation is planned for all production schedules during initial startup. Subsequent manpower scheduling is adjusted to manufacturing capability and capacity consistent with rate production agreements.
- The drawing release system must be controlled and disciplined.
- The manufacturing flow must minimize tooling changes and machine adjustments and ensure that alternate flow plans have been developed.
- A mechanism must be established that ensures the delivery of critical items with long lead time 4–6 weeks before required.
- All new equipment or processes that will be used to produce the item must be identified.

Risk: Manpower and Personnel. Product development and support systems must be designed with as complete an understanding as possible of user manpower and personnel skill profiles. A mismatched yields reduced field reliability, increased equipment training, technical manual costs, and redesign as problems in these areas are discovered during demonstration tests and early fielding. Discovery of increased skill and training requirements late in the acquisition process creates a difficult catch-up problem and often leads to poor system performance.

Prevention

- Manpower and skill requirements must be based on formal analysis of previous experience on comparable systems and maintenance concepts.
- Manpower cost factors used in design and support trade-off analyses. Must take into account costs to train or replace experienced personnel, as well as the true overhead costs.

Risk: Training, Materials, and Equipment. On some programs, training requirements are not addressed adequately, resulting in great difficulty in operation and support of the hardware. Training programs, materials, and equipment such as simulators may be more complex and costly than the hardware they support. Delivery of effective training materials and equipment depends on the understanding of final production design configuration, maintenance concepts, and skill levels of personnel to be trained. On many programs, training materials and equipment delivery schedules are overly ambitious. The results include poor training, inaccuracies in technical content of materials, and costly redesign and modification of training equipment:

Prevention

- Contractors must be provided with clear descriptions of user personnel qualifications and current training programs of comparable systems, to be used in prime hardware and training systems design and development.
- On-the-job training capability must be incorporated in the prime equipment design as a method to reduce the need for additional training equipment.

17.16 DEPENDENCIES BETWEEN RISKS

PMBOK® Guide, 4th Edition
11.5.3.1 Risk Response
  Planning—Risk Registers
  (updates)

If project managers had unlimited funding, they could generally identify a multitude of risk events, both significant and insignificant. With a large number of possible risk events, it is impossible to address each and every situation, and thus it may be necessary to prioritize risks.

Assume that the project manager categorizes the risks according to the project's time, cost, and performance constraints as illustrated in Figure 17–11. According to the figure, the project manager should focus his efforts on reducing the schedule-related risks. However, it must be recognized that even if schedule has the highest priority, you may also have to start work on cost and technical performance-related issues at the same time, but the schedule-related issues may have the greatest resources applied.

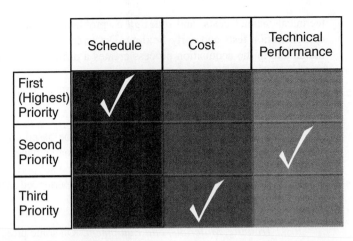

FIGURE 17–11. Prioritization of risks.

The prioritization of risks could be established by either the project manager or the project sponsor, or even by the customer. The prioritization of risks can also be industry specific, or even country specific as shown in Figure 17–12. It is highly unlikely that any project management methodology would dictate the prioritization of risks. A well-thought-out risk analysis methodology *does* dictate, or at least reveal, the priority of risks, but then project management input may change the resulting priority. It is simply impossible to develop standardization in this area such that the application could be uniformly applied to each and every project.

The prioritization of risks for an individual project is a good starting point and could work well if most risks were not interrelated. We know from trade-off analysis that

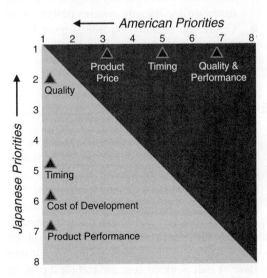

FIGURE 17–12. Ordering of trade-offs (Note: Lower priorities more often undergo trade-offs.).

changes to a schedule may induce changes in cost and/or performance. The changes may not occur in both dimensions because this depends on the objective functions and market constraints of the buyer and seller. Therefore, even though schedules have the highest priority in Figure 17–11, risk response to the schedule risk events may cause immediate evaluation of the technical performance risk events. Yes, risks are interrelated.

The interdependencies between risks can also be seen from Table 17–12. The first column identifies certain actions that the project manager can take in pursuit of the possible benefits listed in column 2. Each of these possible benefits, in turn, can cause additional risks, as shown in column 3. In other words, risk mitigation strategies that are designed to take advantage of a possible benefit could create another risk event that is more severe. As an example, working overtime could save you $15,000 by compressing the schedule. But if the employees make more mistakes on overtime, retesting may be required, additional materials may need to be purchased, and a schedule slippage could occur, thus causing a loss of $100,000. Therefore, is it worth risking a loss of $100,000 to save $15,000?

To answer this question, we can use the concept of expected value, assuming we can determine the probabilities associated with mistakes being made and the cost of the mistakes. Without any knowledge of these probabilities, the actions taken to achieve the possible benefits would be dependent upon the project manager's tolerance for risk.

Most project management professionals seem to agree that the most serious risks, and the ones about which we seem to know the least, are the technical risks. The worst situation is to have multiple technical risks that interact in an unpredictable or unknown manner. As an example, you are managing a new product development project. Marketing has provided you with two technical characteristics that would make the product highly desirable in the marketplace.

The exact relationship between these two characteristics is unknown. However, your technical subject matter experts have prepared the curve shown in Figure 17–13. According to the curve, the two characteristics may end up moving in opposite directions. In other words, maximizing one characteristic may require degradation in the second characteristic.

Working with marketing, you prepare the specification limits according to characteristic B in Figure 17–13. Because these characteristics interact in often unknown ways, the specification limit on characteristic B may force characteristic A into a region that would

TABLE 17–12. RISK INTERDEPENDENCIES

Action	Possible Benefit	Risk
• Work overtime	• Schedule compression	• More mistakes; higher cost and longer schedule
• Add resources	• Schedule compression	• Higher cost and learning curve shift
• Parallel work	• Schedule compression	• Rework and higher costs
• Reduce scope	• Schedule compression and lower cost	• Unhappy customer and no follow-on work
• Hire low-cost resources	• Lower cost	• More mistakes and longer time period
• Outsource critical work	• Lower cost and schedule compression	• Contractor possesses critical knowledge at your expense

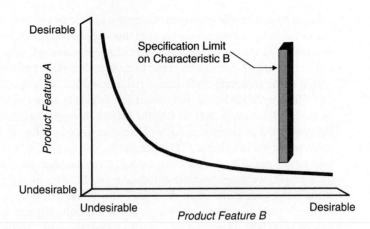

FIGURE 17–13. Interacting risks.

make the product less desirable to the ultimate consumer. Figure 17–13 is a utility representation of product feature A versus product feature B, and the curve is Pareto optimal—meaning that you cannot have more product feature A without having less product feature B.

Although project management methodologies provide a framework for risk management and the development of a risk management plan, it is highly unlikely that any methodology would be sophisticated enough to account for the identification of technical dependency risks. The time and cost associated with the identification, analysis and handling of technical risk dependencies could severely tax the project financially.

As companies become successful in project management, risk management evolves into a structured process that is performed continuously throughout the life cycle of the project. The four most common factors supporting the need for continuous risk management are how long the project lasts, how much money is at stake, the degree of developmental maturity, and the interdependencies between the different risks. For example, consider Boeing's aircraft projects where designing and delivering a new plane might require ten years and a financial investment of more than $5 billion.

Table 17–13 shows the characteristics of risks at Boeing. The table does not mean to imply that risks are mutually exclusive of each other. New technologies can appease customers, but production risks increase because the learning curve is lengthened with new technology compared to accepted technology. The learning curve can be lengthened further when features are custom-designed for individual customers. In addition, the loss of suppliers over the life of a plane can affect the level of technical and production risk. The relationships among these risks require the use of a risk management matrix and continuous risk assessment.

Another critical interdependency is the relationship between change management and risk management, both of which are part of the singular project management methodology. Each risk management strategy can result in changes that generate additional risks. Risks and changes go hand in hand, which is one of the reasons companies usually integrate risk

TABLE 17–13. RISK CATEGORIES AT BOEING

Type of Risk	Risk Description	Risk Handling Strategy
Financial	Up-front funding and payback period based upon number of planes sold	• Funding by life-cycle phases • Continuous financial risk management • Sharing risks with subcontractors • Risk reevaluation based upon sales commitments
Market	Forecasting customers' expectations on cost, configuration, and amenities based upon a 30–40 year life of a plane	• Close customer contact and input • Willingness to custom-design per customer • Develop a baseline design that allows for customization
Technical	Because of the long lifetime for a plane, must forecast technology and its impact on cost, safety, reliability, and maintainability	• A structured change management process • Using proven designs and technology rather than unproven designs and high risk technology • Parallel product improvement and new product development processes
Production	Coordination of manufacturing and assembly of a large number of subcontractors without impacting cost, schedule, quality, or safety	• Close working relationships with subcontractors • A structured change management process • Lessons learned from other new airplane programs • Use of learning curves

management and change management into a singular methodology. Table 17–14 shows the relationship between managed and unmanaged changes. If changes are unmanaged, then more time and money are needed to perform risk management, which often takes on the appearance and behavior of crisis management. And what makes the situation even worse is that higher salaried employees and additional time are required to assess the additional risks resulting from unmanaged changes. Managed changes, on the other hand, allow for a lower cost risk management plan to be developed.

TABLE 17–14. UNMANAGED VERSUS MANAGED CHANGES

	Where Time Is Invested	How Energy Is Invested	Which Resources Are Used
Unmanaged Change	• Back-end	• Rework • Enforcement • Compliance • Supervision	• Senior management and key players only
Managed Change	• Front-end	• Education • Communication • Planning • Improvements • Value added	• Stakeholders (internal) • Suppliers • Customers

Project management methodologies, no matter how good, cannot accurately define the dependencies between risks. It is usually the responsibility of the project team to make these determinations.

17.17 THE IMPACT OF RISK HANDLING MEASURES

Most project management methodologies include risk management, which can be used to:

- Create an understanding of the potential risks and their effects
- Provide an early warning system when the risk event is imminent
- Provide clear guidance on how to manage and contain the risk event, if possible
- Restore the system/process after the risk event occurs
- Provide a means for escape and rescue should all attempts fail

Some guidance in risk management is necessary because each stakeholder could have a different tolerance for risk. Risk and safety system policies, procedures, and guidelines exist primarily for the lower three levels in Figure 17–14. The customer's tolerance for risk could be significantly greater or less than the company's tolerance. Also, based upon the project's requirements, any given project could be willing to accept significantly more or less risk than the organizational procedures normally allow.

The project management methodology may very well dictate the magnitude of the risk handling measures to be undertaken. The risk handling measures for risk assumption may be significantly more complex than measures for avoidance. Figure 17–15 shows the extent of the risk handling strategy versus the magnitude of the risks. As the magnitude of the risk increases, an overreaction may occur that places undue pressure on the risk management process and the project management methodology. The cost of maintaining these risk handling measures should not overly burden the project. Excessive risk management

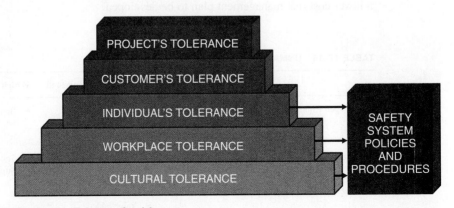

FIGURE 17–14. Tolerance for risk.

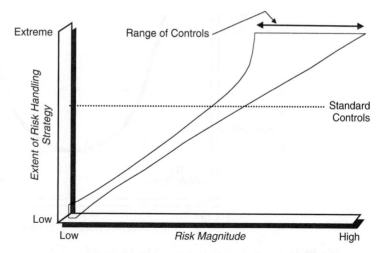

FIGURE 17–15. Risk handling measures.

procedures may require that the project manager spend more time and money than appropriate.

If an organization goes overboard in its investment in risk management, the results can be devastating, as shown in Figure 17–16. Overinvestment in risk management could lead to financial disaster if the project's risk events do not call for substantial measures or expenses. However, underinvestment in risk management for a project with numerous and complex risk events could lead to heavy losses and damages, possibly leading to project failure. Some sort of parity position is needed.

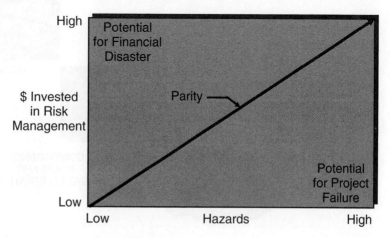

FIGURE 17–16. Investment in risk management.

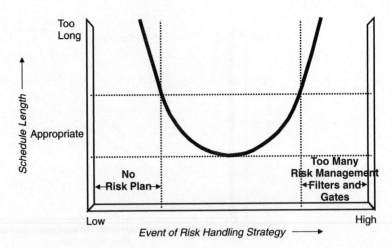

FIGURE 17–17. Risk handling versus schedule length.

Determining the proper amount of risk control measures is not easy. This can be seen from Figure 17–17, which illustrates the impact on the schedule constraint. If too few risk handling measures are in place, or if there simply is no risk handling plan, the result may be an elongated schedule due to ineffective risk handling measures. If excessive risk handling measures are in place, such as too many filters and gates, the schedule can likewise

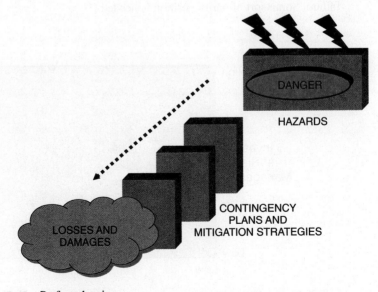

FIGURE 17–18. Perfect planning.

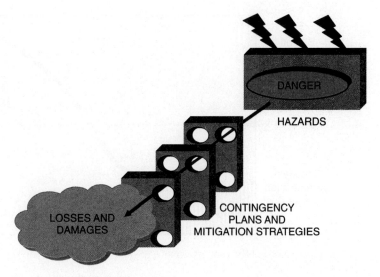

FIGURE 17–19. Imperfect planning.

be elongated because the workers are spending too much time on contingency planning. The same can be said for a risk management process with excessive risk reporting, documentation, and risk management meetings (i.e., too many gates). This results in very slow progress. A proper balance is needed.

Similarly, investing in risk management is not a guarantee that losses and damages will be prevented. Figure 17–18 illustrates perfect planning for risk management. The organization prepares a primary and possibly secondary risk handling plan for each potential hazard. Unfortunately, real-world planning is often imperfect, as shown in Figure 17–19, and some losses and damages may still occur, even for known risk issues.

17.18 RISK AND CONCURRENT ENGINEERING

PMBOK® Guide, 4th Edition
11.6 Risk Monitoring and Control

Most companies desire to get to the marketplace in a timely manner because the rewards for being the first-to-market can be huge in both profitability and market share. Getting to the marketplace quickly often entails using concurrent engineering, or overlapping activities. The critical question is, "How much overlapping can we incur before we get diminishing returns?"

The risks involved with overlapping activities are shown in Figure 17–20. Overlapping activities can lead to schedule compression and lower costs. However, too much overlapping can lead to excessive rework and unanticipated problems that can generate significant schedule slippages and cost overruns. Finding the optimal overlapping point that increases benefits while decreasing rework is difficult.

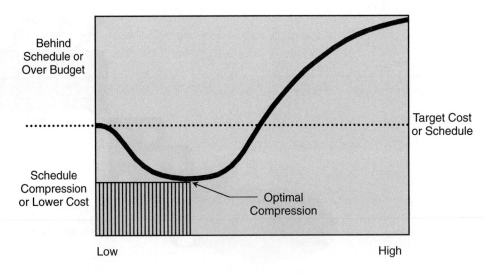

FIGURE 17–20. Overlapping risks.

Although there may exist numerous reasons for the rework, two common problems are:

● Combining new technology development and product development technology
● An insufficient test and evaluation program

To illustrate why these problems occur, consider a situation where the sales and marketing force promises the marketplace a new product with advanced technology that hasn't yet been developed. To compress the schedule, the product development team begins designing the product without knowing whether or not (and when) the technology can be developed. Production teams are asked to develop manufacturing plans without having any drawings. This results in massive changes when the product final reaches production.

There are three questions that need to be continuously addressed:

● Can the new technology be developed?
● Can we demonstrate the new technology within the product?
● Can the product then be manufactured within the time, cost, and performance (i.e., reliability) constraints?

Simultaneous development of technologies and products has become commonplace. To decrease the risks of rework, there should be a demonstration that the technology can work as expected. Leading firms that use concurrent engineering do not include a new technology in a product until the technology reaches a prescribed level of maturity. They have disciplined processes that match requirements with technological capability before product development is launched. These companies have learned the hard lesson of not committing to new products that outstrip their technological know-how. These practices stem from their recognition that resolving technology problems after product development begins can result in a tenfold cost increase; resolving these problems in production could increase costs by a hundredfold.

Some commonly accepted practices to reduce risks include:

- Flexibility in both the resources provided and the product's performance requirements to allow for uncertainties of technical progress
- Disciplined paths for technology to be included in products, with strong gatekeepers to decide when to allow it into a product development program
- High standards for judging the maturity of the technology
- The imposition of strict product development cycle times
- Rules concerning how much innovation can be accepted on a product before the next generation must be launched (these rules are sometimes referred to as technology readiness levels)

Collectively, these factors create a healthy environment for developing technology and making good decisions on what to include in a product.

Overlapping activities can be very risky if problems are discovered late in the cycle. One common mistake is to begin manufacturing before a sufficient quantity of engineering drawings is available for review. This normally is the responsibility of systems integration personnel. Systems integration should conclude with a critical design review of engineering drawings and confirmation that the system's design will meet requirements—a key knowledge point. It should also result in firm cost and schedule targets and a final set of requirements for the current version of the product. Decision-makers should insist on a mature design, supported with complete engineering drawings, before proceeding to even limited production. Having such knowledge at this point greatly contributes to product success and decreases costly rework.

As an example, Boeing had released over 90 percent of the engineering drawings on its 777-200 airplane halfway through its product development program. This allowed Boeing to have near certainty that the design for the 777-200 airplane would meet requirements. On the other hand, a different program had released only about half of its engineering drawings at approximately the same point in development. The other program encountered numerous technical problems in testing that resulted in redesigns, cost increases, and schedule delays.

Companies intent on decreasing the risks of concurrent engineering have found ways to employ testing in a manner that avoids late-cycle churns, yet enables them to efficiently yield products in less time, with higher performance, and at a lower cost. Generally, these practices are prompted by problems—and late-cycle churn—encountered on earlier products. Both Boeing and Intel were hurt by new products in which testing found significant problems late in development or in production that may have been preventable. Boeing absorbed cost increases in one line of aircraft and delivered it late to the first customer; Intel had to replace more than a million microprocessors that contained a minor, but nevertheless well-publicized, flaw. On subsequent products, these firms were able to reduce such problems by changing their approach to testing and evaluation and were able to deliver more sophisticated products on time, within budget, and with high quality.

Boeing encountered significant difficulties late in the development of its 747-400 airliner, which delayed its delivery to the customer and increased costs. When the 747-400 was delivered to United Airlines in 1990, Boeing had to assign 300 engineers to solve problems that testing had not revealed earlier. The resulting delivery delays and initial service problems irritated the customer and embarrassed Boeing. Boeing officials stated that this experience prompted the company to alter its test approach on subsequent aircraft, culminating with the 777-200 program of the mid-1990s. According to company officials, the 777-200 testing was the most extensive conducted on any Boeing commercial aircraft. As a result, Boeing delivered a Federal Aviation Administration–certified, service-ready 777-200 aircraft at initial delivery and reduced change, error, and rework by more than 60 percent.

A hallmark of the 777-200's success was the extended-range twin-engine certification for transoceanic flight it received from the Federal Aviation Administration on the first aircraft. This certification is significant because it normally takes about two years of actual operational service before the Federal Aviation Administration grants extended range certification. In the case of the 777-200, the testing and evaluation effort provided enough confidence in the aircraft's performance to forego the operational service requirements.

Intel has also employed testing to reduce late-cycle churn on its new microprocessors. According to Intel officials, the company learned this lesson the hard way—by inadvertently releasing the initial Pentium® microprocessor with a defect. After the release, Intel discovered a flaw in one of the Pentium® microprocessor's higher level mathematical functions. Using analytical techniques, Intel concluded that this flaw would not significantly affect the general public because it would occur only very rarely. Intel, however, miscalculated the effect on the consumer and was forced to replace more than a million microprocessors at a cost of about $500 million. Intel underwent a significant corporate change in its test approach to ensure that bugs like this did not "escape" to the public again. As a result, the performance of subsequent microprocessors, like the Pentium® Pro and Pentium® III microprocessors, has significantly improved. Despite adopting a much more rigorous testing and evaluation approach, Intel did not increase the amount of time it took to develop new, more sophisticated microprocessors. In fact, Intel's rate of product release increased over time.[20]

17.19 STUDYING TIPS FOR THE PMI® PROJECT MANAGEMENT CERTIFICATION EXAM

This section is applicable as a review of the principles to support the knowledge areas and domain groups in the PMBOK® Guide. This chapter addresses:

● Risk Management
● Planning
● Execution

20. *A More Constructive Test Approach Is Key to Better Weapon System Outcomes,* Best Practice Series, GAO/NSIAD-00-199, Government Accounting Office, July 2000, pp. 23–25.

- Controlling
- Professional Responsibility

Understanding the following principles is beneficial if the reader is using this text to study for the PMP® Certification Exam:

- What is meant by a risk
- Components of a risk
- That risk management is performed throughout the project and involves possibly the entire team
- Types of risks
- What is meant by one's tolerance for risk
- Sources of a risk
- What is meant by a risk event
- Components of a risk management plan
- Risk gathering techniques such as the Delphi technique and brainstorming
- Quantitative risk analysis such as expected value and Monte Carlo simulation
- Qualitative risk assessment
- What is meant by decision trees
- Risk response modes

The following multiple-choice questions will be helpful in reviewing the principles of this chapter:

1. The two major components of a risk are:
   A. Time and cost
   B. Uncertainty and damage
   C. Quality and time
   D. Cost and decision-making circumstances

2. Risk management is normally performed by:
   A. Developing contingency plans
   B. Asking the customer for help
   C. Asking the sponsor for help
   D. Developing work-around situations

3. Future outcomes that provide favorable opportunities are called:
   A. Favorable risks
   B. Opportunities
   C. Contingencies
   D. Surprises

4. The cause of a risk event is usually referred to as:
   A. An opportunity
   B. A hazard
   C. An outcome
   D. An unwanted surprise

5. If there is a 40 percent chance of making $100,000 and a 60 percent chance of losing $150,000, then the expected monetary outcome is:
   A. $50,000
   B. −$50,000

C. $90,000
D. −$90,000

6. Assumption, mitigation, and transfer are examples of risk:
   A. Contingencies
   B. Uncertainties
   C. Expectations
   D. Responses

7. In which life-cycle phase would project uncertainty be the greatest?
   A. Initiation
   B. Planning
   C. Execution
   D. Closure

8. In which life-cycle phase would the financial risks of a project be the greatest?
   A. Initiation
   B. Planning
   C. Execution
   D. Closure

9. Identifying a risk as high, moderate, or low would be an example of which risk assessment?
   A. Go-for-broke
   B. Adverse
   C. Qualitative
   D. Quantitative

10. Monte Carlo simulation is an example of which risk assessment?
    A. Go-for-broke
    B. Adverse
    C. Qualitative
    D. Quantitative

11. Which of the following is *not* a valid reason for managing a risk?
    A. Minimizing the risk's likelihood
    B. Minimizing the risk's unfavorable consequences
    C. Maximizing the probability of the risk's favorable consequences
    D. Providing a late-as-possible warning system

12. Which of the following is generally *not* part of overall risk management?
    A. Defining the roles and responsibilities of the team members
    B. Establishing a risk reporting format
    C. Select of the project manager
    D. Risk scoring and interpretation

13. A technique for risk evaluation that uses a questionnaire, a series of rounds, and reports submitted in confidence and then circulated with the source unidentified is called:
    A. The Delphi technique
    B. The work group
    C. Unsolicited team responses
    D. A risk management team

14. Risk symptoms or early warning signs are called:
    A. Vectors
    B. Triggers
    C. Pre-events
    D. Contingency events

15. Which of the following is *not* a risk quantification tool or technique?
    A. Interviewing
    B. Decision tree analysis
    C. Objective setting
    D. Simulation

16. A technique that depicts interactions among decisions and associated events is called:
    A. Decision tree analysis
    B. Earned value measurement system
    C. Network scheduling system
    D. Payoff matrix

17. Varying one risk driver at a time, either in small increments or from optimistic to pessimistic estimates while keeping all other drivers fixed, is called:
    A. Decision tree analysis
    B. Sensitivity analysis
    C. Network analysis
    D. Earned value analysis

18. A risk response strategy that generally reduces the probability or impact of the event without altering the project's objectives is called:
    A. Avoidance
    B. Acceptance
    C. Mitigation
    D. Transfer

19. Earned value measurement is an example of:
    A. Risk communication planning
    B. Risk identification planning
    C. Risk response
    D. Risk monitoring and control

20. The difference between being proactive and reactive is the development of a:
    A. Payoff table
    B. Range of probabilities
    C. Range of payoffs
    D. Contingency plan

ANSWERS

1. B
2. A
3. B

4. B
5. B
6. D
7. A
8. D
9. C
10. D
11. D
12. C
13. A
14. B
15. C
16. A
17. B
18. C
19. D
20. D

PROBLEMS

17–1 You have $1,000,000 worth of equipment at the job site and wish to minimize your risk of direct property damage by taking out an insurance policy. The insurance company provides you with its statistical data as shown below:

Type of Damage	Probability (%)	Amount of Damage (Loss) (%)
Total	0.02	100
Medium	0.08	40
Low	0.10	20
No Damage	99.8	0

If the insurance company uses expected value to calculate premiums, then how much would you expect the premium to be, assuming the insurance company adds on $300 for handling and profit?

17–2 You have been asked to use the expected-value model to assess the risk in developing a new product. Each strategy requires a different sum of money to be invested and produces a different profit payoff as shown below:

Strategy	States of Nature		
	Complete Failure	Partial Success	Total Success
S_1	<$50K>	<30K>	70K
S_2	<80K>	20K	40K
S_3	<70K>	0	50K
S_4	<200K>	<50K>	150K
S_5	0	0	0

Assume that the probabilities for each state are 30 percent, 50 percent, and 20 percent, respectively.

   a. Using the concept of expected value, what risk (i.e., strategy) should be taken?
   b. If the project manager adopts a go-for-broke attitude, what strategy should be selected?
   c. If the project manager is a pessimist and does not have the option of strategy S_5, what risk would be taken?
   d. Would your answer to part c change if strategy S_5 were an option?

17–3 Your company has asked you to determine the financial risks of manufacturing 6,000 units of a product rather than purchasing them from a vendor at $66.50 per unit. The production line will handle exactly 6,000 units and requires a one-time setup cost of $50,000. The production cost is $60/unit.

Your manufacturing personnel inform you that some of the units may be defective, as shown below:

% defective	0	1	2	3	4
probability of occurrence (%)	40	30	20	6	4

Defective items must be removed and replaced at a cost of $145/defective unit. However, 100 percent of units purchased from vendors are defect-free.

Construct a payoff table, and using the expected-value model, determine the financial risk and whether the make or buy option is best.

17–4 Below are four categories of risk and ways that a company is currently handling the risks. According to Section 17.11, which risk handling options are being used? More than one answer may apply.

   a. A company is handling its high R&D financial risk by taking on partners and hiring subcontractors. The partners/subcontractors are expected to invest some of their own funds in the R&D effort in exchange for sole-source, long-term production contracts if the product undergoes successful commercialization.
   b. A company has decided to handle its marketing risks by offering a family of products to its customer base. Different features exist for each product offered.
   c. A company has product lines with a life expectancy of ten years or more. The company is handling its technical risks by performing extensive testing on new

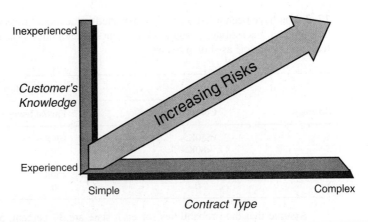

FIGURE 17–21. Future risks.

components and performing parallel technical development efforts for downstream enhancements.

    d. A company has large manufacturing costs for its high-tech products. The company will not begin production until it has a firm commitment for a certain quantity. The company uses learning curves and project management to control its costs.

17–5 A telecommunications firm believes that the majority of its income over the next ten years will come from organizations outside of the United States. More specifically, the income will come from third world nations that may have very little understanding or experience in project management. The company prepared Figure 17–21. What causes the increasing risks in Figure 17–21?

17–6 In the 1970s and 1980s, military organizations took the lead in developing ways to assess total program risk. One approach was to develop a rigorous process for identifying specific technical risk at the functional level and translating this detailed information through several steps. In this way, it was believed that risks could easily be monitored and corrected, as shown in Figure 17–22. Why is this method not being supported today?

17–7 As an example of the situation in Problem 17–6, Figure 17–23 shows risk categories at the program, subsystem, and functional levels. Starting at the bottom, data are developed for five engineering indicators and rated according to "high," "medium," or "low" risk. Results of this assessment are then summarized for each subsystem to provide a system overview. This is often considered a template risk analysis method. What are the advantages and disadvantages of this approach? Why is this method not used extensively today?

17–8 With the explosion of computer hardware and software during the 1970s and 1980s, companies began developing models to assess the technical risk for the computer hardware and software effort. One such model is discussed in this problem. Although some people contend that there may still exist applicable use for this model, others argue that the model is obsolete and flawed with respect to current thinking. After reading the paragraphs below, explain why the model may have limited use today for technical risk management.

    Previously, we showed that risk quantification could be found by use of an expected-value calculation. However, there are more sophisticated approaches that involve templates combined

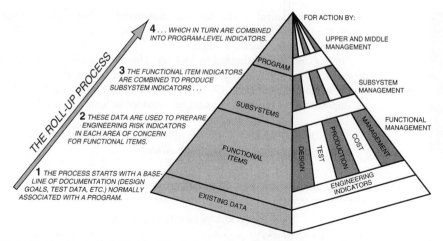

FIGURE 17–22. Technical risk identification at appropriate management levels (ONAS P 4855-X).

with the expected-value model. Here, we can develop mathematical expressions for failure and risk for *specific* types of projects.

Risk can be simply modeled as the interaction of two variables: probability of failure (P_f) and the effect or consequence of the failure (C_f). Consequences may be measured in terms of technical performance, cost, or schedule. A simple model can be used to highlight areas where the probability of failure (P_f) is high (even if there is a low probability of occurrence). Mathematically, this model can be expressed as the union of two sets, P_f and C_f. Table 17–15

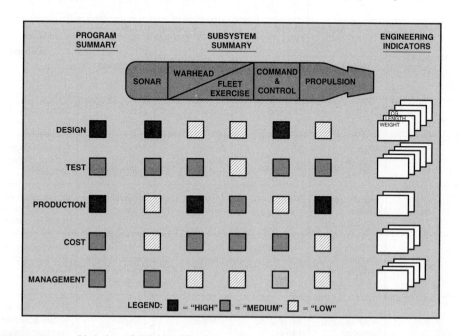

FIGURE 17–23. Variation of risk identification products with management level (ONAS P4855-X).

shows a mathematical model for risk assessment on hardware–software projects. In other words, the risk factor (defined as $P_f \times C_f$) will be largest where both P_f and C_f are large, and may be high if either factor is large.

In this case, P_f is estimated by looking at hardware and software maturity, complexity, and dependency on interfacing items. The probability of failure, P_f, is then quantified from ratings similar to the factors in Table 17–15. C_f is calculated by looking at the technical, cost, and schedule implications of failure. For example, consider an item with the following characteristics:

- Uses off-the-shelf hardware with minor modifications to software database
- Is based on simply designed hardware

TABLE 17–15.   A MATHEMATICAL MODEL FOR RISK ASSESSMENT

$$(1)\ \text{Risk Factor} = P_f + C_f - P_f \bullet C_f$$

$(2)\ P_f = a \bullet P_{M_{hw}} + b \bullet P_{M_{sw}} + c \bullet P_{C_{hw}} + d \bullet P_{C_{sw}} + e \bullet P_D$

where:

$P_{M_{hw}}$ = Probability of failure due to degree of hardware maturity

$P_{M_{sw}}$ = Probability of failure due to degree of software maturity

$P_{C_{hw}}$ = Probability of failure due to degree of hardware complexity

$P_{C_{sw}}$ = Probability of failure due to degree of software complexity

P_D = Probability of failure due to dependency on other items

and where: a, b, c, d, and e are weighting factors whose sum equals one.

$(3)\ C_f = f \bullet C_t + g \bullet C_c + h \bullet C_s$

where:

C_t = Consequence of failure due to technical factors

C_c = Consequence of failure due to changes in cost

C_s = Consequence of failure due to changes in schedule

and where: f, g, and h are weighting factors whose sum equals one.

Magnitude	Maturity Factor (P_M)		Complexity Factor (P_C)		Dependency Factor (P_D)
	Hardware $P_{M_{hw}}$	Software $P_{M_{sw}}$	Hardware $P_{C_{hw}}$	Software $P_{C_{sw}}$	
0.1	Existing	Existing	Simple design	Simple design	Independent of existing system, facility, or associate contractor
0.3	Minor redesign	Minor redesign	Minor increases in complexity	Minor increases in complexity	Schedule dependent on existing system, facility, or associate contractor
0.5	Major change feasible	Major change feasible	Moderate increase	Moderate increase	Performance dependent on existing system performance, facility, or associate contractor
0.7	Technology available, complex design	New software, similar to existing	Significant increase	Significant increase/major increase in # of modules	Schedule dependent on new system schedule, facility, or associate contractor
0.9	State of art, some research complete	State of art, never done before	Extremely complex	Extremely complex	Performance dependent on new system schedule, facility, or associate contractor

(continues)

TABLE 17–15. A MATHEMATICAL MODEL FOR RISK ASSESSMENT (*Continued*)

Magnitude	Technical Factor (C_t)	Cost Factor (C_c)	Schedule Factor (C_s)
0.1 (low)	Minimal or no consequences, unimportant	Budget estimates not exceeded, some transfer of money	Negligible impact on program, slight development schedule change compensated by available schedule slack
0.3 (minor)	Small reduction in technical performance	Cost estimates exceed budget by 1 to 5 percent	Minor slip in schedule (less than 1 month), some adjustment in milestones required
0.5 (moderate)	Some reduction in technical performance	Cost estimates increased by 5 to 20 percent	Small slip in schedule
0.7 (significant)	Significant degradation in technical performance	Cost estimates increased by 20 to 50 percent	Development schedule slip in excess of 3 months
0.9 (high)	Technical goals cannot be achieved	Cost estimates increased in excess of 50 percent	Large schedule slip that affects segment milestones or has possible effect on system milestones

- Requires software of somewhat minor increase in complexity
- Involves a new database to be developed by a subcontractor

Using Table 17–15, the probability of failure, P_f, would be calculated as follows:

Assume that the weighting factors for *a, b, c, d,* and *e* are 20 percent, 10 percent, 40 percent, 10 percent, and 20 percent, respectively.

$$
\begin{array}{llll}
P_M \text{ (hardware)} & = 0.1 & \quad 0.2\, P_M \text{ (h)} & = 0.02 \\
P_M \text{ (software)} & = 0.3 & \quad 0.1\, P_M \text{ (s)} & = 0.03 \\
P_C \text{ (hardware)} & = 0.1 & \quad 0.4\, P_C \text{ (h)} & = 0.04 \\
P_C \text{ (software)} & = 0.3 & \quad 0.1\, P_C \text{ (s)} & = 0.03 \\
P_D & = 0.9 & \quad 0.2\, P_D & = \underline{0.18} \\
& & & = 0.30
\end{array}
$$

Then, assuming the weighting factors shown in equation (2) of Table 17–15 are as indicated above, the P_f on this item would be 0.30.

If the consequence of the item's failure because of technical factors would cause some problems of a correctable nature, but correction would result in an 8 percent cost increase and two-month schedule slip, the consequence of failure, C_f, would be calculated from Table 17–15 as follows:

$$
\begin{array}{llll}
C_t = 0.3 & \quad 0.4\, C_t = 0.12 \\
C_c = 0.5 & \quad 0.5\, C_c = 0.25 \\
C_s = 0.5 & \quad 0.1\, C_s = \underline{0.12} \\
& \quad 0.42
\end{array}
$$

Then C_f for this item [assuming that the weighting factors in equation (3) of Table 17–15 are as indicated above] would be 0.42.

From equation (1) of Table 17–14, the risk factor would be

$$0.30 + 0.42 - (0.30)(0.42) = 0.594$$

In other words, the risk associated with this item is medium. Because most of the risk associated with this example arises from software changes, in particular the use of a subcontractor in this area, we can conclude that the risk can be reduced when the computer software developer is held "accountable for work quality and is subject to both incentives and penalties during all phases of the system life cycle."

Similar risk analyses would be performed for all other items and a risk factor would be obtained for each identified risk area. Risk areas would then be prioritized according to source of the risk (for example, are other items exhibiting excessive risk due to subcontractor software development?).

17–9 Figure 17–24 shows a probability-impact (or risk mapping) matrix that is frequently used as part of the risk analysis prioritization process. Here, ordinal probability-of-occurrence and consequence-of-occurrence risk scales are used (5, 4, 3, 2, 1 correspond to E, D, C, B, A since the actual scale coefficients are unknown). In this figure, L represents a low risk, which would generally be a risk acceptable to the project manager. The letter M represents a moderate risk, which will likely require a risk response. The letter H represents a high risk, which will need one or more risk responses. What are the advantages of using a high–moderate–low (or red–yellow–green) risk designation as opposed to assigning quantitative numbers to each cell and risk level (e.g., $5 \times 5 = 25$ to $1 \times 1 = 1$)?

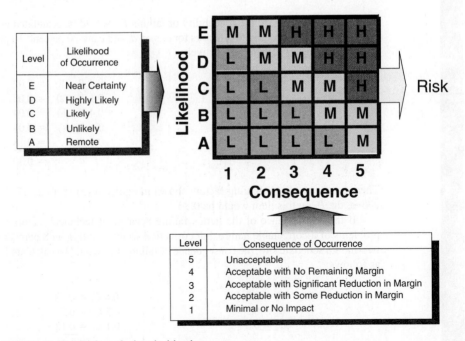

FIGURE 17–24. Risk analysis prioritization.

TELOXY ENGINEERING (A)

Teloxy Engineering has received a one-time contract to design and build 10,000 units of a new product. During the proposal process, management felt that the new product could be designed and manufactured at a low cost. One of the ingredients necessary to build the product was a small component that could be purchased for $60 in the marketplace, including quantity discounts. Accordingly, management budgeted $650,000 for the purchasing and handling of 10,000 components plus scrap.

During the design stage, your engineering team informs you that the final design will require a somewhat higher-grade component that sells for $72 with quantity discounts. The new price is substantially higher than you had budgeted for. This will create a cost overrun.

You meet with your manufacturing team to see if they can manufacture the component at a cheaper price than buying it from the outside. Your manufacturing team informs you that they can produce a maximum of 10,000 units, just enough to fulfill your contract. The setup cost will be $100,000 and the raw material cost is $40 per component. Since Teloxy has never manufactured this product before, manufacturing expects the following defects:

Percent defective	0	10	20	30	40
Probability of occurrence	10	20	30	25	15

All defective parts must be removed and repaired at a cost of $120 per part.

1. Using expected value, is it economically better to make or buy the component?
2. Strategically thinking, why might management opt for other than the most economical choice?

TELOXY ENGINEERING (B)

Your manufacturing team informs you that they have found a way to increase the size of the manufacturing run from 10,000 to 18,000 units in increments of 2000 units. However, the setup cost will be $150,000 rather than $100,000 for all production runs greater than 10,000 units and defects will cost the same $120 for removal and repair.

1. Calculate the economic feasibility of make or buy.
2. Should the probability of defects change if we produce 18,000 units as opposed to 10,000 units?
3. Would your answer to question 1 change if Teloxy management believes that follow-on contracts will be forthcoming? What would happen if the probability of defects changes to 15 percent, 25 percent, 40 percent, 15 percent, and 5 percent due to learning-curve efficiencies?

Learning Curves

Related Case Studies (from Kerzner/*Project Management Case Studies*, 3rd Edition)	Related Workbook Exercises (from Kerzner/*Project Management Workbook and PMP®/CAPM® Exam Study Guide*, 10th Edition)	PMBOK® Guide, 4th Edition, Reference Section for the PMP® Certification Exam
None	• Multiple Choice Exam	• Time Management • Cost Management

18.0 INTRODUCTION

PMBOK® Guide, 4th Edition
6.4 Activity Duration Estimates

Competitive bidding has become an integral part of the project management responsibility in many industries. A multitude of estimating techniques are available in such fields as construction, aerospace, and defense to assist project managers in arriving at a competitive bid. If the final bid is too high, the company may not be competitive. If the bid is too low, the company may have to incur the cost of the overrun out of its own pocket. For a small firm, this overrun could lead to financial disaster.

Perhaps the most difficult projects to estimate are those that involve the development and manufacturing of a large quantity of units. As an example, a company is asked to bid on the development and manufacture of 15,000 components. The company is able to develop a cost for the manufacture of its first unit, but what will be the cost for the 10th, 100th, 1,000th, or 10,000th unit? Obviously, the production cost of each successive unit should be less than the previous unit, but by how much? Fortunately there exist highly accurate estimating techniques referred to as "learning" or "experience" curves.

18.1 GENERAL THEORY

Experience curves are based on the old adage that practice makes perfect. A product can always be manufactured better and in a shorter time period not only the second time, but each succeeding time. This concept is highly applicable to labor-intensive projects, such as those in manufacturing where labor forecasting has been a tedious and time-consuming effort.

It wasn't until the 1960s that the true implications of experience curves became evident. Personnel from the Boston Consulting Group showed that each time cumulative production doubled, the total manufacturing time and cost fell by a *constant* and *predictable* amount. Furthermore, the Boston Consulting Group showed that this effect extended to a variety of industries such as chemicals, metals, and electronic components.

Today's executives often measure the profitability of a corporation as a function of market share. As market share increases, profitability will increase, more because of lower production costs than increased margins. This is the experience curve effect. Large market shares allow companies to build large manufacturing plants so that the fixed capital costs are spread over more units, thus lowering the unit cost. This increase in efficiency is referred to as *economies of scale* and may be the main reason why large manufacturing organizations may be more efficient than smaller ones.

Capital equipment costs follow the rule of six-tenths power of capacity. As an example, consider a plant that has the capacity of producing 35,000 units each year. The plant's construction cost was $10 million. If the company wishes to build a new plant with a capacity of 70,000 units, what will the construction cost be?

$$\frac{\$ \text{ new}}{\$ \text{ old}} = \left(\frac{70,000}{35,000}\right)^{0.6}$$

Solving for $ new, we find that the new plant will cost approximately $15 million, or one and one-half times the cost of the old plant. (For a more accurate determination, the costs must be adjusted for inflation.)

18.2 THE LEARNING CURVE CONCEPT

Learning curves stipulate that manufacturing man-hours (specifically direct labor) will decline each time a company doubles its output. Typically, learning curves produce a cost and time savings of 10 to 30 percent each time a company's experience at producing a product doubles. As an example, consider the data shown in Table 18–1, which represents a company operating on a 75 percent learning curve. The time for the second unit is 75 percent of the time of the first unit. The time for the fortieth unit is 75 percent of the time for the twentieth unit. The time for the 800th unit is 75 percent of the time for the 400th unit. Likewise, we can *forecast* the time for the 1,000th unit as being 75 percent of the time for the 500th unit. In this example, the time decreased by a fixed amount of 25 percent. Theoretically, this decrease could occur indefinitely.

TABLE 18–1. CUMULATIVE PRODUCTION AND LABOR-HOUR DATA

Cumulative Production	Hours This Unit	Cumulative Total Hours
1	812	812
2	609	1,421
10	312	4,538
12	289	5,127
15	264	5,943
20	234	7,169
40	176	11,142
60	148	14,343
75	135	16,459
100	120	19,631
150	101	25,116
200	90	29,880
250	82	34,170
300	76	38,117
400	68	45,267
500	62	51,704
600	57	57,622
700	54	63,147
800	51	68,349
840	50	70,354

In Table 18–1, we could have replaced the man-hours per production unit with the cost per production unit. It is more common to use man-hours because exact costs are either not always known or not publicly disclosed by the firm. Also, the use of costs implies the added complexity of considering escalation factors on salary, cost of living adjustments, and possibly the time value of money. For projects under a year or two, costs are often used instead of man-hours.

These types of costs are often referred to as value-added costs, and can also appear in the form of lower freight and procurement costs through bulk quantities. The value-added costs are actually cost savings for both the customer and contractor.

The learning curve was adapted from the historical observation that individuals performing repetitive tasks exhibit an improvement in performance as the task is repeated a number of times. Empirical studies of this phenomenon yielded three conclusions on which the current theory and practice are based:

- The time required to perform a task decreases as the task is repeated.
- The amount of improvement decreases as more units are produced.
- The rate of improvement has sufficient consistency to allow its use as a prediction tool.

The consistency in improvement has been found to exist in the form of a constant percentage reduction in time required over successive doubled quantities of units produced.

It's important to recognize the significance of using the learning curve for manufacturing projects. Consider a project where 75 percent of the total direct labor is in assembly (such as aircraft assembly) and the remaining 25 percent is machine work. With direct labor, learning improvements are possible, whereas with machine work, output may be restricted due to the performance of the machine. In the above example, with 75 percent direct labor and 25 percent machine work, a company may find itself performing on an 80 percent learning curve. But, if the direct labor were 25 percent and the machine work were 75 percent, then the company may find itself on a 90 percent learning curve.

18.3 GRAPHIC REPRESENTATION

Figure 18–1 shows the learning curve plotted from the data in Table 18–1. The horizontal axis represents the total number of units produced. The vertical axis represents the total labor hours (or cost) for each unit. The labor-hour graph in Figure 18–1 represents a hyperbola when drawn on ordinary graph paper (i.e., rectangular coordinates). The curve shows that the difference or amount of labor-hour reduction is *not* consistent. Rather, it declines by a continuously diminishing amount as the quantities are doubled. But the rate of change or decline has been found to be a constant percentage of the prior cost, because the decline in the base figure is proportionate to the decline in the amount of change. To illustrate this, we can use the data in Table 18–1, which was used to construct Figure 18–1. In doubling production from the first to the second unit, a reduction of 203 hours occurs. In doubling

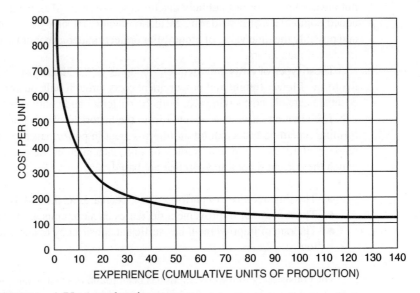

FIGURE 18–1. A 75 percent learning curve.

from 100 to 200 units, a reduction of 30 hours occurs. However in both cases, the percentage decrease was 25 percent. Again, in going from 400 to 800 units, a 25 percent reduction of 17 hours results. We can therefore conclude that, as more units are produced, the rate of change remains constant but the magnitude of the change diminishes.

When the data from Figure 18–1 are plotted on log-log paper, the result is a straight line, which represents the learning curve as shown in Figure 18–2.

There are two fundamental models of the learning curve in general use; the unit curve and the cumulative average curve. Both are shown in Figure 18–2. The unit curve focuses on the hours or cost involved in specific units of production. The theory can be stated as follows: As the total quantity of units produced doubles, the cost per unit decreases by some constant percentage. The constant percentage by which the costs of doubled quantities decrease is called the rate of learning.

The "slope" of the learning curve is related to the rate of learning. It is the difference between 100 percent and the rate of learning. For example, if the hours between doubled quantities are reduced by 20 percent (rate of learning), it would be described as a curve with an 80 percent slope.

To plot a straight line, one must know either two points or one point and the slope of the line. Generally speaking, the latter is more common. The question is whether the company knows the man-hours for the first unit or uses a projected number of man-hours for a target or standard unit to be used for pricing purposes.

The cumulative average curve in Figure 18–2 can be obtained from columns 1 and 3 in Table 18–1. Dividing column 3 by column 1, we find that the average hours for the first

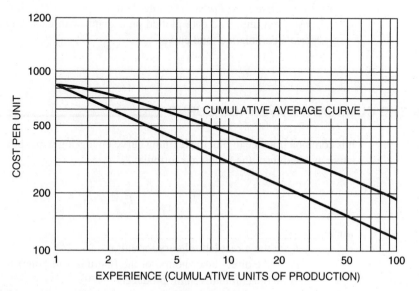

FIGURE 18–2. Logarithmic plot of a 75 percent learning curve.

100 units is 196 hours. For 200 units, the average is 149 hours. This becomes important in determining the cost for a manufacturing project.

18.4 KEY WORDS ASSOCIATED WITH LEARNING CURVES

Understanding a few key phrases will help in utilizing learning curve theory:

- *Slope of the curve.* A percentage figure that represents the steepness (constant rate of improvement) of the curve. Using the unit curve theory, this percentage represents the value (e.g., hours or cost) at a doubled production quantity in relation to the previous quantity. For example, with an experience curve having 80 percent slope, the value of unit two is 80 percent of the value of unit one, the value of unit four is 80 percent of the value at unit two, the value at unit 1000 is 80 percent of the value of unit 500, and so on.
- *Unit one.* The first unit of product actually completed during a production run. This is not to be confused with a unit produced in any reproduction phase of the overall acquisition program.
- *Cumulative average hours.* The average hours expended per unit for all units produced through any given unit. When illustrated on a graph by a line drawn through each successive unit, the values form a cumulative average curve.
- *Unit hours.* The total direct labor hours expended to complete any specific unit. When a line is drawn on a graph through the values for each successive unit, the values form a unit curve.
- *Cumulative total hours.* The total hours expended for all units produced through any given unit. The data plotted on a graph with each point connected by a line form a cumulative total curve.

18.5 THE CUMULATIVE AVERAGE CURVE

It is common practice to plot the learning curve on log-log paper but to calculate the cumulative average from the following formula:

$$T_x = T_1 X^{-K}$$

where

T_x = the direct labor hours for unit n
T_1 = the direct labor hours for the first unit (unit one)
X = the cumulative unit produced
$-K$ = a factor derived from the slope of the experience curve

Typical values for the exponent K are:

Learning curve %	K
100	0.0
95	0.074
90	0.152
85	0.235
80	0.322
75	0.415
70	0.515

As an example, consider a situation where the first unit requires 812 hours and the company is performing on a 75 percent learning curve. The man-hours required for the 250th unit would be:

$$T_{250} = (812)(250)^{-0.415}$$

$$= 82 \text{ hours}$$

This agrees with the data in Table 18–1.

Sometimes companies do not know the time for the first unit. Instead, they assume a target unit and accompanying target man-hours. As an example, consider a company that assumes that the standard for performance will be the 100th unit, which is targeted for 120 man-hours, and performs on a 75 percent learning curve. Solving for T_1 we have:

$$T_1 = T_x X^{-K}$$

$$= (120)(100)^{0.415}$$

$$= 811 \text{ hours}$$

This is in approximate agreement with the data in Table 18–1. The cumulative average number of labor hours can be *approximated* from the expression

$$T_c = \frac{T_1 X^{-K}}{1-K}$$

where T_c = cumulative average labor hours for the Xth unit.

$$X = \text{cumulative units produced}$$
$$T_1 = \text{direct labor hours for first unit}$$

For the 250th unit,

$$T_c = \frac{(812)(250)^{-0.415}}{1-0.415}$$

$$= 135 \text{ hours}$$

TABLE 18–2. SAMPLE COST REDUCTIONS DUE TO INCREASED EXPERIENCE

Ratio of Old Experience to New Experience	Experience Curve					
	70%	75%	80%	85%	90%	95%
1.1	5%	4%	3%	2%	1%	1%
1.25	11	9	7	5	4	2
1.5	19	15	12	9	6	3
1.75	25	21	16	12	8	4
2.0	30	25	20	15	10	5
2.5	38	32	26	19	13	7
3.0	43	37	30	23	15	8
4.0	51	44	36	28	19	10
6.0	60	52	44	34	24	12
8.0	66	58	49	39	27	14
16.0	76	68	59	48	34	19

Source: Derek F. Abell and John S. Hammond, *Strategic Market Planning.* Reprinted by permission of Pearson Education (Upper Saddle River, NJ., © 1979), p. 109.

From Table 18–1, the cumulative average for the 250th unit is 34,170 man-hours divided by 250, or 137 hours. We must remember that the above expression is merely an approximation. Significant errors can occur using this expression for fewer than 100 units. For large values of X, the error becomes insignificant.

It is possible to use the learning curve equation to develop Table 18–2, which shows typical cost reductions due to increased experience. Suppose that the production level is quadrupled and you are performing on an 80 percent learning curve. Using Table 18–2, the costs will be reduced by 36 percent.

18.6 SOURCES OF EXPERIENCE

There are several factors that contribute to the learning curve phenomenon. None of the factors perform entirely independently, but are interrelated through a complex network. However, for simplicity's sake, these factors will be sorted out for discussion purposes.

- *Labor efficiency.* This is the most common factor, which says that we learn more each time we repeat a task. As we learn, the time and cost of performing the task should diminish. As the employee learns the task, less managerial supervision is required, waste and inefficiency can be reduced or even eliminated, and productivity will increase.

Unfortunately, labor efficiency does not occur automatically. Personnel management policies in the area of *workforce stability* and *worker compensation* are of vital importance. As workers mature and become more efficient, it becomes increasingly important to maintain this pool of skilled labor. Loss of a contract or interruption between contracts could force employees to seek employment elsewhere. In certain industries, like aerospace and defense, engineers are often regarded as migratory workers moving from contract to contract and company to company.

Upturns and downturns in the economy can have a serious impact on maintaining experience curves. During downturns in the economy, people work more slowly, trying to preserve their jobs. Eventually the company is forced into a position of having to reassign people to other activities or to lay people off. During upturns in the economy, massive training programs may be needed in order to accelerate the rate of learning.

If an employee is expected to get the job done in a shorter period of time, then the employee expects to be adequately compensated. Wage incentives can produce either a positive or negative effect based on how they are applied. Learning curves and productivity can become a bargaining tool by labor as it negotiates for greater pay.

Fixed compensation plans generally do not motivate workers to produce more. If an employee is expected to produce more at a lower cost, then the employee expects to receive part of the cost savings as either added compensation or fringe benefits.

The learning effect goes beyond the labor directly involved in manufacturing. Maintenance personnel, supervisors, and persons in other line and staff manufacturing positions also increase their productivity, as do people in marketing, sales, administration, and other functions.

- *Work specialization and methods improvements.*[1] Specialization increases worker proficiency at a given task. Consider what happens when two workers, who formerly did both parts of a two-stage operation, each specialize in a single stage. Each worker now handles twice as many items and accumulates experience twice as fast on the more specialized task. Redesign of work operations (methods) can also result in greater efficiency.
- *New production processes.* Process innovations and improvements can be an important source of cost reductions, especially in capital-intensive industries. The low-labor-content semiconductor industry, for instance, achieves experience curves at 70 percent to 80 percent from improved production technology by devoting a large percentage of its research and development to process improvements. Similar process improvements have been observed in refineries, nuclear power plants, and steel mills, to mention a few.
- *Getting better performance from production equipment.* When first designed, a piece of production equipment may have a conservatively rated output. Experience may reveal innovative ways of increasing its output. For instance, capacity of a fluid catalytic cracking unit typically "grows" by about 50 percent over a ten-year period.[2]
- *Changes in the resource mix.* As experience accumulates, a producer can often incorporate different or less expensive resources in the operation. For instance, less skilled workers can replace skilled workers or automation can replace labor.

---

1. The next six elements are from Derek F. Abell and John S. Hammond, *Strategic Market Planning,* © 1979, pp. 112–113. Reprinted by permission of Pearson Education, Inc., Upper Saddle River, NJ.

2. Reprinted by permission of *Harvard Business Review.* From Winfred B. Hirschmann, "Profit from the Learning Curve," *Harvard Business Review,* 42, no. 1 (January–February 1964), p. 125. Copyright © 1964 by the Harvard Business School Publishing Corporation; all rights reserved.

- *Product standardization.* Standardization allows the replication of tasks necessary for worker learning. Even when flexibility and/or a wider product line are important marketing considerations, standardization can be achieved by modularization. For example, by making just a few types of engines, transmissions, chassis, seats, body styles, and so on, an auto manufacturer can achieve experience effects due to specialization in each part. These in turn can be assembled into a wide variety of models.
- *Product redesign.* As experience is gained with a product, both the manufacturer and customers gain a clear understanding of its performance requirements. This understanding allows the product to be redesigned to conserve material, allow greater efficiency in manufacture, and substitute less costly materials and resources, while at the same time improving performance on relevant dimensions. The change from wooden to brass works of clocks in the early 1800s is a good example; so are the new designs and substitution of plastic, synthetic fiber, and rubber for leather in ski boots.
- *Incentives and disincentives.* Compensation plans and other sources of experience can be both incentives and disincentives. Incentives can change the slope of the learning curve, as shown in Figure 18–3. This is referred to as a "toe-down" learning curve where a more favorable learning process can occur. In Figure 18–4, we have a "toe-up," or "scallop," learning curve, which is the result of disincentives. After the toe-up occurs, the learning curve may have a new slope that was not as favorable as the original slope. According to Hirschmann,[3]

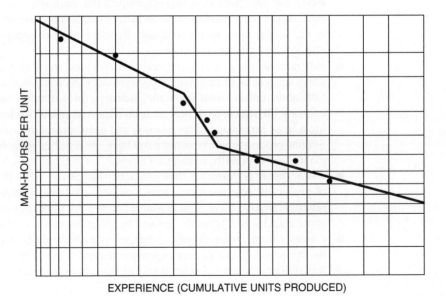

EXPERIENCE (CUMULATIVE UNITS PRODUCED)

FIGURE 18–3. A "toe-down" learning curve.

3. Reprinted by permission of *Harvard Business Review.* From Winfred B. Hirschmann, "Profit from the Learning Curve," *Harvard Business Review,* 42, no. 1 (January–February 1964), p. 126. Copyright © 1964 by the Harvard Business School Publishing Corporation; all rights reserved.

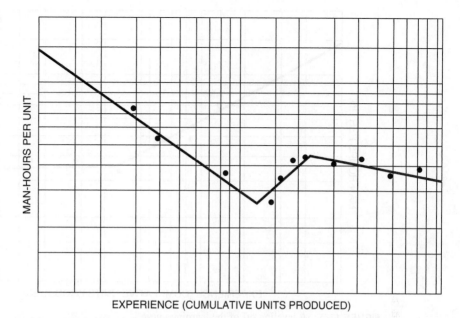

FIGURE 18–4. A "toe-up" learning curve.

A rise in the curve can occur in the middle of a contract too, owing to a substantial interruption (such as that caused by introducing changes in a model, by moving operations to a new building, or by halting operations for a while so that forgetting occurs). Shortly after operations recommence and skill in handling changes is acquired, the curve declines rapidly to approach the old slope. Such a break in the curve occurs frequently enough to have acquired the descriptive term "scallop." In fact, if, instead of merely a change being made, a new model is introduced, or a new type of item is put into production, the scallop occurs initially and the curve essentially starts again. Thus, the direct labor input reverts back to what it had been when the first item of the preceding type was put into production (assuming that the two items were of similar type and configuration).

Worker dissatisfaction can also create a leveling off of the learning curve, as shown in Figure 18–5. This leveling off can also occur as a result of inefficiencies due to closing out of a production line or transferring workers to other activities at the end of a contract.

18.7 DEVELOPING SLOPE MEASURES

Research by the Stanford Research Institute revealed that many different slopes were experienced by different manufacturers, sometimes on similar manufacturing programs. In fact, manufacturing data collected from the World War II aircraft manufacturing industry had slopes ranging from 69.7 percent to almost 100 percent. These slopes averaged 80 percent,

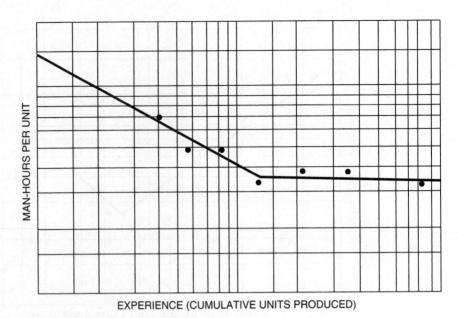

FIGURE 18–5. A leveling off of the learning curve.

giving rise to an industry average curve of 80 percent. Other research has developed measures for other industries, such as 95.6 percent for a sample of 162 electronics programs. Unfortunately, this industry average curve is frequently misapplied by practitioners who use it as a standard or norm. When estimating slopes without the benefit of data from the plant of the manufacturer, it is better to use learning curve slopes from similar items at the manufacturer's plant rather than the industry average.

The analyst needs to know the slope of the learning curve for a number of reasons. One is to facilitate communication, because it is part of the language of the learning curve theory. The steeper the slope (lower the percent), the more rapidly the resource requirements (hours) will decline as production increases. Accordingly, the slope of the learning curve is usually an issue in production contract negotiation. The slope of the learning curve is also needed to project follow-on costs, using either learning tables or a computer. Also, a given slope may be established as a standard based on reliable historical experience. Learning curves developed from actual experience on current production can then be compared against this standard slope to determine whether the improvement on a particular contract is or is not reasonable.

18.8 UNIT COSTS AND USE OF MIDPOINTS

The use of the learning curve is dependent on the methods of recording costs that companies employ. An accounting or statistical record system must be devised by a company so that data are available for learning curve purposes. Otherwise, it may be impossible to construct a learning curve. Costs, such as labor hours per unit or dollars per unit, must be identified with the unit of product. It is preferable to use labor hours rather than dollars, because the latter contain an additional variable—the effect of inflation or deflation (both wage-rate and material cost changes)—that the former does not contain. In any event, the record system must have definite cutoff points for such costs permitting identification of the costs with the units involved. Most companies use a lot-release system, whereby costs are accumulated on a job order in which the number of units completed are specified and the costs are cut off at the completion of the number of units. In this case, however, the costs are usually equated with equivalent units rather than actual units. Because the job order system is commonly used, the unit cost is not the actual cost per unit in the lot. This means that when lots are plotted on graph paper, the unit value corresponding with the average cost value must be found.

18.9 SELECTION OF LEARNING CURVES

Existing experience curves, by definition, reflect past experience. Trend lines are developed from accumulated data plotted on logarithmic paper (preferably) and "smoothed out" to portray the curve. The type of curve may represent one of several concepts. The data may have been accumulated by product, process, department, or by other functional or organizational segregations, depending on the needs of the user. But whichever experience curve concept or method of data accumulation is selected for use, based on suitability to the experience pattern, the data should be applied consistently in order to render meaningful information to management. Consistency in curve concept and data accumulation cannot be overemphasized, because existing experience curves play a major role in determining the project experience curve for a new item or product.

When selecting the proper curve for a new production item when only one point of data is available and the slope is unknown, the following, in decreasing order of magnitude, should be considered.

- Similarity between the new item and an item or items previously produced.
- Physical comparisons
  - Addition or deletion of processes and components
  - Differences in material, if any
  - Effect of engineering changes in items previously produced
- Duration of time since a similar item was produced
  - Condition of tooling and equipment
  - Personnel turnover
  - Changes in working conditions or morale
- Other comparable factors between similar items
  - Delivery schedules
  - Availability of material and components
  - Personnel turnover during production cycle of item previously produced
  - Comparison of actual production data with previously extrapolated or theoretical curves to identify deviations

It is feasible to assign weights to these factors as well as to any other factors that are of a comparable nature in an attempt to quantify differences between items. These factors are again historical in nature and only comparison of several existing curves and their actuals would reveal the importance of these factors.

If at least two points of data are available, the slope of the curve may be determined. Naturally the distance between these two points must be considered when evaluating the reliability of the slope. The availability of additional points of data will enhance the reliability of the curve. Regardless of the number of points and the assumed reliability of the slope, comparisons with similar items are considered the most desirable approach and should be made whenever possible.

A value for unit one may be arrived at either by accumulation of data or statistical derivation. When production is underway, available data can be readily plotted, and the curve may be extrapolated to a desired unit. However, if production has yet to be started, actual unit-one data would not be available, and a theoretical unit-one value would have to be developed. This may be accomplished in one of three ways:

- A statistically derived relationship between the preproduction unit hours and first unit hours can be applied to the actual hours from the preproduction phase.
- A cost estimating relationship (CER) for first-unit cost based on physical or performance parameters can be used to develop a first-unit cost estimate.
- The slope and the point at which the curve and the labor standard value converge are known. In this case, a unit-one value can be determined. This is accomplished by dividing the labor standard by the appropriate unit value.

18.10 FOLLOW-ON ORDERS

Once the initial experience curve has been developed for either the initial order or production run, the values through the last unit on the cumulative average and unit curves can be determined. Follow-on orders and continuations of production runs, which are considered extensions of the original orders or runs, are plotted as extensions on the appropriate curve. However, the cumulative average value through the final point of the extended curve is not the cumulative average for the follow-on portion of that curve. It is the cumulative average for both portions of the curve, assuming no break in production. Thus estimating the cost for the follow-on effort only requires evaluation of the differences between cumulative average costs for the initial run and the follow-on. Likewise, the last-unit value for both portions of the unit curve would represent the last-unit value for the combined curve.

18.11 MANUFACTURING BREAKS

The manufacturing break is the time lapse between the completion of an order or manufacturing run of certain units of equipment and the commencement of a follow-on order or restart of a manufacturing run for identical units. This time lapse disrupts the continuous flow of manufacturing and constitutes a definite cost impact. The time lapse under discussion here pertains to significant periods of time (weeks and months), as opposed to the minutes or hours for personnel allowances, machine delays, power failures, and the like.

It is logical to assume that because the experience curve has a time-cost relationship, a break will affect both time and cost. Therefore, the length of the break becomes as significant as the length of the initial order or manufacturing run. Because the break is quantifiable, the remaining factor to be determined is the cost of this lapse in manufacturing (that is, the additional cost incurred over and above that which would have been incurred had either the initial order or the run continued through the duration of the follow-on order or the restarted run).

When a manufacturer relies on experience curves as management information tools, it can be assumed that the necessary, accurate data for determining the initial curves have been accumulated, recorded, and properly validated. Therefore, if the manufacturer has experienced breaks, the experience curve data for the orders (lots) or runs involved should be available in such form that appropriate curves can be developed.

George Anderlohr suggests a method that assumes loss of learning is dependent on five factors[4]:

- *Manufacturing personnel learning.* In this area, the physical loss of personnel, either through regular movement or layoff, must be determined. The company's personnel records can usually furnish evidence on which to establish this learning loss. The percentage of learning lost by the personnel retained on other plant projects should also be ascertained. These people will lose their physical dexterity and familiarity with the product, and the momentum of repetition.

- *Supervisory learning.* Once again, a percentage of supervisory personnel will be lost as a result of the break in repetition. Management will make a greater effort to retain this higher caliber of personnel, so the physical loss, in the majority of cases, will be far less than in the area of production personnel. However, the supervisory personnel retained will lose their overall familiarity with the job, so that the guidance they can furnish will be reduced. In addition, because of the loss of production personnel, the supervisor will have no knowledge of the new hires and their individual personalities and capabilities.

- *Continuity of productivity.* This relates to the physical positioning of the line, the relationship of one work station to another, and the location of lighting, bins, parts, and tools within the work station. It also includes the position adjustment to optimize the individual's needs. In addition, a major factor affecting this area is the balanced line or the work-in-process buildup. Of all the elements of learning, the greatest initial loss is suffered in this area.

- *Methods.* This area is least affected by a break. As long as the method sheets are kept on file, learning can never be completely lost. However, drastic revisions to the method sheets may be required as a result of a change from soft to hard tooling.

4. George Anderlohr, "What Product Breaks Costs," *Industrial Engineering,* September 1969, pp. 34–36.

- *Special tooling.* New and better tooling is a major contributor to learning. In relating loss in tooling area to learning, the major factors are wear, physical misplacement, and breakage. An additional consideration must be the comparison of short-run, or so-called soft, tooling to long-run, or hard, tooling, and the effect of the transition from soft to hard tooling.

18.12 LEARNING CURVE LIMITATIONS

There are limitations to the use of learning curves, and care must be taken to avoid erroneous conclusions. Typical limitations include:

- The learning curve does not continue forever. The percentage decline in hours/dollars diminishes over time.
- The learning curve knowledge gained on one product may not be extendable to other products unless there exist shared experiences.
- Cost data may not be readily available in order to construct a meaningful learning curve. Other problems can occur if overhead costs are included with the direct labor cost, or if the accounting codes cannot separate work packages sufficiently in order to identify those elements that truly demonstrate experience effects.
- Quantity discounts can distort the costs and the perceived benefits of learning curves.
- Inflation must be expressed in constant dollars. Otherwise, the gains realized from experience may be neutralized.
- Learning curves are most useful on long-term horizons (i.e., years). On short-term horizons, benefits perceived may not be the result of learning curves.
- External influences, such as limitations on materials, patents, or even government regulations, can restrict the benefits of learning curves.
- Constant annual production (i.e., no growth) may have a limiting experience effect after a few years.

18.13 PRICES AND EXPERIENCE

If the competitive marketplace is stable, then as cost decreases as a function of the learning curve experience, prices will decrease similarly. This assumes that profit margins are expressed as a percentage of price rather than in absolute dollar terms. Therefore, the gap between selling price and cost will remain a constant, as shown in Figure 18–6.

Unfortunately, price and cost will most likely follow the relationship shown in Figure 18–7. Companies that use learning curves develop pricing policies based on either an industry average cost or an average cost based on a target production volume. In phase A, new product prices are less than the company cost, because the market would probably be

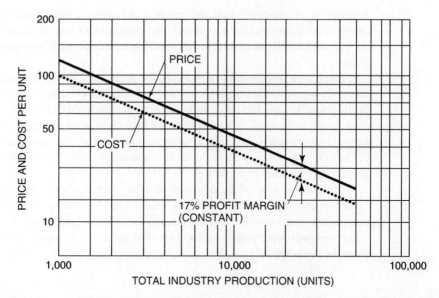

FIGURE 18–6. An idealized price–cost relationship when profit margin is constant. *Source:* Derek F. Abell and John S. Hammond, *Strategic Market Planning,* Prentice-Hall, © 1979, p. 115. Reprinted by permission of Pearson Education, Upper Saddle River, NJ.

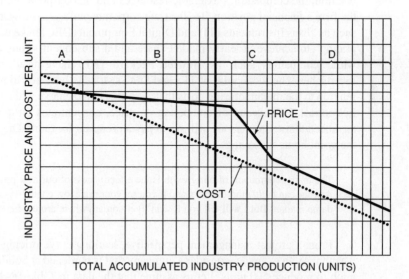

FIGURE 18–7. Typical price–cost relationships. *Source:* Derek F. Abell and John S. Hammond, *Strategic Market Planning,* Prentice-Hall, © 1979, p. 116. Reprinted by permission of Pearson Education, Upper Saddle River, NJ. Adapted from *Perspectives on Experience,* The Boston Consulting Group, 1972, p. 21.

reluctant to purchase the first few items at the *actual production cost.* As the company enters phase B, profits begin to materialize as the experience curve takes hold. Fixed costs are recovered. Price may remain firm because of market strategies adopted by the market leader.

The longer one remains in phase B, the greater the profits. Unfortunately, phase B is relatively unstable. One or more competitors will quickly drop their prices, because if the profit potential were too large, new entrants into the highly profitable marketplace would soon occur. In phase C, prices drop faster than costs, thus forcing a shakeout of the marketplace where marginal producers exit the market. The shakeout phase ends when prices begin to follow industry costs down the experience curve. This is phase D, which represents a stable market condition.

The average cost of the dominant market producers virtually regulates the industry. Whatever learning curve the industry leader uses, the competitors must match it. If the competitors' costs or volume cannot match the industry leader, then the slower rate of cost reductions will force profits to decrease or disappear, thus eliminating these competitors from the marketplace.

18.14 COMPETITIVE WEAPON

Learning curves are a strong competitive weapon, especially in developing a pricing strategy. The actual pricing strategy depends on the product life-cycle stage, the firm's market position, the competitor's available resources and market position, the time horizon, and the firm's financial position. To illustrate corporate philosophy toward pricing, companies such as Texas Instruments (TI) and Digital Equipment (DEC) have used "experience curve pricing" to achieve an early market share and a subsequent strong competitive position, while companies such as Hewlett-Packard (HP) have used completely different approaches. The focal point of TI's and DEC's strategy has been to price a new product in relation to the manufacturing costs that they expect to achieve when the product is mature. In contrast, HP, instead of competing on price, concentrates on developing products so advanced that customers are willing to pay a premium for them. Dr. David Packard drives the point home by saying,

> The main determinant of our growth is the effectiveness of our new product programs. . . . Anyone can build market share, and if you set your price low enough, you can get the whole damn market. But I will tell you it won't get you anywhere around here.[5]

From a project management perspective, learning curve pricing can be a competitive weapon. As an example, consider a company that is burdened at $60/hour and is bidding on a job to produce 500 units. Let us assume that the data in Table 18–1 apply. For 500 units

5. "Hewlett-Packard: When Slower Growth Is Smarter Management," *Business Week,* June 9, 1975, pp. 50–58.

of production, the cumulative total hours are 51,704, giving us an average rate of 103.4 hours per unit. The cost for the job would be 51,704 hours × $60/hour, or $3,102,240. If the target profit is 10 percent, then the final bid should be $3,412,464. This includes a profit of $310,224.

Even though a 10 percent profit is projected, the *actual* profit may be substantially less. Each product is priced out an average of 103.4 hours/unit. The first unit, however, will require 812 hours. The company will *lose* 708.6 hours × $60/hour, or $42,516, on the first unit produced. The 100th unit will require 120 hours, giving us a loss of $996 (i.e., [120 hours − 103.4 hours] × $60/hour). Profit will begin when the 150th unit is produced, because the hours required to produce the 150th unit are less than the average hour per unit of 103.4.

Simply stated, the first 150 units are a drain on cash flow. The cash-flow drain may require the company to "borrow" money to finance operations until the 150th unit is produced, thus lowering the target profit.

18.15 STUDYING TIPS FOR THE PMI® PROJECT MANAGEMENT CERTIFICATION EXAM

This section is applicable as a review of the principles to support the knowledge areas and domain groups in the PMBOK® Guide. This chapter addresses:

- Time Management
- Cost Management

Understanding the following principles is beneficial if the reader is using this text to study for the PMP® Certification Exam:

- What is meant by a learning curve
- Uses of a learning curve
- How learning curves can be used for estimating

The following multiple-choice questions will be helpful in reviewing the principles of this chapter:

1. According to learning curve theory, learning takes place at a fixed rate whenever the production levels:
   A. Increase higher than normal
   B. Increase, but at a lower than normal rate
   C. Double
   D. Quadruple

2. Learning curve theory is most appropriate for estimating which costs?
   A. R&D
   B. Engineering
   C. Marketing
   D. Manufacturing

3. On a 90 percent learning curve, the 100th unit required 80 hours. How many hours would the 200th unit require?
   A. 200
   B. 180
   C. 100
   D. 90

4. Which of the following can be a source of improvement to a learning curve?
   A. New, more efficient production processes
   B. Product redesigns
   C. Higher quality raw materials
   D. All of the above

ANSWERS

   1. C
   2. D
   3. D
   4. D

PROBLEMS

18–1 When a learning curve is plotted on ordinary graph paper, the curve appears to level off. But when the curve is plotted on log-log paper, it appears that the improvements can go on forever. How do you account for the difference? Can the improvements occur indefinitely? If not, what factors could limit continuous improvement?

18–2 A company is performing on an 85 percent learning curve. If the first unit requires 620 hours, how much time will be required for the 300th unit?

18–3 A company working on a 75 percent learning curve has decided that the production standard should be 85 hours of production for the 100th unit. How much time should be required for the first unit? If the first unit requires more hours than you anticipated, does this mean that the learning curve is wrong?

18–4 A company has just received a contract for 700 units of a certain product. The pricing department has predicted that the first unit should require 2,250 hours. The pricing department believes that a 75 percent learning curve is justified. If the actual learning curve is 77 percent, how much money has the company lost? Assume that a fully burdened hour is $65. What percentage error in total hours results from a 2 percent increase in learning curve percentage?

18–5 If the first unit of production requires 1,200 hours and the 150th unit requires 315 hours, then what learning curve is the company performing at?

18–6 A company has decided to bid on a follow-on contract for 500 units of a product. The company has already produced 2,000 units on a 75 percent learning curve. The 2000th unit requires 80 hours of production time. If a fully burdened hour is $80 and the company wishes to generate a 12 percent profit, how much should be bid?

18–7 Referrring to question 18–6, how many units of the follow-on contract must be produced before a profit is realized?

18–8 A manufacturing company wishes to enter a new market. By the end of next year, the market leader will have produced 16,000 units on an 80 percent learning curve, and the year-end price is expected to be $475/unit. Your manufacturing personnel tell you that the first unit will require $7,150 to produce and, with the new technology you have developed, you should be able to perform at a 75 percent learning curve. How many units must you produce and sell over the next year in order to compete with the leader at $475/unit at year end? Is your answer realistic, and what assumptions have you made?

18–9 Rylon Corporation is an assembler of electrical components. The company estimates that for the next year, the demand will be 800 units. The company is performing on an 80 percent learning curve. The company is considering purchasing some assembly machinery to acceler-ate the assembly time. Most assembly activities are 85–90 percent labor intensive. However, with the new machinery, the assembly activities will be only 25–45 percent labor intensive. If the company purchases and installs the new equipment, it will occur after the 200th unit is pro-duced. Therefore, the remaining 600 units will be produced with the new equipment. The 200th unit will require 620 hours of assembly. However, the 201st unit will require only 400 hours of assembly but on a 90 percent learning curve.

  a. Will the new machine shorten product assembly time for all 800 units and, if so, by how many hours?
  b. If the company is burdened by $70 per hour, and the new equipment is depreciated over five years, what is the most money that the company should pay for the new equipment? What assumptions have you made?

Contract Management[1]

Related Case Studies (from Kerzner/*Project Management Case Studies*, 3rd Edition)	Related Workbook Exercises (from Kerzner/Project *Management Workbook and PMP/CAPM Exam Study Guide,* 10th Edition)	PMBOK Guide, 4th Edition, Reference Section for the PMP Certification Exam
None	Multiple Choice Exam Crossword Puzzle on Procurement Management	Procurement Management

19.0 INTRODUCTION

PMBOK® Guide, 4th Edition
Chapter 12 Procurement
  Management

In general, companies provide services or products based on the requirements set forth in invitations for competitive bids issued by the client or the results of direct contract negotiations with the client. One of the most important factors in preparing a proposal and estimating the cost and profit of a project is the type of contract expected. The confidence by which a bid is prepared is usually dependent on how much of a risk the contractor will incur through the

---

1. The title of this chapter has been changed from Procurement Management in the Ninth Edition to Contract Management in this edition. Contract management includes procurement management. Procurement management is the buyer's side of contract management, and sales/proposal management is the seller's side of contract management. All those sellers (contractors) managing project contracts may not find it necessary to use the PMBOK® Guide, Chapter 12, because they may not be procuring anything.

contract. Certain types of contracts provide relief for the contractor since onerous risks[2] exist. The cost must therefore consider how well the contract type covers certain high- and low-risk areas.

Prospective clients are always concerned when, during a competitive bidding process, one bid is much lower than the others. The client may question the validity of the bid and whether the contract can be achieved for the low bid. In cases such as this, the client usually imposes incentive and penalty clauses in the contract for self-protection.

Because of the risk factor, competitors must negotiate not only for the target cost figures but also for the type of contract involved since risk protection is the predominant influential factor. The size and experience of the client's own staff, urgency of completion, availability of qualified contractors, and other factors must be carefully evaluated. The advantages and disadvantages of all basic contractual arrangements must be recognized to select the optimum arrangement for a particular project.

19.1 PROCUREMENT

PMBOK® Guide, 4th Edition
Chapter 12 Introduction
12.1 Plan Purchases and
   Acquisitions

Procurement can be defined as the acquisition of goods or services. Procurement (and contracting) is a process that involves two parties with different objectives who interact on a given market segment. Good procurement practices can increase corporate profitability by taking advantage of quantity discounts, minimizing cash flow problems, and seeking out quality suppliers. Because procurement contributes to profitability, procurement is often centralized, which results in standardized practices and lower paperwork costs.

All procurement strategies are frameworks by which an organization attains its objectives. There are two basic procurement strategies:

- **Corporate Procurement Strategy:** The relationship of specific procurement actions to the corporate strategy. An example of this would be centralized procurement.
- **Project Procurement Strategy:** The relationship of specific procurement actions to the operating environment of the project. An example of this would be when the project manager is allowed to perform sole source procurement without necessarily involving the centralized procurement group, such as purchasing one ounce of a special chemical for an R&D project.

Project procurement strategies can differ from corporate procurement strategies because of constraints, availability of critical resources, and specific customer requirements. Corporate strategies might promote purchasing small quantities from several qualified vendors, whereas project strategies may dictate sole source procurement.

---

2. *Onerous risks* are unfair risks that the contractor may have to bear. Quite often, the contract negotiations may not reach agreement on what is or is not an onerous risk.

Procurement planning usually involves the selection of one of the following as the primary objective:

- Procure all goods/services from a single source.
- Procure all goods/services from multiple sources.
- Procure only a small portion of the goods/services.
- Procure none of the goods/services.

> **PMBOK® Guide, 4th Edition**
> 1.8 Enterprise Environmental
> Factors

Another critical factor is the environment in which procurement must take place. There are two environments: macro and micro. The macro environment includes the general external variables that can influence how and when we do procurement. The PMBOK® Guide refers to this as "Enterprise Environmental Factors." These include recessions, inflation, cost of borrowing money, whether a buyer or seller's market exists, and unemployment. As an example, a foreign corporation had undertaken a large project that involved the hiring of several contractors. Because of the country's high unemployment rate, the decision was made to use only domestic suppliers/contractors and to give first preference to contractors in cities where unemployment was the greatest, even though there were other more qualified suppliers/contractors.

The microenvironment is the internal procurement processes of the firm, especially the policies and procedures imposed by the firm, project, or client in the way that procurement will take place. This includes the procurement/contracting system, which contains four processes according to the PMBOK® Guide, Fourth Edition:

- Plan Procurements
- Conduct Procurements
- Administer Procurements
- Close Procurements

It is important to understand that, in certain environments such as major projects for the Department of Defense (DoD), the contracting process is used as the vehicle for transitioning the project from one life-cycle phase to the next. For example, a contract can be awarded for the design, development, and testing of an advanced jet aircraft engine. The contract is completed when the aircraft engine testing is completed. If the decision is made at the phase gate review to proceed to aircraft engine production, the contracting process will be reinitiated for the new effort. Thus, the above four PMBOK® Guide processes would be repeated for each life-cycle phase. As the project progresses from one phase to the next, and additional project knowledge is acquired through each completed phase, the level of uncertainty (and risk) is reduced. The reduction in project risk allows the use of lower-risk contracts throughout the project life cycle. During higher-risk project phases such as conceptual, development, and testing, cost-type contracts are traditionally used. During the lower-risk project phases such as production and sustainment, fixed-priced contracts are typically used.

It is also important to note that the above four PMBOK® Guide processes focus only on the buyer's side of contract management.

Contract management is defined as "art and science of managing a contractual agreement throughout the contracting process."[3] Since contracts involve at least two parties—the buyer and the seller (contractor), contract management processes are performed by *both* the buyer and seller. The seller's contract management processes, which correspond to the buyer's processes, consist of the following activities[4]:

- **Presales Activity:** The process of identifying prospective and current customers, determining customer's needs and plans, and evaluating the competitive environment.
- **Bid/No Bid Decision-Making:** The process of evaluating the buyer's solicitation, assessing the competitive environment and risks against the opportunities of a potential business deal, and then deciding whether to proceed.
- **Bid/Proposal Preparation:** The process of developing offers in response to a buyer's solicitation or based on perceived buyer needs, for the purpose of persuading the buyer to enter into a contract.
- **Contract Negotiation and Formation:** The process of reaching a common understanding of the nature of the project and negotiating the contract terms and conditions for the purpose of developing a set of shared expectations and understandings.
- **Contract Administration:** The process of ensuring that each party's performance meets contractual requirements.
- **Contract Closeout:** The process of verifying that all administrative matters are concluded on a contract that is otherwise physically complete. This involves completing and settling the contract, including resolving any open items.

As can be seen from the previous discussion, the last two phases of the seller's contract management processes are identical to the buyer's contract management processes. This is because the buyer and seller are both performing the same contract management activities and working off of the same contract document.

19.2 PLAN PROCUREMENTS

PMBOK® Guide, 4th Edition
12.1 Plan Procurements

The first step in the procurement process is the planning for purchases and acquisitions, specifically the development of a procurement plan that states what to procure, when, and how. This process includes the following:

- Defining the need for the project
- Development of the procurement statement of work, specifications, and work breakdown structure

---

3. Gregory A. Garrett and Rene G. Rendon, *Contract Management: Organizational Assessment Tools* (Ashburn, VA: National Contract Management Association, 2005), p. 270.
4. See note 3.

- Preparing a WBS dictionary, if necessary
- Performing a make or buy analysis
- Laying out the major milestones and the timing/schedule
- Determining if long lead procurement is necessary
- Cost estimating, including life-cycle costing
- Determining whether qualified sellers exist
- Identifying the source selection criteria
- Preparing a listing of possible project/procurement risks (i.e., a risk register)
- Developing a procurement plan
- Obtaining authorization and approval to proceed

PMBOK® Guide, 4th Edition
12.1.3.2 Procurement Statements of Work

Previously, in Chapter 11, we discussed the statement of work and the scope statement. There could be separate and different statements of work for each product to be procured. The statement of work (SOW) is a *narrative* description of the work to be accomplished and/or the resources to be supplied. The identification of resources to be supplied has taken on paramount importance during the last 10 years or so. During the 1970s and 1980s, small companies were bidding on megajobs only to subcontract out more than 99 percent of all of the work. Lawsuits were abundant and the solution was to put clauses in the SOW requiring that the contractor identify the names and resumes of the talented *internal* resources that would be committed to the project, including the percentage of their time on the project.

In addition to SOWs, organizations also use statements of objectives (SOOs) for projects that are designed as "performance-based" effort. Performance-based projects are now the preference in the federal government. SOOs are used when the procuring organization wants to leverage the advanced technologies, capabilities, and expertise of the potential contractors in the marketplace. Instead of using a SOW, which describes in specific detail to the contractor *what* work needs to be performed and *how* it should be performed, the SOO only describes the end objectives of the project (*what* are the project's end objectives). In response to a solicitation containing a SOO, the potential contractors develop and propose their own SOW that provides the detailed specifics on *how* they intend to perform the work. The source selection process entails comparing the various contractor-developed SOWs, each contractor applying its own unique technologies, capabilities, and expertise to the project effort. The proposal evaluation process includes making trade-offs between differing levels of proposed performance (as reflected in the contractor SOW), as well as proposed price.

Specifications are written, pictorial, or graphic information that describe, define, or specify the services or items to be procured. There are three types of specifications:

- *Design Specifications:* These detail what is to be done in terms of physical characteristics. The risk of performance is on the buyer.
- *Performance Specifications:* These specify measurable capabilities the end product must achieve in terms of operational characteristics. The risk of performance is on the contractor.
- *Functional Specifications:* This is when the seller describes the end use of the item to stimulate competition among commercial items, at a lower overall cost. This is a subset of the performance specification, and the risk of performance is on the contractor.

There are always options in the way the end item can be obtained. Feasible procurement alternatives include make or buy, lease or buy, buy or rent, and lease or rent. Buying domestic or international is also of critical importance, especially to the United Auto Workers Union. Factors involving the make or buy analysis are shown below:

PMBOK® Guide, 4th Edition
12.1.2.1 Make or Buy Analysis

- The make decision
  - Less costly (but not always!!)
  - Easy integration of operations
  - Utilize existing capacity that is idle
  - Maintain direct control
  - Maintain design/production secrecy
  - Avoid unreliable supplier base
  - Stabilize existing workforce
- The buy decision
  - Less costly (but not always!!)
  - Utilize skills of suppliers
  - Small volume requirement (not cost effective to produce)
  - Having limited capacity or capability
  - Augment existing labor force
  - Maintain multiple sources (qualified vendor list)
  - Indirect control

The lease or rent decision is usually a financial endeavor. Leases are usually longer term than renting. Consider the following example. A company is willing to rent you a piece of equipment at a cost of $100 per day. You can lease the equipment for $60 per day plus a one-time cost of $5000. What is the breakeven point, in days, where leasing and renting are the same?

Let X be the number of days.

$$\$100X = \$5000 + \$60X$$

$$\uparrow \qquad\qquad \uparrow$$

Renting   Leasing

Solving, $X = 125$ days

Therefore, if the firm wishes to use this equipment for more than 125 days, it would be more cost effective to sign a lease agreement rather than a rental agreement.

Procurement planning must address the risks on the contract as well as the risks with procurement. Some companies have project management manuals with sections that specifically address procurement risks using templates. As an example, the following is a partial list of procurement risks as identified in the ABB Project Management Manual[5]:

---

5. Adapted from Harold Kerzner, *Advanced Project Management: Best Practices on Implementation,* 2nd ed. (Hoboken, NJ: Wiley, 2004), pp. 346–348.

- Contract and agreements (penalty/liquidated damages, specifications open to misinterpretation, vague wording, permits/licenses, paperwork requirements)
- Responsibility and liability (force majeure, liability limits for each party, unclear scope limitations)
- Financial (letters of credit, payment plans, inflation, currency exchange, bonds)
- Political (political stability, changes in legislation, import/export restrictions, arbitration laws)
- Warranty (nonstandard requirements, repairs)
- Schedule (unrealistic delivery time, work by others not finished on time, approval process, limitations on available resources)
- Technical and technology (nonstandard solutions, quality assurance regulations, inspections, customer acceptance criteria)
- Resources (availability, skill levels, local versus external)

The procurement plan will address the following questions:

- How much procurement will be necessary?
- Will they be standard or specialized procurement activities?
- Will we make some of the products or purchase all of them?
- Will there be qualified suppliers?
- Will we need to prequalify some of the suppliers?
- Will we use open bidding or bidding from a preferred supplier list?
- How will we manage multiple suppliers?
- Are there items that require long lead procurement?
- What type of contract will be used, considering the contractual risks?
- Will we need different contract types for multiple suppliers?
- What evaluation criteria will be used to score the proposals?

19.3 CONDUCTING THE PROCUREMENTS

PMBOK® Guide, 4th Edition
12.2 Conduct Procurements

Once the requirements are identified and a procurement plan has been prepared, a requisition form for each item to be procured is sent to procurement to begin the procurement or requisition process. The process of conducting the procurements includes:

- Evaluating/confirming specifications (are they current?)
- Confirming qualified sources
- Reviewing past performance of sources
- Reviewing of team or partnership agreements
- Producing the solicitation package

The solicitation package is prepared during the procurements planning process but utilized during the next process, conduct procurements. In most situations, the same solicitation package for each deliverable must be sent to each possible supplier so that the playing field is level. A typical solicitation package would include:

- Bid documents (usually standardized)
- Listing of qualified vendors (expected to bid)
- Proposal evaluation criteria (source selection criteria)
- Bidder conferences
- How change requests will be managed
- Supplier payment plan

Standardized bid documents usually include standard forms for compliance with EEO, affirmative action, OSHA/EPA, minority hiring, and so on. A listing of qualified vendors appears in order to drive down the cost. Quite often, one vendor will not bid on the job because it knows that it cannot submit a lower bid than one of the other vendors. The cost of bidding on a job is an expensive process.

PMBOK® Guide, 4th Edition
12.2.2.1 Bidder Conferences

The solicitation package also describes the manner in which solicitation questions will be addressed, namely bidder conferences. Bidder conferences are used so that no single bidder has more knowledge than others. If a potential bidder has a question concerning the solicitation package, then it *must* wait for the bidders' conference to ask the question so that all bidders will be privileged to the same information. This is particularly important in government contracting. There may be several bidders' conferences between solicitation and award. Project management may or may not be involved in the bidders' conferences, either from the customer's side or the contractor's side. Some companies do not use bidder conferences and allow bidders to send in questions. However, the answer to each question is provided to all bidders.

The solicitation package usually provides bidders with information on how the bids will be evaluated. Contracts are not necessarily awarded to the lowest bidders. Some proposal evaluation scoring models assign points in regard to each of the following, and the company with the greatest number of points may be awarded the contract:

- Understanding of the requirements
- Overall bid price
- Technical superiority
- Management capability
- Previous performance (or references)
- Financial strength (ability to stay in business)
- Intellectual property rights
- Production capacity (based upon existing contracts and potential new contracts)

Bidder conferences are also held as part of debriefing sessions whereby the bidders are informed as to why they did not win the contract. Under some circumstances, bidders who feel that their bid or proposal was not evaluated correctly can submit a "bid protest," which may require a detailed reappraisal of their bid. The bid protest is not necessarily a complaint that the wrong company won the contract, but rather a complaint that their proposal was not evaluated correctly.

19.4 CONDUCT PROCUREMENTS: REQUEST SELLER RESPONSES

PMBOK® Guide, 4th Edition
12.2.2.5 Advertising

Selection of the acquisition method is the critical element to request seller responses. There are three common methods for acquisition:

- Advertising
- Negotiation
- Small purchases (i.e., office supplies)

Advertising is when a company goes out for sealed bids. There are no negotiations. Competitive market forces determine the price and the award goes to the lowest bidder.

Negotiation is when the price is determined through a bargaining process. In such a situation, the customer may go out for a:

PMBOK® Guide, 4th Edition
12.2.1.5 Seller Proposals

- Request for information (RFI)
- Request for quotation (RFQ)
- Request for proposal (RFP)
- Invitation for bids (IFB)

The RFP is the most costly endeavor for the seller. Large proposals may contain separate volumes for cost, technical performance, management history, quality, facilities, subcontractor management, and others. Bidders may be hesitant to spend large sums of money bidding on a contract unless the bidder believes that they have a high probability of winning the contract or will be reimbursed by the buyer for all bidding costs.

As mentioned previously, some companies utilize an invitation for bid (IFB) process. Using the IFB process, only selected companies are allowed to bid. Either all or part of the companies on the buyer's preferred contractor list may be allowed to bid.

In government agencies, IFBs are used in sealed bidding procurements. In government sealed bid procurements, the competing offerors submit priced bids in response to IFBs. These IFBs contain all of the necessary technical documents, specifications, and drawings needed for a bidder to develop a priced offer. Thus, in sealed bid procurement, there are no discussions or negotiations, and the contract is always awarded to the lowest acceptable offer using a firm fixed-priced contract.

19.5 CONDUCT PROCUREMENTS: SELECT SELLERS

PMBOK® Guide, 4th Edition
12.2.3.1 Select Sellers

Part of source selection process includes the application of the evaluation criteria to the contractor's proposals. As previously stated, the proposal evaluation criteria were determined, developed, and included in the solicitation during the plan procurement phase of the contracting process. The evaluation criteria reflect the selected contract award strategy, which is typically either a price-based award strategy or best-value award strategy. The priced-based award strategy is used when the contract will be awarded to the lowest priced, technically acceptable proposal.

The best-value award strategy is used when the contract may be awarded to either the lowest priced, technically acceptable offer or a higher-priced proposal offering a higher level of performance. During a best-value source selection, the procuring organization conducts trade-offs among price, performance, and other nonprice factors to select the proposal that offers the overall best value to the buyer.

While several criteria can be used, the most common are time, cost, expected management team of the project (i.e., quality of assigned resources), and previous performance history. As an example, assume that 100 points maximum can be given to each of the four criteria. The seller that is selected would have the greatest number of total points out of 400 points. Weighing factors can also be applied to each of the four criteria. As an example, previous performance may be worth 200 points, thus giving 500 points as a maximum. Therefore, the lowest price supplier may be downgraded significantly because of past performance and not receive the contract.

Selecting the appropriate seller is not necessarily left exclusively to the evaluation criteria. A negotiation process can be part of the selection process because the buyer may like several of the ideas among the many bidders and then may try to have the preferred seller take on added work at no additional cost to the buyer. The negotiation process also includes inclusion and exclusions. The negotiation process can be competitive or noncompetitive. Noncompetitive processes are called sole-source procurement.

On large contracts, the negotiation process goes well beyond negotiation of the bottom line. Separate negotiations can be made on price, quantity, quality, and timing. Vendor relations are critical during contract negotiations. The integrity of the relationship and previous history can shorten the negotiation process. The three major factors of negotiations are:

- Compromise ability
- Adaptability
- Good faith

Negotiations should be planned for. A typical list of activities would include:

- Develop objectives (i.e., min-max positions)
- Evaluate your opponent
- Define your strategy and tactics
- Gather the facts
- Perform a complete price/cost analysis
- Arrange "hygiene" factors

If you are the buyer, what is the *maximum* you will be willing to pay? If you are the seller, what is the *minimum* you are willing to accept? You must determine what motivates your opponent. Is your opponent interested in profitability, keeping people employed, developing a new technology, or using your name as a reference? This knowledge could certainly affect your strategy and tactics.

Hygiene factors include where the negotiations will take place. In a restaurant? Hotel? Office? Square table or round table? Morning or afternoon? Who faces the windows and who faces the walls?

There should be a postnegotiation critique in order to review what was learned. The first type of postnegotiation critique is internal to your firm. The second type of post-negotiation critique is with all of the losing bidders to explain why they did not win the contract.

Once negotiations are completed, each selected seller will receive a signed contract. Unfortunately there are several types of contracts. The negotiation process also includes the selection of the type of contract, and the final type of contract may be different than what was identified in the solicitation package.

Conclusion: The objective of the conduct procurements process is to negotiate a contract type and price that will result in reasonable contractor risk and provide the contractor with the greatest incentive for efficient and economic performance.

There are some basic contractual terms that should be understood before looking at the various contracts. These include:

- **Agent:** The person or group of people officially authorized to make decisions and represent their firm. This includes signing the contract.
- **Arbitration:** The settling of a dispute by a third party who renders a decision. The third party is not a court of law, and the decision may or may not be legally binding.
- **Breach of Contract:** To violate a law by an act or omission or to break a legal obligation.
- **Contract:** An agreement entered into by two or more parties and the agreement can be enforced in a court of law.
- **Executed Contract:** A contract that has been completed by all concerned parties.
- **Force Majeure Clause:** A provision in a contract that excuses the parties involved from any liability or contractual obligations because of acts of God, wars, terrorism, or other such events.
- **Good Faith:** Honesty and fair dealings between all parties involved in the contract.
- **Infringement:** A violation of someone's legally recognized right.
- **Liquidated Damages:** An amount specified in a contract that stipulates the reasonable estimation of damages that will occur as a result of a breach of contract.
- **Negligence:** The failure to exercise one's activity in such a manner that a reasonable person would do in a similar situation.
- **Noncompete Clause:** A covenant providing restrictions on starting up a competing business or working for a competitor within a specified time period.
- **Nondisclosure Clause:** A covenant providing restrictions on certain proprietary information such that it cannot be disclosed without written permission.
- **Nonconformance:** Performance of work in such a manner that it does not conform to contractual specifications or requirements.
- **Penalty Clause:** An agreement or covenant, identified in financial terms, for failure to perform.
- **Privity of Contract:** The relationship that exists between the buyer and seller of a contract.

- **Termination or Termination Liability:** An agreement between the buyer and seller as to how much money the seller will receive should the project be terminated prior to the scheduled completion date and without all of the contractual deliverable being completed.
- **Truth in Negotiations:** This clause in the contract states that both the buyer and seller have been truthful in the information provided during contract negotiations.
- **Waiver:** An intentional relinquishment of a legal right.
- **Warranty:** A promise, either verbal or written, that certain facts are true as represented.

There are certain basic elements of most contracts:

- **Mutual Agreement:** There must be an offer and acceptance.
- **Consideration:** There must be a down payment.
- **Contract Capability:** The contract is binding only if the contractor has the capability to perform the work.
- **Legal Purpose:** The contract must be for a legal purpose.
- **Form Provided by Law:** The contract must reflect the contractor's legal obligation, or lack of obligation, to deliver end products.

The two most common contract forms are completion contracts and term contracts.

- **Completion Contract:** The contractor is required to deliver a definitive end product. Upon delivery and formal acceptance by the customer, the contract is considered complete, and final payment can be made.
- **Term Contract:** The contract is required to deliver a specific "level of effort," not an end product. The effort is expressed in woman/man-days (months or years) over a specific period of time using specified personnel skill levels and facilities. When the contracted effort is performed, the contractor is under no further obligation. Final payment is made, irrespective of what is actually accomplished technically.

The final contract is usually referred to as a *definitive* contract, which follows normal contracting procedures such as the negotiation of all contractual terms, condition cost, and schedule prior to initiation of performance. Unfortunately, negotiating the contract and preparing it for signatures may require months of preparation. If the customer needs the work to begin immediately or if long-lead procurement is necessary, then the customer may provide the contractor with a *letter contract* or *letter of intent*. The letter contract is a preliminary written instrument authorizing the contractor to begin immediately the manufacture of supplies or the performance of services. The final contract price *may* be negotiated after performance begins, but the contractor may not exceed the "not to exceed" face value of the contract. The definitive contract must still be negotiated.

The type of contract selected is based upon the following:

- Overall degree of cost and schedule risk
- Type and complexity of requirement (technical risk)
- Extent of price competition

- Cost/price analysis
- Urgency of the requirements
- Performance period
- Contractor's responsibility (and risk)
- Contractor's accounting system (is it capable of earned value reporting?)
- Concurrent contracts (will my contract take a back seat to existing work?)
- Extent of subcontracting (how much work will the contractor outsource?)

19.6 TYPES OF CONTRACTS

PMBOK® Guide, 4th Edition
12.1.2.3 Contract Types

Before analyzing the various types of contracts, one should be familiar with the terminology found in them.

- The *target cost* or *estimated cost* is the level of cost that the contractor will most likely obtain under normal performance conditions. The target cost serves as a basis for measuring the true cost at the end of production or development. The target cost may vary for different types of contracts even though the contract objectives are the same. The target cost is the most important variable affecting research and development.
- *Target* or *expected profit* is the profit value that is negotiated for, and set forth, in the contract. The expected profit is usually the largest portion of the total profit.
- *Profit ceiling* and *profit floor* are the maximum and minimum values, respectively, of the total profit. These quantities are often included in contract negotiations.
- *Price ceiling* or *ceiling price* is the amount of money for which the government is responsible. It is usually measured as a given percentage of the target cost, and is generally greater than the target cost.
- *Maximum* and *minimum fees* are percentages of the target cost and establish the outside limits of the contractor's profit.
- The *sharing arrangement* or *formula* gives the cost responsibility of the customer to the cost responsibility of the contractor for each dollar spent. Whether that dollar is an overrun or an underrun dollar, the sharing arrangement has the same impact on the contractor. This sharing arrangement may vary depending on whether the contractor is operating above or below target costs. The *production point* is usually that level of production above which the sharing arrangement commences.
- *Point of total assumption* is the point (cost or price) where the contractor assumes all liability for additional costs.

Because no single form of contract agreement fits every situation or project, companies normally perform work in the United States under a wide variety of contractual arrangements, such as:

- Cost-plus percentage fee
- Cost-plus fixed fee
- Cost-plus guaranteed maximum

- Cost-plus guaranteed maximum and shared savings
- Cost-plus incentive (award fee)
- Cost and cost sharing
- Fixed price or lump sum
- Fixed price with redetermination
- Fixed price incentive fee
- Fixed price with economic price adjustment
- Fixed price incentive with successive targets
- Fixed price for services, material, and labor at cost (purchase orders, blanket agreements)
- Time and material/labor hours only
- Bonus-penalty
- Combinations
- Joint venture

At one end of the range is the *cost-plus,* a fixed-fee type of contract where the company's profit, rather than price, is fixed and the company's responsibility, except for its own negligence, is minimal. At the other end of the range is the *lump sum* or *turnkey* type of contract under which the company has assumed full responsibility, in the form of profit or losses, for timely performance and for all costs under or over the fixed contract price. In between are various types of contracts, such as the guaranteed maximum, incentive types of contracts, and the bonus-penalty type of contract. These contracts provide for varying degrees of cost responsibility and profit depending on the level of performance. Contracts that cover the furnishing of consulting services are generally on a per diem basis at one end of the range and on a fixed-price basis at the other end of the range.

There are generally five types of contracts to consider: fixed-price (FP), cost-plus-fixed-fee (CPFF), or cost-plus-percentage-fee (CPPF), guaranteed maximum-shared savings (GMSS), fixed-price-incentive-fee (FPIF), and cost-plus-incentive-fee (CPIF) contracts. Each type is discussed separately.

- Under a *fixed-price* or *lump-sum contract,* the contractor must carefully estimate the target cost. The contractor is required to perform the work at the negotiated contract value. If the estimated target cost was low, the total profit is reduced and may even vanish. The contractor may not be able to underbid the competitors if the expected cost is overestimated. Thus the contractor assumes a large risk.

   This contract provides maximum protection to the owner for the ultimate cost of the project, but has the disadvantage of requiring a long period for preparation and adjudications of bids. Also, there is the possibility that, because of a lack of knowledge of local conditions, all contractors may necessarily include an excessive amount of contingency. This form of contract should never be considered by the owner unless, at the time bid invitations are issued, the building requirements are known exactly. Changes requested by the owner after award of a contract on a lump sum basis lead to troublesome and sometimes costly extras.

- Traditionally, the *cost-plus-fixed-fee* contract has been employed when it was believed that accurate pricing could not be achieved any other way. In the CPFF contract, the cost may vary but the fee remains firm. Because, in a cost-plus contract, the contractor agrees only to use his best efforts to perform the work, good performance and poor performance are, in effect, rewarded equally. The total dollar profit tends to produce low rates of return, reflecting the small amount of risk that the contractor assumes. The fixed fee is usually a small percentage of the total or true cost. The cost-plus contract requires that the company books be audited.

With this form of contract the engineering-construction contractor bids a fixed dollar fee or profit for the services to be supplied by the contractor, with engineering, materials, and field labor costs to be reimbursed at actual cost. This form of bid can be prepared quickly at a minimal expense to contractor and is a simple bid for the owner to evaluate. Additionally, it has the advantage of establishing incentive to the contractor for quick completion of the job.

If it is a *cost-plus-percentage-fee* contract, it provides maximum flexibility to the owner and permits owner and contractor to work together cooperatively on all technical, commercial, and financial problems. However, it does not provide financial assurance of ultimate cost. Higher building cost may result, although not necessarily so, because of lack of financial incentive to the contractor compared with other forms. The only meaningful incentive that is evident today is the increased competition and prospects for follow-on contracts.

- Under the *guaranteed maximum-share savings* contract, the contractor is paid a fixed fee for his profit and reimbursed for the actual cost of engineering, materials, construction labor, and all other job costs, but only up to the ceiling figure established as the "guaranteed maximum." Savings below the guaranteed maximum are shared between owner and contractor, whereas contractor assumes the responsibility for any overrun beyond the guaranteed maximum price.

This contract form essentially combines the advantages as well as a few of the disadvantages of both lump sum and cost-plus contracts. This is the best form for a negotiated contract because it establishes a maximum price at the earliest possible date and protects the owner against being overcharged, even though the contract is awarded without competitive tenders. The guaranteed maximum-share savings contract is unique in that the owner and contractor share the financial risk and both have a real incentive to complete the project at lowest possible cost.

- *Fixed-price-incentive-fee* contracts are the same as fixed-price contracts except that they have a provision for adjustment of the total profit by a formula that depends on the final total cost at completion of the project and that has been agreed to in advance by both the owner and the contractor. To use this type of contract, the project or contract requirements must be firmly established. This contract provides an incentive to the contractor to reduce costs and therefore increase profit. Both the owner and contractor share in the risk and savings.

- *Cost-plus-incentive-fee* contracts are the same as cost-plus contracts except that they have a provision for adjustment of the fee as determined by a formula that compares the total project costs to the target cost. This formula is agreed to in advance by both the owner and contractor. This contract is usually used for long-duration or R&D-type projects. The company places more risk on the contractor and forces him to plan ahead carefully and strive to keep costs down. Incentive contracts are covered in greater detail in Section 19.7.

- Another type of contract incentives are *award fees*. Whereas incentive fees are *objectively* determined, that is, based on objective calculations comparing actual cost to target costs, actual delivery to target delivery, or actual performance to target performance, award fees are more *subjectively* determined. Award fees are used when it is not feasible or effective to determine objective contract incentives. Award fees are earned when the contractor meets higher (over and above the basic requirements of the contract) levels of performance, quality, timeliness, or responsiveness in performing the contract effort. Award fee contracts include an award fee plan that explains the award fee evaluation criteria for any given time period (typically one year), as well as the total dollar amount of the award fee pool. Typically a contract award fee evaluation board convenes at the end of each award fee period to evaluate the contractor's performance in relation to the award fee criteria established in the award fee plan. The award fee determination official, either the project manager or a level above the project manager, makes the actual determination on the amount of award fee earned by the contractor for that specific period. Award fee provisions can be part of cost or fixed-priced contracts.

  For major services contracts, award term incentives are used as incentives for the contractor to achieve higher levels of performance, quality, timeliness, or responsiveness in performing the services contract effort. Award term is similar to award fee, but instead of awarding the successful contractor additional dollars (fee), the contractor earns additional time (contract performance periods) on the service contract. Thus, instead of ending the final contract period of performance, and then having to recompete for the follow-on contract, the successful contractor is awarded with time extensions (additional periods of performance) to the contract performance period.

Other types of contracts that are not used frequently include:

- The *fixed-price incentive successive targets* contract is an infrequently used contract type. It has been used in the past in acquiring systems with very long lead time requirements where follow-on production contracts must be awarded before design or even production confirmation costs have been confirmed. Pricing data for the follow-on contract is inconclusive. This type of contract can be used in lieu of a letter contract or cost-plus arrangement.
- The *fixed-price with redetermination* contract can be either prospective or retroactive. The prospective type allows for future negotiations of two or more firm, fixed-price contracts at prearranged times. This is often used when future costs and pricing are expected to change significantly. The retroactive FPR contract allows for adjusting contract price after performance has been completed.
- *Cost* (CR) and *cost-sharing* (CS) contracts have limited use. Cost contracts have a "no fee" feature that has limited use except for nonprofit educational institutions conducting research. Cost-sharing contracts are used for basic and applied research where the contractor is expected to benefit from the R&D by transferring knowledge to other parts of the business for commercial gain and to improve the contractor's competitive position.

Table 19–1 identifies the advantages and disadvantages of various contracting methods that are commonly used.

The type of contract that is acceptable to the client and the company is determined by the circumstances of each individual project and the prevailing economic and competitive conditions. Generally, when work is hard to find, clients insist on fixed-price bids. This type of proposal is usually a burden to the contractor because of the proposal costs involved (about 1 percent of the total installed cost of the project), and the higher risk involved in the execution of the project on such a basis.

When there is an upsurge in business, clients are unable to insist on fixed-price bids and more work is awarded on a cost-plus basis. In fact, where a special capacity position exists, or where time is a factor, the client occasionally negotiates a cost-plus contract with only one contractor. Another technique is to award a project on a cost-plus basis with the understanding that the contract will be converted at a later date, when the scope has been better defined and unknowns identified, to another form, such as a lump sum for services. This approach is appealing to both the client and the contractor.

As we mentioned earlier, the client frequently has a standard form of contract that is used as the basis of negotiation or the basis of requests for proposals. A company should review the client's document carefully to assure itself that it understands how the client's document differs from what is its preferred position. Any additional duties or responsibilities assigned to your company merit careful scrutiny if the additional legal consequences and increased financial risks are to be evaluated properly.

It is important that you use an adequate and realistic description of the work to be undertaken and a careful evaluation and pricing of the scope of the work to be performed and the responsibilities and obligations assumed. The preparation of a proposal requires a clear understanding between the client and your company as to the rights, duties, and responsibilities of your company. The proposal defines what it intends to do and can do, what it neither intends doing nor is qualified to undertake, and the manner and basis of its compensation. Thorough analysis of these matters before, not after, submission of the proposal is essential.

19.7 INCENTIVE CONTRACTS

PMBOK® Guide, 4th Edition
12.1.2.3 Contract Types

To alleviate some of the previously mentioned problem areas, clients, especially the government, have been placing incentive objectives into their contracts. The fixed-price-incentive-fee (FPIF) contract is an example of this. The essence of the incentive contract is that it offers a contractor more profit if costs are reduced or performance is improved and less profit if costs are raised or if performance goals are not met. Cost incentives take the form of a sharing formula generally expressed as a ratio. For example, if a 90/10 formula were negotiated, the government would pay for 90 cents and the contractor 10 cents for every dollar above the target cost. Thus it benefits both the contractor and the government to reduce costs, because the contractor must consider that 10 percent of every dollar must be spent by the company. Expected profits can thus be increased by making maximum use of the contractor's managerial skills.

TABLE 19–1. CONTRACT COMPARISON

Contract Type	Advantages	Disadvantages
Cost-plus-fee	• Provides maximum flexibility to owner • Minimizes contractor profits • Minimizes negotiations and preliminary specification costs • Permits quicker start, earlier completion • Permits choice of best-qualified, not lowest-bidding, contractor • Permits use of same contractor from consultation to completion, usually increasing quality and efficiency	• No assurance of actual final cost • No financial incentive to minimize time and cost • Permits specification of high-cost features by owner's staff • Permits excessive design changes by owner's staff increasing time and costs
Guaranteed maximum-share savings	• Provides firm assurance of ultimate cost at earliest possible date • Insures prompt advice to owner of delays and extra costs resulting from changes • Provides incentive for quickest completion • Owner and contractor share financial risk and have mutual incentive for possible savings • Ideal contract to establish owner–contractor cooperation throughout execution of project	• Requires complete auditing by owner's staff • Requires completion of definitive engineering before negotiation of contract
Fixed price/lump sum	• Provides firm assurance of ultimate cost • Insures prompt advice to owner of delays and extra costs resulting from changes • Requires minimum owner follow-up on work • Provides maximum incentive for quickest completion at lowest cost • Involves minimal auditing by owner's staff	• Requires exact knowledge if what is wanted before contract award • Requires substantial time and cost to develop inquiry specs, solicit, and evaluate bids. Delays completion 3–4 months • High bidding costs and risks may reduce qualified bidders • Cost may be increased by excessive contingencies in bids to cover high-risk work
Fixed price for services, material, and labor	• Essentially same as cost-plus-fee contract • Fixes slightly higher percentage of total cost • Eliminates checking and verifying contractor's services	• May encourage reduction of economic studies and detailing of drawings: produce higher costs for operation, construction, maintenance • Other disadvantages same as cost-plus-fee contract
Fixed price for imported goods and services, local costs reimbursable	• Maximum price assured for high percentage of plant costs • Avoids excessive contingencies in bids for unpredictable and highly variable local costs • Permits selection of local suppliers and subcontractors by owner	• Same extended time required for inquiry specs, quotations, and evaluation as fixed lump-sum for complete project • Requires careful definition of items supplied locally to insure comparable bids • No financial incentive to minimize field and local costs

In the FPIF contract, the contractor agrees to perform a service at a given fixed cost. If the total cost is less than the target cost, then the contractor has made a profit according to the incentive-fee formula. If the total cost exceeds the target cost, then the contractor loses money.

Consider the following example, which appears in Figure 19–1. The contractor has a target cost and target profit. However, there is a price ceiling of $11,500, which is the maximum price that the contractor will be paid. If the contractor performs the work below the target cost of $10,000, then additional profit will be made. For example, by performing the work for $9,000, the contractor will receive a profit of $1,150, which is the target profit of $850 plus $300 for 30% of the underrun. The contractor will receive a total price of $10,150.

If the cost exceeds the target cost, then the contractor must pay 30% of the overrun out of the contractor's profits. However, the fixed-price-incentive-fee (FPIF) contract has a point of total assumption. In this example, the point of total assumption is the point where all additional costs are burdened by the contractor. From Figure 19–1, the point of total assumption is when the cost reaches $10,928. At this point, the final price of $11,500 is reached. If the cost continues to increase, then all profits may disappear and the contractor may be forced to pay the majority of the overrun.

When the contract is completed, the contractor submits a statement of costs incurred in the performance of the contract. The costs are audited to determine allowability and

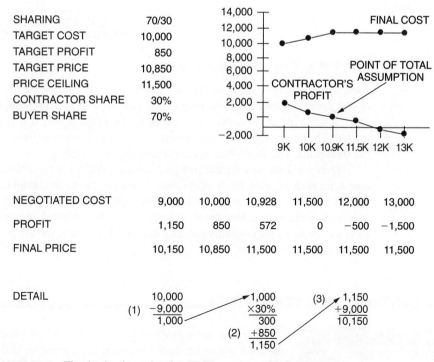

SHARING	70/30
TARGET COST	10,000
TARGET PROFIT	850
TARGET PRICE	10,850
PRICE CEILING	11,500
CONTRACTOR SHARE	30%
BUYER SHARE	70%

NEGOTIATED COST	9,000	10,000	10,928	11,500	12,000	13,000
PROFIT	1,150	850	572	0	−500	−1,500
FINAL PRICE	10,150	10,850	11,500	11,500	11,500	11,500

DETAIL

$$
\begin{array}{ll}
(1) & \begin{array}{r} 10,000 \\ -9,000 \\ \hline 1,000 \end{array}
\end{array}
\qquad
\begin{array}{r} 1,000 \\ \times 30\% \\ \hline 300 \end{array}
\qquad
(2)\ \begin{array}{r} +850 \\ \hline 1,150 \end{array}
\qquad
(3)\ \begin{array}{r} 1,150 \\ +9,000 \\ \hline 10,150 \end{array}
$$

FIGURE 19–1. Fixed-price-incentive-fee (FPIF) contract with firm target.

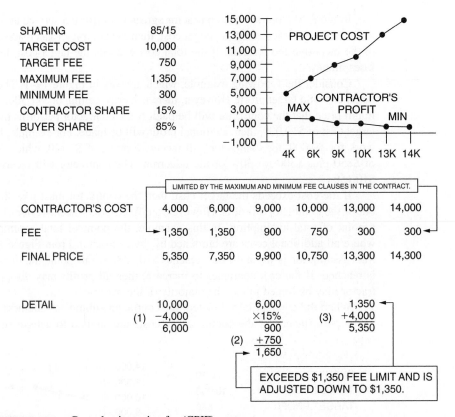

SHARING	85/15
TARGET COST	10,000
TARGET FEE	750
MAXIMUM FEE	1,350
MINIMUM FEE	300
CONTRACTOR SHARE	15%
BUYER SHARE	85%

LIMITED BY THE MAXIMUM AND MINIMUM FEE CLAUSES IN THE CONTRACT.

CONTRACTOR'S COST	4,000	6,000	9,000	10,000	13,000	14,000
FEE	1,350	1,350	900	750	300	300
FINAL PRICE	5,350	7,350	9,900	10,750	13,300	14,300

DETAIL

(1)
$$\begin{array}{r} 10,000 \\ -4,000 \\ \hline 6,000 \end{array}$$

$$\begin{array}{r} 6,000 \\ \times 15\% \\ \hline 900 \end{array}$$

(2)
$$\begin{array}{r} +750 \\ \hline 1,650 \end{array}$$

(3)
$$\begin{array}{r} 1,350 \\ +4,000 \\ \hline 5,350 \end{array}$$

EXCEEDS $1,350 FEE LIMIT AND IS ADJUSTED DOWN TO $1,350.

FIGURE 19–2.   Cost-plus-incentive-fee (CPIF) contract.

questionable charges are removed. This determines the negotiated cost. The negotiated cost is then subtracted from the target cost. This number is then multiplied by the sharing ratio. If the number is positive, it is added to the target profit. If it is negative, it is subtracted. The new number, the final profit, is then added to the negotiated cost to determine the final price. The final price never exceeds the price ceiling.

Figure 19–2 shows a typical cost-plus-incentive-fee (CPIF) contract. In this contract, the contractor is reimbursed 100% of the costs. However, there is a maximum fee (i.e., profit) of $1,350 and a minimum fee of $300. The final allowable profit will vary between the minimum and maximum fee. Because there appears more financial risk for the customer in a CPIF contract, the target fee is usually less than in an FPIF contract, and the contractor's portion of the sharing ratio is smaller.

19.8 CONTRACT TYPE VERSUS RISK

PMBOK® Guide, 4th Edition
12.1.1.5 Risk-Related Contract Decisions

The amount of profit on a contract is most frequently based upon how the risks are to be shared between the contractor and the customer. For example, on a firm-fixed-price contract, the contractor absorbs 100 percent

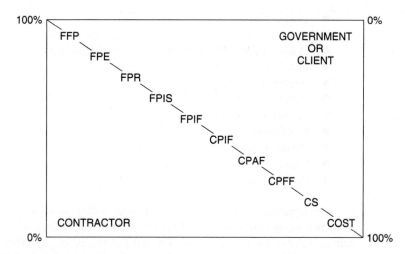

FIGURE 19–3. Contract types and risk types.

of the risks (especially financial) and expects to receive a larger profit than on other types of contracts. On cost, cost-plus, and cost-sharing contracts, the customer absorbs up to 100 percent of the risks and expects the contractor to work for a lower than expected profit margin or perhaps no profit at all.

All other types of contracts may have a risk sharing formula between the customer and the contractor. Figure 19–3 shows the relative degree of risk between the customer and the contractor for a variety of contracts.

19.9 CONTRACT ADMINISTRATION CYCLE

PMBOK® Guide, 4th Edition
12.3 Contract Administration

The contract administrator is responsible for compliance by the seller to the buyer's contractual terms and conditions and to make sure that the final product is fit for use. Contract administrators can shut down a manufacturing plant by allowing the seller to make late deliveries. Although a contract administrator is a member of the project team for project reporting purposes (dotted line reporting), the contract administrator can report to a line function such as corporate legal and may even be an attorney. The functions of the corporate administrator include:

- Change management
- Specification interpretation
- Adherence to quality requirements
- Inspections and audits
- Warranties
- Performance reporting
- Records management

- Contractor (seller) management
- Contractor (seller) performance report card
- Documenting seller's performance (for future source selection teams)
- Production surveillance
- Approval of waivers
- Breach of contract
- Claims administration
- Resolution of disputes
- Payment schedules
- Project termination
- Project closure

The larger the contract, the greater the need for the contract administrator to resolve ambiguity in the contract. Sometimes, large contracts that are prepared by teams of attorneys contain an *order of precedence* clause. The order of precedence specifies that any inconsistency in the solicitation of the contract shall be resolved in a given order of procedure such as:

A. Specifications (first priority)
B. Other instructions (second priority)
C. Other documents, such as exhibits, attachments, appendices, SOW, contract data requirements list (CDRL), etc. (third priority)
D. Contract clauses (fourth priority)
E. The schedule (fifth priority)

Generally speaking, an ambiguous contract will be interpreted against the party who drafted the document. However, there is an offsetting rule called *Patent Ambiguity*. This includes the following:

- The offeror in a "bid" situation is expected to be knowledgeable about ordinary and normal industrial or construction practices pertinent to its work.
- The presumption is made that the offeror has made reasonable and complete review of the contractual documents before preparing and submitting them.
- Failure to notify of patent ambiguity works against the offeror if the claim is later submitted based on ambiguity.

Perhaps the majority of the contract administrator's time is spent handling changes. The following definitions describe the types of changes:

- *Administrative change:* A unilateral contractual change, in writing, that does not affect the substantive rights of the parties (i.e., a change in the paying office or the appropriation funding).
- *Change order:* A written order, signed by the contracting officer, directing the contractor to make a change.
- *Contract modification:* Any written change in the terms of the contract.

- *Undefinitized contractual action:* Any contractual action that authorizes the commencement of work prior to the establishment of a final definitive price.
- *Supplemental agreement:* A contract modification that is accompanied by the mutual action of both parties.
- *Constructive change:* Any effective change to the contract caused by the actions or inaction of personnel in authority, or by circumstances that cause a contractor to perform work differently than required by written contract. The contractor may file a claim for equitable adjustment in the contract.

Typical causes of constructive changes include:

- Defective specification with impossibility of performance
- Erroneous interpretation of contract
- Overinspection of work
- Failure to disclose superior knowledge
- Acceleration of performance
- Late or unsuitable owner or customer furnished property
- Failure to cooperate
- Improperly exercised options
- Misusing proprietary data

Based on the type of contract, terms, and conditions, the customer may have the right to terminate a contract for convenience at any time. However, the customer must compensate the contractor for his preparations and for any completed and accepted work relating to the terminated part of the contract.

The following are reasons for termination for convenience of the customer:

- Elimination of the requirement
- Technological advances in the state-of-the-art
- Budgetary changes
- Related requirements and/or procurements
- Anticipating profits not allowed

The following are reasons for termination for default due to contractor's actions:

- Contractor fails to make delivery on scheduled date.
- Contractor fails to make progress so as to endanger performance of the contract and its terms.
- Contractor fails to perform any other provisions of the contract.

If a contract is terminated due to default, then the contractor may not be entitled to compensation of work in progress but not yet accepted by the customer. The customer may even be entitled to repayment from the contractor of any advances or progress payments applicable to such work. Also, the contractor may be liable for any excess reprocurement costs. However, contractors can seek relief through negotiations, a Board of Contracts Appeals, or Claims Court.

The contract administrator is responsible for performance control. This includes inspection, acceptance, and breach of contract/default. If the goods/services do not comply with the contract, then the contract administrator has the right to:

● Reject the entire shipment
● Accept the entire shipment (barring latent defects)
● Accept part of the shipment

In government contracts, the government has the right to have the goods repaired with the costs charged back to the supplier or fix the goods themselves and charge the cost of repairs to the supplier. If the goods are then acceptable to the government, then the government may reduce the contract amount by an appropriate amount to reflect the reduced value of the contract.

Project managers often do financial closeout once the goods are shipped to the customer. This poses a problem if the goods must be repaired. Billing the cost of repairs against a financially closed out project is called *backcharging*. Most companies do not perform financial closeout until at least 90 days after delivery of goods.

19.10 CONTRACT CLOSURE

The contract administrator is responsible for verification that all of the work performed and deliverables produced are acceptable to the buyer. Contractual closure is then followed up with administrative closure, which includes:

● Documented verification that the output was accepted by the buyer
● Debriefing the seller on their overall performance
● Documenting seller's performance (documentation will be used in future source selections when evaluating contractor's past performance)
● Identifying room for improvement on future contracts
● Archiving all necessary project documentation
● Performing a lessons-learned review
● Identifying best practices

The seller also performs administrative closure once contractual closure is recognized. For the seller, an important subset of administrative closure is financial closure, which is the closing out of all open charge numbers. If financial closure occurs before contractual closure, then the project manager runs the risk that the charge numbers may have to be reopened to account for the cost of repairs or defects. This is referred to as back-charging. Back-charging can be a monumental headache for the project manager especially if the accounting group identified the unused money in the code of accounts as excess profits.

19.11 USING A CHECKLIST

To assist a company in evaluating inquiries and preparing proposals and contracts, a checklist of contract considerations and provisions can be helpful in the evaluation of each proposal and form of contract to insure that appropriate safeguards are incorporated. This checklist is also used for sales letters and brochures that may promise or represent a commercial commitment. Its primary purpose is to remind users of the legal and commercial factors that should be considered in preparing proposals and contracts. Table 19–2 shows the typical major headings that would be considered in a checklist. A key word concept also provides an excellent checklist of the key issues to be considered. It will be useful as a reminder in preparation for contractor-client agreement discussions.

The following contract provisions will minimize risk, and should be included in proposals and contracts:

- Scope of services and description of project
- Contract administration
- Terms of payment
- Client obligation and supplied items
- Warranties and guarantees
- Liability limitation and consequential damages
- Indemnity
- Taxes
- Patent indemnification
- Confidential information
- Termination provisions
- Changes and extras
- Assignments
- Delays, including *force majeure*
- Insurance requirements

TABLE 19–2. TYPICAL MAIN HEADING FOR A CONTRACT PROVISIONS CHECKLIST

I.	Definitions of contract terms
II.	Definition of project scope
III.	Scope of services and work to be performed
IV.	Facilities to be furnished by client (for service company use)
V.	Changes and extras
VI.	Warranties and guarantees
VII.	Compensation to service company
VIII.	Terms of payment
IX.	Definition of fee base (cost of the project)
X.	State sales and/or use taxes
XI.	Taxes (other than sales use taxes)
XII.	Insurance coverages
XIII.	Other contractual provisions (including certain general provisions)
XIV.	Miscellaneous general provisions

- Arbitration
- Escalation (lump sum)
- Time of completion

Because of the variations among proposals and contracts, it is not feasible to prepare material specifically suited for each situation. It is also not practical to establish a standard form of contract or standard provisions to be included in a contract.

However, an increasing number of clients have certain set ideas as to the content of the proposal and contract. Therefore, it would be extremely helpful to develop a standard list and file of draft contract clauses that could be used with some modification for each bid. In addition, because clients occasionally ask for a "typical" contract, the draft clauses can be combined into a "typical" or "draft" contract that can be given to a client. Even though this "typical" contract agreement may not be sufficient for every situation, it can be a starting place. It would also be valuable to maintain a summary of commercially oriented company policies for reference in reviewing a client's contract provisions.

Negotiating for the type of contract is a two-way street. The contractor desires a certain type of contract to reduce risk. The client desires a certain type of contract to reduce costs. Often the client and contractor disagree. It is not uncommon in industry for prospective projects to be canceled because of lack of funds, disagreements in contract negotiations, or changing of priorities.

19.12 PROPOSAL-CONTRACTUAL INTERACTION

It is critical during the proposal preparation stage that contract terms and conditions be reviewed and approved before submission of a proposal to the client. The contracts (legal) representative is responsible for the preparation of the contract portion of the proposal. Generally, contracts with the legal department are handled through or in coordination with the proposal group. The contract representative determines or assists with the following:

- Type of contract
- Required terms and conditions
- Any special requirements
- Cash-flow requirements
- Patent and proprietary data
- Insurance and tax considerations
- Finance and accounting

The sales department, through the proposal group, has the final responsibility for the content and outcome of all proposals and contracts that it handles. However, there are certain aspects that should be reviewed with others who can offer guidance, advice, and assistance to facilitate the effort. In general, contract agreements should be reviewed by the following departments:

- Proposal
- Legal
- Insurance
- Tax
- Project management
- Engineering
- Estimating
- Construction (if required)
- Purchasing (if required)

Responsibility for collecting and editing contract comments rests with the proposal manager. In preparing contract comments, consideration should be given to comments previously submitted to the client for the same form of agreement, and also previous agreements signed with the client.

Contract comments should be reviewed for their substance and ultimate risk to the company. It must be recognized that in most instances, the client is not willing to make a large number of revisions to his proposed form of agreement. The burden of proof that a contract change is required rests with the company; therefore each comment submitted must have a good case behind it.

Occasionally, a company is confronted with a serious contract comment for which it is very difficult to express their position. In such instances, it is better to flag the item for further discussion with the client at the conference table. A good example of this is taxes on cost-plus foreign projects. Normally, when submitting a proposal for such work, a company does not have sufficient definitive information to establish its position relative to how it would like to handle taxes; that is:

- What is the client's position on taxes?
- Will one or two agreements be used for the work? Who will the contracting parties be?
- Time will not permit nor is the cost justifiable for a complete tax assessment.
- Contract procedures have not been established. Would we buy in the name of the company or as agents without liability for the client?

The legal department should be advised of information pertinent to its functions as promptly as possible as negotiations develop. Proposal personnel should also be familiar with the standard contract forms the company uses, its contract terms, and available conditions, including those developed jointly between sales and the legal department, as well as the functions, duties, and responsibilities of the legal department. In addition, key areas that are normally negotiated should be discussed so that proposal personnel have a better understanding of the commercial risks involved and why the company has certain positions.

By the time the client has reviewed the proposal, the company's legal position is fixed commercially if not legally. Therefore, sales and proposal personnel should understand and be prepared to put forward the company's position on commercially significant legal considerations, both in general and on specific issues that arise in connection with a particular project. In this way, sales will be in a position to assert, and sell, the company's position at the appropriate time.

Proposals should send all bid documents, including the client's form of contract, or equivalent information, along with the proposal outline or instructions to the legal department upon receipt of documents from the client. The instructions or outline should indicate the assignment of responsibility and include background information on matters that are pertinent to sales strategy or specific problems such as guarantees, previous experience with client, and so on.

Proposals should discuss briefly with the legal department what is planned by way of the project, the sales effort, and commercial considerations. If there is a kickoff meeting, a representative of the legal department should attend if it is appropriate. The legal department should make a preliminary review of the documents before any such discussion or meeting.

The legal department reviews the documents and prepares a memorandum of comment and any required contract documents, obtaining input where necessary or advisable. If the client has included a contract agreement with the inquiry, the legal department reviews it to see if it has any flaws or is against some set policy of the company. Unless a lesser level of effort is agreed upon, this memorandum will cover all legal issues. This does not necessarily mean that all such issues must be raised with the client.

The purpose of the memorandum is to alert the proposal department to the issues and suggest solutions, usually in the form of contract comments. The memo may make related appropriate commercial suggestions. If required, the legal department will submit a proposed form of contract, joint venture agreement, and so on. Generally, the legal department follows standards that have been worked out with sales and uses standard forms and contract language that were found to be salable in the past and to offer sufficient protection.

At the same time, proposals reviews the documents and advises the legal department of any pertinent issues known by or determined by proposals. This is essential not only because proposals has the final responsibility but also because proposals is responsible for providing information to, and getting comments from, others, such as purchasing, engineering, and estimating.

Proposals reviews and arranges for any other review of the legal department's comments and documents and suggests the final form of comments, contract documents, and other relevant documents including the offer letter. Proposals reviews proposed final forms with the legal department as promptly as possible and prior to any commercial commitment.

Normal practice is to validate proposals for a period of thirty to sixty days following date of submission. Validation of proposals for periods in excess of this period may be required by special circumstances and should be done only with management's concurrence. Occasionally, it is desirable to validate a bid for fewer than thirty days. The validity period is especially important on lump sum bids. On such bids, the validity period must be consistent with validity times of quotations received for major equipment items. If these are not consistent, additional escalation on equipment and materials may have to be included in the lump sum price, and the company's competitive position could thereby be jeopardized.

Occasionally, you may be requested to submit with your proposals a schedule covering hourly rate ranges to reimbursable personnel. For this purpose, you should develop a standard schedule covering hourly rate ranges and average rates for all personnel in the reimbursable category. The hourly rate ranges are based on the lowest-paid person and the highest-paid person in any specific job classification. In this connection, if there are any oddball situations, the effect of such is not included. Average rates are based on the average of all personnel in any given job classification.

One area that is critical to the development of a good contract is the definition of the scope of work covered by the contract. This is of particular importance to the proposal manager, who is responsible for having the proper people prepared for the scope of work description. What is prepared during proposal production most likely governs the contract preparation and eventually becomes part of that contract. The degree to which the project scope of work must be described in a contract depends on the pricing mechanism and contract form used.

A contract priced on a straight per diem basis or on the basis of reimbursement of all costs plus a fee does not normally require a precise description of either the services to be performed or the work to be accomplished.

Usually, a general description is adequate. This, however, is not the case if the contract is priced by other methods, especially fixed price, cost sharing, or guaranteed maximum. For these forms of contracts, it is essential that considerable care be taken to set forth in the contract documents the precise nature of the work to be accomplished as well as the services to be performed.

In the absence of a detailed description of the work prepared by the client, you must be prepared to develop such a description for inclusion in your proposal. When preparing the description of the work for inclusion in the contract documents, the basic premise to be followed must be that the language in the contract will be strictly interpreted during various stages of performance. The proper preparation of the description of the work as well as the evaluation of the requirements demands coordination among sales, administration, cost, and technical personnel both inside and outside the organization. Technical personnel within the organization or technical consultants from outside must inform management whether there is an in-house capability to successfully complete the work. Determination also must be made of whether suitable subcontracts or purchase orders can be awarded. In the major areas, firm commitments should be obtained. Technical projections must be effected relative to a host of problems, including delivery or scheduling requirements, the possibility of changes in the proposed scope of work, client control over the work, quality control, and procedures.

An inadequate or unrealistic description of the work to be undertaken or evaluation of the project requirements marks the beginning of an unhappy contract experience.

19.13 SUMMARY

While it is essential that companies obtain good contracts with a minimum of risk provisions, it is equally important that those contracts be effectively administered. The following guidelines can aid a company in preparing its proposals and contracts and administering operations:

- Use of the checklist in the preparation of all proposals and contracts
- Evaluation of risks by reference to the suggested contract provisions wherever appropriate

- Review by the legal department prior to submission to the client of all major proposals and contracts and of other contracts with questionable provisions
- Appropriate pricing or insuring of risks under the contract
- Improving contract administration at appropriate levels
- Periodic review and updating of the entire contract procedure including basic risk areas, administration, and so on.

19.14 STUDYING TIPS FOR THE PMI® PROJECT MANAGEMENT CERTIFICATION EXAM

This section is applicable as a review of the principles to support the knowledge areas and domain groups in the PMBOK® Guide. This chapter addresses:

- Procurement Management

Understanding the following principles is beneficial if the reader is using this text to study for the PMP® Certification Exam:

- What is meant by procurement planning
- What is meant by solicitation and a solicitation package
- Different types of contracts and relative degree of risk associated with each one
- Role of the contract administrator
- What is meant by contractual closure or closeout

The following multiple-choice questions will be helpful in reviewing the principles of this chapter:

1. The contractual statement-of-work document is:
   A. A nonbinding legal document used to identify the responsibilities of the contractor
   B. A definition of the contracted work for government contracts only
   C. A narrative description of the work/deliverables to be accomplished and/or the resource skills required
   D. A form of specification

2. A written or pictorial document that describes, defines, or specifies the services or items to be procured is:
   A. A specification document
   B. A Gantt chart
   C. A blueprint
   D. A risk management plan

3. The "order of precedence" is:
   A. The document that specifies the order (priority) in which project documents will be used when it becomes necessary to resolve inconsistencies between project documents

   B. The order in which project tasks should be completed
   C. The relationship that project tasks have to one another
   D. The ordered list (by quality) of the screened vendors for a project deliverable

4. In which type of contract arrangement is the contractor *least likely* to want to control costs?
   A. Cost plus percentage of cost
   B. Firm-fixed price
   C. Time and materials
   D. Purchase order

5. In which type of contract arrangement is the contractor *most likely* to want to control costs?
   A. Cost plus percentage of cost
   B. Firm-fixed price
   C. Time and materials
   D. Fixed-price-incentive-fee

6. In which type of contract arrangement is the *contractor* at the most risk of *absorbing all cost overruns?*
   A. Cost plus percentage of cost
   B. Firm-fixed price
   C. Time and materials
   D. Cost-plus-incentive-fee

7. In which type of contract arrangement is the *customer* at the most risk of absorbing excessive cost overruns?
   A. Cost plus percentage of cost
   B. Firm-fixed price
   C. Time and materials
   D. Fixed-price-incentive-fee

8. What is the primary objective the customer's project manager focuses on when selecting a contract type?
   A. Transferring all risk to the contractor
   B. Creating reasonable contractor risk with provisions for efficient and economical performance incentives for the contractor
   C. Retaining all project risk, thus reducing project contract costs
   D. None of the above

9. Which type of contract arrangement is specifically designed to give a contractor relief for inflation or material/labor cost increases on a long-term contract?
   A. Cost plus percentage of cost
   B. Firm-fixed price
   C. Time and materials
   D. Firm-fixed price with economic price adjustment

10. Which of the following is not a factor to consider when selecting a contract type?
   A. Type/complexity of the requirement
   B. Urgency of the requirement
   C. Extent of price competition
   D. All are factors to consider.

11. In a fixed-price-incentive-fee contract, the "point of total assumption" refers to the point in the project cost curve where:
   A. The customer assumes responsibility for every additional dollar that is spent in fulfillment of the contract.

B. The contractor assumes responsibility for every additional dollar that is spent in fulfillment of the contract.

C. The price ceiling is reached after the contractor recovers the target profit.

D. None of the above

12. A written *preliminary* contractual instrument prepared prior to the issuance of a definitive contract that authorizes the contractor to begin work immediately, within certain limitations, is known as a:

A. Definitive contract

B. Preliminary contract

C. Letter contract/letter of intent

D. Purchase order

13. A contract entered into after following normal procedures (i.e., negotiation of terms, conditions, cost, and schedule) but prior to initiation of performance is known as a:

A. Definitive contract

B. Completed contract

C. Letter contract/letter of intent

D. Pricing arrangement

14. Which of the following is not a function of the contract administration activity?

A. Contract change management

B. Specification interpretation

C. Determination of contract breach

D. Selection of the project manager

15. A fixed-price contract is typically sought by the project manager from the customer's organization when:

A. The risk and consequences associated with the contracted task are large and the customer wishes to transfer the risk.

B. The project manager's company is proficient at dealing with the contracted activities.

C. Neither the contractor nor the project manager understand the scope of the task.

D. The project manager's company has excess production capacity.

16. Which of the following are typical actions a customer would take if the customer received nonconforming materials or products and the customer did not have the ability to bring the goods into conformance?

A. Reject the entire shipment but pay the full cost of the contract

B. Accept the entire shipment, no questions asked

C. Accept the shipment on condition that the nonconforming products will be brought into conformance by the vendor at the vendor's expense.

D. Accept the shipment and resell it to a competitor

17. If a project manager requires the use of a piece of equipment, what is the breakeven point where leasing and renting are the same?

Cost Categories	Renting Costs	Leasing Costs
Annual maintenance	$ 0.00	$3,000.00
Daily operation	$ 0.00	$ 70.00
Daily rental	$100.00	$ 0.00

A. 300 days

B. 30 days

C. 100 days

D. 700 days

18. In which type of incentive contract is there a maximum or minimum value established on the profits allowed for the contract?

A. Cost-plus-incentive-fee contract

B. Fixed-price-incentive-fee contract

C. Time-and-material-incentive-fee contract

D. Split-pricing-incentive-fee contract

19. In which type of incentive contract is there a maximum or minimum value established on the final price of the contract?

A. Cost-plus-incentive-fee contract

B. Fixed-price-incentive-fee contract

C. Time-and-material-incentive-fee contract

D. Split-pricing-incentive-fee contract

20. A cost-plus-incentive-fee contract has the following characteristics:

- Sharing ratio: 80/20
- Target cost: $100,000
- Target fee: $12,000
- Maximum fee: $14,000
- Minimum fee: $9,000

How much will the contractor be reimbursed if the cost of performing the work is $95,000?

A. $98,000

B. $100,000

C. $108,000

D. $114,000

21. Using the same data from Problem 20, and the same contract type, how much will the contractor be reimbursed if the cost of performing the work is $85,000?

A. $97,000

B. $99,000

C. $112,000

D. $114,000

22. Using the same data from Problem 20, and the same contract type, how much will the contractor be reimbursed if the cost of performing the work is $120,000?

A. $112,000

B. $119,000

C. $126,000

D. $129,000

23. A fixed-price-incentive-fee contract has the following characteristics:

- Sharing ratio: 70/30
- Target cost: $100,000
- Target fee: $8,000
- Price ceiling: $110,000

How much will the contractor be reimbursed if the cost of performing the work is $90,000?

A. $91,000

B. $101,000

C. $103,000

D. $110,000

24. Using the same data from Problem 23, and the same contract type, how much will the contractor be reimbursed if the cost of performing the work is $102,000?
    A. $104,000
    B. $107,400
    C. $109,400
    D. $110,000

25. Using the same data from Problem 23, and the same contract type, how much will the contractor be reimbursed if the cost of performing the work is $105,000?
    A. $105,000
    B. $106,500
    C. $110,000
    D. $111,500

ANSWERS

1. C
2. A
3. A
4. A
5. B
6. B
7. A
8. B
9. D
10. D
11. B
12. C
13. A
14. D
15. A
16. C
17. C
18. A
19. B
20. C
21. B
22. D
23. B
24. C
25. C

Quality Management[1]

Related Case Studies (from Kerzner/*Project Management Case Studies,* 3rd Edition)	Related Workbook Exercises (from Kerzner/*Project Management Workbook and PMP®/CAPM® Exam Study Guide,* 10th Edition)	PMBOK® Guide, 4th Edition, Reference Section for the PMP® Certification Exam
None	• Constructing Process Charts • Constructing Cause-and-Effect Charts and Pareto Charts • The Diagnosis of Patterns of Process Instability, Part (A): $\bar{X}$ Charts • The Diagnosis of Patterns of Process Instability, Part (B): $\bar{R}$ Charts • Quality Circles • Quality Problems • Multiple Choice Exam • Crossword Puzzle on Quality Management	• Quality Management

1. Appreciation is given to Terry Fischer (PMP) and Dr. Frank Anbari (PMP) for their invaluable assistance in the preparation of this chapter.

20.0 INTRODUCTION

<table>
<tr><td>PMBOK® Guide, 4th Edition
Chapter 8 Quality
8.1.1 Quality Planning Inputs</td><td>During the past twenty years, there has been a revolution in quality. Improvements have occurred not only in product quality, but also in leadership quality and project management quality. The changing views of quality appear in Table 20–1.</td></tr>
</table>

Unfortunately, it takes an economic disaster or a recession to get management to recognize the need for improved quality. Prior to the recession of 1979–1982, Ford, General Motors, and Chrysler viewed each other as the competition rather than the Japanese. Prior to the recession of 1989–1994, high-tech engineering companies never fully recognized the need for shortening product development time and the relationship between project management, total quality management, and concurrent engineering.

The push for higher levels of quality appears to be customer driven. Customers are now demanding:

- Higher performance requirements
- Faster product development
- Higher technology levels
- Materials and processes pushed to the limit
- Lower contractor profit margins
- Fewer defects/rejects

One of the critical factors that can affect quality is market expectations. The variables that affect market expectations include:

- Salability: the balance between quality and cost
- Produceability: the ability to produce the product with available technology and workers, and at an acceptable cost

TABLE 20–1. CHANGING VIEWS OF QUALITY

Past	Present
• Quality is the responsibility of blue-collar workers and direct labor employees working on the floor	• Quality is everyone's responsibility, including white-collar workers, the indirect labor force, and the overhead staff
• Quality defects should be hidden from the customers (and possibly management)	• Defects should be high-lighted and brought to the surface for corrective action
• Quality problems lead to blame, faulty justification, and excuses	• Quality problems lead to cooperative solutions
• Corrections-to-quality problems should be accomplished with minimum documentation	• Documentation is essential for "lessons learned" so that mistakes are not repeated
• Increased quality will increase project costs	• Improved quality saves money and increases business
• Quality is internally focused	• Quality is customer focused
• Quality will not occur without close supervision of people	• People want to produce quality products
• Quality occurs during project execution	• Quality occurs at project initiation and must be planned for within the project

- Social acceptability: the degree of conflict between the product or process and the values of society (i.e., safety, environment)
- Operability: the degree to which a product can be operated safely
- Availability: the probability that the product, when used under given conditions, will perform satisfactorily when called upon
- Reliability: the probability of the product performing without failure under given conditions and for a set period of time
- Maintainability: the ability of the product to be retained in or restored to a performance level when prescribed maintenance is performed

Customer demands are now being handled using total quality management (TQM). Total quality management is an ever-improving system for integrating various organizational elements into the design, development, and manufacturing efforts, providing cost-effective products or services that are fully acceptable to the ultimate customer. Externally, TQM is customer oriented and provides for more meaningful customer satisfaction. Internally, TQM reduces production line bottlenecks and operating costs, thus enhancing product quality while improving organizational morale.

20.1 DEFINITION OF QUALITY

PMBOK® Guide, 4th Edition
Chapter 8 Introduction

Mature organizations readily admit that they cannot accurately define quality. The reason is that quality is defined by the customer. The Kodak definition of quality is those products and services that are perceived to meet or exceed the needs and expectations of the customer at a cost that represents outstanding value. The ISO 9000 definition is "the totality of feature and characteristics of a

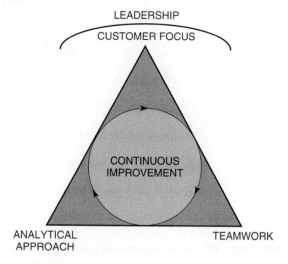

FIGURE 20–1. Kodak's five quality principles.

product or service that bears on its ability to satisfy stated or implied needs." Terms such as fitness for use, customer satisfaction, and zero defects are goals rather than definitions.

Most organizations view quality more as a process than a product. To be more specific, it is a continuously improving process where lessons learned are used to enhance future products and services in order to

- Retain existing customers
- Win back lost customers
- Win new customers

Therefore, companies are developing quality improvement processes. Figure 20–1 shows the five quality principles that support Kodak's quality policy. Figure 20–2 shows a more detailed quality improvement process. These two figures seem to illustrate that organizations

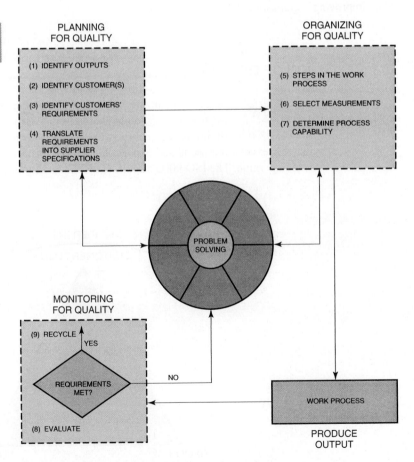

PMBOK® Guide, 4th Edition
8.1 Quality Planning

FIGURE 20–2. The quality improvement process. (Source unknown.)

are placing more emphasis on the quality process than on the quality product and, therefore, are actively pursuing quality improvements through a continuous cycle.

20.2 THE QUALITY MOVEMENT

PMBOK® Guide, 4th Edition
Chapter 8 Introduction
During the past hundred years, the views of quality have changed dramatically. Prior to World War I, quality was viewed predominantly as inspection, sorting out the good items from the bad. Emphasis was on problem identification. Following World War I and up to the early 1950s, emphasis was still on sorting good items from bad. However, *quality control* principles were now emerging in the form of:

- Statistical and mathematical techniques
- Sampling tables
- Process control charts

From the early 1950s to the late 1960s, quality control evolved into quality assurance, with its emphasis on problem avoidance rather than problem detection. Additional quality assurance principles emerged, such as:

- The cost of quality
- Zero-defect programs
- Reliability engineering
- Total quality control

Today, emphasis is being placed on strategic quality management, including such topics as:

- Quality is defined by the customer.
- Quality is linked with profitability on both the market and cost sides.
- Quality has become a competitive weapon.
- Quality is now an integral part of the strategic planning process.
- Quality requires an organization-wide commitment.

Although many experts have contributed to the success of the quality movement, the three most influential contributors are W. Edwards Deming, Joseph M. Juran, and Phillip B. Crosby. Dr. Deming pioneered the use of statistics and sampling methods from 1927 to 1940 at the U.S. Department of Agriculture. During these early years, Dr. Deming was influenced by Dr. Shewhart, and later applied Shewhart's Plan/Do/Check/Act cycle to clerical tasks. Figure 20–3 shows the Deming Cycle for Improvement.

Deming believed that the reason companies were not producing quality products was that management was preoccupied with "today" rather than the future. Deming postulated that 85 percent of all quality problems required management to take the initiative and change the process. Only 15 percent of the quality problems could be controlled by

PMBOK® Guide, 4th Edition

3.1 Project Planning Processes

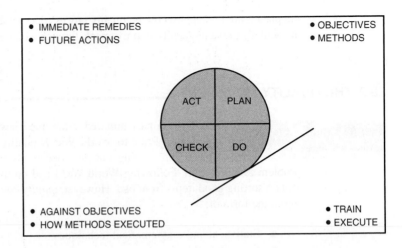

- IMMEDIATE REMEDIES
- FUTURE ACTIONS

- OBJECTIVES
- METHODS

- AGAINST OBJECTIVES
- HOW METHODS EXECUTED

- TRAIN
- EXECUTE

FIGURE 20–3. The Deming Cycle for Improvement.

the workers on the floor. As an example, the workers on the floor were not at fault because of the poor quality of raw materials that resulted from management's decision to seek out the lowest cost suppliers. Management needed to change the purchasing policies and procedures and develop long-term relationships with vendors.

Processes had to be placed under statistical analysis and control to demonstrate the repeatability of quality. Furthermore, the ultimate goals should be a continuous refinement of the processes rather than quotas. Statistical process control charts (SPCs) allowed for the identification of common cause and special (assignable) cause variations. Common cause variations are inherent in any process. They include poor lots of raw material, poor product design, unsuitable work conditions, and equipment that cannot meet the design tolerances. These common causes are *beyond* the control of the workers on the floor and therefore, for improvement to occur, actions by management are necessary.

Special or assignable causes include lack of knowledge by workers, worker mistakes, or workers not paying attention during production. Special causes can be identified by workers on the shop floor and corrected, but management still needs to change the manufacturing process to reduce common cause variability.

Deming contended that workers simply cannot do their best. They had to be shown what constitutes acceptable quality and that continuous improvement is not only possible, but necessary. For this to be accomplished, workers had to be trained in the use of statistical process control charts. Realizing that even training required management's approval, Deming's lectures became more and more focused toward management and what they must do.

Dr. Juran began conducting quality control courses in Japan in 1954, four years after Dr. Deming. Dr. Juran developed his 10 Steps to Quality Improvement (see Table 20–2), as well as the Juran Trilogy: Quality Improvement, Quality Planning, and Quality Control. Juran stressed that the manufacturer's view of quality is adherence to specifications but the customer's view of quality is "fitness for use." Juran defined five attributes of "fitness for use."

TABLE 20–2. VARIOUS APPROACHES TO QUALITY IMPROVEMENT

Deming's 14 Points for Management	Juran's 10 Steps to Quality Improvement	Crosby's 14 Steps to Quality Improvement
1. Create constancy of purpose for improvement of product and service.	1. Build awareness of the need and opportunity for improvement.	1. Make it clear that management is committed to quality.
2. Adopt the new philosophy.	2. Set goals for improvement.	2. Form quality improvement teams with representatives from each department.
3. Cease dependence on inspection to achieve quality.	3. Organize to reach the goals (establish a quality council, identify problems, select projects, appoint teams, designate facilitators).	3. Determine where current and potential quality problems lie.
4. End the practice of awarding business on the basis of price tag alone. Instead, minimize total cost by working with a single supplier.	4. Provide training.	4. Evaluate the cost of quality and explain its use as a management tool.
5. Improve constantly and forever every process for planning, production, and service.	5. Carry out projects to solve problems.	5. Raise the quality awareness and personal concern of all employees.
6. Institute training on the job.	6. Report progress.	6. Take actions to correct problems identified through previous steps.
7. Adopt and institute leadership.	7. Give recognition.	7. Establish a committee for the zero-defects program.
8. Drive out fear.	8. Communicate results.	8. Train supervisors to actively carry out their part of the quality improvement program.
9. Break down barriers between staff areas.	9. Keep score.	9. Hold a "zero-defects day" to let all employees realize that there has been a change.
10. Eliminate slogans, exhortations, and targets for the work force.	10. Maintain momentum by making annual improvement part of the regular systems and processes of the company.	10. Encourage individuals to establish improvement goals for themselves and their groups.
11. Eliminate numerical quotas for the workforce and numerical goals for management.		11. Encourage employees to communicate to management the obstacles they face in attaining their improvement goals.
12. Remove barriers that rob people of workmanship. Eliminate the annual rating or merit system.		12. Recognize and appreciate those who participate.
13. Institute a vigorous program of education and self-improvement for everyone.		13. Establish quality councils to communicate on a regular basis.
14. Put everybody in the company to work to accomplish the transformation.		14. Do it all over again to emphasize that the quality improvement program never ends.

- Quality of design: There may be many grades of quality
- Quality of conformance: Provide the proper training; products that maintain specification tolerances; motivation
- Availability: reliability (i.e., frequency of repairs) and maintainability (i.e., speed or ease of repair)
- Safety: The potential hazards of product use
- Field use: This refers to the way the product will be used by the customer

Dr. Juran also stressed the cost of quality (Section 20.8) and the legal implications of quality. The legal aspects of quality include:

- Criminal liability
- Civil liability
- Appropriate corporate actions
- Warranties

Juran believes that the contractor's view of quality is conformance to specification, whereas the customer's view of quality is fitness for use when delivered and value. Juran also admits that there can exist many grades of quality. The characteristics of quality can be defined as:

- Structural (length, frequency)
- Sensory (taste, beauty, appeal)
- Time-oriented (reliability, maintainability)
- Commercial (warrantee)
- Ethical (courtesy, honesty)

The third major contributor to quality was Phillip B. Crosby. Crosby developed his 14 Steps to Quality Improvement (see Table 20–2) and his Four Absolutes of Quality:

- Quality means conformance to requirements.
- Quality comes from prevention.
- Quality means that the performance standard is "zero defects."
- Quality is measured by the cost of nonconformance.

Crosby found that the cost of not doing things right the first time could be appreciable. In manufacturing, the price of nonconformance averages 40 percent of operating costs.

20.3 COMPARISON OF THE QUALITY PIONEERS

Deming's definition of quality is "continuous improvement." Although variations cannot be entirely eliminated, we can learn more about them and eventually reduce them. The ultimate goal obviously is zero defects, but this error-free work may not be economically feasible or practical.

Juran believes that for quality to improve, we must resolve "sporadic" problems and "chronic" problems. Sporadic problems are short-term problems that generate sudden changes for the worse in quality; techniques exist for identifying and controlling them.

TABLE 20–3. COMPARISON OF THE EXPERTS

	Deming	Juran	Crosby
Definition of quality	Continuous improvement	Fitness for use	Conformance to requirements
Application	Manufacturing-driven companies	Technology-driven companies	People-driven companies
Target audience	Workers	Management	Workers
Emphasis on	Tools/system	Measurement	Motivation (behavioral)
Type of tools	Statistical process control	Analytical, decision-making and cost-of-quality	Minimal use
Use of goals and targets	Not used	Used for breakthrough projects	Posted goals for workers

"Chronic" problems, on the other hand, may require scientific breakthrough to achieve higher levels of quality. Chronic problems exist because workers may not accept change and refuse to admit that there may be a better way of doing things. Solving chronic problems requires breakthrough projects, specific targets usually established on a yearly basis, strong and visible senior management support, and the use of quality experts to lead the company-wide quality improvement programs. Unlike Deming, who avoids the use of targets and quotas, Juran's objective is to get management to accept the habit of an annual quality improvement program based upon well-defined targets.

Juran's method for determining the cost of quality, therefore, suggests that the pursuit of quality will pay for itself only up to a certain point, and beyond that point costs may rise significantly.

Crosby argues that the cost of quality includes only the nonconformance costs, whereas Juran includes both conformance and nonconformance costs. Crosby's argument is that the conformance costs of prevention and appraisal are not really the cost of quality but more so the cost of doing business. Therefore, Crosby argues that quality is free, and the only associated costs of quality should be those of nonconformance. Crosby does not emphasize analytical techniques other than measurement methods for nonconformance costs, and he relies heavily upon motivation and the role of senior management.

Table 20–3 compares the approach to quality of the three experts. Although all three emphasize the need for quality and the importance/role of senior management, each goes about it differently.

20.4 THE TAGUCHI APPROACH[2]

PMBOK® Guide, 4th Edition
8.1.2.5 Design of Experiments

After World War II the allied forces found that the quality of the Japanese telephone system was extremely poor and totally unsuitable for long-term

---

2. Taken from Ranjit Roy, *A Primer on the Taguchi Method* (Dearborn, MI: Society of Manufacturing Engineers, 1990), Chapter 2. Reproduced by permission.

communication purposes. To improve the system, the allied command recommended that Japan establish research facilities similar to the Bell Laboratories in the United States in order to develop a state-of-the-art communication system. The Japanese founded the Electrical Communication Laboratories (ECL) with Dr. Taguchi in charge of improving the R&D productivity and enhancing product quality. He observed that a great deal of time and money was expended in engineering experimentation and testing. Little emphasis was given to the process of creative brainstorming to minimize the expenditure of resources.

Dr. Taguchi started to develop new methods to optimize the process of engineering experimentation. He developed techniques that are now known as the Taguchi Methods. His greatest contribution lies not in the mathematical formulation of the design of experiments, but rather in the accompanying philosophy. His approach is more than a method to lay out experiments. His is a concept that has produced a unique and powerful quality improvement discipline that differs from traditional practices.

These concepts are:

1. Quality should be designed into the product and not inspected into it.
2. Quality is best achieved by minimizing the deviation from a target. The product should be so designed that it is immune to uncontrollable environmental factors.
3. The cost of quality should be measured as a function of deviation from the standard and the losses should be measured system-wide.

Taguchi built on Deming's observation that 85 percent of poor quality is attributable to the manufacturing process and only 15 percent to the worker. Hence, he developed manufacturing systems that were "robust" or insensitive to daily and seasonal variations of environment, machine wear, and other external factors. The three principles were his guides in developing these systems, testing the factors affecting quality production, and specifying product parameters.

Taguchi believed that the better way to improve quality was to design and build it into the product. Quality improvement starts at the very beginning, that is, during the design stages of a product or a process, and continues through the production phase. He proposed an "off-line" strategy for developing quality improvement in place of an attempt to inspect quality into a product on the production line. He observed that poor quality cannot be improved by the process of inspection, screening, and salvaging. No amount of inspection can put quality back into the product; it merely treats a symptom. Therefore, quality concepts should be based upon, and developed around, the philosophy of prevention. The product design must be so robust that it is immune to the influence of uncontrolled environmental factors on the manufacturing processes.

His second concept deals with actual methods of effecting quality. He contended that quality is directly related to deviation of a design parameter from the target value, not to conformance to some fixed specifications. A product may be produced with properties skewed toward one end of an acceptance range yet show shorter life expectancy. However, by specifying a target value for the critical property and developing manufacturing processes to meet the target value with little deviation, the life expectancy may be much improved.

His third concept calls for measuring deviations from a given design parameter in terms of the overall life-cycle costs of the product. These costs would include the cost of scrap, rework, inspection, returns, warranty service calls, and/or product replacement. These costs provide guidance regarding the major parameters to be controlled.

Limitations

The most severe limitation of the Taguchi method is the need for timing with respect to product/process development. The technique can only be effective when applied early in the design of the product/process system. After the design variables are determined and their nominal values are specified, experimental design may not be cost-effective. Also, though the method has wide-ranging applications, there are situations in which classical techniques are better suited; in simulation studies involving factors that vary in a continuous manner, such as the torsional strength of a shaft as a function of its diameter, the Taguchi method may not be a proper choice.

Selecting Design Parameters for Reduced Variation

Taguchi strives to attain quality by reducing the variation around the target. In an effort to reduce variations, he searched for techniques that allow variability to be reduced without necessarily eliminating the causes of variation. Often in an industrial setting, totally removing the causes of variation can be expensive. A no-cost or low-cost solution may be achieved by adjusting the levels and controlling the variation of other factors. This is what Taguchi tries to do through his *parameter design* approach where there is no cost or low cost in reducing variability. Furthermore, the cost savings realized far exceed the cost of additional experiments needed to reduce variations.

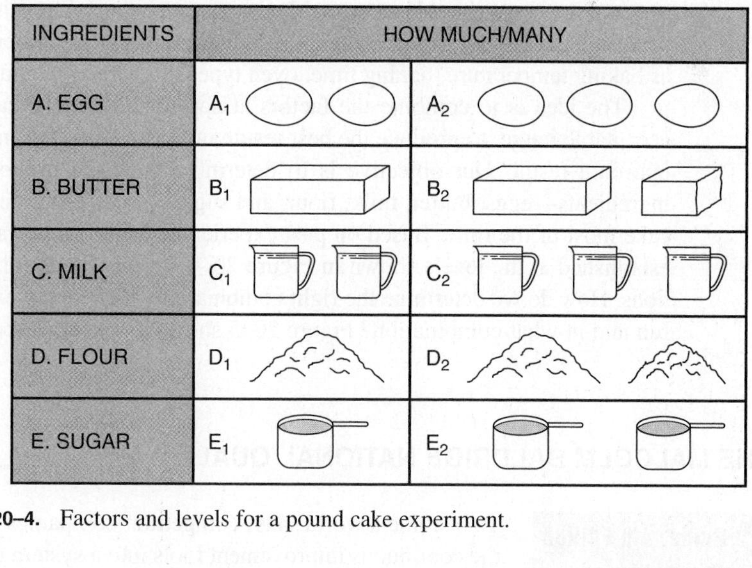

FIGURE 20–4. Factors and levels for a pound cake experiment.

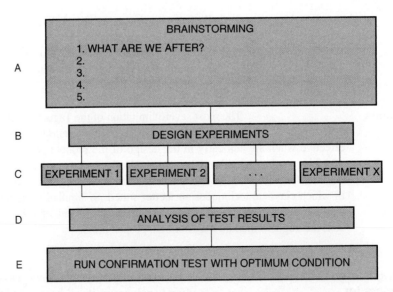

FIGURE 20–5. A Taguchi experiment flow diagram. *Source:* Ranjit Roy, *A Primer on the Taguchi Method* (Dearborn, MI: Society of Manufacturing Engineers, 1990), p. 231. Reproduced by permission.

The Taguchi method is most effective when applied to experiments with multiple factors. But the concept of selecting the proper levels of design factors, and reducing the variation of performance around the optimum/target value, can be easily illustrated through an example.

Consider a baking process. Assume several bakers are given the same ingredients to bake a pound cake, the object being to produce the best-tasting cake. Within limits, they can adjust the amount of ingredients, but they can only use the ingredients provided. They are to make the best cake within available design parameters. Taguchi's approach would be to design an experiment considering all baking ingredients and other influencing factors such as baking temperature, baking time, oven type (if a variable), and so on.

The idea is to combine the factors at appropriate levels, each within the respective acceptable range, to produce the best result and yet exhibit minimum variation around the optimum result. Our objective is to determine the right proportions of the five major ingredients—eggs, butter, milk, flour, and sugar—so that the recipe will produce the best cake most of the time. Based on past experience, the working ranges of these factors are established at the levels shown in Figure 20–4. At this point we face the following questions. How do we determine the right combination? How many experiments do we need to run and in what combination? Figure 20–5 shows a Taguchi experiment flow diagram.

20.5 THE MALCOLM BALDRIGE NATIONAL QUALITY AWARD _____

PMBOK® Guide, 4th Edition
Chapter 8 Introduction

To become a world-class competitor, companies need a model to integrate the continuous improvement tools into a system that involves participative cross-functional implementation. In 1987, this need was recognized at the national level with the establishment of the Malcolm Baldrige National Quality Award. The award is presented to those companies that have achieved a level of world-class competition through quality management of products and services.

The criteria for the award include:

- *The leadership category:* Examines primarily how the senior executives create and sustain a clear and visible quality value system along with a supporting management system to guide all activities of the company. Also examines the senior executives' and the company's leadership and support of quality developments both inside and outside the company.
- *The strategic planning category:* Examines how the company sets strategic directions, and how it determines key action plans. Also examines how the plans are translated into an effective performance management system.
- *The customer and market focus category:* Examines how the company determines requirements and expectations of customers and markets. Also examines how the company enhances relationships with customers and determines their satisfaction.
- *The information and analysis category:* Examines the management and effectiveness of the use of data and information to support key company processes and the company's performance management system.
- *The human resource development and management category:* Examines how the workforce is enabled to develop and utilize its full potential, aligned with the company's objectives. Also examines the company's efforts to build and maintain an environment conducive to performance excellence, full participation, and personal and organizational growth.
- *The process management category:* Examines the key aspects of process management, including customer-focused design, product, and service delivery processes, support processes, and supplier and partnering processes involving all work units. The category examines how key processes are designed, effectively managed, and improved to achieve better performance.
- *The business results category:* Examines the company's performance and improvement in key business areas: customer satisfaction, financial and marketplace performance, human resource, supplier and partner performance, and operational performance. Also examined are performance levels relative to competitors.

Some companies that have been honored with the award include IBM, General Motors, Xerox, Kodak, AT&T, Westinghouse, Federal Express, Ritz-Carlton, Armstrong Building Products, and Motorola. Generally speaking, only two or three companies a year win the award.

20.6 ISO 9000

The International Organization for Standardization (ISO), based in Geneva, Switzerland, is a consortium of approximately 100 of the world's industrial nations. The American

National Standards Institute (ANSI) represents the United States. ISO 9000 is *not* a set of standards for products or services, nor is it specific to any one industry. Instead, it is a quality system standard applicable to any product, service, or process anywhere in the world.

The information included in the ISO 9000 series includes:

ISO 9000:   This defines the key terms and acts as a road map for the other standards within the series.

ISO 9001:   This defines the model for a quality system when a contractor demonstrates the capability to design, produce, and install products or services.

ISO 9002:   This is a quality system model for quality assurance in production and installation.

ISO 9003:   This is a quality system model for quality assurance in final inspection and testing.

ISO 9004:   This provides quality management guidelines for any organization wishing to develop and implement a quality system. Guidelines are also available to determine the extent to which each quality system model is applicable.

There are several myths concerning the ISO 9000 series. First, ISO 9000 is *not* a European standard, although it may be necessary to do business within the European Community. ISO 9000 is based on American quality standards that are still being used. Second, ISO 9000 is *not* a paperwork nightmare. Although documentation is a necessary requirement, the magnitude of the documentation is less than most people believe. Third, becoming ISO 9000 certified does *not* guarantee that your organization will produce quality products or services. Instead, it confirms that the appropriate system is in place.

ISO 9000 is actually a three-part, never-ending cycle including planning, controlling, and documentation. *Planning* is required to ensure that the objectives, goals, authority, and responsibility relationships of each activity are properly defined and understood. *Controlling* is required to ensure that the goals and objectives are met, and that problems are anticipated or averted through proper corrective actions. *Documentation* is used predominantly for feedback on how well the quality management system is performing to satisfy customer's needs and what changes may be necessary.

There always exists the question of how ISO 9000 relates to the Malcolm Baldrige Award. ISO 9000 requirements fall predominantly into the "quality assurance of products and services" section of the Malcolm Baldrige Award. It does touch the other six sections in varying degrees.

ISO 9000 provides minimum requirements needed for certification. The Malcolm Baldrige National Quality Award (MBNQA) tries to identify the "best in class." Organizations wishing to improve quality are encouraged to consider practices of and benchmark against past recipients of the MBNQA as "role models."

The International Organization for Standardization has recently developed the ISO 14000 series standards. ISO 14000 is an evolving series that provides business management with the structure for managing environmental impacts, including the basic management system, performance evaluation, auditing, labeling, and life-cycle assessment.

20.7 QUALITY MANAGEMENT CONCEPTS

PMBOK® Guide, 4th Edition
Chapter 8 Introduction
8.1.1 Quality Planning Inputs

The project manager has the ultimate responsibility for quality management on the project. Quality management has equal priority with cost and schedule management. However, the direct measurement of quality may be the responsibility of the quality assurance department or the assistant project manager for quality. For a labor-intensive project, management support (i.e., the project office) is typically 12–15 percent of the total labor dollars of the project. Approximately 3–5 percent can be attributed to quality management. Therefore, as much as 20–30 percent of all the labor in the project office could easily be attributed to quality management.

From a project manager's perspective, there are six quality management concepts that should exist to support each and every project. They include:

- Quality policy
- Quality objectives
- Quality assurance
- Quality control
- Quality audit
- Quality program plan

Ideally, these six concepts should be embedded within the corporate culture.

Quality Policy

The quality policy is a document that is typically created by quality experts and fully supported by top management. The policy should state the quality objectives, the level of quality acceptable to the organization, and the responsibility of the organization's members for executing the policy and ensuring quality. A quality policy would also include statements by top management pledging its support to the policy. The quality policy is instrumental in creating the organization's reputation and quality image.

Many organizations successfully complete a good quality policy but immediately submarine the good intentions of the policy by delegating the implementation of the policy to lower-level managers. The implementation of the quality policy is the responsibility of top management. Top management must "walk the walk" as well as "talk the talk." Employees will soon see through the ruse of a quality policy that is delegated to middle managers while top executives move onto "more crucial matters that really impact the bottom line."

A good quality policy will:

- Be a statement of principles stating what, not how
- Promote consistency throughout the organization and across projects
- Provide an explanation to outsiders of how the organization views quality
- Provide specific guidelines for important quality matters
- Provide provisions for changing/updating the policy

Quality Objectives   Quality objectives are a part of an organization's quality policy and consist of specific objectives and the time frame for completing them. The quality objectives must be selected carefully. Selecting objectives that are not naturally possible can cause frustration and disillusionment. Examples of acceptable quality objectives might be: to train all members of the organization on the quality policy and objectives before the end of the current fiscal year, to set up baseline measurements of specific processes by the end of the current quarter, to define the responsibility and authority for meeting the organization's quality objectives down to each member of the organization by the end of the current fiscal year, etc.

Good quality objectives should:

- Be obtainable
- Define specific goals
- Be understandable
- State specific deadlines

Quality Assurance   Quality assurance is the collective term for the formal activities and managerial processes that attempt to ensure that products and services

PMBOK® Guide, 4th Edition
8.2 Quality Assurance

meet the required quality level. Quality assurance also includes efforts external to these processes that provide information for improving the internal processes. It is the quality assurance function that attempts to ensure that the project scope, cost, and time functions are fully integrated.

The Project Management Institute Guide to the Body of Knowledge (PMBOK)® refers to quality assurance as the management section of quality management. This is the area where the project manager can have the greatest impact on the quality of his project. The project manager needs to establish the administrative processes and procedures necessary to ensure and, often, prove that the scope statement conforms to the actual requirements of the customer. The project manager must work with his team to determine which processes they will use to ensure that all stakeholders have confidence that the quality activities will be properly performed. All relevant legal and regulatory requirements must also be met.

A good quality assurance system will:

- Identify objectives and standards
- Be multifunctional and prevention oriented
- Plan for collection and use of data in a cycle of continuous improvement
- Plan for the establishment and maintenance of performance measures
- Include quality audits

Quality Control   Quality control is a collective term for activities and techniques, within the process, that are intended to create specific quality characteristics. Such activities include continually monitoring processes, identifying and eliminating problem causes, use of statistical process control to reduce the variability and to increase the efficiency of processes. Quality control certifies that the organization's quality objectives are being met.

The PMBOK® refers to quality control as the technical aspect of quality management. Project team members who have specific technical expertise on the various aspects of the project play an active role in quality control. They set up the technical processes and procedures that ensure that each step of the project provides a quality output from design and development through implementation and maintenance. Each step's output must conform to the overall quality standards and quality plans, thus ensuring that quality is achieved.

A good quality control system will:

- Select what to control
- Set standards that provide the basis for decisions regarding possible corrective action
- Establish the measurement methods used
- Compare the actual results to the quality standards
- Act to bring nonconforming processes and material back to the standard based on the information collected
- Monitor and calibrate measuring devices
- Include detailed documentation for all processes

Quality Audit

PMBOK® Guide, 4th Edition
8.2.2.2 Quality Audit

A quality audit is an independent evaluation performed by qualified personnel that ensures that the project is conforming to the project's quality requirements and is following the established quality procedures and policies.

A good quality audit will ensure that:

- The planned quality for the project will be met.
- The products are safe and fit for use.
- All pertinent laws and regulations are followed.
- Data collection and distribution systems are accurate and adequate.
- Proper corrective action is taken when required.
- Improvement opportunities are identified.

Quality Plan

PMBOK® Guide, 4th Edition
8.1.3.1 Quality Plan

The quality plan is created by the project manager and project team members by breaking down the project objectives into a work breakdown structure. Using a treelike diagramming technique, the project activities are broken down into lower-level activities until specific quality actions can be identified. The project manager then ensures that these actions are documented and implemented in the sequence that will meet the customer's requirements and expectations. This enables the project manager to assure the customer that he has a road map to delivering a quality product or service and therefore will satisfy the customer's needs.

A good quality plan will:

- Identify all of the organization's external and internal customers
- Cause the design of a process that produces the features desired by the customer

- Bring in suppliers early in the process
- Cause the organization to be responsive to changing customer needs
- Prove that the process is working and that quality goals are being met

20.8 THE COST OF QUALITY

PMBOK® Guide, 4th Edition
8.1.2.2 Cost of Quality

To verify that a product or service meets the customer's requirements requires the measurement of the costs of quality. For simplicity's sake, the costs can be classified as "the cost of conformance" and "the cost of non-conformance." Conformance costs include items such as training, indoctrination, verification, validation, testing, maintenance, calibration, and audits. Nonconforming costs include items such as scrap, rework, warranty repairs, product recalls, and complaint handling.

Trying to save a few project dollars by reducing conformance costs could prove disastrous. For example, an American company won a contract as a supplier of Japanese parts. The initial contract called for the delivery of 10,000 parts. During inspection and testing at the customer's (i.e., Japanese) facility, two rejects were discovered. The Japanese returned *all* 10,000 components to the American supplier stating that this batch was not acceptable. In this example, the nonconformance cost could easily be an order of magnitude greater than the conformance cost. The moral is clear: *Build it right the first time.*

Another common method to classify costs includes the following:

- *Prevention costs* are the up-front costs oriented toward the satisfaction of customer's requirements with the first and all succeeding units of product produced without defects. Included in this are typically such costs as design review, training, quality planning, surveys of vendors, suppliers, and subcontractors, process studies, and related preventive activities.
- *Appraisal costs* are costs associated with evaluation of product or process to ascertain how well all of the requirements of the customer have been met. Included in this are typically such costs as inspection of product, lab test, vendor control, in-process testing, and internal–external design reviews.
- *Internal failure costs* are those costs associated with the failure of the processes to make products acceptable to the customer, before leaving the control of the organization. Included in this area are scrap, rework, repair, downtime, defect evaluation, evaluation of scrap, and corrective actions for these internal failures.
- *External failure costs* are those costs associated with the determination by the customer that his requirements have not been satisfied. Included are customer returns and allowances, evaluation of customer complaints, inspection at the customer, and customer visits to resolve quality complaints and necessary corrective action.

Figure 20–6 shows the expected results of the total quality management system on quality costs. Prevention costs are expected to actually rise as more time is spent in prevention activities throughout the organization. As processes improve over the long run,

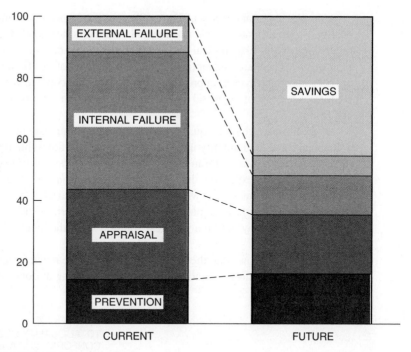

FIGURE 20–6. Total quality cost.

appraisal costs will go down as the need to inspect in quality decreases. The biggest savings will come from the internal failure areas of rework, scrap, reengineering, redo, and so on. The additional time spent in up-front design and development will really pay off here. And, finally, the external costs will also come down as processes yield first-time quality on a regular basis. The improvements will continue to affect the company on a long-term basis in both improved quality and lower costs. Also, as project management matures, there should be further decreases in the cost of both maintaining quality and developing products.

Figure 20–6 shows that prevention costs can increase. This is not always the case. Prevention costs actually decrease without sacrificing the purpose of prevention if we can identify and eliminate the costs associated with waste, such as waste due to

- Rejects of completed work
- Design flaws
- Work in progress
- Improperly instructed manpower
- Excess or noncontributing management (who still charge time to the project)
- Improperly assigned manpower
- Improper utilization of facilities
- Excessive expenses that do not necessarily contribute to the project (i.e., unnecessary meetings, travel, lodgings, etc.)

Another important aspect of Figure 20–6 is that 50 percent or more of the total cost of quality can be attributed to the internal and external failure costs. Complete elimination of failures may seem like an ideal solution but may not be cost-effective. As an example, see Figure 20–7. There are assumptions in the development of this figure. First, the cost of failure (i.e., nonconformance) approaches zero as defects become fewer and fewer. Second, the conformance costs of appraisal and prevention approach infinity as defects become fewer and fewer.

If the ultimate goal of a quality program is to continuously improve quality, then from a financial standpoint, quality improvement may not be advisable if the positive economic return becomes negative. Juran argued that as long as the per unit cost for prevention and appraisal were less expensive than nonconformance costs, resources should be assigned to prevention and appraisal. But when prevention and appraisal costs begin to increase the per unit cost of quality, then the policy should be to maintain quality. The implication here is that zero defects may not be a practical solution since the total cost of quality would not be minimized.

Figure 20–6 shows that the external failure costs are much lower than the internal failure costs. This indicates that most of the failures are being discovered *before* they leave the

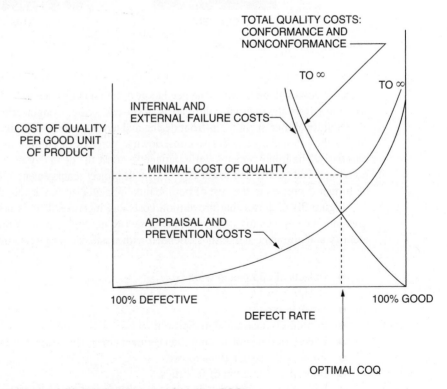

FIGURE 20–7. Minimizing the costs of quality (COQ).

functional areas or plants. This is particularly important if we consider the life-cycle cost model discussed in Section 14.19. We showed that typical life-cycle costs are:

- R&D: 12 percent
- Acquisition: 28 percent
- Operations and support: 60 percent

Since 60 percent of the life-cycle cost occurs *after* the product is put into service, then small increases in the R&D and acquisition areas could generate major cost savings in operation and support due to better design, higher quality, less maintenance, and so forth.

20.9 THE SEVEN QUALITY CONTROL TOOLS[3]

PMBOK® Guide, 4th Edition
8.3 Perform Quality Control

Over the years, statistical methods have become prevalent throughout business, industry, and science. With the availability of advanced, automated systems that collect, tabulate, and analyze data, the practical application of these quantitative methods continues to grow.

More important than the quantitative methods themselves is their impact on the basic philosophy of business. The statistical point of view takes decision-making out of the subjective autocratic decision-making arena by providing the basis for objective decisions based on quantifiable facts. This change provides some very specific benefits:

- Improved process information
- Better communication
- Discussion based on facts
- Consensus for action
- Information for process changes

Statistical process control (SPC) takes advantage of the natural characteristics of any process. All business activities can be described as specific processes with known tolerances and measurable variances. The measurement of these variances and the resulting information provide the basis for continuous process improvement. The tools presented here provide both a graphical and measured representation of process data. The systematic application of these tools empowers business people to control products and processes to become world-class competitors.

The basic tools of statistical process control are data figures, Pareto analysis, cause-and-effect analysis, trend analysis, histograms, scatter diagrams, and process control charts. These basic tools provide for the efficient collection of data, identification of patterns in the data, and measurement of variability. Figure 20–8 shows the relationships among these seven tools and their use for the identification and analysis of improvement opportunities. We will review these tools and discuss their implementation and applications.

---

3. This section is taken from H. K. Jackson and N. L. Frigon, *Achieving the Competitive Edge* (New York: Wiley, 1996), Chapters 6 and 7. Reproduced by permission.

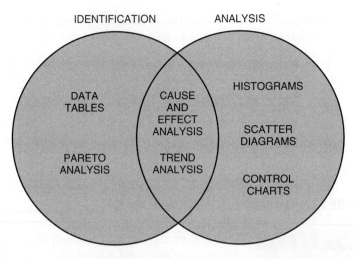

FIGURE 20–8. The seven quality control tools.

Data Tables
Data tables, or data arrays, provide a systematic method for collecting and displaying data. In most cases, data tables are forms designed for the purpose of collecting specific data. These tables are used most frequently where data are available from automated media. They provide a consistent, effective, and economical approach to gathering data, organizing them for analysis, and displaying them for preliminary review. Data tables sometimes take the form of manual check sheets where automated data are not necessary or available. Data figures and check sheets should be designed to minimize the need for complicated entries. Simple-to-understand, straightforward tables are a key to successful data gathering.

Figure 20–9 is an example of an attribute (pass/fail) data figure for the correctness of invoices. From this simple check sheet, several data points become apparent. The total

DEFECT	SUPPLIER				
	A	B	C	D	TOTAL
INCORRECT INVOICE	////	/		//	7
INCORRECT INVENTORY	/////	//	/	/	9
DAMAGED MATERIAL	///		//	///	8
INCORRECT TEST DOCUMENTATION	/	///	////	//	10
TOTAL	13	6	7	8	34

FIGURE 20–9. Check sheet for material receipt and inspection.

number of defects is 34. The highest number of defects is from supplier A, and the most frequent defect is incorrect test documentation. We can subject these data to further analysis by using Pareto analysis, control charts, and other statistical tools.

In this check sheet, the categories represent defects found during the material receipt and inspection function. The following defect categories provide an explanation of the check sheet:

- Incorrect invoices: The invoice does not match the purchase order.
- Incorrect inventory: The inventory of the material does not match the invoice.
- Damaged material: The material received was damaged and rejected.
- Incorrect test documentation: The required supplier test certificate was not received and the material was rejected.

Cause-and-Effect Analysis   After identifying a problem, it is necessary to determine its cause. The cause-and-effect relationship is at times obscure. A considerable amount of analysis often is required to determine the specific cause or causes of the problem.

Cause-and-effect analysis uses diagramming techniques to identify the relationship between an effect and its causes. Cause-and-effect diagrams are also known as fishbone diagrams. Figure 20–10 demonstrates the basic fishbone diagram. Six steps are used to perform a cause-and-effect analysis.

Step 1. Identify the problem. This step often involves the use of other statistical process control tools, such as Pareto analysis, histograms, and control charts, as well as brainstorming. The result is a clear, concise problem statement.

Step 2. Select interdisciplinary brainstorming team. Select an interdisciplinary team, based on the technical, analytical, and management knowledge required to determine the causes of the problem.

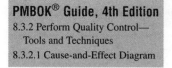

PMBOK® Guide, 4th Edition
8.3.2 Perform Quality Control—
  Tools and Techniques
8.3.2.1 Cause-and-Effect Diagram

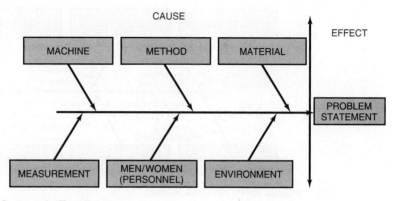

FIGURE 20–10. Cause-and-effect diagram.

Step 3. Draw problem box and prime arrow. The problem contains the problem statement being evaluated for cause and effect. The prime arrow functions as the foundation for their major categories.

Step 4. Specify major categories. Identify the major categories contributing to the problem stated in the problem box. The six basic categories for the primary causes of the problems are most frequently personnel, method, materials, machinery, measurements, and environment, as shown in Figure 20–10. Other categories may be specified, based on the needs of the analysis.

Step 5. Identify defect causes. When you have identified the major causes contributing to the problem, you can determine the causes related to each of the major categories. There are three approaches to this analysis: the random method, the systematic method, and the process analysis method.

    Random method. List all six major causes contributing to the problem at the same time. Identify the possible causes related to each of the categories, as shown in Figure 20–11.

    Systematic method. Focus your analysis on one major category at a time, in descending order of importance. Move to the next most important category only after completing the most important one. This process is diagrammed in Figure 20–12.

    Process analysis method. Identify each sequential step in the process and perform cause-and-effect analysis for each step, one at a time. Figure 20–13 represents this approach.

Step 6. Identify corrective action. Based on (1) the cause-and-effect analysis of the problem and (2) the determination of causes contributing to each major category, identify corrective action. The corrective action analysis is performed in the same manner as the cause-and-effect analysis. The cause-and-effect diagram is simply reversed so that the problem box becomes the corrective action box. Figure 20–14 displays the method for identifying corrective action.

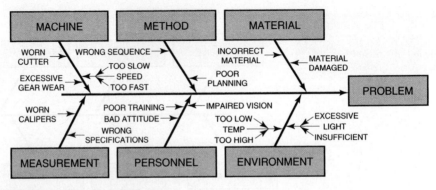

FIGURE 20–11. Random method.

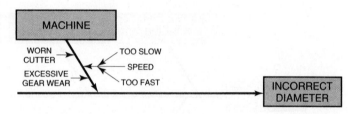

FIGURE 20–12.   Systematic method.

Histogram
A histogram is a graphical representation of data as a frequency distribution. This tool is valuable in evaluating both attribute (pass/fail) and variable (measurement) data. Histograms offer a quick look at the data at a single point in time; they do not display variance or trends over time. A histogram displays how the cumulative data look *today*. It is useful in understanding the relative frequencies (percentages) or frequency (numbers) of the data and how those data are distributed. Figure 20–15 illustrates a histogram of the frequency of defects in a manufacturing process.

Pareto Analysis
A Pareto diagram is a special type of histogram that helps us to identify and prioritize problem areas. The construction of a Pareto diagram may involve data collected from data figures, maintenance data, repair data, parts scrap rates, or other sources. By identifying types of nonconformity from any of these data sources, the Pareto diagram directs attention to the most frequently occurring element.

PMBOK® Guide, 4th Edition
8.3.2.5 Perform Quality Control—
Tools and Techniques

There are three uses and types of Pareto analysis. The basic Pareto analysis identifies the vital few contributors that account for most quality

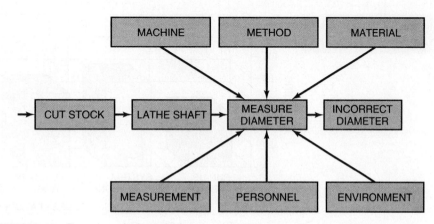

FIGURE 20–13.   Process analysis method.

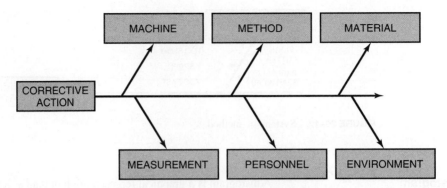

FIGURE 20–14. Identify corrective action.

problems in any system. The comparative Pareto analysis focuses on any number of program options or actions. The weighted Pareto analysis gives a measure of significance to factors that may not appear significant at first—such additional factors as cost, time, and criticality.

The basic Pareto analysis chart provides an evaluation of the most frequent occurrences for any given data set. By applying the Pareto analysis steps to the material receipt and inspection process described in Figure 20–16, we can produce the basic Pareto analysis demonstrated in Figure 20–17. This basic Pareto analysis quantifies and graphs the frequency of occurrence for material receipt and inspection and further identifies the most significant, based on frequency.

A review of this basic Pareto analysis for frequency of occurrences indicates that supplier A is experiencing the most rejections with 38 percent of all the failures.

Pareto analysis diagrams are also used to determine the effect of corrective action, or to analyze the difference between two or more processes and methods. Figure 20–18

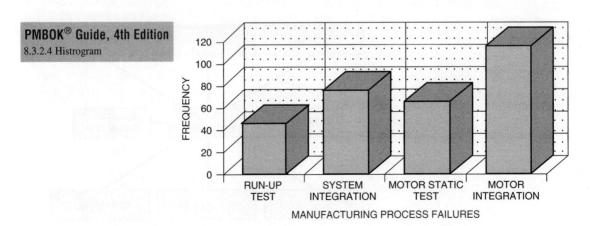

FIGURE 20–15. Histogram for variables.

MATERIAL RECEIPT AND INSPECTION FREQUENCY OF FAILURES			
SUPPLIER	FAILING FREQUENCY	PERCENT FAILING	CUMULATIVE PERCENT
A	13	38	38
B	6	17	55
C	7	20	75
D	9	25	100

FIGURE 20–16. Basic Pareto analysis.

displays the use of this Pareto method to assess the difference in defects after corrective action.

Scatter Diagrams

PMBOK® Guide, 4th Edition
8.3.2.7 Perform Quality Control—
    Tools and Techniques

Another pictorial representation of process control data is the scatter plot or scatter diagram. A scatter diagram organizes data using two variables: an independent variable and a dependent variable. These data are then recorded on a simple graph with *X* and *Y* coordinates showing the relationship between the variables. Figure 20–19 displays the relationship between two of the data elements from solder qualification test scores. The

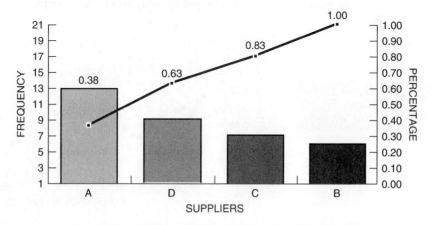

FIGURE 20–17. Basic Pareto analysis.

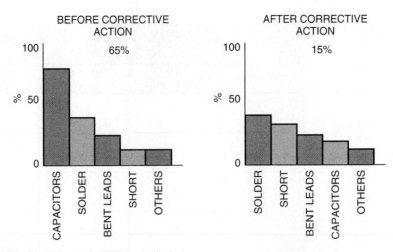

FIGURE 20–18. Comparative Pareto analysis.

independent variable, experience in months, is listed on the *X* axis. The dependent variable is the score, which is recorded on the *Y* axis.

These relationships fall into several categories, as shown in Figure 20–20. In the first scatter plot there is no correlation—the data points are widely scattered with no apparent pattern. The second scatter plot shows a curvilinear correlation demonstrated by the U shape of the graph. The third scatter plot has a negative correlation, as indicated by the downward slope. The final scatter plot has a positive correlation with an upward slope.

From Figure 20–19 we can see that the scatter plot for solder certification testing is somewhat curvilinear. The least and the most experienced employees scored highest, whereas those with an intermediate level of experience did relatively poorly. The next tool, trend analysis, will help clarify and quantify these relationships.

PMBOK® Guide, 4th Edition
8.3.2.7 Perform Quality
  Control—Tools and Techniques

FIGURE 20–19. Solder certification test scores.

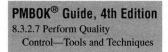

PMBOK® Guide, 4th Edition
8.3.2.7 Perform Quality
    Control—Tools and Techniques

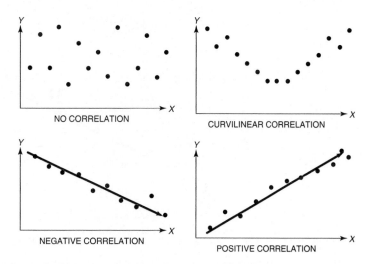

FIGURE 20–20. Scatter plot correlation.

Trend Analysis

Trend analysis is a statistical method for determining the equation that best fits the data in a scatter plot. Trend analysis quantifies the relationships of the data, determines the equation, and measures the fit of the equation to the data. This method is also known as curve fitting or least squares.

Trend analysis can determine optimal operating conditions by providing an equation that describes the relationship between the dependent (output) and independent (input) variables. An example is the data set concerning experience and scores on the solder certification test (see Figure 20–21).

The equation of the regression line, or trend line, provides a clear and understandable measure of the change caused in the output variable by every incremental change of the input or independent variable. Using this principle, we can predict the effect of changes in the process.

One of the most important contributions that can be made by trend analysis is forecasting. Forecasting enables us to predict what is likely to occur in the future. Based on the regression line we can forecast what will happen as the independent variable attains values beyond the existing data.

Control Charts

The use of control charts focuses on the prevention of defects, rather than their detection and rejection. In business, government, and industry, economy and efficiency are always best served by prevention. It costs much more to produce an unsatisfactory product or service than it does to produce a satisfactory one. There are

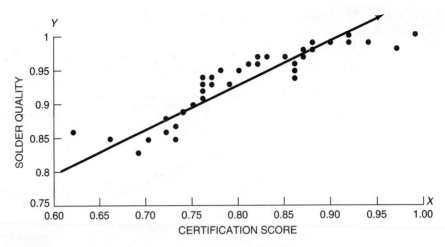

FIGURE 20–21.   Scatter plot solder quality and certification score.

PMBOK® Guide, 4th Edition
8.3.2.2 Perform Quality Control—
Tools and Techniques

many costs associated with producing unsatisfactory goods and services. These costs are in labor, materials, facilities, and the loss of customers. The cost of producing a proper product can be reduced significantly by the application of statistical process control charts.

Control Charts and the Normal Distribution

The construction, use, and interpretation of control charts is based on the normal statistical distribution as indicated in Figure 20–22. The centerline of the control chart represents the average or mean of the data ($\overline{X}$). The upper and lower control limits (UCL and LCL), respectively, represent this mean plus and minus three standard deviations of the data ($\overline{X} \pm 3s$). Either the lowercase s or the Greek letter σ (sigma) represents the standard deviation for control charts.

The normal distribution and its relationship to control charts is represented on the right of the figure. The normal distribution can be described entirely by its mean and standard deviation. The normal distribution is a bell-shaped curve (sometimes called the Gaussian distribution) that is symmetrical about the mean, slopes downward on both sides to infinity, and theoretically has an infinite range. In the normal distribution 99.73 percent of all measurements lie within $\overline{X} + 3s$ and $\overline{X} - 3s$; this is why the limits on control charts are called three-sigma limits.

Companies like Motorola have embarked upon a six-sigma limit rather than a three-sigma limit. The benefit is shown in Table 20–4. With a six-sigma limit, only two defects per billion are allowed. Maintaining a six-sigma limit can be extremely expensive unless the cost can be spread out over, say, 1 billion units produced.

Control chart analysis determines whether the inherent process variability and the process average are at stable levels, whether one or both are out of statistical control (not

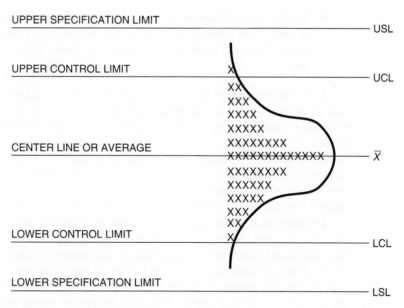

FIGURE 20–22. The control chart and the normal curve.

stable), or whether appropriate action needs to be taken. Another purpose of using control charts is to distinguish between the inherent, random variability of a process and the variability attributed to an assignable cause. The sources of random variability are often referred to as common causes. These are the sources that cannot be changed readily, without significant restructuring of the process. Special cause variability, by contrast, is subject to correction within the process under process control.

- *Common cause variability or variation:* This source of random variation is always present in any process. It is that part of the variability inherent in the process itself. The cause of this variation can be corrected only by a management decision to change the basic process.

TABLE 20–4. ATTRIBUTES OF THE NORMAL (STANDARD) DISTRIBUTION

Specification Range (in ± Sigmas)	Percent within Range	Defective Parts per Billion
1	68.27	317,300,000
2	95.45	45,400,000
3	99.73	2,700,000
4	99.9937	63,000
5	99.999943	57
6	99.9999998	2

● *Special cause variability or variation:* This variation can be controlled at the local or operational level. Special causes are indicated by a point on the control chart that is beyond the control limit or by a persistent trend approaching the control limit.

To use process control measurement data effectively, it is important to understand the concept of variation. No two product or process characteristics are exactly alike, because any process contains many sources of variability. The differences between products may be large, or they may be almost immeasurably small, but they are always present. Some sources of variation in the process can cause immediate differences in the product, such as a change in suppliers or the accuracy of an individual's work. Other sources of variation, such as tool wear, environmental changes, or increased administrative control, tend to cause changes in the product or service only over a longer period of time.

To control and improve a process, we must trace the total variation back to its sources: common cause and special cause variability. Common causes are the many sources of variation that always exist within a process that is in a state of statistical control. Special causes (often called assignable causes) are any factors causing variation that cannot be adequately explained by any single distribution of the process output, as would be the case if the process were in statistical control. Unless all the special causes of variation are identified and corrected, they will continue to affect the process output in unpredictable ways.

The factors that cause the most variability in the process are the main factors found on cause-and-effect analysis charts: people, machines, methodology, materials, measurement, and environment. These causes can either result from special causes or be common causes inherent in the process.

● The theory of control charts suggests that if the source of variation is from chance alone, the process will remain within the three-sigma limits.
● When the process goes out of control, special causes exist. These need to be investigated, and corrective action must be taken.

Control Chart Types

PMBOK® Guide, 4th Edition
8.3.2.2 Perform Quality
Control—Tools and Techniques

Just as there are two types of data, continuous and discrete, there are two types of control charts: variable charts for use with continuous data and attribute charts for use with discrete data. Each type of control chart can be used with specific types of data. Table 20–5 provides a brief overview of the types of control charts and their applications.

Variables Charts. Control charts for variables are powerful tools that we can use when measurements from a process are variable. Examples of variable data are the diameter of a bearing, electrical output, or the torque on a fastener.

As shown in Table 20–5, $\overline{X}$ and R charts are used to measure control processes whose characteristics are continuous variables such as weight, length, ohms, time, or volume. The p and np charts are used to measure and control processes displaying attribute characteristics in a sample. We use p charts when the number of failures is expressed as a fraction, or np charts when the failures are expressed as a number. The c and u charts are used to measure

TABLE 20–5. TYPES OF CONTROL CHARTS AND APPLICATIONS

Variables Charts	Attributes Charts
$\bar{X}$ and R charts: To observe changes in the mean and range (variance) of a process.	p chart: For the fraction of attributes nonconforming or defective in a sample of varying size.
$\bar{X}$ and s charts: For a variable average and standard deviation.	np charts: For the number of attributes nonconforming or defective in a sample of constant size.
$\bar{X}$ and s^2 charts: for a variable average and variance.	c charts: For the number of attributes nonconforming or defects in a single item within a subgroup, lot, or sample area of constant size.
	u charts: For the number of attributes nonconforming or defects in a single item within a subgroup, lot, or sample area of varying size.

the number or portion of defects in a single item. The c control chart is applied when the sample size or area is fixed, and the u chart when the sample size or area is not fixed.

Attribute Charts. Although control charts are most often thought of in terms of variables, there are also versions for attributes. Attribute data have only two values (conforming/nonconforming, pass/fail, go/no-go, present/absent), but they can still be counted, recorded, and analyzed. Some examples are: the presence of a required label, the installation of all required fasteners, the presence of solder drips, or the continuity of an electrical circuit. We also use attribute charts for characteristics that are measurable, if the results are recorded in a simple yes/no fashion, such as the conformance of a shaft diameter when measured on a go/no-go gauge, or the acceptability of threshold margins to a visual or gauge check.

It is possible to use control charts for operations in which attributes are the basis for inspection, in a manner similar to that for variables but with certain differences. If we deal with the fraction rejected out of a sample, the type of control chart used is called a p chart. If we deal with the actual number rejected, the control chart is called an np chart. If articles can have more than one nonconformity, and all are counted for subgroups of fixed size, the control chart is called a c chart. Finally, if the number of nonconformities per unit is the quantity of interest, the control chart is called a u chart.

The power of control charts (Shewhart techniques) lies in their ability to determine if the cause of variation is a special cause that can be affected at the process level, or a common cause that requires a change at the management level. The information from the control chart can then be used to direct the efforts of engineers, technicians, and managers to achieve preventive or corrective action.

The use of statistical control charts is aimed at studying specific ongoing processes in order to keep them in satisfactory control. By contrast, downstream inspection aims to identify defects. In other words, control charts focus on prevention of defects rather than detection and rejection. It seems reasonable, and it has been confirmed in practice, that economy and efficiency are better served by prevention rather than detection.

Control Chart Components

All control charts have certain features in common (Figure 20–23). Each control chart has a centerline, statistical control limits, and the calculated attribute or control data. Some control charts also contain specification limits.

The centerline is a solid (unbroken) line that represents the mean or arithmetic average of the measurements or counts. This line is also referred to as the X bar line ($\bar{X}$). There are two statistical control limits: the upper control limit for values greater than the mean and the lower control limit for values less than the mean.

Specification limits are used when specific parametric requirements exist for a process, product, or operation. These limits usually apply to the data and are the pass/fail criteria for the operation. They differ from statistical control limits in that they are prescribed for a process, rather than resulting from the measurement of the process.

The data element of control charts varies somewhat among variable and attribute control charts. We will discuss specific examples as a part of the discussion on individual control charts.

Control Chart Interpretation

PMBOK® Guide, 4th Edition
8.3.2.2 Perform Quality
Control—Tools and Techniques

There are many possibilities for interpreting various kinds of patterns and shifts on control charts. If properly interpreted, a control chart can tell us much more than whether the process is in or out of control. Experience and training can help extract clues regarding process behavior, such as that shown in Figure 20–24. Statistical guidance is invaluable, but an intimate knowledge of the process being studied is vital in bringing about improvements.

A control chart can tell us when to look for trouble, but it cannot by itself tell us where to look, or what cause will be found. Actually, in many cases, one of the greatest

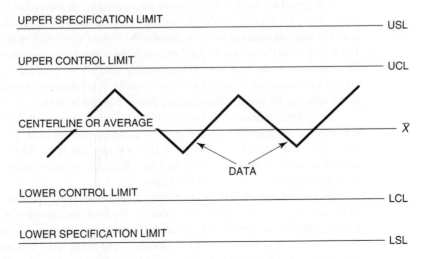

FIGURE 20–23.   Control chart elements.

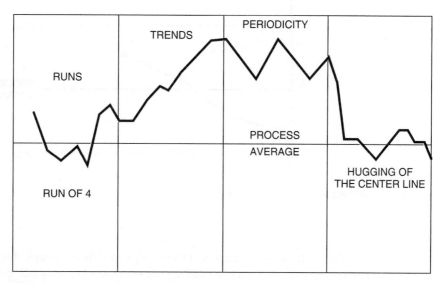

FIGURE 20–24. Control chart interpretation.

benefits from a control chart is that it tells when to leave a process alone. Sometimes the variability is increased unnecessarily when an operator keeps trying to make small corrections, rather than letting the natural range of variability stabilize. The following paragraphs describe some of the ways the underlying distribution patterns can behave or misbehave.

Runs. When several successive points line up on one side of the central line, this pattern is called a run. The number of points in that run is called the length of the run. As a rule of thumb, if the run has a length of seven points, there is an abnormality in the process. Figure 20–25 demonstrates a run.

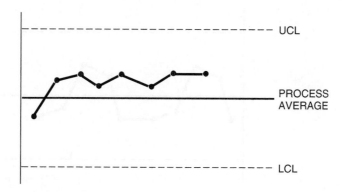

FIGURE 20–25. Process run.

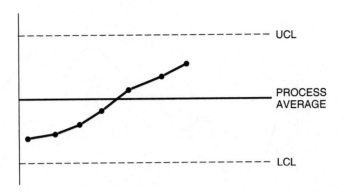

FIGURE 20–26. Control chart trends.

Trends. If there is a continued rise of all in a series of points, this pattern is called a trend. In general, if seven consecutive points continue to rise or fall, there is an abnormality. Often, the points go beyond one of the control limits before reaching seven. Figure 20–26 demonstrates a trend.

Periodicity. Points that show the same pattern of change (rise or fall) over equal intervals denote periodicity. Figure 20–27 demonstrates periodicity.

Hugging the Centerline or Control Limit. Points on the control chart that are close to the central line, or to the control limit, are said to hug the line. Often, in this situation, a different type of data or data from different factors have been mixed into the subgroup. In such cases it is necessary to change the subgrouping, reassemble the data, and redraw the control chart. To decide whether there is hugging of the centerline, draw two lines on the control chart, one between the centerline and the UCL and the other between the centerline and the LCL. If most of the points are between these two lines, there is an abnormality. To see whether there is hugging of one of the control limits, draw a line two-thirds of the

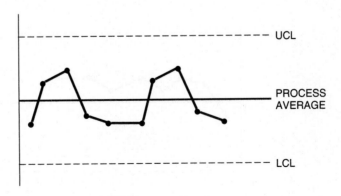

FIGURE 20–27. Control chart periodicity.

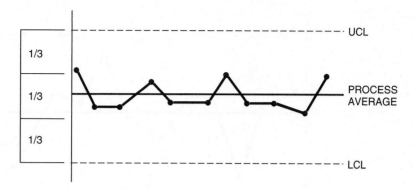

FIGURE 20–28. Hugging the centerline.

distance between the centerline and each of the control lines. There is abnormality if 2 out of 3 points, 3 out of 7 points, or 4 out of 10 points lie within the outer one-third zone. The abnormalities should be evaluated for their cause(s) and the corrective action taken. Figure 20–28 demonstrates data hugging the LCL.

Out of Control. An abnormality exists when data points exceed either the upper or lower control limits. Figure 20–29 illustrates this occurrence.

In Control. No obvious abnormalities appear in the control chart. Figure 20–30 demonstrates this desirable process state.

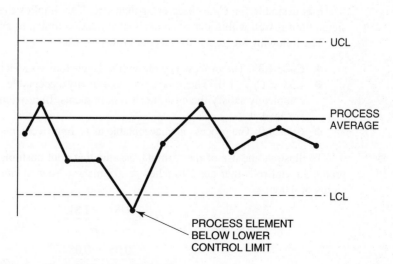

FIGURE 20–29. Control chart out of control.

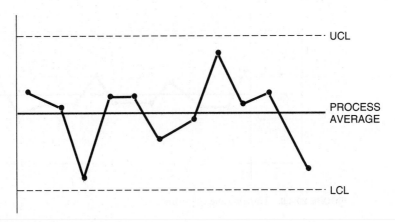

FIGURE 20–30. Process in control.

20.10 PROCESS CAPABILITY (C_P)

Process capability, for a stable manufacturing process, is the ability to produce a product that conforms to design specifications. Because day-to-day variations can occur during manufacturing, process capability is a statement about product uniformity. Process capability, as measured by the quality characteristics of the product of the process, is expressed as the mean value plus or minus three standard deviations. Mathematically:

$$C_P = \frac{\text{USL} - \text{LSL}}{6\sigma}$$

It is desirable for C_P to be greater than one. This implies that the process of three-sigma limit is well within the customer's specification limits, as shown in Figure 20–31. The following are generally accepted rules for C_P:

- $C_P > 1.33$: The process is well within the customer's specifications requirements.
- $1.33 \geq C_P > 1.0$: The process is marginally acceptable. The process may not completely satisfy the customer's requirements. Improvements in process control are needed.
- $C_P \leq 1.0$: The process is unacceptable as is. Improvements are mandatory.

To illustrate the use of the formula, assume that your customer's requirements are to produce metal rods that are 10 inches $\pm$ .05 inches. Your manufacturing process has a sigma of 0.008.

$$C_P = \frac{\text{USL} - \text{LSL}}{6\sigma}$$

$$= \frac{0.05 + 0.05}{6(0.008)}$$

$$= 2.08$$

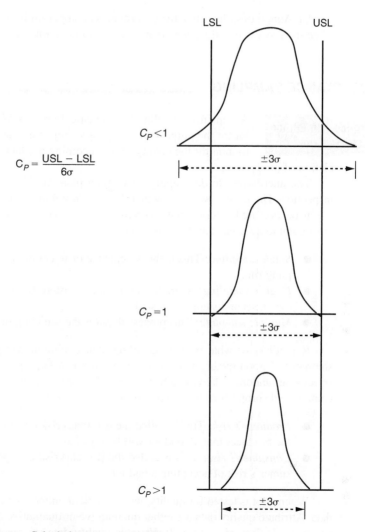

$$C_P = \frac{USL - LSL}{6\sigma}$$

FIGURE 20–31. Calculating process capability.

Looking at Figure 20–31, C_P is the relative spread of the process width within the specification width. Unfortunately, the spread of the process capability, even for very good values, could be poorly positioned within the specification width. The process width could easily be hugging either the USL or LSL. Today, process capability is measured by both C_P and C_{Pk}, where C_{Pk} is the capability index with correction (k) for noncentrality. According to Dr. Frank Anbari, the formula for C_{Pk} can be simplified as:

$$C_{Pk} = \left| \frac{CL - \text{Closest specification limit}}{3\sigma} \right|$$

where CL is the center of the process, that is, its average.

Dr. Anbari postulates that the C_P provides an upper limit for the C_{Pk}, which is reached when the process is fully centered around the nominal dimension.

20.11 ACCEPTANCE SAMPLING

PMBOK® Guide, 4th Edition
8.3 Perform Quality Control

Acceptance sampling is a statistical process of evaluating a portion of a lot for the purpose of accepting or rejecting the entire lot. It is an attempt to monitor the quality of the incoming product or material after the completion of production.

The alternatives to developing a sampling plan would be 100% inspection and 0% inspection. The costs associated with 100% are prohibitive, and the risks associated with 0% inspection are likewise large. Therefore, some sort of compromise is needed. The three most commonly used sampling plans are:

- *Single sampling:* This is the acceptance or rejection of a lot based upon one sampling run.
- *Double sampling:* A small sample size is tested. If the results are not conclusive, then a second sample is tested.
- *Multiple sampling:* This process requires the sampling of several small lots.

Regardless of what type of sampling plan is chosen, sampling errors can occur. A shipment of good-quality items can be rejected if a large portion of defective units are selected at random. Likewise, a bad-quality shipment can be accepted if the tested sample contains a disproportionately large number of quality items. The two major risks are:

- *Producer's risk:* This is called the α (alpha) risk or type I error. This is the risk to the producer that a good lot will be rejected.
- *Consumer's risk:* This is called the β (beta) risk or type II error. This is the consumer's risk of accepting a bad lot.

When a lot is tested for quality, we can look at either "attribute" or "variable" quality data. Attribute quality data are either quantitative or qualitative data for which the product or service is designed and built. Variable quality data are quantitative, continuous measurement processes to either accept or reject the lot. The exact measurement can be either destructive or nondestructive testing.

20.12 IMPLEMENTING SIX SIGMA[4]

PMBOK® Guide, 4th Edition
Chapter 8 Introduction

Six Sigma is a business initiative first espoused by Motorola in the early 1990s. Recent Six Sigma success stories, primarily from the likes of

---

4. Adapted from Forrest W. Breyfogle, III, *Implementing Six Sigma* (New York: Wiley, 1999), pp. 5–7.

General Electric, Sony, AlliedSignal, and Motorola, have captured the attention of Wall Street and have propagated the use of this business strategy. The Six Sigma strategy involves the use of statistical tools within a structured methodology for gaining the knowledge needed to create products and services better, faster, and less expensively than the competition. The repeated, disciplined application of the master strategy on project after project, where the projects are selected based on key business issues, is what drives dollars to the bottom line, resulting in increased profit margins and impressive return on investment from the Six Sigma training. The Six Sigma initiative has typically contributed an average of six figures per project to the bottom line. The Six Sigma project executioners are sometimes called "black belts," "top guns," "change agents," or "trailblazers," depending on the company deploying the strategy. These people are trained in the Six Sigma philosophy and methodology and are expected to accomplish at least four projects annually, which should deliver at least $500,000 annually to the bottom line. A Six Sigma initiative in a company is designed to change the culture through breakthrough improvement by focusing on out-of-the-box thinking in order to achieve aggressive, stretch goals. Ultimately, Six Sigma, if deployed properly, will infuse intellectual capital into a company and produce unprecedented knowledge gains that translate directly into bottom line results.[5]

Former General Electric (GE) CEO Jack Welch described Six Sigma as "the most challenging and potentially rewarding initiative we have ever undertaken at General Electric." The GE 1997 annual report stated that Six Sigma delivered more than $300 million to its operating income. In 1998, they expected to more than double this operating profit impact. GE listed in its annual report the following to exemplify these Six Sigma benefits:

- Medical Systems described how Six Sigma designs have produced a 10-fold increase in the life of CT scanner X-ray tubes—increasing the "uptime" of these machines and the profitability and level of patient care given by hospitals and other health care providers.
- Superabrasives—our industrial diamond business—described how Six Sigma quadrupled its return on investment and, by improving yields, is giving it a full decade's worth of capacity despite growing volume—without spending a nickel on plant and equipment capacity.
- Our railcar leasing business described 62% reduction in turnaround time at its repair shops: an enormous productivity gain for our railroad and shipper customers and for a business that's now two or three times faster than its nearest rival because of Six Sigma improvements. In the next phase across the entire shop network, black belts and green belts, working with their teams, redesigned the overhaul process, resulting in a 50% further reduction in cycle time.
- The plastics business, through rigorous Six Sigma process work, added 300 million pounds of new capacity (equivalent to a "free plant"), saved $400 million in investment and will save another $400 by 2000.[6]

---

5. Information in this paragraph was contributed by J. Kiemele, Ph.D., of Air Academy Associates.

6. 1998 GE Annual Report.

20.13 LEAN SIX SIGMA AND DMAIC[7]

Six Sigma is a quality initiative that was born at Motorola in the 1980s. The primary focus of the Six Sigma process improvement methodology, also known as DMAIC, is to reduce defects that are defined by the customer of the process. This customer can be internal or external. It is whoever is in receipt of the process output. Defects are removed by careful examination from a Six Sigma team made up of cross-functional positions having different lines of sight into the process. The team follows the rigor of the define, measure, analyze, improve, and control (DMAIC) methodology to determine the root cause(s) of the defects. The team uses data and appropriate numerical and graphical analysis tools to raise awareness of process variables generating defects. Data collection and analysis is at the core of Six Sigma. "Extinction by instinct" is the phrase often used to describe intuitive decision-making and performance analysis. It has been known to generate rework, frustration, and ineffective solutions. Six Sigma prescribes disciplined gathering and analysis of data to effectively identify solutions.

Lean manufacturing is another aspect of process improvement derived mostly from the Toyota Production System (TPS). The primary focus of lean is to remove waste and improve process efficiency. Lean is often linked with Six Sigma because both emphasize the importance of minimal process variation. Lean primarily consists of a set of tools designed to assist in the identification and steady elimination of waste (muda), allowing for the improvement of quality as well as cycle time and cost reduction. To solve the problem of waste, lean manufacturing utilizes several tools. These include accelerated DMAIC projects known as kaizen events, cause-and-effect analysis using "five whys" and error proofing with a technique known as poka-yoke.

Kaizen Events. The source of the word *kaizen* is Japanese: *Kai* (take apart) and *Zen* (make good). This is an action-oriented approach to process improvement. Team members devote 3–5 consecutive days to quickly work through the DMAIC methodology in a workshop fashion.

Five Whys. This technique is used to move past symptoms of problems and drill down to the root causes. With every answer comes a new question until you've gotten to the bottom of the problem. Five is a rule of thumb. Sometimes you'll only need three questions, other times it might take seven. The goal is to identify the root cause of process defects and waste.

Poka-Yoke. The source of this technique is Japanese: *Yokeru* (to avoid) and *Poka* (inadvertent errors). There are three main principles of poka-yoke. (1) Make wrong actions more difficult. (2) Make mistakes obvious to the person so that the mistake can be corrected. (3) Detect errors so that downstream consequences can be prevented by stopping the flow or other corrective action. The philosophy behind this technique is that it's good to do things right the first time, but it is even better to make it impossible to do it wrong the first time.

When Six Sigma and lean manufacturing are integrated, the project team utilizes the project management methodology to lead them through the lean Six Sigma toolbox and make dramatic improvements to business processes. The overall goal is to reduce defects that impact the internal and external customer and eliminate waste that impact the cycle times and costs.

---

7. The section was provided by Anne Foley, Director of Six Sigma for the International Institute for Learning.

20.14 QUALITY LEADERSHIP[8]

<table>
<tr><td>**PMBOK® Guide, 4th Edition**
Chapter 9 Human Resources
Management</td><td>Consider for a moment the following seven items:</td></tr>
</table>

- Teamwork
- Strategic integration
- Continuous improvement
- Respect for people
- Customer focus
- Management-by-fact
- Structured problem-solving

Some people contend that these seven items are the principles of project management when, in fact, they are the seven principles of the total quality management program at Sprint. Project management and TQM have close similarity in leadership and team-based decision-making. According to Breyfogle,[9] American managers have often conducted much of their business through an approach that is sometimes called *management by results*. This type of management tends to focus only on the end result, that is, process yield, gross margin, sales dollars, return on investment, and so on. Emphasis is placed on a chain of command with a hierarchy of standards, objectives, controls, and accountability. Objectives are translated into work standards or quotas that guide the performance of employees. Use of these numerical goals can cause short-term thinking, misdirected focus, fear (e.g., of a poor job performance rating), fudging the numbers, internal conflict, and blindness to customer concerns. This type of management is said to be like trying to keep a dog happy by forcibly wagging its tail.

Quality leadership is an alternative that emphasizes results by working on methods. In this type of management, every work process is studied and constantly improved so that the final product or service not only meets but exceeds customer expectations. The principles of quality leadership are customer focus, obsession with quality, effective work structure, control yet freedom (e.g., management in control of employees yet freedom given to employees), unity of purpose, process defect identification, teamwork, and education and training. These principles are more conducive to long-term thinking, correctly directed efforts, and a keen regard for the customer's interest.

Quality leadership does have a positive effect on the return on investment. In 1950, Deming described this chain reaction of getting a greater return on investment as follows: improve quality → decrease costs → improve productivity → decrease prices → increase market share in business → provide jobs → increase return on investment. Quality is not something that can be delegated to others. Management must lead the transformation process.

To give quality leadership, the historical hierarchical management structure needs to be changed to a structure that has a more unified purpose using project teams. A single

---

8. Adapted from Forrest W. Breyfogle, III, *Implementing Six Sigma* (New York: Wiley, 1999), pp. 28–29.

9. Adapted from Forrest W. Breyfogle, III, *Implementing Six Sigma* (New York: Wiley, 1999), pp. 28–29.

person can make a big difference in an organization. However, one person rarely has enough knowledge or experience to understand everything within a process. Major gains in both quality and productivity can often result when a team of people pool their skills, talents, and knowledge.

Teams need to have a systematic plan to improve the process that creates mistakes/defects, breakdowns/delays, inefficiencies, and variation. For a given work environment, management needs to create an atmosphere that supports team effort in all aspects of business. In some organizations, management may need to create a process that describes hierarchical relationships between teams, the flow of directives, how directives are transformed into action and improvements, and the degree of autonomy and responsibility of the teams. The change to quality leadership can be very difficult. It requires dedication and patience to transform an entire organization.

20.15 RESPONSIBILITY FOR QUALITY

Everyone in an organization plays an important role in quality management. In order for an organization to become a quality organization, all levels must actively participate, and, according to Dr. Edwards Deming, the key to successful implementation of quality starts at the top.

Top management must drive fear from the workplace and create an environment where cross-functional cooperation can flourish. The ultimate responsibility for quality in the organization lies in the hands of upper management. It is only with their enthusiastic and unwavering support that quality can thrive in an organization.

The project manager is ultimately responsible for the quality of the project. This is true for the same reason the president of the company is ultimately responsible for quality in a corporation. The project manager selects the procedures and policies for the project and therefore controls the quality. The project manager must create an environment that fosters trust and cooperation among the team members. The project manager must also support the identification and reporting of problems by team members and avoid at all costs a "shoot the messenger" mentality.

The project team members must be trained to identify problems, recommend solutions, and implement the solutions. They must also have the authority to limit further processing when a process is outside of specified limits. In other words, they must be able to halt any activity that is outside of the quality limits set for the project and work toward a resolution of the problem at any point in the project.

20.16 QUALITY CIRCLES

Quality circles are small groups of employees who meet frequently to help resolve company quality problems and provide recommendations to management. Quality circles were initially developed in Japan and have achieved some degree of success in the United States.

The employees involved in quality circles meet frequently either at someone's home or at the plant before the shift begins. The group identifies problems, analyzes data, recommends solutions, and carries out management-approved changes. The success of quality circles is heavily based upon management's willingness to listen to employee recommendations.

The key elements of quality circles include:

- They give a team effort.
- They are completely voluntary.
- Employees are trained in group dynamics, motivation, communications, and problem solving.
- Members rely upon each other for help.
- Management support is active but as needed.
- Creativity is encouraged.
- Management listens to recommendations.

The benefits of quality circles include:

- Improved quality of products and services
- Better organizational communications
- Improved worker performance
- Improved morale

20.17 JUST-IN-TIME MANUFACTURING (JIT)

Just-in-time manufacturing is a process that continuously stresses waste reduction by optimizing the processes and procedures necessary to maintain a manufacturing operation. Part of this process is JIT purchasing or inventory where the materials needed appear just in time for use, thus eliminating costs associated with material handling, storage, paperwork, and even inspection. In order to eliminate inspection, the customer must be convinced that the contractor has adhered to all quality requirements. In other words, JIT inventory pushes quality assurance and quality control for that product down to the contractor's level.

The customer benefits from JIT purchasing by developing long-term relationships with *fewer* suppliers, thus lowering subcontractor management costs. The contractor benefits by having long-term contracts. However, the contractor must agree to special conditions such as on-site inspections by the customer's executives, project manager, or quality team, or even allowing an on-site customer representative at the contractor's location.

JIT purchasing has been widely adopted in Japan, but only marginal success has occurred here in the United States. Table 20–6 shows the relative comparison of American versus Japanese quality practices.

Another part of JIT manufacturing is the identification and continuous reduction of waste. Shigeo Shingo of Toyota Motor Company has identified seven wastes that should be the targets of a continuous improvement process. These appear in Table 20–7.

TABLE 20–6. COMPARATIVE ANALYSIS OF PURCHASING PRACTICE: TRADITIONAL U.S. AND JAPANESE JIT

Purchasing Activity	JIT Purchasing	Traditional Purchasing
Purchase lot size	Purchase in small lots with frequent deliveries	Purchase in large batch size with less frequent deliveries
Selecting supplier	Single source of supply for a given part in nearby geographical area with a long-term contract	Rely on multiple sources of supply for a given part and short-term contracts
Evaluating supplier	Emphasis is placed on product quality, delivery performance, and price, but *no* percentage of reject from supplier is acceptable	Emphasis is placed on product quality, delivery performance, and price but about two percent reject from supplier is acceptable
Receiving inspection	Counting and receiving inspection of incoming parts is reduced and eventually eliminated	Buyer is responsible for receiving, counting, and inspecting all incoming parts
Negotiating and bidding process	Primary objective is to achieve product quality through a long-term contract and fair price	Primary objective is to get the lowest possible price
Determing mode of transportation	Concern for both inbound and outbound freight, and on-time delivery. Delivery schedule left to the buyer	Concern for outbound freight and lower outbound costs. Delivery schedule left to the supplier
Product specification	"Loose" specifications. The buyer relies more on performance specifications than on product design and the supplier is encouraged to be more innovative	"Rigid" specifications. The buyer relies more on design specifications than on product performance and suppliers have less freedom in design specifications
Paperwork	Less formal paperwork. Delivery time and quantity level can be changed by telephone calls	Requires great deal of time and formal paperwork. Changes in delivery date and quantity require purchase orders
Packaging	Small standard containers used to hold exact quantity and to specify the precise specifications	Regular packaging for every part type and part number with no clear specifications on product content

Source: Sang M. Lee and A. Ansari, "Comparative Analysis of Japanese Just-in-Time Purchasing and Traditional Purchasing Systems," *International Journal of Operations and Product Management,* 5, no. 4 (1985), pp. 5–14.

Two new topics are now being discussed as part of JIT manufacturing: value-added manufacturing and stockless production. Value-added manufacturing advocates the elimination of any step in the manufacturing process that does not add value to the product for the customer. Examples include process delays, transporting materials, work-in-process inventories, and excessive paperwork. Stockless production promotes little inventories for

TABLE 20–7. THE SEVEN WASTES

1. *Waste of overproduction.* Eliminate by reducing setup times, synchronizing quantities and timing between processes, compacting layout, visibility, and so forth. Make only what is needed now.
2. *Waste of waiting.* Eliminate through synchronizing work flow as much as possible, and balance uneven loads by flexible workers and equipment.
3. *Waste of transportation.* Establish layouts and locations to make transport and handling unnecessary if possible. Then rationalize transport and material handling that cannot be eliminated.
4. *Waste of processing itself.* First question why this part or product should be made at all, then why each process is necessary. Extend thinking beyond economy of scale or speed.
5. *Waste of stocks.* Reduce by shortening setup times and reducing lead times, by synchronizing work flows and improving work skills, and even by smoothing fluctuations in demand for the product. Reducing all the other wastes reduces the waste of stocks.
6. *Waste of motion.* Study motion for economy and consistency. Economy improves productivity, and consistency improves quality. First improve the motions, then mechanize or automate. Otherwise there is danger of automating waste.
7. *Waste of making defective products.* Develop the production process to prevent defects from being made so as to eliminate inspection. At each process, accept no defects and make no defects. Make processes failsafe to do this. From a quality process comes a quality product—automatically.

Source: R. Hall, *Attaining Manufacturing Excellence.* (Homewood, IL: Dow-Jones-Irwin, 1987), p. 26.

raw materials, work in process, and finished goods. Everything ends up being made to order and then delivered as needed. Waste becomes nonexistent. The practicality and risks of this approach may not be feasible for either the company or the project manager.

20.18 TOTAL QUALITY MANAGEMENT (TQM)[10]

PMBOK® Guide, 4th Edition
8.1.1 Quality Planning

There is no explicit definition of total quality management. Some people define it as providing the customer with quality products at the right time and at the right place. Others define it as meeting or exceeding customer requirements. Internally, TQM can be defined as less variability in the quality of the product and less waste.

Figure 20–32 shows the basic objectives and focus areas of a TQM process. Almost all companies have a primary strategy to obtain TQM, and the selected strategy is usually in place over the long term. The most common primary strategies are listed below. A summary of the seven primary improvement strategies mapped onto 17 corporations is shown in Table 20–8.

Primary strategies:

● Solicit ideas for improvement from employees.
● Encourage and develop teams to identify and solve problems.

10. This section has been adapted from C. Carl Pegels, *Total Quality Management* (Danvers, MA: Boyd & Fraser, 1995), pp. 4–27.

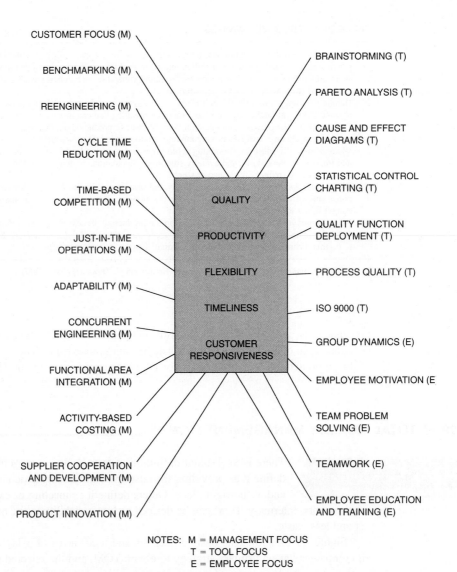

FIGURE 20–32. TQM objectives and focus areas. *Source:* C. Carl Pegels, *Total Quality Management* (Danvers, MA: Boyd & Fraser, 1995), p. 6.

- Encourage team development for performing operations and service activities resulting in participative leadership.
- Benchmark every major activity in the organization to ensure that it is done in the most efficient and effective way.
- Utilize process management techniques to improve customer service and reduce cycle time.

TABLE 20–8. PRIMARY IMPROVEMENT STRATEGIES EMPLOYED BY LISTED CORPORATIONS

	Strategy						
	P1	*P2*	*P3*	*P4*	*P5*	*P6*	*P7*
Asea, Brown, Boveri		X					
AT&T				X			
Cigna						X	
DuPont				X			
Eastman Kodak		X					
Eaton Corp.	X	X	X				
Ford Motor Company	X						
General Motors				X			
Goodyear Tire		X	X				
IBM Rochester							X
ICL Plc							X
Johnson Controls							X
Motorola				X			
New England Corp.					X		
New York Life		X	X				
Pratt and Whitney			X				
Xerox Corp.				X			

Source: C. Carl Pegels, *Total Quality Management* (Danvers, MA: Boyd & Fraser, 1995), p. 21.

TABLE 20–9. SECONDARY IMPROVEMENT STRATEGIES EMPLOYED BY LISTED CORPORATIONS

	Strategy									
	S1	*S2*	*S3*	*S4*	*S5*	*S6*	*S7*	*S8*	*S9*	*S10*
AMP Corp.	X	X							X	X
Asea, Brown, Boveri	X	X								
British Telecom	X						X			
Chrysler Corp.					X				X	X
Coca-Cola									X	
Corning							X			
Eastman Kodak	X									
Eaton Corp.									X	
Fidelity Investment	X			X					X	
Ford Motor Company									X	
Fujitsu Systems	X		X						X	
General Motors					X				X	X
Holiday Inns			X							
IBM Rochester	X			X					X	X
ICL Plc		X								
Johnson Controls	X	X					X			
Motorola								X		
New England Corp.	X									
New York Life	X	X								
Pratt and Whitney						X				
Procter & Gamble		X							X	
The Forum Corp.	X									X
VF Corp.				X						X
Xerox Corp.	X									

Source: C. Carl Pegels, *Total Quality Management* (Danvers, MA: Boyd & Fraser, 1995).

- Develop and train customer staff to be entrepreneurial and innovative in order to find ways to improve customer service.
- Implement improvements so that the organization can qualify as an ISO 9000 supplier.

There also exist secondary strategies that, over the long run, focus on operations and profitability. Typical secondary strategies are shown below, and Table 20–10 identifies the secondary improvement strategies by listed companies.

Secondary strategies:

- Maintain continuous contact with customers; understand and anticipate their needs.
- Develop loyal customers by not only pleasing them but by exceeding their expectations.
- Work closely with suppliers to improve their product/service quality and productivity.
- Utilize information and communication technology to improve customer service.
- Develop the organization into manageable and focused units in order to improve performance.
- Utilize concurrent or simultaneous engineering.

TABLE 20–10. SUMMARY ILLUSTRATIONS OF QUANTIFIED IMPROVEMENTS ACHIEVED

AMP. On-time shipments improved from 65% to 95%, and AMP products have nationwide availability within three days or less on 50% of AMP sales.

Asea, Brown, Boveri. Every improvement goal customers asked for—better delivery, quality responsiveness, and so on—was met.

Chrysler. New vehicles are now being developed in 33 months versus as long as 60 months 10 years ago.

Eaton. Increased sales per employee from $65,000 in 1983 to about $100,000 in 1992.

Fidelity. Handles 200,000 information calls in 4 telephone centers; 1,200 representatives handle 75,000 calls, and the balance is automated.

Ford. Use of 7.25 man-hours of labor per vehicle versus 15 man-hours in 1980; Ford Taurus bumper uses 10 parts compared to 100 parts on similar GM cars.

General Motors. New vehicles are now being developed in 34 months versus 48 months in the 1980s.

IBM Rochester. Defect rates per million are 32 times lower than four years ago and on some products exceed six sigma (3.4 defects per million).

Pratt & Whitney. Defect rate per million was cut in half; a tooling process was shortened from two months to two days; part lead times were reduced by 43%.

VF Corp. Market response system enables 97% in-stock rate for retail stores compared to 70% industry average.

NCR. Checkout terminal was designed in 22 months versus 44 months and contained 85% fewer parts than its predecessor.

AT&T. Redesign of telephone switch computer completed in 18 months versus 36 months; manufacturing defects reduced by 87%.

Deere & Co. Reduced cycle time of some of its products by 60%, saving 30% of usual development costs.

Source: C. Carl Pegels, *Total Quality Management* (Danvers, MA: Boyd & Fraser, 1995), p. 27.

- Encourage, support, and develop employee training and education programs.
- Improve timeliness of all operation cycles (minimize all cycle times).
- Focus on quality, productivity, and profitability.
- Focus on quality, timeliness, and flexibility.

Information about quality improvements is difficult to obtain from corporations. Most firms consider this information confidential and do not like to publish for fear of providing an advantage to their competitors. As a result, the information in Table 20–11 is sketchy. It is simply a snapshot of a limited number of quantitative performance improvements that were achieved by firms as part of their total quality management programs.

One noteworthy achievement is Ford's reduction in man-hours to build a vehicle from 15 to 7.25. Although this took 10 years to achieve, it is still a sterling example of productivity improvement. IBM Rochester, Minnesota's reduction in defects per million by a factor of 32 over a 4-year period is also noteworthy. And the ability of Chrysler and General Motors to reduce their design development times for new vehicles from 60 and 48 months to the current 33 and 34 months, respectively, is an achievement that indicates the return of competitiveness to the U.S. automobile industry.

20.19 STUDYING TIPS FOR THE PMI® PROJECT MANAGEMENT CERTIFICATION EXAM

This section is applicable as a review of the principles to support the knowledge areas and domain groups in the PMBOK® Guide. This chapter addresses:

- Quality Management

Understanding the following principles is beneficial if the reader is using this text to study for the PMP® Certification Exam:

- Contributions by the quality pioneers
- Concept of total quality management (TQM)
- Ddifferences between quality planning, quality assurance, and quality control
- Importance of a quality audit
- Quality control tools
- Concept of cost of quality

The following multiple-choice questions will be helpful in reviewing the principles of this chapter:

1. Which of the following is not part of the generally accepted view of quality today?
   A. Defects should be highlighted and brought to the surface.
   B. We can inspect quality.

    C. Improved quality saves money and increases business.

    D. Quality is customer-focused.

2. In today's view of quality, who defines quality?

    A. Contractors' senior management

    B. Project management

    C. Workers

    D. Customers

3. Which of the following are tools of quality control?

    A. Sampling tables

    B. Process charts

    C. Statistical and mathematical techniques

    D. All of the above

4. Which of the following is true of modern quality management?

    A. Quality is defined by the customer.

    B. Quality has become a competitive weapon.

    C. Quality is now an integral part of strategic planning.

    D. All are true.

5. A company dedicated to quality usually provides training for:

    A. Senior management and project managers

    B. Hourly workers

    C. Salaried workers

    D. All employees

6. Which of the following quality gurus believe "zero-defects" is achievable?

    A. Deming

    B. Juran

    C. Crosby

    D. All of the above

7. What are the components of Juran's Trilogy?

    A. Quality Improvement, Quality Planning, and Quality Control

    B. Quality Improvement, Zero-Defects, and Quality Control

    C. Quality Improvement, Quality Planning, and Pert Charting

    D. Quality Improvement, Quality Inspections and Quality Control

8. Which of the following is not one of Crosby's Four Absolutes of Quality?

    A. Quality means conformance to requirements.

    B. Quality comes from prevention.

    C. Quality is measured by the cost of conformance.

    D. Quality means that the performance standard is "zero-defects."

9. According to Deming, what percentage of the costs of quality is generally attributable to management?

    A. 100%

    B. 85%

    C. 55%

    D 15%

10. Inspection:

    A. Is an appropriate way to ensure quality

    B. Is expensive and time-consuming

     C. Reduces rework and overall costs

     D. Is always effective in stopping defective products from reaching the customer

11. The Taguchi Method philosophies concentrate on improving quality during the:

     A. Conceptual Phase

     B. Design Phase

     C. Implementation Phase

     D. Closure Phase

12. A well-written policy statement on quality will:

     A. Be a statement of how, not what or why

     B. Promote consistency throughout the organization and across projects

     C. Provide an explanation of how customers view quality in their own organizations

     D. Provide provisions for changing the policy only on a yearly basis

13. Quality assurance includes:

     A. Identifying objectives and standards

     B. Conducting quality audits

     C. Planning for continuous collection of data

     D. All of the above

14. What is the order of the four steps in Deming's Cycle for Continuous Improvement?

     A. Plan, do, check, and act

     B. Do, plan, act, and check

     C. Check, do, act, and plan

     D. Act, check, do, and plan

15. Quality audits:

     A. Are unnecessary if you do it right the first time

     B. Must be performed daily for each process

     C. Are expensive and therefore not worth doing

     D. Are necessary for validation that the quality policy is being followed and adhered to

16. Which of the following are typical tools of statistical process control?

     A. Pareto analysis

     B. Cause-and-effect analysis

     C. Process control charts

     D. All of the above

17. Which of the following methods is best suited to identifying the "vital few?"

     A. Pareto analysis

     B. Cause-and-effect analysis

     C. Trend analysis

     D. Process control charts

18. When a process is set up optimally, the upper and lower specification limits typically are:

     A. Set equal to the upper and lower control limits

     B. Set outside the upper and lower control limits

     C. Set inside the upper and lower control limits

     D. Set an equal distance from the mean value

19. The upper and lower control limits are typically set:

     A. One standard deviation from the mean in each direction

     B. 3σ (three sigma) from the mean in each direction

     C. Outside the upper and lower specification limits

     D. To detect and flag when a process may be out of control

20. Which of the following is *not* indicative of today's views of the quality management process applied to a given project?
    A. Defects should be highlighted and brought to the surface.
    B. The ultimate responsibility for quality lies primarily with senior management or sponsor but everyone should be involved.
    C. Quality saves money.
    D. Problem identification leads to cooperative solutions.

21. If the values generated from a process are normally distributed around the mean value, what percentage of the data points generated by the process will *not* fall within plus or minus three standard deviations of the mean?
    A. 99.7%
    B. 95.4%
    C. 68.3%
    D. 0.3%

ANSWERS

1. B
2. D
3. D
4. D
5. D
6. C
7. A
8. C
9. B
10. B
11. B
12. B
13. D
14. A
15. D
16. D
17. A
18. B
19. B
20. B
21. D

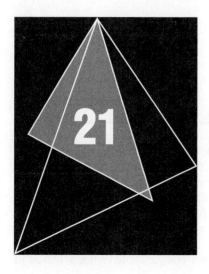

Modern Developments in Project Management

Related Case Studies (from Kerzner/*Project Management Case Studies,* 3rd Edition)	Related Workbook Exercises (from Kerzner/*Project Management Workbook and PMP®/CAPM® Exam Study Guide,* 10th Edition)	PMBOK® Guide, 4th Edition, Reference Section for the PMP® Certification Exam
• Lakes Automotive • Ferris HealthCare, Inc. • Clark Faucet Company	• Project Management Maturity Questionnaire • Multiple Choice Exam	None

21.0 INTRODUCTION

PMBOK® Guide, 4th Edition
PMBOK Chapters 1, 2, and 3 (inclusive)

As more industries accept project management as a way of life, the change in project management practices has taken place at an astounding rate. But what is even more important is the fact that these companies are sharing their accomplishments with other companies during benchmarking activities.

Eight recent interest areas are included in this chapter:

- The project management maturity model (PMMM)
- Developing effective procedural documentation
- Project management methodologies
- Continuous improvement
- Capacity planning
- Competency models

- Managing multiple projects
- End-of-phase review meetings

21.1 THE PROJECT MANAGEMENT MATURITY MODEL (PMMM)

All companies desire excellence in project management. Unfortunately, not all companies recognize that the time frame can be shortened by performing strategic planning for project management. The simple use of project management, even for an extended period of time, does *not* lead to excellence. Instead, it can result in repetitive mistakes and, what's worse, learning from your own mistakes rather than from the mistakes of others.

Companies such as Motorola, Nortel, Ericsson, and Compaq perform strategic planning for project management, and the results are self-explanatory. What Nortel and Ericsson have accomplished from 1992 to 1998, other companies have not achieved in twenty years of using project management.

Strategic planning for project management is unlike other forms of strategic planning in that it is most often performed at the middle-management level, rather than by executive management. Executive management is still involved, mostly in a supporting role, and provides funding together with employee release time for the effort. Executive involvement will be necessary to make sure that whatever is recommended by middle management will not result in unwanted changes to the corporate culture.

Organizations tend to perform strategic planning for new products and services by laying out a well-thought-out plan and then executing the plan with the precision of a surgeon. Unfortunately, strategic planning for project management, if performed at all, is done on a trial-by-fire basis. However, there are models that can be used to assist corporations in performing strategic planning for project management and achieving maturity and excellence in a reasonable period of time.

The foundation for achieving excellence in project management can best be described as the project management maturity model (PMMM), which is comprised of five levels, as shown in Figure 21–1. Each of the five levels represents a different degree of maturity in project management.

- *Level 1—Common Language:* In this level, the organization recognizes the importance of project management and the need for a good understanding of the basic knowledge on project management, along with the accompanying language/terminology.
- *Level 2—Common Processes:* In this level, the organization recognizes that common processes need to be defined and developed such that successes on one project can be repeated on other projects. Also included in this level is the recognition that project management principles can be applied to and support other methodologies employed by the company.
- *Level 3—Singular Methodology:* In this level, the organization recognizes the synergistic effect of combining all corporate methodologies into a singular methodology,

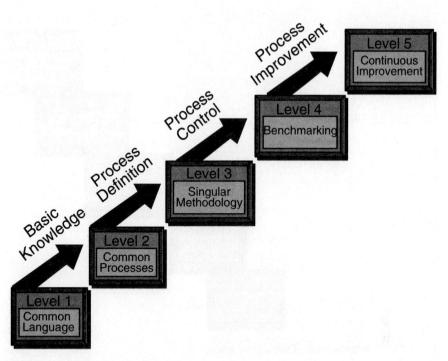

FIGURE 21–1. The five levels of maturity.

the center of which is project management. The synergistic effects also make process control easier with a single methodology than with multiple methodologies.

- *Level 4—Benchmarking:* This level contains the recognition that process improvement is necessary to maintain a competitive advantage. Benchmarking must be performed on a continuous basis. The company must decide whom to benchmark and what to benchmark.

- *Level 5—Continuous Improvement:* In this level, the organization evaluates the information obtained through benchmarking and must then decide whether or not this information will enhance the singular methodology.

When we talk about levels of maturity (and even life-cycle phases), there exists a common misbelief that all work must be accomplished sequentially (i.e., in series). This is not necessarily true. Certain levels can and do overlap. The magnitude of the overlap is based upon the amount of risk the organization is willing to tolerate. For example, a company can begin the development of project management checklists to support the methodology while it is still providing project management training for the workforce. A company can create a center for excellence in project management before benchmarking is undertaken.

Although overlapping does occur, the order in which the phases are completed cannot change. For example, even though Level 1 and Level 2 can overlap, Level 1 *must* still be completed before Level 2 can be completed. Overlapping of several of the levels can take place, as shown in Figure 21–2.

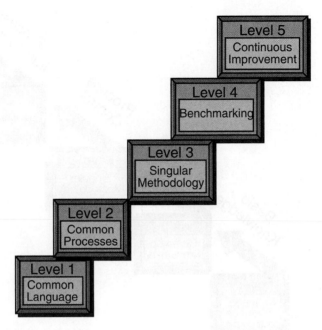

FIGURE 21–2. Overlapping levels.

- *Overlap of Level 1 and Level 2:* This overlap will occur because the organization can begin the development of project management processes either while refinements are being made to the common language or during training.
- *Overlap of Level 3 and Level 4:* This overlap occurs because, while the organization is developing a singular methodology, plans are being made as to the process for improving the methodology.
- *Overlap of Level 4 and Level 5:* As the organization becomes more and more committed to benchmarking and continuous improvement, the speed by which the organization wants changes to be made can cause these two levels to have significant overlap. The feedback from Level 5 back to Level 4 and Level 3, as shown in Figure 21–3, implies that these three levels form a continuous improvement cycle, and it may even be possible for all three of these levels to overlap.

Level 2 and Level 3 generally do not overlap. It may be possible to begin some of the Level 3 work before Level 2 is completed, but this is highly unlikely. Once a company is committed to a singular methodology, work on other methodologies generally terminates. Also, companies can create a Center for Excellence in project management early in the life-cycle process, but will not receive the full benefits until later on.

Risks can be assigned to each level of the PMMM. For simplicity's sake, the risks can be labeled as low, medium, and high. The level of risk is most frequently associated with

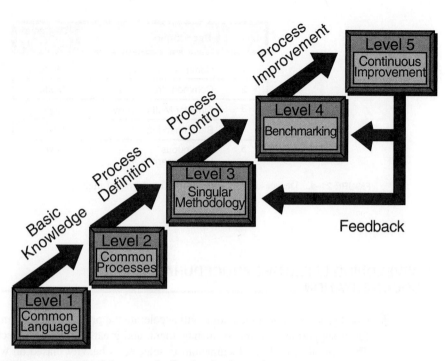

FIGURE 21–3. Feedback between the five levels of maturity.

the impact on the corporate culture. The following definitions can be assigned to these three risks:

- *Low Risk:* Virtually no impact upon the corporate culture, or the corporate culture is dynamic and readily accepts change.
- *Medium Risk:* The organization recognizes that change is necessary but may be unaware of the impact of the change. Multiple-boss reporting would be an example of a medium risk.
- *High Risk:* High risks occur when the organization recognizes that the changes resulting from the implementation of project management will cause a change in the corporate culture. Examples include the creation of project management methodologies, policies, and procedures, as well as decentralization of authority and decision-making.

Level 3 has the highest risk and degree of difficulty for the organization. This is shown in Figure 21–4. Once an organization is committed to Level 3, the time and effort needed to achieve the higher levels of maturity have a low degree of difficulty. Achieving Level 3, however, may require a major shift in the corporate culture.

These types of maturity models will become more common in the future, with generic models being customized for individual companies. These models will assist management in performing strategic planning for excellence in project management.

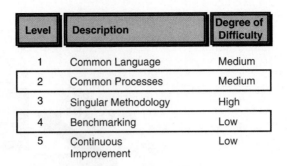

Level	Description	Degree of Difficulty
1	Common Language	Medium
2	Common Processes	Medium
3	Singular Methodology	High
4	Benchmarking	Low
5	Continuous Improvement	Low

FIGURE 21–4. Degrees of difficulty of the five levels of maturity.

21.2 DEVELOPING EFFECTIVE PROCEDURAL DOCUMENTATION

Good procedural documentation will accelerate the project management maturity process, foster support at all levels of management, and greatly improve project communications. The type of procedural documentation selected is heavily biased on whether we wish to manage formally or informally, but it should show how to conduct project-oriented activities and how to communicate in such a multidimensional environment. The project management policies, procedures, forms, and guidelines can provide some of these tools for delineating the process, as well as a format for collecting, processing, and communicating project-related data in an orderly, standardized format. Project planning and tracking, however, involve more than just the generation of paperwork. They require the participation of the entire project team, including support departments, subcontractors, and top management, and this involvement fosters unity. Procedural documents help to:

- Provide guidelines and uniformity
- Encourage useful, but minimum, documentation
- Communicate information clearly and effectively
- Standardize data formats
- Unify project teams
- Provide a basis for analysis
- Ensure document agreements for future reference
- Refuel commitments
- Minimize paperwork
- Minimize conflict and confusion
- Delineate work packages
- Bring new team members on board
- Build an experience track and method for future projects

Done properly, the process of project planning must involve both the performing and the customer organizations. This leads to visibility of the project at various organizational levels, and stimulates interest in the project and the desire for success.

The Challenges

Even though procedural documents can provide all these benefits, management is often reluctant to implement or fully support a formal project management system. Management concerns often center around four issues: overhead burden, start-up delays, stifled creativity, and reduced self-forcing control. First, the introduction of more organizational formality via policies, procedures, and forms might cost money, and additional funding may be needed to support and maintain the system. Second, the system is seen as causing start-up delays by requiring additional project definition before implementation can start. Third and fourth, the system is often perceived as stifling creativity and shifting project control from the responsible individual to an impersonal process. The comment of one project manager may be typical: "My support personnel feel that we spend too much time planning a project up front; it creates a very rigid environment that stifles innovation. The only purpose seems to be establishing a basis for controls against outdated measures and for punishment rather than help in case of a contingency." This comment illustrates the potential misuse of formal project management systems to establish unrealistic controls and penalties for deviations from the program plan rather than to help to find solutions.

How to Make It Work

Few companies have introduced project management procedures with ease. Most have experienced problems ranging from skepticism to sabotage of the procedural system. Many use incremental approaches to develop and implement their project management methodology. Doing this, however, is a multifaceted challenge to management. The problem is seldom one of understanding the techniques involved, such as budgeting and scheduling, but rather is a problem of involving the project team in the process, getting their input, support, and commitment, and establishing a supportive environment.

The procedural guidelines and forms of an established project management methodology can be especially useful during the project planning/definition phase. Not only does project management methodology help to delineate and communicate the four major sets of variables for organizing and managing the project—(1) tasks, (2) timing, (3) resources, and (4) responsibilities—it also helps to define measurable milestones, as well as report and review requirements. This provides project personnel the ability to measure project status and performance and supplies the crucial inputs for controlling the project toward the desired results.

Developing an effective project management methodology takes more than just a set of policies and procedures. It requires the integration of these guidelines and standards into the culture and value system of the organization. Management must lead the overall efforts and foster an environment conducive to teamwork. The greater the team spirit, trust, commitment, and quality of information exchange among team members, the more likely the team will be to develop effective decision-making processes, make individual and group commitments, focus on problem-solving, and operate in a self-forcing, self-correcting control mode.

Established Practices   Although project managers may have the right to establish their own policies and procedures, many companies design project control forms that can be used uniformly on all projects. Project control forms serve two vital purposes by establishing a common framework from which:

- The project manager will communicate with executives, functional managers, functional employees, and clients.
- Executives and the project manager can make meaningful decisions concerning the allocation of resources.

Some large companies with mature project management structures maintain a separate functional unit for forms control. This is quite common in aerospace and defense, but is also becoming common practice in other industries and in some smaller companies.

Large companies with a multitude of different projects do not have the luxury of controlling projects with three or four forms. There are different forms for planning, scheduling, controlling, authorizing work, and so on. It is not uncommon for companies to have 20 to 30 different forms, each dependent upon the type of project, length of project, dollar value, type of customer reporting, and other such arguments. Project managers are often allowed to set up their own administration for the project, which can lead to long-term damage if they each design their own forms for project control.

The best method for limiting the number of forms appears to be the task force concept, where both managers and doers have the opportunity to provide input. This may appear to be a waste of time and money, but in the long run provides large benefits.

To be effective, the following ground rules can be used:

- Task forces should include managers as well as doers.
- Task force members must be willing to accept criticism from other peers, superiors, and especially subordinates who must "live" with these forms.
- Upper-level management should maintain a rather passive (or monitoring) involvement.
- A minimum of signature approvals should be required for each form.
- Forms should be designed so that they can be updated periodically.
- Functional managers and project managers must be dedicated and committed to the use of the forms.

Categorizing the Broad Spectrum of Documents   The dynamic nature of project management and its multifunctional involvement create a need for a multitude of procedural documents to guide a project through the various phases and stages of integration.

Especially for larger organizations, the challenge is not only to provide management guidelines for each project activity, but also to provide a coherent procedural framework within which project leaders from all disciplines can work and communicate with each other. Specifically, each policy or procedure must be consistent with and accommodating to the various other functions that interface with the project over its life cycle. This complexity of intricate relations is illustrated in Figure 21–5.

One simple and effective way of categorizing the broad spectrum of procedural documents is by utilizing the work breakdown concept, as shown in Figure 21–6. Accordingly, the principal procedural categories are defined along the principal project life-cycle phases. Each category is then subdivided into (1) general management guidelines, (2) policies, (3) procedures, (4) forms, and (5) checklists. If necessary, the same concept can be carried forward one additional step to develop policies, procedures, forms, and checklists for the various project and functional sublevels of operation. Although this might be needed for very large programs, an effort should be made to minimize "layering" of policies and procedures to avoid new problems and costs. For most projects, a single document covers all levels of project operations.

As We Mature . . .

As companies become more mature in executing the project management methodology, project management policies and

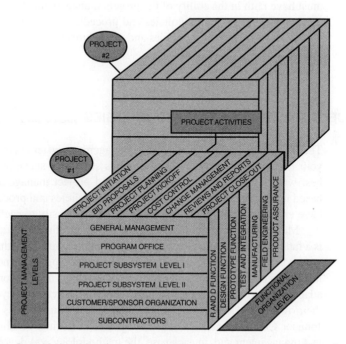

FIGURE 21–5. Interrelationship of project activities with various functional/organizational levels and project management levels.

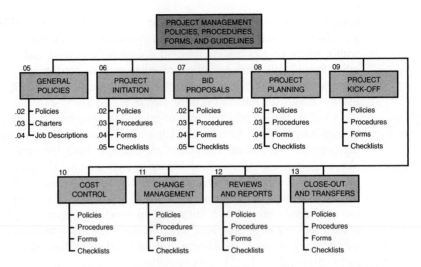

FIGURE 21–6. Categorizing procedural documents within a work breakdown structure.

procedures are disregarded and replaced with guidelines, forms, and checklists. More flexibility is provided the project manager. Unfortunately, this takes time because executives must have faith in the ability of the project management methodology to work without the rigid controls provided by policies and procedures. Yet all companies seem to go through the evolutionary stages of policies and procedures before they get to guidelines, forms, and checklists.

21.3 PROJECT MANAGEMENT METHODOLOGIES

The ultimate purpose of any project management system is to increase the likelihood that your organization will have a continuous stream of successfully managed projects. The best way to achieve this goal is with good project management methodologies that are based upon guidelines and forms rather than policies and procedures. Methodologies must have enough flexibility that they can be adapted easily to each and every project.

Methodologies should be designed to support the corporate culture, not vice versa. It is a fatal mistake to purchase a canned methodology package that mandates that you change your corporate culture to support it. If the methodology does not support the culture, it will not be accepted. What converts any methodology into a world-class methodology is its adaptability to the corporate culture. There is no reason why companies cannot develop their own methodology. Companies such as Compaq Services, Ericsson, Nortel Networks, Johnson Controls, and Motorola are regarded as having world-class methodologies for project management and, in each case, the methodology was developed internally. Developing your own methodology internally to guarantee a fit with the corporate culture usually provides a much greater return on investment than purchasing canned packages that require massive changes.

Even the simplest methodology, if accepted by the organization and used correctly, can increase your chances of success. As an example, Matthew P. LoPiccolo, Director of I.S. Operations for Swagelok Company, describes the process Swagelok went through to develop its methodology:

> We developed our own version of an I.S. project management methodology in the early 90s. We had searched extensively and all we found were a lot of binders that we couldn't see being used effectively. There were just too many procedures and documents. Our answer was a simple checklist system with phase reviews. We called it Checkpoint.
>
> As strategic planning has become more important in our organization, the need for improved project management has risen as well. Project management has found its place as a key tool in executing tactical plans.
>
> As we worked to improve our Checkpoint methodology, we focused on keeping it simple. Our ultimate goal was to transform our methodology into a one-page matrix that was focused on deliverables within each project phase and categorized by key project management areas of responsibility. The key was to create something that would provide guidance in daily project direction and decision making. In order to gain widespread acceptance, the methodology needed to be easy to learn and quick to reference. The true test of its effectiveness is our ability to make decisions and take actions that are driven by the methodology.
>
> We also stayed away from the temptation to buy the solution in the form of a software package. Success is in the application of a practical methodology not in a piece of software. We use various software products as a tool set for scheduling, communicating, effort tracking, and storing project information such as time, budget, issues and lessons learned.

The summary description of the methodology developed by Swagelok is shown in Table 21–1. Swagelok also realized that training and education would be required to support both the methodology and project management in general. Table 21–2 shows the training plan created by Swagelok Company.

21.4 CONTINUOUS IMPROVEMENT

All too often complacency dictates the decision-making process. This is particularly true of organizations that have reached some degree of excellence in project management, become complacent, and then realize too late that they have lost their competitive advantage. This occurs when organizations fail to recognize the importance of continuous improvement.

Figure 21–7 illustrates why there is a need for continuous improvement. As companies begin to mature in project management and reach some degree of excellence, they achieve a sustained competitive advantage. The sustained competitive advantage might

TABLE 21-1. SWAGELOK COMPANY'S CHECKPOINT METHODOLOGY, VERSION 3

Project Management	Assessment	Initiate Define/Plan	Design Specify	Deliver Construct/Integrate	Close Deploy/Transition
Key deliverables	Feasibility report	Project charter Business requirements Technical requirements	Detailed business rqmts. Systems analysis Design prototype	System construction System integration pilot test Implementation plan	Project deliverables evaluation Operational transition Vendor performance report Project performance report
Approval	Feasibility report review Assessment approval	Project approval	Design approval Prototype approval	Construct integrate approval Deployment approval	Project audit Completion approval
Scope	Scope boundaries	Scope/deliverables Benefits/value Assumptions & alternatives Strategic & tactical: impact/ priority/alignment	Change request procedures Issue management procedure	Change management Issue management	Manage delivered value
Human resource	Resource identification	Roles and responsibilities General resource capacity Training requirements Business sponsors	Resource impact & assignment Team training	Resource management Resource performance Knowledge transfer End-customer training	Resource performance evaluation
Time	"Window of opportunity"	Preliminary project schedule Time reporting database	Work breakdown structure Project plan	Execute & monitor plan	Verify activity/completion Close time buckets

938

Cost	Cost projections	Capital budget Operating budget Return on investment	Budget details	Execute & monitor budget	Close cost centers
Procurement	Alternatives evaluation	Hardware Software Consulting services Vendor RFPs	Vendor selection Contract finalization	Purchase hardware & software Vendor performance report	Ongoing maintenance agreements Vendor performance evaluation
Quality	Vendor assessment Quality requirements	Quality plan Previous lessons learned	Test approach Config. management approach Review lessons learned Walkthroughs/reviews	Test plans Test (i.e., unit, integration, system, acceptance)	Process review Post-implementation review Capture to lessons learned
Risk	"Opportunity costs"	Risk assessment	Risk management plan	Risk mitigation	Capture to lessons learned
Technology	Architecture alignment	Architecture requirements	Architecture verification	Technology architecture	Architecture review
Communication	Inter- and intra-program coordination	Communication requirements project site	Progress reports Meetings schedule Project site update	Project site update	Administrative closure

TABLE 21–2. SWAGELOK COMPANY'S TRAINING PLAN

	Training Programs			
Project Management Toolset	**Project Manager**	**Line Manager**	**Project Team Member**	**Executive Managment**
1) Project management concepts	PMP class PMP certification PMO overview	PM 101 PM 102 — Small project management	PM 101	Executive overview
2) Checkpoint (methodology)	Teaching level	Basic understanding	Basic understanding	Overview
3) MS project (scheduling)	Knowledgeable to expect	Basic understanding	Not required	Not required
4) TSP (effort tracking)	Management level	Management level	Time entry	Not required
5) Budget DB (budget tracking)	Management of project budget	Owner of department budget	Not required	Not required
6) SICL DB (issues/changes/lessons management)	Owner	How to view	How to view	Not required
7) Netmosphere (project communication)	Owner, publisher	How to view	How to view	How to view

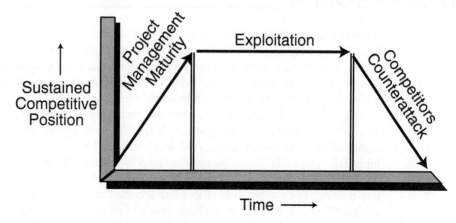

FIGURE 21–7. Why there is a need for continuous improvement.

very well be the single most important strategic objective of the firm. The firm will then begin the exploitation of its sustained competitive advantage.

Unfortunately, the competition is not sitting by idly watching you exploit your sustained competitive advantage. As the competition begins to counterattack, you may lose a large portion, if not all, of your sustained competitive advantage. To remain effective and competitive, the organization must recognize the need for continuous improvement, as shown in Figure 21–8. Continuous improvement allows a firm to maintain its competitive advantage even when the competitors counterattack.

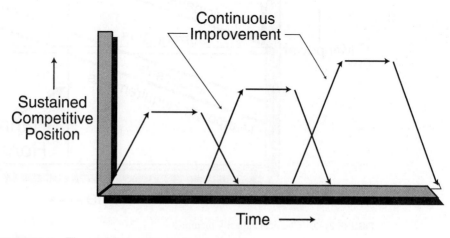

FIGURE 21–8. The need for continuous improvement.

21.5 CAPACITY PLANNING

As companies become excellent in project management, the benefits of performing more work in less time and with fewer resources becomes readily apparent. The question, of course, is how much more work can the organization take on? Companies are now struggling to develop capacity planning models to see how much new work can be undertaken within the existing human and nonhuman constraints.

Figure 21–9 illustrates the classical way that companies perform capacity planning. The approach outlined in this figure holds true for both project- and non–project-driven organizations. The "planning horizon" line indicates the point in time for capacity planning. The "proposals" line indicates the manpower needed for approved internal projects or a percentage (perhaps as much as 100 percent) for all work expected through competitive bidding. The combination of this line and the "manpower requirements" line, when compared against the current staffing, provides us with an indication of capacity. This technique can be effective if performed early enough such that training time is allowed for future manpower shortages.

The limitation to this process for capacity planning is that only human resources are considered. A more realistic method would be to use the method shown in Figure 21–10, which can also be applied to both project-driven and non–project-driven organizations. From Figure 21–10, projects are selected based upon such factors as strategic fit, profitability, who the customer is, and corporate benefits. The objectives for the projects selected are then defined in both business and technical terms, because there can be both business and technical capacity constraints.

The next step is a critical difference between average companies and excellent companies. Capacity constraints are identified from the summation of the schedules and plans.

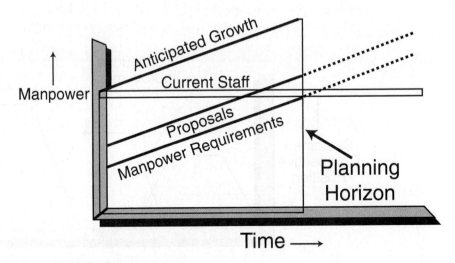

FIGURE 21–9.   Classical capacity planning.

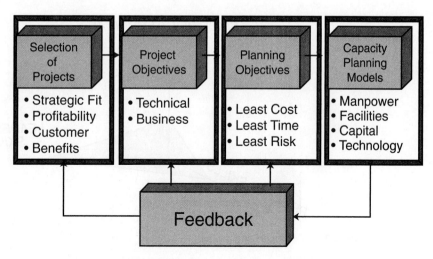

FIGURE 21–10. Improved capacity planning.

In excellent companies, project managers meet with sponsors to determine the objective of the plan, which is different than the objective of the project. Is the objective of the plan to achieve the project's objective with the least cost, least time, or least risk? Typically, only one of these applies, whereas immature organizations believe that all three can be achieved on every project. This, of course, is unrealistic.

The final box in Figure 21–10 is now the determination of the capacity limitations. Previously, we considered only human resource capacity constraints. Now we realize that the critical path of a project can be constrained not only by time but also by available manpower, facilities, cash flow, and even existing technology. It is possible to have multiple critical paths on a project other than those identified by time. Each of these critical paths provides a different dimension to the capacity planning models, and each of these constraints can lead us to a different capacity limitation. As an example, manpower might limit us to taking on only four additional projects. Based upon available facilities, however, we might only be able to undertake two more projects, and based upon available technology, we might be able to undertake only one new project.

21.6 COMPETENCY MODELS

In the twenty-first century, companies will replace job descriptions with competency models. Job descriptions for project management tend to emphasize the deliverables and expectations from the project manager, whereas competency models emphasize the specific skills needed to achieve the deliverables.

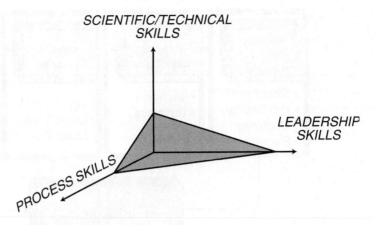

FIGURE 21–11. Competency model.

Figure 21–11 shows the competency model for Eli Lilly. Project managers are expected to have competencies in three broad areas[1]:

- Scientific/technical skills
- Leadership skills
- Process skills

For each of the three broad areas, there are subdivisions or grade levels. A primary advantage of a competency model is that it allows the training department to develop customized project management training programs to satisfy the skill requirements. Without competency models, most training programs are generic rather than customized.

Competency models focus on specialized skills in order to assist the project manager in making more efficient use of his or her time. Figure 21–12, although argumentative, shows that with specialized competency training, project managers can increase their time effectiveness by reducing time robbers and rework.

Competency models make it easier for companies to develop a complete project management curriculum, rather than a singular course. This is shown in Figure 21–13. As companies mature in project management and develop a company-wide core competency model, an internal, custom-designed curriculum will be developed. Companies, especially large ones, will find it necessary to maintain a course architecture specialist on their staff.

---

1. A detailed description of the Eli Lilly competency model and the Ericsson competency model can be found in Harold Kerzner, *Applied Project Management* (New York: Wiley, 1999), pp. 266–283.

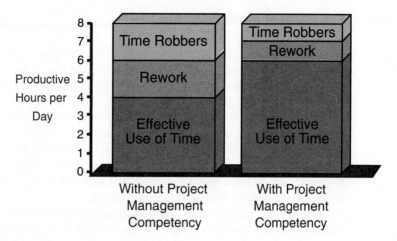

FIGURE 21–12. Core competency analysis.

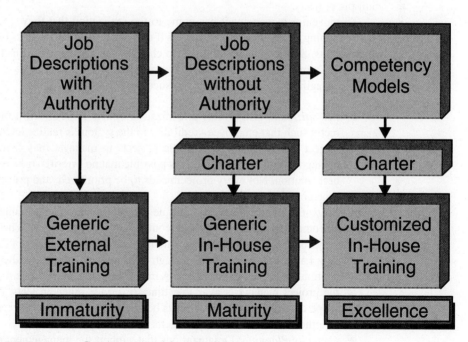

FIGURE 21–13. Competency models and training.

21.7 MANAGING MULTIPLE PROJECTS

As organizations mature in project management, there is a tendency toward having one person manage multiple projects. The initial impetus may come either from the company sponsoring the projects or from project managers themselves. There are several factors supporting the managing of multiple projects. First, the cost of maintaining a full-time project manager on all projects may be prohibitive. The magnitude and risks of each individual project dictate whether a full-time or part-time assignment is necessary. Assigning a project manager full-time on an activity that does not require it is an overmanagement cost. Overmanagement of projects was considered an acceptable practice in the early days of project management because we had little knowledge on how to handle risk management. Today, methods for risk management exist.

Second, line managers are now sharing accountability with project managers for the successful completion of the project. Project managers are now managing at the template levels of the WBS with the line managers accepting accountability for the work packages at the detailed WBS levels. Project managers now spend more of their time integrating work rather than planning and scheduling functional activities. With the line manager accepting more accountability, time may be available for the project manager to manage multiple projects.

Third, senior management has come to the realization that they must provide high-quality training for their project managers if they are to reap the benefits of managing multiple projects. Senior managers must also change the way that they function as sponsors. There are six major areas where the corporation as a whole may have to change in order for the managing of multiple projects to succeed.

- *Prioritization:* If a project prioritization system is in effect, it must be used correctly such that employee credibility in the system is realized. One risk is that the project manager, having multiple projects to manage, may favor those projects having the highest priorities. It is possible that no prioritization system may be the best solution. Not every project needs to be prioritized, and prioritization can be a time-consuming effort.
- *Scope Changes:* Managing multiple projects is almost impossible if the sponsors/customers are allowed to make continuous scope changes. When using multiple projects management, it must be understood that the majority of the scope changes may have to be performed through enhancement projects rather than through a continuous scope change effort. A major scope change on one project could limit the project manager's available time to service other projects. Also, continuous scope changes will almost always be accompanied by reprioritization of projects, a further detriment to the management of multiple projects.
- *Capacity Planning:* Organizations that support the management of multiple projects generally have a tight control on resource scheduling. As a result, the organization must have knowledge of capacity planning, theory of constraints, resource leveling, and resource limited planning.
- *Project Methodology:* Methodologies for project management range from rigid policies and procedures to more informal guidelines and checklists. When managing

multiple projects, the project manager must be granted some degree of freedom. This necessitates guidelines, checklists, and forms. Formal project management practices create excessive paperwork requirements, thus minimizing the opportunities to manage multiple projects. The project size is also critical.

- *Project Initiation:* Managing multiple projects has been going on for almost 40 years. One thing that we have learned is that it can work well as long as the projects are in relatively different life-cycle phases because the demands on the project manager's time are different for each life-cycle phase.
- *Organizational Structures:* If the project manager is to manage multiple projects, then it is highly unlikely that the project manager will be a technical expert in all areas of all projects. Assuming that the accountability is shared with the line managers, the organization will most likely adopt a weak matrix structure.

21.8 END-OF-PHASE REVIEW MEETINGS

For more than 20 years, end-of-phase review meetings were simply an opportunity for executives to "rubber stamp" the project to continue. As only good news was presented the meetings were used to give the executives some degree of comfort concerning project status.

Today, end-of-phase review meetings take on a different dimension. First and foremost, executives are no longer afraid to cancel projects, especially if the objectives have changed, if the objectives are unreachable, or if the resources can be used on other activities that have a greater likelihood of success. Executives now spend more time assessing the risks in the future rather than focusing on accomplishments in the past.

Since project managers are now becoming more business-oriented rather than technically oriented, the project managers are expected to present information on business risks, reassessment of the benefit-to-cost ratio, and any business decisions that could affect the ultimate objectives. Simply stated, the end-of-phase review meetings now focus more on business decisions, rather than on technical decisions.

22 THE BUSINESS OF SCOPE CHANGES

22.0 INTRODUCTION

PMBOK® Guide, 4th Edition
5.4 Verify Scope

Very few projects are ever completed according to the original plan. The changes to the plan result from either increased knowledge, a need for competitiveness, or changing customer/consumer tastes. Once the changes are made, there is almost always an accompanying increase in the budget and/or elongation of the schedule.

The process for recommending and approving scope changes can vary based upon whether or not the client is internal or external to the organization. Scope changes for external clients have long been viewed as a source of added profitability on projects. Years ago, it was common practice on some Department of Defense contracts to underbid the original contract during competitive bidding to assure the award of the contract and then push through large quantities of lucrative scope changes.

External customers were rarely informed of gaps in their statements of work that could lead to scope charges. And even if the statement of work was clearly written, it was often intentionally misinterpreted for the benefit of seeking out scope changes whether or not the scope changes were actually needed. For some companies, scope changes were the prime source of corporate profitability. During competitive bidding, executives would ask the bidding team two critical questions before submitting a bid; (1) What is our cost of doing the work we are committing to? and (2) How much additional work can we push through in scope changes once the contract is awarded to us? Often, the answer to the second question determined the magnitude of the bid.

To make matters even worse, in the early years of project management the Department of Defense requested that the contractors' project managers have a command of technology rather than an understanding of technology. Engineers with advanced degrees were assigned as project managers, and their objective was often to exceed rather than merely meet specifications. This resulted in additional scope changes and often increased the risks in the project.

Another problem that surfaced was the downstream effect of upstream scope changes on large projects involving multiple contractors. When contractors work sequentially, as shown in Figure 22–1, a scope change in an upstream contractor may not have a serious impact on downstream contractors. Contractor B usually needs the output from contractor A to begin work. If contractor A initiates a scope change, the impact on contractor B may be minimal. If there were an impact on contractor B, the customer would incur the added costs. But if the contractors are performing work partially or completely in parallel, as shown in Figure 22–2, the downstream effect can be devastating. A relatively simple decision of an upstream contractor to change over to higher grade or lower grade raw materials can cause major changes in the project plans and scope baselines of downstream contractors.

As an example using Figure 22–2, contractor A was awarded a contract for the development of a liquid fuel rocket. The liquid fuel had already been formulated, tested, and agreed to by the customer. Contractor B was awarded the contract for the construction of the launch site storage and fueling system based upon the agreed-to formulation. Contractor B began preparation of the blueprints for the construction of the fueling system and also began pouring concrete for the foundation. Contractor A continuously exchanged information with contractor B to make sure that no major changes were anticipated. But when contractor A put forth a change request to modify the liquid fuel formulation to extend the range of the rocket, the result was a major scope change in the work already accomplished by contractor B. The new formulation required costly changes to the insulation and the pumps. Simply stated, if the customer were to approve the change request from contractor A,

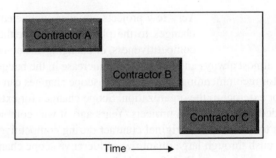

FIGURE 22–1. Sequential contractors.

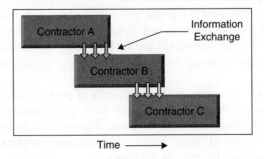

FIGURE 22–2. Overlapping contractors.

then the customer might also need to approve the accompanying change requests from all downstream contractors that would be impacted by contractor A's change request.

22.1 NEED FOR BUSINESS KNOWLEDGE

Using scope changes as a source of revenue has been an acceptable practice for projects funded by customers external to the organization. But for internal customers, there are numerous other reasons for scope changes, as shown in Figure 22–3. For projects that are internal, scope changes must be targeted, and this is the weakest link because it requires business knowledge as well as technical knowledge. As an example, scope changes should not be implemented at the expense of risking exposure to product liability lawsuits or safety issues. Likewise, costly scope changes exclusively for the sake of enhancing image or reputation should be avoided. Also, scope changes should not be implemented if the payback period for the product is drastically extended in order to recover the costs of the scope changes.

Scope changes should be based upon a solid business foundation. Unfortunately, this was not always the case because there existed a difference between project and program management. Historically, project managers were responsible for the development of the product and its accompanying features. Scope changes were often initiated by the project managers based upon technical value rather than business value. For example, developing a very high quality product may seem nice at the time, but there must be customers willing to pay the higher price. It was quite common for project managers to make scope changes that were not a good fit with marketing objectives. The result was often a product that nobody wanted or could afford.

Once the product was developed, it would be turned over to a program manager for commercialization, marketing, and sales. Program managers, functioning like product line managers, would then make all of the business-related decisions including authorization of all of the expensive downstream scope changes. Project managers were technically oriented personnel, whereas program managers were skilled in marketing and sales.

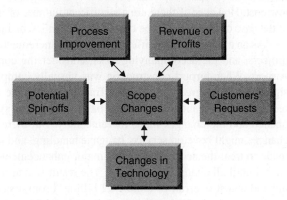

FIGURE 22–3. Factors to consider for scope changes.

Today, project managers are expected to make sound business decisions rather than merely project or technical decisions. But it wasn't that long ago when companies were nonbelievers that project management could benefit the business as a whole. Companies are slowly coming to the realization that they are managing their business by projects. Project managers are now expected to make sound business decisions rather than merely project decisions, and this includes the decision on scope changes.

22.2 TIMING OF SCOPE CHANGES

Everyone seems to understand that the further we go into the project's life-cycle phases, the more costly the scope changes. As we progress through the life-cycle phases, more variables are introduced into the system such that the financial impact of a small scope change can be quite large because of the cost involved with reversing previous decisions. Making scope changes in production are more costly than scope changes in technology development.

For the development of new products, it may take as many as 60 ideas to develop one successful new product. Each idea may undergo numerous scope changes before the idea is officially abandoned. Any scope changes made after capital expenditures are incurred can have a significant impact on total cost and schedule.

Another critical factor involving timing is whether the scope changes are radical scope changes. Radical scope changes require either a breakthrough in technology or the design of an entirely new platform. As an example, a competitor launches a new product that may cause the marketplace to view your product as being obsolete. In order to remain competitive, you may need to consider radical scope changes to remain competitive or to outdo the competition. Radical scope changes focus more so on creativity than execution. Radical scope changes may require a breakthrough in technology accompanied by the consumption of vast resources.

Another timing issue is whether the scope change should be done incrementally or clustered together and approved as enhancement projects. Incremental scope changes are often referred to as scope creep. These incremental changes can be done quickly and at a relatively low cost. However, if there are a significant number of incremental scope changes, such as in the case of perpetual scope creep, the project's schedule can be elongated.

As an example of the difference between incremental scope changes and enhancement projects, and the accompanying risks, consider the construction of Denver International Airport. When airport construction was virtually completed, United Airlines had not yet signed a lease agreement. Before finally signing the lease agreement, United Airlines requested significant changes to the baggage handling system. The airport could have been opened and the changes made incrementally. But there was no telling what impact these changes might have on service, baggage handling, and passenger safety. The decision was made to treat the scope changes as major enhancement changes, thus keeping the airport closed until all changes were made. The result was almost a 2-year slippage in the schedule and a cost overrun of almost $2 billion. From a safety perspective, the decision may have been correct. Today, Denver International Airport is viewed as a success.

Another common example of enhancement projects occurs in information technology (IT). An IT project manager was responsible for the development and implementation of an IT package for inventory control at the company's eight manufacturing plants. The project team included a representative from each of the plants. The representatives were told that they must work together and clearly articulate their requirements at the beginning of the project, and that no scope changes would be allowed once the scope statement was completed. Once the project was initiated, all scope change requests would be accomplished at some time in the future using an enhancement project. The project team understood the project manager's concerns and worked diligently to clearly articulate the requirements.

After the project was nearly 70 percent complete, the plants became concerned that they had not clearly defined their requirements and changes would be necessary. Simply stated, the plants asserted that they could not use the software without the changes being made. The project manager refused to make the scope changes because the resources were already committed to another project that would start immediately after this project was completed. Approving the scope changes would now have a serious impact on the schedules of two projects.

The project manager held her ground and refused to make the incremental scope changes despite the complaints of the plants. Once the project was implemented, the plants admitted that they could still use the existing software, and the enhancement project did not take place until 4 years later. This approach worked well for the project manager, but not all projects have this outcome.

22.3 BUSINESS NEED FOR A SCOPE CHANGE

There must be a valid business purpose for a scope change. This includes the following factors at a minimum:

- An assessment of the customers' needs and the added value that the scope change will provide
- An assessment of the market needs including the time required to make the scope change, the payback period, return on investment, and whether the final product selling price will be overpriced for the market
- An assessment on the impact on the length of the product life cycle
- An assessment on the competition's ability to imitate the scope change
- Is there a product liability associated with the scope change and can it impact our image?

Scope changes can be for existing products or for new products. Support for existing products is usually a defensive scope change designed to penetrate new markets with existing products. Support for new products is usually an offensive scope change designed to provide new products/services to existing customers as well as seeking out new markets.

22.4 RATIONALE FOR NOT APPROVING
A SCOPE CHANGE

Some scope change requests are the result of wishful thinking or the personal whims of management and not necessarily based upon sound business judgment. In such cases, the scope changes may need to be cancelled. Typical rationalization for termination or not approving a scope change includes:

- The cost of the scope change is excessive and the final cost of the deliverable may make us noncompetitive.
- The return on investment may occur too late.
- The competition is too stiff and not worth the risks.
- There are insurmountable obstacles and technical complexity.
- There are legal and regulatory uncertainties.
- The scope change may violate the company's policy on nondisclosure, secrecy, and confidentiality agreements.

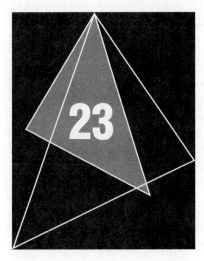

The Project Office

23

23.0 INTRODUCTION

PMBOK® Guide, 4th Edition
1.4.4 PMO

Today, companies are managing their business by projects. The result has been a vast amount of project management information surfacing from all areas of the company. This information focuses on best practices in the project management, the usefulness of an enterprise project management methodology, the benefits of project management, and how project management is improving the profitability of the company. As companies begin to recognize the favorable effect that project management has on performance, all of this project management knowledge is treated as intellectual property. Emphasis is now placed upon achieving professionalism in project management using the project office (PO) concept, where the project management office (PMO) becomes the guardian for the project management intellectual property. The concept of a PO or PMO could very well be the most important project management activity in this decade.

955

23.1 PRESENT-DAY PROJECT OFFICE

The 1990s began with a recession that took a heavy toll on white-collar ranks. Management's desire for efficiency and effectiveness led them to take a hard look at non-traditional management techniques such as project management. Project management began to expand to non–project-driven industries. The benefits of using project management, which were once seen as applicable only to the aerospace, defense, and heavy construction industries, were now recognized as being applicable for other industries.

By the late 1990s, as more of the benefits of project management became apparent, management understood that there might be a significant, favorable impact on the corporate bottom line. This led management to two important conclusions:

- Project management had to be integrated and compatible with the corporate reward systems for sustained project management growth.
- Corporate recognition of project management as a profession was essential in order to maximize performance.

The recognition of project management professionalism led companies to accept PMI's Certification Program as the standard and to recognize the importance of the PO concept. Consideration was being given for all critical activities related to project management to be placed under the supervision of the PO. This included such topics as:

- Standardization in estimating
- Standardization in planning
- Standardization in scheduling
- Standardization in control
- Standardization in reporting
- Clarification of project management roles and responsibilities
- Preparation of job descriptions for project managers
- Preparation of archive data on lessons learned
- Continuous project management benchmarking
- Developing project management templates
- Developing a project management methodology
- Recommending and implementing changes and improvements to the existing project management methodology
- Identifying project management standards
- Identifying best practices in project management
- Performing strategic planning for project management
- Establishing a project management problem-solving hotline
- Coordinating and/or conducting project management training programs
- Transferring knowledge through coaching and mentorship
- Developing a corporate resource capacity/utilization plan
- Assessing risks in projects
- Planning for disaster recovery in projects

- Performing or participating in the portfolio management of projects
- Acting as the guardian for project management intellectual property

With these changes taking place, organizations began changing the name of the PO to the Center of Excellence (COE) in project management. The COE was mainly responsible for providing information to stakeholders rather than actually executing projects or making midcourse corrections to a plan.

23.2 IMPLEMENTATION RISKS

Each activity assigned to the PO brought with it both advantages and disadvantages. The majority of the disadvantages were attributed to the increased levels of resistance to the new responsibilities given to the PO. For simplicity sake, the resistance levels can be classified as low risk, moderate risk, and high risk according to the following definitions:

- *Low Risk:* Easily accepted by the organization with very little shift in the balance of power and authority. Virtually no impact on the corporate culture.
- *Moderate Risk:* Some resistance by the corporate culture and possibly a shift in the balance of power and authority. Resistance levels can be overcome in the near term and with minimal effort.
- *High Risk:* Heavy pockets of resistance exist and a definite shift in some power and authority relationships. Strong executive leadership may be necessary to overcome the resistance.

Not every PO has the same responsibilities. Likewise, the same responsibilities implemented in two POs can have differing degrees of the best interest of the organization.

Evaluating potential implementation risks is critical. It may be easier to gain support for the establishment of a PO by implementing low-risk activities first. The low-risk activities are operational activities to support project management efforts in the near term whereas the high-risk activities are more in line with strategic planning responsibilities and possibly the control of sensitive information. For example, low-risk activities include mentorship, developing standards, and project management training. High-risk activities include capacity planning, benchmarking, and dissemination of information.

Senior managers were now recognizing that project management and the PO had become invaluable assets for senior management as well as for the working levels.

During the past ten years, the benefits for the executive levels of management of using a PO have become apparent. They include:

- Standardization of operations
- Company rather than silo decision-making
- Better capacity planning (i.e., resource allocations)

- Quicker access to higher-quality information
- Elimination or reduction of company silos
- More efficient and effective operations
- Less need for restructuring
- Fewer meetings which rob executives of valuable time
- More realistic prioritization of work
- Development of future general managers

All of the above benefits are either directly or indirectly related to the project management intellectual property. To maintain the project management intellectual property, the PO must maintain the vehicles for capturing the data and then disseminate the data to the various stakeholders. These vehicles include the company project management intranet, project web sites, project databases, and project management information systems. Since much of this information is necessary for both project management and corporate strategic planning, there must exist strategic planning for the PO.

As we entered the twenty-first century, the PO became commonplace in the corporate hierarchy. Although the majority of activities assigned to the PO had not changed, there was now a new mission for the PO: supporting the entire corporation.

The PO was now servicing the corporation, especially the strategic planning activities, rather than focusing on a specific customer. The PO was transformed into a corporate center for control of the project management intellectual property. This was a necessity as the magnitude of project management information grew almost exponentially throughout the organization.

23.3 TYPES OF PROJECT OFFICES

Three types of POs are commonly used in companies:

- *Functional PO:* This type of PO is utilized in one functional area or division of an organization, such as information systems. The major responsibility of this type of PO is to manage a critical resource pool, that is, resource management. The PMO may or may not actually manage projects.
- *Customer Group PO:* This type of PO is for better customer management and customer communications. Common customers or projects are clustered together for better management and customer relations. Multiple customer group POs can exist at the same time and may end up functioning as a temporary organization. In effect, this acts like a company within a company. This type of PMO will have a permanent project manager assigned and managing projects.
- *Corporate (or Strategic) PO:* This type of PO services the entire company and focuses on corporate and strategic issues rather than functional issues. If this type of PMO does management projects, it is for cost reduction efforts.

Companies can champion more than one type of PO at the same time. For example, there can exist both a functional PO and a strategic/corporate PO that work together.

23.4 NETWORKING PROJECT MANAGEMENT OFFICES

Because of political infighting for control of the PMO, many companies have established multiple PMOs all of which are networked together by a "coordinating" PMO. Other companies that are multinational have created regional PMOs that are groupings of project management associates (project managers, team members, etc.) who perform project management duties within specific regional or industry-specific areas. In this case, the primary PMO responsibilities are:

- Promoting the enterprise project management methodology
- Promoting the use of standard project management tools
- Assuring standardization in project execution and delivery
- Maintaining a source of project management subject matter expertise
- Coordinating multinational project management knowledge

23.5 PROJECT MANAGEMENT INFORMATION SYSTEMS

Given the fact that the PO is now the guardian of the project management intellectual property, there must exist processes and tools for capturing this information. This information can be collected through four information systems, as shown in Figure 23–1. Each information system can be updated and managed through the company intranet.

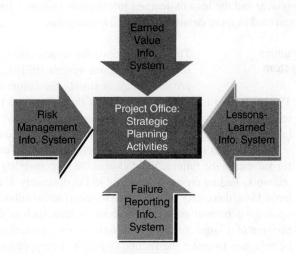

FIGURE 23-1. Project management information systems.

Earned Value Measurement
Information System

The earned value measurement information system is common to almost all project managers. It provides sufficient information to answer two questions:

- Where is the project today?
- Where will the project end up?

This system either captures or calculates the planned and actual value of the work, the actual costs, cost and schedule variances (in hours or dollars and percent), the estimated cost at completion, the estimated time at completion, percent complete, and trends.

The earned value measurement information system is critical for a company that requires readily available information for rapid decision-making. It is easier to make small rather than large changes to a project plan. Therefore, variances from the performance management baseline must be identified quickly such that corrective action can be taken in small increments.

Risk Management
Information System

The second information system provides data on risk management. The risk management information system (RMIS) stores and allows retrieval of risk-related data. It provides data for creating reports and serves as the repository for all current and historical information related to project risk. The information will include risk identification documentation (possibly by using templates), quantitative and qualitative risk assessment documents, contract deliverables if appropriate, and any other risk-related reports. The PMO will use the data from the RMIS to create reports for senior management and retrieve data for the day-to-day management of projects. By using risk management templates, each project will produce a set of standard reports for periodic reporting and have the ability to create ad hoc reports in response to special queries. This information is directly related to the failure reporting information system and the lessons-learned information system. The last two information systems are covered in more detail in the next two sections.

Performance Failure
Information System

The PO may have the responsibility for maintaining the performance failure information system (PFIS). The failure could be a complete project failure or simply the failure of certain tests within the project. The PFIS must identify the cause(s) of the failure and possibly recommendations for the removal of the cause(s). The cause(s) could be identified as coming from problems entirely internal to the organization or from coordinated interactions with subcontractors.

It is the PO's responsibility to develop standards for maintaining the PFIS rather than for validating the failure. Validation is the responsibility of the team members performing the work. Failure reporting can lead to the discovery of additional and more serious problems. First, there may be resistance to report some failures for fear that it may reflect badly upon the personnel associated with the failure, such as the project sponsors. Second, each division of a large company may have its own procedures for recording failures and may be reluctant to make the failure visible in a corporate-wide database. Third, there could exist many different definitions of what is or is not a failure. Fourth, the PO may be at the mercy of others to provide accurate, timely, and complete information.

The failure report must identify the item that failed, symptoms, conditions at the time of the failure, and any other pertinent evidence necessary for corrective action to take place. Failure analysis, which is the systematic analysis of the consequences of the failure on the project, cannot be completed until the causes of the failure have been completely identified. The PO may simply function as the records keeper to standardize a single company-wide format and database for reporting the results of each project. This could be part of the lessons-learned review at the end of each project.

Consider the following example: An aerospace company had two divisions that often competed with one another through competitive bidding on government contracts. Each conducted its own R&D activities and very rarely exchanged data. One of the divisions spent six months working on an R&D project that was finally terminated and labeled as a failure. Shortly thereafter, it was learned that the sister division had worked on exactly the same project a year ago and achieved the same unproductive results. Failure information had not been exchanged, resulting in the waste of critical resources.

Everyone recognizes the necessity for a corporate-wide information system for storing failure data. But there always exists the risk that some will view this as a loss of power. Others may resist for fear that their name will be identified along with the failure. The overall risk with giving this responsibility to the project office is low to moderate.

Lessons-Learned (Postmortem Analysis) Information System Some companies work on a vast number of projects each year, and each of these projects provides valuable information for improving standards, estimating for future bidding, and the way business is conducted. All of this information is intellectual property and must be captured for future use. Lessons-learned reviews are one way to obtain this information.

If intellectual property from projects is to be retained in a centralized location, then the PO must develop expertise in how to conduct a postmortem analysis meeting. At that meeting, four critical questions must be addressed:

- What did we do right?
- What did we do wrong?
- What future recommendations can be made?
- How, when, and to whom should the information be disseminated?

23.6 DISSEMINATION OF INFORMATION

A problem facing most organizations is how to make sure that critical information, such as KPIs (key performance indicators) and CSFs (critical success factors), are known throughout the organization. Intranet lessons-learned databases would be one way to share information. However, a better way might be for the PO to take the lead in preparing lessons-learned case studies at the end of each project. The case studies could then be used in future training programs throughout the organization and be intranet-based.

As an example, a company completed a project quite successfully, and the project team debriefed senior management at the end of the project. The company had made significant breakthroughs in various manufacturing processes used for the project, and senior management wanted to make sure that this new knowledge would be available to all other divisions.

The decision was made to dissolve the team and reassign the people to various divisions throughout the organization. After six months had passed, it became evident that very little knowledge had been passed on to the other divisions. The team was then reassembled and asked to write a lessons-learned case study to be used during project management training programs.

Although this approach worked well, there also exist detrimental consequences that make this approach difficult to implement. Another company had adopted the concept of having to prepare lessons-learned case studies. Although the end result of one of the projects was a success, several costly mistakes were made during the execution of the project due to a lack of knowledge of risk management and poor decision-making. Believing that lessons-learned case studies should include mistakes as well as successes, the PO team preparing the case study included all information.

Despite attempts to disguise the names of the workers that made the critical mistakes, everyone in the organization knew who worked on the project and was able to discover who the employees were. Several of the workers involved in the project filed a grievance with senior management over the disclosure of this information, and the case studies were then removed from training programs. It takes a strong organizational culture to learn from mistakes without retribution to the employees. The risk here may be moderate to high for the PO to administer this activity.

23.7 MENTORING

Project management mentoring is a critical PO activity. Most people seem to agree that the best way to train someone in project management is with on-the-job training. One such way would be for inexperienced project managers to work directly under the guidance of an experienced project manager, especially on large projects. This approach may become costly if the organization does not have a stream of large projects.

Perhaps the better choice would be for the PO to assume a mentoring role whereby inexperienced project managers can seek advice and guidance from the more experienced project managers who report either "solid" or "dotted" to the PO. This approach has three major benefits. First, the line manager to whom the project manager reports administratively may not have the necessary project management knowledge or experience capable of assisting the worker in times of trouble. Second, the project manager may not wish to discuss certain problems with his or her superior for fear of retribution. Third, given the fact the PO may have the responsibility for maintaining lessons-learned files, the project mentoring program could use these files and provide the inexperienced project manager with early warning indicators that potential problems could occur.

The mentoring program could be done on a full-time basis or on an as-needed basis, which is the preferred approach. Full-time mentoring may seem like a good idea, but it includes the risk that the mentor will end up managing the project. The overall risk for PO mentoring is low.

23.8 DEVELOPMENT OF STANDARDS AND TEMPLATES _____

A critical component of any PO is the development of project management standards. Standards foster teamwork by creating a common language. However, developing excessive standards in the form of policies and procedures may be a mistake because it may not be possible to create policies and procedures that cover every possible situation on every possible project. In addition, the time, money, and people required to develop rigid policy and procedure standards would make PO implementation unlikely because of head-count requirements.

Forms and checklists can be prepared in a template format such that the information can be used on a multitude of projects. Templates should be custom-designed for a specific organization rather than copied from another organization that may not have similar types of projects or a similar culture. Reusable templates should be prepared *after* the organization has completed several projects, whether successfully or unsuccessfully, and where lessons-learned information can be used for the development and enhancement of the templates.

There is a danger in providing templates as a replacement for the more formalized standards. First, because templates serve as a guide for a general audience, it may not satisfy the needs of any particular program. Second, there is the risk that some perspective users of the templates, especially inexperienced project managers, may simply adopt the templates "as required, as written" despite the fact that they do not fit his or her program.

The reason for providing templates is *not* to tell the team how to do their job, but to give the project manager and his or her staff a starting point for their own project initiation, planning, execution, control, and closure processes. Templates should stimulate proactive thinking about what has to be done and possibly some ideas on how to do it. Templates and standards often contain significantly more information than most project managers need. However, the templates and standards should be viewed as the key to keeping things simple and the project managers should be able to tailor the templates and standards to suit the needs of the project by focusing on the key critical areas.

Templates and standards should be updated as necessary. Since the PO is most likely responsible for maintaining lessons-learned files and project postmortem analysis, it is only fitting that the PO evaluate these data to seek out key performance indicators which could dictate template enhancements. Standards and templates can be regarded as a low-risk PO activity.

23.9 PROJECT MANAGEMENT BENCHMARKING _____

Perhaps the most interesting and most difficult activity assigned to a PO is benchmarking. Just like mentoring, benchmarking requires the use of experienced project managers. The assigned individuals must know what to look for and what questions to ask, have the ability to recognize a good fit with the company and how to evaluate the data, and what recommendations to make.

Benchmarking is directly related to strategic planning for project management and can have a pronounced effect on the corporate bottom line based on how quickly the changes are implemented. In recent years, companies have discovered that best practices can be benchmarked against organizations not necessarily in your line of business. For example, an aerospace division of a large firm had been using project management for over 30 years. During the early 1990s, the firm had been performing benchmarking studies but *only* against other aerospace firms. Complacency had set in, with the firm believing that it was in equal standing with is competitors in the aerospace field. In the late 1990s, the firm began benchmarking against firms outside of its industry, specifically telecommunications, computers, electronics, and entertainment. Most of these firms had been using project management for less than 5 years and, in that time, had achieved project management performance that exceeded the aerospace firm. Now, the aerospace firm benchmarks against all industries.

Project office networking for benchmarking purposes could very well be in the near future for most firms. Project office networking could span industries and continents. In addition, it may become commonplace even for competitors to share project management knowledge. However, at present, it appears that the majority of project management benchmarking is being performed by organizations whose function is entirely benchmarking. These organizations charge a fee for their services and conduct symposiums for their membership whereby project management best practices data are shared. In addition, they offer database services against which you can compare your organization:

- Other organizations in your industry
- Other organizations in different industry sectors
- Other employee responses within your company
- Other organizations by company size
- Other organizations by project size

Some organizations have a strong resistance to benchmarking. The arguments against benchmarking include:

- It doesn't apply to our company or industry.
- It wasn't invented here.
- We're doing fine! We don't need it.
- Let's leave well enough alone.
- Why fix something that isn't broken?

Because of these concerns, benchmarking may be a high-risk activity because of the fear of what will be found and the recommended changes.

23.10 BUSINESS CASE DEVELOPMENT

One of the best ways for a PO to support the corporate strategic planning function is by becoming expert in business case development. More specifically, this includes expertise in feasibility studies and cost-benefit analysis. In the Scope Management section of the PMBOK® Guide, one of the outputs of the Scope Initiation Process is the identification/appointment of a project manager. This is accomplished after the business case is developed. There are valid arguments for assigning the project manager after the business case is developed:

- The project manager may not be able to contribute to the business case development.
- The project might not be approved and/or funded, and it would be an added cost to have the project manager on board early.
- The project might not be defined well enough to determine at an early stage the best person to be assigned as the project manager.

While these arguments seem to have merit, there is a more serious issue in that the project manager ultimately assigned may not have sufficient knowledge about the assumptions, constraints, and alternatives considered during the business case development. This could lead to a less than optimal project plan. It is wishful thinking to believe that the project charter, which may have been prepared by someone completely separated from the business case development efforts, contains all of the necessary assumptions, alternatives, and constraints.

One of the axioms of project management is that the earlier the project manager is assigned, the better the plan and the greater the commitment to the project. Companies argue that the project manager's contribution is limited during business case development. The reason for this belief is because the project managers have never been trained in how to perform feasibility studies and cost-benefit analysis. These courses are virtually nonexistent in the seminar marketplace as a publicly offered course, but some companies have custom-designed courses specifically for their company.

Business case development often results in a highly optimistic approach with little regard for the schedule and/or the budget. Pressure is then placed upon the project manager to accept arguments and assumptions made during business case development. If the proj-ect fails to meet business case expectations, then the blame is placed upon the project manager.

The PO must develop expertise in feasibility studies, cost-benefit analysis, and business case development. This expertise lends itself quite readily to templates, forms, and checklists. The PO can then become a viable support arm to the sales force in helping them make more realistic promises to the customers and possibly assist in generating additional sales. In the future, the PO might very well become the company experts in feasibility studies and cost-benefit analyses and may eventually conduct customized training for the organization on these subjects. Marketing and sales personnel who traditionally perform these activities may view this as a high risk.

23.11 CUSTOMIZED TRAINING (RELATED TO PROJECT MANAGEMENT)

For years, the training branch of the human resources group had the responsibility of working with trainers and consultants in the design of customized project management training programs. While many of these programs were highly successful, there were many that were viewed as failures. One division of a large company recognized the need for training in project management. The training department went out for competitive bidding and selected a trainer. The training department then added in its own agendas after filtering all of the information concerning the goals and deliverables sought by the division requesting the training. The trainer never communicated directly with the organization requesting the training and simply designed the course around the information presented by the training department. The training program was viewed as a failure and the consultant/trainer was never invited back. Postmortem analysis indicated the following conclusions:

- The training branch (and the requesting organization) never recognized the need to have the trainer meet directly with the requesting organization.
- The training group received input from senior management, unknown to the requesting organization, as to what information they wished to see covered, and the resulting course satisfied nobody's expectation.
- The trainer requested that certain additional information be covered while other information was considered inappropriate and should be deleted. The request fell upon deaf ears.
- The training department informed the trainer that it wanted only lecture, no case studies, and minimal exercises. This was the way it was done in other courses. The participant evaluations complained about lack of exercises and case studies.

While the training group believed that their actions were in the company's best interest, the results were devastating. The trainer was also at fault for allowing this situation to exist.

Successful project management implementation has a positive effect on corporate profitability. Given that this is true, why allow nonexperts to design project management coursework? Even line managers who believe that their organization requires project management knowledge may not know what to stress and what not to stress from the PMBOK® Guide.

The PO has the expertise in designing project management course content. The PO maintains intellectual property on lessons-learned files and project postmortem analysis, giving the PO valuable insight on how to obtain the best return on investment on training dollars. This intellectual property could also be invaluable in assisting line managers in designing courses specific to their organization. This activity is a low risk for the PO.

23.12 MANAGING STAKEHOLDERS

All companies have stakeholders. Apprehension may exist in the minds of some individuals that the PO will become the ultimate project sponsor responsible for all stakeholders. While this may happen in the future, it is highly unlikely that it will occur in the near term.

The PO focuses its attention primarily on internal (organizational) stakeholders. It is not the intent of a PO to replace executive sponsorship. As project management matures within an organization, it is possible that not all projects will require executive sponsorship. In such situations, the PO (and perhaps middle management) may be given the added responsibility of some sponsorship activities, most likely for internal projects.

The PO is a good "starting point" for building and maintaining alliances with key stakeholders. However, the PO's activities are designed to benefit the entire company, and giving the PO sponsorship responsibility may create a conflict of interest for PO personnel. Partnerships with key stakeholders must be built and nurtured, and that requires time. Stakeholder management may rob the PO personnel of valuable time needed for other activities. The overall risk for this activity is low.

23.13 CONTINUOUS IMPROVEMENT

Given the fact that the PO is a repository of the project management intellectual property, the PO may be in the best position to identify continuous improvement opportunities. The PO should not have unilateral authority for implementing the changes, but rather should have the ability to recommend changes. Some organizations maintain a strategic policy board or executive steering committee that, as one of its functions, evaluates PO continuous improvement recommendations.

23.14 CAPACITY PLANNING

Of all of the activities assigned to a PO, the most important activity in the eyes of senior management could very well be capacity planning. For executives to fulfill their responsibility as the architects of the corporate strategic plan, they must know how much additional work the organization can take on as well as, when and where without excessively burdening the existing labor pool. The PO must work closely with senior management on all activities related to portfolio management and project selection. Strategic timing, which is the process of deciding which projects to work on and when, is a critical component of strategic planning.

Senior management could "surf" the company intranet on an as-needed basis to view the status of an individual project without requiring personal contact with the team. But to satisfy the requirements for strategic timing, all projects would need to be combined into a single database that identifies the following:

- Resources committed per time period per functional area
- Total resource pool per functional area
- Available resources per time period per functional area

There may be some argument whether the control of this database should fall under the administration of the PO. The author believes that this should be a PO responsibility because:

- The data would be needed by the PO to support strategic planning efforts and project portfolio management.
- The data would be needed by the PO to determine realistic timing and costs to support competitive bidding efforts.

- The PO may be delegated the responsibility to determine resource skills required to undertake additional work.
- The data will be needed by the PO for upgrades and enhancements to this database and other impacted databases.
- The data may be necessary to perform feasibility studies and cost-benefit analysis.

This activity is a high-risk effort for the PO because line managers may see this as turf infringement.

23.15 RISKS OF USING A PROJECT OFFICE

Risks and rewards go hand in hand. The benefits of a PO can be negated if the risks of maintaining a PO are not effectively managed. Most risks do not appear during the creation of the PO, but more do so well after implementation. These risks include:

- *Headcount:* Once the organization begins recognizing the benefits of using a PO, there is a natural tendency to increase headcount in the PO with the false belief that additional benefits will be forthcoming. While this belief may be valid in some circumstances, the most common result is diminishing returns. As more of the organization becomes knowledgeable in project management, the headcount in the PO should decrease.
- *Burnout:* Employee burnout is always a risk. Using rotational or part-time assignments can minimize the risk. It is not uncommon for people working in a project office to still report "solid" to their line manager and "dotted" to the project office.
- *Excessive Paperwork:* Excessive paperwork costs millions of dollars to prepare and can waste precious time. Project activities work much better when using forms, guidelines, and checklists rather than the more rigid policies and procedures. To do this effectively requires a culture based upon trust, teamwork, cooperation, and effective communications.
- *Organizational Restructuring:* Information is power. Given the fact that the PO performs more work laterally than vertically, there can be power struggles for control of the PO, especially the project managers. Project management and a PO can work quite well within any organizational structure that is based upon trust, teamwork, cooperation, and effective communications.
- *Trying to Service Everyone in the Organization:* The company must establish some criteria for when the PO should be involved. The PO does not necessarily monitor all projects. The most common threshold limits for when to involve the PO include:
  - Dollar value of the project
  - Time duration of the project
  - Amount and complexity of cross-functionality
  - Risks to the company
  - Criticality of the project (i.e., cost reductions)

A critical question facing many executives is "How do we as executives measure the return-on-investment as a result of implementing a project office?" The actual measurement can be described in both qualitative and quantitative terms. Qualitatively, the executives can look at the number of conflicts coming up to the executive levels for resolution. With an effective PO acting as a filter, fewer conflicts should go up to the executive levels. Quantitatively, the executives can look at the following:

- *Progress Reviews:* Without a PO, there may exist multiple scheduling formats, perhaps even a different format for each project. With a PO and standardization, the reviews are quicker and more meaningful.
- *Decision-Making:* Without a PO, decisions are often delayed and emphasis is placed upon action items rather than meaningful decisions. With a PO, meaningful decisions are possible.
- *Wasted Meetings:* Without a PO, executives can spend a great deal of time attending too many and very costly meetings. With a PO and more effective information, the executives can spend less time in meetings and more time dealing with strategic issues rather than operational issues.
- *Quantity of Information:* Without a PO, the executives can end up with too little or too much project information. This may inhibit effective decision-making. With a PO and standardization, executives find it easier to make timely decisions. The prime responsibility of senior management is strategic planning and deployment and worrying about the future of the organization. The prime responsibility of middle-level and lower-level management is to worry about operational issues. The responsibility of the PO is to act as a bridge between all these levels and make it easier for all levels to accomplish their goals and objectives.

MANAGING CRISIS PROJECTS

24.0 INTRODUCTION

Project managers have become accustomed to managing within a structure process such as an enterprise project management methodology. The statement of work had gone through several iterations and was clearly defined. A work breakdown structure existed and everyone understood his or her roles and responsibilities as defined in the responsibility assignment matrix (RAM). All of this took time to do.

This is the environment we all take for granted. Now, let's change the scenario. The president of the company calls you into his office and informs you that several people have just died using one of your company's products. You are being placed in charge of this crisis project. The lobby of the building is swamped with the news media, all of which want to talk to you to hear your plan for addressing the crisis. The president informs you that the media knows you have been assigned as the project manager, and that a news conference has been set up for one hour from now. The president also asserts that he wants to see your plan for managing the crisis no later than 10:00 PM. this evening. Where do you begin? What should you do first? Time is now an extremely inflexible constraint rather than merely a constraint that may be able to be changed. Time does not exist to perform all of the activities you are accustomed to doing. You may need to make hundreds if not thousands of decisions quickly, and many of these are decisions you never thought that you would have to make. This is crisis project management.

24.1 UNDERSTANDING CRISIS MANAGEMENT

The field of crisis management is generally acknowledged to have started in 1982 when seven people died after ingesting Extra-Strength Tylenol capsules that were laced with cyanide. Johnson & Johnson, the parent of Tylenol, handled the situation in such a manner that it became the standard for crisis management.

Today, crises are neither rare nor random. They are part of our everyday lives. Crises cannot always be foreseen or prevented, but when they occur, we must do everything possible to manage them swiftly and effectively. We must also identify lessons learned and best practices so that mistakes are not repeated on future crises that will certainly occur.

Some crises are so well entrenched in our minds that they are continuously referenced in a variety of courses in business schools. Some crises that have become icons in society include:

- Hurricane Katrina
- Mad cow disease
- The Space Shuttle *Challenger* explosion
- The Space Shuttle *Columbia* reentry disaster
- The Tylenol poisonings
- Nestlé's infant formula controversy
- The Union Carbide chemical plant explosion in Bhopal, India
- The Exxon *Valdez* oil spill
- The Chernobyl nuclear disaster
- The Three Mile Island nuclear disaster
- The Russian submarine *Kursk* disaster
- The Enron and Worldcom bankruptcies

Some crises are the result of acts of God or natural disasters. The public is generally forgiving when these occur. Crisis management, however, deals primarily with man-made crises such as product tampering, fraud, and environmental contamination. Unlike natural disasters, these man-made crises are not inevitable, and the general public knows this and is quite unforgiving. When the Exxon *Valdez* oil spill occurred, Exxon refused to face the media for 5 days. Eventually, Exxon blamed the ship's captain for the accident and also attacked the Alaska Department of the Environment for hampering its emergency efforts. Stonewalling the media and assuming a defensive posture created extensive negative publicity for Exxon.

Most companies neither have any processes in place to anticipate these crises, even though they perform risk management activities, nor do they know how to manage them effectively after they occur. When lives are lost because of man-made crises, the unforgiving public becomes extremely critical of the companies responsible for the crises. Corporate reputations are very fragile. Reputations that had taken years to develop can be destroyed in hours or days.

Some people contend that with effective risk management practices, these crises can be prevented. While it is true that looking at the risk triggers can prevent some crises, not all crises can be prevented. However, best practices in crisis management can be developed and implemented such that when a crisis occurs, we can prevent a bad situation from getting worse.

For some time, corporations in specific industries have found it necessary to simulate and analyze worst-case scenarios for their products and services. Product tampering would be an example. These worst-case scenarios have been referred to as contingency plans, emergency plans, or disaster plans. These scenarios are designed around "known unknowns," where at least partial information exists on what events could happen.

Crisis management requires a heads-up approach with a very quick reaction time combined with a concerted effort on the part of possibly all employees. In crisis management, decisions have to be made quickly, often without even partial information and perhaps before the full extent of the damages are known. Events happen so quickly and so unpredictably that it may be impossible to perform any kind of planning. Roles and responsibilities of key individuals may change on a daily basis. There may be very active involvement by a majority of the stakeholders, many of which had previously been silent. Company survival could rest entirely on how well a company manages the crisis.

Crises can occur within any company, irrespective of the size. The larger the company involved in the crisis, the greater the media coverage. Also, crises can occur when things are going extremely well. The management guru, Peter Drucker, noted that companies that have been overwhelmingly successful for a long time tend to become complacent even though the initial assumptions and environmental conditions have changed. Under these conditions, crises are more likely to occur. Drucker called this "the failure of success."

Crisis management is now an integral part of training programs on professional responsibilities for project managers. This encompasses dealing with a multitude of stakeholders, being honest with the news media and the clients, and demonstrating a sincere concern for morality and ethics in project management.

Before discussions on the role of the project manager, it is important to examine the lessons learned from previous crises. What is unfortunate is that most of the lessons learned will come from improper handling of the crisis.

24.2 FORD VERSUS FIRESTONE

Product recalls are costly and embarrassing for the auto industry. Improper handling of a recall can have an adverse effect on consumer confidence and the selling price of the stock. Ford and tire manufacturer Firestone are still suffering from the repercussions of their handling of a product recall in 2000–2001.

In August of 2000, Firestone recalled 6.5 million tires in the United States, primarily because of tread separation problems on Ford Explorers [sports utility vehicles (SUV)]. The problems with the tires had been known several years earlier. In 1997–1998, Saudi Arabia reported tread separation on the SUV Explorer. In August 1999, Firestone replaced the tires in Saudi Arabia. In February 2000, Firestone replaced the tires in Malaysia and Thailand, and in May 2000, the tires were replaced in Venezuela.

Initially, it was believed that the problem might be restricted to countries with hot climates and rough roads. However, by May 2000, the U.S. National Highway Traffic Safety Administration (NHTSA) had received 90 complaints involving 27 injuries and 4 deaths. A U.S. recall of 6.5 million tires took place in August 2000.

Ford and Firestone adopted a unified response concerning the recall. Unfortunately, accidents continued after the recall. Ford then blamed Firestone for flaws in the tires and Firestone blamed Ford for design flaws in the SUV Explorer. The Ford–Firestone relationship quickly deteriorated.

The finger pointing between Ford and Firestone was juicy news for the media. Because neither company was willing to accept responsibility for its actions, probably because of the impending lawsuits, consumer confidence in both companies diminished, as did their stock prices. Consumer sentiment was that financial factors were more important than consumer safety.

Ford's CEO, Jac Nasser, tried to allay consumer fears, but his actions did not support his words. In September of 2000, he refused to testify at the Senate and House Commerce Subcommittee on tire recall stating that he was too busy. In October of 2000, Masatoshi Ono resigned as CEO of Bridgestone, Firestone's parent company. In October of 2001, Jac Nasser resigned. Both executives departed and left behind over 200 lawsuits filed against their companies.

Lessons Learned

1. Early warning signs appeared but were marginally addressed.
2. Each company blamed the other leaving the public with the belief that neither company could be trusted with regard to public safety.
3. Actions must reinforce words; otherwise, the public will become nonbelievers.

24.3 THE AIR FRANCE *CONCORDE* CRASH

On July 25, 2000, an Air France *Concorde* flight crashed on takeoff killing all 109 people on board and 4 people on the ground. Air France immediately grounded its entire *Concorde* fleet pending an accident investigation. In response to media pressure, Air France used its website for press releases, expressed sorrow and condolence from the company, and arranged for some financial consideration to be paid to the relatives of the victims prior to a full legal settlement. The chairman of Air France, Jean-Cyril Spinetta, visited the accident scene the day of the accident and later attended a memorial service for the victims.

Air France's handling of the crisis was characterized by fast and open communication with the media and sensitivity for the relatives of the victims. The selling price of the stock declined rapidly the day of the disaster but made a quick recovery.

British Airways (BA) also flew the *Concorde*, but took a different approach immediately following the accident. BA waited a month before grounding all *Concorde* flights indefinitely, and only after the Civil Aviation Authority announced it would be withdrawing the *Concorde*'s airworthiness certification. Eventually, the airworthiness certification was reinstated, but it took BA's stock significantly longer to recover its decline in price.

Lessons Learned

1. Air France and British Airways took different approaches to the crisis.
2. The Air France chairman showed compassion by visiting the site of the disaster as quickly as possible and attending a memorial service for the victims. British Airways did neither, thus disregarding their social responsibility.

24.4 INTEL AND THE PENTIUM CHIP

Intel, the manufacturer of Pentium chips, suffered an embarrassing moment resulting in a product recall. A mathematics professor, while performing prime number calculations on 10-digit numbers, discovered significant round-off errors using the Pentium chips. Intel believed that the errors were insignificant and would show up only in every few billion calculations. But the mathematician was performing billions of calculations and the errors were now significant.

The professor informed Intel of the problem. Intel refused to take action on the problem, stating that these errors were extremely rare and would affect only a very small percentage of Pentium users. The professor went public with the disclosure of the error.

Suddenly, the small percentage of people discovering the error was not as small as originally thought. Intel still persisted in its belief that the error affected only a small percentage of the population. Intel put the burden of responsibility on the user to show that his or her applications necessitated a replacement chip. Protests from consumers grew stronger. Finally the company agreed to replace all chips, no questions asked, after IBM announced it would no longer use Pentium chips in its personal computers.

Intel created its own public relations nightmare. Its response was slow and insincere. Intel tried to solve the problem solely through technical channels and completely disregarded the human issue of the crisis. Telling people who work in hospitals or air traffic control that there is a flaw in their computer but it is insignificant is not an acceptable response. Intel spent more than a half billion dollars in the recall, significantly more than the cost of an immediate replacement.

Lessons Learned
1. Intel's inability to take immediate responsibility for the crisis and develop a crisis management plan made the situation worse.
2. Intel completely disregarded public opinion.
3. Intel failed to realize that a crisis existed.

24.5 THE RUSSIAN SUBMARINE *KURSK*

In August of 2000, the sinking of the nuclear-powered submarine *Kursk* resulted in the deaths of 118 crewmembers. Perhaps the crew could never have been saved, but the way the crisis was managed was a major debacle for both the Russian Navy and the Russian government.

Instead of providing honest and sincere statements to the media, the Russian Ministry of Defense tried to downplay the crisis by disclosing misleading information, telling the public that the submarine had run aground during a training exercise and that the crew was in no immediate danger. The ministry spread a rumor that there was a collision with a NATO submarine. Finally, the truth came out, and by the time the Russians sought assistance in mounting a rescue mission, it was too late.

Vladimir Putin, Russia's president, received enormous unfavorable publicity for his handling of the crisis. He was vacationing in southern Russia at the time and appeared on Russian television clad in casual clothes, asserting that the situation was under control. He then disappeared from sight for several days, which angered the public and family members of the crew, indicating his lack of desire to be personally involved in the crisis. When he finally visited the *Kursk*'s home base, he was greeted with anger and hostility.

Lessons Learned
1. Lying to the public is unforgivable.
2. Russia failed to disclose the seriousness of the crisis.
3. Russia failed to ask other countries for assistance in a timely manner.
4. Russia demonstrated a lack of social responsibility by refusal to appear at the site of the crisis and showed a lack of compassion for the victims and their families.

24.6 THE TYLENOL POISONINGS[1]

In September 1982, seven people died after taking Extra-Strength Tylenol laced with cyanide. All of the victims were relatively young. These deaths were the first ever to result from what came to be known as product tampering. All seven individuals died within a one-week time period. The symptoms of cyanide poisoning are rapid collapse and coma and are difficult to treat.

On the morning of September 30, 1982, reporters began calling the headquarters of Johnson & Johnson asking about information on Tylenol and Johnson & Johnson's reaction to the deaths. This was the first that Johnson & Johnson had heard about the deaths and the possible link to Tylenol.

From the start, the company found itself entering a closer relationship with the press than it was accustomed to. Johnson & Johnson bitterly recalled an incident 9 years earlier in which the media circulated a misleading report suggesting that some baby powder had been contaminated by asbestos. But in the Tylenol case, Johnson & Johnson opened its doors. For one thing, the company was getting some of its most accurate and up-to-date information about what was going on around the country from the reporters calling in for comment. For another, Johnson & Johnson needed the media to get out as much information to the public as quickly as possible and prevent a panic.

The chairman of Johnson & Johnson was James Burke, 57, a 30-year veteran of Johnson & Johnson. Mr. Burke had to protect the company's image, allay fears that the public may have in the use of Tylenol products, and work with a multitude of stakeholders, including government agencies. Burke decided to be the spokesperson for the media and personally take charge of the crisis project at Johnson & Johnson.

---

1. The complete Tylenol case study is too large to be included here. For additional information and details on the Tylenol crises, see "The Tylenol Tragedies" in Harold Kerzner, *Project Management Case Studies*, 3rd ed. (Hoboken, NJ: Wiley, 2006), pp. 509–536.

Stakeholder Management   Burke had several options available to him on how to handle the crisis. Deciding which option to select would certainly be a difficult decision. Looking over Burke's shoulder were the stakeholders who would be affected by Johnson & Johnson's decision. Among the stakeholders were stockholders, lending institutions, employees, managers, suppliers, government agencies, and the consumers.

CONSUMERS

The consumers had the greatest stake in the crisis because their lives were on the line. The consumers must have confidence in the products they purchase and believe that they were safe to use as directed.

STOCKHOLDERS

The stockholders had a financial interest in the selling price of the stock and the dividends. If the cost of removal and replacement, or in the worst-case scenario of product redesign, were substantial, it could lead to a financial hardship for some investors that were relying on the income.

LENDING INSTITUTIONS

Lending institutions provide loans and lines of credit. If the present and/or future revenue stream is impaired, then the funds available might be reduced and the interest rate charge could increase. The future revenue stream of its products could affect the quality rating of its debt.

GOVERNMENT

The primary concern of the government was in protecting public health. In this regard, government law enforcement agencies were committed to apprehending the murderer. Other government agencies would provide assistance in promoting and designing tamper-resistant packages in an effort to restore consumer confidence.

MANAGEMENT

Company management had the responsibility to protect the image of the company, as well as its profitability. To do this, management must convince the public that management will take whatever steps are necessary to protect the consumer.

EMPLOYEES

Employees have the same concerns as management but are also somewhat worried about possible loss of income or even employment.

Whatever decision Johnson & Johnson selected was certain to displease at least some of the stakeholders. Therefore, how does a company decide which stakeholders' needs are more important? How does a company prioritize stakeholders?

For Jim Burke and the entire strategy committee, the decision was not very difficult—just follow the corporate credo. For more than 45 years, Johnson & Johnson had a corporate credo that clearly stated that the company's first priority is to the users of Johnson & Johnson's products and services. Everyone knew the credo, what it stood for, and the fact that it must be followed. The corporate credo guided the decision-making process, and everyone knew it without having to be told. The credo stated that the priorities, in order, were:

1. To the consumers
2. To the employees
3. To the communities being served
4. To the stockholder

When the crisis had ended, Burke recalled that no meeting had been convened for the first critical decision: to be open with the press and put the consumer's interest first. "Every one of us knew what we had to do," Mr. Burke commented. "There was no need to meet. We had the credo philosophy to guide us."

Accolades
Some people believed that James Burke almost single-handedly saved Tylenol, especially when Wall Street believed that the Tylenol name was dead. Burke had courageously made some decisions against the advice of government agents and some of his own colleagues. He appeared on a variety of talk shows, such as the Phil Donahue Show and 60 Minutes. His open and honest approach to the crisis convinced people that Johnson & Johnson was also a victim. According to Johnson & Johnson spokesman, Bob Andrews, "The American public saw this company was also the victim of an unfortunate incident and gave us our market back."

Both Johnson & Johnson and James Burke received nothing but accolades and support from the media and general public in the way the crisis was handled. A sampling of opinion from newspapers across the United States includes:

- *Wall Street Journal:* "Johnson & Johnson, the parent company that makes Tylenol, set the pattern of industry response. Without being asked, it quickly withdrew Extra-Strength Tylenol from the market at a very considerable expense . . . The company chose to take a large loss rather than expose anyone to further risk. The anti-corporation movement may have trouble squaring that with the devil theories it purveys."
- *Washington Post:* "Though the hysteria and frustration generated by random murder have often obscured the company's actions, Johnson & Johnson has effectively demonstrated how a major business ought to handle a disaster. From the day the deaths were linked to the poisoned Tylenol . . . Johnson & Johnson has succeeded in portraying itself to the public as a company willing to do what's right regardless of cost."
- *Express and News* (San Antonio, Texas): "In spite of the $100 million loss it was facing, the company . . . never put its interests ahead of solving the murders and protecting the public. Such corporate responsibility deserves support."
- *Evening Independent* (St. Petersburg, Florida): "The company has been straightforward and honest since the first news of the possible Tylenol link in the Chicago-area deaths. Some firms would have tried to cover up, lie or say 'no

comment.' Johnson & Johnson knows better. Its first concern was to safeguard the public from further contamination, and the best way to do that was to let people know what had occurred by speaking frankly with the news media."

- *Morning News* (Savannah, Georgia): "Tylenol's makers deserve applause for their valiant attempt to recover from the terrible blow they have suffered."

Lessons Learned

1. On crisis projects, the (executive) project sponsor will be more actively involved and may end up performing as the project manager as well.
2. The project sponsor should function as the corporate spokesperson, responsible for all crisis communications. Strong communication skills are therefore mandatory.
3. Open and honest communications are essential, especially with the media.
4. The company must display a social consciousness as well as a sincere concern for people, especially victims and their families.
5. Managing stakeholders with competing demands is essential.
6. The company, and especially the project sponsor, must maintain a close working relationship with the media.
7. A crisis committee should be formed and composed of the senior-most levels of management.
8. Corporate credos can shorten the response time during a crisis.
9. The company must be willing to seek help from all stakeholders and possibly also government agencies.
10. Corporate social responsibility must be a much higher priority than corporate profitability.
11. The company, specifically the project sponsor, must appear at the scene of the crisis and demonstrate a sincere compassion for the families of those injured.
12. The company must try to prevent a bad situation from getting worse.
13. Manage the crisis as though all information is public knowledge.
14. Act quickly and with sincerity.
15. Assume responsibility for your products and services and your involvement in the crisis.

24.7 NESTLÉ'S MARKETING OF INFANT FORMULA

During a crisis, corporations have an obligation to demonstrate social responsibility, hopefully, without any impact on the profitability of the firm. During the past 40 years, laws, guidelines, and codes have been enacted; not only to show how this social responsibility can be applied, but also to make sure it is done ethically and without any harm to society. But what happens if the guidelines, codes, and laws are nonexistent? What happens if the corporation truly believes that they are doing good service for humanity, but at the same time part of society believes that an injustice has occurred? Such was the case at Nestlé's where corporate management emphatically believed that they were doing a good service for humanity with the distribution of infant formula to Third World nations. However, infant mortality, estimated to be in the thousands, occurred in Third World nations as a result of Nestlé's aggressive marketing campaign.

Nestlé's Infant Formula   Nestlé's entrance into the infant formula industry was derived from a desire to increase profitability and gain a greater market share. Nestlé began developing and marketing its infant formula in the 1920s, when infant formulas were proven to be nutritionally better than condensed milk for the baby and a successful alternative to breastfeeding. Nestlé first entered the Third World markets in Brazil and had expanded over the next 50 years into approximately 20 additional Third World markets. It is important to note that Nestlé never intended that the infant formula serve as a replacement for breast milk, but rather as a supplement.

The product, itself, was of high quality and nutritionally superior to other alternatives in processed products. The product could meet a nutritional need as a supplement to breastfeeding and as a second-best alternative when women were unable or unwilling to breastfeed. Therefore, the product could be socially useful. The product, itself, was not the problem, but rather the misuse of the product. The probability of product misuse was greater for some consumers than others. In this specific case, it was the low-income and low-educated consumers in Third World countries that Nestlé was targeting and who were the most vulnerable.

Product Misuse   The product is nutritiously beneficial for the infant only if it is mixed with water and fed through clean and sterile bottles, and stored in refrigerators. Many consumers in Third World countries could not meet these product requirements. Many of these poor consumers did not have access to clean and drinkable water. They sometimes used the main water supplies, rivers or lakes, which were also used for laundry and bathing. They did not have the funds to afford fuel required for boiling water. Sanitary facilities and disposal systems were often undeveloped and contaminated. An unsanitary bottle or liquid can produce harmful infections to the infant. Some parents in Third World countries were illiterate and not fully able to follow product directions. Mothers did not properly mix formulas or follow correct sanitary procedures. Their lack of financial resources also restricted them from purchasing additional needed formula, causing mothers to dilute the formula mix to make the formula stretch further.

All of these forms of product misuse caused serious and fatal health effects to infants. Consumers in Third World countries were provided with Nestlé products that could not be used properly. The environmental restrictions, along with improper resources and guidance made it a very risky market for Nestlé to target unless consumers were provided with a basic understanding of sanitation and nutrition. Nestlé failed to assure this type of awareness and more importantly, faulted in providing social responsibility.

The Rising Opposition   As word spread about the tragic events of infant formula occurring in Third World countries, many critics started to raise their voices and demand changes in Nestlé's marketing practices. Eventually, a boycott group formed and obtained large-scale support from the medical profession and the clergy. The boycotters never questioned the quality, need, and importance of the infant formula. The boycotters were simply opposed to the aggressive marketing campaign in Third World nations. Nestlé maintained its original position that it was doing a good deed for society, but public opinion sided with the boycotters. Eventually, Nestlé conceded to the demands of the boycotters.

Lessons Learned The following lessons can be learned from the Nestlé crisis:

1. Nestlé's actions were not representative of company demonstrating social responsibility. Its actions may have been legally correct, but they were also morally and ethically incorrect.
2. Nestlé should have used the media to its advantage rather than attacking the media. That made the situation even worse.
3. Nestlé remained in a state of denial over the crisis and refused to accept accountability for its actions. As a result, the media relentlessly looked for "skeletons in the closets," found some, and reported the results to the public.
4. Nestlé assumed that the public was ignorant of the magnitude of the crisis.
5. The longer the crisis remains in the public's eye, the greater the tendency for the company to be portrayed as a villain rather than as a victim.
6. Because of Nestlé's inactions, the size and influence of the boycott grew.
7. Nestlé eventually ran out of options and the corporate image became tarnished because of inactions.
8. Nestlé neglected to realize the importance of demonstrating a concern for people during the crisis.

24.8 THE SPACE SHUTTLE *CHALLENGER* DISASTER[2]

On January 28, 1986, the Space Shuttle *Challenger* lifted off the launch pad at 11:38 AM. Approximately 74 seconds into the flight, the *Challenger* was engulfed in an explosive burn and all communication and telemetry ceased. Seven brave crewmembers lost their lives. Following the accident, significant energy was expended trying to ascertain whether or not the accident was predictable. Controversy arose from the desire to assign, or to avoid, blame. Some publications called it a management failure, specifically in risk management, while others called it a technical failure.

Lessons Learned The following lessons were learned from the *Challenger* disaster:

1. The crisis was created by a poor organizational culture.
2. There were significant early warning signs, which if addressed, could have avoided the crisis. They were ignored.

2. Only summary information is provided in this chapter. For a more in-depth analysis of the case, see "The Space Shuttle *Challenger* Disaster" in Harold Kerzner, *Project Management Case Studies*, 3rd ed. (Hoboken, NJ: Wiley, 2006), pp. 425–474.

3. The chain of command insulated managers and executives from bad news.
4. Management refused to listen to workers who were pleading for help.
5. There was a questionable concern for human life indicated by the pressure to maintain the schedule at all costs.

24.9 THE SPACE SHUTTLE *COLUMBIA* DISASTER[3]

On February 1, 2003, the Space Shuttle *Columbia* began its reentry into the Earth's atmosphere. The shuttle relied upon the heat-resistant materials and the heat shield to protect it from the heat-producing friction encountered during reentry. Unfortunately, a problem occurred and the shuttle disintegrated during reentry into the atmosphere, killing its seven-person crew.

A *Columbia* Accident Investigation Board was convened to address the accident. Seven months later, the board released its findings. The technical cause of the accident was traced to liftoff, where a large piece of fuel tank insulation dislodged and hit and damaged the heat-resistant tiles on the leading edge of *Columbia*'s left wing and punched a hole. The metal components on the shuttle melt at about 2000°F. The heat-resistant ceramic tiles melt at about 3000°F. The tiles prevent the 10,000°F reentry heat from penetrating the vehicle. During reentry, the heat was then able to penetrate the left wing, eventually melting part of the internal structure of the wing causing it to collapse, and resulting in the shuttle tumbling out of control during reentry.

While the dislodged insulation was the technical or physical cause of the accident, the Accident Investigation Board concluded that NASA's culture was equally at fault for the accident, and that NASA's culture was a detriment to safety. These conclusions stated that NASA had relied on past success as a substitute for sound engineering. NASA maintained organizational barriers preventing the communication of critical safety information, and stifled professional differences of opinion.[4] In particular, the board identified attitudes at NASA that were "incompatible with an organization that deals with high-risk technology."[5]

The board also concluded that management of the Space Shuttle Program demonstrated a strong resistance to new information and technologies that may have been able to prevent the disaster. They also failed to develop a simple contingency plan for a reentry emergency. "They were convinced, without study, that nothing could be done about such an emergency. The intellectual curiosity and skepticism that a solid safety culture requires was almost entirely absent."[6]

While these conclusions were damaging to NASA's credibility, there were still more damaging conclusions. Many of the critical issues addressed by *Columbia*'s Accident

---

3. For a more in-depth analysis, see "The Space Shuttle *Columbia* Disaster" in Harold Kerzner, *Project Management Case Studies*, 3rd ed. (Hoboken, NJ: Wiley, 2006), pp. 475–481.
4. See Columbia Accident Investigation Board, *Report Volume 1*, 1 August 2003, 9. Obtainable at <http://www.caib.us/news/report/volume1/chapters.html>.
5. See note 4, Columbia Accident Investigation Board, 8.
6. See note 4, Columbia Accident Investigation Board, 181.

Investigation Board were also identified 17 years earlier by the Presidential Commission investigation the Space Shuttle *Challenger* disaster. Lessons learned from the *Challenger* disaster had not been fully implemented some 17 years later.

Lessons Learned

The following lessons were learned from the *Columbia* disaster:

1. Risk planning was virtually nonexistent.
2. There were no contingency plans for several of the high-risk portions of the space flight.
3. There was a silent safety program in place.
4. There was a poor transfer of knowledge, particularly lessons learned, from the *Challenger* disaster.

24.10 VICTIMS VERSUS VILLAINS

The court of public opinion usually casts the deciding ballot as to whether the company involved in the crisis should be treated as a victim or a villain in the way they handled the crisis. The two determining factors are most often the company's demonstration of corporate social responsibility during the crisis and how well they dealt with the media.

During the Tylenol poisoning, Johnson & Johnson's openness with the media, willingness to accept full responsibility for its products, and rapid response to the crisis irrespective of the cost were certainly viewed favorably by the general public. Johnson & Johnson was viewed as a victim of the crisis. Nestlé, on the other hand, was viewed as a villain despite its belief that it was doing good for humanity with its marketing of the infant formula.

Table 24–1 shows how the general public viewed the company's performance during the crisis. The longer the crisis lasts, the greater the tendency that the company will be portrayed as a villain.

TABLE 24–1   PUBLIC VIEW OF COMPANY PERFORMANCE

Crisis	Public Opinion View
Tylenol poisonings	Victim
Nestlé and the infant formula	Villain
Challenger explosion	Villain
Columbia reentry disaster	Villain
Exxon *Valdez* oil spill	Villain
Russian submarine, *Kursk*	Villain
Ford and Firestone	Villains
Concorde: Air France	Victim
Concorde: British Airways	Villain
Intel and Pentium	Villain

24.11 LIFE-CYCLE PHASES

Crises can be shown to go through the life cycles illustrated in Figure 24–1. Unlike traditional project management life-cycle phases, each of these phases can be measured in hours or days rather than months. Unsuccessful management of any of these phases could lead to a corporate disaster.

Most crises are preceded by early warning signs or risk triggers indicating that a crisis may occur. This is the early warning phase. Typical warning signals might include violations of safety protocols during technology development, warnings from government agencies, public discontent, complaints from customers and warnings/concerns from lower-level employees.

Most companies are poor at risk management, especially at evaluation of early warning signs. Intel, the Nestlé case, and the shuttle disasters were examples of this. Today, project managers are trained in the concepts of risk management, but specifically related to the management of the project, or with the development of the product. Once the product is commercialized, the most serious early warning indicators can appear and, by that time, the project manager may be reassigned to another project. Someone else must then evaluate the early warning sings.

Early warning signs are indicators of potential risks. Time and money is a necessity for evaluation of these indicators, which preclude the ability to evaluate all risks. Therefore, companies must be selective in the risks they consider.

The next life-cycle phase is the understanding of the problem causing the crisis. For example, during the Tylenol poisonings, once the deaths were related to the Tylenol capsules, the first concern was to discover whether the capsules were contaminated during the manufacturing process (i.e., an inside job) or during distribution and sales (i.e., an outside job). Without a fact-based understanding of the crisis, the media can formulate their own cause of the problem and pressure the company to follow the wrong path.

The third life-cycle phase is the damage assessment phase. The magnitude of the damage will usually determine the method of resolution. Underestimating the magnitude of the damage and procrastination can cause the problem to escalate to a point where the cost of correcting the problem can grow by orders of magnitude. Intel found this out the hard way.

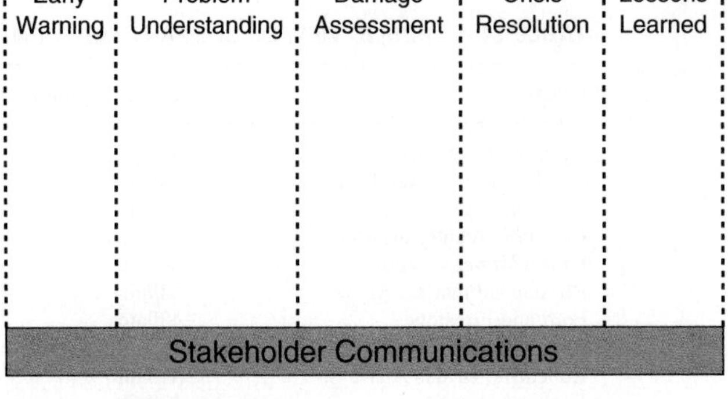

Figure 24–1   Crisis management life-cycle phases.

The crisis resolution stage is where the company announces its approach to resolve the crisis. The way the public views the company's handling of the crisis has the potential to make or break the company.

The final stage, lessons learned, mandates that companies learn from not only their own crises but from how others handled their crises. Learning from the mistakes of others is better than learning from one's own mistakes.

Perhaps the most critical component in Figure 24–1 is stakeholder communications. When a crisis occurs, the assigned project manager may need to communicate with stakeholders that previously were of minor importance, such as the media and government agencies, and all of whom have competing interests. These competing interests mandate that the project managers understand stakeholder needs and objectives and also possess strong communication skills, conflict resolution skills, and negotiation skills.

24.12 PROJECT MANAGEMENT IMPLICATIONS

While it is true that every crisis has its unique characteristics, there is some commonality that can affect project management. Some implications for project management include:

1. **Leader of the Crisis Team:** It is important to understand who will be leading the crisis team. It is quite rare that a project manager will be given the responsibility to manage a crisis team, at least with our definition of a crisis. Many of the decisions that need to be made are not those made by project managers when performing their normal duties. The project sponsor will most likely assume a dual role and be the leader of the project team as well as acting as the sponsor. As in the Tylenol case, it is common for the CEO to assume primary responsibility for managing the crisis team. The leader of the crisis team must have complete authority to commit corporate resources to the project. The project manager, as we know it, will function in an assistant project manager capacity.

2. **The Crisis Committee:** In time of crisis, there should exist a crisis committee composed of the senior-most levels of management. The crisis committee should also have multifunctional membership. Project managers and assistant project managers will then report to the entire membership of the committee rather than to a single sponsor.

3. **Crisis Communications:** The leader of the crisis team will be the primary spokesperson for the crisis and ultimately responsible for all media communications. The media cannot be ignored and has the power to portray the company as either a victim or a villain. The senior-most levels of management, especially those executives with professional communication skills, must perform crisis communication with the media. It is essential that the corporation speak with one voice, accompanied by swiftness, honesty, openness, sincerity and compassion for the victims and their families. Information must not be withheld from the public. Withholding information from the media with the excuse that the information is incomplete may be viewed as stonewalling the media.

4. **Stakeholder Management:** The crisis team must identify all of the parties affected by the crisis. This includes bankers, stockholders, employees, suppliers, customers, top management, government agencies, and the like. Each stakeholder can have a different interest in how the crisis is resolved, such as a financial, medical, environmental, political, or social interest. The crisis team must also be willing to ask for help from external agencies such as the FBI, Environmental Protection Agency, Federal Emergency Management Agency, and the Red Cross. The assistance of these external stakeholders can be invaluable.

5. **Assume Responsibility:** The company must accept responsibility for its actions (or inactions) immediately, and without being coerced into doing so. This will most likely fair well with the media.

6. **Response Time:** In every crisis, there is usually a small window of opportunity where quick and decisive action can limit or even reduce the damages. Another reason for a quick response is because of the media. The longer the company takes to act, the greater the likelihood the media will look upon the company unfavorably.

7. **Compassion:** The respect for people is mandatory. It is essential that the company expresses and demonstrates compassion for all injured parties and their families, irrespective of who was actually at fault for the crisis. The emotions of the victims and their families can be expected to run high. The public expects the company to demonstrate compassion. This also includes being on the scene of the disaster as quickly as possible. Delaying a visit to the crisis scene may be viewed as a lack of compassion or, even worse, that the company is hiding something.

8. **Documentation:** Because of the multitude of legal issues that may be encountered during a crisis, most of the decisions made will need to be clearly documented. The project manager and the associated team members should possess strong writing skills.

9. **Capture Lessons Learned:** Crisis can occur without warning. Companies are expected to capture lesson learned from both internal and external crises. This includes the examination of risk triggers, developing risk management templates, and perhaps even a corporate credo.

The Rise, Fall, and Resurrection of Iridium: A Project Management Perspective

25.0 INTRODUCTION

The Iridium Project was designed to create a worldwide wireless handheld mobile phone system with the ability to communicate anywhere in the world at any time. Executives at Motorola regarded the project as the eighth wonder of the world. But more than a decade later and after investing billions of dollars, Iridium had solved a problem that very few customers needed solved. What went wrong? How did the Iridium Project transform from a leading-edge technical marvel to a multi-billion-dollar blunder? Could the potential catastrophe have been prevented?[1]

> What it looks like now is a multibillion-dollar science project. There are fundamental problems: The handset is big, the service is expensive, and the customers haven't really been identified.
> —Chris Chaney, Analyst, A.G. Edwards, 1999

> There was never a business case for Iridium. There was never market demand. The decision to build Iridium wasn't a rational business decision. It was more of a religious decision. The remarkable thing is that this happened at a big corporation, and that there was not a rational decision-making process in place to pull the plug. Technology for technology's sake may not be a good business case.[2]
> —Herschel Shosteck, Telecommunication Consultant

> Iridium is likely to be some of the most expensive space debris ever.
> —William Kidd, Analyst, C.E. Unterberg, Towbin (now part of Collins Stewart plc)

---

1. Some of the material has been adapted from Sydney Finkelstein and Shade H. Sanford, "Learning from Corporate Mistakes: The Rise and Fall of Iridium," *Organizational Dynamics*, Vol. 29, No. 2, pp. 138–148; 2000. © 2000 by Elsevier Sciences, Inc. Reproduced by permission.
2. Stephanie Paterik, "Iridium Alive and Well," *The Arizona Republic*, April 27, 2005, p. D5.

In 1985, Bary Bertiger, chief engineer in Motorola's strategic electronics division, and his wife Karen were on a vacation in the Bahamas. Karen tried unsuccessfully to make a cellular telephone call back to her home near the Motorola facility in Chandler, Arizona, to close a real-estate transaction. Unsuccessful, she asked her husband why it would not be possible to create a telephone system that would work anywhere in the world, even in remote locations.

At this time, cell technology was in its infancy but was expected to grow at an astounding rate. AT&T projected as many as 40 million subscribers by 2000.[3] Cell technology was based on tower-to-tower transmission as shown in Figure 25–1. Each tower or "gateway" ground station reached a limited geographic area or cell and had to be within the satellite's field of view. Cell phone users likewise had to be near a gateway that would uplink the transmission to a satellite. The satellite would then downlink the signal to another gateway that would connect the transmission to a ground telephone system. This type of communication is often referred to as bent-pipe architecture. Physical barriers between the senders/receivers and the gateways—such as mountains, tunnels, and oceans—created interference problems and therefore limited service to high-density communities. Simply stated, cell phones couldn't leave home. And, if they did, there would be additional "roaming" charges. To make matters worse, every country had its own standards, and some cell phones were inoperable when traveling in other countries.

Communications satellites, in use since the 1960s, were typically geostationary satellites that orbited at altitudes of more than 22,300 miles. At this altitude, three geosynchronous satellites and just a few gateways could cover most of the Earth. But satellites at this altitude meant large phones and annoying quarter-second voice delays. Comsat's Planet 1

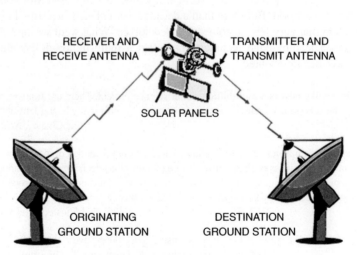

FIGURE 25–1.   Typical satellite communication architecture.

3. Judith Bird, "Cellular Technology in Telephones," *Data Processing*, vol. 27, no. 8, October 1985, p. 37.

phone, for example, weighed in at a computer-case-sized 4.5 pounds. Geosynchronous satellites require signals with a great deal of power. Small mobile phones, with a 1-watt signal, could not work with satellites positioned at this altitude. Increasing the power output of the mobile phones would damage human tissue. The alternative was therefore to move the satellites closer to Earth such that less power would be needed. This would require significantly more satellites the closer we get to Earth and additional gateways. Geosynchronous satellites, which are 100 times further away from Earth than low Earth-orbiting (LEO) satellites, could require almost 10,000 times as much power as LEOs, if everything else were the same.[4]

When Bary Bertiger returned to Motorola, he teamed up with Dr. Raymond Leopold and Kenneth Peterson to see if such a worldwide system could be developed while overcoming all of the limitations of existing cell technology. There was also the problem that LEO satellites would be orbiting the Earth rapidly and going through damaging temperature variations—from the heat of the sun to the cold shadow of Earth.[5] The LEO satellites would most likely need to be replaced every 5 years. Numerous alternative terrestrial designs were discussed and abandoned. In 1987 research began on a constellation of LEO satellites moving in polar orbits that could communicate directly with telephone systems on the ground and with one another.

Iridium's innovation was to use a large constellation of low-orbiting satellites approximately 400–450 miles in altitude. Because Iridium's satellites were closer to Earth, the phones could be much smaller and the voice delay imperceptible. But there were still major technical design problems. With the existing design, a large number of gateways would be required, thus substantially increasing the cost of the system. As they left work one day in 1988, Dr. Leopold proposed a critical design element. The entire system would be inverted whereby the transmission would go from satellite to satellite until the transmission reached the satellite directly above the person who would be receiving the message. With this approach, only one gateway Earth station would be required to connect mobile-to-landline calls to existing land-based telephone systems. This was considered to be the sought-after solution and was immediately written in outline format on a whiteboard in a security guard's office. Thus came forth the idea behind a worldwide wireless handheld mobile phone with the ability to communicate anywhere and anytime.

25.1 NAMING THE PROJECT "IRIDIUM"

Motorola cellular telephone system engineer, Jim Williams, from the Motorola facility near Chicago, suggested the name Iridium. The proposed 77-satellite constellation reminded him of the electrons that encircle the nucleus in the classical Bohr model of the atom. When he consulted the periodic table of the elements to discover which atom had 77

---

4. See note 3.
5. Bruce Gerding, Personal Communications via Satellite: An Overview," *Telecommunications*, vol. 30, no. 2, February 1996, pp. 35, 77.

electrons, he found iridium—a creative name that had a nice ring. Fortunately, the system had not yet been scaled back to 66 satellites, or else he might have suggested the name Dysprosium.

25.2 OBTAINING EXECUTIVE SUPPORT

Initially, Bertiger's colleagues and superiors at Motorola had rejected the Iridium concept because of its cost. Originally, the Iridium concept was considered perfect for the U.S. government. Unfortunately, the era of lucrative government-funded projects was coming to an end, and it was unlikely that the government would fund a project of this magnitude. However, the idea behind the Iridium concept intrigued Durrell Hillis, the general manager of Motorola's Space and Technology Group. Hillis believed that Iridium was workable if it could be developed as a commercial system. Hillis instructed Bertiger and his team to continue working on the Iridium concept but to keep it quiet.

> "I created a bootleg project with secrecy so no one in the company would know about it," Hillis recalls. He was worried that if word leaked out, the ferociously competitive business units at Motorola, all of which had to fight for R&D funds, would smother the project with nay-saying.[6]

After 14 months of rewrites on the commercialized business plan, Hillis and the Iridium team leaders presented the idea to Robert Galvin, Motorola's chairman at the time, who gave approval to go ahead with the project. Robert Galvin, and later his successor and son Christopher Galvin, viewed Iridium as a potential symbol of Motorola's technological prowess and believed that this would become the eighth wonder in the world. In one of the initial meetings, Robert Galvin turned to John Mitchell, Motorola's president and chief operating officer, and said, "If you don't write out a check for this John, I will, out of my own pocket."[7] To the engineers at Motorola, the challenge of launching Iridium's constellation provided considerable motivation. They continued developing the project that resulted in initial service in November 1998 at a total cost of over $5 billion.

25.3 LAUNCHING THE VENTURE

On June 26, 1990, Hillis and his team formally announced the launch of the Iridium Project to the general public. The response was not very pleasing to Motorola with skepticism over

---

6. David S. Bennahum, "The United Nations of Iridium," *Wired*, Issue 6.10, October 1998, p. 194.
7. Quentin Hardy, "How a Wife's Question Led Motorola to Chase a Global Cell-Phone Plan," *Wall Street Journal* (Eastern edition), New York, December 16, 1996, p. A1.

the fact that this would be a new technology, the target markets were too small, the revenue model was questionable, obtaining licenses to operate in 170 countries could be a problem, and the cost of a phone call might be overpriced. Local phone companies that Motorola assumed would buy into the project viewed Iridium as a potential competitor since the Iridium system bypassed traditional landlines. In many countries, Postal Telephone and Telegraph (PTT) operators are state owned and a major source of revenue because of the high profit margins. Another issue was that the Iridium Project was announced before permission was granted by the Federal Communications Commission (FCC) to operate at the desired frequencies.

Both Mitchell and Galvin made it clear that Motorola would not go it alone and absorb the initial financial risk for a hefty price tag of about $3.5 billion. Funds would need to be obtained from public markets and private investors. In order to minimize Motorola's exposure to financial risk, Iridium would need to be set up as a project-financed company. Project financing involves the establishment of a legally independent project company where the providers of funds are repaid out of cash flow and earnings, and where the assets of the unit (and only the unit) are used as collateral for the loans. Debt repayment would come from the project company only, rather than from any other entity. A risk with project financing is that the capital assets may have a limited life. The potential limited life constraint often makes it difficult to get lenders to agree to long-term financial arrangements.

Another critical issue with project financing especially for high-technology projects is that the projects are generally long term. It would be nearly 8 years before service would begin, and in terms of technology, 8 years is an eternity. The Iridium Project was certainly a "bet on the future." And if the project were to fail, the company could be worth nothing after liquidation.

In 1991, Motorola established Iridium Limited Liability Corporation (Iridium LLC) as a separate company. In December of 1991, Iridium promoted Leo Mondale to vice president of Iridium International. Financing the project was still a critical issue. Mondale decided that, instead of having just one gateway, there should be as many as 12 regional gateways that plugged into local, ground-based telephone lines. This would make Iridium a truly global project rather than appear as an American-based project designed to seize market share from state-run telephone companies. This would also make it easier to get regulatory approval to operate in 170 countries. Investors would pay $40 million for the right to own their own regional gateway. As stated by Flower:

> The motive of the investors is clear: They are taking a chance on owning a slice of a de-facto world monopoly. Each of them will not only have a piece of the company, they will own the Iridium gateways and act as the local distributors in their respective home markets. For them it's a game worth playing.[8]

There were political ramifications with selling regional gateways. What if in the future the U.S. government forbids shipment of replacement parts to certain gateways? What if

---

8. Joe Flower, Iridium, *Wired*, Issue 1.05, November, 1993.

sanctions are imposed? What if Iridium were to become a political tool during international diplomacy because of the number of jobs it creates?

In addition to financial incentives, gateway owners were granted seats on the board of directors. As described by David Bennahum, reporter for *Wired*:

> Four times a year, 28 Iridium board members from 17 countries gather to coordinate overall business decisions. They met around the world, shuttling between Moscow, London, Kyoto, Rio de Janeiro, and Rome, surrounded by an entourage of assistants and translators. Resembling a United Nations in miniature, board meetings were conducted with simultaneous translation in Russian, Japanese, Chinese, and English.[9]

The partner with the largest equity share was Motorola. For its contribution of $400 million, Motorola originally received an equity stake of 25 percent, and 6 of 28 seats on Iridium's board. Additionally, Motorola made loan guarantees to Iridium of $750 million, with Iridium holding an option for an additional $350 million loan.

For its part, Iridium agreed to $6.6 billion in long-term contracts with Motorola that included $3.4 billion for satellite design and launch, and $2.9 billion for operations and maintenance. Iridium also exposed Motorola to developing satellite technology that would provide the latter with significant expertise in building satellite communications systems, as well as vast intellectual property.

25.4 THE IRIDIUM SYSTEM[10]

The Iridium system is a satellite-based, wireless personal communications network providing a robust suite of voice features to virtually any destination anywhere on Earth.

The Iridium system comprises three principal components: the satellite network, the ground network, and the Iridium subscriber products including phones and pagers. The design of the Iridium network allows voice and data to be routed virtually anywhere in the world. Voice and data calls are relayed from one satellite to another until they reach the satellite above the Iridium Subscriber Unit (handset) and the signal is relayed back to Earth.

25.5 THE TERRESTRIAL AND SPACE-BASED NETWORK[11]

The Iridium constellation consists of 66 operational satellites and 11 spares orbiting in a constellation of 6 polar planes. Each plane has 11 mission satellites performing as nodes in the telephony network. The remaining 11 satellites orbit as spares ready to replace any

9. See note 6, Bennahum, p. 136.
10. This is the operational version of the Iridium system today taken from the Iridium website, www.Iridium.com.
11. See note 10.

unserviceable satellite. This constellation ensures that every region on the globe is covered by at least one satellite at all times.

The satellites are in a near-polar orbit at an altitude of 485 miles (780 km). They circle the Earth once every 100 minutes traveling at a rate of 16,832 miles per hour. The satellite weight is 1500 pounds. Each satellite is approximately 40 feet in length and 12 feet in width. In addition, each satellite has 48 spot beams, 30 miles in diameter per beam.

Each satellite is cross-linked to four other satellites: two satellites in the same orbital plane and two in an adjacent plane. The ground network is comprised of the system control segment and telephony gateways used to connect into the terrestrial telephone system. The System Control Segment is the central management component for the Iridium system. It provides global operational support and control services for the satellite constellation, delivers satellite-tracking data to the gateways, and performs the termination control function of messaging services. The System Control Segment consists of three main components: four telemetry tracking and control sites, the operational support network, and the satellite network operation center. The primary linkage between the system control segment, the satellites, and the gateways is via K-band feeder links and cross-links throughout the satellite constellation.

Gateways are the terrestrial infrastructure that provides telephony services, messaging, and support to the network operations. The key features of gateways are their support and management of mobile subscribers and the interconnection of the Iridium network to the terrestrial phone system. Gateways also provide network management functions for their own network elements and links.

25.6 PROJECT INITIATION: DEVELOPING THE BUSINESS CASE

For the Iridium Project to be a business success rather than just a technical success, there had to exist an established customer base. Independent studies conducted by A.T. Kearney, Booz, Allen & Hamilton, and Gallup indicated that 34 million people had a demonstrated need for mobile satellite services, with that number expected to grow to 42 million by 2002. Of these 42 million, Iridium anticipated 4.2 million to be satellite-only subscribers, 15.5 million satellite and world terrestrial roaming subscribers, and 22.3 million terrestrial roaming-only subscribers.

A universal necessity in conducting business is ensuring that you are never out of touch. Iridium would provide this unique solution to business with the essential communications tool. This proposition of one phone, one number with the capability to be accessed anywhere, anytime was a message that target markets—the global traveler, the mining, rural, maritime industries, government, disaster relief, and community aid groups—would readily embrace.

Also at the same time of Iridium's conception, there appeared to be another potentially lucrative opportunity in the telecommunications marketplace. When users of mobile or cellular phones crossed international borders, they soon discovered that there existed a lack of common standards, thus making some phones inoperable. Motorola viewed this as an opportunity to create a worldwide standard allowing phones to be used anywhere in the world.

The expected breakeven market for Iridium was estimated between 400,000 and 600,000 customers globally assuming a reasonable usage rate per customer per month. With a launch date for Iridium service established for 1998, Iridium hoped to recover all of its investment within one year. By 2002, Iridium anticipated a customer base of 5 million users. The initial Iridium target market had been the vertical market, those of the industry, government, and world agencies that have defended needs and far-reaching communication requirements. Also important would be both industrial and public sector customers. Often isolated in remote locations outside of cellular coverage, industrial users were expected to use handheld Iridium satellite services to complement or replace their existing radio or satellite communications terminals. The vertical markets for Iridium would include:

- Aviation
- Construction
- Disaster relief/emergency
- Forestry
- Government
- Leisure travel
- Maritime
- Media and entertainment
- Military
- Mining
- Oil and gas
- Utilities

Using its own marketing resources, Iridium appeared to have identified an attractive market segment after having screened over 200,000 people, interviewed 23,000 people from 42 countries, and surveyed over 3000 corporations.

Iridium would also need regional strategic partners, not only for investment purposes and to share the risks but to provide services throughout their territories. The strategic regional partners or gateway operating companies would have exclusive rights to their territories and were obligated to market and sell Iridium services. The gateways would also be responsible for end-user sales, activation and deactivation of Iridium services, account maintenance, and billing.

Iridium would need each country to grant full licenses for access to the Iridium system. Iridium would need to identify the "priority" countries that account for the majority of the business plan.

Because of the number of countries involved in the Iridium network, Iridium would need to establish global Customer Care Centers for support services in all languages. No matter where an Iridium user was located, he or she would have access to a customer service representative in his or her native language. The Customer Care Centers would be strategically located to offer 24-hours-a-day, 7-days-a-week, and 365-days-a-year support.

25.7 THE "HIDDEN" BUSINESS CASE

The decision by Motorola to invest heavily into the Iridium Project may have been driven by a secondary or hidden business case. Over the years, Motorola achieved a reputation of being a first-mover (i.e., first to market). With the Iridium Project, Motorola was poised to capture first-mover advantage in providing global telephone service via low-Earth-orbiting satellites. In addition, even if the Iridium Project never resulted in providing service, Motorola would still have amassed valuable intellectual property that would make Motorola possibly the major player for years to come in satellite communications. There may have also been the desire of Robert and Christopher Galvin to have their names etched in history as the pioneers in satellite communication.

25.8 RISK MANAGEMENT

Good business cases identify the risks that the project must consider. For simplicity sake, the initial risks associated with the Iridium Project could be classified as follows.

Technology Risks   Although Motorola had some technology available for the Iridium Project, there was still the need to develop additional technology, specifically satellite communications technology. The development process was expected to take years and would eventually result in numerous patents.

Mark Gercenstein, Iridium's vice president of operations, explains the system's technological complexity:

> More than 26 completely impossible things had to happen first, and in the right sequence (before we could begin operations)—like getting capital, access to the marketplace, global spectrum, the same frequency band in every country of operations.[12]

While there was still some risk in the development of new technology, Motorola had the reputation of being a high-tech, can-do company. The engineers at Motorola believed that they could bring forth miracles in technology. Motorola also had a reputation for being a first-mover with new ideas and products, and there was no reason to believe that this would not happen on the Iridium Project. There was no competition for Iridium at its inception.

Because the project schedule was more than a decade in duration, there was the risk of technology obsolescence. This required that certain assumptions be made concerning technology a decade downstream. Developing a new product is relatively easy if the

12. Peter Grams and Patrick Zerbib, "Caring for Customers in a Global Marketplace," *Satellite Communications*, October 1998, p. 25.

environment is stable. But in a high-tech environment that is both turbulent and dynamic, it is extremely difficult to determine how customers will perceive and evaluate the product 10 years later.

Development Risks   The satellite communication technology, once developed, had to be manufactured, tested, and installed in the satellites and ground equipment. Even though the technology existed or would exist, there was still the transitional or development risks from engineering to manufacturing to implementation, which would bring with it additional problems that were not contemplated or foreseen.

Financial Risks   The cost of the Iridium Project would most certainly be measured in the billions of dollars. This would include the costs for technology development and implementation, the manufac.ture and launch of satellites, the construction of ground support facilities, marketing, and supervision. Raising money from Wall Street's credit and equity markets was years away. Investors were unlikely to put up the necessary hundreds of millions of dollars on merely an idea or a vision. The technology needed to be developed, and possibly accompanied by the launch of a few satellites before the credit and equity markets would come on board.

Private investors were a possibility, but the greatest source of initial funding would have to come from the members of the Iridium consortium. While sharing the financial risks among the membership seemed appropriate, there was no question that bank loans and lines of credit would be necessary. Since the Iridium Project was basically an idea, the banks would require some form of collateral or guarantee for the loans. Motorola, being the largest stakeholder (and also with the "deepest pockets") would need to guarantee the initial loans.

Marketing Risks   The marketing risks were certainly the greatest risks facing the Iridium membership. Once again, the risks were shared among its membership where each member was expected to sign up customers in its geographic area.

Each consortium member had to aggressively sign up customers for a product that didn't exist yet, no prototypes existed to be shown to the customers, limitations on the equipment were unknown as yet, and significant changes in technology could occur between the time the customer signed up and the time the system was ready for use. Companies that see the need for Iridium today may not see the same need 10 years later.

Motivating the consortium partners to begin marketing immediately would be extremely difficult since marketing material was nonexistent. There was also the very real fear that the consortium membership would be motivated more so by the technology rather than the necessary size of the customer base required.

The risks were interrelated. The financial risks were highly dependent upon the marketing risks. If a sufficient customer base could not be signed up, there could be significant difficulty in raising capital.

25.9 THE COLLECTIVE BELIEF

Although the literature doesn't clearly identify it, there was most likely a collective belief among the workers assigned to the Iridium Project. The collective belief is a fervent, and perhaps blind, desire to achieve that can permeate the entire team, the project sponsor, and even the most senior levels of management. The collective belief can make a rational organization act in an irrational manner.

When a collective belief exists, people are selected based upon their support for the collective belief. Nonbelievers are pressured into supporting the collective belief, and team members are not allowed to challenge the results. As the collective belief grows, both advocates and nonbelievers are trampled. The pressure of the collective belief can outweigh the reality of the results.

There are several characteristics of the collective belief, which is why some large, high-tech projects are often difficult to kill:

- Inability or refusal to recognize failure
- Refusing to see the warning signs
- Seeing only what you want to see
- Fearful of exposing mistakes
- Viewing bad news as a personal failure
- Viewing failure as a sign of weakness
- Viewing failure as damage to one's career
- Viewing failure as damage to one's reputation

25.10 THE EXIT CHAMPION

Project champions do everything possible to make their project successful. But what if the project champions, as well as the project team, have blind faith in the success of the project? What happens if the strongly held convictions and the collective belief disregard the early warning signs of imminent danger? What happens if the collective belief drowns out dissent?

In such cases, an exit champion must be assigned. The exit champion sometimes needs to have some direct involvement in the project in order to have credibility. Exit champions must be willing to put their reputation on the line and possibly face the likelihood of being cast out from the project team. According to Isabelle Royer[13]:

> Sometimes it takes an individual, rather than growing evidence, to shake the collective belief of a project team. If the problem with unbridled enthusiasm starts as an unintended consequence of the legitimate work of a project champion, then what may be needed is a

---

13. Isabelle Royer, "Why Bad Projects Are So Hard to Kill," *Harvard Business Review*, February 2003, p. 11.

countervailing force—an exit champion. These people are more than devil's advocates. Instead of simply raising questions about a project, they seek objective evidence showing that problems in fact exist. This allows them to challenge—or, given the ambiguity of existing data, conceivably even to confirm—the viability of a project. They then take action based on the data.

The larger the project and the greater the financial risk to the firm, the higher up the exit champion should reside. On the Iridium Project, the collective belief originated with Galvin, Motorola's CEO. Therefore, who could possibly function as the exit champion on the Iridium Project? Since it most likely should be someone higher up than Galvin, the exit champion should have been someone on the board of directors or even the entire Iridium board of directors. Unfortunately, the entire Iridium board of directors was also part of the collective belief and shirked their responsibility for oversight on the Iridium Project. In the end, Iridium had no exit champion. Large projects incur large cost overruns and schedule slippages. Making the decision to cancel such a project, once it has started, is very difficult, according to David Davis.[14]

> The difficulty of abandoning a project after several million dollars have been committed to it tends to prevent objective review and recosting. For this reason, ideally an independent management team—one not involved in the projects development—should do the recosting and, if possible, the entire review. . . . If the numbers do not holdup in the review and recosting, the company should abandon the project. The number of bad projects that make it to the operational stage serves as proof that their supporters often balk at this decision.
>
> . . . Senior managers need to create an environment that rewards honesty and courage and provides for more decision making on the part of project managers. Companies must have an atmosphere that encourages projects to succeed, but executives must allow them to fail.

The longer the project, the greater the necessity for the exit champions and project sponsors to make sure that the business plan has "exit ramps" such that the project can be terminated before massive resources are committed and consumed. Unfortunately, when a collective belief exists, exit ramps are purposefully omitted from the project and business plans.

25.11 IRIDIUM'S INFANCY YEARS

By 1992, the Iridium Project attracted such stalwart companies as General Electric, Lockheed, and Raytheon. Some companies wanted to be involved to be part of the satellite technology revolution, while others were afraid of falling behind the technology curve. In any event, Iridium was lining up strategic partners, but slowly.

14. David Davis, "New Projects: Beware of False Economics," *Harvard Business Review*, March–April 1985, pp. 100–101. Copyright © 1985 by the President and Fellows of Harvard College. All rights reserved.

The Iridium Plan, submitted to the FCC in August, 1992, called for a constellation of 66 satellites, expected to be in operation by 1998, and more powerful than originally proposed, thus keeping the project's cost at the previously estimated $3.37 billion. But the Iridium Project, while based up lofty forecasts of available customers, was now attracting other companies competing for FCC approval on similar satellite systems including Loral Corp., TRW Inc., and Hughes Aircraft Co., a unit of General Motors Corp. There were at least nine companies competing for the potential billions of dollars in untapped revenue possible from satellite communications.

Even with the increased competition, Motorola was signing up partners. Motorola had set an internal deadline of December 15, 1992, to find the necessary funding for Iridium. Signed letters of intent were received from the Brazilian government and United Communications Co., of Bangkok, Thailand, to buy 5 percent stakes in the project, each now valued at about $80 million. The terms of the agreement implied that the Iridium consortium would finance the project with roughly 50 percent equity and 50 percent debt.

When the December 15th deadline arrived, Motorola was relatively silent on the signing of funding partners, fueling speculation that it was having trouble. Motorola did admit that the process was time consuming because some investors required government approval before proceeding. Motorola was expected to announce at some point, perhaps in the first half of 1993, whether it was ready to proceed with the next step, namely receiving enough cash from its investors, securing loans, and ordering satellite and group equipment.

As the competition increased, so did the optimism about the potential size of the customer base.

> "We're talking about a business generating billions of dollars in revenue," says John F. Mitchell, Vice Chairman at Motorola. "Do a simple income extrapolation," adds Edward J. Nowacki, a general manager at TRW's Space & Electronics Group, Redondo Beach, Calif., which plans a $1.3 billion, 12-satellite system called Odyssey. "You conclude that even a tiny fraction of the people around the world who can afford our services will make them successful." Mr. Mitchell says that if just 1 % to 1.5 % of the expected 100 million cellular users in the year 2000 become regular users at $3 a minute, Iridium will breakeven. How does he know this? "Marketing studies," which he won't share. TRW's Mr. Nowacki says Odyssey will blanket the Earth with two-way voice communication service priced at "only a slight premium" to cellular. "With two million subscribers we can get a substantial return on our investment," he says. "Loral Qualcomm Satellite Services, Inc. aims to be the 'friendly' satellite by letting phone-company partners use and run its system's ground stations," says Executive Vice President Anthony Navarra. "By the year 2000 there will be 15 million unserved cellular customers in the world," he says.[15]

But while Motorola and other competitors were trying to justify their investment with "inflated market projections" and a desire from the public for faster and clearer reception, financial market analysts were not so benevolent. First, market analysts questioned the size of the customer base that would be willing to pay $3000 or more for a satellite phone in

---

15. John J. Keller, "Telecommunications: Phone Space Race Has Fortune at Stake," *Wall Street Journal* (Eastern edition), New York, January 18, 1993. p. B1.

addition to $3–$7 per minute for a call. Second, the system required a line-of-sight transmission, which meant that the system would not work in buildings or in cars. If a businessman were attending a meeting in Bangkok and needed to call his company, he must exit the building, raise the antenna on his $3000 handset, point the antenna toward the heavens, and then make the call. Third, the low-flying satellites would eventually crash into the Earth's atmosphere every 5–7 years because of atmospheric drag and would need to be replaced. That would most likely result in high capital costs. And fourth, some industry analysts believed that the startup costs would be closer to $6 billion to $10 billion rather than the $3.37 billion estimated by Iridium. In addition, the land-based cellular phone business was expanding in more countries, thus creating another competitive threat for Iridium.

The original business case needed to be reevaluated periodically. But with strong collective beliefs and no exit champions, the fear of a missed opportunity, irrespective of the cost, took center stage.

Reasonably sure that 18 out of 21 investors were on board, Motorola hoped to start launching test satellites in 1996 and begin commercial service by 1998. But critics argued that Iridium might be obsolete by the time it actually starts working.

Eventually, Iridium was able to attract financial support from 19 strategic partners:

- AIG Affiliated Companies
- China Great Wall Industry Corporation (CGWIC)
- Iridium Africa Corporation (based in Cape Town)
- Iridium Canada, Inc.
- Iridium India Telecom Private Ltd, (ITIL)
- Iridium Italia S.p.A.
- Iridium Middle East Corporation
- Iridium SudAmerica Corporation
- Khrunichev State Research and Production Space Center
- Korea Mobile TELECOM
- Lockheed Martin
- Motorola
- Nippon Iridium Corporation
- Pacific Electric Wire & Cable Co. Ltd (PEWC)
- Raytheon
- STET
- Sprint
- Thai Satellite Telecommunications Co., Ltd.
- Verbacom

Seventeen of the strategic partners also participated in gateway operations with the creation of operating companies.

The Iridium board of directors consisted of 28 telecommunications executives. All but one board member was a member of the consortium as well. This made it very difficult for the board to fulfill its oversight obligation effectively given the members' vested/financial interest in the Iridium Project.

In August 1993, Lockheed announced that it would receive $700 million in revenue for satellite construction. Lockheed would build the satellite structure, solar panels, attitude and propulsion systems, along with other parts and engineering support. Motorola and Raytheon Corp. would build the satellite's communications gear and antenna.

In April 1994, McDonnell Douglas Corp. received from Iridium a $400 million contract to launch 40 satellites for Iridium. Other contracts for launch services would be awarded to Russia's Khrunichev Space Center and China's Great Wall Industry Corporation, both members of the consortium. The lower-cost contracts with Russia and China were putting extraordinary pressure on U.S. providers to lower their costs.

Also at the same time, one of Iridium's competitors, the Globalstar system, which was a 48-satellite mobile telephone system led by Loral Corporation, announced that it intended to charge 65 cents per minute in the areas it served. Iridium's critics were arguing that Iridium would be too pricey to attract a high volume of callers.[16]

25.12 DEBT FINANCING

In September 1994, Iridium said that it had completed its equity financing by raising an additional $733.5 million. This brought the total capital committed to Iridium through equity financing to $1.57 billion. The completion of equity financing permitted Iridium to enter into debt financing to build the global wireless satellite network.

In September 1995, Iridium announced that it would be issuing $300 million 10-year senior subordinated discounted notes rated Caa by Moody's and CCC+ by Standard & Poor's, via the investment banker Goldman Sachs Inc. The bonds were considered to be high-risk, high-yield "junk" bonds after investors concluded that the rewards weren't worth the risk.

The rating agencies cited the reasons for the low rating to be yet unproven sophisticated technology and the fact that a significant portion of the system's hardware would be located in space. But there were other serious concerns:

- The ultimate cost of the Iridium Project would be more like $6 billion or higher rather than $3.5 billion, and it was unlikely that Iridium would recover that cost.
- Iridium would be hemorrhaging cash for several more years before service would begin.
- The optimistic number of potential customers for satellite phones may not choose the Iridium system.
- The number of competitors had increased since the Iridium concept was first developed.
- If Iridium defaulted on its debt, the investors could lay claim to Iridium's assets. But what would investors do with more than 66 satellites in space, waiting to disintegrate upon reentering the atmosphere?

---

16. Jeff Cole, "McDonnell Douglas Said to Get Contract to Launch 40 Satellites for Iridium Plan," *Wall Street Journal* (Eastern edition), New York, April 12, 1994, p. A4.

Iridium was set up as "project financing" in which case, if a default occurred, only the assets of Iridium could be attached. With project financing, the consortium's investors would be held harmless for any debt incurred from the stock and bond markets and could simply walk away from Iridium. These risks associated with project financing were well understood by those that invested in the equity and credit markets.

Goldman Sachs & Co., the lead underwriter for the securities offering, determined that for the bond issue to be completed successfully, there would need to exist a completion guarantee from investors with deep pockets, such as Motorola. Goldman Sachs cited a recent $400 million offering by one of Iridium's competitors, Globalstar, which had a guarantee from the managing general partner, Loral Corp.[17]

Because of the concern by investors, Iridium withdrew its planned $300 million debt offering. Also, Globalstar, even with its loan guarantee, eventually withdrew its $400 million offering. Investors wanted both an equity position in Iridium and a 20 percent return. Additionally, Iridium would need to go back to its original 17-member consortium and arrange for internal financing.

In February 1996, Iridium had raised an additional $315 million from the 17-member consortium and private investors. In August 1996, Iridium had secured a $750 million credit line with 62 banks co-arranged by Chase Securities Inc., a unit of Chase Manhattan Corp., and the investment banking division of Barclays Bank PLC. The credit line was oversubscribed by more than double its original goal because the line of credit was backed by a financial guarantee by Motorola and its AAA credit rating. Because of the guarantee by Motorola, the lending rate was slightly more than the 5.5 percent baseline international commercial lending rate and significantly lower than the rate in the $300 million bond offering that was eventually recalled.

Despite this initial success, Iridium still faced financial hurdles. By the end of 1996, Iridium planned on raising more than $2.65 billion from investors. It was estimated that more than 300 banks around the globe would be involved, and that this would be the largest private debt placement ever. Iridium believed that this debt placement campaign might not be that difficult since the launch date for Iridium services was getting closer.

25.13 THE M-STAR PROJECT

In October 1996, Motorola announced that it was working on a new project dubbed M-Star, which would be a $6.1 billion network of 72 low-orbit satellites capable of worldwide voice, video, and high-speed data links targeted at the international community. The project was separate from the Iridium venture and was expected to take 4 years to complete after FCC approval. According to Bary Bertiger, now corporate vice president and general manager of Motorola's satellite communications group, "Unlike Iridium, Motorola

---

17. Quentin Hardy, "Iridium Pulls $300 Million Bond Offer; Analysts Cite Concerns About Projects," *Wall Street Journal* (Eastern edition), New York, September 22, 1995, p. A5.

has no plans to detach M-Star as a separate entity. We won't fund it ourselves, but we will have fewer partners than in Iridium."[18]

The M-Star Project raised some eyebrows in the investment community. Iridium employed 2000 people but M-Star had only 80. The Iridium Project generated almost 1100 patents for Motorola, and that intellectual property would most likely be transferred to M-Star. Also, Motorola had three contracts with Iridium for construction and operation of the global communication system providing for approximately $6.5 billion in payments to Motorola over a 10-year period that began in 1993. Was M-Star being developed at the expense of Iridium? Could M-Star replace Iridium? What would happen to the existing 17-member consortium at Iridium if Motorola were to withdraw its support in lieu of its own internal competitive system?

25.14 A NEW CEO

In 1996, Iridium began forming a very strong top management team with the hiring of Dr. Edward Staiano as CEO and vice chairman. Prior to joining Iridium in 1996, Staiano had worked for Motorola for 23 years, during which time he developed a reputation for being hard-nosed and unforgiving. During his final 11 years with Motorola, Staiano led the company's General Systems Sector to record growth levels. In 1995, the division accounted for approximately 40 percent of Motorola's total sales of $27 billion. In leaving Motorola's payroll for Iridium's, Staiano gave up a $1.3 million per year contract with Motorola for a $500,000 base salary plus 750,000 Iridium stock options that vested over a 5-year period. Staiano commented,

> I was spending 40 percent to 50 percent of my time (at Motorola) on Iridium anyway . . .
> If I can make Iridium's dream come true, I'll make a significant amount of money.[19]

25.15 SATELLITE LAUNCHES

At 11:28 AM on a Friday morning the second week of January 1997, a Delta 2 rocket carrying a Global Positioning System (GPS) exploded upon launch, scattering debris above its Cape Canaveral launch pad. The launch, which was originally scheduled for the third quarter of 1996, would certainly have an impact on Iridium's schedule while an industry board composed of representatives from McDonnell-Douglas and the Air Force determined the cause of the explosion. Other launches had already been delayed for a variety of technical reasons.

---

18. Quentin Hardy, "Motorola Is Plotting New Satellite Project—M-Star Would Be Faster Than the Iridium System, Pitched to Global Firms," *Wall Street Journal* (Eastern edition), New York, October 14, 1996, p. B4.
19. Quentin Hardy, Staiano Is Leaving Motorola to Lead Firm's Iridium Global Satellite Project," *Wall Street Journal* (Eastern edition), New York, December 10, 1996, p. B8.

In May of 1997, after six failed tries, the first five Iridium satellites were launched. Iridium still believed that the target date for launch of service, September 1998, was still achievable but that all slack in the schedule had been eliminated due to the earlier failures.

By this time, Motorola had amassed tremendous knowledge on how to mass-produce satellites. As described by Bennahum:

> The Iridium constellation was built on an assembly line, with all the attendant reduction in risk and cost that comes from doing something over and over until it is no longer an art but a process. At the peak of this undertaking, instead of taking 18 to 36 months to build one satellite, the production lines disgorged a finished bird every four and a half days, sealed it in a container, and placed it on the flatbed of an idling truck that drove it to California or Arizona, where a waiting Boeing 747 carried it to a launchpad in the mountains of Taiyuan, China, or on the steppes of Baikonur in Kazakhstan.[20]

25.16 AN INITIAL PUBLIC OFFERING (IPO)

Iridium was burning cash at the rate of $100 million per month. Iridium filed a preliminary document with the Security and Exchange Commission (SEC) for an initial public offering (IPO) of 10 million shares to be offered at $19–$21 a share. Because of the launch delays, the IPO was delayed.

In June of 1997, after the first five satellites were placed in orbit, Iridium filed for an IPO of 12 million shares priced at $20 per share. This would cover about 3 months of operating expenses including satellite purchases and launch costs. The majority of the money would go to Motorola.

25.17 SIGNING UP CUSTOMERS

The reality of the Iridium concept was now at hand. All that was left to do was to sign up 500,000–600,000 customers, as predicted, to use the service. Iridium set aside $180 million for a marketing campaign including advertising, public relations, and worldwide, direct mail effort. Part of the advertising campaign included direct mail translated into 13 languages, ads on television and on airlines, airport booths, and Internet web pages.

How to market Iridium was a challenge. People would certainly hate the phone. According to John Windolph, executive director of marketing communications at Iridium, "It's huge! It will scare people. It is like a brick-size device with an antenna like a stout bread stick. If we had a campaign that featured our product, we'd lose." The decision was

---

20. See note 6, Bennahum, 1998.

to focus on the fears of being out of touch. Thus the marketing campaign began. But Iridium still did not have a clear picture of who would subscribe to the system. An executive earning $700,000 would probably purchase the bulky phone, have his or her assistant carry the phone in his or her briefcase, be reimbursed by his company for the use of the phone, and pay $3–$7 per minute for calls, also a business expense. But are there 600,000 executives worldwide that need the service?

There were several other critical questions that needed to be addressed. How do we hide or downplay the $3400 purchase price of the handset and the usage cost of $7 per minute? How do we avoid discussions about competitors that are offering similar services at a lower cost? With operating licenses in about 180 countries, do we advertise in all of them? Do we take out ads in *Oil and Gas Daily*? Do we advertise in girlie magazines? Do we use full-page or double-page spreads?

Iridium had to rely heavily upon its "gateway" partners for marketing and sales support. Iridium itself would not be able to reach the entire potential audience. Would the gateway partners provide the required marketing and sales support? Do the gateway partners know how to sell the Iridium system and the associated products?

The answer to these questions appeared quickly.

> Over a matter of weeks, more than one million sales inquiries poured into Iridium's sales offices. They were forwarded to Iridium's partners—and many of them promptly disappeared, say several Iridium insiders. With no marketing channels and precious few sales people in place, most global partners were unable to follow up on the inquiries. A mountain of hot sales tips soon went cold.[21]

25.18 IRIDIUM'S RAPID ASCENT

On November 1, 1998, the Iridium system was officially launched. It was truly a remarkable feat that the 11-year project was finally launched, just a little more than a month late.

> After 11 years of hard work, we are proud to announce that we are open for business. Iridium will open up the world of business, commerce, disaster relief and humanitarian assistance with our first-of-its-kind global communications service . . . The potential use of Iridium products is boundless. Business people who travel the globe and want to stay in touch with home and office, industries that operate in remote areas—all will find Iridium to be the answer to their communications needs."[22]

On November 2, 1998, Iridium began providing service. With the Iridium system finally up and running, most financial analysts issued "buy" recommendations for Iridium

21. Leslie Cauley, "Losses in Space—Iridium's Downfall: The Marketing Took a Back Seat to Science—Motorola and Partners Spent Billions on Satellite Links for a Phone Few Wanted," *Wall Street Journal* (Eastern edition), New York, August 18, 1999. p. A1.
22. Excerpts from the Iridium press release, November 1, 1998.

stock with expected yearly revenues of $6–$7 billion within 5 years. On January 25, 1999, Iridium held a news conference to discuss its earnings for the fourth quarter of 1998. Ed Staiano, CEO of Iridium announced:

> In the fourth quarter of 1998, Iridium made history as we became the first truly global mobile telephone company. Today, a single wireless network, the Iridium Network, covers the planet. And we have moved into 1999 with an aggressive strategy to put a large number of customers on our system, and quickly transform Iridium from a technological event to a revenue generator. We think the prospects for doing this are excellent. Our system is performing at a level beyond expectations.
>
> Financing is now in place through projected cash flow positives. Customer interest remains very high and a number of potentially large customers have now evaluated our service and have given it very high ratings. With all of this going for us, we are in position to sell the service and that is precisely where we are focusing the bulk of our efforts.[23]

At the same conference call, Roy Grant, CFO of Iridium, added:

> Last week Iridium raised approximately $250 million through a very successful 7.5 million-share public offering. This offering had three major benefits. It provided $250 million of cash to our balance sheet. It increased our public float to approximately 20 million shares. And it freed up restrictions placed on $300 million of the $350 million of Motorola guarantees. These restrictions were placed on that particular level of guarantees by our bankers in our $800 million secured credit facility.
>
> With this $250 million, combined with the $350 million of additional guarantees from Motorola, this means we have approximately $600 million of funds in excess of what we need to break cash flow breakeven. This provides a significant contingency for the company.[24]

December, 1998   In order to make its products and services known to travelers, Iridium agreed to acquire Claircom Corporation from AT&T and Rogers Cantel Mobile Communications for about $65 million. Claircom provided in-flight telephone systems for U.S. planes as well as equipment for international carriers. The purchase of Claircom would be a marketing boost for Iridium.

The problems with large, long-term technology projects were now appearing in the literature. As described by Bennahum:

> "This system does not let you do what a lot of wired people want to do," cautions Professor Heather Hudson, who runs the telecommunications program at the University of

---

23. Excerpts from the Iridium conference call, January 25, 1999.
24. See note 23.

San Francisco and studies the business of wireless communications. "Nineteen-nineties technologies are changing so fast that it is hard to keep up. Iridium is designed from a 1908s perspective of a global cellular system. Since then, the Internet has grown and cellular telephony is much more pervasive. There are many more opportunities for roaming than were assumed in 1989. So there are fewer businesspeople who need to look for an alternative to a cell phone while they are on the road."[25]

Additionally, toward the late 1990s, some industry observers felt that Motorola had additional incentive to ensure that Iridium succeeded, irrespective of the costs—namely, protecting its reputation. Between 1994 and 1997, Motorola had suffered slowing sales growth, a decline in net income, and declining margins. Moreover, the company had experienced several previous business mishaps, including a failure to anticipate the cellular industry's switch to digital cell phones, which played a major role in Motorola's more than 50 percent share-price decline in 1998.

25.19 IRIDIUM'S RAPID DESCENT

It took more than a decade for the Iridium Project to ascend and only a few months for descent. In the first week of March, almost 5 weeks after the January teleconference, Iridium's financial woes began to surface. Iridium had expected 200,000 subscribers by the end of 1998 and additional subscribers at a rate of 40,000 per month. Iridium's bond covenants stated a target of 27,000 subscribers by the end of March. Failure to meet such a small target could send investor confidence spiraling downward. Iridium had only 10,000 subscribers. The market that was out there 10 years ago was not the market that was there today. Also, 10 years ago there was little competition for Iridium.

Iridium cited the main cause of the shortfall in subscriptions as being shortages of phones, glitches in some of the technology, software problems, and, most important, a lack of trained sales channels. Iridium found out that it had to train a sales staff and that Iridium itself would have to sell the product, not its distributors. The investor community did not appear pleased with the sales problem that should have been addressed years ago, not 4 months into commercial service.

Iridium's advertising campaign was dubbed "Calling Planet Earth" and promised that you had the freedom to communicate anytime and anywhere. This was not exactly true because the system could not work within buildings or even cars. Furthermore, Iridium underestimated the amount of time subscribers would require to examine and test the system before signing on. In some cases, this would be 6 months.

Many people blamed marketing and sales for Iridium's rapid descent:

> True, Iridium committed so many marketing and sales mistakes that its experiences could form the basis of a textbook on how not to sell a product. Its phones started out costing

---

25. See note 6, Bennehum, 1998.

$3,000, were the size of a brick, and didn't work as promised. They weren't available in stores when Iridium ran a $180 million advertising campaign. And Iridium's prices, which ranged from $3.00 to $7.50 a call, were out of this world.[26]

Iridium's business plan was flawed. With service beginning on November 2, 1998, it was unlikely that 27,000 subscribers would be on board by March of 1999 given the time required to test the product. The original business plan required that the consortium market and sell the product prior to the onset of service. But selling the service from just a brochure was almost impossible. Subscribers want to touch the phone, use it, and test it prior to committing to a subscription.

Iridium announced that it was entering into negotiations with its lenders to alter the terms of an $800 million secured credit agreement due to the weaker-than-expected subscriber and revenue numbers. Covenants on the credit agreement included the following[27]:

Date	Cumulative Cash Revenue ($ Millions)	Cumulative Accrued Revenue ($ Millions)	Number of Satellite Phone Subscribers	Number of System Subscribers[*]
March 31, 1999	$ 4	$ 30	27,000	52,000
June 30, 1999	50	150	88,000	213,000
Sept. 30, 1999	220	470	173,000	454,000

[*]Total system subscribers include users of Iridium's phone, fax, and paging services.

The stock, which had traded as high as almost $73 per share, was now at approximately $20 per share. And, in yet another setback, the CFO, Roy T. Grant, resigned.

April, 1999

Iridium's CEO, Ed Staiano, resigned at the April 22 board meeting. Sources believed that Staiano resigned when the board nixed his plan requesting additional funds to develop Iridium's own marketing and distribution team rather than relying on its strategic partners. Sources also stated another issue in that Staiano had cut costs to the barebones at Iridium but could not get Motorola to reduce its lucrative $500 million service contract with Iridium. Some people believed that Staiano wanted to reduce the Motorola service contract by up to 50 percent. John Richardson, the CEO of Iridium Africa Corp., was assigned as interim CEO. Richardson's expertise was in corporate restructuring. For the quarter ending March, Iridium said it had a net loss of $505.4 million, or $3.45 a share. The stock fell to $15.62 per share. Iridium managed to attract just 10,294 subscribers 5 months after commercial rollout.

---

26. James Surowieckipp, "The Latest Satellite Startup Lifts Off. Will It Too Explode?" *Fortune Magazine*, October 25, 1999, pp. 237–254.
27. Iridium World Communications Ltd., 1998 Annual Report.

One of Richardson's first tasks was to revamp Iridium's marketing strategy. Iridium was unsure as to what business it was in. According to Richardson:

> The message about what this product was and where it was supposed to go changed from meeting to meeting. . . . One day, we'd talk about cellular applications, the next day it was a satellite product. When we launch in November, I'm not sure we had a clear idea of what we wanted to be.[28]

May, 1999   Iridium officially announced that it did not expect to meet its targets specified under the $800 million loan agreement. Lenders granted Iridium a 2-month extension. The stock dropped to $10.44 per share, party due to a comment by Motorola that it might withdraw from the ailing venture.

Wall Street began talking about the possibility of bankruptcy. But Iridium stated that it was revamping its business plan and by month's end hoped to have chartered a new course for its financing. Iridium also stated in a regulatory filing that it was uncertain whether it would have enough cash to complete the agreement to purchase Claircom Communications Group Inc., an in-flight telephone service provider, for the promised $65 million in cash and debt.

Iridium had received extensions on debt payments because the lending community knew that it was no small feat transforming from a project plan to an operating business. Another reason why the banks and creditors were willing to grant extensions was because bankruptcy was not a viable alternative. The equity partners owned all of the Earth stations, all distribution, and all regulatory licenses. If the banks and creditors forced Iridium into bankruptcy, they could end up owning a satellite constellation that could not talk to the ground or gateways.

June, 1999   Iridium received an additional 30-day extension beyond the 2-month extension it had already received. Iridium was given until June 30 to make a $90 million bond payment. Iridium began laying off 15 percent of its 550-employee workforce including two senior officers. The stock had now sunk to $6 per share and the bonds were selling at 19 cents on the dollar.

> We did all of the difficult stuff well, like building the network, and did all of the no-brainer stuff at the end poorly.[29]
>
> —John Richardson, CEO, Iridium

Iridium's major mistake was a premature launch for a product that wasn't ready. People became so obsessed with the technical grandeur of the project that they missed fatal marketing traps . . . Iridium's international structure has proven almost impossible to manage: the

---

28. Carleen Hawn, "High Wireless Act," *Forbes*, June 14, 1999, pp. 60–62.
29. See note 28.

28 members of the board speak multiple languages, turning meetings into mini-U.N. conferences complete with headsets translating the proceedings into five languages.[30]

—John Richardson, CEO, Iridium

. . . We're a classic MBA case study in how not to introduce a product. First we created a marvelous technological achievement. Then we asked how to make money on it.

—John Richardson, CEO, Iridium

Iridium was doing everything possible to avoid bankruptcy. Time was what Iridium needed. Some industrial customers would take 6–9 months to try out a new product, but would be reluctant to subscribe if it appeared that Iridium would be out of business in 6 months. In addition, Iridium's competitors were lowering their prices significantly, putting further pressure on Iridium. Richardson then began providing price reductions of up to 65 percent off of the original price for some of Iridium's products and services.

July, 1999   The banks and investors agreed to give Iridium yet a third extension to August 11 to meet its financial covenants. Everyone seemed to understand that the restructuring effort was much broader than originally contemplated.

Motorola, Iridium's largest investor and general contractor, admitted that the project may have to be shut down and liquidated as part of bankruptcy proceedings unless a restructuring agreement could be reached. Motorola also stated that if bankruptcy occurred, Motorola would continue to maintain the satellite network, but for a designated time period only.

Iridium had asked its consortium investors and contractors to come up with more money. But to many consortium members, it looked like they would be throwing good money after bad. Several partners made it clear that they would simply walk away from Iridium rather than providing additional funding. That could have a far-reaching effect on the service at some locations. Therefore, all partners had to be involved in the restructuring. Wall Street analysts expected Iridium to be allowed to repay its cash payments on its debt over several years or offer debt holders an equity position in Iridium. It was highly unlikely that Iridium's satellites orbiting the Earth would be auctioned off in bankruptcy court.

August, 1999   On August 12, Iridium filed for bankruptcy protection. This was like having "a dagger stuck in their heart" for a company that a few years earlier had predicted financial breakeven in just the first year of operations. This was one of the 20 largest bankruptcy filings up to this time. The stock, which had been trading as little as $3 per share, was suspended from the NASDAQ on August 13, 1999. Iridium's phone calls had been reduced to around $1.40–$3 per minute and the handsets were reduced to $1500 per unit.

---

30. Leslie Cauley, "Losses in Space—Iridium's Downfall: The Marketing Took a Back Seat to Science," *Wall Street Journal* (Eastern edition), New York, August 18, 1999, p. A1.

There was little hope for Iridium. Both the business plan and the technical plan were flawed. The business plan for Iridium seemed like it came out of the film *Field of Dreams* where an Iowa corn farmer was compelled to build a baseball field in the middle of a corn crop. A mysterious voice in his head said, "Build it and they will come." In the film, he did, and they came. While this made for a good plot for a Hollywood movie, it made a horrible business plan.

> If you build Iridium, people may come. But what is more likely is, if you build something cheaper, people will come to that first.
> —Herschel Shosteck, Telecommunication Consultant, 1992

The technical plan was designed to build the holy grail of telecommunications. Unfortunately, after spending billions, the need for the technology changed over time. The engineers that designed the system, many of whom had worked previously on military projects, lacked an understanding of the word "affordability" and the need for marketing a system to more than just one customer, namely the Department of Defense.

> Satellite systems are always far behind the technology curve. Iridium was completely lacking the ability to keep up with Internet time.[31]
> —Bruce Egan, Senior Fellow at Columbia University's Institute for Tele-Information

September, 1999   Leo Mondale resigned as Iridium's chief financial officer. Analysts believed that Mondale's resignation was the result of a successful restructuring no longer being possible. According to one analyst, "If they (Iridium) were close (to a restructuring plan), they wouldn't be bringing in a whole new team."

25.20 THE IRIDIUM "FLU"

The bankruptcy of Iridium was having a flu-like effect on the entire industry. ICO Global Communications, one of Iridium's major competitors, also filed for bankruptcy protection just 2 weeks after the Iridium filing. ICO failed to raise $500 million it sought from public-rights offerings that had already been extended twice. Another competitor, the Globalstar Satellite Communications System, was still financially sound. Anthony Navarro, globalstar chief operating officer stated: "They (Iridium) set everybody's expectations way too high."[32]

---

31. Stephanie Paterik, "Iridium Alive and Well," *The Arizona Republic*, April 27, 2005, p. D5.
32. Quentin Hardy, "Surviving Iridium," *Forbes*, September 6, 1999. pp. 216–217.

25.21 SEARCHING FOR A WHITE KNIGHT

Iridium desperately needed a qualified bidder who would function as a white knight. It was up to the federal bankruptcy court to determine whether someone was a qualified bidder. A qualified bidder was required to submit a refundable cash deposit or letter of credit issued by a respected bank that would equal the greater of $10 million or 10 percent of the value of the amount bid to take control of Iridium.

According to bankruptcy court filing, Iridium was generating revenue of $1.5 million per month. On December 9, 1999, Motorola agreed to a $20 million cash infusion for Iridium. Iridium desperately needed a white knight quickly or it could run out of cash by February 15, 2000. With a monthly operating cost of $10 million, and a staggering cost of $300 million every few years for satellite replenishment, it was questionable if anyone could make a successful business from Iridium's assets because of asset specificity.

The cellular phone entrepreneur Craig McCaw planned on a short-term cash infusion while he considered a much larger investment to rescue Iridium. He was also leading a group of investors who pledged $1.2 billion to rescue the ICO satellite system that filed for bankruptcy protection shortly after the Iridium filing.[33]

Several supposedly white knights came forth, but Craig McCaw's group was regarded as the only credible candidate. Although McCaw's proposed restructuring plan was not fully disclosed, it was expected that Motorola's involvement would be that of a minority stakeholder. Also, under the restructuring plan, Motorola would reduce its monthly fee for operating and maintaining the Iridium system from $45 million to $8.8 million.[34]

25.22 THE DEFINITION OF FAILURE (OCTOBER, 1999)

The Iridium network was an engineering marvel. Motorola's never-say-die attitude created technical miracles and overcame NASA-level technical problems. Iridium overcame global political issues, international regulatory snafus, and a range of other geopolitical issues on seven continents. The Iridium system was, in fact, what Motorola's Galvin called the eighth wonder of the world.

But did the bankruptcy indicate a failure for Motorola? Absolutely not! Motorola collected $3.65 billion in Iridium contracts. Assuming $750 million in profit from these contracts, Motorola's net loss on Iridium was about $1.25 billion. Simply stated, Motorola spent $1.25 billion for a project that would have cost them perhaps as much as $5 billion out of their own pocket had they wished to develop the technology themselves. Iridium provided Motorola with more than 1000 patents in building satellite communication systems. Iridium allowed Motorola to amass a leadership position in the global satellite industry.

---

33. "Craig McCaw Plans Cash Infusion to Support Cash-Hungry Iridium," *Wall Street Journal* (Eastern edition), New York, February 7, 2000, p. 1.
34. "Iridium Set to Get $75 Million from Investors Led by McCaw," *Wall Street Journal* (Eastern edition), New York, February 10, 2000, p. 1.

Motorola was also signed up as the prime contractor to build the 288-satellite "Internet in the Sky," dubbed the Teledesic Project. Backers of the Teledesic Project, which had a price tag of $15 billion to transmit data, video, and voice, included Boeing, Microsoft's Chairman Bill Gates, and cellular magnate Craig McCaw. Iridium had enhanced Motorola's reputation for decades to come.

Motorola stated that it had no intention of providing additional funding to ailing Iridium, unless of course other consortium members followed suit. Several members of the consortium stated that they would not provide any additional investment and were considering liquidating their involvement in Iridium.[35]

In March 2000 McCaw withdrew his offer to bail out Iridium even at a deep discount, asserting that his efforts would be spent on salvaging the ICO satellite system instead. This, in effect, signed Iridium's death warrant. One of the reasons for McCaw's reluctance to rescue Iridium may have been the discontent by some of the investors who would have been completely left out as part of the restructuring effort, thus losing perhaps their entire investment.

25.23 THE SATELLITE DEORBITING PLAN

With the withdrawal of McCaw's financing, Iridium notified the U.S. Bankruptcy Court that Iridium had not been able to attract a qualified buyer by the deadline assigned by the court. Iridium would terminate its commercial service after 11:59 PM on March 17, 2000, and that it would begin the process of liquidating its assets.

Immediately following the Iridium announcement, Motorola issued the following press release[36]:

> Motorola will maintain the Iridium satellite system for a limited period of time while the deorbiting plan is being finalized. During this period, we also will continue to work with the subscribers in remote locations to obtain alternative communications. However, the continuation of limited Iridium service during this time will depend on whether the individual gateway companies, which are separate operating companies, remain open.
>
> In order to support those customers who purchased Iridium service directly from Motorola, Customer Support Call Centers and a website that are available 24 hours a day, seven days a week have been established by Motorola. Included in the information for customers is a list of alternative satellite communications services.

The deorbiting plan would likely take 2 years to complete at a cost of $50–$70 million. This would include all 66 satellites and the other 22 satellites in space serving as spare or decommissioned failures. Iridium would most likely deorbit the satellites four at a time by firing their thrusters to drop them into the atmosphere where they would burn up.

35. Scott Thurm, "Motorola Inc., McCaw Shift Iridium Tactics," *Wall Street Journal* (Eastern edition), New York, February 18, 2000, p. 1.
36. Motorola press release, August, 1999.

25.24 IRIDIUM IS RESCUED FOR $25 MILLION

In November 2000 a group of investors led by an airline executive won bankruptcy court approval to form Iridium Satellite Corporation and purchase all remaining assets of failed Iridium Corporation. The purchase was at a fire-sale price of $25 million, which was less than a penny on the dollar. As part of the proposed sale, Motorola would turn over responsibility for operating the system to Boeing. Although Motorola would retain a 2 percent stake in the new system, Motorola would have no further obligations to operate, maintain, or decommission the constellation.

Almost immediately after the announcement, Iridium Satellite was awarded a $72 million contract from the Defense Information Systems Agency, which is part of the Department of Defense (DoD).

> Iridium will not only add to our existing capability, it will provide a commercial alternative to our purely military systems. This may enable real civil/military dual use, keep us closer to leading edge technologically, and provide a real alternative for the future.[37]
> —Dave Oliver, Principal Deputy Undersecretary of Defense for Acquisition

Iridium had been rescued from the brink of extinction. As part of the agreement, the newly formed company acquired all of the assets of the original Iridium and its subsidiaries. This included the satellite constellation, the terrestrial network, Iridium real estate, and the intellectual property originally developed by Iridium. Because of the new company's significantly reduced cost structure, it was able to develop a workable business model based upon a targeted market for Iridium's products and services.

> "Everyone thinks the Iridium satellites crashed and burned, but they're all still up there."[38]
> —Weldon Knape, World Communication center (WCC) Chief Executive Officer, April 27, 2005

A new Iridium phone costs $1495 and is the size of a cordless home phone. Older, larger models start at $699, or you can rent one for about $75 per week. Service costs $1–$1.60 a minute.[39]

25.25 EPILOGUE

February 6, 2006, Iridium satellite declared that 2005 was the best year ever. The company had 142,000 subscribers, which was a 24 percent increase from 2004, and the 2005

---

37. "DoD Awards $72 Million to Revamp Iridium," *Satellite Today*, Potomac: December 7, 2000, vol. 3, Iss. 227, p. 1.

38. Stephanie Paterik, "Iridium Alive and Well," *The Arizona Republic*, April 27, 2005.

39. See note 38.

revenue was 55 percent greater than in 2004. According to Carmen Lloyd, Iridium's CEO, "Iridium is on an exceptionally strong financial foundation with a business model that is self-funding."[40]

For the year ending 2006, Iridium had $212 million in sales and $54 million in profit. Iridium had 180,000 subscribers and a forecasted growth rate of 14–20 percent per year. Iridium had changed its business model, focusing on sales and marketing first and hype second. This allowed them to reach out to new customers and new markets.[41]

25.26 SHAREHOLDER LAWSUITS

The benefit to Motorola, potentially at the expense of Iridium and its investors, did not go unnoticed. At least 20 investor groups filed suit against Motorola and Iridium, citing:

- Motorola milked Iridium and used the partners' money to finance its own foray into satellite communication technology.
- By using Iridium, Motorola ensured that its reputation would not be tarnished if the project failed.
- Most of the money raised through the IPOs went to Motorola for designing most of the satellite and ground-station hardware and software.
- Iridium used the proceeds of its $1.45 billion in bonds, with interest rates from 10.875 to 14 percent, mainly to pay Motorola for satellites.
- Defendants falsely reported achievable subscriber numbers and revenue figures.
- Defendants failed to disclose the seriousness of technical issues.
- Defendants failed to disclose delays in handset deliveries.
- Defendants violated covenants between itself and its lenders.
- Defendants delayed disclosure of information, provided misleading information, and artificially inflated Iridium's stock price.
- Defendants took advantage of the artificially inflated price to sell significant amounts of their own holdings for millions of dollars in personal profit.

25.27 THE BANKRUPTCY COURT RULING

On September 4, 2007, after almost 10 months, the Bankruptcy Court in Manhattan ruled in favor or Motorola and irritated the burned creditors that had hoped to get a $3.7 billion judgment against Motorola. The judge ruled that even though the capital markets were "terribly wrong" about Iridium's hopes for huge profits, Iridium was "solvent" during the

---

40. Iridium Press Release, February 6, 2006.
41. Adapted from Reena Jana, "Companies Known for Inventive Tech Were Dubbed the Next Big Thing and Then Disappeared. Now They're Back and Growing," *Business Week*, Innovation, April 10, 2007.

critical period when it successfully raised rather impressive amounts of debt and equity in the capital markets.

The court said that even though financial experts now know that Iridium was a hopeless one-way cash flow, flawed technology project, and doomed business model, Iridium was solvent at the critical period of fundraising. Even when the bad news began to appear, Iridium's investors and underwriters still believed that Iridium had the potential to become a viable enterprise.

The day after the court ruling, newspapers reported that Iridium LLC, the now privately held company, was preparing to raise about $500 million in a private equity offering to be followed by an IPO within the next year or two.

25.28 AUTOPSY[42]

There were several reasons for Iridium's collapse:

CELLULAR BUILD-OUT DRAMATICALLY REDUCED THE TARGET MARKET'S NEED FOR IRIDIUM'S SERVICE

Iridium knew its phones would be too large and too expensive to compete with cellular service, forcing the company to play in areas where cellular was unavailable. With this constraint in mind, Iridium sought a target market by focusing on international business executives who frequently traveled to remote areas where cellular phone service wasn't available. Although this market plan predated the rise of cell phones, Iridium remained focused on the business traveler group through the launch of its service. As late as 1998, CEO Staiano predicted Iridium would have 500,000 subscribers by the end of 1999.

One of the main problems with Iridium's offering was that terrestrial cellular had spread faster than the company had originally expected. In the end, cellular was available. Due to Iridium's elaborate technology, the concept-to-development time was 11 years—during this period, cellular networks spread to cover the overwhelming majority of Europe and even migrated to developing countries such as China and Brazil. In short, Iridium's marketing plan targeted a segment—business travelers—whose needs were increasingly being met by cell phones that offered significantly better value than Iridium.

IRIDIUM'S TECHNOLOGICAL LIMITATIONS AND DESIGN STIFLED ADOPTION

Because Iridium's technology depended on line-of-sight between the phone antenna and the orbiting satellite, subscribers were unable to use the phone inside moving cars, inside buildings, and in many urban areas. Moreover, even in open fields users had to align the phone just right in order to get a good connection. As a top industry consultant said to us in an interview, "you can't expect a CEO traveling on business in Bangkok to leave a building, walk outside on a street corner, and pull out a $3000 phone." Additionally, Iridium lacked

42. Sydney Finkelstein and Shade H. Sanford, "Learning from Corporate Mistakes: The Rise and Fall of Iridium," *Organizational Dynamics*, vol. 29, no. 2, pp. 138–148; 2000, © 2000 by Elsevier Sciences, Inc. Reproduced by permission.

adequate data capabilities, an increasingly important feature for business users. Making matters worse were annoyances such as the fact that battery recharging in remote areas required special solar-powered accessories. These limitations made the phone a tough sell to Iridium's target market of high-level traveling businesspeople.

The design of Iridium's phone also hampered adoption. In November 1997, John Windolph, Iridium director of marketing communications, described the handset in the following manner: "It's huge! It will scare people. If we had a campaign that featured our product, we'd lose." Yet a year later Iridium went forward with essentially the same product. The handset, although smaller than competitor Comsat's Planet 1, was still literally the size of a brick.

POOR OPERATIONAL EXECUTION PLAGUED IRIDIUM

Manufacturing problems also caused Iridium's launch to stumble out of the gate. Management launched the service before enough phones were available from one of its two main suppliers, Kyocera, which was experiencing software problems at the time. Ironically, this manufacturing bottleneck meant that Iridium couldn't even get phones to the few subscribers that actually wanted one. The decision to launch service in November 1998, in spite of the manufacturing problems, was made by CEO Staiano, although not without opposition. As one report put it, "[John Richardson] claimed to be vociferous in board meetings, arguing against the November launch. Neither the service, nor the service providers, were ready. Supply difficulties meant that there were few phones available in the market."

IRIDIUM'S PARTNERS DID NOT PROVIDE ADEQUATE SALES AND MARKETING SUPPORT

Although at first Motorola had difficulty attracting investors for Iridium, by 1994 Iridium LLC had partnerships with 18 companies including Sprint, Raytheon, Lockheed Martin, and a variety of companies from China, the Middle East, Africa, India, and Russia. In exchange for investments of $3.7 billion, the partners received equity and seats on Iridium LLC's board of directors. In 1998, 27 of the 28 directors on Indium's board were either Iridium employees or directly appointed by Iridium's partners.

Iridium's partners would ultimately control marketing, pricing, and distribution when the service came on line. Iridium's revenues came from wholesale rates for its phone service. Unfortunately for Iridium, its partners, outside the United States in particular, delayed setting up marketing teams and distribution channels. "The gateways were very often huge telecoms," said Stephane Chard, chief analyst at Euroconsult, a Paris-based research firm. "To them, Iridium was a tiny thing." So tiny, in fact, that Iridium's partners failed to build sales teams, create marketing plans, or set up distribution channels for their individual countries. As the *Wall Street Journal* reported, "with less than six months to go before the launch of the service, time became critical . . . Most partners didn't reveal they were behind schedule."

25.29 FINANCIAL IMPACT OF THE BANKRUPTCY _____

At the time of the bankruptcy, equity investments in Iridium totaled approximately $2 billion. Most analysts, however, considered the stock worthless. Iridium's stock price, which had

IPOed at $20 per share in June 1997, and reached an all time high of $72.19 in May 1998, had plummeted to $3.06 per share by the time Iridium declared bankruptcy in August 1999. Moreover, the NASDAQ exchange reacted to the bankruptcy news by immediately halting trading of the stock, and actually delisted Iridium in November 1999. Iridium's partners—who had also made investments by building ground stations, assembling management teams, and marketing Iridium services—were left with little to show for their equity. Iridium's bondholders didn't fare much better than its equity holders. After Iridium declared bankruptcy, its $1.5 billion in bonds were trading for around 15 cents on the dollar as the company entered restructuring talks with its creditors.

25.30 WHAT REALLY WENT WRONG?

Iridium will go down in history as one of the most significant business failures of the 1990s. That its technology was breathtakingly elegant and innovative is without question. Indeed, Motorola and Iridium leaders showed great vision in directing the development and launch of an incredibly complex constellation of satellites. Equally as amazing, however, was the manner in which these same leaders led Iridium into bankruptcy by supporting an untenable business plan.

Over the past several years, there have been perhaps thousands of articles written about Iridium's failure to attract customers and its resulting bankruptcy. Conventional wisdom often argues that Iridium was simply caught off guard by the spread of terrestrial cellular. By focusing almost strictly on what happened, such an analysis provides little in the way of valuable learning. A more interesting question is why Iridium's failure happened—namely why did the company continue to press forward with an increasingly flawed business plan.

Three Forces Combined to Create Iridium's Failure

Three forces combined to create Iridium's business failure. First, an "escalating commitment," particularly among Motorola executives who pushed the project forward in spite of known and potentially fatal technology and market problems. Second, for personal and professional reasons Iridium's CEO was unwilling to cut losses and abandon the project. And third, Iridium's board was structured in a way that prevented it from performing its role of corporate governance.

PROBLEM 1: ESCALATING COMMITMENT

During the 11 years that passed between Indium's initial concept to its actual development, its business plan eroded. First, the gradual build-out of cellular dramatically shrank Iridium's target market—international executives who regularly traveled to areas not covered by terrestrial cellular. Second, it became apparent over time that Iridium's phones would have significant design, operational, and cost problems that would further limit usage.

Motorola's decision to push Iridium forward in spite of a deeply flawed business plan is a classic example of the pitfalls of "escalating commitment." The theory behind escalating commitment is based in part on the "sunk cost fallacy"—making decisions based on the size of previous investments rather than on the size of the expected return. People tend to escalate their commitment to a project when they (a) believe that future gains are available, (b) believe they can turn a project around, (c) are publicly committed or identified with the project, and (d) can recover a large part of their investment if the project fails.

Motorola's involvement in the Iridium Project met all four of these conditions. In spite of known problems, top executives maintained blind faith in Iridium. To say that Iridium's top management was unaware of Iridium's potential problems would be wholly inaccurate. In fact, Iridium's prospectus written in 1998 listed 25 full pages of risks including:

- A highly leveraged capital structure
- Design limitations—including phone size
- Service limitations—including severe degradation in cars, buildings, and urban areas
- High handset and service pricing
- The build-out of cellular networks
- A lack of control over partners' marketing efforts

During Iridium's long concept-to-development time, there is little evidence to suggest that Motorola or Iridium made any appreciable progress in addressing any of these risks. Yet Iridium went forward, single-mindedly concentrating on satellite design and launch while discounting the challenges in sales and marketing the phones. The belief that innovative technology would eventually attract customers, in fact, was deeply ingrained in Motorola's culture.

Indeed, Motorola's history was replete with examples of spectacular innovations that had brought the company success and notoriety. In the 1930s, Paul Gavin developed the first affordable car radio. In the 1940s, Motorola rose to preeminence when it developed the first handheld two-way radio, which was used by the Army Signal Corps during World War II. In the 1950s, Motorola manufactured the first portable television sets. In the 1969, Neil Armstrong's first words from the Moon were sent by a transponder designed and manufactured by the company. In the 1970s and 1980s, Motorola enjoyed success by developing and manufacturing microprocessors and cellular phones.

By the time it developed the concept for Iridium in the early 1990s, Motorola had experienced over 60 years of success in bringing often startling new technology to consumers around the world. Out of this success, however, came a certain arrogance and biased faith in the company's own technology. Just as Motorola believed in the mid-1990s that cellular customers would be slow to switch from Motorola's analog phones to digital phones produced by Ericsson and Nokia, their faith in Iridium and its technology was unshakable.

PROBLEM 2: STAIANO'S LEADERSHIP WAS A DOUBLE-EDGED SWORD

Dr. Edward Staiano became CEO of Iridium in late 1996—before the company had launched most of its satellites. During his previous tenure with Motorola, Staiano had developed a reputation as intimidating and demanding—imposing in both stature, at 6 feet 4 inches, and in temperament. Staiano combined his leadership style with an old Motorola ethic that argued leaders had a responsibility to support their projects. Staiano also had significant financial incentives to push the project forward, rather than cutting losses and moving on. In both 1997 and 1998, he received 750,000 Iridium stock options that vested over a 5-year period. Indeed, this fact didn't escape Staiano's attention when he took the CEO position in late 1996, stating: "If I can make Indium's dream come true, I'll make a significant amount of money."

Ironically, the demanding leadership style, commitment to the project at hand, and financial incentives that made Ed Staiano such an attractive leader for a startup company such as Iridium turned out to be a double-edged sword. Indeed, these same characteristics also made him unwilling to abandon a project with a failed business plan and obsolete technology.

PROBLEM 3: INDIUM'S BOARD DID NOT PROVIDE ADEQUATE CORPORATE GOVERNANCE

In 1997, Iridium's board had 28 directors—27 of whom were either Iridium employees or directors designated by Iridium's partners. The composition, not to mention size, of Iridium's board created two major problems. First, the board lacked the insight of outside directors who could have provided a diversity of expertise and objective viewpoints. Second, the fact that most of the board was comprised of partner appointees made it difficult for Iridium to apply pressure to its partners in key situations—such as when many partners were slow to set up the necessary sales and marketing infrastructure prior to service launch. In the end, Iridium's board failed to provide proper corporate oversight and limited Iridium's ability to work with its partners effectively.

25.31 LESSONS LEARNED

Executives Should Evaluate Projects Such as Iridium as Real Options

Projects with long concept-to-development times pose unique problems for executives. These projects may seem like good investments during initial concept development; but by the time the actual product or service comes on line, both the competitive landscape and the company's ability to provide the service or product have often changed significantly.

To deal with long concept-to-development times, executives should evaluate these projects as real options. A simple model would be a two-stage project. The first stage is strategic in nature and provides the opportunity for a further investment and increased return in the second stage. When the initial stage is complete, however, the company must reevaluate the expected return of future investments based on a better understanding of the product/service and the competitive landscape.

Iridium is a textbook example of a project that would have benefited from this type of analysis. The Iridium project itself essentially consisted of two stages. During stage one (1987–1996), Motorola developed the technology behind Iridium. During stage two (1996–1999), Motorola built and launched the satellites—and the majority of Iridium's costs occurred during this part of the project.

Investment in R&D for Iridium Was Appropriate—Follow-on Investment Was Not

Looking back, it would be unfair to assert that the initial decision to invest in R&D for Iridium was a mistake. In the late 1980s, Iridium appeared to have a sound business plan. Travel among business executives was increasing and terrestrial cellular networks didn't cover many of their destinations. It was certainly not unreasonable to foresee a large demand for a wireless phone that had no geographic boundaries. In turn, the investment in R&D was reasonable as it provided the option to deploy (or not deploy) the complex Iridium satellite system 9 years later.

By 1996, however, when Iridium had to make the decision of whether to invest in building and launching satellites, much had changed. Not only had the growth in cellular networks drastically eroded Iridium's target market, but Iridium's own technology was never able to overcome key design, cost, and operational problems. Put simply, Iridium didn't have a viable business plan. Armed with this additional insight, a reasonable evaluation of the project would have precluded further investment.

Executives Must Build Option Value Assessments into Their Business Plans

The key to using the option value approach is to include it in the business plan. Specifically, executives must specify a priori when they will reevaluate the project and its merits. During this evaluation, the company should objectively evaluate updated market data and its own ability to satisfy changing customer demands. The board of directors plays a key role in this process by making sure that inertia doesn't carry a failed project beyond its useful life. This is particularly important when company executives have ancillary reasons, such as concerns about personal reputation or compensation, to press forward in spite of a flawed business plan.

Top executives were publicly committed to, and identified with, Iridium. Just as important as its financial investment in Iridium was Motorola's psychological investment in the project. Motorola's chairmen, Robert Galvin and later his son Christopher Galvin, publicly expressed support for Iridium and looked to it as an example of Motorola's technological might. Indeed, it was Robert Galvin, Motorola's chairman at the time, who first gave Bary Bertiger approval to go ahead with Iridium, after Bertiger's superiors had rejected the project as being too costly. In the end, both Galvins staked much of Motorola's reputation on Iridium's success, and the project provided Motorola and the rest of its partners with a great deal of cachet.

Costs of Risky Projects Can Be Reduced via Opportunities for Contracting and Learning

Motorola did gain important benefits from its relationship with Iridium. In fact, Motorola signed $6.6 billion in contracts to design, launch, and operate Iridium's 66 satellites and manufacture a portion of its handsets. David Copperstein of Forrester Research described Motorola's deal with Iridium as "a pretty crafty way of creating a no-lose situation." Other analysts were less complimentary: "That contract [Motorola's $50 million a month agreement with Iridium to provide operational satellite support] is absurdly lucrative for Motorola," said Armand Mussey, an analyst who followed the industry for Bank of America Securities, "Iridium needs to cut that by half."

These contracts—while lucrative—also gave Motorola an incentive to push Iridium forward regardless of its business plan. Even if Iridium failed, Motorola would still generate significant new revenues along the way. In quantifying the importance of Motorola's contracts with Iridium, in May 1999 Wojtek Uzdelewicz of SG Cowen estimated that Motorola had already earned and collected $750 million in profits from its dealings with the company. Based on these offsetting profits, he placed Motorola's total exposure in Iridium to be between $1.0 and $1.15 billion—much less than many observers realized.

Further, Iridium would ultimately expose Motorola to developing satellite technology and the patent protection that came with it. This exposure came at a time when Motorola was interested in entering the satellite communications industry beyond Iridium, in projects such as Craig McCaw's Teledesic—a $9 billion project consisting of a complex constellation of LEO satellites designed to provide global high-speed Internet access.

Strategic Leadership of CEOs and Boards Can Make, or Break, Strategic Initiatives

In an era where executive compensation is dominated by stock options, the Iridium story should give pause to those who see only the benefits of options-based pay. Financial incentives are extremely powerful, and companies that rely on them for motivation must be particularly careful to consider both intended and unintended consequences. Would CEO Staiano have been more attentive to the numerous warning signs with Iridium if stock options didn't play such a large role in his compensation package? The heavy emphasis on options gave Staiano an incentive to persist with the Iridium strategy; it was the only opportunity he had to make the options pay.

The lessons of the board of directors at Iridium are just as stark. Surely few boards can operate with 28 members, most representing different constituencies surely holding different goals. That all but one board member was a member of the Iridium consortium similarly speaks volumes about the vigilance of the board in fulfilling its oversight function. Actually, this type of board, consisting as it does of representatives of investors, is becoming more common in high-technology startups. Companies such as General Magic, Excite At Home, and Net2Phone have all had multiple investors, typically represented on the board, and not always agreeing on strategic direction. In fact, General Magic's development of a personal digital assistant was severely hampered by its dependence on investors such as Apple, Sony, IBM, and AT&T. With Iridium, the magnitude of the ancillary contractual benefits Motorola derived from Iridium appear rather out-sized given Iridium's financial condition. An effective board should be simultaneously vigilant and supportive, a tall order for an insider-dominated, multiple-investor board.

25.32 CONCLUSION

What is fascinating about studying cases like Iridium is that what look like seemingly incomprehensible blunders are really windows into the world of managerial decision-making, warts and all. In-depth examinations of strategy in action can highlight how such processes as escalating commitment are real drivers of managerial action. When organizations stumble, observers often wonder why the company, or the top management, did something so "dumb." Much more challenging is to start the analysis by assuming that management is both competent and intelligent and then ask, why did they stumble? The answers one gets with this approach tend to be at once both more interesting and revealing. Students of strategy and organization can surely benefit from such a probing analysis.

APPENDIX A

Solutions to the Project Management Conflict Exercise

Part One: Facing the Conflict

After reading the answers that follow, record your score on line 1 of the worksheet on page 314.

A. Although many project managers and functional managers negotiate by "returning" favors, this custom is not highly recommended. The department manager might feel some degree of indebtedness at first, but will surely become defensive in follow-on projects in which you are involved, and might even get the idea that this will be the only way that he will be able to deal with you in the future. If this was your choice, allow one point on line 1.

B. Threats can only lead to disaster. This is a surefire way of ending a potentially good arrangement before it starts. Allow no points if you selected this as your solution.

C. If you say nothing, then you accept full responsibility and accountability for the schedule delay and increased costs. You have done nothing to open communications with the department manager. This could lead into additional conflicts on future projects. Enter two points on line 1 if this was your choice.

D. Requesting upper-level management to step in at this point can only complicate the situation. Executives prefer to step in only as a last resort. Upper-level management will probably ask to talk to the department manager first. Allow two points on line 1 if this was your choice.

E. Although he might become defensive upon receiving your memo, it will become difficult for him to avoid your request for help. The question, of course, is when he will give you this help. Allow eight points on line 1 if you made this choice.

F. Trying to force your solution on the department manager will severely threaten him and provide the basis for additional conflict. Good project managers will always try to predict emotional reactions to whatever decisions they might be forced to make. For this choice, allow two points on line 1 of the worksheet.

G. Making an appointment for a later point in time will give both parties a chance to cool off and think out the situation further. He will probably find it difficult to refuse your request for help and will be forced to think about it between now and the appointment. Allow ten points for this choice.

1025

H. An immediate discussion will tend to open communications or keep communication open. This will be advantageous. However, it can also be a disadvantage if emotions are running high and sufficient time has not been given to the selection of alternatives. Allow six points on line 1 if this was your choice.

I. Forcing the solution your way will obviously alienate the department manager. The fact that you do intend to honor his request at a later time might give him some relief especially if he understands your problem and the potential impact of his decision on other departments. Allow three points on line 1 for this choice.

Part Two: Understanding Emotions

Using the scoring table shown on page 1027, determine your total score. Record your total in the appropriate box on line 2 of the worksheet on page 314. There are no "absolutely" correct answers to this problem, merely what appears to be the "most" right.

Part Three: Establishing Communications

A. Although your explanations may be acceptable and accountability for excess costs may be blamed on the department manager, you have not made any attempt to open communications with the department manager. Further conflicts appear inevitable. If this was your choice, allow a score of zero on line 3 of the worksheet.

B. You are offering the department manager no choice but to elevate the conflict. He probably has not had any time to think about changing his requirements and it is extremely doubtful that he will give in to you since you have now backed him into a corner. Allow zero points on line 3 of the worksheet.

C. Threatening him may get him to change his mind, but will certainly create deteriorating working relationships both on this project and any others that will require that you interface with his department. Allow zero points if this was your choice.

D. Sending him a memo requesting a meeting at a later date will give him and you a chance to cool down but might not improve your bargaining position. The department manager might now have plenty of time to reassure himself that he was right because you probably aren't under such a terrible time constraint as you led him to believe if you can wait several days to see him again. Allow four points on line 3 of the worksheet if this was your choice.

E. You're heading in the right direction trying to open communications. Unfortunately, you may further aggravate him by telling him that he lost his cool and should have apologized to you when all along you may have been the one who lost your cool. Expressing regret as part of your opening remarks would benefit the situation. Allow six points on line 3 of the worksheet.

F. Postponing the problem cannot help you. The department manager might consider the problem resolved because he hasn't heard from you. The confrontation should not be postponed. Your choice has merit in that you are attempting to open up a channel for communications. Allow four points on line 3 if this was your choice.

G. Expressing regret and seeking immediate resolution is the best approach. Hopefully, the department manager will now understand the importance of this conflict and the need for urgency. Allow ten points on line 3 of the worksheet.

	Reaction	Personal or Group Score
A. I've given you my answer. See the general manager if you're not happy.	Hostile or Withdrawing	4
B. I understand your problem. Let's do it your way.	Accepting	4
C. I understand your problem, but I'm doing what is best for my department.	Defensive or Hostile	4
D. Let's discuss the problem. Perhaps there are alternatives.	Cooperative	4
E. Let me explain to you why we need the new requirements.	Cooperative or Defensive	4
F. See my section supervisors. It was their recommendation.	Withdrawing	4
G. New managers are supposed to come up with new and better ways, aren't they?	Hostile or Defensive	4
	Total: Personal	
	Total: Group	

Part Four: Conflict Resolution

Use the table shown on page 1028 to determine your total points. Enter this total on line 4 of the worksheet on page 314.

Part Five: Understanding Your Choices

A. Although you may have "legal" justification to force the solution your way, you should consider the emotional impact on the organization as a result of alienating the department manager. Allow two points on line 5 of the worksheet.

B. Accepting the new requirements would be an easy way out if you are willing to explain the increased costs and schedule delays to the other participants. This would certainly please the department manager, and might even give him the impression that he has a power position and can always resolve problems in this fashion. Allow four points on line 5 of your worksheet.

C. If this situation cannot be resolved at your level, you have no choice but to request upper-level management to step in. At this point you must be pretty sure that a compromise is all but impossible and are willing to accept a go-for-broke position. Enter ten points on line 5 of the worksheet if this was your choice.

	Mode	Personal or Group Score
A. The requirements are my decision and we're doing it my way.	Forcing	4
B. I've thought about it and you're right. We'll do it your way.	Withdrawal or Smoothing	4
C. Let's discuss the problem. Perhaps there are alternatives.	Compromise or Confrontation	4
D. Let me explain why we need the new requirements.	Smoothing, Confrontation, or Forcing	4
E. See my section supervisors; they're handling it now.	Withdrawal	4
F. I've looked over the problem and I might be able to ease up on some of the requirements.	Smoothing or Compromise	4
	Total: Personal	
	Total: Group	

D. Asking other managers to plead your case for you is not a good situation. Hopefully upper-level management will solicit their opinions when deciding on how to resolve the conflict. Enter six points on line 5 if this was your choice, and hope that the functional managers do not threaten him by ganging up on him.

Part Six: Interpersonal Influences

A. Threatening the employees with penalty power will probably have no effect at all because your conflict is with the department manager, who at this time probably could care less about your evaluation of his people. Allow zero points on line 6 of the worksheet if you selected this choice.

B. Offering rewards will probably induce people toward your way of thinking provided that they feel that you can keep your promises. Promotions and increased responsibilities are functional responsibilities, not those of a project manager. Performance evaluation might be effective if the department manager values your judgment. In this situation it is doubtful that he will. Allow no points for this answer and record the results on line 6 of the worksheet.

C. Expert power, once established, is an effective means of obtaining functional respect provided that it is used for a relatively short period of time. For long-term efforts, expert power can easily create conflicts between project and functional managers. In this situation, although relatively short term, the department manager probably will not consider you as an expert, and this might carry on down to his functional subordinates. Allow six points on line 6 of the worksheet if this was your choice.

D. Work challenge is the best means of obtaining support and in many situations can overcome personality clashes and disagreements. Unfortunately, the problem occurred because of complaints by the functional personnel and it is therefore unlikely that work challenge would be effective here. Allow eight points on line 6 of the worksheet if this was your choice.

E. People who work in a project environment should respect the project manager because of the authority delegated to him from the upper levels of management. But this does not mean that they will follow his directions. When in doubt, employees tend to follow the direction of the person who signs their evaluation form, namely, the department manager. However, the project manager has the formal authority to "force" the line manager to adhere to the original project plan. This should be done only as a last resort, and here, it looks as though it may be the only alternative. Allow ten points if this was your answer and record the result on line 6 of the worksheet.

F. Referent power cannot be achieved overnight. Furthermore, if the department manager feels that you are trying to compete with him for the friendship of his subordinates, additional conflicts can result. Allow two points on line 6 of the worksheet if this was your choice.

APPENDIX B
Solution to Leadership Exercise

Situation 1
A. This technique may work if you have proven leadership credentials. Since three of these people have not worked for you before, some action is necessary.
B. The team should already be somewhat motivated and reinforcement will help. Team building must begin by showing employees how they will benefit. This is usually the best approach on long-term projects. (5 points)
C. This is the best approach if the employees already understand the project. In this case, however, you may be expecting too much out of the employees this soon. (3 points)
D. This approach is too strong at this time, since emphasis should be on team building. On long-term projects, people should be given the opportunity to know one another first. (2 points)

Situation 2
A. Do nothing. Don't overreact. This may improve productivity without damaging morale. See the impact on the team first. If the other members accept Tom as the informal leader, because he has worked for you previously, the results can be very favorable. (5 points)
B. This may cause the team to believe that a problem exists when, in fact, it does not.
C. This is duplication of effort and may reflect on your ability as a leader. Productivity may be impaired. (2 points)
D. This is a hasty decision and may cause Tom to overreact and become less productive. (3 points)

Situation 3
A. You may be burdening the team by allowing them to struggle. Motivation may be affected and frustration will result. (1 point)
B. Team members expect the project manager to be supportive and to have ideas. This will reinforce your relationship with the team. (5 points)
C. This approach is reasonable as long as your involvement is minimum. You must allow the team to evolve without expecting continuous guidance. (4 points)
D. This action is premature and can prevent future creativity. The team may allow you to do it all.

Situation 4   A. If, in fact, the problem does exist, action must be taken. These types of problems do not go away by themselves.

B. This will escalate the problem and may make it worse. It could demonstrate your support for good relations with your team, but could also backfire. (1 point)

C. Private meetings should allow you to reassess the situation and strengthen employee relations on a one-on-one basis. You should be able to assess the magnitude of the problem. (5 points)

D. This is a hasty decision. Changing the team's schedules may worsen the morale problem. This situation requires replanning, not a strong hand. (2 points)

Situation 5   A. Crisis management does not work in project management. Why delay until a crisis occurs and then waste time having to replan?

B. This situation may require your immediate attention. Sympathizing with your team may not help if they are looking toward you for leadership. (2 points)

C. This is the proper balance: participative management and contingency planning. This balance is crucial for these situations. (5 points)

D. This may seriously escalate the problem unless you have evidence that performance is substandard. (1 point)

Situation 6   A. Problems should be uncovered and brought to the surface for solution. It is true that this problem may go away, or that Bob simply does not recognize that his performance is substandard.

B. Immediate feedback is best. Bob must know your assessment of his performance. This shows your interest in helping him improve. (5 points)

C. This is not a team problem. Why ask the team to do your work? Direct contact is best.

D. As above, this is your problem, not that of the team. You may wish to ask for their input, but do not ask them to perform your job.

Situation 7   A. George must be hurting to finish the other project. George probably needs a little more time to develop a quality report. Let him do it. (5 points)

B. Threatening George may not be the best situation because he already understands the problem. Motivation by threatening normally is not good. (3 points)

C. The other team members should not be burdened with this unless it is a team effort.

D. As above, this burden should not be placed on other team members unless, of course, they volunteer.

Situation 8   A. Doing nothing in time of crisis is the worst decision that can be made. This may frustrate the team to a point where everything that you have built up may be destroyed.

B. The problem is the schedule slippage, not morale. In this case, it is unlikely that they are related.

C. Group decision making can work but may be difficult under tight time constraints. Productivity may not be related to the schedule slippage. (3 points)

D. This is the time when the team looks to you for strong leadership. No matter how good the team is, they may not be able to solve all of the problems. (5 points)

Situation 9 A. A pat on the back will not hurt. People need to know when they are doing well.
  B. Positive reinforcement is a good idea, but perhaps not through monetary rewards. (3 points)
  C. You have given the team positive reinforcement and have returned authority/responsibility to them for phase III. (5 points)
  D. Your team has demonstrated the ability to handle authority and responsibility except for this crisis. Dominant leadership is not necessary on a continuous basis.

Situation 10 A. The best approach. All is well. (5 points)
  B. Why disturb a good working relationship and a healthy working environment? Your efforts may be counterproductive.
  C. If the team members have done their job, they have already looked for contingencies. Why make them feel that you still want to be in control? However, if they have not reviewed the phase III schedule, this step may be necessary. (3 points)
  D. Why disturb the team? You may convince them that something is wrong or about to happen.

Situation 11 A. You cannot assume a passive role when the customer identifies a problem. You must be prepared to help. The customer's problems usually end up being your problems. (3 points)
  B. The customer is not coming into your company to discuss productivity.
  C. This places a tremendous burden on the team, especially since it is the first meeting. They need guidance.
  D. Customer information exchange meetings are *your* responsibility and should not be delegated. You are the focal point of information. This requires strong leadership, especially during a crisis. (5 points)

Situation 12 A. A passive role by you may leave the team with the impression that there is no urgency.
  B. Team members are motivated and have control of the project. They should be able to handle this by themselves. Positive reinforcement will help. (5 points)
  C. This approach might work but could be counterproductive if employees feel that you question their abilities. (4 points)
  D. Do not exert strong leadership when the team has already shown its ability to make good group decisions.

Situation 13 A. This is the worst approach and may cause the loss of both the existing and follow-on work.
  B. This may result in overconfidence and could be disastrous if a follow-on effort does not occur.
  C. This could be very demoralizing for the team, because members may view the existing program as about to be canceled. (3 points)
  D. This should be entirely the responsibility of the project manager. There are situations where information may have to be withheld, at least temporarily. (5 points)

Situation 14   A. This is an ideal way to destroy the project-functional interface.
    B. This consumes a lot of time, since each team member may have a different opinion. (3 points)
    C. This is the best approach, since the team may know the functional personnel better than you do. (5 points)
    D. It is highly unlikely that you can accomplish this.

Situation 15   A. This is the easiest solution, but the most dangerous if it burdens the rest of the team with extra work. (3 points)
    B. The decision should be yours, not your team's. You are avoiding your responsibility.
    C. Consulting with the team will gain support for your decision. It is highly likely that the team will want Carol to have this chance. (5 points)
    D. This could cause a demoralizing environment on the project. If Carol becomes irritable, so could other team members.

Situation 16   A. This is the best choice. You are at the mercy of the line manager. He may ease up some if not disturbed. (5 points)
    B. This is fruitless. They have obviously tried this already and were unsuccessful. Asking them to do it again could be frustrating. Remember, the brick wall has been there for two years already. (3 points)
    C. This will probably be a wasted meeting. Brick walls are generally not permeable.
    D. This will thicken the brick wall and may cause your team's relationship with the line manager to deteriorate. This should be used as a last resort *only* if status information cannot be found any other way. (2 points)

Situation 17   A. This is a poor assumption. Carol may not have talked to him or may simply have given him her side of the project.
    B. The new man is still isolated from the other team members. You may be creating two project teams. (3 points)
    C. This may make the new man uncomfortable and feel that the project is regimented through meetings. (2 points)
    D. New members feel more comfortable one-on-one, rather than having a team gang up on them. Briefings should be made by the team, since project termination and phaseout will be a team effort. (5 points)

Situation 18   A. This demonstrates your lack of concern for the growth of your employees. This is a poor choice.
    B. This is a personal decision between you and the employee. As long as his performance will not be affected, he should be allowed to attend. (5 points)
    C. This is not necessarily a problem open for discussion. You may wish to informally seek the team's opinion. (2 points)
    D. This approach is reasonable but may cause other team members to feel that you are showing favoritism and simply want their consensus.

Situation 19   A. This is the best choice. Your employees are in total control. Do nothing. You must assume that the employees have already received feedback. (5 points)
   B. The employees have probably been counseled already by your team and their own functional manager. Your efforts can only alienate them. (1 point)
   C. Your team already has the situation under control. Asking them for contingency plans at this point may have a detrimental effect. They may have already developed contingency plans. (2 points)
   D. A strong leadership role now may alienate your team.

Situation 20   A. A poor choice. You, the project manager, are totally accountable for all information provided to the customer.
   B. Positive reinforcement may be beneficial, but does nothing to guarantee the quality of the report. Your people may get overcreative and provide superfluous information.
   C. Soliciting their input has some merit, but the responsibility here is actually yours. (3 points)
   D. Some degree of leadership is needed for all reports. Project teams tend to become diffused during report writing unless guided. (5 points)

APPENDIX C
Dorale Products Case Studies

DORALE PRODUCTS (A)

PMBOK® GUIDE AREA	**INTEGRATION MANAGEMENT**
	SCOPE MANAGEMENT
SUBJECT AREA	**DEFINING A PROJECT**

Background Dorale Products was undergoing favorable growing pains. Business was good. New product development was viewed as the driving force for the company's future growth. The company was now spending significantly more money for new product development, yet the number of new products reaching the market place was significantly less than in prior years. Also, some of the products reaching the market place were taking longer than expected to recover their R&D costs, while others became obsolete too quickly.

Management recognized that some sort of structured decision-making process had to be put in place whereby management could either cancel a project early before massive resources were committed or redirect efforts to different objectives. David Mathews was assigned as the project manager in charge of developing a new product development (project management) methodology for Dorale Products.

David understood the benefits of a project management methodology, especially as a structured decision-making process. It would serve as a template or a repetitive process such that project success could be incurred over and over again. The methodology would contain sections for project scope definition, planning, scheduling, and monitoring and control. There would also be a section on the role of the project manager, line managers, and executive sponsors.

To make the project management methodology easy to use and adaptable to all projects, the methodology would be constructed using forms, guidelines, templates, and checklists rather than the more rigid policies and procedures. This would certainly lower the cost of using the methodology and make it easier to adapt to a multitude of projects. The project managers could then decide whether to implement the methodology on an informal basis or on a more formal basis.

The first draft of the new methodology was completed and ready for review by the vice president (VP) of operations who had been assigned as the project sponsor. After a review of the methodology, a meeting was held between the sponsor and the project manager (PM).

The Meeting

VP: "I have read over the methodology. Is it your expectation that the methodology should be used on every project?"

PM: "We could probably justify using the methodology on every project. This would give us a really good structured decision-making process."

VP: "Using the methodology is costly and perhaps not all projects should require the use of the methodology. I can rationalize the use of the methodology on a $500,000 project. But what if the project is only $25,000 or $50,000? What if the project is 30 days in length rather than our usual 6- to 12-month effort?"

PM: "I guess we need to define the threshold limits on when project management should be used."

VP: "I have a concern that we should define not only when to use project management but also what a project is. If an activity remains entirely in one functional area, is it still a project according to your definition? Should we also define a threshold limit on how many functional departments must be involved before we define an activity as a project?"

PM: "I'll go back to the drawing board and get back to you in a week or so."

Questions

1. What is a reasonable definition of a project?
2. Is every activity a project or should there be a minimum number of functional boundaries that need to be crossed? If so, how many boundaries?
3. How do we determine when project management should be used and when an activity can be handled effectively by one functional group without the use of project management?
4. Do all projects need project management?
5. Since the use of a formal project management methodology requires time and money, what should be "reasonable" threshold limits for its use?

DORALE PRODUCTS (B)

PMBOK® GUIDE AREA	**INTEGRATION MANAGEMENT**
	SCOPE MANAGEMENT
SUBJECT AREA	**DEFINING A PROGRAM**

Background Dorale Products had just developed a project management methodology for the development of new products. Although the methodology was designed exclusively for new product development, the vice president of operations believed that other applications for the methodology would be possible. A meeting was held between the project manager responsible for the development of the methodology and the vice president of operations.

The Meeting

VP: "The company has invested significant time and money in the development of this methodology. It would be a shame if the methodology could not be applied elsewhere in the organization. As an example, there has to be commonality between new product development and information systems projects. Can we use this methodology, or part of it, for both new product development and information systems development?"

PM: "I'm not sure we can do that. The requirements for information systems projects are different, as are the life-cycle phases. A common project management methodology would have to be highly generic to be applicable to all types of projects."

VP: "Are you telling me that we will need to invest more time and more money to develop a family of methodologies?"

PM: "The methodology we've developed can be applied to all of our activities except information technology efforts. All of our projects are similar, or in the same domain group, except for IT. The IT people may require their own methodology, and I can understand their rationale for wanting it this way."

VP: "I assume from your comments that our existing methodology applies equally as well to programs as it does to projects. After all, isn't a program just a continuation of a project?"

PM: "I did not consider applying our methodology to programs as well as projects. Let me think about this, and I will get back to you."

Questions

1. Does it seem practical to have both a project management methodology and a systems development methodology in use concurrently?
2. What is the definition of a program? How does it differ from the definition of a project?
3. Does the project management methodology apply equally well to programs as it does to projects?

DORALE PRODUCTS (C) _____

PMBOK® GUIDE AREA   **INTEGRATION MANAGEMENT**
                            SCOPE MANAGEMENT

SUBJECT AREA   **PROJECT MANAGEMENT APPLICATIONS**

Background   Dorale Products has just completed the development of a project management methodol-
                 ogy. Although the methodology was to be used for new project development, there was
                 hope that the methodology could be applied to other products as well.

The Meeting VP: "Have we restricted ourselves on what type of projects we can use our
                       methodology?"

                 PM: "The answer is both yes and no! Every activity in the company, regardless of the
                       functional area, can be regarded as a project. But not all projects require the use
                       of the methodology or even project management."

                 VP: "When we had these conversations months ago at the onset of this development
                       process, you convinced me that we were managing our business by projects. Are
                       you now changing your mind?"

                 PM: "Not at all. The main skill requirement for our project managers is integration
                       management. The greater the integration requirements, the greater the need for
                       project management."

                 VP: "Now I'm really confused. First you tell me that all projects need project man-
                       agement, and now you say that not all projects need the use of a methodology.
                       What am I missing here?"

Questions   1. Should all projects require the use of the principles of project management?
                 2. What type of projects should or should not require the use of a project management
                    methodology?
                 3. How does the magnitude of the integration requirements affect your answer to the pre-
                    vious question?
                 4. What conclusions can be made about the applications of project management?

DORALE PRODUCTS (D) _____

PMBOK® GUIDE AREA   **INTEGRATION MANAGEMENT**
                            SCOPE MANAGEMENT

SUBJECT AREA   **PROJECT MANAGEMENT PROCESSES**

Background   Dorale Products developed a methodology for the management of projects. A vice presi-
                 dent was assigned as the project sponsor to oversee the development of the project

management methodology. It was now time for the sponsor to introduce the methodology to the executive levels of management. The vice president met with his project manager to prepare the handouts for the executive committee briefing.

The Meeting

VP: "I have looked over the methodology and am concerned that I cannot easily recognize the structure to the methodology. If I cannot identify the structure, then how can I effectively make a presentation to other executives?"

PM: "Good methodologies should be based upon guidelines, forms, and checklists, rather than policies and procedures. We must have this flexibility to adapt the methodology to a multitude of projects."

VP: "I agree with you. But there must still be some overall structure to the project management process."

PM: "Integration management involves three process areas, namely, the integration of the development of the plan, the integration of the execution of the plan, and the integration of changes to the plan. Our methodology is broken down into life-cycle phases, and these three integrative processes are included in each life-cycle phase, though not specifically addressed. I have tried to use the principles of the PMBOK®."

VP: "Let me look at the methodology again and see if I can relate it to what you've said."

Questions

1. Is the project manager correct in his definition of the integration management process areas?
2. Can it be difficult to identify these process areas in each life-cycle phase? If so, then what can we do to make them more visible?
3. What should the vice president say in his presentation about the structure of the methodology?

DORALE PRODUCTS (E) ⎯⎯⎯⎯⎯⎯⎯⎯⎯⎯⎯⎯⎯⎯⎯⎯⎯⎯⎯⎯

PMBOK® GUIDE AREA   **INTEGRATION MANAGEMENT**
     SCOPE MANAGEMENT

SUBJECT AREA   **LIFE-CYCLE PHASES**

Background   The vice president made his presentation to the other senior officers concerning the methodology. Emphasis was placed upon the 10 life-cycle phases. The other executives had several questions concerning the use of 10 life-cycle phases. The vice president returned to the project manager for another meeting.

The Meeting VP: "The other executives have concerns that 10 life-cycle phases are too many. You have 10 end-of-phase gate reviews which require that most of our executives attend. That seems excessive."

PM: "I agree. The more I think about it, the more I believe that 10 are too many. I'll be spending most of my time planning for gate review meetings, rather than managing the project."

VP: "Another concern of our executives was their role or responsibility at the gate review meetings. The methodology is unclear in this regard."

PM: "Once again, I must agree with you. We should have an established criterion for what constitutes passing the gate reviews."

Questions 1. What are the primary benefits for using life-cycle phases? Are there disadvantages as well?"
2. How many life-cycle phases are appropriate for a methodology?
3. What is the danger of having too many gate review meetings?
4. Who determines what information should be presented at each gate review meeting?
5. What questions should the information at the gate review meeting be prepared to answer?

DORALE PRODUCTS (F)

PMBOK® GUIDE AREA   **INTEGRATION MANAGEMENT**
                              SCOPE MANAGEMENT

SUBJECT AREA   **DEFINING SUCCESS**

Background When the executive committee made the final review of the project management methodology, they identified a lack of understanding of what would constitute project success. The recommendation was to establish some type of criteria that would identify project success.

The Meeting VP: "We have a problem with the identification of success on a project. We need more clarification."

PM: "I assumed that meeting the deliverables specified by the customer constituted success."

VP: "What if we meet only 92 percent of the specification? Is that a success or a failure? What if we overrun our new product development process but bring in more new customers? What if the project basically fails but we develop a good customer relationship during that process?"

PM: "I understand what you are saying. Perhaps we should identify both primary and secondary contributions to success."

Questions
1. What is the standard definition of success (i.e., primary factors)? How does this relate to the triple constraint?
2. What would be examples of secondary success factors?
3. What would be a reasonable definition of project failure?
4. Should these definitions and factors be included in a project management methodology?
5. Are there any risks with inserting the primary and secondary success factors into the methodology?

DORALE PRODUCTS (G)

PMBOK® GUIDE AREA
INTEGRATION MANAGEMENT
SCOPE MANAGEMENT
HUMAN RESOURCES MANAGEMENT

SUBJECT AREA
ROLE OF THE EXECUTIVE

Background Although senior management seemed somewhat pleased with the new methodology, there was some concern that the role of senior management was ill-defined. The vice president felt that this needed to be addressed quickly so that other executives would understand that they have a vital role in the project management process.

The Meeting VP: "Many of our executives are not knowledgeable in project management and need some guidance on how to function as a project sponsor. Without this role clarification, some sponsors might be "invisible" while others may tend to be too actively involved. We need a balance."

PM: "I understand your concerns and agree that some role description is needed. However, I don't see how the role description will prevent someone from becoming invisible or overbearing."

VP: "That's true, but we still need a starting point. We may need to teach them how to function as a sponsor."

PM: "If the sponsor can change based upon which life-cycle phase we are in, then we should delineate the role of the sponsor per phase."

VP: "That is a good point. Let's also make sure we define the role of the sponsor at the gate review meetings."

Questions
1. What should be the primary role for the sponsor?
2. Will the role change based upon life-cycle phases?
3. Is it advisable for the sponsor to change based upon the life-cycle phase?
4. Will role delineation in the methodology force the sponsor to perform as expected?
5. What should be the sponsor's role during gate review meetings?

DORALE PRODUCTS (H)

PMBOK® GUIDE AREA	**INTEGRATION MANAGEMENT**
	SCOPE MANAGEMENT
	HUMAN RESOURCES MANAGEMENT
SUBJECT AREA	**ROLE OF LINE MANAGERS**

Background   The project management methodology was finally beginning to take shape. However, even though the basic structure of the methodology was in place, there were still gaps that had to be filled in. One of these gaps was a well-defined role for the line managers.

The Meeting   VP: "From what I've read about project management, it is very difficult at first to get line managers to effectively support projects. I want our line managers to become fully committed to project management as quickly as possible."

PM: "I agree with you! It's not good for a line manager to assign people to a project and then take no interest in the project at all."

VP: "I believe the line managers have the power to make or break a project. Simply stated, we need them to share in the accountability after they assign resources."

PM: "I'm not exactly sure how to do that. There is no way that I as a project manager can force a line manager to share accountability with me for the project's success or failure."

VP: "I know this will be difficult at first, but I believe it can be done. The methodology should define the expectations that the executives have on the role of the line managers in each life-cycle phase as well as the working relationships in each phase. See if you can get some of our line managers to help you in this regard."

PM: "On most of our projects, the technical direction to the employees is still provided by the line managers, even after the employee is assigned. Most of our project managers have an understanding of technology, not a command of technology. However, we do have some projects where the technical know-how resides with the project manager, who must then provide daily technical supervision. How do I cover both bases in the design of the methodology?"

VP: "It seems to me that in one situation the project manager would be negotiating with the line manager for deliverables, and in the second situation the negotiation would be for specific people. I'm sure you'll find a way to incorporate this into the methodology."

Questions   1. Should a methodology also include staffing policies? If so, what would be an example of a staffing policy?

2. When should a project manager negotiate for people, and when should the project manager negotiate for deliverables?

3. Should a staffing policy also distinguish between full-time and part-time assignments?

4. How should a company handle a situation where the line managers refuse to support project management, even though it is defined as part of the methodology?
5. Should staffing policies and the role of line management be defined in terms of policies and procedures or simply guidelines?

DORALE PRODUCTS (I)

PMBOK® GUIDE AREA
INTEGRATION MANAGEMENT
SCOPE MANAGEMENT
HUMAN RESOURCES MANAGEMENT

SUBJECT AREA
INTERPERSONAL SKILLS FOR PROJECT MANAGERS

Background With the role of the line manager and senior manager somewhat defined, Dorale believed that only individuals with specialized, interpersonal skills would become the best project managers. The company contemplated the preparation of a list of "universal" skills necessary to function as a project manager.

The Meeting VP: "I would like to see a list of desired personal characteristics for project managers included in our methodology. Surely this can be done."

PM: "I think we can define knowledge areas more easily than interpersonal skills. It is easier for us to decide whether or not the project manager needs a command of technology or understanding of technology by looking at the requirements of the project. But interpersonal skills are more complicated."

VP: "I don't understand why. Please explain!"

PM: "We appoint project managers to manage deliverables, not people. Our line managers are providing significantly more daily direction to the assigned workers than do our project managers."

VP: "Are you telling me that project managers do not require any management skills or interpersonal skills while managing a project?"

PM: "That's not really what I'm saying. I just believe that the skills needed to be a project manager are probably significantly different than the skills needed to be a line manager."

VP: I agree! See what kind of list you can develop."

Questions 1. What types of interpersonal skills are needed to be an effective project manager?
2. How do the interpersonal skills of a project manager differ from the skills needed to be an effective line manager?
3. Is your answer to the first two questions dependent upon the fact that in project management multiple-boss reporting is required?
4. Should the list that you have created be dependent upon whether or not the project manager has wage and salary responsibility (or input) for the team members?

5. Why is it often difficult for experienced line managers to become full-time project managers? (Or, in some cases, even part-time project managers?)

6. Some project managers have a command of technology while others have an understanding of technology. Can this command or understanding of technology influence the interpersonal skills needed to be a project manager?

7. Can the interpersonal skills requirements change if the project manager focuses on deliverables rather than people?

8. Can someone with very strong technical skills also have undesirable project management interpersonal skills?

9. Should a project management methodology identify the desired interpersonal skills of a project manager or should it be done on a project-by-project basis only?

DORALE PRODUCTS (J)

PMBOK® GUIDE AREA	**INTEGRATION MANAGEMENT** **SCOPE MANAGEMENT** **HUMAN RESOURCES MANAGEMENT**
SUBJECT AREA	**PROJECT STAFFING POLICIES AND PROCEDURES**

Background   Dorale expected conflicts to arise over the staffing of projects. There was some concern over whether or not a project management methodology should contain policies and procedures for project staffing.

The Meeting   VP: "We need some sort of direction in our methodology for the staffing of projects. If we do not have policies and procedures in this regard, then there is no guarantee that the project manager will receive adequate and timely resources."

PM: "I'm not sure I know how to do this. Right now, we are advocating that our project managers negotiate with the line managers for deliverables, rather than people. It is then the responsibility of the line manager to provide adequate resources to get the job done."

VP: "I agree with you, but we still need direction. Project managers must make it clear what the job specifically requires so that the line manager provides the right resources. I do not want to get into a conflict situation where the project manager blames the line manager for not providing the right resources and the line manager blames the project manager for improperly defining the scope."

PM: "That seems more like accepting accountability than staffing."

VP: "Perhaps so, but it is related to staffing. I want the line managers to provide the projects with personnel with the qualification levels necessary to meet the budgetary limits. We cannot afford to have projects that are loaded with the highest salaried workers."

PM: "That's a good idea. It might also be advisable to have some policy that mandates that the project managers release the assigned workers at their earliest convenience so that they can be picked up on other projects."

Questions

1. Is it appropriate for a project management methodology to contain policies and procedures on project staffing?
2. Should staffing policies and procedures be directed to project managers, line managers, or both?
3. Should project sponsors be involved in decisions affecting project staffing? If so, what specifically is their involvement and for the staffing of which positions?
4. How do you develop a policy that "forces" a project manager to release people to other projects, assuming they are no longer required on the existing project?
5. Is project staffing an "accountability" decision?
6. Is it the responsibility of the project manager or line manager to adequately define the skill level required to complete a task?
7. Should staffing policies be applied to full-time personnel, part-time personnel, or both?

DORALE PRODUCTS (K)

PMBOK® GUIDE AREA **INTEGRATION MANAGEMENT**
SCOPE MANAGEMENT
HUMAN RESOURCES MANAGEMENT

SUBJECT AREA **THE PROJECT/PROGRAM OFFICE**
Background

The methodology developed by Dorale focused on relatively small projects with time durations of less than 18 months. Could the same methodology be used on large projects?

The Meeting

VP: "Most of our projects have manpower requirements of 10–20 people with time durations of 18 months or less. Last week, at the executive committee meeting, we approved several large projects that may run for 3 years or more and require more than 40 people full time. How will we manage these projects?"

PM: "I assume you are talking about projects that will be managed by a project office rather than simply by a project manager."

VP: "On large projects, the project manager is more of a project office manager than a project manager. Shouldn't our methodology also discuss the role of a project office and a project office manager?"

Questions
1. What criteria should exist in deciding when to use a project office as opposed to just a project manager?
2. Are the integration management responsibilities of a project office manager different than those of a project manager?
3. What is a project office?
4. What is the role of a project office manager?
5. Can the members of a project office be part-time or must they be full-time?
6. If employees are assigned full-time to a project office, can they still report administratively to their line managers?
7. Can the assigned project office employees be full-time and yet the project manager is part-time?
8. Can project staffing policies be defined for a project office or is it more project-specific?

Solutions to the Dorale Products Case Studies

CASE STUDY (A)

1. A project is a unique activity, with a well defined objective with constraints, that consumes resources, and is generally multi-functional. The project usually provides a unique product service or deliverable.
2. Generally, there is no minimum number of boundaries that need to be crossed.
3. Usually this is based upon the amount of integration required. The greater the amount of integration, the greater the need for project management.
4. All projects could benefit from the use of project management, but on some very small projects, project management may not be necessary.
5. Reasonable thresholds for the use of the project management methodology are based upon dollar value, risk, duration, and number of functional boundaries crossed.

CASE STUDY (B)

1. In many companies, one enterprise project methodology may be impractical. There may be one methodology for developing a unique product or service, and another one for systems development.
2. A program is usually longer in duration than a project and is comprised of several projects.
3. Project management methodologies apply equally to both programs and projects.

CASE STUDY (C) ────────────────────────────────────

1. All projects should use the principles of project management but may not need to use the project management methodology.
2. Projects that do not require the methodology are those that are of short duration, low dollar value and stay within one functional department.
3. Methodologies are generally required for all projects that necessitate large scale integration. However, if the cost associated with the use of the methodology is low, or the methodology is not complex, then it could be argued that the methodology should be used on all projects.
4. This is a valid argument that the principles of project management should be applied to all projects, irrespective of constraints.

CASE STUDY (D) ────────────────────────────────────

1. The project manager is partially correct in his definition of the integration management process. The project manager's definition is aligned more so with the 2000 PMBOK® Guide rather than the 2004 version.
2. It can be difficult to identify these processes in each life cycle phase. However, a good project management methodology will solve this problem.
3. A good project management methodology is based upon forms, guidelines, templates, and checklists, and is applicable to a multitude of projects. The more structure that is added into the methodology, the more control one has, but this may lead to the detrimental result of limiting the flexibility that project teams need to have for one methodology that can be adapted to a multitude of projects.

CASE STUDY (E) ────────────────────────────────────

1. The primary benefits are standardization and control of the process. The disadvantage occurs when this is done with policies and procedures rather than forms, guidelines, templates, and checklists.
2. Most good methodologies have no more than five or six life cycle phases.
3. With too many gate review meetings, the project manager spends most of his/her time managing the gate review meetings rather than managing the project.
4. The stakeholders that are in attendance at the gate review meetings determine what information should be presented. Templates and checklists can be established for the gate review meetings as well as the stages.
5. At a minimum, the questions addressed should include: (1) Where are we today?, (2) Where will we end up?, and (3) What special problems exist?

CASE STUDY (F)

1. The standard definition of success is within time, cost, scope or quality, and accepted by the customer.
2. Secondary success factors might include profitability and follow-on work.
3. It is more difficult to define failure as opposed to success. People believe that failure is an unsatisfied customer. Others believe that failure is a project which, when completed, provided no value or learning.
4. Absolutely, but they can be modifies to fit a particular project or the needs of a particular sponsor.
5. Lack of flexibility may be the result.

CASE STUDY (G)

1. The primary role of the sponsor is to help the PM resolve problems that may be beyond the control of the PM.
2. The role of the sponsor can and will change based upon the life cycle phase.
3. There are two schools of thought; some believe that the same person should remain as sponsor for the duration of the project while others believe that the sponsor can change based upon the life cycle phase. There are advantages and disadvantages of both approaches, and it is often based upon the type of project and the importance of the customer.
4. Not necessarily, but is it a good starting point in explaining to to new sponsors their role and responsibility.
5. Verify that the current phase has been completed correctly and authorize initiation of the next phase.

CASE STUDY (H)

1. Good methodologies identify staffing policies. As an example, a project manager may have the right to identify the skill level desired by the workers, but this may be open for negotiations.
2. Project managers that possess a command of technology normally negotiate for people whereas project managers without a command of technology negotiate for deliverables.
3. This question is argumentative because it may involve an argument over effort versus duration. The line manager may carry more weight in this regard than the project manager since this may very well be based upon the availability of personnel.
4. This is why project sponsors exist; to act as a referee when there are disagreements and to make sure that line management support exists.
5. Guidelines are always better than policies and procedures, at least in the eyes of the author.

CASE STUDY (I)

1. Core skills include decision-making, communications, conflict resolution, negotiations, mentorship, facilitation, and leadership without having authority.
2. Line management skills often focus on superior-subordinate relationships whereas PM skills focus on team-building where the people on the team are not necessarily under the control of the PM (and may actually be superior in rank to the PM).
3. Multiple-boss report is also a concern because the control and supervision of the worker may be spread across several individuals.
4. Wage and salary administration is an important factor. If the PM has this responsibility, the workers will adapt to the PM because he/she has an influence over their performance review and salary. Without this responsibility, the PM may be forced to adapt to the workers rather than vice-versa.
5. Line managers are accustomed to managing with authority whereas project managers are not.
6. When a PM has a command of technology, he/she may align closer with the skills of a line manager rather than a PM.
7. PMs usually negotiate for deliverables when they do not have a command of technology and this can influence the interpersonal skills needed for a particular project.
8. Yes.
9. The identification should be in general terms only so that it may be applicable to a multitude of projects.

CASE STUDY (J)

1. Yes, but in general terms only.
2. Both, in order to minimize conflicts.
3. Sponsors usually take an active role in selection of the PM, but take a passive role in functional staffing so as not to usurp the authority of their line managers.
4. There is no really effective to do this other than by closing out some of the functional charge numbers.
5. Yes, if mandated by senior management.
6. The PM can request any skill level desired, but the final decision almost always rests with the line manager.
7. Both.

CASE STUDY (K)

1. The size of the project, duration, risk, and importance of the customer.
2. They are the same and may even be more detailed.
3. A project management team.

4. The role of the PM is to coordinate and integrate the activities of the project management team.
5. They can be part-time or full-time based upon the needs of the project.
6. Yes. An example of this may be the quality specialist assigned to the project.
7. Yes.
8. Policies can be established to staffing of a project office team, but it may be company-specific or client specific.

APPENDIX E

Alignment of the PMBOK® Guide to the Text

This appendix cross-lists the PMBOK® Guide 4th Edition sections with this textbook. Not every section in the PMBOK® Guide is addressed in this textbook, only the major categories.

PMBOK® Guide	Page
1.0	927
1.2	2
1.3	2
1.4.2	55, 148
1.4.3	24
1.4.4	169, 191
1.5	49
1.6	4, 8, 191, 287, 385, 415, 472, 955
1.8	841
2.0	66, 68, 927
2.1	418, 474, 611
2.1.1	66, 511
2.1.2	68
2.1.3	68
2.2	22, 955
2.3	12, 174, 384
2.3.2	18
2.3.5	955
2.3.6	12
2.3.8	12
2.4	27, 119
2.4.1	76

PMBOK® Guide	Page
10.2	234, 235
10.2.2	244
10.2.2.2	642
10.3	235, 242
10.3.1.2	556
10.3.3	238
10.4.2	235
10.5.3.1	678
11.1	743, 753
11.2	602, 746, 755, 797
11.3	761, 766
11.4	761, 771
11.4.2	772
11.4.2.2	618, 619
11.5	782
11.5.3.1	793
11.6	788
12.1	840, 842
12.1.1.5	858
12.1.2.1	844
12.1.2.3	851, 855
12.1.3.2	426, 431, 442, 843
12.2	845
12.2.1.5	847
12.2.2.1	846
12.2.2.5	847
12.2.3.1	847, 850
12.3	735, 859
12.3.1.3	599
12.4	863
12.4.2.1	479

This section of the appendix cross-lists some figures.

PMBOK® Guide	Page
Figure 2-8	106, 117
Figure 2-9	106, 117
Figure 2-10	106, 117
Figure 5-6	436, 438, 638, 639
Figure 6-6	503

There are several ways that one can study for the PMI®'s PMP® Examination. The author recommends the following approach:

1. Read over a specific area of knowledge chapter in the PMBOK® Guide.
2. Then, read over the chapter(s) in this text that correspond to that area of knowledge
3. Then re-read the area of knowledge in the PMBOK® Guide for a second time. Usually things fall into place better after the second reading of the PMBOK® Guide.
4. Now it is time to measure what you have learned. Answer the multiple choice questions at the end of the chapter(s) where the area of knowledge information was found. Collect whatever practice questions you can find, such as with the workbook that can accompany this text. The more question you answer, the more prepared you will be to pass the exam.

Some of the sources mentioned previously for practice questions can dramatically help the learning process. For example, the software provided by the International Institute for Learning allows the user to:

● Test on all questions in an area of knowledge
● Test on all questions with a domain area, such as project initiation
● Test on all questions related to a specific PMBOK® Guide section such as Section 3.2.

Author Index

Subject Index _____